DICTIONARY

OF

NATIONAL BIOGRAPHY

1981–1985

THE

DICTIONARY

OF

NATIONAL BIOGRAPHY

1981–1985

EDITED BY
LORD BLAKE

AND

C. S. NICHOLLS

With an Index covering the years 1901–1985
in one alphabetical series

OXFORD UNIVERSITY PRESS
1990

Oxford University Press, Walton Street, Oxford OX2 6DP
Oxford New York Toronto
Delhi Bombay Calcutta Madras Karachi
Petaling Jaya Singapore Hong Kong Tokyo
Nairobi Dar es Salaam Cape Town
Melbourne Auckland
and associated companies in
Berlin Ibadan

Oxford is a trade mark of Oxford University Press

Published in the United States
by Oxford University Press, New York

British Library Cataloguing in Publication Data
The Dictionary of National Biography.
1981–1985: with an index covering the years 1901–1985
in one alphabetical series.
1. Great Britain — Biographies — Collections
I. Blake, Robert Blake, 1916– II. Nicholls, C. S.
(Christine Stephanie)
920'.041
ISBN 0–19–865210–0

Library of Congress Cataloging in Publication Data
The dictionary of national biography, 1981–1985: with an
index covering the years 1901–1985 in one alphabetical series/edited by
Lord Blake and C.S. Nicholls.
p.   cm.
1. Great Britain — Biography — Dictionaries.   I. Blake, Robert, 1916–
II. Nicholls, C. S. (Christine Stephanie)
920.041—dc20   DA28.D59   1990   89–39766
ISBN 0–19–865210–0

Text processing by the Oxford Text System and
Wyvern Typesetting Ltd.
Printed in Great Britain by
Courier International Ltd.
Tiptree, Essex

PREFATORY NOTE

THE 380 individuals noticed in this Supplement died between 1 January 1981 and 31 December 1985. They are mostly British subjects but also include some who were originally British and then took other nationalities and some who spent a significant part of their career in Britain but did not become British subjects. There are, in addition, a few outstanding citizens of the Commonwealth.

This is the first time that the Dictionary has used a five-year, rather than a ten-year span for a supplementary volume. The experiment has been tried because there was public pressure for the Supplements to appear more frequently and new technology made the prospect feasible. It has certainly been easier to handle half the number of people (there were 748, 745, and 754 in the three previous decennial Supplements) and a distinct advantage has emerged: in the case of many entrants a five-year period makes it possible to capture the personal knowledge of colleagues who under the ten-year rule might already themselves have joined the entrants as candidates for inclusion.

It has long been the practice of DNB editors to ask people who knew or worked with the entrants to write the articles. They have a perspective on their subject which a historian writing from a distance would lack. Far from setting up memorial shrines, these contributors are generally rigorous in their assessment of their subjects. The editors are very grateful to them, for they realize how much work is required to encapsulate a person in 750–1,000 words (only in exceptional cases are more allocated).

Their essays display a variety of human endeavour which will inform, delight, and sometimes startle the reader. They tell us, for example, that Arthur Askey as a boy sang solo in the presence of the archbishops of Canterbury and York when the Lady chapel of Liverpool Cathedral was consecrated. Gertrude Caton-Thompson, the archaeologist, slept in an empty tomb at Abydos with cobras for company and a pistol under her pillow to ward off prowling hyenas. David Cecil, sixth Marquess of Exeter, who ran as a sprinter in the 1924, 1928, and 1932 Olympics, the last while he was MP for Peterborough, had a similar speed of speech; when he visited the offices of the Hansard shorthand writers to correct his speech proofs, he improved the speeches. Sir Christopher Cox, the compulsive talker, would say 'I want to pick your brains for five minutes' and would then talk uninterruptedly himself for an hour. Charles Douglas-Home, who had a firm attachment to spectacularly shabby clothes, was not only editor of *The Times* but also a fellow of the Royal College of Music. The cricketer Percy Fender, who made the fastest ever hundred (since equalled), was also a wine merchant and a regular writer for the *Evening News*; he was the first man to use a typewriter in the press box, an innovation not well received. Philip Hall, the mathematician, who cared nothing for hot water or central heating, spoke poetry beautifully, not only in English. Antony Head (Viscount Head), soldier and politician, rode in the Grand National, gained an able bodied seaman's certificate, became minister of defence, and was an expert ornithologist and entomologist, rising early when high commissioner for Lagos in pursuit of rare varieties of butterfly. Christmas Humphreys, the barrister, became a Buddhist, wrote four books of poetry, loved Chinese art, and was convinced that the Earl of Oxford wrote the plays attributed to Shakespeare. John Pringle, professor of zoology at Oxford, warm-hearted behind a cold exterior, was once compared with a *bombe surprise* turned inside out.

The entrants are drawn from many walks of life: the categories run from A (Actor),

v

through M (Malefactor), to Z (Zoologist). Among politicians are George Brown (Lord George-Brown) and R. A. Butler (Lord Butler of Saffron Walden); among actors Dame Flora Robson, Sir Michael Redgrave, and Sir Ralph Richardson; among broadcasters (for the first time radio personnel appear in considerable numbers and television is beginning to feature) Stuart Hibberd, Alvar Lidell, and Roy Plomley; among writers Sir John Betjeman, Philip Larkin, Robert Graves, Sir William Empson, and J. B. Priestley; among musicians Sir William Walton, Sir Adrian Boult, and Herbert Howells (who all died within a fortnight of each other); among the armed services Sir Claude Auchinleck, Sir Douglas Bader, Sir Arthur Harris, and Lord Fraser of the North Cape; among journalists James Cameron and Edward Crankshaw; among astronomers Sir Martin Ryle; among physicists Paul Dirac; among artists Ben Nicholson; among educationists Sir Robert Birley; among spies Anthony Blunt and Donald Maclean; and among Commonwealth leaders Indira Gandhi.

The oldest person in the volume is Sir Robert Mayer (1879–1985), the patron of music, who was sent to England from Germany in 1896 and who celebrated his hundredth birthday with the publication of an autobiography confidently entitled *My First Hundred Years*. He is closely followed by Sir George Schuster (1881–1982), founder of Atlantic College, who on his ninety-eighth birthday went to London to be measured for two new suits and who was still writing to headmasters on educational matters in his 102nd year, and H. C. Barnard (1884–1985), professor of education, who denied that there was any merit in being old and wrote in his old age a splendid biographical account of his Victorian childhood *Were Those The Days?* The youngest persons in the book are Michael Hailwood, world champion motor-cycle rider, who ironically died in a traffic accident, and Shiva Naipaul, the writer.

The editors would like to express their heartfelt thanks to the experts in many fields who generously helped them select the entrants to this Supplement. They are busy and important people who nevertheless gave their advice freely. Without their assistance this book would be the poorer. We also owe a great debt of gratitude to Mrs Joan Goodier and Mrs Lesley Jenkins who coped cheerfully and efficiently with secretarial matters and the new technology. The Oxford University Press is always a stalwart supporter of this Dictionary, which we hope is a credit to it, and the Bodleian Library grants us facilities without which we would be unable to function. To both institutions we are profoundly grateful.

In conclusion, we have embarked upon a volume which contains biographies of those people who have inadvertently been omitted from the DNB in the past. Gerard Manley Hopkins, whose fame was posthumous, is an example. If any readers have sought and failed to find in the DNB a biography of someone whom they think ought to have been included, please will they write to the address below, giving reasons for their subject's candidature.

BLAKE

C. S. NICHOLLS

Dictionary of National Biography,
Bodleian Library,
Clarendon Building,
Broad Street,
Oxford

15 May 1989

LIST OF CONTRIBUTORS

ADAMS, Sara Trerice:
Adams (M. G. A.).
ALDERSON, Brian Wouldhave:
Challans (Mary Renault).
ALLASON, Rupert William Simon (West, Nigel):
Mitchell (G. R.).
ALLIBONE, Thomas Edward:
Burch; Sykes.
AMBLER, Eric:
Mason.
AMERY, Julian:
Assheton (Clitheroe); Lennox-Boyd (Boyd of Merton).
AMIS, Kingsley:
Betjeman.
ARLOTT, (Leslie Thomas) John:
Voce.
ARMSTRONG OF ILMINSTER, Robert Temple Armstrong, Baron:
Maud (Redcliffe-Maud); Mayer.
ARMSTRONG, Sir Thomas Henry Wait:
Howells; McKie.
ASCHERSON, (Charles) Neal:
Crankshaw.
ATTENBOROUGH, (Ralph) John:
Goudge.
AVERY, Gillian Elise:
Opie.

BACON, Sir Nicholas Hickman Ponsonby, Bart:
Bacon.
BAKER, Richard Douglas James:
Lidell; Plomley.
BALL, Harold William:
White.
BALLHATCHET, Kenneth Arthur:
Spear.
BANNISTER, Sir Roger Gilbert:
Noel-Baker.
BARKER, Theodore Cardwell:
Pilkington.
BARNES, Dame (Alice) Josephine (Mary Taylor):
Wright.
BARNETT, Correlli Douglas:
Roskill.
BARR, James:
Caird.
BARROS, James:
Gouzenko.
BAXTER, (James) Neil:
Coia.
BAXTER-WRIGHT, Keith Stanley:
Burton.
BEESON, Trevor Randall:
Collins (L. J.).

BELL, Alan Scott:
Moncreiffe of that Ilk.
BELL, George Douglas Hutton:
Engledow.
BELOFF, Nora:
Ward (Jackson of Lodsworth).
BERRY, Francis:
Knight.
BETTS, Alan Osborn:
Amoroso.
BIGGS-DAVISON, Sir John Alec:
Fraser (H. C. P. J.).
BLAKE, Robert Norman William Blake, Baron:
Bryant; Vaizey.
BOND, Brian James:
Auchinleck.
BOWNESS, Sir Alan and BOWNESS, Sophie:
Nicholson.
BOYD, Sir John McFarlane:
Duffy.
BRIDGE, Antony Cyprian:
Wolfenden.
BROWN, (John) David:
Bush.
BROWN, Sir John Gilbert Newton:
Blackwell.
BROWNLIE, Ian:
Fitzmaurice; Waldock.
BRYANS, Dame Anne Margaret:
Pery (Limerick).
BURGES, (Norman) Alan:
Watt.
BUTLER, George:
Halliday.

CARR, Sir (Albert) Raymond (Maillard):
Somerset (Beaufort).
CARVER, (Richard) Michael (Power) Carver, Baron:
O'Connor; Ritchie.
CAZALET, Edward Stephen:
Douglas-Home.
CECIL, Robert:
Maclean.
CHADWICK, Sir Henry:
Dykes Bower.
CHADWICK, (William) Owen:
Neill.
CHARLTON, John Fraser:
Smallwood.
CHARNOCK, Henry:
Deacon.
CLAPHAM, Arthur Roy:
Godwin.
CLIFFORD, Paul Rowntree:
Goodall.

CLIVE, Nigel David:
Oldfield; Rennie.
COLVILLE, Sir John Rupert:
Head.
COMPTON-HALL, (Patrick) Richard:
Miers.
CONNON, Bryan:
Nichols.
COOKE, Alexander Macdougall:
Witts.
COOPER, Leo and HARTMAN, Thomas
Robert:
Marshall-Cornwall.
CORK, Richard Graham:
Penrose.
CROSS, Sir Kenneth Brian Boyd:
Elmhirst.
CRUICKSHANK, Charles Greig:
Sweet-Escott.
CUNNINGHAM, Angela Joan:
Demant.
CURRAN, Sir Samuel Crowe:
Dee.

DALITZ, Richard Henry:
Dirac.
DALTON, (Henry) James (Martin):
Lutyens.
DAVIE, Donald Alfred:
Bunting.
DAVIES, Richard Gareth:
Richards (O. W.).
DAVIES, Robert William:
Carr.
DAVIS, Thomas Beven:
Kirkman.
DE LANGE, Nicholas Robert Michael:
Parkes.
DENBIGH, Kenneth George:
Danckwerts.
DENISON, (John) Michael (Terence
Wellesley):
Bridge.
DENNISTON, Robin Alastair:
Welchman.
DERBYSHIRE, Sir Andrew George:
Johnson-Marshall.
DEVLIN, Patrick Arthur Devlin, Baron:
Hodson.
DOUGLAS-HOME, William:
Johnson (C.).
DRAPER, Gerald Irving Anthony Dare:
Russell (Russell of Liverpool).
DRURY, Sir (Victor William) Michael:
Gillie.
DRYDEN, Colin John:
Chapman.
DUGDALE, Sir William Stratford:
Boyd-Rochfort.
DUTHIE, Robert Buchan:
Charnley.

ELLIOT OF HARWOOD, Katharine Elliot,
Baroness:
Davidson (Northchurch).
ESHER, Lionel Gordon Baliol Brett, Baron:
Sheppard.
EVANS, Sydney Hall:
Abbott.

FAITHFULL, Lucy Faithfull, Baroness:
Younghusband.
FALKNER, Sir (Donald) Keith:
Baillie.
FARRAR-HOCKLEY, Sir Anthony Heritage:
Gale.
FAWKES, Richard Brian:
Keating.
FERMOR, Patrick Michael Leigh:
Toynbee; Winn (St Oswald).
FINNISTON, Sir (Harold) Montague:
McCance.
FISCHER, David Hackett:
Barraclough.
FONTEYN, Dame Margot (Dame Margot
Fonteyn de Arias):
Lopokova (Keynes, L.).
FOOT, Michael:
Cameron (M. J. W.).
FOOT, Michael Richard Daniell:
Montagu (E. E. S.); Sporborg.
FORD, Colin John:
Brandt.
FORD, Sir Edward William Spencer:
Adeane.
FORD, Sir (Richard) Brinsley:
Gwynne-Jones.
FRASER, Sir Ian James:
Warburg.
FRASER OF KILMORACK, (Richard) Michael
Fraser, Baron:
Redmayne.

GARDNER, John Linton:
Jacob.
GARLICK, Kenneth John:
Ellis.
GELDER, Michael Graham:
Hill.
GILMOUR, Sir Ian Hedworth John Little, Bart.:
Butler (Butler of Saffron Walden).
GOLDBERG, Sir Abraham:
McNee.
GOODBODY, Douglas Maurice:
Milford.
GOODY, John Rankine:
Fortes.
GORE, Frederick John Pym:
Fitton.
GORING, Marius:
Robson.
GOWING, Margaret Mary:
Hinton (Hinton of Bankside).

GREENHILL OF HARROW, Denis Arthur
Greenhill, Baron:
Gore-Booth; Scott (R. H.).
GREY, Dame Beryl:
Dolin.
GRIEVE, William Percival:
Lindsay.
GRUFFYDD, (Robert) Geraint:
Foster (I. L.).
GUEST, Ivor Forbes:
Rambert.
HALLS, Wilfred Douglas:
Lauwerys.
HALSEY, Albert Henry:
Roseveare.
HANAK, Harry:
Seton-Watson.
HARCOURT, Geoffrey Colin:
Robinson (J. V.); Sraffa.
HARRIS, James (Gordon Shute):
Hillier (H. G. K.).
HARRIS, Nicholas James Watson:
Hailwood.
HART, Jenifer Margaret:
Corbett Ashby.
HART-DAVIS, Sir Rupert Charles:
Fulford.
HARTSHORNE, John Norman:
Harland.
HARTMAN, Thomas Robert and COOPER,
Leo:
Marshall-Cornwall.
HASTINGS, Max Macdonald:
Fraser (Fraser of North Cape); Moorehead.
HAWORTH, Lionel:
Hooker.
HAYTER, Sir William Goodenough:
Cox.
HEATH, Edward Richard George:
Boyle (Boyle of Handsworth).
HEMINGFORD. See HERBERT
HENNIKER, John Patrick Edward Chandos
Henniker-Major, Baron:
Wates.
HEPWORTH, John Vord:
Dixey.
HERBERT, (Dennis) Nicholas, Baron
Hemingford:
Clark (W. D.).
HERRMANN, Frank:
Wilson.
HEUSTON, Robert Francis Vere:
Widgery.
HEYMAN, Jacques:
Baker (J. F.).
HILL, (John) Richard:
Macintyre.
HILLS, Richard ('Dick') Michael:
Morecambe.
HODGES, Sheila:
Cronin.

HODGSON, Sir Maurice Arthur Eric:
Beeching.
HODSON, Henry Vincent:
Caroe.
HOOSON, (Hugh) Emlyn Hooson, Baron:
Ashley.
HORNE, Michael Rex:
Kerensky.
HOSKINS, Percy Kellick:
Adams (J. B.).
HOYOS, Sir (Fabriciano) Alexander:
Adams (J. M. G. M.).
HUDD, Roy:
Sarony.
HUDDLESTON, (Ernest Urban) Trevor:
Scott (G. M.).
HUGHES, Glyn Tegai:
Lloyd (Richard Llewellyn).
HUGHES, (John) Trevor:
Russell (D. S.).
HULTON, Paul Hope:
Hermes.
HUTCHISON, Sidney Charles:
Gross.
HUXLEY, Thomas:
Polunin.

IMISON, Richard Geoffrey:
Gielgud.
INGRAMS, Richard Reid:
Cockburn.

JACOBS, David Lewis:
Fury; Monro.
JAGO, Thomas:
Joseph.
JAHODA, Marie:
Freud.
JAMES, Eric Arthur:
Robinson (J. A. T.).
JARRETT, Sir Clifford George:
Lang.
JEFFES, James Henry Elliston:
Richardson (F. D.).
JENKINS, Roy Harris:
Ormsby Gore (Harlech).
JOHNSON, Hugh Eric Allan:
Mitchell (J. A. H.).
JOHNSON, Nevil:
Vickers.
JOHNSON, Paul Bede:
Priestley.
JOLL, James Bysse:
Storry.
JONES, Emrys Lloyd:
Bennet.
JONES, Sir Ewart Ray Herbert:
Johnson (A. W.); Spinks.
JONES, (Henry) John (Franklin):
Carritt.

JONES, Sir James Duncan:
Sharp.
JONES, James Larkin ('Jack'):
Tewson.
JUDGE, Harry George:
Beloe.

KEATING, Henry Reymond Fitzwalter:
Marsh.
KEITH OF CASTLEACRE, Kenneth Alexander
Keith, Baron:
Strong.
KENYON, Sir George Henry:
Armitage.
KING, Roger John Benjamin:
Marrian.
KIRK-GREENE, Anthony Hamilton Millard:
Robertson.
KIRWAN, Sir (Archibald) Laurence (Patrick):
Caton-Thompson.
KITSON, Michael William Lely:
Blunt.
KNIGHT, Sir Arthur William:
Shonfield.
KNIGHT, Bernard Henry:
Simpson.
KORNBERG, Sir Hans Leo:
Krebs.
KOSINSKI, Dorothy Mary:
Cooper (A. W. D.).
KUHN, Heinrich Gerhard:
Jackson (D. A.).

LAING, Sir (John) Maurice:
Mitchell (G. W.).
LAMB, Richard Anthony:
Horrocks.
LAWSON JOHNSTON, Hugh de Beauchamp:
Pitman.
LAWTON, Sir Frederick Horace:
Humphreys.
LEACH, Sir Edmund Ronald:
Richards (A. I.).
LEE, Sir (Henry) Desmond (Pritchard):
Guthrie.
LEITCH, Sir George:
Wansbrough-Jones.
LEVIN, (Henry) Bernard:
West.
LEVY, Mervyn Montague:
Hillier (T. P.).
LEWIS, Dan:
Darlington.
LEYSER, Karl Joseph:
Wallace-Hadrill.
LING, Roger John:
Ward-Perkins.
LLOYD-JONES, Sir (Peter) Hugh (Jefferd):
Lobel.
LOCK, Stephen Penford:
Clegg.

LOVELL, Sir (Alfred Charles) Bernard:
Husband; Ryle.
LOW, Rachael:
Rotha.
LOWE, Ian Harlowe:
Cardew.
LUBBOCK, Roger John:
FitzGibbon.
LUCAS, Cyril Edward:
Hardy.
LUCAS, Percy Belgrave:
Bader.
LUSTY, Sir Robert Frith:
Swinnerton.
LYON, Peter Hazelip:
MacDonald.

MACANDREW, Hugh Holmes:
MacTaggart.
MACRAE, Norman Alastair Duncan:
Tyerman.
McCRUM, Michael William:
fforde.
MacKINNON, Donald MacKenzie:
Farmer.
McKITTERICK, David John:
Keynes.
MACLAGAN, Michael:
Lascelles; Norrington; Prittie.
MACLEAN, Alan Duart:
Johnson (Snow).
MACLURE, (John) Stuart:
Savage.
McVEAGH, Diana Mary:
Neel.
McWHIRTER, Norris Dewar:
Cecil (Exeter, and Burghley).
MADDEN, (Albert) Frederick:
Perham.
MALPAS, James Spencer:
Bodley Scott.
MANN, William Somervell:
Curzon.
MIALL, (Rowland) Leonard:
Bartlett; Hibberd; Mitchell (L. S. F.).
MIDWINTER, Eric Clare:
Cooper (T. F.); Holloway.
MILLER, Edward:
Postan.
MILLER, Peter Lamont:
Pringle (J. W. S.).
MOLLISON, Patrick Loudon:
Race.
MONTAGUE, Michael Jacob:
Shanks.
MORGAN, Kenneth Owen:
Greenwood (Greenwood of Rossendale).
MORIARTY, Denis Edmund Hugh:
Clifton-Taylor.
MORLEY, Sheridan Robert:
Niven; Richardson (R. D.).

MORRISON, Victor William:
Hastings.
MOTT, Sir Nevill Francis:
Heitler; Massey.
MURDOCH, Richard Bernard:
Askey.

NICHOLAS, (John Keiran) Barry (Moylan):
Lawson.
NICOLL, Douglas Robertson:
Cooper (J. E. S.).
NICHOLLS, Christine Stephanie:
Boot.
NICHOLSON, Marjorie:
Citrine.
NICHOLSON, Matthew Archibald:
Birley.
NICOLSON, Nigel:
Bagnold (Jones).
NIMMO, Derek:
Le Mesurier.
NORMAN, Richard Oswald Chandler:
Waters.
NORTH, Peter Machin:
Morris.
NUGENT OF GUILDFORD, George Richard
Hodges Nugent, Baron:
Amory.
NUTTING, Sir (Harold) Anthony, Bart.:
Trevelyan.

OATLEY, Sir Charles William:
Eastwood.
OGSTON, Alexander George:
Peters.
OSBORNE, Charles Thomas:
Boult.
OXBURY, Harold Frederick:
Hargrave; Lyons (W.); McEvoy; Mannin; Masters; Matthews; Schwartz; Williams.

PARTRIDGE, Francis Catherine:
Garnett.
PATTINSON, (William) Derek:
Scott (J. A. G.).
PAUL, Geoffrey David:
Janner.
PERRY, Samuel Victor:
Porter.
PHILLIPS, Charles Garrett:
Liddell.
PHIPPS, Simon Wilton:
Batten.
PICKERING, Sir Edward Davies:
Barnetson; Evans (T. M.).
PINKER, Robert Arthur:
Marshall.
PIPER, Sir David Towry:
Clark (K. M.); Waterhouse.
POPE, Michael Douglas:
Lewis.

POWELL, (John) Enoch:
Birch (Rhyl).
PRAGNELL, Anthony William:
Fraser (R. B.).
PROBERT, Henry Austin:
Cameron (Cameron of Balhousie); Pike.

RACE, Stephen Russell:
Kennedy.
RAWLINSON OF EWELL, Peter Anthony Grayson
Rawlinson, Baron:
Diplock.
READ, John:
Butler.
REASON, (Bruce), John:
Wakefield (Wakefield of Kendal).
RICHARDS, Denis George:
Harris; Lloyd (H. P.).
RICHARDS, Sir James Maude:
Gibberd.
RICHARDS, Sir Rex Edward:
Thompson (H. W.).
ROBERTSON, Edwin Hanton:
Phillips.
ROBSON, James Scott:
Davidson (L. S. P.).
ROE, Derek Arthur:
Oakley.
ROSE, Kenneth Vivian:
Alice (Athlone).
ROSEBLADE, James Edward:
Hall.
ROSKILL, Eustace Wentworth Roskill, Baron:
Baker (G. G.).
ROTHSCHILD, Miriam Louisa:
Foster (J. G.).
RUSSELL, Cosmo Rex Ivor:
Curtis.

ST JOHN, John Richard:
Frere.
SAUMAREZ SMITH, William Hanbury:
Paul.
SHACKLETON, Robert Millner:
Watson.
SHELBOURNE, Sir Philip:
Showering.
SHERRIN, Edward George:
Brahms.
SHOENBERG, David:
Kapitza:
SHUKMAN, Harold:
Schapiro.
SILBERSTON, (Zangwill) Aubrey:
Robbins.
SMALLMAN, Raymond Edward:
Raynor.
SMITH, David Burton:
Thomas.
STAFFORD, Godfrey Harry:
Adams (J. B.).

STEEL, Donald MacLennan Arklay:
Crawley; Wethered.
STEWART OF FULHAM, (Robert) Michael
(Maitland) Stewart, Baron:
Brown (George-Brown).
STRAWSON, John Michael:
Cunningham.
STREET, Sarah Caroline Jane:
Boulting.
STREETEN, Paul Patrick:
Balogh.
STRUDWICK, John Philip:
Chambers.
STUART, Alan:
Kendall.
SUMMERSON, Sir John Newenham:
Pevsner.
SUTCLIFFE, Peter Hoyle:
Gérin.
SWANN, Michael Meredith Swann, Baron:
Danielli.
SWANTON, Ernest William:
Fender.
SWIRE, John Anthony:
Keswick.
SYMONS, Julian Gustave:
Grigson.

TATTERSALL, Arthur:
Evans (Evans of Hungershall).
TAYLOR, Gerald:
Nemon.
THOMSON OF MONIFIETH, George Morgan
Thomson, Baron:
Garner.
THRUSH, Brian Arthur:
Sugden.
TIERNEY, (James) Neil:
Walton.
TIZARD, Barbara Patricia:
Pringle (M. L. K.).
TIZARD, Sir (John) Peter (Mills):
Sheldon.
TOBIN, John O'Hara:
Evans (D. G.).
TOOK, Barry:
Emery.
TOWNSHEND, Charles Jeremy Nigel:
Lyons (F. S. L.).
TRAPP, Joseph Burney:
Yates.
TRETHOWAN, Sir (James) Ian (Raley):
McKenzie.
TREWIN, John Courtenay:
Redgrave.
TULLY, (William) Mark:
Gandhi.

VAN OSS, (Adam) Oliver:
Schuster.

VARAH, (Edward) Chad:
Hampson.
VAUGHAN WILLIAMS, (Joan) Ursula (Penton):
Holst.

WAIN, John Barrington:
Empson; Graves; Larkin.
WALTON OF DETCHANT, John Nicholas, Baron:
McAlpine.
WALWYN, Kim Scott:
Ellerman (Bryher).
WATSON, Derek:
Searle.
WAUGH, Auberon Alexander:
Waugh.
WAYMARK, Peter Astley Grosvenor:
Dors.
WEBBER, (Mary) Teresa (Josephine):
Ker.
WEBSTER, Alan Brunskill:
Hunter.
WEEKS, John Reginald:
Llewelyn Davies.
WELCH, John Frederic:
Heffer.
WENDEN, David John:
Montagu (I. G. S.); Warner.
WEST, Nigel (Allason, Rupert William
Simon):
Mitchell (G. R.).
WHEATCROFT, Geoffrey:
Naipaul.
WHITELAW, William Stephen Ian Whitelaw,
Viscount:
Hare (Blakenham).
WHITEMAN, (Elizabeth) Anne (Osborn):
Aldridge.
WHITFIELD, Michael:
Russell (F. S.).
WIGODER, Basil Thomas Wigoder, Baron:
Byers.
WILKINS, Maurice Hugh Frederick:
Randall.
WILLIAMS, Sir David:
John.
WILLIAMS, Peter Orchard:
Boyd.
WILLIAMS, Trevor Illtyd:
Calder (Ritchie-Calder); Tetlow.
WILSON, Raymond:
Barnard.
WINDLESHAM, David James George Hennessy,
Baron:
Collins (N. R.).
WISE, Michael John:
Beaver.
WONTNER, Sir Hugh Walter Kingwell:
D'Oyly Carte.
WYATT OF WEEFORD, Woodrow Lyle Wyatt,
Baron:
Koestler; Wigg.

WYLIE, Shaun:
  Newman.

YOUNG OF DARTINGTON, Michael Young,
  Baron:
  Jackson (B. A.).

ZEC, Donald David:
  Zec.

ZIEGLER, Philip Sandeman:
  Lewin.

DICTIONARY

OF

NATIONAL BIOGRAPHY

(TWENTIETH CENTURY)

PERSONS WHO DIED 1981-1985

ABBOTT, ERIC SYMES (1906-1983), Anglican priest, was born 26 May 1906 at Nottingham, the younger son and second of three children of William Henry Abbott, schoolteacher, and his wife, Mary Symes, also a teacher. A Dame Agnes Mellor scholar at Nottingham High School, Abbott won a further scholarship to Jesus College, Cambridge (of which he became an honorary fellow in 1966), to read classics. A second class in part ii (1928) following a first in part i (1927) was probably to be accounted for by many hours on the river coxing the college 1st VIII, winning the award of a Trial cap, and being president of the university branch of the Student Christian Movement. In 1928 he crossed Jesus Lane to Westcott House and there came under the single most formative influence in his life: B. K. Cunningham. He obtained a third class in part i of the theological tripos in 1929.

Ordained in St Paul's Cathedral to serve at St John's, Smith Square (1930-2), Abbott was then requested by King's College, London, to become chaplain. So began the twenty-five years in which he was to give all his care and almost all of himself to students who were being prepared for ordination. He was also chaplain to Lincoln's Inn in 1935-6. In 1936, at the age of thirty, Abbott accepted the wardenship of Lincoln Theological College with a staff that included a future archbishop of Canterbury. Here he shared his Anglican perception of priesthood with his students, opening up for them a way of interior faith, prayer, and self-discipline appropriate for the demands of parochial pastoral ministry. He was also canon and prebendary of Lincoln Cathedral (1940-60).

World War II was coming to an end when Abbott was invited back to King's College as dean (1945-55). This unique position as dean of the whole college in harness with a lay principal, as well as being head of the department and faculty of theology and warden of the hostel for theological students in Vincent Square, enabled him to influence the post-war revival of the college as a whole, to build up the staff in theology, and to create a postgraduate one-year college at Warminster for the immediate pre-ordination spiritual and pastoral training (Lincoln style) of King's ordinands.

The total burden of this complex responsibility would have been more than enough; but to this public work was added private work of an exacting kind, a seemingly endless sequence of individuals drawn to him by his preaching and personality to seek spiritual guidance. This led to a vast correspondence (the 'apostolate of the post' as he called it). Even on holiday he would spend most mornings and evenings writing letters and the 'famous' postcards. It is not surprising that latent heart trouble which was to dog his later years was aggravated by his years in London. The invitation in 1956 to become the sixth warden of Keble College, Oxford (but the first to be elected by the fellows under a new set of statutes), undoubtedly prolonged his working life. He was made an honorary fellow in 1960.

A colleague writing about the bracing effect of Abbott's four years at Keble comments: 'His main achievement was to give the college confidence in itself. He consolidated the college both in its internal and external relations. Though not an academic he understood and encouraged academic excellence.' His swift intelligence, acute perception, and subtle wit, together with his accessibility, drew senior and junior members alike to consult him.

In 1959 came the invitation to be dean of Westminster. His vision of the Abbey was of 'a great church in which all questing men and women, irrespective of faith and race, would "see Jesus"'. The theme for the celebration in 1965-6 of the 900th anniversary of the founding of the Abbey was 'One People'. Events were planned to relate the Abbey to the needs and aspirations of the modern world. The dean was present at almost all.

Throughout his years as dean, Abbott influenced the form and preparation of services for weddings, memorials, and celebrations of national independence. Three royal weddings prepared and conducted by him further deepened his pastoral relationship with the royal family. His presence, prayer, and preaching, his care for art, music, literature, and drama enhanced the worship and work of the Abbey during fifteen years—years which for him included episodes of debilitating ill health. He became a freeman of the City of Westminster in 1973. In 1974 he accepted the need to retire, but continued his ministry to individuals from his home in Vincent Square.

In 1966 he was appointed KCVO and received an honorary DD from London University. He was unmarried. He died at Haslemere 6 June 1983. The stone that marks his grave in the Abbey carries the inscription: 'Friend and counsellor of many, he loved the Church of England, striving to make this House of Kings a place of pilgrimage and prayer for all peoples. *Pastor Pastorum.*'

[*Eric Symes Abbott: A Portrait*, 1984 (privately printed, obtainable from Westminster Abbey library); private information; personal knowledge.]   SYDNEY EVANS

ABRAHAMS, DORIS CAROLINE (1901–1982), novelist, critic, journalist, and songwriter. [See BRAHMS, CARYL.]

ADAMS, SIR JOHN BERTRAM (1920–1984), engineer, was born 24 May 1920 in Kingston, Surrey, the only son and younger child of John Albert Adams, who had worked for Paquin's, a high-class London couturier, but who had lost his job because he was badly gassed in World War I, and his wife, Emily Searles. Between 1931 and 1936 he attended Eltham College but because of financial hardship in the family he left early and went to work at the Siemens laboratories in Woolwich. He continued his studies at the South-East London Technical Institute and obtained a Higher National Certificate in 1939.

During World War II he worked at the Telecommunications Research Establishment, first at Swanage and then at Malvern. His particular responsibility was the development of microwave radar systems. It was here that his outstanding engineering skills first came to be recognized. In 1945 he moved to the Atomic Energy Research Establishment at Harwell where his exceptional ability again became quickly evident and in 1953 the director, Sir John Cockcroft [q.v.], arranged for him to go to the newly formed high energy physics laboratory, CERN, in Geneva. He went as a junior member of the team formed to build the world's largest particle accelerator and before

he left in 1961 he had served as director-general of this large international laboratory for a year.

Adams left CERN to take charge of Britain's controlled thermonuclear programme at Cockcroft's request. This was the second laboratory that Adams built, essentially from a green field at Culham. Once again he created a splendidly effective research organization. It is somewhat surprising, however, that Adams did little to develop the vision of closer European collaboration in controlled fusion until 1974 when he accepted an invitation to join the Joint European Torus (JET) Scientific and Technical Committee. His support of the programme then went a long way towards securing approval for the project. The director appointed to take charge of the JET project was Hans-Otto Wüster who had worked with Adams during his second period at CERN.

Adams was appointed controller at the Ministry of Technology in the first Labour government of Harold Wilson (later Lord Wilson of Rievaulx) although he remained director of Culham. In 1966 he became member for research of the UK Atomic Energy Authority. Neither of these two appointments suited his talents particularly well and in April 1969 he chose to return to CERN to take charge of a project whose future, to say the least, was very uncertain. The CERN proposal was to build a new giant accelerator on a completely new site somewhere in Europe. The British government initially refused to join this project. In less than two years Adams transformed the plan so that the accelerator could be built on the CERN site at a fraction of the originally estimated cost. Britain then joined the project and it went ahead. The machine was completed in just over five years from approval and was built within the originally estimated cost. On the completion of this project in 1975 Adams was named executive director-general of a unified CERN laboratory. The period from 1976 to 1981 was one of remarkable accelerator development at CERN. When Adams's term of office as director-general ended he remained at CERN but ceased to play any significant role in its affairs. His experience and knowledge were, however, much sought after outside CERN both in Europe and in the United States of America.

Adams's achievements were recognized through the award of many honours and distinctions. They include the following: honorary doctorates of Geneva (1960), Birmingham (1961), Surrey (1966), Strathclyde (1978), and Milan (1980); Röntgen prize of the University of Giessen (1960); Duddell medal of the Physical Society (1961); Royal Society Leverhulme and Royal medals (1972 and 1977); and Faraday medal (Institution of Electrical Engineers,

1977). He was elected FRS in 1963 and a foreign member of the USSR Academy of Sciences in 1982. He was appointed CMG (1962) and knighted (1981). He was a fellow of Wolfson College, Oxford (1966), the Manchester College of Technology, the Institution of Electrical Engineers, and the Institute of Physics.

In 1943 Adams married Renie, daughter of Joseph Warburton, engineer. They had one son and two daughters. Adams died at his home near Geneva 4 March 1984, near the mountains where he loved to walk and ski, well before his active career had come to an end.

[G. H. Stafford in *Biographical Memoirs of Fellows of the Royal Society*, vol. xxxii, 1986; private information; personal knowledge.]

G. H. STAFFORD

ADAMS, JOHN BODKIN (1899–1983), medical practitioner, was born 21 January 1899 in Randalstown, county Antrim, the elder son (there were no daughters) of Samuel Adams JP, watch maker, and his wife, Ellen Bodkin, formerly of Desertmartin, county Tyrone. The younger son was born in 1903 and died of pneumonia in 1916. Shortly after Adams's birth the family moved to Ballinderry Bridge, county Tyrone, where he attended a Methodist day school. The strictly religious family worshipped regularly at a Plymouth Brethren meeting place where the father frequently preached. In 1911 the family moved to Coleraine mainly to permit the elder son to study at a local academy. He qualified in 1917 for Queen's College, Belfast, where he studied anatomy and embryology. Adams qualified MB, BCH, and BAO in 1921. He was an ophthalmic house surgeon and casualty officer in Bristol Royal Infirmary. An advertisement for a 'Christian doctor assistant' subsequently led him to join an Eastbourne general practice. He paid £2,000 for the partnership, which he raised by taking out life insurance and using it to secure a bank overdraft. By relieving his partners of all night calls Adams was soon able to rent a modest house in Upperton Road and invite his mother (who died in 1943) to join him. The family reunion brought a revival of the religious pattern—family prayers in the home and the advent of a Young Crusaders' bible class.

When his partners retired Adams was able to slot into the 'good life' of Eastbourne. He looked for a more suitable residence and in 1930 for £5,000 he acquired Kent Lodge, a Victorian villa later to be valued in the doctor's estate at £100,000.

The charm and ebullience of the soft-spoken, loquacious, teetotal bachelor quickly led him into the homes of the wealthy. By 1955 he had established himself in hunting and fishing circles. Yet he did not ignore his poor patients. On the other hand signs of the doctor's avaricious nature were beginning to appear, and rumours began. In 1935 Mrs Matilda Whitton, a seventy-two-year-old patient, left Adams £3,000. Relatives disputed the will but it was upheld by the courts. The rumours were revived between 1944 and 1955 during which period Adams received fourteen legacies totalling £21,600 from octogenarian patients.

In July 1956 fifty-year-old Mrs Gertrude Hullett, another of Adams's patients, died. Heart-broken at the recent death of her husband, Mrs Hullett had often spoken to her solicitor and friends of her intention to take her own life. Adams was not present when she died and an autopsy established that she had swallowed a massive dose of barbiturates. The inquest jury found that she had committed suicide 'of her own free will', but in winding up the proceedings the coroner announced that the chief constable 'had invoked the aid of Scotland Yard to investigate certain deaths in the neighbourhood'. Mrs Hullett's will disclosed that she had left Adams a Rolls-Royce car and had previously given him a cheque for £1,000 which he had 'specially cleared'.

On 24 November Adams was arrested on a series of minor charges. Soon a further charge was added: 'That he did in November 1950 feloniously and with malice aforethought murder Edith Alice Morrell.' Refused bail, Adams spent the next 111 days in prison. Meanwhile, the bodies of two other patients were exhumed but only one of them was in good enough condition to decide the cause of death, cerebral thrombosis, which was that given by Adams on the death certificate. At the committal proceedings the Crown maintained it was essential that evidence regarding the deaths of Mr and Mrs Hullett should be given and despite defence objections the magistrates agreed—a decision which was later (1967) to lead to a change in the law.

Physically Adams seemed strong enough to withstand the trial but his facial appearance was against him. In good humour and smiling he resembled a genial Mr Pickwick but in sombre, grim moments he tended to appear quite capable of the crimes soon to be laid at his door.

The trial was opened at the Old Bailey by the attorney-general, Sir Reginald Manningham-Buller (later Viscount Dilhorne, q.v.). He first called four nurses who gave their version from memory of the treatment of Mrs Morrell six years earlier. Then the defence, conducted by (Sir) F. Geoffrey Lawrence [q.v.], produced a trump card: the nurses' actual notebooks which gave a very different version. The seams of the Crown's case had started to come apart, and worse was to follow.

The chief medical witness, Arthur Henry Douthwaite, senior physician at Guy's, revised his theories in the witness box after admitting he had not seen the reports of three other doctors in Cheshire who had started Mrs Morrell on a course of heroin and morphia before she came under the care of Adams. Another expert, in answer to the judge, Patrick (later Lord) Devlin, said 'I do not think it is possible absolutely to rule out a sudden catastrophic intervention by some natural cause.' The defence, perhaps wisely, decided not to put the talkative Adams into the witness box. In his summing-up Devlin told the jury that the case for the defence was manifestly a strong one. The jury found Adams not guilty.

On 26 July Adams appeared at Lewes assizes to answer the minor charges which remained on the calendar. He pleaded guilty and was fined £2,400. As a consequence he was struck off the medical register but four years later his name was restored and he returned to a flock of private patients. He died 5 July 1983 in Eastbourne General Hospital after breaking a leg. He left an estate of £402,970, a sum which included substantial libel damages and the sale of Kent Lodge.

[Patrick Devlin, *Easing the Passing*, 1985; Rodney Hallworth and Mark Williams, *Where There's a Will . . .*, 1983; Percy Hoskins, *Two Were Acquitted*, 1984; Sybille Bedford, *The Best We Can Do*, 1958; private information; personal knowledge.]

PERCY HOSKINS

ADAMS, JOHN MICHAEL GEOFFREY MANNINGHAM ('TOM') (1931–1985), prime minister of Barbados, was born 24 September 1931 in St Michael, Barbados, the only child of (Sir) Grantley Herbert Adams [q.v.], later prime minister of the Federation of the West Indies, and his wife, Grace, only daughter of Alexander Thorne, shipping agent, and his wife, Millicent. He was educated at the Ursuline Convent, Harrison College in Barbados, and Magdalen College, Oxford, to which he was admitted after winning the Barbados scholarship in 1951. He obtained third class honours in philosophy, politics, and economics in 1954.

In 1954 he began to study law at Gray's Inn; he was called to the bar in 1959. While a law student, he worked as a producer in the Overseas Service of the BBC where he met Genevieve Turner, who was employed as a secretary. They were married in 1962. Genevieve was the daughter of Philip Turner, CBE, principal assistant solicitor (later solicitor) to the Post Office. There were two sons of the marriage.

Adams worked for six months in the Gray's Inn chambers of (Sir) Dingle Foot [q.v.] and then returned to Barbados in January 1963. After settling his family and setting himself up as a barrister in chambers he began his political career. He was secretary of the Barbados Labour Party from 1965 to 1969. He won a seat in the House in 1966 and was leader of the opposition from 1971 to 1976. He became prime minister and finance minister of Barbados in September 1976 and remained in that office until his death in 1985.

He had a formidable task before him when he entered public life. He had to function under the shadow of his illustrious father's name and, in addition, he was faced with a redoubtable opponent, Errol Barrow, who led the Democratic Labour Party and was then prime minister of Barbados. But at the BBC he had developed the talents that made him the island's most skilful communicator. His oratory was such that he attracted large audiences to meetings of the Assembly. And, while he excelled in parliamentary debate, he soon showed he was no mean opponent on the political platform. In addition, what helped him to win three by-elections and two general elections was his flair for organization.

One of his ambitions was to bring to reality his father's dream of a West Indies, united and self-governing. He was actively involved in the Caribbean Community which sought to attain some of the late Federation's objectives. He was in every sense a Caribbean man. He promptly sent military assistance to St Vincent when the premier, Milton Cato, asked for help in 1979 against a revolt in the Grenadines. And when the premier of Grenada, Maurice Bishop, was killed in 1983 and that island was threatened with a Marxist take-over, Adams joined with the leaders of the Organization of Eastern Caribbean States and persuaded President Ronald Reagan of the USA to assist them in a rescue mission. Adams regretted the decline of British influence.

Adams was tall and handsome and possessed the common touch. He was fond of gardening and philately, bridge and poker. He loved dancing and merry-making, even to the extent of embarrassing his entourage. He drove himself hard both at work and play and slept only four hours a night. In due course his many-sided activities began to take a toll on his health. He made light of the first signs that should have warned him that his heart was beginning to feel the strain. Later he consulted a Harley Street specialist, who advised open-heart surgery late in 1984. Adams decided to postpone any such operation until he had completed pressing business abroad that would help him in the unceasing tasks of social reform and financial stability. However, he collapsed on the afternoon of 11 March 1985, at Ilaro Court, the official residence of the prime minister, and a medical certificate confirmed that his death that day was due to cardiac failure.

[F. A. Hoyos, *Tom Adams*, 1988; private information; personal knowledge.]

ALEXANDER HOYOS

ADAMS, MARY GRACE AGNES (1898-1984), pioneer in broadcasting and television, was born 10 March 1898 at Well House Farm, Hermitage, Berkshire, the elder of two daughters (her sister died aged three) and second of four children of Edward Bloxham Campin, farmer, of Hermitage, and his wife, Catherine Elizabeth Mary, daughter of Edwin Gunter, farmer, of Alveston. Mary's mother brought up the three surviving children in Penarth, Wales, under conditions of great hardship; Mary's father died of consumption in 1910. A scholarship to Godolphin School, Salisbury, led to University College, Cardiff, where Mary Adams gained first class honours in botany (1921). She then became a research scholar and Bathurst student at Newnham College, Cambridge (1921-5), and published papers on cytology.

She held tutorial and lectureship posts at Cambridge for extramural and board of Civil Service studies (1925-30). In 1928 she broadcast a series (published as *Six Talks on Heredity*, 1929), which proved a transition point. Gripped by the educational possibilities of the BBC, she joined its staff in 1930, as adult education officer. Her inspiring flair for teaching, especially the young, and organizational skills had a new outlet. She stayed in sound broadcasting until 1936: under the exactitudes of Sir John (later Lord) Reith [q.v.] she combined information with informality in obligatorily scripted programmes.

She was appointed to the newly established television service in 1936, becoming the first woman television producer. An atmosphere of history-in-the-making, experimentation, and excitement prevailed, though the budget was miniscule. Mary Adams was in charge of education, political material, talks, and culture. She persuaded the eminent—for example, C. E. M. Joad, (Sir) S. Gordon Russell, (Sir) Julian Huxley, and (Sir) John Betjeman [qq.v.]—to appear on this alarming new medium, but with the outbreak of war in 1939 the service was shut down.

From 1939 to 1941 Mary Adams was director of Home Intelligence in the Ministry of Information and the years 1942-5 were spent working in North American Service broadcasting, where she produced morale-boosting programmes—for example, *Transatlantic Quiz*.

Returning to BBC TV in 1946, she was ever breaking new ground, bubbling with ideas, enlisting the services of the distinguished in arts and science for programmes, and cutting red tape. She was head of television talks (1948-54) and assistant to the controller of TV programmes (1954-8). She stimulated others to achieve, while she promoted and produced many programmes (always with a purpose): cooking; gardening; art; intellectual quizzes—for example, *Animal, Vegetable, Mineral?* (1952-9, 1971); material for children, such as *Muffin the Mule* (1946-55) and *Andy Pandy* (1950-69, 1970-6); science, for example *Eye on Research* (1957-61); and world celebrations, such as *The Restless Sphere* with the Duke of Edinburgh in International Geophysical Year (1957). First to make medicine accessible to the public, she shook the medical profession out of its secretive complacency with such series as *A Matter of Life and Death* (1949-51, 1952), *Matters of Medicine* (1952-3, 1960), *The Hurt Mind* (1957), and *Your Life in Their Hands* (1958-64). There were many repeats and revivals of her programmes.

When she retired in 1958 she began a new career. The Consumers' Association, which had been established in 1957, resulted from the deliberations of Mary Adams and Julian Huxley in 1937. She became its deputy chairman (1958-70), and induced a nervous *Which?* magazine to produce the first comparative tests of contraceptives and a reluctant BBC to produce the first consumer programmes giving brand ratings. Simultaneously, she advised the BBC (and became, contrastingly, a member of the Independent Television Authority, 1965-70), wrote for *Punch*, and played far-sighted roles in organizations concerned with Anglo-Chinese understanding, the unmarried mother and her child, British rail design, BMA planning, mental health, womens' groups, telephone users, and eugenics.

With science, as with the arts, she made people sit up, with ideas that were daring, in support of what was new, often with a strong feminist slant, and expressed with animation, even provocation. Spurning notes for public speaking, she arrived well informed, sensed the atmosphere, then, in a style that teased and delighted, injected serious and original proposals. She was a creative persuader at work, yet paradoxically indecisive at home. She was appointed OBE (1953), an associate of Newnham College, Cambridge (1956-69), and a fellow of University College, Cardiff (1983).

She married on 23 February 1925 (Samuel) Vyvyan (Trerice) Adams, the younger son of Samuel Trerice Adams, canon of Ely and rural dean of Cambridge. Her husband was Conservative MP for West Leeds (1931-45), an early determined anti-Nazi, and a radical reformer. Adopted for the Conservative seat of Darwen in 1951, and earmarked for cabinet rank, his untimely drowning on 13 August of that year denied this political resurgence. There was one daughter of the marriage, born in 1936.

In appearance Mary Adams was small, bird-like, always well dressed, with bright blue eyes and, in public, a knowing confidence. Her character, though full of contradictions, was magnetic. She was a socialist, a romantic communist, and could charm with her charisma, spontaneity, and quick informed intelligence. She was a fervent atheist and advocate of humanism and common sense, accepting her stance without subjecting it to analysis. These qualities ensured she was the centre of attention in a social setting, and she involved herself with all the right people. Yet a streak within compelled her to dread, then court, loneliness and it decreed that her marriage was to be less than happy: she could not live in harmony with herself, nor be truly supportive to those close to her who needed help. She had an unexpected, dry put-down humour. A stroke in 1980 affected her memory and she died in University College Hospital, London, 15 May 1984.

[*The Times*, 18 May 1984; information from Nora Wood, Cyril Campin, Ida Hughes-Stanton, and Andrew Armit; personal knowledge.]   SALLY ADAMS

ADEANE, MICHAEL EDWARD, BARON ADEANE (1910–1984), private secretary to Queen Elizabeth II, was born 30 September 1910 in London, the only son of Captain Henry Robert Augustus Adeane, of the Coldstream Guards, who was killed in action in 1914, and his wife, Victoria Eugenie, daughter of Arthur John Bigge (later Lord Stamfordham, q.v.). He was educated at Eton and Magdalene College, Cambridge, where he achieved a second class (first division) in part i of the history tripos (1930) and a first class (second division) in part ii (1931). The college made him an honorary fellow in 1971.

He then joined the Coldstream Guards and from 1934 to 1936 was aide-de-camp to two successive governors-general of Canada, the Earl of Bessborough and Lord Tweedsmuir [qq.v.]. In 1937 he was appointed equerry and assistant private secretary to King George VI and accompanied the King and Queen on their visit to Canada and the United States in the summer of 1939. On the outbreak of war he rejoined his regiment, being promoted to major in 1941. From 1942 to 1943 he was a member of the joint staff mission in Washington with the acting rank of lieutenant-colonel. From 1943 to 1945 he served with the 5th battalion Coldstream Guards as company commander and second-in-command. In the battle of Normandy he had to take over command of the battalion, was wounded in the stomach, and mentioned in dispatches. In 1945 he returned to Buckingham Palace and for the remainder of the reign served as assistant private secretary to King George VI. In 1947 he was a member of the royal party on their visit to South Africa and in 1952, having been seconded to the staff of Princess Elizabeth, was with her in Kenya at the time of her father's death. The new Queen decreed that he should continue as one of her assistant private secretaries, which he did until, on the retirement of Sir Alan Lascelles [q.v.] on 1 January 1954, he became her principal private secretary, retaining this office until his retirement in 1972. He was also keeper of the Queen's archives (1953–72).

The three main duties of the Queen's private secretary are to be the link between the monarch and her ministers, especially her prime ministers, to make the arrangements for her public engagements and for the numerous speeches which she is called upon to make, and to deal with her massive correspondence.

The first twenty years of the reign were marked by demands for an ever-expanding programme of public engagements at home and abroad, and by an intrusive, and not always charitable, scrutiny of the Queen and her family by the media. It was largely due to Adeane that the monarchy was able to adjust to these pressures, while retaining its essential dignity and mystery. Although some judged his advice to be unduly cautious and the speeches which he drafted for the Queen to be lacking in imagination, he was able to avoid the controversies to which a more adventurous private secretary might have exposed a constitutional sovereign. Adeane had to deal with six British prime ministers and with many more from Commonwealth countries as well as with their governors-general. He treated all with equal courtesy and respect and was invariably well briefed on their personalities and policies. These qualities showed to particular advantage during overseas tours. During these years the Queen visited almost every country in the Commonwealth, several of them more than once, and many foreign countries including most of those in western Europe. Adeane was responsible for the arrangements for all these visits. He also performed a notable service to the royal family by his compelling evidence to the select committee of the House of Commons on the civil list in 1971 outlining the Queen's workload and commitments. This led later to the civil list being reviewed annually and submitted to Parliament in the same way as a departmental budget.

Professor Harold Laski [q.v.] wrote in 1942: 'The Secretary to the Monarch occupies to the

Crown much the same position that the Crown itself in our system occupies to the Government; he must advise and encourage and warn . . . The Royal Secretary walks on a tight-rope below which he is never unaware that an abyss is yawning . . . A bad Private Secretary, who was rash or indiscreet or untrustworthy might easily make the system of constitutional monarchy unworkable' (*Fortnightly Review*, vol. clviii, July–Dec. 1942). By these criteria, Adeane was highly successful.

In his style of work, he closely resembled his grandfather, Lord Stamfordham, of whom it was said (*DNB 1931–40*) that he was 'a man of persistent industry, making it his practice to finish the day's work within the day'. Like him too, he was 'regarded by his colleagues with a love which perhaps never wholly cast out fear'. His wisdom, sense of humour, and discretion endeared him to most other members of the royal family and their households, to whom his advice was always available. But he could also be severe on any lack of tact or competence. He was a popular member of several dining clubs, and although he enjoyed conversation, he would never gossip about the royal family. A totally concentrated listener, he rarely came away from such occasions without useful information which he stored in a capacious and accurate memory.

Adeane was at heart a countryman, a fine but unassuming shot and a skilful fisherman. He was an enthusiastic gardener, on the roof of his house in Windsor Castle and in the small gardens of the house in Chelsea and the cottage in Aberdeenshire which he made his homes in his retirement. He also painted in water-colours and was a voracious reader of biography, history, and Victorian novels, especially those of Anthony Trollope [q.v.]. Modest, even spartan in his personal life, he nevertheless appreciated good food and wine and liked to smoke a good cigar. He was entirely free from social, religious, and racial prejudice.

On his retirement in 1972 he acquired several directorships, including those of Phoenix Assurance, the Diners Club, the Banque Belge, and the Royal Bank of Canada. He was also appointed chairman of the Royal Commission on Historical Monuments and served as the Queen's representative on the board of the British Library. He was a fellow of the Society of Antiquaries and a governor of Wellington College. As a member of the House of Lords he sat on the cross-benches but spoke rarely.

His honours came in a steady progression after the war until his retirement. He was appointed MVO in 1946, CB in 1947, KCVO in 1951, KCB in 1955, GCVO in 1962, GCB in 1968, and, on his retirement in 1972, received the Royal Victorian Chain. He was made a privy councillor in 1953 and a life peer in 1972.

In 1939 he married Helen, elder daughter of Richard Chetwynd-Stapleton, stockbroker, of Headlands, Berkhamsted. They had one daughter, who died in 1953, and one son, Edward, who became private secretary to the Prince of Wales (1979–85). Adeane died in Aberdeen 30 April 1984 of heart failure after enjoying two days' fishing in the Dee. It was characteristic of his modesty that, by his special request, no thanksgiving service was held in his memory. A portrait of him by David Poole, commissioned by the Queen, hangs at Buckingham Palace.

[*The Times*, 2 May 1984; personal knowledge.]   EDWARD FORD

ALDRIDGE, JOHN ARTHUR MALCOLM (1905–1983), painter and gardener, was born near Woolwich 26 July 1905, the second of three sons (there were no daughters) of Major John Barttelot Aldridge, DSO, of the Royal Field Artillery, and his wife, Margaret Jessica Goddard, the daughter of a Leicester architect. His father died when he was three, and his mother subsequently married again and had a daughter. He won scholarships both to Uppingham and Corpus Christi College, Oxford, where he obtained a second in classical honour moderations (1926) and a third in *literae humaniores* (1928). At Oxford his diverse activities included both rugby and the Opera Club, but his interest in painting was already strong; with the intention of teaching himself to paint, he moved to Hammersmith in 1928. His work came to the attention of Ben Nicholson [q.v.], who in 1930 invited him to exhibit in London with the Seven and Five Society. In 1933 he held his first one-man show, and a year later some of his paintings were chosen for the Venice Biennale.

During the 1930s Aldridge was closely associated with a group which included Robert Graves [q.v.], (John) Norman Cameron, Laura Riding, Len Lye, and Lucie Brown, whom he was later to marry. Both Deyá, Majorca, which Graves made his home, and the Place House, Great Bardfield, Essex, bought by Aldridge in 1933, were centres for vigorous literary and artistic activity; these included productions of the Seizin Press. Aldridge notably illustrated Laura Riding's *The Life of the Dead* (1934), and also designed the dust-jackets for most of Graves's novels. Towards the end of the decade he began to design wallpapers, as did Edward Bawden, then also living in Great Bardfield. In 1939 Aldridge's name was included among those artists whom the third Viscount Esher proposed should be exempted from military service.

Between 1941 and 1945 he served in the Royal Army Service Corps and, after being commissioned, in the Intelligence Corps. During his time in North Africa and Italy, he made many drawings and water-colours, some of which are a valuable record of army and civilian life in wartime. He did distinguished work as an interpreter of air photographs.

After demobilization, he returned to Great Bardfield and resumed both his painting in oils and the care of his garden. In 1949 (Sir) William Coldstream, recently appointed head of the Slade School of Fine Art, invited him to become a part-time member of staff at the School. For the next twenty-one years, until his retirement in 1970, and although without any formal art school training himself, Aldridge proved a successful and much-liked teacher. In 1948 he first exhibited at the Royal Academy; he was elected ARA in 1954, and RA in 1963. He played an active part in the Academy's affairs, serving several times on the hanging committee. Locally, he was one of the principal organizers of an interesting experiment in the 1950s, when a number of painters and designers living in or near Great Bardfield exhibited their work in their own houses.

As a painter, Aldridge was primarily interested in landscape and vernacular buildings, particularly in Essex, abroad in Italy and France, and in Majorca, on visits to Robert Graves. He also produced some noteworthy still-life paintings, and a few portraits, including one of Robert Graves, which was later acquired by the National Portrait Gallery. His early work made a considerable impression; it had a freshness and directness, and the description 'stark and wiry' suits it well. This economical formality began to change during the 1930s and, after the war, he became more concerned with detailed, accurate representation, perhaps in part because he had little liking for the then fashionable abstract and decorative painting. His interest in gardening and painting ran closely together; as he wrote in 1959, he saw the development of a garden as 'a process which combines selection, precision and an understanding of the nature and possibilities of the materials in a way which is analogous to painting'. His friendship with the classical architect Raymond Erith [q.v.] was an important influence on him; a common passion for gardening drew him to near neighbours such as Sir Cedric Morris and John Nash [q.v.].

Aldridge was a gentle, friendly man, of great charm, with a scholarly interest in painting, architecture, and gardening. In appearance he was tall and spare. He was a generous host in his beautiful Tudor house, in itself a work of art which owed much to the taste of his first wife, Lucie. It was there that he died 3 May 1983. He was married twice; in 1940, to (Cecilia) Lucie (Leeds) Brown, daughter of Isaac Eb-enezer Leeds Saunders, farmer, of Clayhithe, near Cambridge, and in 1970, after the dissolution of his first marriage the same year, to Margareta ('Gretl') Anna Maria Cameron, widow of the poet (John) Norman Cameron and daughter of Dr Friederich Viktor Bajardi, Hofrat at Graz. She died early in 1983. There were no children of either marriage.

[Great Bardfield exhibition catalogue, 1959; *Times Literary Supplement*, 2 January 1987, pp. 12-13; private information; personal knowledge.]   ANNE WHITEMAN

ALICE MARY VICTORIA AUGUSTA PAULINE (1883-1981), Princess of Great Britain and Ireland and Countess of Athlone, was born at Windsor Castle 25 February 1883, the elder child and only daughter of Prince Leopold George Duncan Albert, first Duke of Albany [q.v.], Queen Victoria's fourth and youngest son, and his wife, Princess Helene Friederike Auguste of Waldeck-Pyrmont. Her father died of haemophilia little more than a year after her birth and she was brought up by her mother at Claremont House, near Esher.

In 1904 Princess Alice was married to the younger brother of the future Queen Mary, Prince Alexander of Teck, whose full names were Alexander Augustus Frederick William Alfred George. A serving officer in the British Army, he abandoned his German title in 1917, adopted the family name of Cambridge, and was created Earl of Athlone [q.v.].

Princess Alice's lifelong vivacity concealed anxiety and sorrow. Her brother Prince Charles Edward, second Duke of Albany, had at the age of fifteen been taken away from Eton to be brought up in Germany as heir to his uncle, the reigning Duke of Coburg. He entered on his unfortunate inheritance in 1900, became a general in the German Army, fought for his adopted country during Word War I, and was deposed in 1918. In the following year he was stripped of his British dukedom of Albany and later became a fervent supporter of the Nazi regime. These events naturally distressed Princess Alice, whose heart was torn between patriotism and affection for an only brother.

As wife of the governor-general of South Africa in 1923-31 and of Canada in 1940-6, Princess Alice proved a memorable proconsul in her own right: graceful, sympathetic, and perpetually amused. But tragedy struck again in 1928. Her son, Rupert Alexander George Augustus, Viscount Trematon (born 1907), had inherited the haemophilia of his grandfather, Prince Leopold. He died of injuries in a motoring accident from which others might have recovered. A younger son, Maurice Francis George, had died in 1910 before he was six months old. There was also one daughter of the marriage, Lady May Helen Emma Cambridge (born 1906), who married a soldier, (Sir) Henry Abel Smith, governor of Queensland 1958-66.

From marriage until 1923, the Princess and her husband lived in Henry III tower, Windsor Castle. Later they had an apartment in Kensington Palace with a country place at Brantridge Park, Sussex. Lord Athlone's death in 1957 dissolved a partnership of more than half a century but did not deflect his widow from a way of life both industrious and convivial. Well into her tenth decade, she remained an active patron of many institutions; the Royal School of Needlework and the Women's Transport Service (FANY) earned her particular interest. Princess Alice's leisure hours were no less productive, and she would continue to knit even while walking up a mountain at Balmoral.

A sense of adventure as well as of thrift led Princess Alice to travel about London by bus. For many years she similarly crossed the Atlantic each winter in a banana boat, combining her duties as chancellor (1950-71) of the University of the West Indies with a holiday in Jamaica. Several times she revisited South Africa and made the long journey to stay with her son-in-law and daughter in Australia.

Although below middle height, Princess Alice had a patrician presence, with aquiline features, observant eyes, and a stylish sense of fashion. She was an engaging talker and needed little prompting to recall life at Windsor under Queen Victoria, whose unsuspected laughter still rang in the ears of her last surviving granddaughter almost a century later. Not all her memories were benign. She never forgave W. E. Gladstone for having cheated her family of a whole year's civil list when her father died a few days before the start of the fiscal year; or Sir Winston Churchill for filling her drawing-room with pungent cigar smoke during the Quebec conference of 1943. Some of these recollections she confided to an entertaining volume of memoirs, *For My Grandchildren* (1966). Her views on public affairs were emphatic and not always predictable. When a colonial governor of radical bent expressed his belief in universal suffrage, she replied: 'Foot, I have never heard such balderdash in my life.' Yet she was the first member of the royal family publicly to advocate birth control; and like her cousin King George V, did not harbour a trace of colour prejudice.

Princess Alice, the last surviving member of the Royal Order of Victoria and Albert, was also appointed GBE in 1937 and GCVO in 1948. She had many honorary degrees. She died 3 January 1981 at Kensington Palace in her ninety-eighth year. After a funeral service in St George's chapel, Windsor, her remains were buried at Frogmore.

[Princess Alice, *For My Grandchildren*, 1966; Theo Aronson, *Princess Alice*, 1981; personal knowledge.]   KENNETH ROSE

AMOROSO, EMMANUEL CIPRIAN (1901-1982), veterinary embryologist and endocrinologist, was born in Port of Spain, Trinidad, 16 September 1901, the third child in the family of eight sons (one of them adopted) and three daughters of Thomas Amoroso, bookkeeper and later estate owner, and his wife, Juliana Centeno. He was educated at St Mary's College in Trinidad. He overcame both prejudice and poverty to obtain an education and finally received medical training in Dublin where he had a distinguished academic career at University College. At first he supported himself by selling newspapers outside the main railway station and later by teaching anatomy. He won several prizes and in the final medical examination he achieved the highest marks ever attained. He graduated B.Sc. in 1926 and MB, B.Ch., BAO (Dubl.) in 1929. Yet he found time to become a useful amateur heavyweight boxer. In 1930 he was awarded a travelling studentship which enabled him to undertake postgraduate studies in Berlin and Freiburg. To ensure that he could read scientific publications in the original text and to avoid any bias introduced by another translator, 'Amo', as he was by then universally known, began to learn all the European languages. He was to become a highly competent linguist. After his period of study in Germany he moved to University College, London, where his interest in experimental embryology was excited and he obtained his Ph.D. degree, the first of a bewildering array of higher degrees and fellowships from many parts of the world.

In 1934 he joined the Royal Veterinary College as a senior assistant in charge of histology and embryology. In 1948 he was appointed to the chair of physiology at the college and for a while also acted as professor of anatomy. In 1950 he became professor of veterinary physiology at London University. In 1957 he was elected FRS; he also held fellowships of the Royal College of Physicians (1966), Royal College of Surgeons (1960), Royal College of Obstetricians and Gynaecologists (1965), and Royal College of Pathologists (1973). He became an honorary associate of the Royal College of Veterinary Surgeons (1959) and of the British Veterinary Association. In 1964 he was awarded the MD of the National University of Ireland. He held numerous honorary doctorates.

Amoroso's interests and research ranged over an extraordinarily diverse group of subjects including cancer and the first studies of the seried veins of the neck of the giraffe. However, his greatest contribution to knowledge lay in the field of reproductive biology, notably in studies of the structure and function of the placenta. He published many papers and stimulating reviews but much of his thought, philosophy, and work was encapsulated in his outstanding chapter in volume two of *Marshall's Physiology of Reproduction* (3rd edn., ed. A. S. Parkes, 1952). He was awarded the Mary Marshall medal of the Society of Fertility and the Carl Hartman medal of the Society for the Study of Reproduction. It was, however, the presentation of the Dale medal by the Society for Endocrinology at the symposium on 29 September 1981 to honour his eightieth birthday which gave him particular pleasure and delighted his many friends and colleagues.

Amoroso was a practical supporter of the learned societies with which he concerned himself and several benefited from his financial acumen. As a chairman of scientific meetings he was outstanding. He possessed the ability to distil complex arguments and to identify key facts from which he presented a lucid and elegant summary. His mastery of the English language was complete and generations of students were captivated by the literary flair of this man who was slightly larger than life with a touch of flamboyance—cigar, pocket handkerchief, and bow-tie—who referred to himself as an Afro-Saxon. He was an inspired teacher as well as research worker and set many others on the road to scientific success. Yet he had time for the less successful and did not forget the difficulties he faced in his early life or the help he received. It was an indication of the impact he made on students and on veterinary preclinical teaching that twenty years after his retirement and six years after his death a new award for outstanding veterinary preclinical research should be designated the Amoroso award.

On his retirement in 1968 the University of London conferred on him the title of emeritus professor and the Royal Veterinary College awarded him its highest honour by making him a fellow. In 1969 Amoroso was appointed CBE and in 1977 he received the Trinity Cross, the highest national award of Trinidad and Tobago. Retirement only marked a further stage in his veterinary career and was followed by a series of visiting professorships and other posts in Santiago, Sydney, Nairobi, and Guelph and by a long period as special professor in the University of Nottingham. He was slowed only by incipient heart disease and continued to work at the Institute of Animal Physiology at Babraham before he died 30 October 1982 at the home of some friends in Leeds.

In 1936 he married 'Peter' Pole, who left him soon afterwards. They had no children. Amoroso's deeply held Catholic beliefs made him unable to divorce his wife, who remained a financial burden to him for most of his life. After his marriage failed he remained a private and rather lonely person.

[Peter Jewell in *Journal of Zoology*, vol. cc, 1983, pp. 1–4; Lord Zuckerman in *Munk's Roll*, vol. vii, 1984; R. V. Short in *Biographical Memoirs of Fellows of the Royal Society*, vol. xxxi, 1985; private information; personal knowledge.]   A. O. BETTS

AMORY, DERICK HEATHCOAT, fourth baronet, and first VISCOUNT AMORY (1899–1981), industrialist and politician, was born 26 December 1899 in London, the second of four sons (there were no daughters) of Sir Ian Murray Heathcoat Heathcoat-Amory, second baronet, and his wife, Alexandra Georgina, daughter of Vice-Admiral Henry George Seymour. He was educated at Eton and Christ Church, Oxford, where he obtained a third class in modern history in 1921. He entered the family silk and textile business, John Heathcoat & Co., a large employer in his native Tiverton, subsequently becoming an extremely successful managing director.

At this stage in his life, 'Derry' Amory, as he was always known to his friends, showed no interest in the national political scene but in 1932 he was elected to the Devon county council, where he served for twenty years, in particular as chairman of the education committee. Already his deep and abiding interest in young people was beginning to play an important part in his life, and he served as county commissioner for Boy Scouts for Devon from 1930 to 1945. This interest was to flower again after he retired from the political scene, when he was chairman of Voluntary Service Overseas (1964–75) and president of the London Federation of Boys' Clubs from 1963 until he died. It was typical of Amory that his interest was expressed not only in filling with distinction the top national positions, but also at the personal level in giving weekend trips to boys from these clubs, in his small yacht where he taught them to crew, and which he sailed himself with such enthusiasm.

In World War II Amory became a lieutenant-colonel in the Royal Artillery (general staff). In the latter part of the war he was charged with the training of paratroopers for the Arnhem operation. Although many years older than the paratroopers, he insisted on going into action with them and was severely wounded. This incident is reminiscent of an occasion in pre-war days when he was managing director of the family business, and steeplejacks were repairing the main factory chimney. Feeling that he should make a personal inspection of the chimney, Amory climbed the ladders to the top. After an anxious moment when he stuck, negotiating the overhang at the rim, he heaved himself on to the top, and descended to safety by ropes down the inside. He was a man with intrepid nerve and a dedicated sense of duty. Few were admitted to his inner thoughts. Behind his genial manner and engaging sense of humour lay personal disciplines of steel.

In 1945 he was elected Conservative MP for Tiverton, stepping into the vacancy left by his cousin who had been killed in action and moved more by a sense of family obligation than a personal inclination for Westminster. After a quiet start in opposition from 1945 to 1951, ministerial ascent was rapid. He was minister of pensions in 1951, minister of state at the Board of Trade in 1953 (when he was admitted to the Privy Council), minister of agriculture (1954–8), and chancellor of the Exchequer (1958–60). He retired in 1960. He was particularly successful as minister of agriculture, having been appointed by Sir Winston Churchill to clear up the mess after the Crichel Down affair. It was entirely in character that in a few months Amory restored confidence and a sense of direction without asking any questions about the débâcle. The 1957 Agriculture Act was very much his personal vision, and a particularly sound piece of agricultural policy. The complaints of farmers are traditional and on one occasion when answering questions in a critical House, he was asked whether it would not be an improvement to the farming scene if he resigned; he replied, with his usual equanimity, thanking his attacker for this constructive suggestion, which he would pass on to the prime minister. The whole House roared with laughter.

It was natural that he should be chosen once again, in 1958, for a rescue operation when Peter (later Lord) Thorneycroft, chancellor of the Exchequer, and his Treasury ministers resigned—a major calamity for the government which Harold Macmillan (later the Earl of Stockton) dismissed as 'a little local difficulty'. With only two months to go before budget day, Amory stepped in and made a resounding success of the budget. When he retired from Westminster in 1960 there was universal surprise and regret at losing such a stalwart, successful, and well-liked minister; there had been a significant body of opinion which saw him as the best qualified successor to Macmillan. In the event, this was never a possibility; Amory had made up his mind to retire from Westminster before he was sixty because he did not approve of old men in political life, on the grounds that they became increasingly remote from the mass of the nation.

On retirement he was created a viscount and went on to fill many distinguished positions, including those of high commissioner for Canada (1961–3), chairman of the Medical Re-

search Council (1960–1 and 1965–9), president of the County Councils' Association, pro-chancellor (1966–72) and later chancellor (from 1972) of Exeter University, prime warden of the Goldsmiths' Company (1971–2), high steward of the borough of South Molton (1960–74), chairman of Exeter Cathedral Appeal, and deputy lieutenant for Devon (1962). He was appointed GCMG (1961) and KG (1968). He was honorary FRCS and had honorary degrees from Exeter (1959), McGill (1961), and Oxford (DCL, 1974). He succeeded his brother as fourth baronet in 1972.

Amory died 20 January 1981 at his home near Tiverton. He was unmarried and the viscountcy became extinct. The baronetcy was inherited by his brother William (born 1901, died 1982) and subsequently by William's elder son Ian (born 1942).

[*The Times*, 12 January 1981; personal knowledge.]   G. R. H. NUGENT

ARMITAGE, SIR ARTHUR LLEWELLYN (1916–1984), lawyer and academic, was born 1 August 1916 at Marsden in Yorkshire, the elder son (there were no daughters) of Kenyon Armitage, master draper, of Oldham, and his wife, Lucy Amelia Beaumont. From Oldham Hulme Grammar School he obtained a foundation scholarship at Queens' College, Cambridge, achieving a first in both parts of the law tripos (1935 and 1936) and being awarded the George Lang prize for Roman law. After two years at Yale University on a Commonwealth Fund fellowship he was called to the bar by the Inner Temple in 1940. From 1940 to 1945 he served in the King's Royal Rifle Corps and the Second Army (as a temporary major). In 1945, preferring the academic world to practice at the bar, he accepted an immediate offer of a fellowship at Queens'.

In 1958 when senior tutor he was elected president of Queens' and in 1965–7 he was vice-chancellor of Cambridge University. He was young to be president of Queens' at the age of forty-two but was a great success there. During his last years he obtained for the college the large Cripps benefaction. This was but the culmination of twelve years of dedication to his college, its pupils, and its tutors and fellows. An example was his work for the department of criminal science leading to the establishment of an independent Institute of Criminology. His loyalty to Queens' was absolute without undermining his duty to Cambridge University as a whole. At the same time he wrote steadily on legal matters, being joint editor of *Clerk and Lindsell on Torts* (1954, 1961, 1969, and 1975) and of the *Case Book on Criminal Law* (1952, 1958, and 1964). He was deputy-chairman of quarter-sessions in Cambridge (1963–71).

From 1970 to 1980 Armitage was vice-chancellor of Manchester University. By 1970, at the end of the period of expansion decreed by Lord Robbins [q.v.], Manchester was the largest of the non-federated universities (that is, those excluding London and Wales). It was therefore natural for others to look to it for leadership and Armitage provided it. He rapidly grasped the important difference between Oxbridge and the established provincial universities—namely their councils' insistence on a strong lay majority in their business and administration. He knew that this influence if nurtured and continued would prove to be the academics' true source of freedom and independence. To this end he took steps to bring council and senate into closer working collaboration leading to much more rapid and fluent decision making and action at critical times. He played a leading role in the higher educational field of this period and linked it with the wider world of government and of business.

Over a period of twenty-five years Armitage was the chairman of important national committees relating to industrial pay and conditions of training and employment in education, the Civil Service, the law, and agriculture, and the effect of lorries on people and the environment. He was a member of the university grants committee (1967–70), the Unesco committee on West Indian universities (1964), and the grants to students committee (1961–5). In 1974–6 he was chairman of the committee of vice-chancellors and principals. He was also chairman of the association of Commonwealth universities, the inter-university council for higher education overseas, and the governors of the Leys School, Cambridge (from 1971).

Armitage was a big man—in energy, enthusiasm, compassion, and authority. He possessed magnanimity and fought his battles in an optimistic spirit. He was gregarious and clubbable—a member of a golf four which consisted entirely of knights. His energy spilled over in constant movement—in lectures as on a social occasion his immediate companions were in danger from his enthusiastic dancing lunges, backwards or sideways. His bushy eyebrows set off his penetrating but friendly glances in private conversation. He was very funny, with a great sense of the absurd. This helped to calm frayed nerves in many difficult situations.

Armitage received honorary degrees from Manchester (1970), Belfast (1980), Liverpool (1981), and Birmingham (1981), and the order of Andres Bello (first class) from Venezuela (1968). In 1969 he became an honorary bencher of the Inner Temple. He was knighted in 1975.

In 1940 he married Joan Kenyon, daughter of Harold Marcroft, yarn agent, of Oldham. They had two daughters. Armitage died 1 February 1984 at his home, Rowley Lodge, Kermincham, Cheshire.

[Private information; personal knowledge.]
GEORGE KENYON

ASHBY, DAME MARGERY IRENE CORBETT (1882–1981), feminist and internationalist. [See CORBETT ASHBY.]

ASHLEY, LAURA (1925–1985), dress designer, interior decorator, and entrepreneur, was born 7 September 1925 at Dowlais, near Merthyr Tydfil, south Wales, the eldest of the four children (two daughters and two sons) of Stanley Lewis Mountney, a civil servant, of Raleigh Avenue, Wallington, Surrey, and his wife, Margaret Elizabeth Davis. She was educated at Elmwood School, Croydon, and at the Aberdare Secretarial School, before becoming a secretary in the City of London at the age of sixteen. In the latter years of World War II she served in the WRNS. After the war she worked at the London headquarters of the Women's Institute.

During the war she had met Bernard Albert, son of Albert Ashley, grocer. In 1949 they were married; they had two sons and two daughters. In the early 1950s Laura designed tea towels which Bernard printed on a printing machine he had designed and developed in an attic in Pimlico. This was followed by the opening of a small mill in Kent to produce a greater range of products but a disastrous flood brought this particular phase of enterprise to an end. In the early 1960s Laura persuaded Bernard that they should seek to develop in her native Wales and the two, with their three eldest children, explored mid-Wales and fell in love with it. They determined to start afresh in Montgomeryshire and in 1963 began in a small way at Machynlleth. Soon afterwards they moved to the old railway station at the village of Carno, which became the headquarters of the large international empire of Laura Ashley plc. They developed in an area that had virtually no industrial background, recruiting their work force from women whose sewing and cutting experience was entirely domestic and men whose knowledge of machines was largely confined to those of the farmyard. But what was initially lacking in skill was more than made up for by wholehearted enthusiasm, ingenuity, and co-operation.

The Ashleys regarded their work force as an extension of the family and their friendly, concerned approach ensured great loyalty and support; the factories worked a four-and-a-half day week only as Laura believed that people on repetitive tasks, in particular, should have time for leisure and their families. The area was suffering rural depopulation, which government grants aimed to stem. The Ashleys took full advantage of this financial support. Jobs were provided at all levels—from the factory floor to the boardroom—in what was once regarded as a remote if beautiful area. Retail subsidiaries of the company later sold from the Ashleys' shops in four continents.

The Ashleys had complementary personalities and talents and were each other's greatest critics. Bernard, a man of immense energy, drive, and flair was also a skilled engineer, designer, and printer. Restless and volatile by temperament, he found in the outwardly calm and composed Laura the ideal rudder to direct his energies. He was knighted in 1987. Laura was essentially a 'revivalist' in her approach, drawing her inspiration from nature and ideas of the past. She was never a designer in the formal sense: rather, she would frequently describe a design for other people to implement, or she would produce an old design, the result of her many researches, and would demonstrate how to change it slightly or modify its colouring. She had her own design philosophy which resulted in a projected lifestyle which had a profound influence on the attitudes of her time. Her childhood recollections of periods of time spent with relations living close to the Welsh rural scene bordering the area of Dowlais influenced her greatly. She believed that most people yearned for a more natural lifestyle than had come to be accepted in modern industrial and urban society. In her clothes, furnishings, and interior decorations she put forward a style which was simpler, kindlier, and more romantic than her contemporaries projected. She revived many of the discarded designs of the eighteenth and nineteenth centuries.

Laura Ashley was an active partner in an enterprise which contradicted the trends of the times. When the mini skirt was in full flower she advocated the maxi skirt, which she considered infinitely more attractive. In an age when designers were emphasizing the more savage and tough side of human nature, she pointed to its pretty, peaceful, and generous side. Personally she was nearly always dressed in a skirt and blouse.

Although there were aspects of a large international company, with its annual turnover of £130 million, which Laura Ashley did not relish, the resourcefulness and realism of her business attitude must not be underestimated. The Ashleys moved abroad to develop the business in Europe. They lived in a French château in Picardy and had a town house in Brussels. Essentially Laura Ashley was a very private person with a profound belief in Christian values and the family as the bases of civilized life. She died 17 September 1985 in hospital in Cov-

entry, following head injuries sustained in a fall down the stairs of her daughter's Cotswolds house.

[Iain Gale and Susan Irvine, *Laura Ashley Style*, 1987; private information; personal knowledge.]   EMLYN HOOSON

ASKEY, ARTHUR BOWDEN (1900–1982), actor and entertainer, was born 6 June 1900 at 19 Moses Street, Liverpool, the elder child and only son of Samuel Askey, secretary of the firm Sugar Products, of Liverpool, and his wife, Betsy Bowden, of Knutsford, Cheshire. Six months after his birth the family moved to 90 Rosslyn Street. He was educated at St Michael's Council School and the Liverpool Institute. His social life was mostly centred round the church. Apart from Sunday services there were Sunday school, the Band of Hope, Scripture Union, and choir practice. He started in the choir as a probationer and was paid three pence a month. Eventually he became head boy. His most memorable singing performance was when the Lady chapel of Liverpool Cathedral was consecrated. He sang solo in the presence of the archbishops of Canterbury and York.

As a schoolboy his summer holidays were spent at Rhyl in north Wales. It was there that he first became addicted to the light entertainment business. A pierrot troupe called 'The Jovial Sisters' performed three times a day on a small wooden stage on the sands. Askey would arrive well before each performance and sit on the sands as near to the stage as he could get. He saw every performance and at the end of the holiday he knew all the words and music of every item. All this remained in his memory and was to be of great use later in his career.

At the age of sixteen Askey went to work at the Liverpool education offices at a salary of £10 a month. After the outbreak of war in 1914 he was often asked to sing to wounded soldiers and apart from solos would sometimes sing duets with Tommy Handley [q.v.], another Liverpudlian destined for show business stardom. In June 1918 he became a private in the Welch Regiment. His army career was short-lived because when peace came in November he was demobilized and returned to the education office.

He soon became fascinated with concert parties. As often as possible he would cross the Mersey to the Olympian Gardens at Rock Ferry where touring shows appeared each week. He memorized his favourite items, especially the jokes, and was secretly determined one day to join a concert party. Later he spent his summer holidays at Douglas in the Isle of Man where there were no less than four concert parties and again he was able to add to his repertoire of jokes and comic songs. He also entered and won prizes at talent competitions. He started his own amateur concert party called 'The Filberts' and joined another as the pianist.

His yearning to become a professional entertainer caused him to write for interviews with theatrical agents in London. When these met with no success he resigned himself to a settled career at the Liverpool education office. In 1924 Askey landed his first professional job as a comedian in a concert party called 'Song Salad'. He was offered a thirty-week tour at £6. 10s. a week. As his salary at work was £3 a week he decided to accept the offer and leave the education office, even though this entailed the extra expense of living accommodation. The concert party was a natural stepping-stone to pantomimes, in several of which Askey appeared with considerable success.

Askey now felt he could afford a wife and in March 1925 he married (Elizabeth) May (died 1974), daughter of Walter Swash, publican. They had one daughter, Anthea. Askey was soon to be introduced to a branch of the entertainment business which was to be his main source of income for the next twelve years—after-dinner entertaining. There were Masonic lodges, city companies, staff dinners, Central Hall concerts, and Sunday League shows. The fee was anything from two to five guineas and sometimes he would appear at as many as three in one night. After several such engagements he was offered occasional weeks touring the music halls but as the salary would have been much less than he could earn with his Masonics, he turned down the offers.

He and his wife then moved to a flat in Golders Green. In 1938 Askey joined Powis Pinder's 'Sunshine' concert party at Shanklin, Isle of Wight, where he was a resounding success for the next eight years. There he was discovered by the BBC who engaged him for a new radio show called 'Band Waggon', in which his partner was Richard Murdoch. The show, first broadcast in January 1938, was an enormous success. The broadcasts were followed by a tour of music halls and performances in the London Palladium in 1939. Richard Murdoch then joined the Royal Air Force, bringing 'Band Waggon' to an end. He was, however, given leave to appear with Askey in four films, *Band Waggon*, *Charlie's Big-hearted Aunt*, *The Ghost Train*, and *I Thank You*.

After 'Band Waggon' Askey appeared in a number of successful musical shows in London's West End: *The Love Racket* (1943), *Follow The Girls* (1945), *The Kid from Stratford* (1948), *Bet Your Life* (1952), and *The Love Match* (1953). He also appeared in many more films and was now an established star in most forms of entertainment. His later films included *Back Room Boy*, *King Arthur was a Gentleman*, *Miss London Limited*, and *Bees in Paradise*. Amongst these engagements Askey managed to fit in a very successful tour of Australia (1949–50). In 1969 he was appointed OBE and in 1981 CBE.

Later in his life he slowed down a little but was never short of offers for films, concerts, television, and radio programmes. His popularity was such that he appeared in no less than ten Royal Command Performances—in 1946, 1948, 1952, 1954, 1955, 1957, 1968, 1972, 1978, and 1980. He is also the only celebrity to have been twice the subject of television's *This is Your Life*.

Arthur Askey stood just five feet three and was able to use his lack of inches for much humour. He always wore glasses and in his earlier and prime professional days had red hair which his daughter inherited. He was always smartly dressed except when portraying a character. He was not a dancer though one critic wrote 'He didn't dance but looked as if he could.' He was a 'busy' little comedian given to extravagant gestures and darting movements, sometimes carried to excess. This exuberance prevented him from coming to immediate terms with television but he later conquered this shortcoming. He was at his best in a large theatre production such as pantomime and musical comedy. He later learned what many comedians lack—repose. His rapport with the public was such that few comic performers can have achieved greater popularity.

After his eightieth birthday his health began to fail but he remained cheerful and full of humour. His last public appearance was in pantomime at the Richmond Theatre in 1981. He died in St Thomas's Hospital 16 November 1982. His autobiography, *Before Your Very Eyes*, was published in 1975.

[Arthur Askey, *Before Your Very Eyes*, 1975 (autobiography); personal knowledge.]

RICHARD MURDOCH

ASSHETON, RALPH, second baronet, and first BARON CLITHEROE (1901–1984), politician and businessman, was born at Downham 24 February 1901, the second of four children and only son of Sir Ralph Cockayne Assheton, first baronet, of Downham Hall, Clitheroe, and his wife, Mildred Estelle Sybella, daughter of John Henry Master, JP, of Montrose House, Petersham, Surrey. His family could trace its descent for more than a thousand years and had sent more than twenty of its members to the House of Commons since 1324. They were British landowners who, though never entering the highest ranks of nobility, had survived the Wars of the Roses and the English civil war, retaining and enlarging their estates. Assheton was educated at Eton and Christ Church, Oxford, where he obtained a second class in modern history in 1923. He was called to the bar at the Inner Temple (1925). He soon, however, felt the pull of the City and became a member of a firm of stockbrokers in 1927. Assheton was also a devout Anglican and represented the diocese of Blackburn in the Church Assembly from 1930 to 1950. He was also high steward of Westminster from 1962 to his death.

He had been an active Conservative at university and at a by-election in 1934 entered the House of Commons as member for Rushcliffe, Nottinghamshire. He soon made his mark by his knowledge of economics and finance. He was appointed parliamentary private secretary to W. D. Ormsby-Gore (later Lord Harlech, q.v.), then minister of works and later at the Colonial Office (1936–8). This led to his being appointed a member of the royal commission on the West Indies (1938–9). On the outbreak of war Assheton was appointed parliamentary secretary to the Ministry of Labour and National Service, a post which he held until 1942, despite Sir Winston Churchill succeeding Neville Chamberlain as premier. In 1942 he was first promoted to parliamentary secretary to the Ministry of Supply and, from December, to financial secretary to the Treasury, 'the mounting block to the cabinet'. Assheton proved his worth dealing with the immense problems connected with wartime finance. He was sworn of the Privy Council in 1944.

In 1944, with a general election looming, Churchill asked him to become chairman of the Conservative and Unionist Party organization. During the war years the Conservative Party had declined in strength; there had been 400 agents in 1939, but by the time Assheton took over there were only 100. He fought hard touring the constituencies to revive Conservative philosophy very largely on a private enterprise and orthodox financial basis. None the less the Conservatives were defeated in 1945 and Assheton lost his own seat. He was, however, elected in the same year as Conservative member for the City of London and returned to the House of Commons as a front-bench speaker on financial and economic affairs. He also became chairman of the Public Accounts Committee (1948–50). When the City of London was disfranchised in 1950 he won back the Labour-held seat of Blackburn West for the Conservatives. He held the seat in 1951 and was generally expected to be included in the cabinet. In fact he was only offered the junior post of postmaster-general which he declined. He accordingly returned to the back-benches where he was chairman of the select committee on nationalized industries. Assheton was a strong supporter of the Suez group and in 1954 voted against the decision to withdraw from the Suez canal zone. In 1955 he retired from Parliament and accepted a hereditary peerage as first Baron Clitheroe. He succeeded to a baronetcy in the same year.

After this he devoted his career to business and held a large number of directorships, the most important of which were Borax Consolidated, the Mercantile Investment Trust, and a joint deputy chairmanship of the National Westminster Bank. He was a council member of the Duchy of Lancaster (1956–77) and was appointed KCVO in 1977. He continued to take a considerable interest in international affairs and more particularly those of Central Africa. He was a director of Tanganyika Consolidated and made a strong speech in the House of Lords urging support for Moise Tshombe's Katanga administration which the United Nations was seeking to integrate into the new Congo Republic (later Zaire). He showed himself a diligent landlord in Lancashire and took a keen interest in local affairs. He was lord lieutenant for Lancashire in 1971–6.

Assheton was very much what Sir Winston Churchill once described as 'an English worthy'. He had a keen intellect and great experience of financial and economic matters but he lacked the more theatrical gifts of politicians. His economic opinions were of an orthodoxy unfashionable until Margaret Thatcher's administration in 1979. He has been well described as 'the right man in the right place at the wrong time'.

In 1924 Assheton married Sylvia Benita Frances, daughter of Frederick William Hotham, sixth Baron Hotham. They had two sons and two daughters; one daughter died at birth. Assheton died 18 September 1984 at Downham Hall, and was succeeded in the baronetcy and the barony by his elder son, Ralph John (born 1929).

[Personal knowledge.]   JULIAN AMERY

ATHLONE, COUNTESS OF (1883–1981), Princess of Great Britain and Ireland. [See ALICE MARY VICTORIA AUGUSTA PAULINE.]

AUCHINLECK, SIR CLAUDE JOHN EYRE (1884–1981), field-marshal, was born at Aldershot 21 June 1884, the elder son and eldest of four children of Colonel John Claude Alexander Auchinleck, of the Royal Artillery, who died when Claude was eight, and his wife, Mary Eleanor, daughter of John Eyre. As the son of a deceased officer he qualified for a foundation place at Wellington College but his mother was hard pressed to pay even the reduced fee of £10 a year. Accustomed to hardship at home or with his mother's relations in Ireland, Auchinleck acquired a lifelong indifference to personal comfort. He passed into Sandhurst just high enough on the list to secure a posting to the Indian Army where in 1904 he was commissioned into the 62nd Punjabis. He displayed a marked aptitude for learning local languages and soon developed a close rapport with the ordinary soldiers of the Indian Army.

On the outbreak of World War I Auchinleck as a captain accompanied his regiment to Egypt where he first saw action in repelling the Turkish attack on the Suez canal in February 1915. From 1916 to the end of the war he served in Mesopotamia, taking part in the desperate attempts to relieve Kut-el-amara and at one point commanding his depleted battalion. Promoted brigade-major to the 52nd brigade, he took part in the victorious advance to Baghdad in operations characterized by appalling conditions, defective medical supplies, and faulty tactics. Though appointed to the DSO (1917) and thrice mentioned in dispatches, his promotion after 1918 was slow. In 1921 he married Jessie, daughter of Alexander Stewart, civil engineer, of Kinloch Rannoch, Perthshire.

He passed at the Staff College, Quetta, in 1919, attended the inaugural course at the Imperial Defence College in 1927, and, as a full colonel, returned to Quetta as an instructor from 1930 to 1932. In 1933 he took over command of the Peshawar brigade which was immediately involved in operations against the upper Mohmands, followed by another campaign in 1935. Auchinleck displayed professional competence and skill in improvization in one of the most difficult forms of warfare; in consequence he was promoted major-general and in 1936 became deputy chief of the general staff in India. In 1938-9 he was a key member of the Chatfield committee on the modernization of the Indian Army. Here he evinced his progressive belief in the feasibility of replacing British officers by Indians.

Early in 1940 he was posted to England to prepare IV Corps for eventual dispatch to France but, as an expert in mountain warfare, was precipitately sent to Norway, after the disastrous opening of the campaign, to take command of the land forces in the Narvik area. Auchinleck's insistence on the provision of essential supplies, artillery, and air cover irritated (Sir) Winston Churchill, then first lord of the Admiralty, and was symptomatic of their relations. The general's forebodings were justified in that under-equipping and tragi-comic mismanagement continued to dog operations which were overshadowed from the outset by the German attack in the west on 10 May. Narvik was duly retaken, only to be speedily abandoned. Auchinleck's report to the war cabinet, stating that the French contingent had impressed him more than the British, was not well received.

Nevertheless he was at once instructed to form V Corps for the defence of southern England against invasion, and in July 1940 he was promoted GOC Southern Command where he experienced considerable friction with his successor at V Corps, Lieutenant-General B. L. Montgomery (later Viscount Montgomery of Alamein, q.v.). In November, when the immediate threat of invasion had passed, Auchinleck was promoted to full general and appointed commander-in-chief in India. This was perhaps a surprising transfer from an active to a secondary, support theatre, but India's military importance was increasing. It was the potential source of many divisions equipped from local manufacturing resources which were badly needed in the Middle East and the Far East; it might become a target for Japanese aggression; and its internal unrest needed handling by a respected Indian Army veteran. Auchinleck won Churchill's approval by his prompt dispatch of a force to help put down the rebellion of Rashid Ali in Iraq. The prime minister decided 'the Auk' had the necessary drive after all, as well as the other professional and personal attributes which he admired, and that he must take over as C-in-C Middle East where Sir A. P. (later Earl) Wavell [q.v.], in his opinion, was too exhausted to achieve the desperately needed victory. Consequently, when Operation Battleaxe failed in June 1941, Churchill ordered Wavell and Auchinleck to change places.

Auchinleck had been warned by successive CIGSs, Sir John Dill and Sir Alan Brooke (later Viscount Alanbrooke) [qq.v.], that Churchill would expect an early offensive and quick results, and if he could not provide them he must explain why very diplomatically. Unfortunately Auchinleck was a bluff, uncompromising soldier, unable or unwilling to learn diplomatic finesse. After a brisk exchange of signals he bluntly told Churchill: 'I must repeat that to launch an offensive with the inadequate means at present at our disposal is not, in my opinion, a justifiable operation of war.' To all Churchill's relentless pressure and cajolery he returned a bleak factual explanation of the desert army's unreadiness for battle. From London, the reinforcements in troops, tanks, and guns seemed more than adequate (particularly in the light of the enemy's deficiencies as revealed by Ultra Intelligence); but in Auchinleck's judgement the troops required training and acclimatization, while the tanks and guns were inferior to the enemy's.

On this occasion Auchinleck got his way. Only on 18 November did Eighth Army open Operation Crusader by advancing round the southern flank of the Axis forces' position on the Italian frontier. General Sir Alan Cunningham [q.v.] intended to drive towards the besieged Tobruk so forcing Rommel to defend his communications on ground not of his own choosing whereupon he would be defeated by the superior numbers of the British armour. An extraordinarily confused situation developed in which tanks and armoured cars milled about the desert desperately trying to retain cohesion and restore order in an unusually dense 'fog of war'. Units of the Axis forces and Eighth Army became interspersed and layered upon one another in a remarkable fashion. As the battle appeared to swing against him with Rommel's bold dash towards the Egyptian frontier, Cunningham lost his nerve and began to talk of retreat. Warned of this situation, Auchinleck flew up to the front, restored confidence, and ordered the offensive to be renewed. Shortly afterwards he relieved Cunningham, replacing him with his deputy chief of staff from Cairo, (Sir) Neil Ritchie [q.v.], initially as a temporary measure to avoid changing the Corps commanders.

The immediate outcome was an impressive victory. Tobruk was relieved and Rommel forced to retreat into Cyrenaica with the loss of 20,000 prisoners and masses of *matériel*. In February 1942 however, Rommel, having received reinforcements, unexpectedly counterattacked, driving the Eighth Army out of Benghazi and back to the Gazala line with its right flank on the sea about thirty-five miles west of Tobruk.

As in the previous year, Auchinleck was placed under increasing pressure from Churchill to mount a new offensive, now with the urgent need to contribute to the relief of Malta by seizing Axis-held airfields. In these circumstances Auchinleck's adamant refusal to fly to London to explain his case that offensive operations before 1 June might risk defeat and the loss of Egypt was unwise. Churchill and the war cabinet were unsympathetic: Auchinleck was instructed to attack before 1 June only to be pre-empted by Rommel on 26 May.

Auchinleck offered another hostage to fortune in his retention of Ritchie as Eighth Army commander despite private warnings that he lacked the quality and experience to stand up to Rommel, and also lacked the confidence of his Corps commanders, W. H. E. Gott [q.v.] and C. W. M. N. (later Lord) Norrie, both senior to him and more experienced in command. Auchinleck in effect got the worst of both worlds: he expressed confidence in Ritchie yet tried to control him at long distance by feeding him detailed advice before and during the Gazala battle. More serious still, as conclusively shown by Field-Marshal Lord Carver, Ritchie was correct in his diagnosis of Rommel's in-

tended line of attack round the south of the British defences, and Auchinleck wrong, despite his access to Ultra. As a consequence of Auchinleck's misjudgement, Norrie's armoured brigades were not concentrated to check and defeat Rommel's initial attack as they might otherwise have been. Auchinleck's Ultra intelligence, which revealed Rommel's severe logistic problems and shortage of tanks and fuel, also caused him to urge Ritchie to seize the initiative for which he was later to be blamed as over-optimistic.

The outcome of the Gazala battle was a clear-cut defeat. Tobruk surrendered and the Eighth Army was driven all the way back to the line of El Alamein. Auchinleck at this late stage in the battle (25 June) removed Ritchie and assumed command himself. Despite Auchinleck's excellent intelligence of Rommel's weaknesses, dispositions, and intentions, there is little evidence that he displayed exceptional skill in his execution of counterattacks, which all petered out ineffectually. Nevertheless his resolution and calm contributed to the defensive success. By mid-July Rommel's advance had been decisively halted and his diaries reveal that he had come close to total defeat. By early August Churchill had decided to remove Auchinleck, replacing him with Montgomery, news which Auchinleck heard on 8 August.

Montgomery's sweeping criticisms of his predecessor sought to deny Auchinleck any credit either for checking Rommel or bequeathing a feasible defensive plan. By the end of July the defensive layout of minefields and field fortifications was indeed well in hand. Montgomery was certainly unjust in telling Churchill a few days after he had taken command that Auchinleck had no plan and in the event of a heavy attack intended to retreat to the Nile delta.

On the other hand Montgomery's buoyant personality did inject new confidence into a tired army, and he did make significant changes to the plans he inherited. In particular he realized immediately that if all available reserves were called up from the delta he would be able to man a continuous line of defence as far south as Alam Halfa. On that ridge he would reinforce his left flank and threaten any enemy attempt to turn it by a counterattack. All contingency plans for insuring against failure were cancelled. Ironically, Auchinleck was dismissed by Churchill for refusing to be pressured into a premature offensive, whereas Montgomery took several additional weeks to prepare his offensive which began on 23 October.

Auchinleck was hurt by the manner of his removal and refused relegation to the command in Persia and Iraq. Consequently he spent nearly a year idle in India before again becoming commander-in-chief there on Wavell's promotion to viceroy in June 1943.

Though deprived of responsibility for the conduct of operations against the Japanese, Auchinleck bore the burden of the vast expansion of the Indian Army and its war industry. He had to provide bases, troops, and supplies for the Burma campaign, and also assist the Chinese and American forces in that theatre. His immense prestige in the Indian Army was a great asset in this task. He was one of the few British generals to gain the respect of the American 'Vinegar Joe' Stilwell but, unlike Wavell, was too conventional a soldier to be sympathetic to the unorthodox methods and exorbitant demands on Indian resources, of Orde Wingate [q.v.].

In 1945 Auchinleck was appointed GCB, and the following year he was promoted field-marshal, but these honours were overshadowed by a personal tragedy. His wife left him for another senior officer at GHQ and they were divorced in 1946. They had no children. Jessie had found her husband's long office hours and dedication to work tiresome and ultimately intolerable. Her departure left a void in the field-marshal's life which he filled by working harder than ever. His last phase in India proved tragic in an even deeper sense. Auchinleck had hoped that India, and hence his cherished Indian Army, would remain united, but he quickly realized that partition of the country and the army were inevitable. He knew that considerable time would be needed for the peaceful reconstitution of two dominion armies without the steadying influence of British officers, the great majority of whom would wish to retire. Partition in 1948 would have been difficult enough, but the sudden decision of the new viceroy, Viscount (later Earl) Mountbatten of Burma [q.v.] to bring forward independence to 15 August 1947, made Auchinleck's task impossible. The Indian Army itself was deeply affected by religious conflict and rapidly disintegrated as an organized force. The meagre British military and police forces were powerless to prevent the massacres that accompanied the partition of the Punjab.

After independence Auchinleck decided to stay on as supreme commander of the Indian and Pakistan forces. He was openly criticized by some Indian leaders as being too favourable to Pakistan, and beneath a polite façade his relations with Mountbatten were increasingly strained. In September 1947 Mountbatten asked him to resign and he left India on 1 December in a mood of bitterness and despair after forty-four years' service. He refused a peerage lest the honour be associated with the events that he considered to be dishonourable.

After retiring from the army, Auchinleck led an active life, frequently revisiting India on business. His numerous appointments included being a governor of Wellington College (1946-59), president of the London Federation of Boys' Clubs (1949-55), vice-president of the Forces Help Society and Lord Roberts's Workshops, and chairman of the Armed Forces Art Society (1950-67). He characteristically refused to join in the battle of the memoirs, but in 1967 presented his personal papers to Manchester University, convinced that history would eventually do him justice.

Auchinleck had a most attractive personality; he detested all forms of pomp, display, and self-advertisement; but his simple tastes in personal matters did not inhibit him as a generous and entertaining host. His former private secretary, Major-General Shahid Hamid, has described Auchinleck in his element among his Indian soldiers: 'There is nothing dominating or overbearing in his character at all . . . His mental energy, enthusiasm, good temper and driving power are immense . . . He has become a legend in his lifetime.'

In 1968, when over eighty, Auchinleck left England to live in an unpretentious flat in Marrakesh. There he received visitors and spent his time painting, fishing, and walking. He had received honorary LLDs from Aberdeen and St Andrews universities in 1948 and from Manchester in 1970. He had many other honours, among which were OBE (1919), CB (1934), CSI (1936), GCIE (1940), and GCB (1945). He was among the ten field-marshals of World War II honoured by a memorial in the crypt of St Paul's in November 1976. He died in his Marrakesh home 23 March 1981.

[The Times and Daily Telegraph, 25 March 1981; Listener, 4 July 1974 and 16 April 1981; John Connell (John Henry Robertson), Auchinleck, 1959; Roger Parkinson, The Auk, 1977; Philip Warner, Auchinleck: the Lonely Soldier, 1981; Michael Carver, Dilemmas of the Desert War, 1986; Shahid Hamid, Disastrous Twilight, 1986.]   BRIAN BOND

B

BACON, SIR EDMUND CASTELL, thirteenth and fourteenth baronet (1903-1982), premier baronet of England and landowner, was born 18 March 1903 at Raveningham Hall, the only son and fifth of six children of Sir Nicholas Henry Bacon, twelfth and thirteenth baronet, and his wife, Constance Alice, younger daughter of Alexander Samuel Leslie-Melville, of Branston Hall, Lincolnshire. He was descended from Sir Nicholas Bacon [q.v.], lord keeper to Queen Elizabeth I. Born into a family where public service was a way of life, he did not flinch from the responsibilities which he later inherited and also created for himself. Educated at Eton and Trinity College, Cambridge, he subsequently studied farming and estate management.

During World War II he commanded the 55 (Suffolk Yeomanry) Anti-Tank Regiment of the Royal Artillery in Normandy and Belgium. He was mentioned in dispatches and appointed OBE in 1945. In 1947 his father died leaving him 4,000 acres in Norfolk, and 10,000 in Lincolnshire, a unique collection of English watercolours, and the collection of John Staniforth Beckett with its Dutch landscapes. He succeeded also to his father's titles, becoming thirteenth baronet of Redgrave (created in 1611) and fourteenth baronet of Mildenhall, created in 1627.

In 1949 he was appointed lord lieutenant of Norfolk, a post he held until 1978. His twenty-nine years' loyal and industrious service to the sovereign and to the county of Norfolk ended with the successful Silver Jubilee appeal in 1977. Norfolk was one of the few counties which exceeded its target.

In 1953 the eastern counties suffered from the severest of weather and tidal surges, Norfolk taking the fullest brunt of them. Bacon energetically organized the forces of relief and headed an appeal for a county relief fund. No other county raised such a large sum in proportion to its population and nowhere else were the funds so promptly distributed to the sufferers. It was this practical sense of administration which later allowed Bacon to spearhead years of crucial refurbishment in Norwich Cathedral after 1956, when he became high steward (until 1979). His religious beliefs were deeply held but they were more a positive moral conviction than an exact allegiance to one particular doctrine of Christianity. When he was made an honorary freeman of Norwich, he said that nothing had given him so much satisfaction as the part he played safeguarding the structure of the cathedral for the next generation.

His interests in agriculture were put to the test in the managing of his own substantial estates. These enterprises and a family property company in London, held since Elizabethan days, allowed him to develop a shrewd business sense which brought him national as well as local appointments. He was president of Eastern Counties Farmers, the country's second largest co-operative, until 1973. As chairman of the British Sugar Corporation during an eleven-year period of rationalization (1957-68), he was able to introduce company policy without the conflicts within the labour force which became so prevalent in the next decade. His most demanding job was as chairman in 1966-71 of the Agricultural National Economic Development Committee ('NEDDY'), whose purpose was to stabilize agricultural markets. Towards the end of his life he was awarded the Royal Agricultural Society's gold medal for distinguished services to agriculture.

He was a director of Lloyds Bank and president of the Norfolk branch of the Magistrates' Association for twenty-nine years. He played an active part in the founding of the University of East Anglia, being pro-chancellor from 1964 to 1973. He was awarded an honorary DCL there in 1969. It was during the period of student unrest and demonstrations in the late 1960s when his skill as chairman, presiding over critical meetings calmly and steadfastly, was most needed. This was also the era when Rhodesia dominated the media and the students' attention. During that time his great friend, Sir Humphrey Gibbs, as governor, had been confined to his residence since the unilateral declaration of independence. Bacon's work in Rhodesia, in supporting his friend during the difficult weeks in 1969, has never been much publicized. He disliked any glamour associated with himself unless it benefited the cause with which he was involved.

In 1965 he was appointed KBE and in 1970 he was among four new Knights of the Garter, the first baronet ever to be so honoured by the Queen.

Although he lived and looked like a country squire, and was happiest dressed in baggy, very old cavalry-twill trousers, he was a versatile man with wide interests. Hugely built, with hunched shoulders and sharp blue eyes, he endeared himself to people by his simple charm. There was always a kind greeting for those he met; he seemed never to forget a face among the many he encountered in his public offices.

Throughout his life the demands upon his time and energy had a deleterious effect upon his health. He suffered two heart attacks, but did not spare himself. In 1936 he married Priscilla Dora, daughter of Colonel (Sir) Charles Edward Ponsonby, later baronet and MP for Sevenoaks. They had one son and four daughters. Bacon died 30 September 1982 at Gainsborough, Lincolnshire. He was succeeded in the baronetcies by his son, Nicholas Hickman Ponsonby (born 1953).

[*The Times*, 2 October 1982; *Eastern Daily Press*, 2 October 1982; private information; personal knowledge.]   NICHOLAS BACON

BADER, SIR DOUGLAS ROBERT STEUART (1910-1982), Royal Air Force officer, was born at St John's Wood, London, 21 February 1910, the younger child and younger son of Frederick Roberts Bader, a civil engineer working in India, and his wife, Jessie McKenzie. He was educated at St Edward's School, Oxford, where he was a scholar, and at the Royal Air Force College, Cranwell, where he was a prize cadet. He finished second in the contest for the sword of honour at Cranwell and his confidential report described him as 'plucky, capable, headstrong'.

Bader was commissioned in August 1930 and was then posted as a pilot officer to No. 23 Fighter Squadron at RAF station, Kenley. He was an exceptional pilot and was selected, with another officer from the squadron, to fly the pair for the RAF at the Hendon air display in 1931 before a crowd of 175,000. As a young officer, he was good looking and charming. He was also determined and dogmatic, and could be thoroughly 'difficult'. He was, however, a natural leader, fearless and always eager for a challenge. While he could be brusque and impatient, he was socially at ease in any company. He was intensely loyal to the causes he cared about and to his friends.

Bader was twenty-one when on 14 December 1931 he crashed on Woodley airfield, near Reading. Both legs had to be amputated at the Royal Berkshire Hospital where his life was saved. He was later transferred to the RAF Hospital at Uxbridge. Six months after his operations, he was walking unaided on his artificial legs. 'I will never use a stick', he said.

He was discharged from the RAF in the spring of 1933. That summer he became a clerk in the Asiatic Petroleum Company (later Shell Petroleum) in the City and on 5 October 1933 married Olive Thelma Exley, daughter of Lieutenant-Colonel Ivo Arthyr Exley Edwards, RAF (retired), and Olive Maud Amy Addison (née Donaldson), secretly at Hampstead register office. Four years later, on 5 October 1937, the two were formally married at St Mary Abbots church, Kensington. There were no children of the marriage.

Bader was re-engaged by the RAF in No-

vember 1939, two months after the outbreak of World War II and on 7 February 1940 he was posted to No. 19 Fighter Squadron at Duxford, near Cambridge, as a flying officer. Within six weeks he was appointed to command 'A' Flight in No. 222 Squadron. As a flight lieutenant, he saw action with the squadron at Dunkirk. Promotion continued and on 24 June 1940 he was posted to command No. 242 (Canadian) Squadron at Coltishall, in Norfolk. He led the nearby Duxford wing with this unit with signal success throughout the Battle of Britain, being appointed to the DSO on 13 September and awarded the DFC a month later.

Bader's advocacy of his much misunderstood 'Big Wing' tactics served to fuel the controversy which existed between the air officers commanding Nos. 11 and 12 Groups of Fighter Command—(Sir) Keith Park and (Sir) Trafford Leigh-Mallory [qq.v.]—and between the C-in-C, Sir Hugh (later Lord) Dowding [q.v.], and the deputy chief of the air staff, Sholto Douglas (later Lord Douglas of Kirtleside, q.v.). The controversy was further inflamed by Leigh-Mallory's decision to ask Bader to accompany him to the high-level, tactical conference held at the Air Ministry on 17 October 1940 with Sholto Douglas in the chair. In March 1941 Bader became the first wing commander flying at Tangmere, in Sussex, leading his three Spitfire squadrons with notable success in the frequent offensive operations over northern France. His aggressive leadership was recognized with the award of a bar both to the DSO and the DFC. The French croix de guerre and the Legion of Honour followed. There were three mentions in dispatches. With his official score of twenty-three enemy aircraft destroyed, Bader was shot down on 9 August 1941 near St Omer in the Pas de Calais and became a prisoner of war until he was released from Colditz in April 1945. He made repeated attempts to escape, refusing repatriation on the grounds that he expected to return to combat.

After the armistice, the 'legless ace' was promoted to group captain and posted to command the North Weald sector in Essex. From here he led the victory fly-past over London on 15 September 1945. He retired from the RAF six months later and in July 1946 rejoined the Shell Company, eventually to become managing director of Shell Aircraft Ltd. (1958-69). He flew himself to many parts of the world, often taking his wife with him. An outstanding games player before his accident, he played golf on his tin legs to a handicap of four.

Paul Brickhill's biography of Bader, *Reach for the Sky* (1954) was followed by the film bearing the same title with Kenneth More playing Bader (1956). This brought the Battle of Britain pilot world-wide fame. Notwithstanding this, his unsung work for the disabled continued apace and in 1956 he was appointed CBE.

Bader, who lived mainly in London, retired from Shell in 1969 and from 1972 to 1978 was a member of the Civil Aviation Authority. He accepted several non-executive directorships, maintained his long-established connections with Fleet Street, and continued with his numerous public speaking engagements. Latterly, his principal business base was as a consultant to Aircraft Equipment International at Ascot.

Bader's wife died in London 24 January 1971 after a long illness. Two years later, on 3 January 1973, he married Mrs Joan Eileen Murray, daughter of Horace Hipkiss, steel mill owner. In the same year was published *Fight for the Sky* (1973), his story of the Hurricane and the Spitfire. Knighted in 1976 for service to the public and the disabled, Bader died in London 4 September 1982, while being driven home through Chiswick after speaking at a dinner in Guildhall. He was FRAeS (1976), an honorary D.Sc. of the Queen's University, Belfast (1976), and DL of Greater London (1977).

[P. B. ('Laddie') Lucas, *Flying Colours* (biography), 1981; private information; personal knowledge.]   P. B. LUCAS

BAGNOLD, ENID ALGERINE, LADY JONES (1889–1981), novelist and playwright, was born in Rochester, Kent, 27 October 1889, the elder child and only daughter of Major (later Colonel) Arthur Henry Bagnold, an inventive military engineer, and his wife, Ethel, daughter of William Henry Alger who inherited a large chemical and fertilizer factory at Cattedown and became mayor of Plymouth. She spent three years of her childhood in Jamaica, where her father was posted, then at Woolwich, and she went to Prior's Field school, Godalming, run by the stimulating Mrs Leonard Huxley.

Enid Bagnold was a tomboyish, dramatic, outdoor, beautiful girl who soon escaped the conventionally respectable life of her parents by taking a flat in Chelsea, and for several years studied art under Walter Sickert [q.v.]. She made friends easily among the marginally Bohemian set, including Henri Gaudier-Brzeska, who sculpted her. In 1912 she met J. T. ('Frank') Harris [q.v.], then aged fifty-six, who employed her as a journalist on his magazine *Hearth and Home*, and with whom she had her first love affair.

In World War I she worked as a nurse in a London hospital, and described her experiences in *A Diary Without Dates* (1918, 2nd edn. 1978), imagining herself writing to Antoine Bibesco, the Romanian diplomat whom she loved for many years with an unrequited passion. Within half an hour of the book's publication she was sacked by her matron for indiscipline, but it had an instantaneous success, H. G. Wells [q.v.] describing it as one of the two most human books written about the war.

Enid Bagnold thus achieved fame while still in her twenties, and her ambition never slackened. Her vitality, humour, audacity, and grace made her an exhilarating companion. She was ebulliently communicative, in talk as in writing, as lavish with words as a pianist is with notes, loving the inexhaustible variety of human experience as much as the language which expressed it. She was, however, too fond of the great and grand to be taken seriously by the literary establishment (apart from Sir C. O. Desmond MacCarthy, q.v., one of her closest friends), and in the view of one critic she remained 'a brilliantly erratic amateur'. Her literary strength lay in her gift for narrative, dialogue, and domestic scenes, for which she drew extensively on her own experience and the characters of her friends.

In 1920 she married Sir G. Roderick Jones [q.v.], chairman and managing director of Reuters. 'Their partnership', wrote her biographer Anne Sebba, 'was marked by loyalty, not fidelity, respect but not passion', but for their children, three sons and a daughter, Enid Bagnold's affection was unstinting. In London the Joneses entertained generously, and it was there, in a study designed for her by Sir Edwin Lutyens [q.v.], and at North End House, Rottingdean, their Sussex home, that she wrote joyfully but intermittently, having no need to write for money.

Her first novel *The Happy Foreigner* (1920), describing her experiences as a VAD in France, was followed by *Serena Blandish* (1924), about a poor girl in search of a rich marriage. Then came a book for children *Alice and Thomas and Jane* (1930), and in 1935 she published her best known book, *National Velvet*, the story of a butcher's daughter who gains a piebald horse in a raffle and, disguised as a boy, wins the Grand National. In 1944 it was made into a hugely successful film, with the youthful Elizabeth Taylor in the starring role. Her next novel, which she and many others considered her best, was *The Squire* (1938), a daringly outspoken description of the relationship between a mother (unmistakably herself) and the child to whom she gives birth.

In 1941 she wrote her first play, *Lottie Dundass*, about a stage-struck typist who dies of a heart attack after her first and only performance. It was an instant success in America and England, and Enid Bagnold thereafter devoted her talent to the theatre, apart from one more novel, *The Loved and Envied* (1951), a story of high life clearly based on her friend Lady Diana Cooper, and her *Enid Bagnold's Autobiography* (*from 1889*) (1969). As a dramatist she had several failures, like *Gertie* (1952), *The Last Joke* (1960), and most disastrously, *Call me Jacky* (1967), but in compensation she enjoyed one major triumph with *The Chalk Garden* (1955) which ran in London for two years and was described by Kenneth Tynan [q.v.] as possibly the finest English comedy since those of William Congreve [q.v.], and one critical success, particularly in America, with *The Chinese Prime Minister* (1964). She involved herself intensively (too intensively for many of the actors and directors) in the production of all her plays, and continued writing till late in her eighties, when she reworked *Call me Jacky* into *A Matter of Gravity* (1975). Her last published books were a selection of her poems (1978) and *Letters to Frank Harris and Other Friends* (1980).

Her husband died in 1962, and the last years of her life, apart from the devotion of her children, became increasingly lonely. She needed constant injections to dull the pain of her arthritis, but she remained gallantly active, describing herself, characteristically and truthfully, as 'an old lady masking a sort of everlasting girl inside'. She died in London 31 March 1981.

She was appointed CBE in 1976. Her bust by Gaudier-Brzeska is owned by her granddaughter, Viscountess Astor.

[Enid Bagnold, *Enid Bagnold's Autobiography* (*from 1889*), 1969, 2nd edn. 1985; Anne Sebba, *Enid Bagnold*, 1986; personal knowledge.]   NIGEL NICOLSON

BAILLIE, DAME ISOBEL (1895-1983), Scottish soprano, was born 9 March 1895 at Hawick, Roxburghshire, the youngest in the family of one son and three daughters of Martin Baillie, master baker, of Hawick, and his wife Isabella, daughter of Robert Douglas, weaver, of Selkirk. She was christened Isabella Douglas and was educated at Princes Road board school and, with a scholarship, at Manchester High School. At a very early age it was noted that 'her voice is something different'. She studied first with Jean Sadler-Fogg and later with Guglielmo Somma. At the age of fifteen she made her first public appearance, for a fee of seven shillings and sixpence, and in the same year sang her first *Messiah* performance.

(Sir) H. Hamilton Harty [q.v.] took great interest in her career and gave her her first Hallé concert (1921). On his advice she studied with Somma in Milan (1925-6). Later Harty wrote to her, 'When you die I want someone to write on your grave, "She was a sweet singer who always respected Music".' He also advised her to change her professional name from Bella to Isobel. Her career developed rapidly in oratorio

and in concert following her first London season in 1923 and she became the most sought after soprano in England for Handel, Haydn, Brahms, and Elgar oratorios. She was the first British artist to sing in the Hollywood Bowl in California (1933). In 1937 she sang in Gluck's *Orphée* at Covent Garden and in 1940 she gave many performances of Gounod's *Faust* in New Zealand, but she did not consider opera her true *métier*. She sang in every corner of Britain and appeared 51 times with the Royal Choral Society (which included 33 performances of *Messiah*). She sang for twenty-six consecutive years with the Hallé orchestra, and for fifteen consecutive years at the Three Choirs Festival. Between 1924 and 1974 she recorded 166 songs and arias covering 100 sessions, starting in the acoustic (horn) era and concluding in the stereo age.

At the height of her career Richard Capell, music critic, wrote: 'The character of Isobel Baillie's singing and her fine technique will be indicated if it is said that her performance in Brahms's *Requiem* has hardly been matched in her time. The trying tessitura of "Ye now are sorrowful" becomes apparently negligible and the term "angelic" has sometimes been applied to suggest the effect, not so much personal as brightly and serenely spiritual made here by her soaring and equable tones.' In Handel's *Messiah* her declamation of the Christmas music, the clarity and vitality of 'Rejoice Greatly', the serenity of 'I Know That My Redeemer Liveth' were acclaimed by all who heard her. *The Times* obituary referred to 'her gift for silken legato and silvery tone' and the *Telegraph* to 'that clear beautifully enunciated, entirely individual voice, capable of surprising richness'. Toscanini described her singing 'Right in the middle of the note and none of this', waving his hands in and out as if playing a concertina, indicating excessive vibrato.

Her red-gold hair and lovely complexion were outstanding physical traits. These together with complete dedication to her 'job', as she called it, captivated the audience. She delighted them by the care she gave to her own appearance for she felt that if people came to hear her she should look her best and *Never Sing Louder than Lovely*, the title of her autobiography (1982).

Sir George Dyson [q.v.], impressed with her performance as the Wife of Bath in his 'The Canterbury Pilgrims' persuaded her to teach at the Royal College of Music (1955-7 and 1961-4). She was a visiting professor at Cornell University (1960-1) and later for many years at the Manchester College of Music. She was described as a teacher 'of rare genius and contagious enthusiasm'. After her retirement from the concert scene she travelled the world giving lecture-recitals. At the age of eighty-five she returned to lecture at Cornell University 'with very impressive vocal examples'. She sang her way through life and at the end quoted Hilaire Belloc [q.v.]: 'It is the best of all trades to make songs, and the second best to sing them.'

She was appointed CBE in 1951 and DBE in 1978. She became honorary RCM (1962) and honorary RAM (1971), and also had honorary degrees from Manchester (MA, 1950) and Salford (D.Lit., 1977).

In 1917 she married Henry Leonard Wrigley, who worked in the cloth trade, when he was on leave from the trenches in Flanders. They had met as fellow artists in concerts in Manchester before World War I. They had one daughter. She lived in Selbourne, Hampshire, until her husband died in 1957 and later in St John's Wood, London. Finally she returned to Manchester to be among old friends. She died there 24 September 1983. There is a water-colour portrait of her (artist unknown) as Marguerite in *Faust*, in the possession of her daughter.

[*The Times*, 26 September 1983; *Daily Telegraph*, 21 November 1983; Richard Capell in *Grove's Dictionary of Music and Musicians*, 5th edn., 1954 (ed. Eric Blom); private information; personal knowledge.]

KEITH FALKNER

BAKER, SIR GEORGE GILLESPIE (1910–1984), judge, was born in Stirling, Scotland, 25 April 1910, the only child of Captain John Kilgour Baker, of Stirling, and his wife, Jane Gillespie. Baker's mother, whom he did not remember, died in 1914. His father, then a captain in the Cameron Highlanders, was killed in 1918. He had remarried shortly before his death. Baker's upbringing thus became the task of relatives who though devoted were unable to afford him the advantages of normal family life. His childhood was lonely. This gave him a lasting sensitivity to the needs of children, a quality which served him well in later years when it became his lot to deal with the child casualties of broken marriages. He was in due course sent to Glasgow Academy and thence to Strathallan School in Perthshire. For his further education he moved south to Oxford, but he retained throughout his life a gentle Scottish accent which led to his nickname of 'Scottie'. He became a senior Hulme scholar at Brasenose College and a pupil of the redoubtable W. T. S. Stallybrass [q.v.], later principal of Brasenose. He obtained a first class in jurisprudence (1930) and a second class in the BCL (1931). He attempted but failed to obtain a fellowship at All Souls.

Baker had always wanted a career at the bar. He joined the Middle Temple, of which he was

elected a Harmsworth scholar, and was called to the bar in 1932. His resources other than his own abilities were limited. Work in London was then scarce. He joined the Oxford circuit and there sought and gradually acquired a small, good general practice.

But, as with almost all his generation, the prospects of steady professional advancement were shattered in 1939 by the outbreak of war. Baker, though by this time married and a father, at once volunteered and joined the Royal West Kent Regiment as a private. He was later commissioned into the Cameronians. He served successively in Africa, Italy, and Germany, his services winning him a military OBE in 1945. After an unsuccessful foray into politics as Conservative candidate for Southall in the 1945 election and a short spell of service at Nuremberg on the war crimes executive, he returned to the uncertainties of post-war practice at the bar, by this time the father of three sons. His determination never faltered. His work returned and indeed increased so rapidly, again largely on circuit, that he was well justified in taking silk in 1952. Two small recorderships had already come his way, Bridgnorth and Smethwick. Then in 1952 he became recorder of Wolverhampton, an important appointment.

In 1954 he was elected leader of the Oxford circuit, a post which he greatly valued, discharged with great tact and skill, and held until his appointment to the High Court bench (and a knighthood) in 1961. By that time his judicial qualities had manifested themselves not only in his recorderships but in his chairmanships of successive government committees between 1956 and 1961. His practice had lain mainly in the fields of criminal and general common law work. He had done some but not a great deal of divorce work. His appointment to the High Court bench came as no surprise but his assignment to the Probate, Divorce, and Admiralty division did. His qualities, thus already manifested, soon justified the appointment. He was patient, courteous, sensitive, and above all decisive. But as with many of his judicial contemporaries in that division he found the state of the law cruel and the hard-fought, bitter, defended divorce cases of the time exceedingly unattractive. He welcomed the changes in the law of divorce in 1969, especially the abolition of the concept of matrimonial offence. He also welcomed the changes made in 1970 regarding the redistribution of matrimonial property upon divorce. Little did he then realize that it would fall to him as president of the newly created Family division to secure the smooth transition from the old law to the new.

His appointment to that post in 1971 came more as a surprise to him than to his friends though there were other contenders. At the same time he was sworn of the Privy Council. He proved a superb administrator and with the support of a younger generation of judges soon welded the new division into an effective and efficient unit. Though technically he had no control over divorce work done in the county courts, in fact through the deep respect which he commanded he was able to ensure greater efficiency in the way that work was done there. He, like others, was disappointed in his ambition to see the creation of a unified Family Court and also by the continued bitterness with which disputes over custody, access, and the division of matrimonial property continued to be conducted.

He retired in 1979. He said he did so because he was tired. That this was true was not surprising. He in fact retired to nurse his wife Jessie Raphael McCall whom he had married in 1935, and to whom he showed lasting devotion. She was the daughter of Thomas Scott Findlay, nurseryman, of Glasgow. There were three sons of the marriage (one of whom became a High Court judge) which was one of great happiness. Baker was deeply devoted to his children and grandchildren. The breakdown in his wife's health some years before his retirement cast a great and continuous burden on him until her death in 1983. Thereafter he sought to mitigate his loneliness by accepting a government invitation to investigate the operation of the emergency laws in Northern Ireland, a task not without danger to himself but which he discharged with characteristic thoroughness, courageously criticizing some aspects of that policy.

Baker's deep personal integrity was founded upon his staunch Presbyterian faith. He sometimes seemed shy and diffident but he had a heart of gold. Amidst all his activities he found time for innumerable interests outside the law, especially in the field of education. His college made him an honorary fellow in 1966. He became a governor of Strathallan School, of Epsom College, and of Wycombe Abbey. He was an active bencher (1961) of the Middle Temple of which he became treasurer in 1976. His recreations included golf and latterly fishing. But the heavy burdens he had carried for so long eventually took their toll. Early in June 1984 he suffered a stroke and he died 13 June 1984 at Mount Vernon Hospital, Northwood, Middlesex.

[*The Times*, 14 June 1984; private information; personal knowledge.]   ROSKILL

BAKER, JOHN FLEETWOOD, BARON BAKER (1901-1985), civil engineer, was born 19 March 1901 at Liscard, Cheshire, the younger child and only son of Joseph William Baker, etcher and water-colour painter, whose house was in Liscard, and his wife, Emily Carole Fleetwood. Baker was educated at Rossall School, from which he won an open mathematical scholarship to Clare College, Cambridge, where he read mechanical sciences in 1920-3. He was placed in the first class in the tripos (1923).

In 1923 jobs in engineering were difficult to find, and it was only in January 1924 that Baker gained employment, by accident, as an assistant to Professor A. J. Sutton Pippard [q.v.] in an investigation into the structural problems of airships. The work went well, and a year later Baker transferred to the Royal Aircraft Works, Cardington, as a technical assistant in the design department; he resigned in 1926 to join Pippard as an assistant lecturer at University College, Cardiff, but he continued to spend time at Cardington. The work was concerned essentially with the determination of elastic stresses in the main transverse space-frames of airships, and Baker's publications from this period show the great complexity of the investigation, and the drudgery (which was later obviated by the use of electronic computers). These few years may be viewed as an apprenticeship served by Baker; a later equivalent would be the concentrated study of a research student in his attainment of a Ph.D.

The next eight years, 1928-36, saw Baker make a first start in his life's work, although it was, as it turned out, a false one. The steel industry set up the steel structures research committee in the late 1920s to try to bring some order into the multitude of conflicting regulations governing the design of steel-framed buildings. The committee had many eminent members from industry, government, and academe, and Baker was appointed as technical officer to the committee, although he was prevented by a severe illness (probably tuberculosis) from taking up his post until January 1930. Three volumes (1931, 1934, and 1936) record the theoretical and experimental findings of the committee, and many of the papers in them were written by Baker. His contributions were recognized immediately, and he was awarded the Telford gold medal of the Institution of Civil Engineers in 1932, and appointed to the professorship of civil engineering at Bristol in 1933.

However, at the end of this work Baker realized that the elastic method of structural analysis could never serve as the basis of a rational design method for steel structures. The actual elastic state of a structure is extremely sensitive to accidental imperfections, and cannot, in any real sense, be predicted. What can be predicted with great accuracy, however, is the collapse load of a ductile structure, and Baker's work was directed from 1936 onwards to the development of the 'plastic' method of design.

Although progress may have seemed slow, a mere dozen years saw the appropriate British standard altered (in 1948) to permit design by plastic collapse methods. A spectacular application of the new ideas had been made in the meantime to the design of the Morrison air-raid shelter (for which Baker received official acknowledgement from the royal commission on awards to inventors); Baker was scientific adviser to the Ministry of Home Security from 1939, and was concerned with many aspects of air-raid precautions, as he described in his *Enterprise versus Bureaucracy* (1978).

In 1943 Baker moved to Cambridge University as a fellow of Clare College and professor of mechanical sciences and head of the engineering department, where he built up a substantial research team working on problems of structural design. At the same time he completely revised the educational programme of the department, and proposed the construction of new buildings, a doubling of the teaching staff, and a great expansion of research activity in all fields of engineering (including management studies). He achieved all these aims comfortably. He was concerned also with more general administration of the university, and found time to serve on the council of the Institution of Civil Engineers, on the University Grants Committee, and on other national bodies. He wrote a massive two-volume account of elastic and plastic methods of design *The Steel Skeleton* (1954, 1956) and, later, a two-volume text *Plastic Design of Frames* (1969, 1971). To this Dictionary he contributed the notices of Sir Charles Inglis and Andrew Robertson. He retired from his chair in 1968.

Baker's pioneering work on the plastic theory of structures was recognized academically by the award of eight medals (including the Royal medal of the Royal Society in 1970), by twelve honorary degrees, by election as FRS in 1956 and as a founder F.Eng. in 1976, by honorary fellowship of the Institution of Mechanical Engineers, and by honorary membership of the Royal Institute of British Architects. He was a fellow and vice-president of the Institution of Civil Engineers and fellow of the Institute of Welding. He was appointed OBE in 1941, knighted in 1961, and made a life peer in 1977. Baker married in 1928 Fiona Mary MacAlister (died 1979), the daughter of John Walker, cotton broker in Liverpool. There were two daughters. Baker died in Cambridge 9 September 1985.

[J. Heyman in *Biographical Memoirs of Fellows of the Royal Society*, vol. xxxiii, 1987; private information; personal knowledge.]

JACQUES HEYMAN

BAKER, PHILIP JOHN NOEL-, BARON NOEL-BAKER (1889-1982), Labour politician and winner of the Nobel peace prize. [See NOEL-BAKER.]

BALOGH, THOMAS, BARON BALOGH (1905-1985), political economist, was born in Budapest 2 November 1905, the elder son and elder child of Emil Balogh, director of the Transport Board, Budapest, and his wife, Eva, daughter of Professor Bernard Levy, of Berlin and Budapest. He was educated in the Modelgymnasium in Budapest, from which so many other distinguished emigré Hungarians graduated. After studying law and economics at the universities of Budapest and Berlin, he went in 1928 to America for two years as a Rockefeller fellow at Harvard University.

He served in the research departments of the Banque de France, the Reichsbank, and the Federal Reserve before he went to England. J. M. (later Lord) Keynes [q.v.], who once said he could hear more of what was going on from Balogh in an hour or two than he himself could pick up during several days in London, published his first article in the *Economic Journal* and helped him to his first job in England with the banking firm of O. T. Falk & Co. in 1932.

While a lecturer at University College, London, between 1934 and 1940, he wrote for the National Institute of Economic and Social Research his *Studies in Financial Organization* (1947). He was naturalized in 1938. Against high-level opposition he went, in 1939, to Balliol College, Oxford, as a lecturer. In 1945 he was elected to a fellowship, which he held until 1973, and in 1960 became a university reader. He was one of the founding members of the Institute of Statistics at Oxford.

In the 1950s Balogh turned his attention increasingly to the problems of the underdeveloped countries. As adviser to the Food and Agriculture Organization of the United Nations (1957-9) he used an afforestation project to design a series of ambitious development plans for the countries round the Mediterranean. Planting trees remained one of his hobbies. The demand of Dominic Mintoff (then prime minister of Malta) for integration with the United Kingdom appealed particularly to Balogh's economic philosophy, and he was deeply disappointed when the negotiations failed.

Harold Wilson (later Lord Wilson of Rievaulx) worked with Balogh throughout the 1950s and early 1960s and in particular on the preparation for the 1964 election. One of Balogh's lines of argument was that a Labour government would be heavily committed to a policy of faster growth, sustained by a strong incomes policy and supported by more state intervention in industry. He thought that the

Treasury would not be capable of carrying out such a policy and this led to the call for a separate Ministry of Expansion or Planning. These ideas were the origin of the Department of Economic Affairs.

After Labour's victory in October 1964, Balogh was brought into the Cabinet Office as adviser on economic affairs, with special reference to external economic policy. Some of Balogh's greatest contributions in Whitehall stemmed more from his uncanny knowledge of everything that was going on than from contributions to committee work, where his criticisms, though scintillating, were too radical. When Wilson's government was defeated in June 1970 he retained his close personal contacts with Harold Wilson, spent more time in Oxford at both Balliol and Queen Elizabeth House, and returned to writing.

Balogh retired from his Oxford readership in 1973 and became a fellow emeritus of Balliol, and a Leverhulme fellow (1973-6). But he was about to enter a new career. In 1974 he was made a minister of state at the Department of Energy and, as a life peer, which he had been created in 1968, its spokesman in the House of Lords. The theme of his maiden speech was the need for a tougher incomes policy, while investment and innovation were carried out, and for stricter foreign exchange controls. He played a key role in the creation of the British National Oil Corporation and became its deputy chairman (1976-8).

He was awarded an honorary doctorate by his old university, Budapest, in 1979, and an honorary degree by York University, Toronto. He married in 1945 Penelope (died 1975), daughter of the Revd Henry Bernard Tower, vicar of Swinbrook, Oxfordshire, and widow of Oliver Gatty. She became a distinguished psychotherapist and they had two sons and a daughter (there was also a daughter from the previous marriage). This marriage was dissolved in 1970. In the same year he married Catherine, daughter of Arthur Cole, barrister, Lincoln's Inn. She was a psychologist and prolific author, particularly of children's books, and previously the wife of (Charles) Anthony Storr, psychiatrist, by whom she had three daughters.

Balogh had a flamboyant mind and considerable moral courage. He was neither a systematic thinker nor a popular figure, indeed he seemed to court hostility. To his friends he showed unbounded loyalty. He died 20 January 1985 at his home in Hampstead, London.

[Private information; personal knowledge.]
PAUL STREETEN

BARNARD, HOWARD CLIVE (1884-1985), educational historian and professor of education in the University of Reading, was born 7 June 1884 at 11 King Edward Street in the heart of the City of London, the only child of Howard Barnard, journalist, and his wife, Sarah Ann Haggis. During the eighteenth century his ancestors established a firm of silversmiths in the City of London, and so became associated with the Goldsmiths' Company. Barnard was sixth in a direct line going back to about 1760 to become 'by patrimony' a member of this Company and a freeman of the City. He was educated at University College School, London, when it was still in Gower Street and at Brasenose College, Oxford, where he was a senior Hulme scholar. After taking third classes in both classical honour moderations (1905) and *literae humaniores* (1907) he was awarded a postgraduate scholarship for a further four years of study. This enabled him to pursue his interest in geography at the London School of Economics and to become a wandering scholar for a year, primarily at the universities of Heidelberg and Caen, where he perfected his German and French. Back at Oxford, he took a diploma in education with distinction and completed a B.Litt. thesis on 'The Educational Theory and Practice of the Port-Royalists', which was published two years later by the Cambridge University Press as *The Little Schools of Port-Royal* (1913). At twenty-nine, Barnard had embarked on what was to be a prolific literary career.

Meanwhile, his career as a schoolmaster had got under way in 1911. At Manchester Grammar School, Chatham House School in Ramsgate, and Bradford Grammar School, he served as assistant master, being variously responsible for teaching German, English, and Geography. Then in 1925 he moved to the headmastership of Gillingham Grammar School, guiding it through a rapid expansion during his twelve years there. In 1937, at a time when it was unusual for a practising teacher to take up a professorship, he moved to the chair of education at Reading; but by then Barnard's distinction as a scholar ran parallel with his distinction as teacher and headmaster.

As scholar and historian, he had been given stimulus and encouragement by his association with King's College, London, where, by a liberal interpretation of the term, he considered himself a part-time 'advanced student'. It was through his connection with King's College, and in particular with Professors John Dover Wilson [q.v.] and J. W. Adamson, with both of whom he struck up a close friendship, that he became an MA (education) and D.Litt. of London University.

In 1918 Barnard brought out *The Port-Royalists in Education*, and in 1922 *The French Tradition in Education*, which remains a classic. It covers the period from the Renaissance to the eve of the revolution, exploring in detail the work of educationists and educational institutions, and impresses by its range and depth alike. Four further books, *Madame de Maintenon and Saint-Cyr* (1934), *Girls at School under the Ancien Régime* (1954), *Fénelon on Education* (1966), and *Education and the French Revolution* (1969) testify to his unflagging interest in French education and are standard works on the subject. Other notable books are *Principles and Practice of Geography Teaching* (1933), *A Short History of English Education, from 1760 to 1944* (1947), a textbook well known to generations of students, and a splendid autobiographical account of his Victorian childhood, *Were Those The Days?* (1970).

Ease of scholarship and an enviable limpidity of style characterized both his writing and lecturing, and Barnard was warmly admired by students and colleagues alike. By nature he was modest, detached, slightly impersonal. He lived to be the 'grand old man' of the university, but typically denied that there was any merit in being old.

For more than thirty years after his retirement he regularly returned to the university to play the organ in the Great Hall. In the same hall, Reading University conferred an honorary D.Litt. on Barnard in 1974; and it was there that family, students, friends, and colleagues held a commemorative service shortly after he died in a nursing home near Godalming 12 September 1985 at the revered age of 101.

In 1910 Barnard married Edith Gwendolen, daughter of John Wish, a civil servant. They had a son and a daughter.

[*Were Those The Days?*, 1970 (autobiography); Peter Gosden, 'Present in the Past', *Times Educational Supplement*, 1 June 1984; *The Times*, 20 September 1985; Raymond Wilson, 'One Hundred Years, Teacher and Scholar', *Comparative Education*, 20 March 1984; private information; personal knowledge.]   RAYMOND WILSON

BARNETSON, WILLIAM DENHOLM, BARON BARNETSON (1917-1981), newspaper proprietor and television chairman, was born in Edinburgh 21 March 1917, the eldest child in the family of three sons and one daughter of William Barnetson, estate agent, of London and Edinburgh, and his wife, Ella Grigor, daughter of Dr Moir, medical practitioner, of Buccleuch Place, Edinburgh. He was educated at the Royal High School, Edinburgh, and at Edinburgh University. Study there was interrupted by the nineteen-year-old Barnetson's decision to become a free-lance war correspondent in the Spanish civil war. From Spain he returned to university and then in the space of one week in July 1940 he married Joan Fairley, daughter of William Fairley Davidson, publican, of Edinburgh, and Agustina Bjarnarson of Iceland; took his MA degree; and began service in the army.

Throughout World War II he served in Anti-Aircraft Command reaching the rank of major. As the war came to an end he was seconded for special duty on the reorganization of West German newspaper and book publishing in the British zone. He was responsible for helping to launch *Die Welt* and for choosing as a suitable person to be its subsequent publisher, an ambitious young German—Axel Springer, who was to become one of West Germany's most influential newspaper proprietors. Returning to Edinburgh, he joined the *Edinburgh Evening News* and from 1948 to 1961 was successively leader writer, editor, and general manager—combining these duties with extra-mural lecturing at the university, and speaking and debating on radio and television.

It was Barnetson's habit to go into his office early each morning to thump out his leaders, articles, and speeches on his own typewriter—a habit he never forsook. On one such occasion, there walked into the deserted office a passenger, fresh from the night sleeper from King's Cross, who announced himself as Harley Drayton [q.v.]. This was the financier who already ruled a formidable commercial empire and was about to expand his interests in the field of publishing. In Barnetson, Drayton recognized a man of integrity, a working journalist capable of moving into the management of a newspaper empire. A famous partnership had been born.

In 1962 Barnetson was made a director of United Newspapers. In the next four years he drew up a strategic plan for the morning, evening, and weekly provincial papers United already owned in Preston, Leeds, Doncaster, Northampton, and elsewhere. On Drayton's death in 1966, Barnetson succeeded him as chairman and, within three years, doubled the size of United by the acquisition of Yorkshire Post Newspapers, as well as acquiring *Punch* magazine. He recognized the value to his companies of good contacts with the professional and commercial worlds, and in the 1970s the Bill Barnetson lunches at the Savoy, attended by royalty, archbishops, cabinet ministers, diplomats, financiers, academics, and fellow newspapermen, became an established part of the London social scene.

Meanwhile, his interests and his directorships continued to grow. From 1976 he was deputy chairman of British Electric Traction; he sat on the boards of Hill Samuel, Trusthouse Forte, Argus Press, and the Press Association. From 1973 he was a member of *The Times* Trust; when Atlantic Richfield acquired the *Observer* in 1976 he became chairman, and from 1979 he was chairman of Thames Television. At one time, reference books listed forty-two organizations with which he was associated.

One of his outstanding contributions to the newspaper world was achieved in the years 1968-79 when he was chairman of Reuters. During this time the agency's turnover rose from £6 million to £73 million and the foundations were laid for Reuters's technological development from a simple news agency into the world-wide business and economic information service that caters for almost every financial market in the world. Reuters brought out to the full the visionary, the shrewd manager, and the leader, in Barnetson.

Amid all this activity, he found time to spend weekends with his wife and family of one son and three daughters, at their home 'Broom' in Crowborough, east Sussex, where he could enjoy his books and his gardening. But there were increasing demands on his energies: the Open University, Press Council, English National Opera, Queen's Silver Jubilee Appeal, and St Bride's church in Fleet Street, received his personal support and encouragement. And through it all, the thumping of the typewriter keys and the puffing at a king-size briar pipe, went on. Ceaseless activity seemed to suit him. For a somewhat bulky Scot he was surprisingly light of step and this lent a pronounced sense of cheerful urgency to his movements. His working philosophy in dealing with all his many interests was simple: 'I leave the man on the spot to get on with the job. But he must know that I am available here all the time.'

Barnetson was knighted in 1972 and made a life peer in 1975. He died in Westminster Hospital 12 March 1981.

[*The Times* and *Daily Telegraph*, 13 March 1981; Guy Schofield, *The Men that Carry the News*, 1975; personal knowledge.]

EDWARD PICKERING

BARRACLOUGH, GEOFFREY (1908-1984), historian, was born in Bradford, Yorkshire, 10 May 1908, the eldest of three children (all sons) of Walter Barraclough, a prosperous wool merchant of that city, and his wife, Edith Mary Brayshaw. He was educated at Ackworth (1917-19), Colwall (1919-21), 'which made a lifelong rebel of me', Bootham School in York (1921-4), and Bradford Grammar School (1924-5). At seventeen he went to work in the family firm of O. S. Daniel & Co., selling essential oils in Europe.

'Whatever essential oils may have been', he wrote, 'I soon discovered that they were not essential to me.' In 1926 he won a scholarship to Oriel College, Oxford, where he developed an interest in medieval history, winning first class honours in the history schools in 1929. Graduate work followed at Munich (1929-31), where he took his new bride, Marjorie Gardiner, who bore him one son (Alan, born 1930).

In 1931-3 he moved to the English School at Rome and did research on the papacy. His first books, *Public Notaries and the Papal Curia* (1934) and *Papal Provisions* (1935) were intended to correct what Barraclough regarded as the provincialism of English medievalists.

In 1934 Barraclough won a fellowship at Merton College, Oxford, and in 1936 moved to St John's College, Cambridge. He also published *Medieval Germany* (2 vols., 1938), and became editor of *Studies in Medieval History*, designed to introduce English readers to German scholarship. By the age of twenty-nine he was the leading authority on German medieval history in the English-speaking world.

When World War II began, Barraclough joined the 'Enigma' project at Bletchley Park, first as an employee of the Foreign Office, and later as a squadron leader in the Royal Air Force. In Bletchley's hut 3 he became 'first of the watch', presiding over the horseshoe table where Luftwaffe dispatches were read during the Battle of Britain. In 1941 he was sent to the Mediterranean, and in 1944 was seconded to France. At the war's end he was in the Air Ministry at London.

During the war he wrote *The Origins of Modern Germany* (1946), which offered a historian's long view of the German problem. The book was widely read, and the policy it recommended was adopted by Britain and the United States.

In 1945 Barraclough became professor of medieval history at Liverpool. He divorced his first wife and married Diana Russell-Clarke, who bore him two sons, Nigel (born 1947) and Giles (born 1948). Diana was the daughter of Edward Russell-Clarke, barrister, of the Inner Temple. The new family moved to an Elizabethan house eighteen miles from Liverpool, and lived there from 1945 to 1956, a decade of happiness and peace in Barraclough's life. At Liverpool he founded a programme in archival administration, where a generation of British archivists was trained. He published *The Mediaeval Empire* (1950), *Cheshire Charters* (1957), and *The Earldom and County Palatine of Chester* (1953). Barraclough also developed a new interest in applying history to present problems, and contributed to the *Listener*, *New Statesman*, *Observer*, and *Guardian*. A collection of essays appeared as *History in a Changing World* (1955). He was also active in the Historical Association, becoming president, 1964-7.

In 1956 Barraclough was offered the Stevenson research chair at the University of London formerly held by Arnold Toynbee [q.v.], and the directorship of the Royal Institute of International Affairs. His task was to publish an annual *Survey of International Affairs*, a labour which he disliked. In 1962 he quit his job, and returned to St John's College, Cambridge, where he wrote his most successful book, *An Introduction to Contemporary History* (1964).

Increasingly restless, Barraclough took flight to America, teaching at Texas (1965) and San Diego (1965-8). He was married a third time, to Gwendolyn Lambert, and in 1968 moved to Brandeis. There he recovered stability in his life, publishing *The Medieval Papacy* (1968). There were no children of this marriage.

In 1970 Barraclough was called to Oxford as Chichele professor of modern history and a fellow of All Souls. There, things went badly again. Barraclough felt confined by the curriculum, oppressed by the ambience of All Souls, and disliked by colleagues who expected him to resume a narrow speciality in medieval history.

After two years of strife, Barraclough abruptly left Oxford and returned to Brandeis, remaining there from 1972 to 1981, his most prolific period. He published *The Crucible of Europe* (1976) a history of the ninth and tenth centuries; *Management in a Changing Economy* (1976), an application of history to economic planning; *Main Trends in History* (1976), a survey commissioned by Unesco; *The Times Atlas of World History* (1978); *Turning Points in World History* (1979); *The Christian World* (1981); and *From Agadir to Armageddon* (1982). He wrote abundantly for the *Nation*, *New Republic*, and *New York Review of Books*.

Barraclough was lionized in Boston, and became a famous figure in American universities. He dressed in a wool Tattersall shirt, an old tweed jacket, drab necktie, ancient Oxford bags, and gleaming British walking shoes. He walked rapidly with a curious rolling gait, his bald head cocked oddly to one side, and his round ruddy face veiled by blue clouds of smoke from a battered pipe.

Honours cascaded upon him, from everywhere but Britain. 'His reputation in Europe and the United States . . . was enormous', an English colleague wrote, but 'British colleagues, especially in his later years, tended to view his achievements rather more coolly.' His contribution was to enlarge the historian's territory beyond the boundaries of the nation state, and to expand the temporal context of history from the study of the past to the study of change. His work inspired young scholars throughout the world.

In 1981 Barraclough retired from Brandeis, and moved to Williams College, and in 1982 he taught at Munich. There he fell severely ill of lung cancer. He returned to his second wife Diana Barraclough in Burford, Oxfordshire, and resumed a project on Cheshire charters which he had put aside thirty years before. As he neared the end of that last labour, he died at Sylvester House, Burford, 26 December 1984.

[Manuscript autobiography owned by Diana Barraclough; Brandeis faculty archives; A. J. P. Taylor, *A Personal History*, 1983; private information; personal knowledge.]

DAVID HACKETT FISCHER

BARTHOLOMEW, (JOHN) ERIC (1926-1984), actor and comedian. [See MORECAMBE, ERIC.]

BARTLETT, (CHARLES) VERNON (OLDFELD) (1894-1983), journalist, author, and broadcaster, was born 30 April 1894 at Westbury, Wiltshire, the second of the three children and only son of Thomas Oldfeld Bartlett, bank manager, of Swanage, and his wife, Beatrice Mary Jecks. He was educated at Blundell's School at Tiverton but psoriasis forced him to leave early. He was sent to live abroad and acquired a fluent mastery of the main European languages that later stood him in good stead as a foreign correspondent.

In World War I Bartlett was in hospital recovering from a wound when his regiment (the Dorset) suffered an early poison gas attack. Later in 1915 he was the only man pulled out alive when his dug-out received a direct hit. He was invalided out of the army with a burning conviction that further world wars must be prevented. He found work in Fleet Street, joining the *Daily Mail* as a general reporter. In 1917 he married Marguerite Eleonore, the Belgian refugee daughter of Henri van Bemden, a well-to-do food merchant of Antwerp. They had two sons, the younger of whom died in 1970.

Also in 1917 Bartlett joined Reuters which later sent him to cover the Paris peace conference, and in 1919 he became a foreign correspondent for *The Times*, reporting the turbulent post-war diplomatic developments from Geneva, Berlin, Rome, and Warsaw. With his international experience and his repugnance of war it was a natural step for him to become, in 1922, the director of the London office of the League of Nations, a post he held until 1932.

Bartlett was an early broadcaster, at first occasionally about the work of the League, and later analysing general international developments in a weekly series *The Way of the World*. He became an outstanding communicator to whom a rising generation owed its interest in foreign affairs. The BBC *Yearbook* for 1931 noted: 'By his unobtrusive charm and his respect for truth and a certain quality of fair play which is inherent in his work, Mr Bartlett has made of international affairs a subject of general interest.'

That quality of fair play sometimes got him into trouble. In October 1933 Nazi Germany walked out of the disarmament conference at Geneva and left the League of Nations. Bartlett was far from pro-Nazi, but broadcasting on the BBC, which had employed him full-time since October 1932, he made the immediate comment: 'I believe the British would have acted in much the same way as Germany has acted if they had been in the same position.'

This remark provoked an immediate political storm. There were violent protests from the prime minister, J. Ramsay MacDonald, and other MPs, as well as in the press. Bartlett himself said he received thousands of letters, the overwhelming majority in support. The BBC told him it would be easier for him to continue broadcasting if he were not on the staff, adding that the politics of any newspaper he joined might affect the frequency of his broadcast invitations. Bartlett therefore resigned, rejecting a lucrative offer from the Labour *Daily Herald* in favour of a smaller one from the more moderate *News Chronicle*. Yet, by what he regarded as cowardly treatment, it was several years before he was asked to broadcast again.

For the next two decades he travelled widely as the *News Chronicle*'s diplomatic correspondent. In 1938, after Chamberlain at Munich yielded to Hitler's demands over Czechoslovakia, he stood as an Independent Progressive candidate opposed to the appeasement policy in the by-election at Bridgwater, then a safe Conservative seat. With no party machine to support him, though he had a number of keen volunteers, he won Bridgwater, and held it until 1950. He contributed usefully to debates on foreign affairs, but the role of an independent member was perhaps too lonely for one so gregarious by nature.

During World War II Bartlett's broadcasting skill was put to good use. He spoke three times a week to America and often in French and German. His *Postscripts*, after the 9 o'clock home news, were wise, compassionate talks, often including touches of sardonic humour, and they did much to sustain domestic morale. For a time he served as British press attaché in Moscow. At one Kremlin dinner, emboldened by vodka, he responded to a toast to the press by declaring that without a free press there could not be a free people. Stalin commented 'That young man talks too much.'

After retiring from the *News Chronicle* in 1954 Bartlett worked as a journalist in Singapore and in 1956 he was appointed CBE. In 1961 he moved to a farm in Tuscany where he produced wine and continued to write books— twenty-eight altogether. His first had been an autobiographical novel (1929) about calf love in pre-war Berlin. Others were about foreign affairs and his travels in South-East Asia, Africa, and Europe, and especially about his beloved Tuscany. An autobiography, written when he was forty-three, provides revealing insights into European diplomacy between the wars.

Bartlett's wife died in 1966 but three years later he made a happy second marriage to Eleanor Needham ('Jo'), daughter of Lieutenant-Colonel Theodore Francis Ritchie DSO, Royal Army Medical Corps, and widow of Walter Menzel, of the International Committee of the Red Cross. Bartlett died at Yeovil 18 January 1983.

[Vernon Bartlett, *Calf Love*, 1929, *This is My Life*, 1937, *I Know What I Liked*, 1974; BBC *Yearbook*, 1931; *Daily Telegraph*, 17 October 1933; Asa Briggs, *Governing the BBC*, 1979; BBC Written Archives Centre; personal knowledge.]   LEONARD MIALL

BATTEN, EDITH MARY (1905–1985), social work pioneer and educationist, was born in London 8 February 1905, the third of the four children and elder daughter of Charles William Batten, general draper, of High Street, Wandsworth, London, and his wife, Edith Wallis, daughter of Robert Black, of Chelsea, London. The family moved to Southport in 1916, where 'Mollie' Batten was educated at the Southport High School for Girls. In 1925 she gained a B.Sc. in chemistry at the University of Liverpool.

Returning to London in 1925, where her father died, Mollie Batten had to find work and took a clerical job in an office in Piccadilly, later moving to McVitie & Price's Hendon factory, where she became an assistant to the personnel officer. From here she was dismissed due to a disagreement with management over the treatment of a worker. In 1928 she became secretary of the Invalid Children's Aid Association branch in West Ham. Residing in the St Helen's Settlement, she became a member of the Church of England, being prepared for confirmation by the vicar of Barnet, the Revd Leslie Hunter [q.v.]. During this time she attended evening classes at the London School of Economics and gained a B.Sc. (Econ.) degree in 1932.

Mollie Batten was appointed warden of the Birmingham Settlement in 1933 and soon became a nationally recognized pioneer in the training of social workers and youth workers. She helped to rebuild and expand the Settlement's premises and was responsible for changes in its constitution, which brought the warden and staff representation on to its council. In 1937-8 she was a JP for the City of Birmingham.

In 1938 she became organizing secretary of the British Association of Residential Settlements. The Hire Purchase Act (1938) owed much to evidence she had collected. She had become a member of the Ministry of Labour's Factory and Welfare Advisory Board in 1938 and in 1940 she joined the Ministry itself. She had a particular concern for the welfare of workers outside working hours. In 1942 she was appointed to the Manpower Services Board and was involved in the call-up of women in the London area. Offered a permanent Civil Service appointment, she nevertheless resigned in 1947 and went to St Anne's College, Oxford, where she read for a theology degree, gaining a second class honours BA in 1949 (MA, 1953).

This and her past experience splendidly equipped her to become principal of William Temple College in 1950. The college, which she moved from Hawarden, Flintshire, to Rugby in 1954, was a memorial to the archbishop and was initially to be a theological college for women. Under Mollie Batten's imaginative guidance, and with the unflagging support of her old friend Leslie Hunter, now bishop of Sheffield and chairman of governors, it became a unique centre for the study of the secular realities of modern society in the light of Christian faith. There were resident students on two and three year courses for the Cambridge diploma and certificate in religious knowledge. But large numbers of men and women came on short courses from industry, social work, education, administration, and the Civil Service. Mollie Batten's wide field of contacts in these areas enabled her to bring in a steady flow of distinguished speakers to stimulate penetrating yet practical thought on the topics in hand. To avoid a previous tendency to heavy cigarette smoking she took to smoking a small pipe while listening to lectures. Sadly, the churches failed to perceive the value of the college or to support it in any way, although the growing number of industrial chaplains found in it and in its principal an important resource. Engagements with the World Council of Churches, with the Church in Canada and in Hong Kong, and contacts in Germany nevertheless demonstrate an impressive reputation beyond these shores.

Mollie Batten retired to London in 1966 but continued in part-time research work for the Church Assembly's Board for Social Responsibility. Moving to Midhurst in 1969 she played an active part in the diocese of Chichester and the neighbouring diocese of Portsmouth, presiding in the latter over a working party on the South Hampshire development plan. She was sometime chairman of the Chichester constituency Labour Party.

She was a lifelong supporter of the Anglican Group for the Ordination of Women. Her vigorous intellect, innovative spirit, warmth of character, and infectious enjoyment of life evoked exceptional responses of loyalty and affection. She was appointed OBE in 1948. She died at Midhurst 27 January 1985. She was unmarried.

[Personal knowledge.]   S. W. PHIPPS

BEAUFORT, tenth DUKE OF (1900-1984), leading figure of equestrian activities. [See SOMERSET, HENRY HUGH ARTHUR FITZROY.]

BEAVER, STANLEY HENRY (1907-1984), geographer, was born 11 August 1907 at Willesden, London, the second son and fifth child in the family of two sons and three daughters of William Henry Beaver, post office letter sorter, and his wife, Hypatia Florence Hobbs. He was educated at Kilburn Grammar School and at University College, London, where he gained a first class honours degree in geography (1928) and the Morris prize in geology. After a course of teacher training he was appointed in 1929 to the staff of the London School of Economics and Political Science. He quickly earned a high reputation as an economic geographer and collaborated with (Sir) L. Dudley Stamp [q.v.] in writing the successful *The British Isles* (1933). Beaver's studies concentrated on the mineral, manufacturing, and transport industries. He contributed *Yorkshire, West Riding* (1941) and *Northamptonshire* (1943) to Stamp's *Land Utilization Survey of Britain*.

After the outbreak of war Beaver's knowledge proved valuable to the geographical section of the Naval Intelligence Division of the Admiralty: he worked especially on southeastern Europe and on railway matters and contributed to the Geographical Handbooks. In 1943 he joined the Ministry of Town and Country Planning in the research division working on mineral problems and derelict land. He developed the collection of air photographs. He continued to advise the ministry after his return to academic life as Sir Ernest Cassel reader in economic geography at LSE in 1946. As exemplified by his *Derelict Land in the Black Country* (1946), based on field survey, and by his work on minerals (notably ironstone) and planning, Beaver made contributions of lasting policy significance. His studies of sand and gravel resources were particularly influential. In 1950 Beaver was appointed to a foundation chair in the University College of North Staffordshire at Keele and threw his energies wholeheartedly into the development of a new university. He founded a climatological station and produced (with E. M. Shaw) *The Climate of Keele* (1971). He wrote on the evolution of the industrial landscape of the Potteries.

Field-work which threw light on the relationships between environment, economy, technical skills, and social conditions was of the essence of Beaver's life as a geographer. He had been early attracted to the work of the Le Play Society and took part in field surveys, for example in Bulgaria, Albania, and Norway. He joined societies to serve them, not for recognition. A founder member of the Institute of British Geographers, he was president in 1964. The Geographical Association elected him to its presidency in 1967 and he was twice president of Section E of the British Association for the Advancement of Science (1961, 1981). He served on the council of the Royal Geographical Society (1943–7) and received its Murchison award in 1962. He was chairman for many years of the Dudley Stamp Memorial Trust. He was elected an honorary fellow of LSE.

Beaver was short, spare, and energetic. His mind was quick: he did not enjoy long committee discussions. He loved railways and knew his Bradshaw. He was a kind man, though when principles were at stake he could speak sharply. He had what was described as a 'pungent humour'. He enjoyed teaching and students responded to his enthusiasm. A central theme was the environmental consequences of technical change. He linked a grasp of geological conditions, detailed knowledge of places, reading in economic history, and an understanding of industrial technologies to analyse the revolution in the British landscape that had resulted from industrial change. Technical development, he argued, did not lead to mastery over nature but rather to increasingly interdependent relationships between societies and environments.

In 1933 he married Elsie, daughter of William Wallace Rogers, chief traveller for the Jaeger Company. They had a son and three daughters. Beaver died 10 November 1984 at his home in Eccleshall, Staffordshire.

[*The Times*, 14 November 1984; *Geographical Journal*, vol. cli, July 1985; *Sociological Review*, August 1985; *LSE Magazine*, vol. lxix, June 1985; A. D. M. Phillips and B. J. Turton (eds.), *Environment, Man and Economic Change: Essays Presented to S. H. Beaver*, 1975; *Transactions of the Institute of British Geographers*, New Series, vol. x, 1985; private information; personal knowledge.]

M. J. WISE

BEECHING, RICHARD, BARON BEECHING (1913–1985), businessman and physicist, was born 21 April 1913 at Sheerness, Kent, the second of four sons (there were no daughters) of Hubert Josiah Beeching, journalist, of Sheerness, and his wife, Annie Twigg, schoolmistress. Educated at Maidstone Grammar School and the Imperial College of Science and Technology, London, he took a first class in his B.Sc. (physics, 1934) and later a Ph.D. (1937). He became a fellow of the college in 1964.

In 1936 he joined the Fuel Research Station at Greenwich and the following year moved to the Mond Nickel Company. In 1943 he was seconded, at his request, to the armaments design department of the Ministry of Supply at Fort Halstead and it was here that his outstanding analytical skills were first given full scope. He became deputy chief engineer of armaments design for all three Services after only three years, at the age of thirty-four. He refused to accept any Service demand for a weapon design if it was based solely on tradition. He probed the fundamentals, and although as a physicist among engineers his appointment was at first greeted with scepticism this quickly changed to respect, then to awe and enthusiasm for his analytical approach.

The department was under the direction of Sir F. Ewart Smith who asked Beeching to join ICI after World War II. In 1948 he became a member of the company's technical department at the Millbank headquarters. He became prominently involved in the early development of man-made fibres and was a member of the Terylene Council which later became the board of ICI fibres division. In 1953 he was appointed a vice-president of ICI (Canada) and supervised the construction of a fibres plant at Millhaven, Ontario. In 1955 he returned to become chairman of what was then the metals division of ICI. Some found this appointment startling but those who worked with him quickly recognized a talent particularly needed in what was a traditional organization in a somewhat archaic industry. His performance as chairman was a great stimulus. Two years later he was promoted to the ICI board as technical director.

In 1960 A. Ernest (later Lord) Marples [q.v.], the Minister of Transport, determined to apply modern management practices to the railways. He set up a committee of inquiry to which Beeching was ICI's nominee, an appointment which was to lead to the role for which he is best known. In 1961 he was asked to join the British Transport Commission as chairman-designate of the new British Railways Board, with a brief to reorganize Britain's entire railway system. He became chairman of British Railways in 1963 and, characteristically, his approach was first to establish the facts.

The railway map he included in the reshaping report showed by the thickness of the lines on it how much traffic flowed on each route. A few thick lines were surrounded by a spider's web of lines which were little used. He proposed the closure of 5,000 miles of track, 2,000 stations, and the shedding of 70,000 staff. A change of government prevented him from completing this plan but a great part of it was accomplished in a short time. It is for this work that he is best remembered. He did not enjoy the image of the 'Beeching axe' but he recognized that much of the railway system could never be properly used and he was not deflected from his closure programme by its unpopularity with the public and trade unionists.

He was created a life peer in 1965 and in the same year returned to ICI as a deputy chairman. He has said his return was a mistake, but it was certainly not from the company's point of view. He established ICI's corporate planning department, and continued to exercise a profound influence on the company.

In 1966 he was appointed chairman of the royal commission on assizes and quarter-sessions. The committee rapidly cut through the complexities of the system of administering criminal justice and the tangles of existing practice. After only three years the Beeching report was accepted by the Labour government which had initiated it, as it was by the Conservative government which followed. It was implemented by the Courts Act of 1971. Beeching will be remembered as one of the outstanding legal reformers of the twentieth century.

In 1970, after leaving ICI, he joined the board of Redland Ltd. He became chairman in May of that year and remained in an executive capacity until 1977. He continued as a director until March 1984 and subsequently retained his links with the company as a special adviser on research and development. During his fifteen years' association with Redland he saw the company grow, as he put it, from 'a big small company to a small big one'—a typically modest statement about a company whose sales had grown to over £1 billion a year.

He joined the board of Furness Withy in 1972 and was chairman from 1973 until ill health forced him to give up the position in 1975. His appointment from outside this company was a complete break with tradition and although he was not a fit man it was with his guidance that plans were laid for successful diversification from an over-reliance on shipping.

His analytical skills were formidable and he applied them with great effect to problems of remarkable diversity. Once when asked what he would most like to do in his remaining years he suggested a really thoroughgoing examination of the entire system of government, but he was under no illusion that he would be asked to do it. 'Governments', he said, 'don't like delegating that sort of thing.'

He was concerned at the suspicion often attaching to top management in large companies, which he believed reflected widely held misconceptions, 'particularly amongst those who enjoy idealism untrammelled by knowledge'. He was a kind man with a quiet sense of humour, happier in small discussions and with little time for small talk.

He held many other directorships and presidencies. Among his awards were honorary degrees of the National University of Ireland (D.Sc., 1967) and of the University of London (LLD, 1966). In 1938 he married Ella Margaret, daughter of William John Tiley, engineer, of Maidstone. There were no children. Beeching died at Queen Victoria Hospital, East Grinstead, 23 March 1985.

[*The Times*, 25 March 1985; private information; personal knowledge.]

MAURICE HODGSON

BELOE, ROBERT (1905–1984), educationist and lay secretary to the archbishop of Canterbury, was born in Winchester 12 May 1905, the eldest of the four sons (there were no daughters) of the Revd Robert Douglas Beloe, the headmaster of Bradfield College, and his wife, (Margaret) Clarissa, the daughter of the Revd Prebendary John Trant Bramson, of Winchester College. Throughout his life his career was marked by the twin interests of education and the Church of England. He was educated at Winchester and at Hertford College, Oxford, where he obtained a second class honours degree in modern history in 1927.

There followed a brief apprenticeship in teaching—at his father's old school, at Eton, and (as some preparation for the very different realities of popular education) at an elementary school in Reading. In 1931 he joined the ranks of educational administration as a trainee in Kent, and was three years later promoted to Surrey, to which the central twenty-five years of his career were devoted. In 1939 he became deputy education officer, and one year later chief education officer.

He was therefore well placed to become a central member of that significant group of chief education officers who were responsible for putting into effect, in the years after World War II, the provisions of the Education Act of 1944 designed by R. A. Butler (later Lord Butler of Saffron Walden, q.v.). The two decades during which Beloe presided over the provision of public education in Surrey represented the high-water mark of the influence of enlightened administrators working amicably with elected councillors. Beloe was among the first to make detailed plans for the expansion of secondary education, which required both the abolition of all fees for secondary school pupils and the provision of secondary education for all pupils over the age of eleven.

The urgent requirements of expansion and building, the traditional respect in which the county grammar school had been held since the Act of 1902, and the prevailing orthodoxies inherited from the 1930s made it inevitable that the majority of secondary school pupils would be directed into secondary modern schools while the more academic minority should enjoy the undoubtedly superior provision of the grammar schools. Surrey was rich in such schools. Beloe's own preference was for a form of common or comprehensive schooling for pupils between the ages of eleven and thirteen, at which point more definitive choices could more appropriately be made. Although this preference did not prevail, he did ensure—as educational expectations and performance rose—that good opportunities for some of the pupils excluded from the grammar schools were provided within the secondary modern schools. In social and economic terms, Surrey was a well-favoured area, and the development of the bilateral school (essentially a secondary modern school with a grammar school stream within it) was a distinctive mark of Beloe's years in that county.

Such a development, in a county warily suspicious of innovation, furnished a natural bridge to the more dramatic growth of comprehensive schooling which was to be so pronounced a feature of the educational history of the 1960s. Beloe himself had little interest in some of the more ambitious claims made for comprehensive education, and little sympathy with their ideological roots. His concern was for the individual (pupil or teacher or colleague) and with the extension to him or her of the advantages enjoyed by those who came, like himself, from a privileged background.

One measure of these advantages in post-war Britain was access to an appropriate public examination attesting educational success. Beloe's introduction of the GCE (General Certificate of Education, as it had been constituted in 1951) into the secondary modern schools of Surrey was an index of the determination of parents, teachers, and pupils to enlarge the scope of that examination. It was, nevertheless, acknowledged that the GCE had weaknesses, which became more obvious as it was opened to less academic pupils.

Beloe had been a member since 1947 of the Secondary Schools Examination Council, the national body then responsible for policy on such matters, and it was therefore particularly appropriate that in 1958 he should be appointed chairman of a committee of that council charged with reviewing the provision of school ex-

aminations. The Beloe report of 1960 enjoys a place in educational history and led to the introduction five years later of the Certificate of Secondary Education (CSE). That examination was designed to meet the needs of pupils across a wide range of ability, to be determined by the curriculum provided in each school, to be controlled very largely by teachers, and to provide an acceptable alternative to the GCE. Its success and popularity created problems—inherent in the provision of two separate examinations for sixteen-year-old pupils—and in 1988 the two examinations were merged in the General Certificate of Secondary Education (GCSE). The CSE, the principal national monument to Beloe's career, represented an important transition and epitomized many of the liberal values of the decade in which it was born.

In 1959 Beloe, to the surprise of many, was appointed as the (first) lay secretary to the archbishop of Canterbury. In that discreet capacity he served for ten years two archbishops, G. F. Fisher (later Lord Fisher of Lambeth, q.v.) and A. M. Ramsey (later Lord Ramsey of Canterbury). He was appointed CBE in 1960 and received in 1979 an honorary DCL from the University of Kent. He was a member of many committees and other bodies, including the royal commission on marriage and divorce (1951–5). His recreations included travel and gardening.

In 1933 he married Amy, daughter of Captain Sir Frank Stanley Di Rose, second baronet, of the 10th Hussars, who was killed in action in 1914. They had one son and two daughters. Beloe died at his home in Richmond, Surrey, 26 April 1984.

[*The Times*, 30 April 1984; *Education*, 4 May 1984; private information; personal knowledge.]   H. G. JUDGE

BENNETT, JACK ARTHUR WALTER (1911–1981), medieval literary scholar, was born 28 February 1911 at Auckland, New Zealand, the elder son and elder child of Ernest Bennett, foreman shoemaker, and his wife, Alexandra Corrall. Both parents were born in Leicester in the 1880s; Ernest worked in shoe manufacturing, and emigrated to New Zealand in 1907; Alexandra followed in 1908. Bennett's younger brother Norman was born in 1916. Bennett was educated at Mount Albert Grammar School and Auckland University College. The most important influence on his work while he was at Auckland was that of P. S. Ardern; to his inspiringly rigorous and sceptical yet generously imaginative teaching Bennett was always to acknowledge a profound debt. While undergraduates, Bennett and his friend James

Bertram founded and edited a controversial literary magazine called *Open Windows*, intended to introduce New Zealand readers to the larger world of literary modernism.

In 1933 Bennett left for Merton College, Oxford, with a postgraduate scholarship awarded after he had taken a first class in the Auckland MA examination. He completed the Oxford BA English course (medieval and philological option) in 1935, with a first class; and at once, with Kenneth Sisam as his supervisor, embarked on a doctoral thesis on 'Old English and Old Norse studies in England from the time of Francis Junius till the end of the eighteenth century'. He finished the thesis and took the D.Phil. degree in 1938; in the same year he was elected to a junior research fellowship at The Queen's College, Oxford.

The outbreak of war in 1939 found Bennett in America. He stayed in the United States until September 1945, working for the Ministry of Information, becoming head of the research division, British Information Services. He returned to Queen's College late in 1945 and took up teaching for the English faculty. In 1947 he was elected to a tutorial fellowship at Magdalen College, where his teaching colleague in English was C. S. Lewis [q.v.]. He now began to help C. T. Onions (he wrote for this Dictionary the notices of both Lewis and Onions), who was also at Magdalen, to edit his journal *Medium Ævum*; he was to succeed Onions as sole editor in 1956 and to continue until 1980. In 1964 he succeeded C. S. Lewis as professor of medieval and renaissance English at Cambridge; in the same year he became a fellow of Magdalene College, and in 1968 was made keeper of the Old Library at Magdalene.

Bennett's first substantial publications were editions of medieval texts: *The Knight's Tale* (1954), and *Devotional Pieces in Verse and Prose from MS Arundel 285 and MS Harleian 6919* (1955). Also published in 1955 was his edition (with H. R. Trevor-Roper) of *The Poems of Richard Corbett*. His three books on Chaucer occupied him intermittently for the following twenty years. *The Parlement of Foules: An Interpretation* appeared in 1957, *Chaucer's Book of Fame* in 1968, and (based on the Alexander lectures given at Toronto in 1970-1) *Chaucer at Oxford and at Cambridge* in 1974. His inaugural lecture at Cambridge, *The Humane Medievalist* (1965), is a vivid statement of professional faith as well as a warm tribute to C. S. Lewis. *Early Middle English Verse and Prose*, a volume of annotated selections which Bennett edited with G. V. Smithers, appeared in 1966.

During the 1950s, 60s and early 70s Bennett was also much involved as general editor of the Clarendon Medieval and Tudor series of annotated texts, to which he himself contributed

two of the later volumes: *Gower* (1968) and *Piers Plowman* (1972). From 1972 to 1976 he published four sets of *Supplementary Notes on 'Sir Gawain and the Green Knight'* and in 1980 a short book *Essays on Gibbon* (this was not such a radical departure as it might seem, since his interest in Gibbon went back to his doctoral thesis over forty years before). Bennett's last completed book, *Poetry of the Passion* (published in 1982), examined poetic treatments of Christ's Passion from the Old English *Dream of the Rood* to *The Anathemata* of David Jones [q.v.]. Also in 1982 appeared a volume of Bennett's uncollected essays; edited by Piero Boitani and published in Rome, it was called *The Humane Medievalist and Other Essays*. Bennett left unfinished at his death a large work which had engaged him for many years—the medieval volume intended for The Oxford History of English Literature. Entitled *Middle English Literature*, this was completed and edited by Douglas Gray and published in 1986. *Medieval Studies for J. A. W. Bennett*, a volume of essays by pupils and friends, edited by Peter Heyworth and intended for his seventieth birthday, appeared too late for him to see it in 1981.

Although Bennett was unquestionably one of Britain's leading literary medievalists, his interests were by no means confined to the Middle Ages. The seventeenth century was a favourite period, and he was surprisingly expert in American literature. Erudite, imaginative, sensitive, his work as a whole was nourished by his curious and highly independent reading in all periods and by a wide and discriminating experience of the arts, especially painting. Though physically frail-seeming, he had a strong constitution which saw him through several periods of illness, and though himself gentle and sweet-natured, essentially a private person, he was also something of a fighter, always committed (sometimes fiercely) to one or another cause, however unpopular. He was a devout Roman Catholic. Through his personal distinction and the originality of his scholarship and learning, he left a deep mark on many of his pupils and graduate students.

Bennett was made a fellow of the British Academy in 1971. In 1976 he held a visiting fellowship at the Australian National University, and was elected a corresponding fellow of the Medieval Academy of Arts and Sciences. In 1978 he was elected to an emeritus fellowship of the Leverhulme Trust, and in 1979 was awarded the Sir Israel Gollancz prize of the British Academy.

Bennett was twice married. His first marriage (1937) to an Oxford pupil, Edith Bannister, was annulled in 1949. In 1951 he married Gwyneth Mary, daughter of Archibald John Nicholas, civil servant. They had two sons. Bennett had long suffered from cardiac asthma; some months after his wife's death in 1980, he was on his way to New Zealand and had interrupted his journey at Los Angeles when he died suddenly 29 January 1981.

[Norman Davis in *Proceedings* of the British Academy, vol. lxviii, 1982; *Magdalen College Record*, 1979; private information; personal knowledge.] EMRYS JONES

BETJEMAN, SIR JOHN (1906-1984), poet, writer on architecture, and broadcaster, was born 28 August 1906 at 52 Parliament Hill Mansions, north London, the only child of Ernest Betjemann, a furniture manufacturer, and his wife, Mabel Bessie Dawson. The family name, of Dutch or German origin, can be traced back to an immigration in the late eighteenth century. The poet adopted his style of it about the age of twenty-one.

He attended the Dragon School, Oxford, and Marlborough College and was active at both in school theatricals and in various forms of writing. He entered Magdalen College, Oxford, in 1925 but was rusticated three years later after failing in divinity. To his father's deep disappointment he declined to enter the family business, becoming successively a preparatory school master (at Thorpe House School, Gerrard's Cross, and at Heddon Court, Cockfosters, Hertfordshire), assistant editor of the *Architectural Review* in 1930, and film critic of the London *Evening Standard* in 1933. Shortly after his marriage that year he moved to a farmhouse in Uffington, in the Vale of White Horse, Berkshire, where his wife was able to keep horses.

His first two collections of poems, *Mount Zion* (1931) and *Continual Dew* (1937), showed a poet already fully formed, with the impeccable ear, delight in skill, and assured mastery of a wide range of tones and themes that so distinguished all his subsequent work in verse. In these early volumes, as later, Betjeman moved with perfect assurance from light pieces, *vers de société*, satirical sketches of muscular padres or philistine businessmen (as in the famously ferocious tirade 'Slough') to sombre reflections on the impermanence of all human things. In a remarkable variety of metres and manners the poems make an equally clear-cut impression on the reader, never drifting into obscurity and never once tainted with the modernism then fashionable. Here too he gave glimpses of the world of gas-lit Victorian churches and railway stations, of grim provincial cities and leafy suburbs that he was to make his own, not forgetting the grimmer contemporary developments, shopping arcades, and bogus Tudor bars, that he saw effacing it and strove to resist.

These concerns are reflected in his publication of 1933, *Ghastly Good Taste*, subtitled 'a depressing story of the rise and fall of English architecture', which attracted more immediate attention than either of his first books of poems. In it he attacked not only modern or modernistic trends but also the other extreme of unthinking antiquarianism, nor had he any time for the safely conventional. While still at school he had become interested in Victorian architecture, thoroughly unmodish as it was at the time. His writings on the subject over the years led to a revival in appreciation of the buildings of that era and paved the way for the eventual founding of the successful Victorian Society. Further afield, he showed among other things his fondness for provincial architecture in his contributions to the Shell series of English county guides, of which the most notable is that on Cornwall, another enthusiasm acquired in boyhood. He had joined the publicity department of Shell in the mid-1930s.

Betjeman's poetical career had begun to flourish with the appearance in 1940 of *Old Lights for New Chancels* and continued with *New Bats in Old Belfries* (1945) and *A Few Late Chrysanthemums* (1954). Many of the poems in these three volumes became classics of their time, including 'Pot Pourri from a Surrey Garden', 'A Subaltern's Love-song' and 'How to Get On in Society'. His *Collected Poems* came out in 1958 and went through many impressions. *Summoned by Bells* is dated 1960, a blank-verse poem of some 2,000 lines that gives an account of his early life up to schoolmastering days with characteristic animation, humour, sadness, and abundance of detail.

Both these volumes were widely successful, the first edition of the *Collected Poems* selling over 100,000 copies. Betjeman's poetry has continued to enjoy a popularity unknown in this country since the days of Rudyard Kipling and A. E. Housman [qq.v.]. No doubt it was poems like the three mentioned above and the more obviously quaint period pieces that made an immediate appeal. Nor should one underestimate the sheer relief and delight to be felt at the appearance of a poetry of contemporary date that was easy to follow and yielded the almost forgotten pleasures of rhyme and metre expertly handled. Nevertheless it may not be instantly obvious how so strongly personal a poet, one given moreover to evoking characters and places that might seem outside general interest, should have proved so welcome. He is full of nasty jolts for the squeamish too.

The answer must lie in the closeness of the concerns of Betjeman's poetry to the ordinary day-to-day experience of his readers, something else that had been far to seek in the work of his contemporaries. For all the delight in the past, it is the past as seen from and against the present; for all the cherished eccentricities—as such hardly repugnant to British taste anyway—the subject is ourselves and our own world. The point was well made by Philip Larkin [q.v.], the friend and admirer who best understood his work: 'He offers us something we cannot find in any other writer—a gaiety, a sense of the ridiculous, an affection for human beings and how and where they live, a vivid and vivacious portrait of mid-twentieth-century English social life' (Philip Larkin, 'It Could Only Happen in England', 1971, in *Required Writing*, 1983, pp. 204–18).

In World War II Betjeman volunteered for the RAF but was rejected and joined the films division of the Ministry of Information. He then became UK press attaché in Dublin (1941–3) to Sir John (later Lord) Maffey [q.v.] and subsequently worked in P branch (a secret department) in the Admiralty, Bath. In 1945 he moved to Farnborough and in 1951 to Wantage where his wife opened a tea shop, King Alfred's Kitchen. By the mid-1950s his main income came from book reviewing, broadcasting, and his poems. He pursued a highly successful career as a broadcaster, and with the help of the image he projected through television, engaging, diffident, exuberant, often launched on some architectural or decorative enthusiasm, he became a celebrated and much-loved figure in national life. He used this position to further zealously the defence of many buildings threatened with demolition. He was able to save many of these, from St Pancras station to Sweeting's fish restaurant in the City of London, though the Euston Arch was lost despite his vigorous campaign. Appropriately, it was at St Pancras, naming a British Rail locomotive after himself, that he was to make his last public appearance on 24 June 1983.

In later years Betjeman continued his work in poetry, publishing *High and Low* in 1966 and *A Nip in the Air* in 1974. The contents of these two volumes reveal no loss of energy; indeed, poems like 'On Leaving Wantage 1972' embody a melancholy, even a tragic, power he had never surpassed. All the same, apart from the ebulliently satirical 'Executive', almost none of them have achieved much individual popularity. They were incorporated entire in the fourth edition of the *Collected Poems* in 1979. Those in *Uncollected Poems* (1982) were such as the poet was content should remain in that state and are unlikely to gain him many new readers, though lovers of his work would not be without any of them. To this Dictionary he contributed the notice of Lord Berners.

No account of Betjeman's life could fail to stress his devoted adherence to the Anglican Church, not only for the sake of its buildings,

its liturgy, and its worshippers but for its faith. Expressions of doubt and the fear of old age and death are strong and memorable in his poetry, but 'Church of England thoughts' are pervasive too, and one of its chief attractions, seldom given proper weight, has been the sense of an undemonstrative but deep Christian belief of a kind able to contain the harsh, ugly, absurd realities of present-day existence.

John Betjeman was a sociable man, one who loved company and valued it the more for being also a shy man. Although he was renowned for his youthful gregariousness and was endlessly affable with all manner of people, his was a life rich in intimacy. Latterly he was partial to small gatherings and old friends and a sufficiency of wine. His expression in repose was timid, perhaps not altogether at ease, and even at the best of times it was possible to surprise on his face a look of great dejection. But all this was blown away in an instant by laughter of a totality that warmed all who knew him. His presence, like his work in verse and prose, was full of the enjoyment he felt and gave.

He was chosen as poet laureate to universal acclaim in 1972. He received many distinctions besides, being awarded the Duff Cooper memorial prize, the Foyle poetry prize, and in 1960 the Queen's medal for poetry. In that year too he was appointed CBE, in 1968 he was elected a Companion of Literature by the Royal Society of Literature, and in 1969 he was knighted. He was an honorary fellow of his old college, Magdalen (1975), and also of Keble College, Oxford (1972). He had honorary degrees from Oxford, Reading, Birmingham, Exeter, City University, Liverpool, Hull, and Trinity College, Dublin. He was also honorary ARIBA.

In 1933 he married Penelope Valentine Hester, only daughter of Field-Marshal Sir Philip Walhouse Chetwode, first Baron Chetwode [q.v.], OM, commander-in-chief in India at the time. In latter years they were amicably separated and Betjeman was cared for by his friend Lady Elizabeth Cavendish, sister of the Duke of Devonshire. Lady Betjeman, a writer of travel books (as Penelope Chetwode) and a devotee of Indian culture, died in 1986. There were a son and a daughter of the marriage. From the mid-1970s Betjeman had suffered increasingly from the onset of Parkinson's disease and he died at Treen, his home in Trebetherick, Cornwall, 19 May 1984. He is buried in nearby St Enodoc churchyard.

[*The Times*, 21 May 1984; Bevis Hillier, *John Betjeman: a Life in Pictures*, 1984, and *Young Betjeman*, 1988; personal knowledge.]

KINGSLEY AMIS

BIRCH, (EVELYN) NIGEL (CHETWODE), BARON RHYL (1906–1981), economist and politician, was born in London 18 November 1906, one of two sons (there were no daughters) of (General Sir) (James Frederick) Noel Birch, GBE, KCB, KCMG, and his wife Florence Hyacinthe, the daughter of Sir George Chetwode, sixth baronet, and sister of (Field-Marshal Lord) Chetwode [q.v.]. Educated at Eton, he became a partner in the stockbroking firm of Cohen Laming Hoare. A skilful operator, especially in the gilt-edged market, he acquired at an early age a sufficient fortune to assure him independence. He joined the Territorial Army (King's Royal Rifle Corps) before 1939 and served in World War II on the general staff in Britain and the Mediterranean theatre, attaining the rank of lieutenant-colonel in 1944.

He was elected to Parliament for Flintshire in 1945 and quickly made a formidable reputation on the Conservative opposition benches for his mordantly witty interventions and speeches and for his advocacy of strictly honest public finance and his critique of the prevalent Keynesian trend. On forming his government, in November 1951 (Sir) Winston Churchill appointed him—some believed as a result of mistaken identity—not, as was expected, to a financial post but as parliamentary under-secretary at the Air Ministry and subsequently at the Ministry of Defence (from the end of February 1952).

He attained ministerial rank in October 1954 as minister of works where he enjoyed, as he said, 'gardening with a staff of three thousand' and deserved to be remembered for his insistence upon retaining the trees in Park Lane when it was widened. He served in the government of Sir Anthony Eden (later the Earl of Avon) as secretary of state for air from December 1955; but the first appointment in his special field came only when Harold Macmillan (later the Earl of Stockton) made him economic secretary to the Treasury—technically a demotion—in January 1957.

Birch supported and encouraged the chancellor of the Exchequer Peter (later Lord) Thorneycroft in evolving a theory of inflation which was the recognizable forerunner of the monetarism of later decades, and when Thorneycroft failed to obtain from his cabinet colleagues the restraint in public expenditure he considered essential, Birch joined him and the financial secretary (J. Enoch Powell) in the triple resignation on 6 January 1958 which Macmillan, in a jaunty phrase which became famous, dismissed as 'a little local difficulty'.

Though he never held office again, Birch exercised great influence on economic and financial questions and continued to castigate what he regarded as the lax and inflationary proclivities of Macmillan's government. When Macmillan was about to be toppled by the scandal of the 'Profumo affair', Birch demanded his resignation in a philippic long famous for its closing phrase from Robert Browning's 'The Lost Leader', 'Never glad confident morning again', which went down in parliamentary memory along with his exclamation to a full House on the announcement of the resignation of E. H. J. N. (later Lord) Dalton [q.v.]: 'They've shot our fox!' The failing eyesight which increasingly inhibited Birch's other activities did not interfere with his stream of contributions to economic debate, which, though meticulously prepared, were invariably delivered without a note.

Birch, who once accused the Treasury of 'the reckless courage of a mouse at bay', was representative of a distinctive element in the House of Commons after World War II. He combined with personal courage, independence, and integrity an understanding of public finance and economic affairs acquired and perfected in the practical life of the City. He was appointed OBE in 1945 and admitted to the Privy Council in 1955.

Having continued from 1950 to represent West Flint, where he resided at Saithailwyd ('seven hearths') near Holywell, he went to the House of Lords in 1970 as a life peer, Lord Rhyl. He married in 1950 Esmé Consuelo Helen, OBE, daughter of Frederic Glyn, fourth Baron Wolverton. There were no children. He died 8 March 1981 at Swanmore, Hampshire.

[Personal knowledge.]   J. ENOCH POWELL

BIRLEY, SIR ROBERT (1903-1982), headmaster of Eton and Charterhouse, and educational administrator, was born 14 July 1903 at Midnapore, the only son and elder child of Leonard Birley, CSI, CIE, of the Indian Civil Service (Bengal), and his first wife, Grace Dalgleish, younger daugher of Maxwell Smyth, indigo planter, formerly of Moffat, Dumfriesshire. His mother died when he was three and he was brought up by his paternal grandparents in Bournemouth. He was educated at Rugby (1917-22) where he specialized in history and soon developed a taste for exploring its byways and oddities, which was stimulated by a tour of Europe between school and university, roaming about Italy, Germany, and Czechoslovakia. By a series of chances he had generally escaped external examinations and this was to colour his approach to education later on. He went to Balliol College, Oxford (1922-5, Brackenbury scholar in history), where he won the Gladstone memorial prize (1924) and obtained first class honours in modern history (1925). Balliol nurtured the strong idealism in Birley's character.

On leaving Oxford he joined the staff at Eton (1926-35), where he made an immediate impact as an inspiring history teacher. He also startled some of the staff by putting the case for the strikers in the general strike of 1926 and organizing relief for the unemployed of Slough in 1931. He developed a lifelong absorption in the magnificent college library from which he obtained much of his large store of often esoteric knowledge. He taught mainly the cleverer boys, introducing them to adult modes of thought and awakening their aesthetic sense. In all he made such an impression that on the retirement of C. A. Alington [q.v.] as head master in 1933, Birley was seriously suggested by some as his successor. But the post went to an older man, (Sir) Claude Elliott [q.v.], and soon Birley was made the headmaster of Charterhouse (1935-47).

As a young liberal-minded headmaster he had a stimulating effect in the school, and brought it into closer touch with the world of affairs. He believed that a headmaster should do as much teaching as possible. In addition to the contact with the pupils that this afforded, Birley and his wife encouraged individual boys, particularly those whom life was not treating well, with private hospitality, both in the school and at their holiday home in Somerset. Soon Charterhouse had to be guided through the war and Birley stoutly resisted all pressures to remove it from Godalming.

Birley's public spirit made him a prominent member of the Headmasters' Conference. In 1942 he was appointed as its member on the government committee set up under Lord Fleming [q.v.] to bridge the gap between private and state education. He contributed much to the final report and at the end of the war his reputation was such that he was appointed educational adviser to the military governor, Control Commission for Germany (1947-9). In Germany Birley worked for reconciliation by bringing people together. He must be given a generous share of the credit for rebuilding the German educational system. He also played a leading part in setting up the Königswinter conferences, the first of which was held in 1950, by which Anglo-German understanding was strengthened and sustained.

In 1949 Birley became head master at Eton, where he remained until 1963. During his first term there he gave the Reith lectures, which were not a success. Historian that he was, he had a deep understanding of the historical role of Eton and a subtle appreciation of its special character. To this Dictionary he contributed the notice of a former provost of Eton, Sir C. H. K. Marten. Birley was not a reforming head master but under him respect grew for the intellect and the arts, and the teaching of science was improved and modernized. He realized the need for technologists in the second half of the twentieth century, and he felt that ideally they should be literate, go to church, and know Greek.

Birley soon became the most influential public school headmaster of his generation and he occupied the chair of the Headmasters' Conference for four years. He saw the main purpose of education as the encouragement of independent thought. The leisurely Eton education of his day, with its late specialization and freedom from the tyranny of the A-level syllabus, admirably suited this approach. However, towards the end of Birley's time some parents grew restive and there was a growing demand for paper qualifications, which he was unable to resist.

The next phase of his life was as remarkable as it was unexpected. On a visit to South Africa Birley's crusading zeal was aroused by the political and racial constraints in that country. He left Eton to become visiting professor of education at the University of the Witwatersrand (1964-7). Tirelessly and single-mindedly he and his wife worked in their favourite cause, freedom of the spirit against repression and injustice. Their presence heartened liberals throughout South Africa and they made innumerable friends among both Whites and Blacks.

Birley filled one more important post on his return from South Africa, that of professor and head of the department of social science and the humanities at the City University, London (1967-71).

Six feet six inches in height, clumsy and unathletic, with a shy smile and an increasing stoop, Birley was an imposing and compelling figure, particularly in his cassock and bands as head master of Eton. Incapable of small talk, he would launch into the serious topic of the moment with excited earnestness, livened by a keen sense of the absurd and a mischievousness which never lapsed into irresponsibility. Anyone so sure of his own rectitude could not fail to have his detractors, but his natural goodness and lofty ideals won the respect of most and the devotion of many. His effect on the moral and intellectual climate of his time was immense.

Birley was given many honours in his life, including honorary doctorates (one of them an Oxford DCL, 1972) at eight universities. He was appointed CMG (1950) and KCMG (1967) but not to the honour to which his talents and sense of service were perhaps best suited, namely a life peerage. He became an honorary fellow of Balliol in 1969. He died 22 July 1982 at Somerton in Somerset.

In 1930 Birley married Elinor Margaret, daughter of Eustace Corrie Frere, FRIBA, architect. They had two daughters, Julia (born 1931, died 1978) and Rachael (born 1935).

[Arthur Hearnden, *Red Robert, a Life of Robert Birley*, 1984; David Astor and Mary Benson (eds.), *Robert Birley 1903-1982*, 1985; Arthur Hearnden (ed.), *The British in Germany*, 1978; Eton College *Chronicle*, September 1963 and September 1982; personal knowledge.]   M. A. NICHOLSON

BLACKWELL, SIR BASIL HENRY (1889-1984), bookseller and publisher, was born 29 May 1889, at Blackwell's bookshop, 51 Broad Street, Oxford, the only son and younger child of Benjamin Henry Blackwell, bookseller, and his wife, Lydia ('Lilla'), daughter of John Taylor, a Norfolk farmer. Basil's grandfather, Benjamin Harris Blackwell, had first become a bookseller in Oxford in 1845, but he died prematurely when his elder son, Benjamin Henry, was only six. However, on New Year's day 1879 Benjamin Henry opened his shop at 50 Broad Street, and by the time Basil was born, the business was well established.

Blackwell was educated at Magdalen College School, Oxford, and in 1907 he achieved a postmastership at Merton College, Oxford. He received a third class in classical honour moderations (1909) and a second in *literae humaniores* (1911). On graduation he went to work for two years in the London office of the Oxford University Press in order to learn about publishing. On New Year's day 1913 he joined his father and in September established the publications department separately from the bookshop. In the next few years (he was rejected for military service because of defective eyesight) he laid the foundation for the publishing firm of Basil Blackwell & Mott (founded in 1922) which encouraged many young Oxford poets. Blackwell and Adrian Mott were also co-directors of the Shakespeare Head Press, which they saved from bankruptcy, and which allowed them to indulge in fine book production.

However, in 1924 Blackwell's father died and he became chairman of B. H. Blackwell Limited. This meant that he had to concentrate on the main business of selling new books. This he did reluctantly, as he preferred publishing to bookselling and selling antiquarian books to new books. Almost immediately on assuming complete responsibility for the business, his attention was drawn to the existence of a ring of collusive bidders in the antiquarian book auctions. Blackwell, jealous of the trade's reputation, was determined to oppose the ring in the auction rooms, and did so successfully. In 1925-6 he was elected president of the Antiquarian Booksellers' Association. Collusive bidding at auctions is difficult to control and

years later, in 1955, there was a further outbreak. Blackwell had the matter raised vigorously in Parliament, and, as a result of the publicity, the Antiquarian Booksellers' Association made a new and even more stringent rule against the practice.

Throughout his life he was to serve his profession well and to exercise considerable influence in the book trade of the United Kingdom. From 1934 to 1936 he was president of the Booksellers' Association. At the time the president of the Publishers' Association was (Sir) Stanley Unwin [q.v.] and this was fortunate as both were in sympathy with the view that booksellers and publishers, so often mutually suspicious, should be brought together in co-operation. At the 1932 Booksellers' Association conference Blackwell, at Unwin's request, had succeeded in persuading booksellers to support the book token scheme, originally proposed by a publisher in 1928, and treated at first with suspicion by booksellers. In 1934 Blackwell invited Unwin to be joint host to an informal weekend conference of fifty booksellers and publishers in Ripon Hall, Oxford, the beneficial influence of which lasted for many years. At the end of his period of office Blackwell's reputation for approachability, concern for the welfare of the trade, and enthusiasm was such that he was affectionately nicknamed the Gaffer, and was universally so called for the rest of his life.

By 1938 the bookselling business was flourishing and he established a second publishing business, Blackwell Scientific Publications Limited. Contrary to the experience of World War I, and to expectations, the sales of books greatly increased in World War II, and in 1945 B. H. Blackwell Limited and its two associated publishing companies were well set. Blackwell's eldest son, Richard Blackwell [q.v.], returned from the war and in the course of the next twenty years gradually assumed responsibility. Blackwell finally resigned the chairmanship after forty-five years in 1969 to become president, and handed on a business vastly increased, both in reputation and in size, although much of the credit for the post-war expansion overseas belonged to his son, Richard.

For the rest of his life to within a week of his death, he continued to attend all board meetings as president, to peruse all accounting statements, and to visit the three main offices. He instinctively supported the attitudes and conventions of a previous age, working long hours in his office and supervising every detail. He believed in paternalism and it was his settled intention that Blackwell's should remain independent and controlled only by those members of the family actively concerned with its management. However, he was always ready to listen to the views, even when unwelcome, of others, and he was invariably generous in the credit he gave for the firm's success to those who served him over the years. He would say that it was the Blackwells' good luck that they could command such ability and loyalty, but in fact, it was due to his personal qualities. He was courteous, sympathetic, and liberal, although he did not allow these qualities to influence his shrewd commercial judgement. He had the reputation of being one of the best read people in the country. He wrote no books, but his letters and speeches were a delight. To this Dictionary he contributed the notice of J. G. Wilson.

In 1941 Blackwell became a JP for Oxford. In 1956 he was knighted. In 1959 he was elected an honorary fellow of Merton College. He was president of the Classical Association (1964-5), and of the English Association (1969-70). In 1979, in his ninetieth year, and on the hundredth anniversary of the founding of B. H. Blackwell Ltd., the University of Oxford conferred on him an honorary DCL. He was also an honorary LLD of Manchester University (1965). But the distinction he cherished most was that of honorary freeman of the City of Oxford, conferred upon him in his eightieth year.

In 1914 he married Marion Christine (died 1977), daughter of John Soans, schoolmaster, of Ramsgate. They had two sons (the elder died in 1980) and three daughters. Blackwell died at Tubney House 9 April 1984. A portrait in oils by John Ward hangs in the boardroom at Beaver House, a large new building constructed by Blackwell's near Oxford station.

[A. L. P. Norrington, *Blackwell's 1879-1979*, 1983; personal knowledge.]   JOHN BROWN

BLAKENHAM, VISCOUNT (1911-1982), politician and farmer. [See HARE, JOHN HUGH.]

BLUNT, ANTHONY FREDERICK (1907-1983), art historian and communist spy, was born in Bournemouth, Hampshire, 26 September 1907, the third and youngest son (there were no daughters) of the Revd Arthur Stanley Vaughan Blunt and his wife, Hilda Violet, daughter of Henry Master of the Madras Civil Service. His father was a kinsman of the poet, anti-imperialist, and libertine, Wilfrid Scawen Blunt [q.v.], his mother a friend of the future Queen Mary; both these connections for Blunt's future career. After a childhood acquiring a lasting enthusiasm for French art and architecture while his father was chaplain of the British embassy in Paris, he went to Marlborough College; his artistic interests were further stim-

ulated by his eldest brother, Wilfrid, a future art master. Going up to Trinity College, Cambridge, on a scholarship, Blunt graduated there with a second in part i of the mathematical tripos (1927) and a first in both parts (1928 and 1930) of the modern languages tripos (French and German). In 1932 he was elected a fellow of the college on the strength of a dissertation on artistic theory in Italy and France during the Renaissance and seventeenth century. By now he was already writing for the *Cambridge Review* and within a year was contributing articles and reviews on modern art to the *Spectator* and *Listener*. At first he championed the modern movement, which for him was a product of the School of Paris, but later, influenced by the Marxism which he had first espoused in 1934, he just as fiercely, if temporarily, attacked modernism as irrelevant to the contemporary political struggle.

While still an undergraduate he was invited to join the Apostles. The values of this exclusive Cambridge society between the wars (derived in part from the teaching of the philosopher, G. E. Moore, q.v.) have been summed up as the cult of the intellect for its own sake, belief in freedom of thought and expression irrespective of the conclusions to which this freedom might lead, and the denial of all moral restraints other than loyalty to friends. An influential minority of the society's members were, moreover, like Blunt himself, homosexual, and, at a time when homosexual acts were still illegal in Britain, he seems to have relished the resulting atmosphere of secrecy and intrigue. Where he stood out from most of his contemporaries was in his phenomenal intellectual energy, powers of concentration, and capacity for self-discipline — qualities he retained into old age. Endowed with charm, vitality, and good looks, he lived an active social life and also travelled widely during the 1930s on the Continent. In 1936-7 he resigned his Cambridge fellowship, joined the staff of the Warburg Institute, lectured on baroque art at the Courtauld Institute, and allowed himself to be drawn into working for the Russian secret intelligence service by the charming, scandalous Guy Burgess. Much ink has since been spilt over the identity of the individual who recruited spies for the Russians at Cambridge in the 1930s, but in Blunt's case his fondness for Burgess, to whom he remained devoted until the latter's death in Moscow in 1963, is probably sufficient explanation. His decision seems to have been at once emotional and cold-blooded. There is no doubt of his hatred of Fascism at the time but whether he was ever a convinced communist is far from clear. Yet his sense of professional dedication would have made him as capable a spy as he was an art historian.

After a futile spell in France in the Field Security Police, he joined the British counter-intelligence service, MI5, in 1940 and remained with it in London throughout World War II. He is known to have given his Soviet controllers every detail of the service's organization and the names of all its personnel and he presumably also gave away any military secrets to which he had access. How much damage he actually did to the British war effort is hard to say; it may have been rather slight. But there was a real danger that the information he supplied might have been leaked to the Germans, so causing the deaths of Allied agents in occupied Europe, or, alternatively, used by the Russians in preparing policies hostile to the British and Americans. At the end of the war in 1945, Blunt left MI5 and thereafter had no more secrets to impart, though he remained in touch with Burgess until the latter's defection to Moscow in 1951 and he continued occasionally to see another of the Cambridge spies, H. A. R. ('Kim') Philby. In May 1951 Philby told Blunt that the security authorities were planning to arrest Donald Maclean [q.v.]. When Blunt passed on the warning both Maclean and Burgess were able to escape to Moscow.

All this went on concurrently with the development of Blunt's career as an art historian. Already before the war he saw that the refugee scholars at the Warburg Institute had brought with them from Hamburg both an intellectual rigour and a soundly-based historical method that were new to the study of the art of the past in Britain. In 1937–9 he published scholarly articles in the Warburg *Journal* on such diverse topics as 'The Hypnerotomachia Poliphili in 17th-century France' and 'Blake's *Ancient of Days*' and began his great work on the seventeenth-century French painter, Nicolas Poussin, characteristically with an article showing that Poussin's 'Notes on Painting' were not original but were largely copied from obscure ancient and Renaissance literary sources. Blunt also helped to establish friendly relations between the Warburg and Courtauld Institutes, becoming deputy director of the latter in 1939. In 1940 most of his fellowship dissertation was published as *Artistic Theory in Italy 1450–1600*; written with his customary lucidity and stylistic grace, it remains a useful introduction to its subject. During the war he wrote further articles in periodicals and a book on the French architect, François Mansart (1941); in 1945 he published a catalogue of the French drawings in the royal collection at Windsor Castle.

In the same year, 1945, to the puzzlement of his friends, who knew of his political sympathies though not of his activities as a spy, he accepted appointment as surveyor of the King's (after 1952 the Queen's) Pictures. One of his motives in taking the job may well have been to deflect suspicion away from himself in the event that any of his fellow conspirators was caught—for who in authority, he would have calculated, would think of doubting the loyalty of a senior royal servant? On the other hand, his activity, soon after appointment, in helping to rescue on behalf of George VI from a castle in Germany what are now said to be compromising letters from the Duke of Windsor quite possibly had no sinister implications so far as Blunt was concerned. At all events, he gave every sign of enjoying the post of surveyor, which he retained until 1972. While the day-to-day work was left to his deputy and eventual successor, (Sir) Oliver Millar, Blunt took the major decisions, including that of opening the Queen's Gallery, Buckingham Palace, in 1962. (For his services he was appointed CVO in 1947 and KCVO in 1956.) In 1947 he had become director of the Courtauld Institute of Art and professor of the history of art in the University of London. Thenceforth the Institute was to be his home (he had a flat at the top of the building, designed by Robert Adam, in Portman Square) and the centre of his life. In almost every sense he was a superb director. He had a natural authority, an infectious enthusiasm for his subject, and a winning way with students and younger colleagues. Teaching more by example than by precept, he inspired those around him to give of their best, and it was under him that the Courtauld, whose staff and student numbers more than doubled during his time, earned the position, which it had had in theory since its foundation in 1931, of being the principal centre for the training of art historians in Britain.

The first phase of his scholarly career was crowned by a masterly survey in the Pelican History of Art series, *Art and Architecture in France 1500–1700* (1953). Lucid, penetrating, and comprehensive, this is still the best study of its subject and is perhaps Blunt's single most successful book. The next dozen years were spent mainly working on Poussin, an artist for whose intellectual power, self-discipline, and personal reticence he had a natural sympathy. His erudite monograph on Poussin, based on a thorough study of the artist's ideas and including a catalogue of the paintings, appeared in 1966–7. Afterwards Blunt turned his attention as a scholar chiefly to Italian baroque architecture, on which he also wrote several books. To this Dictionary he contributed the notices of Frederick Antal and Tomás Harris. He retired from the Courtauld and the university in 1974, covered with British and French academic honours, including honorary D.Litts. of Bristol (1961), Durham (1963), and

Oxford (1971), and the Legion of Honour (1958).

Yet all this time he was at risk of exposure as a former spy. For many years he successfully resisted interrogation by the security services but in 1964, after the FBI had found a witness prepared to testify that Blunt had tried to recruit him during the 1930s, he made a secret confession in return for a promise of immunity from prosecution. In the later 1970s the pressure mounted again, as a result of investigations by independent writers on espionage relying on information leaked by former security officers. On 15 November 1979, the prime minister confirmed in the House of Commons that Blunt had been an agent of, and talent spotter for, Russian intelligence before and during World War II, although she added that there was insufficient evidence on which criminal charges could be brought. His knighthood was annulled, as was the honorary fellowship he had held at Trinity College since 1967, and immediately the press, radio, and television began a campaign of vilification. There was also much discussion by intellectuals in the serious press not only of Blunt but of the whole phenomenon of the Cambridge spies who had put belief in communism above loyalty to country in the 1930s and 1940s.

Undoubtedly some of the agitation was motivated by class hatred, and it is a striking fact that both Blunt's own actions and the treatment of him not only by the public but also by officials were pervaded at every turn by the class divisions in British society. More immediately, his career can perhaps best be explained by the fatal conjunction in him of his own outstanding gifts and his desire to be at once part of the establishment and against it; or, as an acquaintance put it, 'The trouble with Anthony was that he wanted both to run with the hare and hunt with the hounds.' He died of a heart attack, in the London flat to which he had retired near the Courtauld, 26 March 1983. He was unmarried.

[E. K. Waterhouse, introduction to *Essays in Renaissance and Baroque Art presented to Anthony Blunt*, 1967; Anthony Blunt, essay in *Studio International*, 1972; Andrew Boyle, *The Climate of Treason*, 1979; Barrie Penrose and Simon Freeman, *Conspiracy of Silence*, 1966; Peter Wright, *Spycatcher*, 1987; John Costello, *Mask of Treachery*, 1988; personal knowledge.]   MICHAEL KITSON

BODLEY SCOTT, SIR RONALD (1906-1982), consultant physician, was born in Bournemouth 10 September 1906, the eldest of six sons (there were no daughters) of Maitland Bodley Scott, a general practitioner in Bournemouth, and his wife, Alice Hilda Durancé

George. He was educated at Marlborough College and Brasenose College, Oxford, where he obtained a second class in natural science (physiology) in 1928. He had been the only son to decide to do medicine, and entered St Bartholomew's Hospital, London, to complete his clinical studies, graduating BM, B.Ch. in 1931.

Once qualified, Bodley Scott served a short period as house physician at St Bartholomew's Hospital, and then left to join the family practice in Bournemouth. He had evidently made such an impression as a student and house physician that he was invited to return to Bart's, first as chief assistant to A. E. Gow, and then as first assistant on the medical unit. This was a recognized route for those who were going on to consulting practice. During this time he obtained his MRCP (Lond. 1933), and developed an interest in clinical haematology. He obtained an Oxford DM in 1937, using the newly developed technique of bone marrow aspiration. His studies with this method led to his describing with A. H. T. Robb-Smith the clinical entity of histiocytic medullary reticulosis in the *Lancet* in 1939.

World War II started not long after he had obtained a consultant appointment at Woolwich Memorial Hospital (1936-71). He entered the Royal Army Medical Corps, and in 1941 joined a very distinguished group of medical men who were to work in the Middle East, a very important theatre of war, where he remained for four years, achieved the rank of lieutenant-colonel, and was responsible for the medical division in No. 63 and No. 43 general hospitals. Even though hard pressed by military medical duties, he still found time to lecture for the British Council, and to examine in medicine in Cairo. He was elected FRCP (Lond.) in 1943.

At the end of the war Bodley Scott returned to London. He was one of the first of five new assistant physicians to be appointed at Bart's (1946) and he rapidly became an internationally recognized figure in the management of leukaemia and lymphoma. He introduced drug treatment for these invariably fatal illnesses, and was one of the first to use nitrogen mustard and other chemotherapeutic agents in Britain. He remained a shrewd general physician, and his penetrating intelligence, great clinical skill, and inordinate capacity for hard work made him one of the pre-eminent physicians of his generation. It was quite remarkable that someone so close to the bedside and concerned with clinical matters could appreciate the value of pathology and the science of medicine. He strove to bring the laboratory and the clinic together, and aspiring young clinicians with good research projects could always count on his support.

In 1949 he was appointed physician to the

household of King George VI and from 1952 to 1973 he was physician to Queen Elizabeth II. He was appointed KCVO in 1964 and GCVO in 1973. He was made consultant physician to British Railways (Eastern Region) in 1957; honorary consultant in haematology to the army from 1957; consultant physician to the Florence Nightingale Hospital in 1958; consultant physician to the King Edward VII Hospital for Officers in 1963, and in the same year honorary consultant to the Royal Navy. Two years later he became honorary consultant to the Ministry of Defence.

As a teacher he excelled in a postgraduate setting. His achievements in his speciality were acknowledged by invitation to give the Lettsomian lecture of the Medical Society of London in 1957, the Langdon-Brown lecture in 1957, the Croonian lecture in 1970, and the most prestigious lecture offered by the Royal College of Physicians, the Harveian oration, in 1976. To appreciate Bodley Scott's breadth of vision it is necessary to turn to the Lettsomian lecture on leukaemia. It must be remembered that at that time no adult patient survived acute leukaemia. He insisted that 'although the outlook was bleak, the use of active drugs must be attempted in order to achieve a remission of the disease, for a nihilistic approach would create an impenetrable barrier against therapeutic advance'.

At the Royal College of Physicians, he was successively councillor (1963-6), censor (1970), and vice-president (1972). He was president of the Medical Society of London (1965-6), president of the British Society of Haematology (1966-7), president of the section of medicine of the Royal Society of Medicine (1967-8). He was made a member of the court of assistants of the Society of Apothecaries of London in 1964, and became master in 1974. He served as a member of council of the Imperial Cancer Research Fund from 1968, and the British Heart Foundation from 1975, eventually becoming chairman of the latter. A chair in cardiology, tenable at St Bartholomew's Hospital, was named after him.

The success of his edition of *Price's Textbook of Medicine* (12th edn. 1978), and of the *Medical Annual*, which he edited from 1979, owed much to his encyclopaedic knowledge of medicine and clear, concise style. He published his last book *Cancer—the Facts* in 1979.

Bodley Scott combined a remarkable ability in clinical medicine with a deep awareness of the need for innovation and experiment. He was a shrewd observer of his fellow men, and behind a reserved, almost shy manner he appreciated their foibles and commented on them with an astringent wit. Despite his reserve, he could speak forcibly and with effect if he detected

humbug, deceit, or injustice. Those who were privileged to have his friendship found a great source of kindness and support. He had a unique ability to inspire loyalty in all who worked for him. A new unit for the care of patients with cancer at St Bartholomew's Hospital is to be named in memory of him.

In 1931 he married (Edith) Daphne (died 1977), daughter of Lieutenant-Colonel Edward McCarthy, of the Royal Marine Artillery. In 1980 Bodley Scott married Jessie, the widow of Alexander Page Gaston, of Sevenoaks, and daughter of Thomas Mutch, farmer in Aberdeenshire. She survived their car accident near Parma, Italy, on 12 May 1982, in which Bodley Scott died.

[*Lancet*, vol. i, 1982, p. 1195; *British Medical Journal*, vol. cclxxxiv, p. 1567; *The Times*, 13 May 1982; 'Medicine in the Twentieth Century' in *The Royal Hospital of St Bartholomew*, ed. V. C. Medvei and J. L. Thornton, 1974; J. S. Malpas in *Munk's Roll*, vol. vii, 1984; personal knowledge.]

J. S. MALPAS

BOOT, HENRY ALBERT HOWARD (1917-1983), physicist, was born 29 July 1917 in Hall Green, Birmingham, the elder child and only son of Henry James Boot, electrical engineer at the firm of Bellings, and his wife, Ruby May Beeson. Always known as 'Harry' he was educated at King Edward's High School, Birmingham, and at Birmingham University, where he studied physics and obtained a B.Sc. in 1938. He had begun work for his Ph.D. (which he obtained in 1941) when war broke out.

(Sir) Mark Oliphant, professor of physics at Birmingham University, had recently been to Stanford University where he had encountered the klystron, an interesting possibility for a high-power generator of microwave radiation for use in radar. He put Boot, (Sir) J. T. Randall [q.v.], and James Sayers to work on the problem, having acquired an Admiralty contract to develop microwave generators (centimetric wave transmitters). Boot and Randall preferred the magnetron to the klystron and by late February 1940 they had constructed a new type of cavity magnetron, with a radiation wavelength of 9.8 cm. By May 1940 an experimental radar set using a pulsed 10 cm. cavity magnetron had been built at the Telecommunications Research Laboratory, Swanage, and by September 1940 a submarine periscope could be detected at a range of seven miles.

The cavity magnetron—later improved by James Sayers who developed the technique of strapping—crucially influenced the outcome of the war, for centimetric radar enabled a precise radar beam and small lightweight radar transmitters to be used. At first Bomber Command aircraft used the transmitters in night raids, and soon anti-aircraft units, convoy escorts, and night fighters all had the equipment. Thousands of cavity magnetrons had been manufactured by the end of World War II in 1945. Boot and Randall were given a prize of £50 from the Royal Society of Arts 'for improving the safety of life at sea' in 1945 but when they applied to the royal commission for awards to inventors they and James Sayers received £36,000 in 1949.

In 1943 the Birmingham physics department returned to the study of atomic physics (for atomic bombs) and Boot moved for a time to British Thomson–Houston at Rugby to continue the development of very high-power magnetrons. He rejoined the Birmingham department (to help build the cyclotron) in 1945 as a Nuffield research fellow. In 1948 he entered government service as a principal scientific officer with the Royal Navy Scientific Service in the Services Electronics Research Laboratories (SERL) at Baldock in Hertfordshire. He was appointed senior principal scientific officer in 1954 and remained at Baldock until his retirement in 1977. During his latter years in Birmingham and at Baldock he continued his work on microwaves and magnetrons and further researched on plasma physics, controlled thermonuclear fusion, lasers, masers, and infra-red viewing devices. Boot had exceptional success in designing, constructing, and operating powerful electrical devices, beginning as a schoolboy with generators of X-rays and of high voltages. He was awarded the John Price Wetherill medal of the Franklin Institute (1958) and the John Scott award (with J. T. Randall, 1959).

Boot and his wife lived for thirty years in a thatched cottage at Rushden near Cambridge, with five acres of land (although Boot disliked gardening) and a garage full of fifty assorted magnetrons. His great love was sailing; he kept two boats at Salcombe in Devon. He was a quiet, modest, and tactful man, neat in his dress and habits of work. He had a friendly and discerning manner, together with a mildly sardonic sense of humour.

In 1948 he married Penelope May, daughter of Luke Herrington, engineer. They had two sons. Boot died 8 February 1983 in the Hope Nursing Home, Cambridge. One of the prototype cavity magnetrons built by GEC is in the Institution of Electrical Engineers and several from Boot's own collection were given to Hatfield Polytechnic.

[H. A. H. Boot and J. T. Randall, 'Historical Notes on the Cavity Magnetron', Institute of Electrical and Electronics Engineers' *Transactions on Electron Devices*, vol. ED-23, no. 7, July 1976; A. L. Norberg, transcript of an interview with H. A. H. Boot, Bancroft Library, University of California, Berkeley, 1979; information from P. B. Moon, M. H. F. Wilkins, and Mrs Penelope Boot.]

C. S. Nicholls

BOOTH, PAUL HENRY GORE-, Baron Gore-Booth (1909–1984), diplomat. [See Gore-Booth.]

BOULT, Sir ADRIAN CEDRIC (1889–1983), orchestra conductor, was born 8 April 1889 in Chester, the only son and younger child of Cedric Randal Boult, JP, oil merchant, and his wife, Katherine Florence Barman. The family were Unitarians. He was educated at Westminster School, and at Christ Church (of which he was made an honorary Student in 1940), Oxford, where he was president of the University Musical Club in 1910, and took a pass degree in 1912. After studying under the distinguished German conductor Arthur Nikisch at the Leipzig Conservatorium in 1912–13, he sat his B.Mus. examination at Oxford in 1913, receiving his degree in 1914. He achieved his Oxford D.Mus. in 1921.

Boult's talent for music had revealed itself at a remarkably early age. At sixteen months he was able to pick out tunes on the piano, and by his seventh birthday he had begun to compose. There was, therefore, never any doubt about his choice of profession. At the beginning of 1914 he joined the music staff of the Royal Opera House, Covent Garden, where he participated in the first British performances of Richard Wagner's *Parsifal* in February and March, playing the off-stage bells. As a young man he suffered from a heart condition which rendered him unfit for active service during World War I. He helped to drill recruits in Cheshire for two years, worked in the war office in 1916–18, and found time to organize concerts in Liverpool with a small orchestra drawn from the ranks of the Liverpool Philharmonic Society. This led to his being invited to conduct the full orchestra at a concert in Liverpool in January 1916. The programme of this, his professional début as a conductor, included works by Bach, Haydn, Liszt, and the contemporary composers Sir C. Hubert H. Parry [q.v.] and Arthur de Greef.

In 1918, at the invitation of its composer, Boult conducted the first performance of *The Planets* by Gustav Holst [q.v.] at a concert in the Queen's Hall, London. In the following year he joined the teaching staff of the Royal College of Music, where he remained until 1930, continuing to accept engagements as a conductor in London. His first experience as a conductor of opera was gained with the British National Opera Company, and in 1926 he rejoined the Covent Garden company as a staff conductor. He was also at this time the musical director of the City of Birmingham Symphony Orchestra (1924–30) and, from 1928 to 1931, conductor of the Bach Choir, London.

The most important phase of Boult's career began when he was invited to succeed Percy Pitt as director of music of the BBC at the beginning of 1930. In addition to his other administrative duties, this involved him in recruiting players for and becoming chief conductor of the newly formed BBC Symphony Orchestra which during the following years he developed into a first-class ensemble. From the beginning the orchestra gave public concerts as well as broadcasting from the BBC studios, and Boult took it on tour in Europe with great success in 1935 and 1936, giving concerts in Brussels, Paris, Zurich, Vienna, and Budapest.

Boult had by this time become well known abroad, having been invited to conduct the Vienna Philharmonic Orchestra in Vienna for the first time in 1933, and having later conducted in Salzburg, New York, and Boston. Nor did he lose contact with the world of opera, his performances of *Die Walküre* at Covent Garden in 1931 and *Fidelio* at Sadler's Wells Theatre in 1930 being considered outstanding. He also introduced much new music in his concerts with the BBC Symphony Orchestra, giving the first performances in England of Alban Berg's *Wozzeck* in 1934, and Busoni's *Doktor Faust* in 1937, perhaps his most notable and memorable operatic achievements.

Relinquishing the position of music director of the BBC in 1942, Boult became associate conductor of the Promenade concerts, and continued as conductor of the BBC Symphony Orchestra until 1950 when, having reached the age of sixty, he was retired by the BBC (the wife of whose new director of music Boult had married) and immediately became musical director of the London Philharmonic Orchestra with which he toured West Germany in 1951 and the Soviet Union in 1956. Although, in the following year, he announced his retirement from the London Philharmonic, he continued to make a number of guest appearances with orchestras at home and abroad, in Europe and the United States, and was able to devote a large part of his time to recording many of the works in his vast repertory, especially the music of Sir Edward Elgar and Ralph Vaughan Williams [qq.v.]. He conducted the music at the coronations of George VI and Elizabeth II.

In 1959, the year of his seventieth birthday, Boult was offered and accepted the presidency of the Royal Scottish Academy of Music, in succession to Vaughan Williams. In the same year he became musical director of the City of Birmingham Symphony Orchestra for the second time (until 1960), and he returned to the Royal College of Music to teach from 1962 to 1966. Among the many honours he received were his knighthood in 1937, the gold medal of the Royal Philharmonic Society in 1944, and the Harvard medal in 1956. He was made a Companion of Honour in 1969. He had honorary degrees from six universities, including Cambridge (Mus.D., 1953) and Oxford (D.Litt., 1979).

One of the leading British musicians of his time, Boult was the least demonstrative of conductors on the concert platform, obtaining his effects by meticulous rehearsal, impeccable musicianship, and a natural authority. A tall man of erect, almost military bearing, Boult was taciturn by nature. However, his courteous manner could occasionally, at rehearsals, give way to storms of violent temper. He was always concerned to present the music as the composer conceived it, and was reluctant to impose his own personality upon a work in the name of interpretation. He excelled in the nineteenth-century classics as well as in the music of his British contemporaries, and was the author of two excellent books on conducting, *The Point of the Stick* (1920) and *Thoughts on Conducting* (1963), as well as a fascinating volume of memoirs, *My Own Trumpet* (1973).

In 1933 he married Ann Mary Grace, daughter of Captain Francis Alan Richard Bowles, RN, JP, of Dully House, Sittingbourne, Kent, and mother of four children from a previous marriage to Sir (James) Steuart Wilson [q.v.]. There were no children of this marriage. Boult died in a London nursing home on 22 February 1983.

[Adrian Cedric Boult, *My Own Trumpet*, 1973; Michael Kennedy, *Adrian Boult*, 1987; Ronald Crichton in *The New Grove Dictionary of Music and Musicians*, 1980 (ed. Stanley Sadie); personal knowledge.]

CHARLES OSBORNE

BOULTING, JOHN (1913–1985), film producer and director, was born in Bray, Berkshire, 21 November 1913, the identical twin brother of Roy and son of Walter Arthur Boulting, financial consultant, of Hove, Sussex, and his wife, Rose Bennett. John and Roy had two brothers—one older than themselves (Peter Cotes, an actor)—and one younger who died aged eight. Boulting was educated at Reading School where he was captain of the rugby team, secretary of the dramatic society, and a leading actor. Interested in politics and the local Labour Party, he was a member of the school debating society. John and Roy formed one of the first cinematograph societies in a public school. In 1933 John began work at Ace Films, a small film distribution company owned by a school friend's father in Wardour Street, London, where he discovered his flair for film business.

Early in 1937 Boulting volunteered as a front-line ambulance driver for the International Brigade in Spain. On his return he formed Charter Films with Roy in November 1937. Their first film, *Ripe Earth* (1937), was a documentary, but they soon branched out into features, their first critical success being *Consider Your Verdict* (1938). After *Trunk Crime* (1939), a moderately successful thriller, and *Inquest* (1939), John produced *Pastor Hall* (1940), one of the first British anti-Nazi films, and *The Dawn Guard* (1941), a documentary for the Ministry of Information. Early in 1941 John joined the RAF as an AC2 flight mechanic but obtained special leave in 1942 to produce *Thunder Rock*. He became a flight lieutenant in the RAF Film Unit where he directed his first film *Journey Together* (1944).

After the war the Boulting brothers worked as independent producers. In 1947 John directed his second film, *Brighton Rock*, the only film of his work approved of by Graham Greene, and produced *Fame is the Spur*, a film about a Labour politician played by (Sir) Michael Redgrave [q.v.]. Boulting's preoccupation with social and political issues continued with *The Guinea Pig* (1949), about a lower middle class schoolboy (played by (Sir) Richard Attenborough) who attends a public school. The third film John Boulting directed, *Seven Days to Noon* (1950), was the first to deal with the moral implications of the atom bomb, winning an Academy award for best original story.

In 1951 he directed *The Magic Box*, which was chosen as the film industry's contribution to the Festival of Britain. As an independent producer, he resented the Rank Organization's stranglehold over film distribution and exhibition. In the following decade he championed the cause of the independent producer and publicly denounced the combines' emphasis on flamboyance and spectacle associated with 'gala' premières.

In the 1950s the Boulting brothers turned to comedy and John became a Liberal. He produced *Brothers in Law* (1957) and *Carlton-Browne of the FO* (1958), and directed *Private's Progress* (1956), *Lucky Jim* (1958), *I'm All Right Jack* (1960), and *Heavens Above!* (1963), films that satirized respectively the legal profession, the Civil Service, the army, universities, trade unions, and the church. Although *I'm All Right Jack* won British Oscars for best screenplay and best British actor (for the performance of Peter Sellers, q.v., as a shop steward), the film

brought the Boultings into conflict with the film technicians' trade union.

In the 1960s the Boultings did not play a significant part in the 'New Wave' British cinema except perhaps with *The Family Way* (1966). The last film directed by John Boulting was *Rotten to the Core* (1965) but he went on to produce *Twisted Nerve* (1968), *There's a Girl in My Soup* (1970), and *Soft Beds, Hard Battles* (1974). He continued to be interested in film industry politics, particularly the challenge posed by television. He forced television companies to pay higher fees to film producers for television showings of feature films. He was managing director of British Lion and Shepperton Studios (1967-72).

John Boulting's life was dominated by films, his most creative period being the 1940s. He enjoyed cricket, tennis, and horse riding. His political views clearly influenced his work and while his films were celebrating the individual against authority he was an active local Liberal. He was married four times—to Veronica, daughter of John Craig Nelson Davidson, barrister, in 1938; to Jacqueline Helen, daughter of Richard Chilver Robert Rice Allerton, a broker at Lloyd's, in 1952; to Ann Marion, daughter of Alan Ware, of the Royal Marines, in 1972; and to Anne Josephine, daughter of Frank Flynn, sales manager, in 1977. By his first wife he had two sons and by his second three daughters. He also had another son. Boulting died 17 June 1985 at his home in Sunningdale, Berkshire.

[*The Times*, 19 June 1985; British Film Institute Information Department; private information.]   SARAH STREET

BOWER, SIR JOHN DYKES (1905-1981), cathedral organist. [See DYKES BOWER.]

BOYD OF MERTON, first VISCOUNT (1904-1983), politician. [See LENNOX-BOYD, ALAN TINDAL.]

BOYD, SIR JOHN SMITH KNOX (1891-1981), physician, was born 18 September 1891 at Largs, Ayrshire, the second of three sons (there were no daughters) of John Knox Boyd, an agent in the Royal Bank of Scotland, and his wife, Margaret Wilson Smith. He was educated at the local school in Largs and then entered Glasgow University in 1908 to study medicine; he graduated MB, Ch.B. in 1913. After house appointments at Glasgow Royal Infirmary he became a ship's surgeon and in 1914 applied for a commission in the Royal Army Medical Corps. By December he was at Ypres. From France he moved to Salonika in 1916 as a medical officer to the Divisional Engineers. In this post he travelled widely through Mesopotamia making his first acquaintance with tropical diseases. By 1917, after some bacteriological training, he was in charge of a mobile laboratory where he worked on the treatment of malaria and studied the prevalent dysentery. He became a pathologist at Salonika in September 1918 but was invalided home in December with 'Spanish influenza'.

Having obtained a regular commission in the Royal Army Medical Corps in 1920 he was appointed to take charge of the brigade laboratory at Nasirabad in Rajputana and later the district laboratory in Mhow, Central Provinces. He returned to the Royal Army Medical College in 1923 where he became demonstrator then assistant professor of pathology. He obtained the DPH diploma at Cambridge in 1924. Back in India in 1929 with the rank of major (1926) he was in charge of laboratories in Bangalore and then Poona. In 1932 he was appointed assistant director of hygiene and pathology at Army HQ Simla. In 1936 he returned to Millbank in charge of the vaccine laboratory. He was awarded the Leishmann medal in 1937 and promoted to lieutenant-colonel in 1938.

After the outbreak of World War II, in August 1940 he was appointed in charge of pathology in the Middle East and in November 1943 became deputy-director of pathology to 21 Army Group. He was promoted colonel in 1944 and brigadier the following year. In 1945 he became director of pathology on the War Office staff. He put the vaccine department of the RAM College on a war footing in 1939 and after its transfer to Tidworth it produced sufficient material for all the needs of the services. During the war he organized a blood transfusion service for the Middle East forces and pioneered the preferential use of whole blood for transfusion in casualties with severe blood loss. During this time he built up some forty laboratories in that zone. He was mentioned in dispatches (1941).

He left the RAMC in 1946 and became director of the Wellcome Laboratories of Tropical Medicine where he remained until 1955. In that year he became a Wellcome trustee and served in this capacity until 1966, becoming deputy chairman in 1965 and a consultant to the Trust until 1968.

Boyd was especially interested in the dysenteric diseases. He studied the difference between 'smooth' and 'rough' colonies of dysenteric bacterial strains, showing that the rough strain contained the group antigen common to all flexner types while the smooth strain lacked this group antigen but possessed its own specific surface antigen. This property made it possible to separate dysenteric bacilli into two groups called subsequently flexneri and boydii. Later

in the Middle East he was responsible for the first trials of sulphaguanidine in the treatment of dysentery. His work on malaria included the first studies on the synthetic anti-malarials, and his studies of typhus in India in 1916 showed that most cases of this disease were transmitted by the mite and flea and not by the tick as had been suspected. Boyd also had a special interest in bacteriophage. His experience with these 'bacterial viruses' led him to formulate a theory for the long-lasting immunity that occurs after recovery from yellow fever, based on the bacteriophage model.

He was a member or chairman of many committees and was president of the Royal Society of Tropical Medicine and Hygiene (1957-9). He was awarded the Manson medal in 1968 and was elected MRCP in 1950, and FRCP and FRS in 1951. He obtained his Glasgow MD in 1948. He was an honorary FRCPE (1960) and FRSM (1965). He was a FRCPath. (1968), an honorary LLD, Glasgow (1957), and an honorary D.Sc., Salford (1969). He was appointed OBE in 1942 and knighted in 1958. To this Dictionary he contributed the notices of Sir Neil Hamilton Fairley and Sir William MacArthur.

Boyd was a deeply honest man of military bearing with a determination to see that his objectives were achieved. During his latter years he was a very formidable figure, but once he made a friend his loyalty was unshakeable. He was a keen golfer and bird watcher. In 1918 he married Elizabeth, daughter of John Edgar, a Dumfriesshire station master. She died in 1956 after many years as a chronic invalid and in 1957 he married his secretary, (Ellen) Mary (Harvey) Bennett (died 1968), daughter of Denis Harvey Murphy, company director, of Northwood, Middlesex. There were no children of either marriage. Boyd died 10 June 1981 at Northwood, Middlesex.

[L. G. Goodwin in *Biographical Memoirs of Fellows of the Royal Society*, vol. xxviii, 1982; diaries and a personal summary of Boyd's work held in Royal Army Medical College, Millbank; L. G. Goodwin in *Munk's Roll*, vol. vii, 1983; personal knowledge.]

P. O. WILLIAMS

BOYD-ROCHFORT, SIR CECIL CHARLES (1887-1983), racehorse trainer, was born 16 April 1887 at the family home, Middleton Park, county Meath, the third son and seventh of eight children of Major Rochfort Hamilton Boyd, of the 15th Hussars (who assumed the additional name of Rochfort in 1888), and his wife, Florence Louisa, daughter of Richard Hemming, of Bentley Manor, Worcestershire.

After schooling at Eton, in 1906 he became a pupil trainer with H. S. Perse at Grateley in Wiltshire. In 1908 he became assistant trainer to Captain R. H. Dewhurst at Newmarket. He left Dewhurst in 1912 to become racing manager to Sir Ernest Cassel [q.v.]. He joined the Scots Guards for the war of 1914-18, was wounded on the Somme, and received the croix de guerre. A captain on demobilization, he returned to Newmarket and to managing Cassel's racing interests at Moulton Paddocks. A chance meeting led to Boyd-Rochfort being invited to manage Marshall Field's horses as well.

On Cassel's death Boyd-Rochfort bought Freemason Lodge at Newmarket and commenced training for the season of 1923. His principal owners were Field and the Dowager Lady Nunburnholme. His first season yielded nineteen successes including the winner of the Newmarket July Cup. Subsequent seasons brought a steady flow of winners but his first winner of a 'listed' race was Royal Minstrel who won the Eclipse Stakes in 1929.

The arrival of the Americans, Messrs William Woodward and Joseph Widener, as owners in 1930 marked a turning point in Boyd-Rochfort's career. The first classic success came in 1933 when Woodward's Brown Betty won the 1,000 guineas. Boyd-Rochfort was now on the crest of the wave and in 1937 he became the leading trainer for the first of five times. In 1927 he had bought the filly Double Life for Lady Zia Wernher and in 1937 he had the satisfaction of training Double Life's son, Precipitation, to win the Ascot Gold Cup. He trained many winners from this family culminating in the victory of Meld in the Oaks of 1955.

The outbreak of war in 1939 forced Boyd-Rochfort to reduce his string to twenty-five horses. However, in 1943 Boyd-Rochfort commenced training the horses from the private stable of King George VI. In 1946 he trained Hypericum, the King's only home bred classic winner. The victories of well bred fillies such as Avila, Angelola, and Above Board laid the foundation of Queen Elizabeth's many successes in the 1950s and 1960s.

On her accession the Queen decided to continue keeping her home bred horses at Freemason Lodge. Aureole was expected to win the coronation year Derby for her in 1953, but finished in second place to Pinza. The 1950s were golden years for the stable, which culminated in Boyd-Rochfort winning both the Derby with Parthia and the King George VI stakes with Alcide in 1959. Thereafter the success rate of the then septuagenarian trainer declined but he continued to turn out top class winners until his retirement in 1968.

Boyd-Rochfort was an excellent judge of a yearling and he liked to have a free hand in the selection of the horses he was to train. He had a deep and detailed knowledge of the Stud Book and consequently the studs of his owners usually flourished following his shrewd original purchases of their foundation mares. While he was always ready to spend money, he could be equally successful when operating within rigid financial controls. He was a punctilious correspondent and kept his owners regularly informed of the progress of their horses. Owners rarely quit the stable.

Boyd-Rochfort was a large man of infinite courtesy and impeccable manners, inspiring respect rather than affection. He ran a stable where good judgement, regularity, good feeding, attention to detail, and patience were the principal ingredients of success. A strict employer, he obtained and kept the best stable staff. He was a master of the art of training stayers. His skill with fillies was particularly noteworthy and P. T. Beasley, his second stable jockey, rated him the best trainer of fillies of his generation. He won thirteen classic races and 1,156 others which earned prize money to the total of £1,651,514.

Boyd-Rochfort was appointed CVO in 1952 and KCVO in 1968, the year in which he retired at the end of the season. In 1944 he married Elizabeth Rohays Mary, daughter of Major-General Sir James Lauderdale Gilbert Burnett, of Leys, thirteenth baronet, and widow of Captain the Hon. Henry Cecil. They had one son. Boyd-Rochfort died at Kilnahard Castle, county Cavan, 17 March 1983.

[B. W. R. Curling, *The Captain*, 1970; Roger Francis Mortimer (ed.), *Biographical Encyclopaedia of British Flat Racing*, 1978; *The Times*, 19 March 1983; *European Racehorse*, June 1983; private information; personal knowledge.]   WILLIAM DUGDALE

BOYLE, EDWARD CHARLES GURNEY, third baronet, and BARON BOYLE OF HANDSWORTH (1923-1981), politician, was born at 63 Queen's Gate, Kensington, 31 August 1923, the elder son and eldest of three children of Sir Edward Boyle, second baronet, barrister, and his wife, Beatrice, daughter of Henry Greig, of Belvedere House, Kent. He was the grandson of Sir Edward Boyle [q.v.], Conservative MP for Taunton in 1906-9, who was created the first baronet in 1904. He was educated at Eton where he was captain of Oppidans, editor of the *Eton Chronicle*, and president of the Political Society. Journalism and politics remained two of his abiding interests. He succeeded his father as baronet in 1945. Towards the end of World War II he served for a short period in the Foreign Office, after which he went up to Oxford as a scholar of Christ Church where he read history. He played an active part in Conservative undergraduate politics and was elected president of the Union in the summer of 1948. Even at that time he was considered to be exceptionally mature, a charming and persuasive, though not a rhetorical, speaker, and one who was likely to make his mark in the outside world. He left Oxford in 1949 with a third class degree, disappointing to him and at first sight somewhat surprising. His intellectual characteristics, however, comprised a wide breadth of interest, a remarkable store of information, and a phenomenal memory for everything which he encountered, but his was not a mind full of innovative ideas or of penetrating analysis. His strength lay in his deep-seated and moderate convictions which guided him all through his life.

It was, no doubt, because he was so impressively grown-up that he was selected to fight a by-election in the Perry Bar division of Birmingham in 1948 whilst still at Oxford. Having lost the by-election he fought the same seat in the 1950 general election when he again lost. Meantime he had become a journalist and was assistant editor of the *National Review* under John Grigg. Later in that year he was elected to the House of Commons for the Handsworth division of Birmingham at the age of twenty-seven, the youngest member in the House. His ability and steadfastness were soon recognized by his appointment the following year as parliamentary private secretary to the under-secretary for air. In 1954 he received his first government post as parliamentary secretary to the Ministry of Supply, after which the whole of his parliamentary life until he left the House of Commons in 1970 was spent on the front bench either in government or in opposition, with one short exception after his resignation at the time of Suez.

In 1955 he became economic secretary to the Treasury and from then on it was economic affairs and education which he enjoyed in politics above all else. A heavy responsibility was placed upon him by the budget of R. A. Butler (later Lord Butler of Saffron Walden, q.v.) in the autumn of 1955. Butler himself was overstrained by the illness and death of his wife. His emergency budget had been whittled down by the cabinet to the point where it was doubtful whether it was worth the trouble it caused. No one, however, had the strength to call a halt to it. Boyle defended it tirelessly in the Commons, in particular the proposition that it was possible to reduce inflation by increasing indirect taxation, thus putting up prices, a proposition which became colloquially known as 'Boyle's Law'.

When Sir Anthony Eden (later the Earl of Avon) announced his decision to use British forces in Egypt, Boyle resigned, though without great fuss. He thought the policy was both dishonourable and doomed to failure. When Harold Macmillan (later the Earl of Stockton) became prime minister in January 1957 he began to heal the wounds in the Conservative Party caused by the Suez adventure by inviting both Julian Amery and Edward Boyle to take office in his government. Boyle became parliamentary secretary to the Ministry of Education, but returned to the Treasury as financial secretary after the general election of October 1959. There he strongly supported proposals for indicative planning and an overall incomes policy. In 1962 he was appointed minister of education and became a member of the cabinet. He was successful in expanding and bringing up to date the educational system. Always an admirer of Butler, he continued to develop the approach set out in the Butler Education Act of 1944.

When the responsibility for science was moved from the lord president of the Council to the enlarged Department of Education and Science in 1964, Quintin Hogg (later Lord Hailsham of St Marylebone) became secretary of state for the new department, and Boyle was made minister of state with special responsibility for higher education. He retained his seat in the cabinet and in no way resented his change of status at the department, for the new structure was one that he had himself urged upon the government.

After the defeat of the Conservative Party in the general election of 1964, Sir Alec Douglas-Home (later Lord Home of the Hirsel) made Boyle shadow home secretary. Here he was unhappy, feeling that his own moderate views on home affairs were all too often in conflict with the more right-wing views of many members of his party. He became opposition front bench spokesman for education and deputy chairman of the party's advisory committee on policy a few months later in February 1965. As leader of the Conservative Party, Edward Heath made a number of attempts in 1965 to persuade Boyle to move to other front-bench positions to enable him to widen his experience in preparation for the highest offices in government. However, he repeatedly refused and in 1970 retired from Parliament in order to devote more time to what had now become his overwhelming interest, education, and in a role where he could express his views and implement his policies without constant interference from those taking part in the political battle. He became vice-chancellor of the University of Leeds in the same year.

Whilst still in Parliament in the second half of the sixties, he had engaged in many additional activities, becoming a director of Penguin Books (1965), a member of the Fulton committee on the Civil Service (1966–8), and pro-chancellor of the University of Sussex (1965-70). After he moved to Leeds his main activity outside the university was as chairman of the top salaries review board. Here he was noted for the fairness and firmness with which he dealt with the intractable problems which arose between governments and the public services during the 1970s. He felt deeply that the constant attacks on public servants from the press and politicians were unjustified and deeply damaging to the national interest. He went to great pains to ensure that those working in the public service received the honourable recognition which was their due, expressed not only in words but in their remuneration.

The early years of Boyle's vice-chancellorship of Leeds University covered that difficult period during the first half of the 1970s when the majority of students on both sides of the Atlantic were opposed to the continuation of the war in Vietnam and, in particular, to those governments which appeared to be supporting it or conniving at it. Boyle's diplomatic touch, administrative skills, concern for the welfare of both teachers and students alike, and above all his energy and humanity in keeping in continuous contact with all aspects of university life enabled him to maintain a reasonable stability where many others failed. Everyone knew that he cared little about himself and his personal position, but that he was determined to maintain the standards of his university in the interests of the future of the students for whom he was responsible. In 1977-9 he was chairman of the committee of vice-chancellors and principals. He found these dignitaries more difficult to handle. He was most happy when he was moving among his own students in his own university.

Boyle was a genial host and a stimulating conversationalist. He had a great love of music, which he shared with Edward Heath, who visited him on the evening before his death. Although he did not play any instrument himself, he possessed a fine collection of gramophone records and a profound knowledge of music and musicians. He was a regular opera goer, favouring particularly Glyndebourne in the summer. He was an admirer of Gabriel Fauré about whom he had collected papers for a book which, alas, was never written. To this Dictionary he contributed the notice of Reginald Maudling. In his early days he was a high Anglican but over his last twenty-five years he moved further and further towards agnosticism.

Boyle, who received honorary degrees from the universities of Leeds (LLD, 1965), Southampton (LLD, 1965), Aston (D.Sc., 1966), Bath (LLD, 1968), Heriot-Watt (D.Litt., 1977), Hull (D.Litt., 1978), and Sussex (LLD, 1972) was also an honorary freeman of the Clothworkers' Company (1975) and a freeman of the City of London, a charter fellow of the College of Preceptors, and an honorary fellow of the Royal College of Surgeons of England (1976). Boyle was admitted to the Privy Council in July 1962 and received a life peerage in 1970. He became a Companion of Honour at a special investiture ceremony at Buckingham Palace on 30 June 1981, shortly before his death.

He died in the vice-chancellor's lodge at Leeds on Monday 28 September 1981 after a prolonged illness, at the early age of fifty-eight. He was unmarried and was succeeded in the baronetcy by his brother, Richard Gurney Boyle (born 1930).

[Private information; personal knowledge.]

EDWARD HEATH

BRAHMS, CARYL (1901-1982), novelist, critic, journalist, and songwriter, was born Doris Caroline Abrahams in Croydon, Surrey, 8 December 1901, the only child of Henry Abrahams, a merchant, and his wife, Pearl, one of the twenty-one children of Moses and Sultana Levi who arrived in England from Constantinople, probably in 1873. She was educated privately, at Minerva College in Stonygate, near Leicester, and at the Royal Academy of Music, where she failed her LRAM. At the Academy, already an embryo critic, she did not care to listen to the noise she made when playing the piano. She began to write light verse for a student magazine and then for the *Evening Standard*. At this time she adopted her pseudonym so that her parents who envisaged a more domestic future for her would be unaware of her literary activities.

Ballet class as a child and her later exposure to the Diaghilev Ballet in the South of France encouraged her to apply to Viscountess Rhondda [q.v.] to write ballet criticism for *Time and Tide*. She also wrote on opera and the theatre for the same newspaper, the *Daily Telegraph*, and many others. In 1930 Gollancz published her volume of children's verse, *The Moon on my Left*.

At this time she met S. J. Simon (Secha Jascha Skidelsky), a White Russian student of agriculture, an international bridge player, and an inspired humorist. In the early thirties they contributed captions for 'Musso, the Home Page Dog' cartoons by (Sir) David Low [q.v.] in the *Evening Standard*. In 1936, during which Caryl Brahms also edited a popular primer, *Footnotes to the Ballet*, they collaborated on a novel, *A Bullet in the Ballet* (1937), their classic comedy thriller which introduced the Stroganoff Ballet Company. Another Stroganoff novel, *Casino for Sale*, followed in 1938. *The Elephant is White* (1939) again deployed a cast of eccentric Russian émigrés. In 1940 they hit on a new and original vein of wild, anachronistic, historical humour with *Don't, Mr Disraeli!* (1940), a novel set, 'not in the Victorian Age but in its literature'. They brought a similar approach to the Elizabethan Age in *No Bed for Bacon* (1941). Their other novels were *No Nightingales* (1944), *Six Curtains for Stroganova* (1945), *Titania has a Mother* (1944), and *Trottie True* (1946). A stage version of *A Bullet in the Ballet*, starring Léonide Massine and Irina Baronova, foundered in Blackpool.

In 1948 they had just begun their last collaboration, *You Were There* (1950), when Simon died suddenly. Caryl Brahms finished the book alone, continued to write fiction, and increased her criticism of theatre and opera as well as ballet. Her collected book of theatre notices, *The Rest of the Evening's My Own*, was published in 1964. She followed it with *Gilbert and Sullivan* (1975) and with *Reflections in a Lake* (1976), a study of Chekhov's four great plays.

In 1954 she had met Ned Sherrin, who asked permission to adapt *No Bed for Bacon* as a stage musical. Seeing that he challenged no comparisons with Simon, she suggested a collaboration. The production of the musical at the Bristol Old Vic was not a success but it laid the foundation of a partnership which over the next twenty-eight years produced seven books, many radio and television scripts, and several plays and musicals for the theatre including *I Gotta Shoe* (1962-3), *Sing a Rude Song* (1970), *Liberty Ranch* (1972), *Nickelby and Me* (1975), *The Spoils of Poynton* (1968, from Henry James, q.v.), and latterly *Beecham* (1980) and *The Mitford Girls* (1981), both of which had respectable West End runs.

In her sixties, her enthusiasm for ballet waned. She became a devotee of show-jumping and the All-England course at Hickstead took the place of Covent Garden in her life as she argued that the horses moved more gracefully than contemporary ballerinas. She also began to write songs for the BBC television programme *That Was The Week That Was* (1962) and, as a lyric writer, she won an Ivor Novello award (1966) for the title song of *Not So Much a Programme More a Way of Life*. Her song compilations were a feature of *Side by Side by Sondheim* (1976) and the TV series *Song by Song* (1979-80). The latter also became a book (1984). In 1978 she was appointed a governor of the National Theatre.

Physically her aspect seemed not to change for the last three decades of her life, until a noticeable enfeeblement in the last two years. She was dark, tiny, with a very prominent nose on which she permanently perched large dark forbidding spectacles. Her pouter pigeon figure and thrust-forward chin matched her combative approach to life. She sparred energetically with colleagues and bank managers and conscientiously encouraged young artistes. Her work with Simon will surely survive as an example of the most sensitive and innovative comic-fiction during the middle years of the century. She died 5 December 1982 at her home, 3 Cambridge Gate, Regent's Park, London. She was unmarried.

[*The Times*, 6 December 1982; Caryl Brahms and Ned Sherrin, *Too Dirty for the Windmill, a Memoir*, 1986; Caryl Brahms, 'Palookas in Peril' (unpublished diary); personal knowledge.]   NED SHERRIN

BRANDT, HERMANN WILHELM ('BILL') (1904-1983), photographer, was born 3 May 1904 in Hamburg (though he seems to have encouraged the belief that he was born in London), the second of four sons (there were no daughters) of Ludwig Walter Brandt, merchant, and his wife, Lili Merck. His school-days were unhappy and at sixteen he contracted tuberculosis. After six years at a Swiss sanatorium, he went to Vienna in search of a cure by psychoanalysis. But his lungs were already clear.

In 1928, while learning photography at a Vienna portrait studio, Brandt took Ezra Pound's portrait. Pound, impressed, introduced him to the most fashionable art photographer of the day, Man Ray, in whose Paris studio Brandt became a student for three months in 1929-30. He learned little directly from Ray but through him Brandt met many members of the Paris art community, notably the Surrealists. Brandt's photographs of this period include carefully composed city scenes as well as attempts to capture movement—a common preoccupation of photographers at the time, able to exploit the first-ever generation of 35 mm. cameras. Brandt himself started with a camera which took small glass negatives, and he soon began to use this for 'candid' shots of people unaware they were being photographed.

Brandt met his first wife, the Hungarian Eva Boros, in Vienna. She was the daughter of Joseph Boros. Married in Barcelona in 1932, they settled in London, where Brandt had some pictures published in the *News Chronicle*. His first book, *The English at Home* (1936), is a coldly analytical foreigner's view of the English class structure. His second, *A Night in London* (1938), is equally cool. Using mainly family and friends as models, the photographs are posed, theatrical, and menacing. After reading *English Journey* (1934) by J. B. Priestley [q.v.], Brandt turned his critical eye on unemployment and poverty in the north of England, also working for *Weekly Illustrated*, *Lilliput*, *Picture Post*, and *Harper's Bazaar*. In World War II he was commissioned to record important buildings and bomb damage. *Lilliput* published his powerful and eerie photographs of conditions in the air-raid shelters alongside Henry Moore's equally disturbing drawings. At this time he also returned to portraits, sombre likenesses of artists, actors, and writers as 'denizens of black-out and of the dark, of those shadowy moments when faces emerge under lamplight but bodies remain dark' (Alan Ross). Reading widely in contemporary and historical literature, he began to visit the places where authors lived, or which they had described. This led to another book, *Literary Britain* (1951).

By now Brandt had acquired a rather ancient wooden Kodak camera with a very small aperture lens. This he used for the pictures in *Perspective of Nudes* (1961). Placing his models among antique furnishings in his flat, or in landscape (especially the seashore), Brandt distorted their forms through the wide angle lens, making them almost abstract, as sculptural as torsos by Sir Jacob Epstein [q.v.] but only just recognizably human. The brooding mystery of these and all Brandt's later photographs owed a great deal to his increasingly contrasty printing style, with few tones between harsh black and stark white. He was moving away from photography's surface record of reality towards poetic self-expression.

Later, Brandt began to photograph inanimate objects in an equally abstract way, especially the stones and seashore flotsam hitherto associated with his nudes; he also experimented with colour. Continuing to take and publish pictures to the end of his life, he was the first photographer to be given a 'one-man show' at the Hayward Gallery in London (in 1970; ironically, the exhibition was first mounted at New York's Museum of Modern Art); his prints were sold by the fashionable London and New York dealers, Marlborough Fine Art.

Soft-spoken, slight, courteous, and with the elusive smile of a Cheshire cat, Brandt seemed almost too unworldly for a professional photographer. He avoided publicity, preferred not to discuss his work ('I am an instinctive, not an intellectual photographer') and scarcely ever consented to be interviewed on radio or television. The Royal College of Art, London, gave him an honorary degree in 1977, and the Royal Photographic Society made him an honorary fellow in 1980. He became RDI in 1978.

Brandt's books, not universally well received at the time of publication, later were among the most prized works of photo-poetry in the history of the medium. The mysterious intensity with which he endowed the commonplace and everyday is unmatched, though many have striven to capture through the lens an imagined landscape of such vivid and awesome intensity. But Brandt alone mastered 'the spell that charges the commonplace with beauty'.

Later Brandt married Marjorie (died 1971), daughter of Henry James Becket. In 1972 he married Noya, daughter of Ivan Leznover, merchant. Brandt died 21 December 1983 in London. He was never naturalized.

[Cyril Connolly and Mark Haworth-Booth, *Bill Brandt, Shadow of Light*, 1977; Michael Hiley, *Bill Brandt: Nudes 1945–80*, 1980; Alan Ross, *Bill Brandt: Portraits*, 1982; Mark Haworth-Booth and David Mellor, *Bill Brandt Behind the Camera*, 1985; personal knowledge.]   COLIN FORD

BRIDGE, (STEPHEN HENRY) PETER (1925–1982), theatrical impresario, was born in Wimbledon 5 May 1925, the only son and elder child of Stephen Henry Howard Bridge, stockbroker, of Wimbledon, and his wife, Ella Mary Twine. He left Bryanston School when sixteen (his mother and sister having been killed by enemy action in 1940 and his father dying in 1943) and was briefly an actor before joining the RAF. He served for five years, landing in Normandy on D-day + 6. He was later in photographic intelligence in Burma and ended as personal assistant to the air officer commanding Hong Kong and air aide-de-camp to the governor.

On demobilization he became assistant manager for Lord Tedder [q.v.] in the RAF Malcolm Clubs; but by 1948 he had begun his life's work as an impresario, with *Set to Partners* by Diana Morgan, and plays by Leslie Sands and Val Gielgud [q.v.]. None of these was particularly successful; and as he now had a wife and soon a family to support, and in part to assuage his unquenchable curiosity and enthusiasm for every aspect of theatre, he became, first, assistant manager to Alec Clunes [q.v.] during his prestigious reign at the Arts Theatre; then briefly managed the Winter Garden and co-presented Christopher Fry's *The Firstborn*.

Next he was a dynamic employee of Lord Gifford's theatre ticket agency Ashton & Mitchell, drumming up new business throughout the country, with a special emphasis on US Air Force bases. There followed two years as itinerant critic for Keith Prowse, assessing the West End prospects of touring productions—a job and a relationship which was to bear much fruit at the peak of his career.

In 1955 with the birth of Independent Television he was appointed director of sport for Associated Rediffusion. He was—astonishingly, for when had he had the time?—one of the country's six professional lawn tennis referees (not to be confused with mere umpires) and produced ITV coverage of Wimbledon for two years.

But the call of the theatre was too strong. In 1957 he returned to management with *The Queen and the Welshman* by Rosemary Ann Sisson. This *succès d'estime* was to be the first of some sixty Bridge productions in the next thirteen years, of which a dozen were considerable box office successes and another dozen were artistically distinguished—a higher percentage of success than perhaps it sounds. A subjective selection would place in the first category: *Guilty Party* (1961), *Difference of Opinion* (1963), *Six of One*—revue (1963), *An Ideal Husband* and *Say Who You Are* (1965), *Wait Until Dark* (1966) and two early plays by Alan Ayckbourn—*Relatively Speaking* (1967) and *How the Other Half Loves* (1969). These eight achieved from 300 to 800 performances each, when 250 was considered a long run. In the 'artistic' category were Hugh Leonard's *Stephen D* (1963), Giles Cooper's *Happy Family* and Arbuzov's *The Promise* (1967), two all-star G. B. Shaw [q.v.] revivals *Too True to be Good* (1965) and *Getting Married* (1967), and a daring American import *Boys in the Band* (1969).

Not surprisingly this West End record, to which must be added six productions in New York, five in Toronto, and one in Australia, took its toll. Bridge had burned himself out physically and emotionally, and for the last decade of his life his managerial output was drastically reduced, but never his enthusiasm. His final production was a revival in 1981 of *Dangerous Corner* by J. B. Priestley [q.v.], graced by an on-stage appearance of the author on the opening night.

A man of the widest theatrical taste, Peter Bridge had an equally all-embracing love of theatre folk. When he employed the idols of his boyhood—the Hulberts, Celia Johnson [qq.v.], Roger Livesey, Margaret Lockwood, and many others—he was giving star-struck thanks for past pleasures as well as showing confidence in their abiding drawing power; but the discovery of some unknown writer (Alan Ayckbourn in Scarborough), some youthful talent at LAMDA, some esoteric offering of The Fringe would arouse in him, if not in his backers, a joy quite as intense.

He was a man without enemies and held in deep affection by the theatre world—something unique for an employer. His shortcomings flowed from his virtues; he could be over optimistic, over generous, over loyal. He was an obsessive round-the-clock communicator—his media being the theatre and the telephone. Good reviews, ideas for new productions, box-office returns, theatre news and rumours—all would be retailed at length to his friends whether they were in bed, hosting a lunch party, or about to go on stage. It was a small price to pay for having such an ally.

He married in 1948 Roslyn Mary, daughter of Douglas Seymour Foster, of independent means. In a happy and close-knit family there were three sons, of whom Andrew became an internationally known theatre lighting designer. Bridge died at his home in Highgate 24 November 1982.

[*The Times*, 27 November 1982; notes supplied by Roslyn Bridge; Ian Carmichael's memorial service address; Michael Denison, *Double Act*, 1985; personal knowledge.]

MICHAEL DENISON

BROWN, GEORGE ALFRED, BARON GEORGE-BROWN (1914-1985), politician, was born 2 September 1914 in Peabody Buildings, Lambeth, London, the elder son and eldest of the four children of George Brown, grocer's packer and later van driver, of London, and his wife, Rosina Harriett Mason. He was educated at Gray Street elementary school and West Square Central School in Southwark. He became a choirboy at the church of St Andrew's By-the-Wardrobe, and in later life remembered Father Sankey, the priest there, and some of his schoolteachers as people who had greatly influenced him. Although his family would not have ranked among the very poorest in London he was, from an early age, acquainted with hardship. This experience, combined with the strong trade-union loyalty of his father and the spiritual element contributed by school and church, produced a man who was to become a champion of the underprivileged and a figure in world politics.

Leaving school at fifteen, Brown became a fur salesman for the John Lewis partnership and then secured a job with the Transport and General Workers' Union. In his spare time he was politically active, especially in the Labour Party League of Youth. In 1937 he married Sophia, a book sewer, daughter of Solomon Levene, bookbinder, and his wife, two outstanding figures in East London Labour politics. There were two daughters of this marriage, and Brown became a grandfather in his early fifties. In 1939 he made his first speech at a Labour Party conference, attacking Sir R. Stafford Cripps [q.v.]. His trade-union work during the war brought him into touch with agricultural workers and George Dallas, a veteran of the Agricultural Workers' Union, helped him to become parliamentary candidate for Belper, for which constituency he was elected in 1945. He became at once parliamentary private secretary to George Isaacs [q.v.], the minister of labour, and in 1947 was taken on, still as PPS, by E. H. J. N. (later Lord) Dalton [q.v.], chancellor of the Exchequer. While in this position he became involved in a plot to put Ernest Bevin [q.v.] in the place of C. R. (later Earl) Attlee as prime minister. The plot failed—Bevin himself would have none of it—but in October 1947 Attlee promoted Brown to be a junior minister of agriculture and fisheries. In 1951 the resignation of Aneurin Bevan [q.v.] produced several ministerial changes and George Brown became minister of works in April and a privy councillor. A few months later, however, the defeat of the Labour government moved him to the opposition benches.

The years of opposition, 1951-64, were hard on Labour MPs. Parliamentary salaries were inadequate, and strife within the party created an atmosphere of discouragement. Despite his rapid rise to the front rank, Brown considered leaving Parliament and resuming trade-union work, but financial help from Cecil King of the *Daily Mirror* made it possible for him to stay at Westminster. He was elected to the shadow cabinet and became spokesman on defence. His ability as an administrator and a debater was widely recognized. His style was sometimes elevated by passionate belief, sometimes degraded by ill-temper, so that he accumulated both admirers and enemies. His famous quarrel with the Russian leaders Khrushchev and Bulganin when they were the Labour Party's guests at dinner can be regarded as a sturdy defence of democracy, or an ill-timed discourtesy, or both. In 1960, on the death of Aneurin Bevan, he was elected deputy leader of the Labour Party and firmly supported the leader, Hugh Gaitskell [q.v.], in the struggle against unilateral nuclear disarmament. On Gaitskell's death in 1963 Brown was deeply disappointed at being defeated (by 144 votes to 103 in the final ballot) by Harold Wilson (later Lord Wilson of Rievaulx) in the election for the leadership: but he set to work to co-operate with Wilson in making plans for the expected Labour government.

Accordingly, Labour's victory in 1964 meant that George Brown became secretary of state for economic affairs, charged with the task of creating a new department which would plan the nation's economy. He was successful in recruiting people of great ability, and a national plan was drawn up which the National Economic Development Council was, with difficulty, persuaded to accept. It was buttressed by a declaration of intent on productivity, prices, and incomes, in which employers and trade unions—again, persuaded with difficulty—accepted the view that incomes and prices depend on productivity and could not be left to conflicting bargaining power. Regional planning councils were set up to work out the application of the plan throughout the country.

In 1966 a sharp balance of payments crisis obliged the government to choose between devaluation of the pound or a stern package of deflationary measures. Brown recognized that deflation would have meant the destruction of much of the work of the new department, but he was overruled by the cabinet. His decision to leave the government, however, was met by an appeal from over 100 Labour MPs and, after piloting through Parliament a measure for the statutory control of prices and incomes, he moved to the Foreign Office in August 1966.

The nineteen months which Brown spent as foreign secretary were packed with difficult problems. He had always followed closely the problems of the Middle East, and, after the 1967 war, was able, with the help of Lord Caradon, to draft the Security Council resolution 242 which set out the principles on which a settlement might be based; but the contending parties refused to put it into practice. He had also taken a keen interest in the approach of the United Kingdom to the European Economic Community; he and Harold Wilson visited all the Community countries, seeking agreement, but were frustrated by General de Gaulle's veto. At one point he seemed close to securing agreement between the USA and the USSR on Vietnam, but this also failed at the last moment. In March 1968 he came into conflict with Harold Wilson, whose conduct of the government seemed to him too autocratic, and the long-suffering prime minister accepted his resignation from the cabinet, which Brown had offered once too often.

His memoirs, entitled *In My Way*, published in 1971, help us to see what he sought to do in the two high offices which he held. In both, he was disappointed in his main objective: but it is noteworthy that innovations which he made in the relations between government and industry, and in the organization of the Foreign Office, have left their mark. His ability was universally recognized, but his explosive temperament, often aggravated by alcohol, hampered his performance.

After his resignation Brown continued, as deputy leader, to work in the Labour Party, and he toured the country vigorously in the 1970 election campaign. It is possible that he did this to the neglect of his own constituency and so contributed to his defeat at Belper. He was then created a life peer, as Lord George-Brown, having changed his surname by deed poll, but his interest in politics declined. In 1972 the University of Milan awarded him the Biancamano prize for his work for Europe, and his efforts for peace in the Middle East were recognized by conferment of the Order of the Cedar of Lebanon (1971). He took up a number of business appointments which involved frequent visits to the Middle East. In 1976 he left the Labour Party after a disagreement on the question of the 'closed shop' and founded a social democratic organization; this however was overshadowed by the Social Democratic Party. By 1980 his health had begun to deteriorate, and he and his family saw less of each other. He went to live in a Cornish village, and, after a lengthy illness, died 2 June 1985 in the Duchy Hospital, Truro, of a liver complaint. Despite the disappointments of his last years he will be remembered as one whose faith and energy did much to raise the Labour Party to influence and power.

[*The Times*, 4 June 1985; George Brown, *In My Way*, 1971 (autobiography); private information; personal knowledge.]

MICHAEL STEWART

BRYANT, SIR ARTHUR WYNNE MORGAN (1899–1985), historical writer, was born 18 February 1899 at Dersingham, a village on the royal Sandringham estate, the elder son (there were no daughters) of (Sir) Francis Morgan Bryant, then chief clerk to the Prince of Wales, and his wife, May, elder daughter of H. W. Edmunds, of Edgbaston. Sir Francis later held various offices in the royal secretariat and became registrar of the Royal Victorian Order. Bryant was brought up in a house adjoining the wall of Buckingham Palace gardens, close to the Royal Mews. It was a world of protocol, pomp, and pageantry. His feeling for a picturesque and glorious English past never left him. He was educated at Pelham House, Sandgate, and Harrow School. He was intended for the army but in 1916 won an exhibition at Pembroke College, Cambridge, which however he did not take up, joining the Royal Flying Corps in 1917 and becoming a pilot officer. His war experience deeply affected him. In January 1919 he went up to Oxford at The Queen's College where he had family connections, to study modern history. He obtained in 1920 (using the surname Morgan-Bryant) a distinction in the shortened honours course for ex-servicemen.

Bryant developed a strong social conscience and a deep belief in the importance of education as a bridge between the 'two nations'. It was this that later led him to become a popular historian. Meanwhile he taught at a London County Council school combining the post with regular attendance at debutante dances. Being tall, dark, and good-looking he often persuaded his dancing partners to help him teach 'ragged' children at the Dickens Library in Somers Town. He was called to the bar by the Inner Temple in 1923, but in the same year accepted the post of principal of the Cambridge School of Arts, Crafts, and Technology. In 1925, the year of his marriage into the old-established Shakerley family of Somerford Park, Cheshire, he moved to become a lecturer in history for the Oxford University delegacy for extra-mural studies, a post he held till 1936. He also acted as educational adviser to the Bonar Law College at Ashridge and in 1929 published his first book, *The Spirit of Conservatism*, for the benefit of Ashridge students.

At this stage several careers were open to him—the bar, Tory politics, drama (he produced historical pageants in both Cambridgeshire and Oxfordshire). His interest in history had been stimulated by cataloguing the voluminous Shakerley archives, and in 1929 a publishing friend asked him to write a new life of Charles II. At the suggestion of Professor Wallace Notestein of Yale he abandoned strict chronology and began with the King's escape from Worcester, weaving in later the details of his early life. This dramatic opening caused the Book Society to make it their October (1931) choice and hence a bestseller. His success rightly convinced Bryant that he could live by his pen. He did so for the rest of his life and lived very well. *King Charles II* (1931) was not only readable but a work of scholarship. It was followed by Bryant's most important contribution to English history, the three volumes (1933, 1935, and 1938) of the life of Samuel Pepys [q.v.] which John Kenyon correctly described as 'one of the great historical biographies in the language'.

Bryant had an almost manic energy. He wrote over forty books and sold over two million copies published largely by Collins. In 1936 he succeeded G. K. Chesterton [q.v.] as writer of 'Our Note Book' in the *Illustrated London News* and continued to do so till his death. His output in the magazine has been reckoned at 2,783,000 words. His secretary, Pamela Street, daughter of A. G. Street [q.v.], wrote a remarkable account of his whirlwind activity into the 1970s— the chaos, confusion, telephone calls, unanswered letters, perpetual rush to achieve deadlines, which characterized life at his house in Rutland Gate.

He divided his time between London society—he was a member of six London clubs as well as the MCC—and the country where he had for a time a farm and was involved in both agriculture and forestry, but London was his world. He was a notable figure in Grillion's, an old parliamentary-cum-literary dining club of which he was joint secretary till his death.

After *Pepys* his books tended to be colourful and readable rather than closely researched, and to jar on younger historians. His best-selling volumes written in World War II incurred some criticism—*English Saga* (1940), *The Years of Endurance, 1793-1802* (1942), *Years of Victory, 1802-1812* (1944). The morale-raising parallel between Napoleon and Hitler does not convince. Partly for these reasons, though one suspects another—envy of popular success—Bryant never received the highest academic honours. After 1945 he produced one work of major significance and research—his two-volume edition, with commentary, of Lord Alanbrooke's diaries (*The Turn of the Tide*, 1957, and *Triumph in the West*, 1959). They created a storm because of criticisms of Sir Winston Churchill, then a sacrosanct figure, but remain essential reading. For much of the rest of his life he was engaged in broad outline histories of England. The first of three new volumes, *Set in a Silver Sea* (1984), appeared before he died, the second (*Freedom's Own Island*, 1986, with a chapter by John Kenyon) posthumously, and there is a third to come. He was a passionate believer in 'communication' and has some claim to be the Lord Macaulay and G. M. Trevelyan [qq.v.] of his day.

Bryant was a Tory patriot paternalist of the old order. He seldom went abroad and, in spite of his partly Welsh ancestry, was an intense English nationalist. He strongly opposed entry into the EEC. But he did not expect to win nor did he repine at defeat.

He was appointed CBE in 1949 and CH in 1967. He was knighted in 1954. He held honorary degrees from the universities of Edinburgh, St Andrews, and New Brunswick. He married first in 1924 Sylvia Mary (died 1950), daughter of Sir Walter Geoffrey Shakerley, third baronet. The marriage was dissolved in 1939. He married secondly in 1941 Anne Elaine, daughter of Bertram Brooke, HH Tuan Muda of Sarawak. This marriage was dissolved in 1976. There were no children. In July 1980 he announced his engagement to (Frances) Laura, widow of the tenth Duke of Marlborough, but the marriage did not take place. Bryant died in Salisbury 22 January 1985.

[*The Times*, 24 January 1985; John Kenyon in *Observer*, 18 February 1979; Pamela Street, *Arthur Bryant, Portrait of a Historian*, 1979; private information; personal knowledge.]   BLAKE

BRYHER (1894-1983), writer and private philanthropist. [See ELLERMAN, (ANNIE) WINIFRED.]

BUNTING, BASIL (1900-1985), poet and translator, was born 1 March 1900 in Scotswood on Tyne, Northumberland, the only son and elder child of Thomas Lowe Bunting, MD, physician and research scientist, and his wife,

Annie Cheesman, daughter of a mining engineer. He was educated at Newcastle Royal Grammar School and Ackworth School, Yorkshire, and then at Leighton Park, Berkshire. In or about April 1918 he was arrested for refusing military conscription on Quaker principles, and spent up to six months in prison. In 1920 he matriculated at the London School of Economics, but in 1922 he left without graduating.

By 1923 he was in Paris, had met Ezra Pound, and was employed by Ford Madox Ford [q.v.] to assist with the *Transatlantic Review*. In 1924 he followed Pound to Rapallo, and sailed along the Tyrrhenian coast. By the next year he was back in London, making for the first time a decent livelihood as music critic for *Outlook* and other magazines. In 1925 he made himself known to T. S. Eliot [q.v.]. Of the poetry written in those years the most substantial piece to survive is the so-called 'Sonata', 'Villon' (1925). After the demise of *Outlook*, a subsidy from the philanthropic American Margaret de Silver enabled Bunting to return briefly to Northumberland, thereafter (1929) to travel in Germany and return to Rapallo, and also briefly to visit the USA.

In 1930 he married Marian, daughter of Howard Leander Culver, businessman and owner of a shoe store; they had two daughters and a son (died 1982). He returned with his bride to Rapallo, where he met W. B. Yeats [q.v.] who briefly recorded their acquaintance. Pound and Louis Zukofsky represented Bunting in anthologies, and in March 1930 appeared in Milan his *Redimiculum Matellarum*. In 1933 penury forced him from Italy to the Canary Islands where he stayed for three years. From the Rapallo years survives 'Chomei at Toyama', and from the time in the Canaries 'The Well of Lycopolis'. His wife left him in 1936, whereupon Bunting returned to England and maintained himself by sailing and fishing off the coast.

On the outbreak of World War II the erstwhile conscientious objector was eager for war service and in 1940-1 served with the RAF, some of the time at sea. On the strength of the classical Persian he had learned at Rapallo in order to read Persian poetry, Bunting prevailed on the Air Ministry to send him as an interpreter to Iran. His very happy life there, and on a motorized convoy from Basra to Tunis, figures, with other of his wartime experiences, in 'The Spoils' (1951). Meanwhile had appeared, in Galveston, Texas, *Poems: 1950*. After the war Bunting was back in Persia, working for the British embassy and subsequently as *Times* correspondent.

In 1948 he married Sima, daughter of Kambar Alladadian, who worked in an oil company; they had a son and a daughter. After 1952 Bunting lived in Northumberland, eking out a pittance by menial journalism for local newspapers. He was forgotten as a poet, and writing no poems, until 1963, when the young prentice-poet Tom Pickard sought him out. That much recognition was all he needed to incite him to his masterpiece, *Briggflatts* (1966). Such fame as he ever achieved came thereafter: in 1971 he was made an honorary D.Litt. of the University of Newcastle, and he was president of the Poetry Society (1972-6) and of Northern Arts (1973-6). But Bunting never escaped penury, and as an old man raised money by poetry readings and stints of teaching in the USA, where he enjoyed, as a poet, fame such as is still accorded him only grudgingly in Britain.

It was Bunting's misfortune that he appeared incorrigibly a modernist in a period when it suited English taste to think that modernism in poetry was a Franco-American aberration which English poets and their readers had blessedly escaped. But he was not programmatically a modernist at all; though his association with Yeats and Pound and Eliot is not accidental, and though he freely acknowledged his debt to exotic sources in ancient Persian and Arabic, his masters in poetry (and in ethics) were Lucretius and Horace (he translated both, with wit and passion); Dante, Villon, and Malherbe; Sir Thomas Wyatt [q.v.] and very notably, as consciously a north-country poet like himself, William Wordsworth [q.v.]. His criticism, still for the most part uncollected, reveals a powerful consistency among these seemingly random or idiosyncratic enthusiasms. As befits a poet who was so much a musician, the peculiar glory of Bunting's verse is its scrupulous playing off of vowel against vowel, consonant against consonant, and quantity against accent, to achieve effects seldom predictably mellifluous but always expressive and often very beautiful. He died in Hexham, Northumberland, 19 April 1985.

[C. F. Terrell (ed.), *Basil Bunting. Man and Poet*, 1981; *The Times*, 19 April 1985.]
DONALD DAVIE

BURCH, CECIL REGINALD (1901-1983), industrial and academic scientist, was born in Oxford 12 May 1901, the fifth and youngest child and third son of George James Burch FRS, professor of physics at the University College, Reading, and his wife, Constance Emily, daughter of Walter Jeffries. His mother ran a young ladies' finishing school at Norham Hall, Oxford, which was strongly patronized by girls from the Continent and so virtually collapsed when war broke out in August 1914, leaving her and her family in straitened circumstances. Her husband had died in March 1914. Burch attended the Dragon School, Oxford, from which he obtained a classical scholarship to Oundle, which, generously, charged his mother no fees. In 1919 he won a senior scholarship, followed by a Ewelme scholarship, to Gonville and Caius College, Cambridge, and was there joined by his older brother who had spent two years on war work after leaving Winchester. His eldest brother was killed in World War I. Burch gained second classes in both parts of the natural sciences tripos (1921 and 1922).

Burch was then offered a college apprenticeship in the Metropolitan-Vickers Electrical Co. of Trafford Park, Manchester. He greatly enjoyed working in the large-machines part of the factory where he learned mechanical skills as well as a choice 'works' vocabulary with which he delighted shocking his friends all his life. As an apprentice in the research department he invented complex electronic circuits and published his first paper on radio atmospherics. At that time the department was the home of the BBC Manchester station 2ZY and Burch operated the power transmitter. With a colleague he operated the first trans-Atlantic transmission, put out by the BBC, to and from Westinghouse broadcasting station. With the same colleague he developed the science of induction heating using very high frequencies, first melting precious metals and then steel; they wrote a classic work on induction heating and 'taught Sheffield' its use for making tons of alloy steels.

Burch was given the task of impregnating insulating materials with 'transformer oil' in a vacuum but found that at low pressures some fractions of the oil distilled leaving a waxy residue. He realized that some of the fractions had very low vapour pressures and might therefore be suitable as the operating fluid in a diffusion pump in place of mercury and thus avoid the necessity of using a liquid-air trap above the pump. He patented the process which he called evaporative distillation and named the low-vapour-pressure products 'apiezon' oils, which came to be used all over the world. The uranium separation plant at Oak Ridge, Tennessee, where the atomic bomb material was made during World War II, could not have been run without Burch's pumps and oil. The Dutch Shell Co. took over production of the apiezon oils and distilled several tons per day. In 1943 the Physical Society awarded Burch the Duddell medal for his pioneering work and he was elected FRS in 1944. He also distilled cod-liver oil, co-operating with the BDH Co. and it was found that vitamin A distilled in his unique apparatus without decomposition; the process was fully developed and is now widely used in the pharmaceutical industry.

Burch had always been interested in optics and had mastered the technique of grinding huge areas of cast iron and speculum metal to a high degree of flatness. When his brother, with whom he had been working and living, died in 1933, he went with a Leverhulme scholarship to Imperial College to produce lenses and mirrors of spherical and aspherical shapes, figured to an accuracy of better than one millionth of an inch over several inches, ultimately improving the Newton 30-inch telescope mirror in Cambridge far beyond its original shape, a task which involved hand grinding all night for many weeks when vibration from London traffic was at its lowest. After taking his Ph.D. he was offered in 1936 a research associateship in the physics department at Bristol University where he spent the rest of his working life (from 1944 until 1966 as a research fellow). Burch decided to make a microscope with aspheric reflecting surfaces at an aperture of 0.55 (an enormously wide aperture; our cameras seldom exceed 2.5). With extremely advanced mathematics he designed the system to be without spherical aberration and primary coma. His success was widely proclaimed and he was inundated with requests for microscopes. Against his will and with the help of a precision engineering firm in Bristol he undertook to make ten. Burch could drive himself very hard but he was not a manager of men and overworked himself in tackling the countless problems of manufacture so that in 1953 he collapsed and suffered from an ulcerated stomach. He blamed a 'lack of self-knowledge' but to work such long hours he dosed himself with dextroamphetamine with amine-oxidase inhibitors to keep awake. At the same time he was supervising research students to whom he gave his time most generously.

For convalescence he went to Cornwall, got interested in tin mining, learned to 'pan' for tin and gold, and invented a mineral classifier of novel helicoidal shape which he patented; in time thousands of tons of tin 'trailings' were passed through his separators and he successfully directed some of his research students into the mining industry. The Royal Society awarded him the Rumford medal (1954) and he was appointed CBE (1958).

Burch was the friend of many: the workmen on the big lathes in Trafford Park, the laboratory assistants who were fascinated by his talk and his manual dexterity, his scientific colleagues who saluted him as a classicist, physicist, and mathematician of outstanding ability. In 1937 he married Enid Grace, daughter of Owen Henry Morice. They had one daughter. His wife was a lecturer in education at Bristol University. Burch died at his home in Bristol 19 July 1983.

[T. E. Allibone in *Biographical Memoirs of Fellows of the Royal Society*, vol. xxx, 1984; scientific papers and a tape-recording in the library of the University of Bristol; personal knowledge.]   T. E. ALLIBONE

BURGHLEY, BARON, sixth MARQUESS OF EXETER, (1905-1981), athlete and parliamentarian. [See CECIL, DAVID GEORGE BROWNLOW.]

BURTON, RICHARD (1925-1984), actor, was born Richard Walter Jenkins 10 November 1925 at Pontrhydyfen, a small Welsh village in the Rhondda Valley, four miles from Port Talbot, the son of Richard Jenkins, a miner, and his wife, Edith Maud Thomas, who had worked as a barmaid at the Miners Arms public house in the village. He was their sixth son, the twelfth of their thirteen children. His mother died in October 1927, and he was brought up by his eldest sister, Cecilia, and her husband, Elfed. The family spoke both Welsh and English and Richard was able to speak Welsh for the rest of his life. He was educated at Eastern Primary School, Port Talbot, and Port Talbot Secondary School. At fifteen he left school to work in the men's outfitting department at the local Co-operative store. Bored, he joined a youth club and experienced the exhilaration of amateur dramatics. Wanting to play rugby, he became a cadet in the local Air Training Corps, where one of the officers was Philip H. Burton, the senior English teacher at the Secondary School and a theatre lover. In appearance young Richard Jenkins was of medium height with fine, wide, blue eyes. Sturdily built, he had the body of a rugby half-back, long and solid in the trunk, but with short legs. He was troubled by boils and his skin was pitted by acne. Nevertheless he was considered extremely attractive. Convinced that in education lay escape from his job, he now focused all his charm on Philip Burton and in September 1941 was readmitted to the Secondary School. He moved into the teacher's lodgings and in 1943, after matriculating, legally renounced his own surname and became Richard Burton.

He made his début in London as Glan in Emlyn Williams's play, *The Druids' Rest*, on 26 January 1944. When the play closed he was called up. On a special six-month wartime course at Exeter College, Oxford, he read English, while also undergoing RAF training. His tutor, Nevill Coghill [q.v.], a gifted amateur director, was captivated by him. Casting Burton as Angelo in his Oxford University Dramatic Society production of *Measure for Measure*, Coghill proclaimed him 'a genius'. Demobilized in 1947, Burton returned to the theatre. In 1948, filming *The Last Days of Dolwyn*, he met the actress Sybil Williams from Ferndale—also a Welsh mining village—where she had been brought up by aunts after the death of her parents. Her father was an under-manager in a coal mine. Sybil and Burton were married in 1949 and had two daughters: Kate (1957) and Jessica (1959).

In 1949, in *The Lady's Not For Burning*, the stillness and simplicity that would become Burton's trademark attracted considerable attention. After his Prince Hal at Stratford-upon-Avon in 1951, his future was assured. His first Hollywood film, *My Cousin Rachel* (1953), also brought him the first of seven unsuccessful Academy award nominations and won him the coveted lead role in *The Robe* (1953). In 1953 at the Old Vic he played his first *Hamlet*. The voice, beautifully modulated, and the physical presence so controlled, created an impression of sensitivity combined with a startling virility. In 1955 his *Henry V* won him the *Evening Standard* Best Actor award. When he alternated the roles of Othello and Iago in 1956 (with John Neville) his reputation as a classical actor seemed unassailable.

Burton was notorious for his romantic exploits behind the scenes. He also had a reputation as both a compelling story-teller in the Welsh tradition and as a fierce drinker. Asked where his ambition lay next—Macbeth perhaps, or Lear?—no one took his reply seriously: 'I want to be a millionaire.' However, when the season finished he settled in Switzerland; he would never appear on the London stage again. In 1960 he played King Arthur in the musical *Camelot* in New York. In 1961 he arrived in Rome to play Mark Antony in the film *Cleopatra* (1963) starring Elizabeth Taylor (the English-born daughter of Francis Taylor, art dealer, who had moved to America on the outbreak of World War II). Burton separated from his wife.

During the next thirteen years he made over twenty films, few of which pleased him or the critics, but as the spy in *The Spy Who Came In From The Cold* (1966) he was superb and *Who's Afraid of Virginia Woolf?* (1966), with Elizabeth Taylor, was justifiably acclaimed. He had married Elizabeth Taylor in Canada in 1964 (his divorce was finalized that year), *en route* to New York with *Hamlet*. He was her fifth husband; she had two sons and a daughter by previous marriages. In 1966 he returned to Oxford to raise money for the OUDS, appearing in the title role of *Dr Faustus* with Elizabeth Taylor as Helen of Troy. They gave their services free but the critics savaged him. When he said he yearned to be an academic, an honorary fellowship at St Peter's College, Oxford, was arranged (1973) but the realities of a don's life made him abandon the experiment.

Burton was appointed CBE in 1970. His drinking was now addictive and in 1974 his

marriage to Elizabeth Taylor was dissolved; they remarried a year later in Botswana and divorced again in 1976. He married Susan Hunt, the English daughter of Frederick Miller, lawyer, and ex-wife of racing driver James Hunt, while playing in *Equus* on Broadway (1976). *Equus* impressed the critics and was filmed.

In 1980, recreating his role in *Camelot*, he collapsed in Los Angeles and underwent surgery on his spine. Fighting alcoholism, he filmed *Wagner*. His marriage was dissolved in 1982 and in 1983 he married Sally Hay, an English continuity girl he had met while making *Wagner*. She was the daughter of Jack Hay, motoring correspondent for the *Birmingham Post*. In his last film, *Ellis Island*, Burton played the father of his real-life daughter, actress Kate Burton. He died of a cerebral haemorrhage 5 August 1984 in hospital at Geneva, and was buried at Celigny, Switzerland, where he lived.

Burton had made his début when the London stage was dominated by actors of flamboyant lyricism—(Sir) Michael Redgrave [q.v.], (Sir) John Gielgud, and Laurence (later Lord) Olivier. His sheer sexuality had confounded and excited critical opinion. Now the obituaries deplored his failure to fulfil expectations, but these were expectations other people had predicted for him. He had done what he wanted with his life: he had achieved fame and riches and experienced passion. Above all he had escaped from the steelworks, the mines, and the Co-op, and a life of stultifying mediocrity that as a young Welsh boy must have once seemed his inevitable destiny.

[Paul Ferris, *Richard Burton*, 1981; Penny Junor, *Burton*, 1985; J. Cottrell and Fergus Cashin, *Richard Burton*, 1971; personal knowledge.]   KEITH BAXTER

BUSH, ERIC WHELER (1899-1985), naval officer, was born 12 August 1899 at Simla, the younger son and younger child of the Revd Herbert Wheler Bush, chaplain to the forces, and his wife, Edith Cornelia, daughter of Dr George Cardew, inspector-general of the Indian Medical Service. Mrs Bush returned to England in 1908 with her two sons, leaving her husband as the principal of the Lawrence Memorial School, Murree, until he too returned, to become vicar of Bathford, in 1912. Eric Bush was educated in England at Stoke House, Stoke Poges, and on 10 May 1912 entered the Royal Naval College, Osborne, as a naval cadet.

After two years' general education, 'Blake term', of which he was a member, proceeded to the RN College, Dartmouth, but the course was interrupted at the end of the first term by the mobilization of the fleet, on 1 August 1914. Bush was appointed to the armoured cruiser *Bacchante*, the flagship of the 7th cruiser squadron, and on 28 August 1914 was present at the battle of the Heligoland Bight. After service in the North Sea and English Channel, the cruiser joined the Mediterranean Fleet and was part of the Dardanelles expeditionary force. On 25 April 1915 Midshipman Bush (promoted in the previous October), commanded a picket-boat which towed boats of the first assault wave to land on 'Anzac' beach and for his services on this day and during the month which followed he was awarded the DSC, besides receiving the first of four mentions in dispatches. In August Bush was similarly engaged during the Suvla landings.

In March 1916 he joined the new battleship *Revenge*, in which he remained until the end of the war, seeing action only at the battle of Jutland. A sub-lieutenant since July 1917, he was one of the 370 naval officers—Kipling's 'gentlemen tired of the sea'—whose education had been interrupted and who were sent to Cambridge University in January 1919 for two terms. By his own account, his main achievement appears to have been the hoisting of a large white ensign on the lantern above his college, Trinity.

Bush's inter-war career was typical for the period, with service aboard a destroyer in the Baltic and Home Fleet being followed by commissions in the East Indies, where he qualified as an interpreter in Hindustani in 1924, and on the China station, where, in 1932, he joined his first command, the Yangtse gunboat *Ladybird*. Between 'sea jobs' he served as a training officer, his infectious enthusiasm for the Service producing an exceptional 'term'—two of whose members were to win VCs—attended the RN staff course and, from 1934 to 1936, occupied the Naval Intelligence Division's Japanese desk.

At the end of 1936 Commander Bush was appointed as executive officer of the Mediterranean Fleet cruiser *Devonshire*. Observation of the fighting during the Spanish civil war and evacuation of refugees gave the Royal Navy much useful experience, but Bush also found time to get married, on 20 May 1938, during a visit to Cannes to Mollie Noël, daughter of Colonel Brian Watts DSO, of the Royal Army Medical Corps. They had two sons.

Promoted in June 1939, he became captain, auxiliary patrols, on the outbreak of war commanding the variegated collection of minor warships and conscripted fishing vessels which closed the straits of Dover to German submarines and dealt with the early magnetic mines. His tireless efforts during the Dunkirk evac-uation, at which he was responsible for the La Panne beaches, earned him the first of three DSOs. In June 1941 he returned to sea, in command of the anti-aircraft cruiser *Euryalus*, in which he remained until September 1943, taking part in all the more notable eastern Mediterranean actions during the period and the invasions of Sicily and Salerno.

Acquaintance with amphibious operations was renewed by his appointment in command of the Force Sword assault group, which he trained and led for the Normandy invasion, where he was responsible for landing 8th Infantry brigade at Ouistreham, on the exposed eastern flank of the assault area. After brief command of the battleship *Malaya* in the autumn of 1944, he returned to combined operations as the chief staff officer of Force W, the amphibious component of the South-East Asia Command. He commanded the assaults on Akyab and the Arakan coast operations and was largely responsible for the planning and execution of the invasions of Ramree island and Rangoon and for the unopposed re-occupation of Malaya and Singapore.

From late 1945 until June 1948, Bush commanded HMS *Ganges*, the boy seamen's training establishment at Shotley, near Harwich. This was to be his last naval appointment, for he was not selected for promotion to flag rank and was retired from the Service, shortly before his forty-ninth birthday.

An enthusiast for all matters maritime and dedicated to the encouragement of youth, Bush became the secretary of the Sea Cadet Council (1948–59). Towards the end of this time he wrote and had published (in 1958) his autobiography, *Bless Our Ship*. His character and personality show through this modest account of a substantial naval career: patriotic without a trace of jingoism or bigotry, his determination and dedication to duty were tempered by kindliness, a keen sense of humour, and love for his wife and two sons, both of whom followed him into the Royal Navy.

Leaving the Sea Cadet Council, he next became the general manager of the Red Ensign Club, in Stepney, which he ran until 1964. Even in retirement he continued his association with the sea, being engaged for several years by the British-India Steam Navigation Co. as a liaison officer and lecturer accompanying educational cruises. So well received were his lectures on the Gallipoli expedition, he was persuaded to write an excellent account of the campaign (*Gallipoli*, 1957) based on meticulous research as well as his own experience. Prior to this, he had compiled two anthologies of poetry and prose, one nautical (*The Flowers of the Sea*, 1962) and the other military (*Salute the Soldier*, 1966). Bush finally retired from the sea to Tunbridge Wells, Kent, where he died 17 June 1985.

[*The Times*, 20 June 1985; Eric Bush, *Bless Our Ship* (autobiography), 1958; *Navy Lists*; family information.]   DAVID BROWN

BUTLER, REGINALD COTTERELL, (1913-1981), sculptor, was born 28 April 1913 at Buntingford, Hertfordshire, the only child of Frederick William Butler and his wife, Edith Barltrop, the master and matron of the Buntingford workhouse. His father was a distant relative of the Irish poet, William Butler Yeats [q.v.]. His grandfather had been a gardener at Chatsworth under Sir Joseph Paxton [q.v.]. His mother was of Anglo-French extraction and a distant relative of the poet George Crabbe [q.v.]. From the age of fourteen he was educated at the Hertford Grammar School. He learned a great deal about the crafts and handling of tools in his father's Buntingford community.

'Reg' Butler (as he preferred to be known) entered architectural practice locally in 1933 and was sufficiently successful to be elected ARIBA in 1937. He was a lecturer in architecture at the Architectural Association School (1937-9), and continued with architectural journalism throughout the war and until 1951. During World War II, as a conscientious objector, he worked as a blacksmith in Sussex. He was interested in sculpture from 1937 onwards, initially being influenced by African primitive art and Henry Moore. His first one-man show was at the Hanover Gallery, London, in 1949. He was awarded the first Gregory fellowship in sculpture at the University of Leeds (1951-3). In 1951 he also became a lecturer at the Slade School of Fine Art, London, eventually becoming head of the department until 1980. The Arts Council of Great Britain and the Greater London Council commissioned work from him for the Festival of Britain in 1951. In 1952 the British Council invited him to take part in an exhibition at the British Pavilion at the Venice Biennale.

He was awarded the grand prix in an international competition for a monument to 'The Unknown Political Prisoner' in 1953, for a sculpture intended to be placed on the Russian-German frontier in Berlin. The sculpture was never built and the maquette was destroyed by a Hungarian refugee when it was exhibited at the Tate Gallery. Butler was able to make a second small model.

Butler's earlier work shows an acute awareness of trends in Britain and Europe after the war and has close affinities to sources as diverse as Henry Moore, Graham Sutherland [q.v.], and Francis Bacon as well as to Picasso, Gonzalez, Richier, Giacometti, and at a later stage,

Balthus and Bellmer. This suggests a certain eclecticism in his nature, but it is fair to say that he was equal to the sources of his inspiration, and at times excelled them. He enjoyed argument and was a stimulating and respected teacher. He loved fast cars and was excited by modern technology and by science fiction. He was one of a number of sculptors working with forged and welded metals in the 1950s. Critics tended to find the work harsh and threatening, reflecting a mood of post-war anxiety which Sir Herbert E. Read [q.v.] summed up as 'the Geometry of Fear'.

In the mid-1950s Butler's work turned from spiky biomorphic metaphors towards a more realist concern with the female figure. Increasingly its erotic nature suggested that the sculptor was searching to invent a series of modern Venuses. In his William Townshend lecture at University College, London, in November 1980, Butler talked at length about his admiration of Stone Age fertility figures. His own last series of doll-like figures, owing something to Indian and Japanese inspiration, seemed to relate twentieth-century sexual fantasies to an ancient tradition. The bronze casts were painted with a sugar-almond surface and were not unlike the treatment of female flesh in the paintings of Cranach. Heads and limbs were made in interchangeable units. The eyes were made of painted resin and covered with a glassy lens. Human hair was implanted in the skulls, but not in the pubic areas. Combining both lust and compassion, Butler's creations achieved the potency of their primitive ancestors.

Butler continued to be a prominent and controversial figure whose work was admired throughout Europe and in America. He is represented at the Tate Gallery in London, the Museum of Modern Art in New York, and the Hirshhorn Collection in Washington, as well as in many other museums. In 1965 he was elected to the Académie Royale des Sciences, des Lettres, et des Beaux-Arts de Belgique. A posthumous memorial exhibition took place at the Tate Gallery in November 1983.

In 1938 Butler married (Mary) Joan ('Jo'), daughter of Robert Child, farmer. They had no children. By his friend Rosemary, a sculptor, daughter of Matthew Young, doctor of medicine, he had two daughters. Butler died 23 October 1981 at Berkhamsted, Hertfordshire, where he had lived since 1953.

[BBC film, 1958; Tate Gallery catalogue, 1983; personal knowledge.]   JOHN READ

BUTLER, RICHARD AUSTEN, BARON BUTLER OF SAFFRON WALDEN, (1902-1982), politician, was born at Attock Serai in the Punjab, India, 9 December 1902, the eldest of a family

Butler, R. A.

of two sons and two daughters of (Sir) Montagu Sherard Dawes Butler [q.v.] and his wife, Anne Gertrude Smith. His father, who had passed top into the Indian Civil Service, was a member of a remarkable academic dynasty (since 1794) of Cambridge dons, which included a master of Trinity, two headmasters of Harrow, and one of Haileybury. He later became governor of the Central Provinces and, finally, of the Isle of Man. His mother, warm, sympathetic, and encouraging, and to whom Butler was always devoted, was one of ten talented children of George Smith, CIE, a Scottish teacher, journalist, and editor in India. She was the sister of Sir George Adam Smith [q.v.].

When Butler was six, he fell from his pony and broke his right arm in three places, an injury which was aggravated by a hot-water bottle burn. The arm never fully recovered, and successful games playing was thus ruled out though he became a keen shot. Returning to be educated in England, Butler attended the Wick preparatory school at Hove. Having rebelled against going to Harrow because of a surfeit of Butlers there and having failed a scholarship for Eton, Butler (by now known as 'Rab' as his father had intended) went to Marlborough. After a final year learning modern languages which were better taught than the classics he had earlier endured, Butler went to France to improve his French with the Diplomatic Service in mind. He won an exhibition to Pembroke College, Cambridge—the money was needed—which after a first class in the modern and medieval languages tripos (1923) was converted into a scholarship. He became secretary of the Union as a Conservative. An unsuccessful love affair and a mainly nervous collapse did not stop him becoming president of the Union (1924). In his fourth year Butler gained a first in history (1925) and a fellowship at Corpus Christi College.

While an undergraduate he had met Sydney Elizabeth Courtauld, a capable, strong-minded girl, who became his wife in April 1926. Her father, Samuel Courtauld [q.v.], an industrialist, settled £5,000 a year on Butler for life tax free. This financial independence enabled him to decide on a parliamentary career, though his father told him that strong personal executive decisions were not his forte and he should aim for the speakership. While the honeymooners went round the world, the Courtauld family secured for them a fairly safe seat, Saffron Walden in Essex, and on their return Butler was duly selected without the complication of competing candidates. He had a comfortable victory in the general election of 1929 and held the seat until his retirement in 1965. Before the election he had become private secretary to Sir Samuel Hoare (later Viscount Templewood,

q.v.), and he soon became known to the party hierarchy. His first notable public act was a sharp exchange in *The Times* with Harold Macmillan (later the Earl of Stockton), who was advised to seek 'a pastime more suited for his talents' than politics.

In the national government in 1931 Hoare became India secretary and Butler his parliamentary private secretary. At the second Round Table conference, Butler was deeply impressed by M. K. Gandhi [q.v.], the current hate figure of many Conservatives and of his father. After a tour of India, Butler became Hoare's under-secretary in September 1932. His support of constitutional reform and knowledge of the Indian scene made him a natural choice, even though he had been in Parliament only three and a half years and was easily the youngest member of the government. India was the issue on which (Sir) Winston Churchill was challenging Stanley Baldwin (later Earl Baldwin of Bewdley), and in the Commons Butler compared himself to 'the miserable animal', a bait 'in the form of a bullock or calf tied to a tree awaiting the arrival of the Lord of the Forest'. Yet he was never devoured by Churchill and proved himself Hoare's able lieutenant in defending the India Bill during the fierce two-and-a-half-year war waged against it by the Conservative right wing.

The Butlers had since 1928 lived in the constituency first at Broxted and then at Stansted Hall, Halstead, where their three sons and a daughter were largely brought up, and where in 1935 Baldwin came for the weekend and Churchill was invited. They also had a flat in Wood Street, London, until they moved to 3 Smith Square in 1938. They entertained generously in both London and the country.

Neville Chamberlain's accession to the premiership in May 1937 brought Butler a welcome release from the India Office but not a department of his own. However, his stint as parliamentary secretary at the Ministry of Labour gave him a useful acquaintance with the depressed areas and with mass unemployment. After nine months he went to the Foreign Office as under-secretary of state in February 1938. With the foreign secretary, the first Earl of Halifax [q.v.], in the House of Lords he was once again prominent—in the long run, indeed, too prominent. The policy of appeasement cut across the Conservative Party much more deeply than India or unemployment, and, when Churchill took over, Butler was on the wrong side of the divide. Appeasement was held against him in a way it was not against those more minor supporters of the Munich agreement, Lord Dunglass (later Lord Home of the Hirsel) and Quintin Hogg (later Lord Hailsham of St Marylebone).

Butler was an enthusiastic Chamberlainite and like Chamberlain regarded Munich not as a means of buying time but as a way of settling differences with Adolf Hitler. He was disposed, however, to interpret Benito Mussolini's invasion of Albania as a general threat to the Balkans, until Chamberlain told him not to be silly and to go home to bed. Butler remained an appeaser down to the outbreak of war, opposing the Polish alliance signed on 25 August 1939 because it would have 'a bad psychological effect on Hitler'. After Chamberlain's fall he, together with Alec Dunglass and two friends, drank to 'the King over the water' and described Churchill as 'the greatest political adventurer of modern times'.

Despite his conspicuous identification with the *ancien régime*, Butler survived Churchill's reconstruction of the government in May 1940. 'I wish you to go on', Churchill told him, 'with your delicate manner of answering parliamentary questions without giving anything away'; the prime minister also expressed appreciation of having been asked to 'Butler's private residence'. The Foreign Office was now a backwater, whose calm was only disturbed by Butler's imprudent conversation about peace with the Swedish minister in June 1940, which Churchill thought might indicate a lukewarm attitude to the war if not defeatism. Bombed out of both Smith Square and his father-in-law's house, Butler went for a time to stay in Belgrave Square with (Sir) Henry Channon, his parliamentary private secretary since 1938.

Butler remained at the Foreign Office against his wishes when Sir Anthony Eden (later the Earl of Avon), whom he did not admire, succeeded Halifax in December 1940. But in July 1941 after nine years as an under-secretary he became president of the Board of Education. Even further removed from the war than the Foreign Office, education was nevertheless a political minefield and had seen no major reform since 1902. Ignoring Churchill's warnings not to stir up either party politics or religious controversy, Butler decided on comprehensive reform. Although in the end he had to exclude the public schools, every child was given the right to free secondary education and, to make that right a reality for the poor, provision was made for the expansion of both nursery and further education and for the raising of the school leaving age. All Butler's formidable diplomatic and political skills were needed to secure the agreement of the churches and the acquiescence of Churchill. The 1944 Education Act was Butler's greatest legislative achievement and was deservedly called after him.

Butler became chairman of the Conservative Party's post-war problems central committee in 1941, and in November 1943 he joined the go-

vernment's reconstruction committee. The only leading Conservative clear-sighted enough to oppose an early election, he became minister of labour in Churchill's 'caretaker' government in May 1945. After the electoral defeat in July—Butler's own majority fell to 1,158—Churchill made him chairman both of the Conservative Research Department and of the high-powered industrial policy committee. From these two positions Butler exerted the major influence in reshaping Conservative policy, and, even more than Macmillan, was chiefly responsible for the civilized conservatism of the post-war party. In 1947 the industrial policy committee produced the *Industrial Charter*, which, Butler later wrote, was 'an assurance, that in the interests of efficiency, full employment, and social security, modern Conservatism would maintain strong central guidance over the operation of the economy'. Mass unemployment was to be a thing of the past; as Butler put it, those who advocated 'creating pools of unemployment should be thrown into them and made to swim'. The right wing regarded Butler's efforts as 'pink socialism', a recurring charge under various names in his later career. He himself believed that, without the rejection of unemployment and the acceptance of the Welfare State, the spectre of the thirties would not be exorcized and the Conservative Party would remain in opposition.

Contrary to the general expectation and his own, Butler became chancellor of the Exchequer in October 1951 and inherited the usual economic crisis. He tackled it by import controls and the resurrection of monetary policy. The cabinet rejected, however, his plan for a floating exchange rate, a decision which Butler both then and later regarded as a fundamental mistake. Butler's first two budgets were popular and successful, expansion and the promotion of enterprise being his general themes, and such was his standing that in September 1952 in the absence of both Churchill and Eden he was left in charge of the government. The same happened for a longer period in the summer of 1953 when, with Eden ill in Boston, Churchill was felled by a stroke. The gravity of Churchill's illness, concealed by his entourage, was known to Butler; this was perhaps the first occasion on which he could have become prime minister had he striven for the job. He had no such thoughts and ran the government well. Since Marlborough, painting had been Butler's chief hobby; after the war he occasionally painted with Churchill, once being commanded by him to 'take the mountains', while his leader would 'take the sea'. Butler thought their paintings were of about the same standard.

At the Treasury Butler, who was one of the two best post-war chancellors, had two special difficulties. Sir Walter Monckton (later Viscount Monckton of Brenchley, q.v.) had been made minister of labour by Churchill to conciliate the unions, and conciliation entailed conceding excessive wage claims, sometimes in concert with the prime minister and without consulting the chancellor. The second was the Conservatives' pledge to build 300,000 houses a year, which Macmillan, the minister of housing, never allowed the chancellor or the cabinet to forget. In consequence, too many of the nation's resources went into the housing drive. In 1954 Butler's third budget was, as he said, a 'carry-on affair' with few changes, but later in the year he predicted the doubling of the country's standard of living within twenty-five years.

In December 1954 his wife died after a long and painful illness. His grief and the loss of her influence as well as the effects of three gruelling years affected Butler's political judgement. His troubles were in any event growing: inflation and balance of payments difficulties necessitated a 'stop', and in February 1955 Butler raised the bank rate and brought back hire-purchase restrictions. Nevertheless he produced an electioneering budget, taking 6d. off income tax. That was his first mistake. After the election Eden invited him to give up the Treasury, but Butler refused, which was his second mistake. A run on the pound compelled an autumn budget whose unimaginativeness underlined the errors of its predecessor—his third mistake. In December 1955 Eden decided to replace Butler with Macmillan, who showed by his stipulated terms that he was determined also to replace Butler as Eden's heir apparent. Butler consented to become merely lord privy seal and leader of the House—his fourth and biggest mistake. He needed a change, but ministerial power in British politics rests with the big departments and for Butler to allow himself to be left without a department was a gratuitous act of unilateral disarmament.

Though Macmillan was to the left of him on economics, there was no issue on which Butler was, in the eyes of the Conservative Party, seen to be right wing. Many Conservatives saw him as a 'Butskellite'. Hence he was always more popular in the country than in his own party. His appearance was not charismatic, with his damaged arm, his sad, irregular features, and his clothes, described by Channon as 'truly tragic'. But behind it there was a Rolls-Royce mind and a sharp sardonic wit which he enjoyed exerting at the expense of his colleagues. He was the master of many types of ambiguity—'my determination is to support the prime minister in all his difficulties' or 'there is no one whose farewell dinner I would rather have attended'—and occasionally the cause of ambiguity in others. His famous saying that Eden was 'the best prime minister we have' was put to him as a

question to which he rashly assented. Butler had a strong vein of innocence, rare in sophisticated politicians. He was also abnormally good-natured and inspired great affection.

Butler was ill when President Nasser of Egypt nationalized the Suez Canal Company in 1956 and was in no danger of being infected by the collective reaction. He missed the first cabinet meeting at which the fatal route to Port Said was mapped and he was not included in the Egypt committee that Eden set up, though he occasionally attended it. His freedom from departmental responsibilities would for once have been an advantage, but cool, detached advice was not what Eden wanted. Over Suez Butler's predicament was acute. Far too intelligent to accept Eden's likening of Nasser to Mussolini, he had nevertheless an 'appeasing' past to live down. Believing that party and public opinion required action of some sort, Butler also believed that Britain should act in accordance with international law.

Hence Butler was in a similar position to John Foster Dulles, the American secretary of state, and was driven to similar deviousness: as the international position altered, different expedients had to be produced to prevent Eden launching an attack on Egypt. But what was permissible in Dulles, trying to divert an ally from folly, looked less so in the cabinet's nominal number two seeking to restrain his leader, sick and unbalanced though Eden was. Butler would probably have done better to state his position unequivocally or to keep quiet or to resign; doubts were not enough. Even so, if he had succeeded, as his phrase went, in keeping Eden 'in a political strait-jacket' he would have done the prime minister and the country a great service. But by October Butler had run out of strait-jackets, and he used the wrong tactics for defeating the Anglo-French-Israeli plan. Instead of joining with Monckton in direct opposition to a grubby conspiracy which was bound to fail, Butler implausibly advocated an open attack on Egypt by the three countries which would have been scarcely less disastrous. After the UN had voted for an emergency force and an Israeli-Egyptian cease-fire seemed imminent, Butler tried to prevent the Anglo-French invasion as it was by then redundant; and when two days later Eden told the cabinet that a cease-fire was essential, Butler like Macmillan strongly supported him.

Butler's deviousness over Suez was honesty itself compared with the duplicity of Eden and some colleagues; and he was more consistent than Macmillan whose fire-eating bellicosity first drove Eden on towards destruction and who then suddenly demanded peace. Yet Butler ended up by pleasing virtually no one, and his varying indiscretions to different back-bench groups gave the impression that he was not playing the game. Others were playing a deeper one.

Eden's retreat to the West Indies to recuperate left Butler to do the salvage work at the head of a weak and divided government. Butler was at his best but gained no credit for limiting damage that he had not caused. Instead he incurred odium for unpopular though necessary decisions, made at Macmillan's insistence, over Britain's unconditional withdrawal from Egypt. In consequence, when Eden finally resigned in January 1957, Butler had no chance of succeeding him. The cabinet voted overwhelmingly for Macmillan, and back-bench soundings gave a similar result. Churchill, too, recommended Macmillan. Eden gave no advice to the Queen: he disliked both men although he preferred Butler. Butler took his defeat well. Macmillan refused him the Foreign Office, and Butler did not insist, accepting the Home Office while remaining leader of the House. At least he now had a department. He also, as under Churchill and Eden, had the government to run from time to time. When Macmillan in 1958 went on his Commonwealth tour after settling his 'little local difficulties' over the resignation of his entire Treasury team in January, Butler was left, as he said, 'to hold the baby'. As usual he held it well, and this time was popular. As home secretary he was a reformer, which was less popular.

After the October 1959 election Butler became chairman of the Conservative Party in addition to being home secretary and leader of the House. Other than demonstrating that there was almost no limit to his capacity for transacting public business—at which he was indeed the unrivalled master—there was little point in Butler's new job. It was in any case scarcely compatible with his existing ones. His leadership of the House entailed trying to get on with the opposition in the Commons, while his chairmanship of the party entailed attacking the opposition in the country. Further, as home secretary, Butler was intent on penal reform, while many of his party faithful were intent on the return of flogging. However Butler was always adept at squaring circles, and he squared those three. Much more important to him than the acquisition of offices was his wedding, in the presence of the couple's ten children, in October 1959 to a relative by marriage of his late wife, Mollie, widow of Augustine Courtauld [q.v.], polar explorer, and daughter of Frank Douglas Montgomerie, of Castle Hedingham, Essex. The marriage was strikingly happy and gave Butler renewed strength. He was an outstanding home secretary, making few mistakes in handling a notoriously tricky department and initiating much useful legislation. He beat the

flogging lobby and passed a major Criminal Justice Act; he reformed the laws of gambling, public houses, prostitution, and charities; and also passed in 1962 the Bill to curb immigration which had been prepared by Churchill's government and successively deferred.

In October 1961 Butler lost two of his offices, retaining only the Home Office, and was made overseer of the common market negotiations which in practice meant little. In March 1962 Macmillan, tired of the squabbling between the Colonial and Commonwealth Offices, formed a new central Africa department and persuaded Butler to take charge of it. This was a real job, if a thankless one; characteristically, Butler merely added it to his other one. But in the cabinet massacre of July 1962 he lost the Home Office and was left with his central African responsibilities with the honorific title of 'first secretary of state' plus the intimation that he would be serving as deputy prime minister. Macmillan was thus able both to heap burdens on to the good-natured Butler and to strip him of them again almost at will. For nearly all his long parliamentary career Butler had been a minister: this gave him a unique experience of administration but made him too addicted to Whitehall ever to think of withdrawing. He had, too, the character and quality of a great public servant.

Macmillan weakened his government by banishing Butler from the home front. Yet the government gained in Africa. At the Victoria Falls conference in July 1963 Butler achieved the seemingly impossible feat of an orderly dissolution of the Central African Federation without conceding full independence to Southern Rhodesia.

Butler made no attempt to take advantage of Macmillan's considerable troubles in the first half of 1963, and the prime minister's revived fortunes had persuaded him to fight the next election, when his prostate operation altered that decision. Butler was yet again asked to deputize. Yet Macmillan was determined to prevent Butler succeeding him and played an unprecedented part in choosing his own successor. At first he supported Hailsham and then switched to Home. Even more important, he devised a procedure under which he kept control of events. In acquiescing, the leading cabinet ministers, Butler above all, were markedly trusting or negligent. And after fudged consultations with cabinet ministers by the lord chancellor, Lord Dilhorne [q.v.], who produced an idiosyncratic reading of the results, and with MPs by the whips, some of whom knew the answer they wanted and went on till they got it, and after some apparent refining of the figures by the chief whip, Sir Martin (later Lord) Redmayne [q.v.], Macmillan adjudged Home the winner.

This decision was leaked on 12 October 1963, the day before Macmillan was to see the Queen. That evening a meeting of cabinet ministers at Enoch Powell's house telephoned Butler urging him to fight. Hailsham did the same very strongly. Butler's response was merely to ask the lord chancellor the next morning to call a meeting of all the leading candidates. Home felt like withdrawing, but was dissuaded by Macmillan who ignored the opposition to his 'compromise' choice and did not change his intended advice to the Queen. Shortly afterwards Home was on his way to the palace where he was asked to see if he could form a government. Even then Butler could have prevailed: both Hailsham and Reginald Maudling [q.v.] had agreed to serve under him, he had much cabinet support, and his wife was urging him on. But his heart was not in the fight, and after reserving his position he became foreign secretary on 20 October. Perhaps, as his father had long ago told him, he could not take strong personal executive decisions. Perhaps, like his old chief in 1940, Halifax, he did not really want the job. More likely he was inhibited by fears of splitting the party; and Home had been a friend since their Chamberlain days. Whatever the truth, his forbearance did not help the Conservatives. The supporters of both Butler and Hailsham thought their man would win the election of 16 October 1964, and both were probably right. Home just lost it. In his farewell message to the party conference, Macmillan hailed the coming into existence of 'the party of our dreams', which accepted a 'pragmatic and sensible compromise between the extremes of collectivism and individualism'; at the very same time he was blocking the man who was at least as responsible as himself for the existence of such a party, thus ensuring that the dream was short-lived. The 1964 election was crucial. A Conservative victory would have consolidated such a party and probably produced a Labour realignment. Defeat led to the later polarization of the parties and an abandonment of Macmillan's 'compromise'.

The rest, politically, was for Butler anticlimax. The job he had wanted in 1957 and 1960 no longer presented much of a challenge. He ran the Foreign Office easily, but had no opportunity or inclination to do anything of note. Had the Conservatives won the election, he would not have been reappointed. Butler was given no part in the election preparations and only a bit part in the election itself, though he gave one rather unfortunate interview. After the election he lost his chairmanship of the Conservative Research Department. Home offered him an earldom which he refused; in 1965 the new prime minister, Harold Wilson (later Lord Wilson of Rievaulx), offered him the mastership of Trinity College, Cambridge, which he accepted. He then accepted a life peerage in 1965 and took his seat on the cross-benches. Butler was the first non-Trinity man to become master for 250 years, and his appointment was at first not wholly welcome in the college. Nevertheless he and his wife were pre-eminently successful there, and in 1972 91 out of 118 fellows present voted for the maximum extension of Butler's term of office. In 1971 he published his autobiography. Lively, wise, and relatively accurate, *The Art of the Possible* was a strong contrast to the multi-volume efforts of Eden and Macmillan and was one of the very few political autobiographies to enhance its author's reputation. This was followed in 1977 by *The Conservatives*, a history of the party, which Butler edited and introduced. In the same year he retired from Trinity. To this Dictionary he contributed the notice of Sir Lionel Fox.

His son, Adam, was a member of the 1979 Conservative government, but Butler like Macmillan had no great liking for the new Conservative regime. In February 1980 he defeated in the Lords the government's proposal to allow local authorities to charge for school transport, which he saw as a breach of the 1944 Act's promise to provide free secondary education for all. Butler's portrait was painted by Margaret Foreman for the National Portrait Gallery, where he was last seen in public. He finished *The Art of Memory* (1982) which was little more than a footnote to its predecessor and was published after his death. He died 8 March 1982 at his home in Great Yeldham, Halstead, Essex.

Butler was sworn of the Privy Council in 1939, and was appointed CH in 1954 and KG in 1971. He was awarded honorary degrees by thirteen universities (including Oxford and Cambridge, both in 1962), and elected an honorary fellow of Pembroke College, Cambridge, in 1941, Corpus Christi College, Cambridge, in 1952, and St Antony's College, Oxford, in 1957. He was rector of Glasgow University (1956-9), high steward, Cambridge University (1958-66), chancellor of Sheffield University (1960-78), chancellor of Essex University from 1962, and high steward, City of Cambridge, from 1963. He was president of the Modern Language Association and of the National Association of Mental Health from 1946, and of the Royal Society of Literature from 1951. He was given the freedom of Saffron Walden in 1954.

[R. A. Butler, *The Art of the Possible*, 1971, and *The Art of Memory*, 1982 (autobiographies); Anthony Howard, *Rab*, 1987; Molly Butler, *August and Rab*, 1987; Robert Rhodes James (ed.), *Chips, The Diaries of Sir*

Henry Channon, 1967; John R. Colville, *The Fringes of Power*, 1985; private information; personal knowledge.]   IAN GILMOUR

BYERS, (CHARLES) FRANK, BARON BYERS (1915–1984), Liberal politician, was born in Liverpool 24 July 1915, the only son (there were two younger daughters) of Charles Cecil Byers, a Lloyd's underwriter, vice-chairman of United Molasses Ltd., and one-time Liberal parliamentary candidate, and his wife, Florence May, daughter of James Fairclough, of Northenden, Cheshire.

Educated at The Hall School, Hampstead, and Westminster School, Byers was captain of football and athletics and princeps oppidanorum at the latter. Going up to Christ Church, Oxford, in October 1934, he emerged four years later with a third class honours degree in politics, philosophy, and economics. He was awarded his blue and international colours as a 220-yard hurdler, holding the British Universities' record for over twelve years. At Oxford his interest in politics developed. He was a Liberal by temperament, instinct, and upbringing and became chairman of the Union of Liberal Students and the University Liberal Club (1937), often recalling with a wry chuckle that his treasurer was one Harold Wilson (later Lord Wilson of Rievaulx, q.v.).

Leaving Oxford in 1938 he joined Gray's Inn, but his studies ended when on 2 September 1939, the day after the outbreak of the war, and only a few weeks after his marriage, he enlisted in the Royal Artillery. In due course he became a lieutenant-colonel on the staff of the Eighth Army. His courage and determination, which were natural attributes, were recognized when he was appointed OBE in 1944, mentioned three times in dispatches, awarded the croix de guerre with palms, and created a chevalier of the Legion of Honour.

Byers had been adopted as prospective Liberal parliamentary candidate for the North Dorset constituency at the age of twenty-two, and returned to England in July 1945 for the general election. He won a remarkable and unexpected victory, being one of only twelve Liberals in the Commons. He was at once conspicuous for his political flair, judgement, and administrative ability, and became chief whip in 1946 after only a few months. However, largely because of boundary changes he narrowly lost his seat in 1950 and failed to regain it in 1951. Already a part-time director of the Rio Tinto Company (later RTZ), he now became full time, organizing the company's world-wide exploration programme. By 1959 he was in a position to resume his political work, and he became for many years the dominant figure in the Liberal Party as director of its election campaigns, chairman (1950–2 and 1965–7), president, general fund-raiser, and peripatetic speaker. He was responsible almost single-handed for the survival of the party in a difficult period. He contested, without success, a by-election in Bolton in 1960.

When in 1964 the Liberal Party was offered its first two life peerages, Byers was an obvious choice. In 1967 he became leader of the Liberal peers, a position which he held unchallenged until his death. He was appointed a privy councillor in 1972, and became chairman of the Company Pension Information Centre in 1973 and part-time consultant at Marks & Spencer in 1977. He was chairman of the Anglo-Israel Association, a member of the committee on privacy (1970–2) led by (Sir) Kenneth Younger [q.v.], chaired a far-reaching report on the organization of British athletics (1968), and was enthusiastically involved in a host of voluntary activities where he was able to demonstrate his passionate concern for people as individuals, particularly the young, the sick, and the deprived.

For twenty years Byers was an outstanding and widely admired party leader in the House of Lords. Lean, wiry, red-haired, and pugnacious, his carefully prepared speeches were appreciated for their logic, forceful delivery, and conciseness of argument. His Liberalism was both caring and practical, as is shown by his moving contributions to the debates on the Immigration Act in 1971. He had an impish and irreverent sense of humour. His occasional impatience was no more than a reflection of his quickness of mind, as was the brusque bark of 'Byers' with which he answered the telephone. He had a profound and sympathetic knowledge of parliamentary customs and procedures, and made a distinctive contribution to the all-party talks in 1968 on the reform of the upper house.

On 15 July 1939 he married another Liberal stalwart, Joan Elizabeth, daughter of William Oliver, company director of Spicers Ltd., of Alfriston, Wayside, Golders Green. They had one son and three daughters. Byers suffered heart attacks in 1973 and again in 1978, but remained fully active. On 6 February 1984 he had his third and fatal attack when he was working in his room at the House of Lords. It is said that members do not die in the House, because coroners have no jurisdiction in a royal palace, but 'on the way to hospital'. It was the only time Byers broke a parliamentary convention.

[Private information; personal knowledge.]
                                        WIGODER

C

CAIRD, GEORGE BRADFORD (1917–1984), biblical scholar, was born in Springfield, Wandsworth, 19 July 1917, the youngest of three children and only son of George Caird, engineer, from Dundee but then in London because of war work, and his wife, Esther Love Bradford. The family home was afterwards in Birmingham, where Caird attended King Edward's School (1929–36) and went on to Peterhouse, Cambridge. He took a first class in both parts of the classical tripos (1938 and 1939), with distinction in Greek and Latin verse. From Cambridge he went on to study for the Congregational ministry at Mansfield College, Oxford (1939–43), and at the same time he did postgraduate research, which gained him his D.Phil. in 1944. From 1943 to 1946 he was minister of Highgate Congregational Church in London, and the instincts and styles of the preacher remained with him throughout his life. In 1945 he married Viola Mary ('Mollie'), daughter of Ezra Benjamin Newport, schoolmaster, of Reigate, and, after their move to Canada which soon followed, they had three sons and one daughter. Caird never tired of talking of his children, all of whom came to be professionally noted, and of his grandchildren.

Caird's career specifically as an academic scholar of the Bible began when he went to St Stephen's College, Edmonton, Alberta, an institution of the United Church of Canada, to teach Old Testament (1946–50). In 1950 he became the first professor of New Testament in the newly formed faculty of divinity at McGill University, Montreal, and he stayed there until 1959; in addition he was (1955–9) principal of the theological college of the United Church in Montreal. He greatly admired the United Church, which fitted well with his spiritual and theological tendencies. But Oxford was his spiritual home, and Mansfield and English Congregationalism drew him back: he became tutor in theology at Mansfield in 1959 and principal in 1970. His academic duties were complemented by service to the church: he was an observer at the second Vatican Council and in 1975–6 was moderator of the General Assembly of the recently formed United Reformed Church.

In Oxford theology Caird's comprehensiveness of scope, command of language and evidence, and deep theological emphasis, combined with his excellence as a lecturer, quickly established him as a central figure. He was appointed to a readership in 1969, and in 1977 became Dean Ireland's professor of the Exegesis of Holy Scripture, a position which carried with it a professorial fellowship at The Queen's College. As the senior person in New Testament studies in Oxford Caird exercised a deep influence and was greatly admired by many of those who heard him.

His books included: *The Truth of the Gospel* (1950), a general account of the whole range of Christian doctrine; a commentary on the Books of Samuel in *The Interpreter's Bible* (1953); *The Apostolic Age* (1955); *Principalities and Powers* (1956); commentaries on St Luke (1963), on Revelation (1966), on Paul's Letters from Prison (1976); and *The Language and Imagery of the Bible* (1980), which received wide notice and was awarded the Collins Religious Book award in 1982. At the time of his death he was preparing a Theology of the New Testament. Among other scholarly achievements may be mentioned his work on the Septuagint: he was Grinfield lecturer on the Septuagint at Oxford for four years from 1961; his work as joint editor of the *Journal of Theological Studies*, from 1977; and his interest in biblical translation, as shown by his warm support for the New English Bible, for which he served on the Apocrypha panel from 1961. To this Dictionary he contributed the notices of Leslie Weatherhead and Samuel Hooke.

His academic distinction was recognized by honorary doctorates of divinity from St Stephen's College, Edmonton (1959), the Diocesan (Anglican) College in Montreal (1959), and Aberdeen University (1966). In 1966 he also gained Oxford's own doctorate of divinity. In 1973 he was elected a fellow of the British Academy.

Caird's thinking combined a moderate criticism, a somewhat conservative theological position, and a strong sense for linguistic nuances and literary values. His handling of the Bible followed critical lines but rejected scepticism; he was judicious and fair but also quite combative in controversy, and his opposition to Bultmannian positions was marked. He greatly emphasized the historical Jesus, while denying that the quest for him would lead away from theological values. Biblical authority was central in all his thinking, yet he completely rejected the fundamentalist understanding of scripture. Religion for him filled the whole of life and affected attitudes to sickness and health, life and death, peace and war (he was a committed pacifist), and such matters as the situation in South Africa. He particularly stressed the element of metaphor and myth in human talk about the divine, and thought that much distortion had arisen because interpreters had taken as literal expressions that were meant from the beginning to be literary figures.

Music was important to him, and he wrote several hymns, some of which were included in standard hymnals. Precision in language was central to his personality; he relaxed in joy among family and friends. In the year in which he would have retired from his chair, he suddenly died in his home at Letcombe Regis, Oxfordshire, 21 April 1984.

[J. Barr in *Proceedings* of the British Academy, vol. lxxi, 1985; D. A. Sykes and H. Chadwick in Mansfield College *Magazine*, no. 186; private information; personal knowledge.]   JAMES BARR

CALDER, PETER RITCHIE, BARON RITCHIE-CALDER (1906-1982), author and journalist, was born 1 July 1906 at 6 Newmonthill, Forfar, Angus, the youngest in the family of one daughter and three sons of David Lindsay Calder, linen worker (later works manager), and his wife, Georgina, daughter of John Ritchie, master mason. He was educated at Forfar Academy, which he left at the age of sixteen.

He began his career in 1922 as a police court reporter with the *Dundee Courier*. He then worked with the D. C. Thomson Press before joining the *Daily News* (1926-30), the *Daily Chronicle* (1930), and the *Daily Herald* (1930-41). During these years his range of interests broadened considerably. Politically he was, and always remained, a dedicated socialist and member of the Labour Party, and the social problems and political strife of the inter-war years provided many subjects for perceptive record and comment. Then a chance assignment to cover a science story opened up a completely new field for him. He realized that application of new scientific knowledge could powerfully augment the conventional approach to social problems. A timely stimulus to this new interest was provided by acquaintanceship—which grew into close and lasting friendship—with John (later Lord) Boyd Orr [q.v.], who was making a name for himself in the field of animal and human nutrition. Seeing malnutrition as a global problem, Calder devoted much effort to publicizing Boyd Orr's ideas and activities. This he did not only through his own newspaper connections but also through documentary films, then establishing themselves as a powerful means of influencing public opinion. Boyd Orr's book *Food, Health and Income* was published in 1936 and Calder played an important part in producing a film version entitled *Enough to Eat?*

The outbreak of war in 1939 inevitably diverted much of his activity into other fields. During the blitz his vivid reports helped to make the world aware that temporarily the front line of the war was in London. In 1941 he was appointed to the Political Warfare Executive of the Foreign Office, a service for which he was appointed CBE in 1945. One of his first tasks during this time was to assist, with Boyd Orr, in the production of a film, *World of Plenty*, to show the world how Britain was using food sent from the USA.

The immediate post-war years were very favourable to the resumption of his career, initially as science editor of the *News Chronicle* (1945-56) and as a member of the editorial staff of the *New Statesman*. The new Labour government was committed to a radical programme of social reform in which a key element was to be the harnessing of science to national needs. On a wider front, the newly created agencies of the United Nations sought to achieve similar ends internationally. In recording and promoting all these activities he found himself involved to an extent that severely taxed even his seemingly boundless enthusiasm and energy. He clearly understood that progress could be made only if public opinion was informed and favourable.

Despite the euphoric view of science created by great wartime achievements such as atomic energy, radar, and penicillin, the post-war media paid little more than lip-service to science. To help to alleviate this, he took a leading role in establishing the Association of British Science Writers, of which he was the first chairman (1949-55), and in which he always took great pride. In 1960 he was awarded the Kalinga award for science writing.

After the war he undertook many assignments for UN agencies, often on the initiative of Boyd Orr, who became the first director-general of the Food and Agriculture Organization and invited him to attend the Famine conference in Washington (1946) as a special adviser. He went on arduous missions to study the utilization of human resources in North Africa, the Congo (1960), and South-East Asia (1962).

His exceptional knowledge of international affairs led the University of Edinburgh to appoint him to the Montague Burton professorship of international relations (1961-7). This was a remarkable tribute to a man who had had no formal higher education. His fluent and easy manner, and varied reminiscences, appealed to the students. The appointment opened many academic doors and he was awarded honorary doctorates by the Open University (1975) and York University in Ontario (1976). He was senior fellow, Center for the Study of Democratic Institutions, Santa Barbara, California (1972-5).

Calder was passionately devoted to the cause of peace, particularly in the post-war years when there was so much uncertainty about where atomic weapons would ultimately take us. He was president of the British Peace Council and a leader of the Campaign for Nuclear Disarmament. He wrote several books and to this Dictionary contributed the notices of the cartoonist Vicky (Victor Weisz) and Lord Francis-Williams. In 1966 he was created a life peer as Baron Ritchie-Calder and three years later was made chairman of the Metrication Board, retiring in 1972.

In 1927 he married Mabel Jane Forbes, daughter of David McKail DPH, FRFPS, of Glasgow; they had three sons and two daughters. He died 31 January 1982 at Western General Hospital, Edinburgh.

[*The Times*, 2, 6, and 12 February 1982; personal knowledge.] TREVOR I. WILLIAMS

CAMERON, (MARK) JAMES (WALTER) (1911–1985), journalist and author, was born in Battersea, London, 17 June 1911, the elder son (there were no daughters) of William Ernest Cameron, barrister-at-law and novelist, and his wife, Margaret Douglas Robertson. His Scottish links always remained unshaken. His grandfather was a Highland divine, whose Gaelic ecclesiastical writings won him local renown, and his father was a lawyer turned author under the name of Mark Allerton. He was educated, on his own testimony, 'erratically, at a variety of small schools, mostly in France', and then started real work, filling paste pots, at the age of sixteen for fifteen shillings a week, in the offices of D. C. Thomson's *Red Star Weekly* in Dundee.

Escaping to the *Scottish Daily Express* in 1938 he was able to afford to marry his first love, Eleanor ("Elma") Mara Murray, an artist. She was the daughter of George John Murray, a surgeon. It was an idyllic but cruelly brief affair, cut short by her death in childbirth in 1940. A daughter was born. Turned down for the armed forces because of 'organic cardiac disease', Cameron moved in 1939 to the *Daily Express* in Fleet Street, subbing other people's copy. He met Elizabeth O'Conor working in the art room of the newspaper, and they were married at the Chelsea registry office on 25 November 1943. She was the daughter of Stanley Punshon Marris and the former wife of Denis Armor O'Conor, by whom she had a son. She was also to have a son by Cameron.

As the war ended, Cameron stumbled on his real trade. One assignment sent him to India at the moment of the independence negotiations; he started to comprehend the moral impetus behind the Asian and the worldwide revolt against imperialism. He had Jawaharlal Nehru and M. K. Gandhi [qq.v.] for his tutors. Another assignment took him to Bikini when the American atom bomb was exploded; his allegiance to the Campaign for Nuclear Disarmament, launched much later, was born in that forgotten summer of 1946. A third assignment took him to expose the brutality and corruption of Syngman Rhee's regime in South Korea. And back at home he chose his own allies with some delicacy. When the *Evening Standard* engaged in what he justly believed to be a shameful slur against the Labour Minister, E. John St Loe Strachey [q.v.] in 1950, he resigned from the Beaverbrook newspapers altogether, and when (Sir) Edward Hulton, the owner of *Picture Post*, fired its talented editor, (Sir) H. T. Hopkinson, for daring to attempt to publish Cameron's Korean dispatches, he resigned again in sympathy (1951).

Soon thereafter his genius flowed in the truly great and truly liberal *News Chronicle* of those times. Many contributed to the triumph: Sir Gerald Barry [q.v.], his editor; Tom Baistow, his foreign editor; and Vicky [q.v.], the cartoonist, soon to become the closest friend of all. His passion and his wit and his readiness to fit every incident into the worldwide scene were all part of his charm. His matchless integrity was part of it too, and yet he could wear his armour without a hint of pride or piety. He could raise journalism to the highest level of literature, like a Swift or a Hazlitt.

But the liberal hour did not last; the Liberal financiers killed off the *News Chronicle* in 1960. Other wars, requiring investigation, had broken out, and Cameron could turn to other instruments than his pen. To persuade the world that the North Vietnamese were made of flesh and blood, in 1965 he staked all the fortune he had not got, and all his precious time and energy, to get in and out of Hanoi with a camera crew. His film, *Western Eyewitness*, was shown on BBC television and was also bought by several overseas stations.

Most of his journalistic accomplishments were recited in what he called his 'experiment in biography', *Point of Departure*, published in 1967. Another title, *An Indian Summer*, which appeared in October 1974, offered even more enticing prospects. After the dissolution of his previous marriage in 1969, in January 1971 he married Moneesha Sarkar, airline receptionist, the daughter of Kandrachanda Ayuppa Appachod, coffee planter. She had a daughter and a son by a previous marriage. She took him on honeymoon to her native land, which ended in a car crash. A bundle of skin, bones, wood, and wire was somehow trans-shipped back to London and there stuck together by the National Health Service and Cameron's wife.

All such accidents, susceptibilities, memories, and broken hopes and dreams were woven together into a single thread in *An Indian Summer*. It is a book about ancient and modern India, brave, beautiful, astringent, withering, and, just occasionally, savage; an Anglo-Indian classic fit to take its place beside *A Passage to India* by E. M. Forster [q.v.]. Cameron was appointed CBE in 1979. In 1965 he won the Granada journalist of the year award. He had honorary degrees from the universities of Lancaster (D.Litt. 1970), Bradford (LLD 1977), and Essex (D.Univ. 1978). He died at his London home, 3 Eton College Road, 26 January 1985.

[*The Times*, 28 January and 8 February 1985; James Cameron, *Touch of the Sun*, 1950, *Mandarin Red*, 1955, *1914*, 1959, *The African Revolution*, 1961, *1916*, 1962, *Witness*, 1966, *Point of Departure*, 1967, *What a Way to Run the Tribe*, 1968, *An Indian Summer*, 1974, and *Cameron in The Guardian*, 1985; personal knowledge.]   MICHAEL FOOT

CAMERON, NEIL, BARON CAMERON OF BALHOUSIE (1920-1985), marshal of the Royal Air Force, was born in Perth 8 July 1920, the only son and younger child of Neil Cameron, an inspector of the poor, and his wife, Isabella Stewart. His father died in the year he was born and he and his sister were brought up by his mother and grandfather in Perth. Having attended Perth Academy he worked in the Royal Bank of Scotland. After joining the RAF Volunteer Reserve in May 1939 he was called up on the outbreak of war and qualified as a pilot. He joined No. 17 Squadron towards the end of the Battle of Britain, was commissioned in 1941, flew in Russia with No. 151 Wing, and in 1942-3 served with the Desert Air Force in No. 213 Squadron. In 1944-5, now a squadron leader, he commanded No. 258 Squadron and flew Hurricanes and Thunderbolts in Burma, earning the DFC (1944) and DSO (1945) for his outstanding leadership.

Awarded a permanent commission after the war, he instructed at the School of Air Support, Old Sarum, attended the Staff College, and in 1949 went to the Air Ministry, where he became seriously ill with sub-acute bacterial endocarditis. He was never again allowed a full flying category, making his eventual rise to the top of the RAF particularly remarkable. A spell instructing at the Staff College (1953-6) enabled him to deepen his thinking and begin writing about air power, and he saw something of the academic world while commanding London University Air Squadron (1956-8). He then served as personal staff officer to the chief of the air staff (1958-60), commanded RAF Abingdon (1960-2), attended the Imperial Defence College (1963), went to the Supreme Headquarters Allied Powers in Europe (1964), and in 1965—now an air commodore—became assistant commandant at Cranwell. There ensued

four years in the Ministry of Defence working in Denis Healey's programme evaluation group and as assistant chief of staff (policy), when he showed the ability to take an overall defence view but in the process aroused suspicions among his single-service contemporaries, and he subsequently saw little chance of further advancement. Nevertheless after serving at Headquarters Air Support Command and RAF Germany he was promoted to air marshal in 1973 to become AOC No. 46 Group. Ten months later he became air member for personnel, where—at a time of major cuts in RAF strength—he combined his essential humanity with the ability to take hard decisions, notably in the matter of redundancy.

He became chief of the air staff in 1976 but had hardly had time to make much impression before the sudden death of Sir Andrew Humphrey [q.v.] led to his becoming chief of the defence staff (1977-9) and marshal of the Royal Air Force (1977). As CAS he had already emphasized the need for better communication within the RAF and deeper thinking about air power, and as CDS he was determined to argue the defence case in public debate; he held strong views on what he saw as a dangerous growth in Soviet military power and his much publicized reference in China to the Russians as 'an enemy' led to their calling him 'a drunken hare' and to a minor political storm at home. Of the domestic issues he faced the most difficult was service pay, where he led his colleagues in confronting the government and winning the battle for the military salary.

After handing over as CDS in 1979 he was appointed principal of King's College, London, in August 1980, giving him the opportunity to show his leadership qualities in the academic environment that had eluded him in his youth. Here, despite failing health, he played a major part in the restructuring of London University and in particular the merger between King's, Chelsea, and Queen Elizabeth colleges. He was created a life peer in 1983 and a Knight of the Thistle the same year. He had been appointed CBE (1967), CB (1971), KCB (1975), and GCB (1976). He received an honorary LLD from Dundee in 1981.

His many wider interests included the RAF Rugby Football Union, the RAF Club, the RAF Museum, the Trident Trust, and the British Atlantic Committee, and underlying all else was a deep Christian faith rooted in his experience when ill in the 1950s and quietly but sincerely demonstrated through his support for St Clement Danes and organizations such as the Officers' Christian Union. A man of great honesty, integrity, and forthrightness, who was widely respected, he saw no difficulty in combining a firm military stance based on a belief in nuclear deterrence with his strongly held Christian convictions.

In 1947 he married Patricia Louise, daughter of Major Edward Asprey, a civil engineer. They had a son and daughter. Cameron died in the Middlesex Hospital 29 January 1985.

[Neil Cameron, *In the Midst of Things*, 1986; official records; personal knowledge.]

HENRY A. PROBERT

CARDEW, MICHAEL AMBROSE (1901-1983), potter, was born 26 May 1901 at Wimbledon, Surrey, the third son and fourth child in the family of five sons and one daughter of Arthur Cardew, a civil servant, and his wife, Alexandra Rhoda, daughter of George William Kitchin, dean of Winchester and of Durham [q.v.]. He was educated at King's College School, Wimbledon, and in 1919 obtained a scholarship to Exeter College, Oxford. He obtained a second in classical honour moderations (1921) and a third class in *literae humaniores* (1923).

From childhood he was surrounded by ceramics which filled his parents' houses in Wimbledon and at Saunton, north Devon. It was the rural pottery made by Edwin Beer Fishley to which he was deeply attracted and he used to visit him at Fremington. When Fishley died in 1911, Cardew felt a sense of deprivation from which came the realization that if pots of the generous warm nature of Fishley's were to be obtained, he would have to make them himself. In the summer of 1921 he persuaded Fishley's grandson to give him lessons in throwing. Few though these were, they confirmed Cardew's love for the craft and precipitated a crisis in his academic work. He was faced with the ultimatum: either to give up potting or to surrender his scholarship. As he had long wanted to study philosophy, Cardew reluctantly agreed to put potting aside. The study of Plato and Aristotle, and the discipline of the school, left their mark. To his innate mental qualities—an alertness and ability to master and express his thoughts in clear and vigorous language—Cardew added a knowledge of method, and 'how to be capable of going on learning'.

In January 1923 he went to St Ives to meet Bernard Leach [q.v.] and Shoji Hamada. Leach has described how 'he strode in, nose and brow straight, handsome as a young Greek god, eyes flashing blue, hair waving gold, and within an hour announced that this was where he wanted to work'. In July he was back. Leach found that his first and best pupil 'possessed a sense of form, the potter's prime gift'. Here he also met Katharine Pleydell-Bouverie, Norah Braden, and Tsuranosuke Matsubayashi.

In 1926 Cardew decided that he must be self-supporting and that he wanted to make slipware rather than the stoneware produced by Leach. Through (Sir) S. Gordon Russell [q.v.] he found the disused pottery at Greet, near Winchcombe in Gloucestershire. Two years later, he had confidence in his powers as a thrower, held his first exhibition in London, and was making a monthly average of 350 golden-brown pots. They were cheap and made to be used. The total value of what he made in 1928 was £293 12s. 6d. He was producing the most genuine lead-glazed slipware since the eighteenth century. He always lived simply, with few material comforts. He loved the music of Handel and Mozart, which he played on clarinet and recorder. In 1933 he married Mary-Ellen ('Mariel'), the daughter of Baron Russell, journalist and contributor to *Printer's Ink*. They had three sons: Seth, who was to follow his father as a potter; Cornelius (died 1981), composer; and Ennis, accountant. In 1939 they bought the inn at Wenford Bridge at St Breward in Cornwall, converting it into a pottery. It remained their home despite long periods of absence and separation.

Cardew was appointed pottery instructor at Achimota College in the Gold Coast (Ghana) in 1942 with responsibility for the tileworks at Alajo. Here he began to make stoneware. Three demanding years were followed by his unsuccessful endeavour to set up the Volta Pottery at Vume Dugame. Despite poor raw materials, disasters in the kilns, and ill health, he produced pots among 'the most beautiful to come from the hands of any modern potter' (George Wingfield-Digby, *The Work of the Modern Potter in England*, 1952). He returned to Wenford Bridge in 1948. Two years later he was appointed pottery officer for the Nigerian government. After travelling extensively and learning Hausa, he set up a pottery training centre at Abuja in northern Nigeria where he worked for fourteen years, absorbing and using the incised native styles on stoneware bowls, casseroles, jars, jugs, stools, and teapots, because, as at Winchcombe, his pots were always intended for use.

He retired to Wenford Bridge in 1965, the year in which he was appointed MBE (a CBE followed in 1981). He also received an honorary doctorate from the Royal College of Art in 1982. His hard apprenticeship in Ghana, together with his capacity for continuing to learn, bore fruit in his book *Pioneer Pottery* (1969). In addition to scientific analyses, this contains the essence of Cardew's experience and thought, showing him to have been a philosopher as well as a potter. It also reveals his gifts as a teacher which found expression in the tours and demonstrations which he gave in the USA, Canada, and Australia. While he never taught in a formal sense, Cardew taught many by example. Until his death in Truro 11 February 1983, he retained the energy and force of character which found expression in his pots. These can best be seen in the collections of the University of Wales at Aberystwyth, the Crafts Study Centre at the Holburne of Menstrie Museum in Bath, the City Museum and Art Gallery at Stoke-on-Trent, and the York City Art Gallery. He is also represented in the national collections, at the Victoria and Albert Museum, the British Museum, and the university museums at Oxford and Cambridge.

[Michael Cardew, *Pioneer Pottery*, 1969, and *A Pioneer Potter* (autobiography), 1988; Garth Clark, *Michael Cardew*, 1978; Tessa Sidey, *Michael Cardew and Pupils*, 1983; private information.]   IAN LOWE

CAROE, SIR OLAF KIRKPATRICK KRUNSE (1892-1981), Indian civil servant, was born in London 15 November 1892, the eldest in the family of two sons and one daughter of William Douglas Caröe [q.v.], architect and son of the Danish consul in Liverpool, and his wife, Grace Desborough, daughter of John Rendall, barrister, of London. From Winchester College he won a demyship at Magdalen College, Oxford, entering in 1911. He obtained a second class in classical honour moderations in 1913. The outbreak of war in 1914 drew him straight into the army, where he rose to the rank of captain in the 4th battalion of the Queen's Regiment.

In 1919 he joined the Indian Civil Service, one of five officers selected for their war records. After four years in the Punjab he was transferred to the Political Service and posted to the North-West Frontier Province. There he spent much of his Indian career, first as a district officer, then as chief secretary to government (1933-4) and finally as governor. While district commissioner in Peshawar (1930-2) he was involved in an incident which his Congress critics recalled when they sought his dismissal in 1947: troops fired on a menacing crowd, killing nine people. No inquiry was ordered, and Caroe was vigorously exculpated by his superiors.

Meanwhile he had also served in Political Department posts in the Persian Gulf, Waziristan, and Baluchistan. In 1934 he was appointed deputy secretary of the External Affairs Department of the Government of India, and in 1939 its secretary, responsible directly to the viceroy. Besides the perennial frontier troubles, World War II witnessed grave external problems arising from the war itself, and important progress in India's direct representation in other countries and international bodies; all those matters fell within Caroe's charge. Appointed CIE in 1932, CSI in 1941, he became a Knight Commander of those orders in 1944 and 1945.

The governorship of the Frontier Province, to which he was appointed in March 1946, was a fitting climax to such a career, but its duration was unhappily curtailed. An Indian National Congress government held office in the NWFP. Ardent for a united India, it had gaoled many supporters of the pro-partition Muslim League. Fresh elections or a referendum were deemed essential by the governor and the viceroy, Viscount (later Earl) Mountbatten of Burma [q.v.]. Caroe was accused by Khan Sahib [q.v.], the provincial premier, and Congress national leaders of interference with his ministers and partiality for the Muslim League. The 'interferences' had been proper exercises of his constitutional powers. Lord Mountbatten wrote to him in June 1947: 'I have a high opinion of your capacity, integrity and selfless devotion to duty under great strain.' The charge of inter-party bias was equally misplaced—'totally unfounded', wrote the viceroy. A provincial governor was not the constitutional tool of a majority administration; he could not, as they did, treat the opposition as enemies of government. Indeed Khan Sahib held the governor in great respect, and he and his family remained Caroe's friends long after the latter's retirement.

Nevertheless Mountbatten had decided that Caroe must be replaced before presiding as governor over a referendum on the Pakistan issue. Lord Ismay [q.v.], the viceroy's chief of staff, had reported Caroe to appear tired. Slightly built, lean of face, and somewhat highly strung, Caroe may have seemed more harassed than with his renowned energy, courage, and balance he really was. He denied being weary and made no offer of a resignation which Mountbatten shrank from demanding. Eventually, however, under strong pressure he preconcerted with the viceroy an exchange of letters, dated 13-17 June 1947, wherein he sought permission to go on leave 'for about two months'.

Caroe's hope of return or re-employment after independence proved vain. In retirement at his home in Sussex, he remained active in many spheres, as president of the Tibet Society of the United Kingdom, chairman of the Overseas League, and author and contributor to the press. His well-received books included *Wells of Power* (1951), on the politics of Middle Eastern oil, *Soviet Empire* (1953), fruit largely of his knowledge of Russia's southward expansionism, and *The Pathans* (1958), a study of the people whom he had long served and loved. With Sir Evelyn Howell he translated the poems of Khushhal (1963). To this Dictionary he contributed the notices of Sir Girja Bajpai and Sir George Cunningham. Oxford honoured him with a

D.Litt. in 1960. He married in 1920 Frances Marion ('Kitty') (died 1969), daughter of Atherton Gwillym Rawstorne, bishop of Whalley. They had two sons. Caroe died 23 November 1981 at his home at Steyning, Sussex.

[Nicholas Mansergh (ed. in chief) and Penderel Moon (ed.), *The Transfer of Power* (documents), HMSO, 12 vols., 1970–83; Register of Magdalen College, Oxford; India and Burma Office Lists; *The Times*, 25 November 1981; private information; personal knowledge.]   H. V. HODSON

CARR, EDWARD HALLETT (1892–1982), historian, diplomat, journalist, and essayist, was born 28 June 1892 in London, the eldest of three children and elder son of Francis Parker Carr, manufacturer, and his wife, (Elizabeth) Jessie Hallett. Educated at Merchant Taylors' School, he was awarded a Craven scholarship at Trinity College, Cambridge, and took a first class in both parts of the classical tripos (1915 and 1916). He served in the Foreign Office from 1916 to 1936. A member of the British delegation to the Paris peace conference in 1919, he was appointed CBE in 1920. He then served successively in Riga, Geneva, and London, rising to the rank of first secretary.

While in Riga in the 1920s he learned Russian and became fascinated with Russian culture and the Russian nineteenth-century intelligentsia, and its challenge to the conventional liberal world-view in which he had been brought up. This led him to the spare-time vocation of literary biographer; he completed in quick succession *Dostoevsky (1821-1881): a New Biography* (1931), *The Romantic Exiles: a Nineteenth Century Portrait Gallery* (1933), dealing with Alexander Herzen and his circle, *Karl Marx: a Study of Fanaticism* (1934), which he later described as a 'foolish book', and *Michael Bakunin* (1937).

In 1936 Carr left the Foreign Office on his appointment as first Woodrow Wilson professor of international politics at the University College of Wales, Aberystwyth, a post he held until 1947. His subsequent publications, notably *The Twenty Years' Crisis, 1919-1939: an Introduction to the Study of International Relations* (1939), were grounded both in his practical experience in international affairs and in a sternly realistic approach to politics. He rejected the prevailing doctrine of the harmony of interests, arguing that international relations are primarily based on hard bargaining between conflicting interests in which the stronger power prevails.

In the early 1930s Carr's initial curiosity about the Bolshevik revolution gave way to enthusiasm about Soviet planning as the answer to the anarchy of capitalism. The mounting political repression in the Soviet Union temporarily undermined his enthusiasm; but during World War II Russian endurance and the eventual victory over Fascism persuaded him that the Soviet system presented a challenge which the western world must meet by combining the best elements of planning and mass democracy with its own individualist traditions. As the influential assistant editor of *The Times* from 1941 to 1946 he advocated continued co-operation with the Soviet Union and the establishment of a new social and economic order in Britain and western Europe. His views were trenchantly expressed in a series of books and lectures, notably *Conditions of Peace* (1942), *The Soviet Impact on the Western World* (1946), and his most mature reflections, *The New Society* (1951). In 1953–5 he was a tutor in politics at Balliol College, Oxford, and from 1955 was a fellow of Trinity College, Cambridge.

In the last winter of war, at the age of fifty-two, Carr decided to write *A History of Soviet Russia*, which was published in fourteen volumes between 1950 and 1978 (vols. ix and x with R. W. Davies); it was his masterpiece and is summarized in his *The Russian Revolution from Lenin to Stalin (1917-1929)* (1979). This enterprise was a natural consequence of his view of the contemporary world, and was facilitated by his considerable knowledge of Russian culture and Soviet affairs. He set out to write a history 'not of the events of the revolution . . . but of the political, social and economic order which emerged from it'. The *History* deals with the first twelve formative years after the revolution, taking the story to the eve of the forced collectivization of peasant agriculture which followed Stalin's consolidation of power. The focus of the work is the emergence of the Soviet state; chapters in successive volumes display the establishment of the principal parts of the edifice—the Communist Party, the soviets, the army, the law, the political police, and the economic and planning institutions. But this analysis is placed firmly in the context of the internal social forces which shaped and were eventually transformed by the Soviet state, and of Soviet external relations.

In his George Macaulay Trevelyan lectures delivered at Cambridge in January–March 1961, published as *What is History?* (1961, 2nd edn. 1986), Carr argued that history is the product of the historians, and the historian the product of his society. No historian can entirely escape from the influences of his time; indeed, a historian who does not have bees buzzing in his bonnet is a 'dull dog'. For Carr, the grand theme of the first twelve years of Soviet history is the rise of a system in which every element, and the whole of society, were subordinated to the goal of transforming a peasant society from above into a modern world power. But he was a great historical craftsman; and his account gives much weight to local and personal factors, and to the complexities and convolutions of the transition. The honesty and care with which Carr handled the evidence enabled his critics to draw from his pages the evidence with which they sought to refute him.

After he completed his *History*, Carr embarked at the age of eighty-five on a history of the Communist International from 1930 to its dissolution in 1943. A first volume, *The Twilight of Comintern, 1930-1935*, was published in 1982, but he did not live to complete the work; one section of the second volume, *The Comintern and the Spanish Civil War* (1984), was published posthumously under the editorship of Tamara Deutscher.

Carr combined his austere analysis of past and present with a passionate belief in human progress; he was a Utopian as well as a realist. He was a quiet man, slightly aloof in manner, but with a ready sense of humour. Devoted to his work, in his later years his imposing height was reduced by a characteristic historian's stoop, and he almost always wore the casual clothes of a working historian. Increasingly a dissident, he was not entirely at home either with the masters of society, whose outlook he rejected, or with their opponents. But he formed long-standing friendships with many scholars and public figures, and willingly helped and encouraged younger scholars. He was elected a fellow of the British Academy in 1956 and ten years later became an honorary fellow of Balliol. He had honorary degrees from Manchester (1964), Groningen (1964), Cambridge (1967), and Sussex (1970).

Carr's first marriage, in 1925, was to Anne Eva Rowe (died 1961), daughter of Thomas Henshell Ward, farmer; they separated in 1945. From 1946 to 1964 he lived with Joyce Marion Forde, formerly wife of (Cyril) Daryll Forde, FBA, anthropologist. She was the daughter of Henry Walter Stock, businessman. His second marriage, in 1966, was to (Catherine) Betty (Abigail) Behrens, historian, who survived him. She was the daughter of Noel Edward Behrens, civil servant and merchant banker. He had one son by his first marriage. Carr died in Cambridge 3 November 1982.

[R. W. Davies in *Proceedings* of the British Academy, vol. lxix, 1983; E. H. Carr papers (University of Birmingham); private information; personal knowledge.]

R. W. DAVIES

CARRITT, (HUGH) DAVID (GRAHAM) (1927-1982), art historian, critic, and dealer in pictures, was born 15 April 1927 in London, the only son of Reginald Graham Carritt, a lecturer in music, and his wife, Christian Norah Begg. He had two sisters, one older, the other his twin. Arriving at preparatory school, he was told he might have one magazine subscription. He asked for *Country Life* which he had been enjoying in bound volumes since he was five or six, in his grandfather's library. At about the same age a mimic and performing flair began to show itself. In the nursery he used to play Maurice Chevalier to his elder sister's curveless Mae West. He was sent to Rugby when he was twelve. Starved of beautiful things to look at, he wrote to owners of collections within bicycle range, asking if he might come and see what they had. They were led by his letters to expect a schoolmaster, and were surprised at being confronted with a schoolboy.

He won an open scholarship to Christ Church, Oxford, where his career ended with a third-class degree in modern history (1948). That result is inexplicable. He already knew so much, and wrote so well; and indeed he did not entirely neglect his academic work. While at Oxford he caught the attention of L. Benedict Nicolson [q.v.], editor of the *Burlington Magazine*. Nicolson took his protégé to visit Bernhard Berenson in Florence. 'The finest eye of his generation', was Berenson's considered judgement.

Always a Londoner, Carritt settled first in a flat over a dairy in Mayfair, working as a private dealer and writing for the *Burlington Magazine*—but also for the *Evening Standard* and the *Spectator*. In 1964 he joined Christie's old master paintings department where he played a large part in reviving the fortunes of the firm; and in 1970 he set up David Carritt Ltd., owned by the international art-investment company of Artemis. Here, in his later years, he mounted a number of small and very personal exhibitions.

For a quarter of a century, between 1952 when he discovered a Caravaggio in a retired naval officer's house, and 1977 when he saw that a picture catalogued as by Carle van Loo in Lord Rosebery's Mentmore collection was in fact an early Fragonard, Carritt dazzled the art world and intrigued the general public with a series of detective feats remarkable for their number, variety, and importance. 'Connoisseurs agree', stated the *New York Times*, 'that he had no peer, and has no successor.'

Carritt himself took most pleasure in his discovery of a Rogier van der Weyden portrait at the cottage of Joan, Lady Baird. This beautiful picture was not just lost; it was unrecorded. It is historically important, a milestone in European portraiture. Its attribution to Rogier entailed much pacing round the Flemish rooms of the National Gallery. He also attributed the Tiepolo allegory fixed to a ceiling of the Egyptian embassy in London, scanned by countless pairs of eyes before his comprehended it. There was a fictional flavour to his discovery of five large and very dirty Guardi canvases rolled up like linoleum in a Dublin shed. In 1956, still in his twenties, Carritt gave informed people cause to admire his eye and brain when a photograph in the Courtauld library recalled to him a drawing in Hamburg, on the basis of which he attributed to Dürer a painting held to be by a minor Veronese master.

But the Mentmore Fragonard, his last coup, will perhaps be longest remembered. It carries the stamp of his boldness as well as his flair. The black and white photograph in the catalogue, available to French specialists the world over, warned Carritt, himself no specialist, that this was not a van Loo, and that the subject of the picture was misdescribed. Fortunately the family had refused him permission to look at the collection before the sale. Otherwise, as always, he would have felt obliged to tell them the truth.

It is sad that he published little, for his writing is a model of clarity and grace, with the authority but not weight habitual to sound scholarship. We can hope for more, since two sets of papers on Chardin and Fragonard are extant, as is the text of lectures given in the United States. Carritt was a lavish spender of effort in other men's causes. Pierre Rosenberg, for example, in the catalogue of the Paris Chardin exhibition of 1979, acknowledged his very significant contribution to that great enterprise. In the earlier days, when he practised regularly, he was an excellent pianist, and his knowledge of the baroque and Viennese classics was thorough. He was a witty and indiscreet talker. Friends relished his cooking. Colleagues agreed that he was no businessman. Friends and colleagues united in admiration of his courage in the grip of cancer. Almost to the end he managed to be sociable. His parties for children were among the most memorable given by this lovable and talented man. He died 3 August 1982 at his Mount Street flat in London. He never married.

[*The Times*, 4 and 9 August 1982; *New York Times*, 5 August 1982; *Art & Auction*, October 1982; *Christie's House Magazine*, December 1982; private information; personal knowledge.]   JOHN JONES

CARTE, DAME BRIDGET D'OYLY (1908–1985), theatrical manager and proprietor of the D'Oyly Carte Opera Company. [See D'OYLY CARTE.]

CATON-THOMPSON,   GERTRUDE (1888–1985), archaeologist and authority on African prehistory, was born in London 1 February 1888, the younger child and only daughter of William Caton Thompson, of Lancashire–Yorkshire stock, and his wife, Ethel Gertrude, daughter of William Bousfield Page, surgeon, of Carlisle. Her father, a barrister and head of the legal branch of the London and North Western Railway, died when she was five years old. Her mother remarried in 1900, to George Moore, general practitioner, of Maidenhead, a widower with four sons and a daughter. Her father had ensured her financial independence and, after being educated privately and at the Links School, Eastbourne, until World War I she enjoyed a social life in London and the country, and included one auspicious visit to Egypt among her travels abroad.

In 1917 she was employed by the Ministry of Shipping and promoted to a senior secretarial post in which she attended the Paris peace conference. She declined a permanent appointment in the Civil Service, and in 1921, aged thirty-three and with none of the usual qualifications, she began her archaeological studies under the Egyptologist (Sir) W. M. Flinders Petrie [q.v.] at University College, London, joining his excavations at Abydos in upper Egypt that winter. The next year was spent at Newnham College, Cambridge (with which she remained associated for the rest of her life), and in attending university courses. She returned to Egypt in 1924 and joined Petrie and Guy Brunton at Qau. While they concentrated their excavations on predynastic cemeteries she had concluded, well ahead of her time, that settlement sites would be more informative, and in characteristically independent fashion she embarked on her own excavations on the site of a predynastic village at Hemamieh. There she made the first discovery of remains, well stratified, of the very early Badarian civilization. With Guy Brunton she wrote *The Badarian Civilization* (1928).

In 1925 she turned to north-western Egypt and the desert margins of Lake Fayum, accompanied by the Oxford geologist Miss Elinor Gardner to assist in an attempt to correlate lake levels with archaeological stratification. In three Fayum seasons they discovered two unknown neolithic cultures which proved later to be related to the Khartoum neolithic. They published their findings in *The Desert Fayum* (1934). Her next assignment followed an invitation in 1929 from the British Association for the Advancement of Science to investigate the great monumental ruins at Zimbabwe in southern Africa. She confirmed the conclusion reached by David Randall-MacIver [q.v.] in 1905 that they belonged to an indigenous African culture and were not, as widely believed, of oriental origin. She was also able to date the ruins back to the eighth or ninth century AD and to produce evidence of Zimbabwe's links with Indian Ocean trade.

Having completed her Zimbabwe report with her usual speed (*The Zimbabwe Culture*, 1931), she and Elinor Gardner embarked in 1930 on an extended campaign of excavations on prehistoric sites at Kharga oasis. These excavations, the first on Saharan oasis sites, inaugurated a far-ranging programme of research on the palaeolithic of North Africa and led to her book *Kharga Oasis in Pre-history* (1952).

Her last excavations, in 1937, were the only ones outside Africa apart from some field-work in Malta in her student days. These were at Hureidha in the Hadhramaut, southern Arabia, where she excavated the Moon Temple and tombs of the fifth and fourth centuries BC. Carried out in a region then rarely visited by western, let alone female, travellers, they were the first scientific excavations in southern Arabia. Again she was accompanied by Elinor Gardner. A third, less compatible member of the party was the writer and traveller (Dame) Freya Stark. *The Tombs and Moon Temple of Hureidha, Hadramaut* appeared in 1944.

She retired from field-work after World War II and from her home in Cambridge pursued her research activities and visited excavations in East Africa. She never sought or accepted a professorship. In 1961, already in her seventies, she became a founder member of the British School of History and Archaeology in East Africa (later the British Institute in Eastern Africa), served on its council for ten years and was later elected an honorary member. An honorary fellowship of Newnham College, Cambridge, and an honorary Litt.D. (1954) of that university were among academic honours bestowed on her. She was elected a fellow of the British Academy in 1944 and not long before her death a fellow of University College, London.

Professionally, Gertrude Caton-Thompson was a formidable personality, a trenchant critic, adamant in academic argument, and an indefatigable worker. She was an intrepid traveller and at Abydos, it is said, slept in an emptied tomb with cobras for company and a pistol under her pillow to ward off prowling hyenas. Although by nature quiet and retiring she was excellent company and her dry humour and laconic, incisive comments on people and events became a rich source of anecdote in Cambridge circles. She was a generous benefactor, especially of the National Trust for the purchase of land near where she lived, Court Farm, Broadway, Worcestershire; a home happily shared with two former Cambridge colleagues, the de Navarros, for thirty years. She died there

18 April 1985, aged ninety-seven. She was unmarried.

[Gertrude Caton-Thompson, *Mixed Memoirs*, 1983; Grahame Clark in *Proceedings* of the British Academy, vol. lxxi, 1985; *Newnham College Roll*, 1986.]   L. P. KIRWAN

CECIL, DAVID GEORGE BROWNLOW, sixth MARQUESS OF EXETER and BARON BURGHLEY (1905-1981), athlete and parliamentarian, was born in the family's vast 350-year-old ancestral home at Burghley, Stamford, Lincolnshire, 9 February 1905, the second child and elder son in the family of four children of the fifth Marquess, William Thomas Brownlow Cecil (1876-1956) and his wife, Myra Rowena Sibell, daughter of William Thomas Orde-Powlett, fourth Baron Bolton. Lord Burghley, as he was to be for more than fifty years, was educated at Eton and from 1923 at Magdalene College, Cambridge, where he studied engineering and obtained a BA in 1926. He earned selection in Britain's 1924 Olympic team but ran poorly. However in 1925 this striking blond-haired, aquiline-featured young patrician leaped into prominence as the world's preeminent hurdler.

He won both the high and low hurdles event in the Inter-Varsity Sports against Oxford in 1925, 1926, and 1927 and became president of the Cambridge University Athletics Club in his last year. The time of 42.5 seconds, recorded in Burghley's diary after he ran round the Great Court of Trinity College, Cambridge, before the twenty-third of the twenty-four noon-time chimes, was still unsurpassed after half a century. He went on to represent Great Britain in eleven full international matches and to win eight Amateur Athletic Association titles and three British Empire titles. He set English native records for the 120 yards high and 220 yards low hurdles with 14.5 and 24.7 seconds. In 1927 he very briefly held the world's 440 yards hurdles record.

His greatest hour came in the 400 metres hurdles final at the 1928 Olympic Games in Amsterdam where, by deliberately running for third place in the semi-final he arrived quite fresh for the six-man final. He and Britain's T. C. Livingstone-Learmonth drew the disadvantageous outside fifth and sixth lanes. Ranged against them were the world record holder and defending champion F. Morgan Taylor (US), Frank J. Cuhel (US), the Italian Luigi Facelli, and Sweden's Sten Pettersson. Burghley went off seemingly too fast but reached the tenth and final flight of the all-wooden barriers just ahead of Cuhel. His finish was described as 'obstinate' and at the tape he won by a long yard in the Olympic record time of 53.4 seconds with Cuhel second and Morgan Taylor third, both in 53.6 seconds. In that era before track suits he donned his greatcoat and confined his comment to 'The Americans are frightfully good losers.' In the fashion of the day his father made no comment whatsoever but touchingly his mother was later found to have kept a secret scrap-book of her son's many triumphs on the track.

Burghley's enjoyment of amateur athletics was such that four years later he defended his Olympic title in Los Angeles. By this time he was married, a father, and an MP. In the general election of 1931 he had won the Labour-held seat of Peterborough with a majority of 12,434. Under the California sun he ran his fastest ever time of 52.2 seconds but still finished behind his fellow Cambridge blue Robert Tisdall (Ireland) (51.7 seconds) and the two Americans Glen Hardin and his old rival Morgan Taylor. In the 4 × 400 metres relay he ran an outstanding third stage for Great Britain in 46.7 seconds, so adding a silver medal to his collection, behind only America's world record-breaking quartet. He was the British Olympic captain in 1932 and 1936.

From 1933 to 1937 Burghley was chairman of the Junior Imperial League (president, 1939). In the general election of 1935 he held his seat with a reduced majority. In the House of Commons Burghley and Alan Lennox-Boyd (later Viscount Boyd of Merton, q.v.) were the two fastest speakers and cherished their visits to the office of the Hansard shorthand writers where they had to be permitted to improve rather than merely correct their speech proofs. Burghley, though a lieutenant in the Grenadier Guards until 1929 (when he resigned and obtained seats on the boards of a number of companies), was unable to take up combatant service in 1939 due to a persistent leg injury. He was appointed a staff captain Tank Supply in 1940, and became assistant director with the rank of lieutenant-colonel in 1942. In August 1943 he left Parliament to become governor of Bermuda where he captivated many American and other Allied visiting dignitaries. He was created KCMG in the same year.

Burghley was chairman of the organizing and executive committee of the fourteenth summer Olympic Games held in Wembley stadium in London in 1948. King George VI adamantly refused to attend and the prime minister, C. R. (later Earl) Attlee, was obliged to intervene since Olympic protocol required that the head of state of the host nation declare the Games open in person. Predictably the King's sense of duty prevailed when on 29 July he performed the ceremony. Burghley, who had only two, instead of the customary six years, in which to organize these first post-war games, had worked tirelessly in surmounting all the difficulties in an era of rationing, bomb damage, and national austerity. The success of this quadrennial festival of what was then amateur sport was a credit to his zeal. Some of his team were less richly honoured for their heroic efforts to make the Games a British success than might have been appropriate had not an honour of any kind been withheld from their chairman. Burghley was president of the AAA (1936–76), the International Amateur Athletic Federation (1946–76), and the British Olympic Association (1966–77, chairman 1936–66). His fervent, some thought naïve, belief that amateur and Olympic sport were a palliative in international strife brought him into conflict over the award in 1974 by the International Olympic committee of the 1980 or twentieth Games to Moscow. He was obdurate in the face of mounting entreaties after the invasion of Afghanistan that Britain should pull out. Supported by Lords Killanin and Luke, his defiance of the prime minister's well-known wishes resulted in his being ostracized by those who thought British participation would also demoralize the four millions imprisoned in Soviet Gulags and be an insult to the captive peoples of eastern Europe.

Despite hip replacement operations Burghley hunted with vigour, latterly with the Burghley Hunt up to 1967. He had succeeded to the marquessate in 1956. He often gave priority to local affairs over debates in the House of Lords because, as he said, 'They are of more importance to Mrs Buggins because it is her roof which is leaking.' In 1961 he was elected mayor of Stamford. He was honorary FRCS and had an honorary LLD from St Andrews University (of which he was rector from 1949 to 1952).

In 1929 Burghley married Mary Theresa, fourth daughter of John Charles Montagu-Douglas-Scott, seventh Duke of Buccleuch and ninth Duke of Queensberry, at a time when the announcement of her engagement to Prince Henry, later Duke of Gloucester [q.v.], was regarded as imminent in royal circles. They had a son and three daughters but the son died of tubercular meningitis at the age of thirteen months in 1934. The marriage was dissolved in 1946 and in the same year Burghley married Diana Mary (died 1982), widow of Lieutenant-Colonel David Walter Arthur William Forbes and daughter of the Hon. Arnold Henderson. The only child of his second marriage, Lady Victoria Leatham, was to become a vigorous and enterprising chatelaine of Burghley House where he died 21 October 1981. He was succeeded in the marquessate by his brother, (William) Martin (Alleyne) (1909–1988).

[*The Times*, 23 October 1981; personal knowledge.]   Norris McWhirter

CHALLANS, (EILEEN) MARY (1905-1983), writer under the name of MARY RENAULT, was born 4 September 1905 at 49 Plashet Road, Forest Gate, West Ham, the elder daughter (there were no sons) of Frank Challans, a physician, and his wife, Clementine Mary Newsome Baxter. Frank Challans was of Huguenot stock and his wife is said to be descended from the seventeenth-century Presbyterian divine, Richard Baxter [q.v.]. After attending a coeducational preparatory school and then Clifton High School, Bristol, Mary Challans went on to St Hugh's College, Oxford, where she read English, receiving a third class degree in 1928.

Her passion for literature and history had been aroused early on through her school, where she had found 'a huge collection of books', and she seems never to have doubted that her vocation was to be a writer. (She had attempted a first novel, about cowboys, at the age of eight—and she was amused to make the heroine of *The Friendly Young Ladies*, 1944, a writer of westerns.) She recognized a need for experience, however, and after a period spent in clerical work she went to the Radcliffe Infirmary, Oxford, where she trained as a nurse, being registered in 1937. Hospitals and medical personnel often figured in her early novels, the first of which to be accepted for publication being an unexceptionable romance, *Purposes of Love* (1939; published in the United States as *Promise of Love*). Because she was uncertain of the reaction of the nursing authorities to this extramural activity she chose to use the name Renault, which she drew from her reading in medieval French.

During World War II Mary Renault continued her work as a nurse, in the course of which she met Julie Mullard, with whom she was to live for the rest of her life (a glimpse of that friendship is also perhaps discernible in *The Friendly Young Ladies*). Novel writing was not entirely ruled out by her work, however, and when, after the war, she published her fourth book, *Return to Night* (1947) it made sufficient impression to win her that year's Metro-Goldwyn-Mayer prize. The option to film it was never taken up, but the £40,000 prize money gave Mary Renault full independence to pursue her career as a professional writer. She and Julie Mullard emigrated to South Africa in 1948, and after a spell in Durban they established a permanent home for themselves at Camps Bay, Cape Town.

In her six novels up to *The Charioteer* (1953), Mary Renault had devoted herself to exploring 'the purposes of love' with an increasing degree of subtlety—and, indeed, *The Charioteer* was a remarkable study for its time of the homosexual relationships between a group of servicemen lodged at Bristol during the Battle of Britain. Through several of these books in a contemporary setting however there ran an undercurrent of references to Classical antiquity which can now be seen as contributing to Mary Renault's direct imaginative assault on the civilization of Ancient Greece which began with *The Last of the Wine* (1956). Here, and in seven succeeding novels, she focused upon clusters of facts for which there was archaeological or historical evidence and then widened her vision to create a credible context within which her, often episodic, narratives could run their courses. Thus, in *The King Must Die* (1958), her most celebrated novel, she gives to Theseus the telling of his own story: dramatic, plangent, but also rationalized through an interpretation of the finds at Knossos and through the author's knowledge of physiology. As with her other historical novels, Mary Renault permits readers to examine or question her sources by substantiating them in a discursive 'Author's Note'.

There has been some dispute, especially among Classical scholars, over the validity of Mary Renault's reconstructions in these books, especially since, like one of her own characters, she worked from 'a translation, for she knew no Greek'. Argument has centred, as might be anticipated, on her interpretation of Alexander's aims and motives through such novels as *Fire From Heaven* (1970), *The Persian Boy* (1972), and her last book, *Funeral Games* (1981), and through her study *The Nature of Alexander* (1975). Nevertheless, given the impossibility of reaching any conclusive picture, and given Mary Renault's own zealous application to her sources, the ultimate justification of her work lies in its power to evoke the enthusiasm of her readers for her subject and to hold at bay what she once called 'an unregenerate Cro Magnon' in modern man, who wishes 'the whole world to grow grey with his egalitarian breath'.

During her years in South Africa Mary Renault, who was a member of the Progressive Party, became closely involved with movements defending freedom of speech and publication, working through the PEN Club, of which she became national president. She travelled widely in Africa and the Mediterranean, but although she was elected a fellow of the Royal Society of Literature she never returned to Britain. She died in Cape Town 13 December 1983.

[*The Times*, 14 and 21 December 1983; *Books and Bookmen*, March 1984; Mary Renault, 'History in Fiction', *Times Literary Supplement*, 23 March 1973; private information.] BRIAN ALDERSON

CHAMBERS, SIR (STANLEY) PAUL (1904-1981), civil servant and industrialist, was born in Edmonton, London, 2 April 1904, the sixth of the eight children (six boys and two girls) of Philip Joseph Chambers, clerk (later company secretary and finally cigar merchant), and his wife, Catherine Emily Abbott. He was educated at the City of London School and the London School of Economics, acquiring the degrees of B.Com. in 1928 and M.Sc.Econ. in 1934.

In 1927 he joined the Inland Revenue Tax Inspectorate. He served in Leeds and in London. In 1935 he was selected for secondment to serve on the Indian income tax inquiry committee. In 1937 he was appointed income tax adviser to the government of India, and he instituted a scheme for the deduction of income tax at source from salaries and wages.

On the outbreak of war in 1939 income tax in Britain needed to be sharply raised, which would create difficulty in paying it in two half-yearly lump sums. Chambers was therefore recalled and in 1940 appointed assistant secretary to set up a deduction scheme. With a small committee he did so. Each deduction was a proportion of the tax assessed for the previous year. The trade unions soon began agitating for deductions to be related to current pay. This at first appeared impossible, and Chambers wrote a white paper saying so, but the agitation continued. Meanwhile, in 1942, Chambers was promoted to membership of the Board of Inland Revenue. Eventually A. G. T. Shingler, a principal inspector of taxes, devised the cumulative principle which became, and still is, the basis of the PAYE (pay as you earn) system. After working out the detailed operation of the scheme, Chambers succeeded in winning the consent of the rest of the department, ministers, trade unions, and the employers' organizations, and in 1943 it took effect. Another achievement of his Inland Revenue days was the negotiation with the USA of the first double taxation agreement made by Britain. In 1945 Chambers was seconded to the Control Commission for Germany as finance director of the British element. His success in this post helped prepare the ground for the German 'economic miracle'.

In 1947 he resigned from the Civil Service and was appointed a director of ICI Ltd. In 1948 he became finance director, in 1952 deputy chairman, and from 1960 to 1968 he was chairman. He did much to modernize the organization and public image of the company, and took it into Europe and America. In 1961 he launched a take-over bid for Courtaulds Ltd., but was eventually beaten off. He was a director from 1951 to 1974 of the National Provincial, later the National Westminster Bank, and, on leaving ICI, he was appointed chairman of three insurance companies, the Royal, the London and Lancashire, and the Liverpool and London and Globe. He was also

a part-time member of the National Coal Board from 1956 to 1960. Between 1951 and 1972 he served, mostly as chairman, on committees reviewing the organization of the Customs and Excise Department, departmental records, London Transport, and the British Medical Association. He also served various terms as president of the National Institute of Economic and Social Research, the British Shippers' Council, the Institute of Directors, the Royal Statistical Society, and the Advertising Association. In his later days he entered the academic world as vice-president of the Liverpool School of Tropical Medicine (1969-74), the first treasurer of the Open University (1969-75), and pro-chancellor of the University of Kent (1972-8).

Chambers's enormous energy and drive showed in his rapid movements and speech and his quick grasp of detail. He was always ready to delegate work; he pushed people hard, but his generous appreciation of their efforts, his great personal charm, and his obvious mastery won him their support and affection. He was a keen gardener, and this included major construction work. He enjoyed music and bridge, but he read little.

He was appointed CIE in 1941, CB in 1944, and KBE in 1965. He was awarded an honorary fellowship at the London School of Economics, and honorary degrees by the universities of Bristol (1963), Liverpool (1967), and Bradford (1967). The Open University gave him an honorary degree in 1975.

In 1926 he married Dorothy Alice Marion, daughter of Thomas Gill Baltershell Copp, printer. They had no children. The marriage was dissolved in 1955, and in that year Chambers married Mrs Edith Pollack, second daughter of Robert Phillips Lamb, accountant. They had two daughters. Chambers died in Kenton, London, 23 December 1981.

[*The Times*, 29 December 1981 and 5 January 1982; *Financial Times*, 29 December 1981; private information; personal knowledge.]

J. P. STRUDWICK

CHAPMAN, (ANTHONY) COLIN (BRUCE) (1928-1982), racing car designer and motor manufacturer, was born at Richmond, Surrey, 19 May 1928, the only son of Stanley Frank Kennedy-Chapman, licensed caterer, of Muswell Hill, and his wife, (Lillian) Mary Bruce. Educated at the Stationers' School, Hornsey, he then gained a B.Sc. (Eng.) at University College, London.

For a technological buccaneer and the most innovative of Formula 1 designers who spent much of his life doing battle with authority, he made a surprisingly placid start to a turbulent professional life. Reading civil engineering because he thought it would be easier, he specialized in structural engineering. Combined with a lifelong interest in aeronautics, this stood him in good stead in producing cars well known for their lightness.

Fast cars, flying, and competitiveness in everything he did were the mainsprings of his life. After gaining his private pilot's licence with his university air squadron, Chapman accepted a short service commission with the RAF. But cars were his first love and by 1948 the first Chapman car based on the Austin 7, known as the Lotus Mark 1, was completed.

He soon discovered that his *métier* lay in cars for circuit racing in which he excelled, rather than trials, and in ingenious interpretation of the rules governing their construction. His budding Lotus so dominated the 1951 season of 750 Formula racing that the regulations were changed to close the engine loophole he had exploited. The mould for twenty-five years of motor racing endeavour was set.

Chapman would bring out some world-beating racing car only to have it eclipsed the following season when his rivals had caught up, copied, refined, and developed it while he had lost interest and moved on to his next design. His Formula 1 cars won 72 Grand Prix, achieved 88 pole positions and 63 fastest laps, took the Constructors' championship seven times, were runners-up five times, and provided a seat for five world champion drivers. Only Ferrari approached this record. As with Enzo Ferrari's team, there were also the long lean periods, mainly because Chapman's full attention was elsewhere or he was stubbornly pursuing an avant-garde design that would never be sanctioned.

Ironically, Chapman's first Grand Prix victory, the 1960 Monaco Grand Prix, was achieved not by a works driver but by Stirling Moss who was so impressed by the Lotus-Climax that he asked his sponsor, Rob Walker, to buy him one. But it was James (Jim) Clark [q.v.], the 1963 and 1965 world champion, who was to dominate the 1500cc Formula and establish Lotus as the marque to beat in the 3-litre Formula before his tragic and untimely death on the Hockenheim track (1968).

The 1965 Dutch Grand Prix found Chapman spending a night in the cells of Zandvoort police station after having felled an officious policeman, who had not seen his pass and ordered him off the track before the start of the race. This unfortunate incident, which received much publicity at the time, somehow symbolized Chapman's frequent clashes with the sport's governing body, FISA (Federation Internationale Sporting Automobile): Chapman doing something which he knew he was perfectly entitled to do with authority trying to stop him. But the same year Clark won the Indianapolis 500, as well as the Formula 1 title.

The first year of the 3-litre Formula belonged to the Brabham team with their stopgap but reliable Repco engine. But Chapman was waiting for the Ford–Cosworth DFV engine for his Lotus 49 which won first time out and went on to power another forty-seven Lotus victories. Chapman was shattered by Clark's death and when Mike Spence was killed in practice for the Indianapolis 500 soon after, he was almost ready to quit. Sadly there were to be yet more tragedies in Lotus cars to come.

The talented Austrian Jochen Rindt, the sport's first posthumous world champion, was to die at Monza, as was Ronnie Peterson, of Sweden, one of the great drivers never to win the title. Despite these disasters, Lotus were able to attract top drivers and Emerson Fittipaldi won the Drivers' championship in 1972. But only with the American veteran, Mario Andretti, could Chapman establish something of the rapport he had had with Jim Clark.

In the Lotus 78 Andretti had an almost unfair advantage in his championship winning year of 1978. Chapman had exploited the phenomenon of ground-effect, shaping the undersides of his cars to generate down-force and enclosing them with side skirts. This was such a novel departure that other teams could not rebuild their cars in time that season but it was destined to be Chapman's last great year with eight Lotus victories.

While other teams refined their own ground-effect cars, Chapman moved on to his next revolutionary development, the Type 88 twin chassis car with the driver and essential components suspended in a kind of 'sprung cab'. This was too much and other teams threatened to withdraw from the 1981 Long Beach Grand Prix if the Lotus was allowed to race. The ban on the 88 soured Chapman and even his natural ebullience seemed crushed.

For three years Lotus won no races as sponsorship and other problems mounted. Meanwhile, the Lotus road car business was in serious financial trouble. A world authority on the use of composite materials, pioneered in his Formula 1 cars, Chapman had somehow found time to become entangled in the ill-fated project to build De Lorean sports cars in Northern Ireland. This unfortunate connection did his reputation no good although nothing can detract from his record in Enzo Ferrari's words as a 'subtle visionary . . . able to produce ideas ahead of their time'.

Inspirer of great loyalty as well as dislike, endearing or devious according to the viewpoint, Chapman left an indelible mark on the summit of motor racing. Of medium height with penetrating blue eyes, his mechanics in the early days nicknamed him 'Chunky', but only behind his back. It was said the fortunes of Lotus could be gauged by whether his waistline was going through a thick or thin phase. Few men could have lived long at the pace Chapman set himself: racing driver, Formula 1 designer and team manager, chairman of the Lotus Car Group, boat builder, aircraft pilot, and family man. He was appointed CBE in 1970 and other distinctions included RDI (1979), FRSA (1968), fellowship of University College London, and honorary Dr RCA (1980).

He married in 1954 Hazel Patricia, daughter of Victor Hayesmere Williams, a builder. They had two daughters and a son. Chapman's constitution could not stand up to the load of work and worry to which he subjected it. He died suddenly 16 December 1982 of a heart attack, aged fifty-four, at East Carleton Manor, Norfolk, the neo-colonial style mansion he had—of course—designed himself.

[Gerard Crombac, *Colin Chapman, the Man and his Cars*, 1986; private information; personal knowledge.]   COLIN DRYDEN

CHARNLEY, SIR JOHN (1911–1982), orthopaedic surgeon, was born 29 August 1911 in Bury, Lancashire, the only son and elder child of Arthur Walter Charnley, chemist, and his wife, Lily Hodgson, a nurse. He was educated at Bury Grammar School and Manchester University, where, exceptionally, he graduated B.Sc. in anatomy and physiology (1932) before entering his clinical studies. Three years later he gained his MB, Ch.B. and in 1936 he became FRCS. He held various surgical appointments at Manchester Royal Infirmary and at Salford Royal Hospital, and it was obvious that he had an outstanding future in surgery. He volunteered for service in the Royal Army Medical Corps in May 1940. After service as resident medical officer at the evacuation of Dunkirk he was graded as an orthopaedic surgeon and was posted to the Middle East as a major and officer in charge of No. 2 Orthopaedic Centre, Cairo.

On demobilization he returned to Manchester Royal Infirmary where he was appointed assistant orthopaedic surgeon and then consultant. In 1950 he published *The Closed Treatment of Common Fractures*, which was a uniquely thought-out analysis of manipulation in the treatment of fractures and used Charnley's unusual talent for drawing on engineering concepts. This was a challenge to his colleagues and destroyed many myths and rigidly held concepts. His mechanical ideas and skills were again seen in the development of simple and direct compression arthrodeses tech-

niques for the knee, ankle, and even hip. This resulted in another monograph, *Compression Arthrodesis* (1953). However, compression arthrodesis of the hip never became widely accepted. Recognizing this, Charnley began to work on the concept of having a mobile hip for painful hip conditions due to arthritis. His discoveries were made possible by his outstanding ability in working with materials, his engineering techniques, and his skills at the lathe, which he utilized at his home in Hale.

In 1961 Charnley established a centre for hip surgery at Wrightington Hospital, Wigan, where he had been the visiting orthopaedic surgeon. There he pioneered one of the greatest contributions to world surgery for thousands of patients. At that time very little was known about the science of biological materials, lubrication, elasticity of tissues, or the acceptance of artificial materials within bone and joint tissues. All of these were critically studied to produce the Charnley prosthesis. The prototype which utilized Teflon plastic was a failure, but eventually a new polyethylene material became available. At the same time he designed sterile air operating enclosures, specialized hip instruments, and total body exhaust systems within impervious gowns, all to reduce the infection rates. Slowly he published the successes and failures and subsequent results of this procedure.

Finally his masterly report and monograph appeared in 1970—*Acrylic Cement in Orthopaedic Surgery*. This was followed by *Low Friction Arthroplasty of the Hip* (1979). Orthopaedic surgeons had long been waiting for such a breakthrough, and began to flock to the hip centre in Wrightington from all over the world. Here they saw the master surgeon performing this complex procedure, and were welcomed and shown the techniques and methods to avoid complication and catastrophe.

Charnley held honorary appointments in Manchester University, in the department of mechanical engineering, and in the Institute of Science and Technology (1966). He had many honours and awards, among them an honorary fellowship of the British Orthopaedic Association (1981) and the Lister medal of the Royal College of Surgeons (1975). He received the honorary D.Sc. of Leeds and Queen's University, Belfast (both 1978), and honorary MDs from Liverpool (1975) and Uppsala (1977). He was appointed CBE in 1970 and became the first and only orthopaedic fellow of the Royal Society in 1975. In 1972 Manchester University recognized their outstanding graduate by making him professor of orthopaedic surgery, a post he held until 1976. The citizens of Bury in 1974 made him a freeman of the borough, and he was knighted in 1977.

Charnley was a man of small stature, but had a strongly built physique which expressed his dynamic non-stop activity, both mental and physical. He was incisive in thought, work, and deed, being intolerant of humbug. He had seemed to be a confirmed bachelor and lived with his mother, until on a skiing slope in Switzerland in 1957 he met Jill Margaret Heaver and three months later they were married. She was the daughter of James Cedric Heaver, an officer in Customs and Excise. They had a son and a daughter. Charnley died in Manchester 5 August 1982.

[*Journal of Bone and Joint Surgery*, 1983, vol. lxv B, no. 1; N. W. Nisbet and Sir Michael Woodruff in *Biographical Memoirs of Fellows of the Royal Society*, vol. xxx, 1984; private information; personal knowledge.]

ROBERT B. DUTHIE

CITRINE, WALTER McLENNAN, first BARON CITRINE (1887–1983), trade-union leader and electrician, was born 22 August 1887 at Liverpool, the second of the three sons and the third of the five children of Alfred Citrine (born, as was his father, Francisco Citrini), seaman, of Liverpool, and his wife, Isabella, hospital nurse, daughter of George McLennan, of Arbroath. He left elementary school at twelve and thereafter educated himself with dictionaries, textbooks on electricity, economics, accountancy and short-hand—in which he became expert—novels and plays, and the pamphlets and books on socialism and trade unionism current in the labour movement of the day. He applied this experience when general secretary of the Trades Union Congress, producing practical publications for trade unionists, notably his classic *ABC of Chairmanship* (1939).

Unable to obtain a job as a cabin boy, Citrine worked in a flour mill before securing an electrical apprenticeship. Employed by different contractors, he gained a varied experience in his trade, but did not encounter trade unionism until 1911. He became active in Liverpool in the Electrical Trades Union when the first successful attempts were being made at collective bargaining. In 1914 he was elected the ETU's first full-time district secretary and in 1920 an assistant general secretary, working in Manchester. He was thus thoroughly grounded in organizing and negotiating before starting work at the TUC as assistant general secretary on 20 January 1924. In October 1925 Citrine was recalled from his first visit to Russia to act as general secretary after the death of Fred Bramley. From September 1926 he was general secretary until he joined the new National Coal Board in 1946 with responsibility for education,

training, and welfare. From 1947 to 1957 he was chairman of the Central Electricity Authority, and thereafter served as a part-time member on the Electricity Council, and on the Atomic Energy Authority (1958–62).

Throughout, Citrine displayed his outstanding administrative capacity, beginning with the files of the ETU and culminating in the transfer of 553 municipal and privately owned electricity undertakings to the Central Electricity Authority and its area boards. But he was much more than a great administrator. Frequently the seminal idea in an important development was his, as in the Mond–Turner conference with employers in 1928, after the general strike, and in his work as a member of the West Indies royal commission in 1938–9. His unsuccessful fight as a Labour candidate for Wallasey in the 1918 general election proved to be a youthful diversion. He declined Ramsay MacDonald's offer of a peerage in 1930, and in 1940 when (Sir) Winston Churchill (whose pre-war rearmament campaign he had supported) wanted Citrine in his government, he preferred to remain with the TUC. Always conscious that trade unions and political parties have different constituencies, he aimed to build the TUC into an organization which governments of any complexion would be prepared to consult. He saw this fulfilled, especially in the war years. He strongly resisted Communist Party attempts at disruption through the National Minority Movement and the 'communist solar system', but as president of the International Federation of Trade Unions (1928–45) he visited Russia again in 1935. In 1945 he presided over the World Trade Union conference which preceded the formation of the World Federation of Trade Unions. In 1944 he attended the Philadelphia conference of the International Labour Organization which established guidelines for its post-war work. Despite these constructive achievements, Citrine considered his years in the electricity supply industry after 1947 to be the most creative and happiest period of his life.

Ruthlessly self-disciplined, Citrine could be a martinet, but he was fair-minded and had considerable personal charm. A lighter side was reflected in his cornet playing, community singing, and love of the opera, and an idiosyncratic interest in palmistry. He published several perceptive but pedestrian accounts of his travels, based on his diary, the most important being *I Search for Truth in Russia* (1936). But he put his heart and understanding into two volumes of autobiography, *Men and Work* (1964) and *Two Careers* (1967) which stand as an education in trade unionism and public service and as a moving account of the making of an upright man. To this Dictionary he contributed the notices of Dame Caroline Haslett and Sir Arthur Pugh.

Citrine was appointed KBE (1935) and GBE (1958). He became Baron Citrine in July 1946. In 1940 he was admitted to the Privy Council. From 1939 to 1947 he was a visiting fellow at Nuffield College, Oxford, and in 1955 Manchester University awarded him an honorary LLD. In 1975 the Electrical Electronic Telecommunication and Plumbing Union, of which he was still a member, awarded him its gold badge.

He married 28 March 1914 Dorothy Helen ('Doris') (died 1973), daughter of Edgar Slade, commercial traveller, of Pendleton, Manchester. They had two sons. The elder, Norman Arthur (born 1914), solicitor, wrote the standard work, *Trade Union Law* (1950), and succeeded to the barony when his father died 22 January 1983 at Brixham, Devon.

[*The Times*, 26 January 1983; Lord Citrine, *Men and Work*, 1964, and *Two Careers*, 1967 (autobiographies); TUC *Annual Report*, 1946; personal knowledge.]

MARJORIE NICHOLSON

CLARK, KENNETH MACKENZIE, BARON CLARK (1903-1983), patron and interpreter of the arts, was born in London 13 July 1903, the only child of Kenneth Mackenzie Clark, and his wife, (Margaret) Alice, daughter of James McArthur, of Manchester and formerly of Paisley, a cousin of her husband's on his mother's side. The father was a wealthy Scottish industrialist, sportsman, gambler, and ultimately alcoholic. The wealth derived from the well-known family textile firm, Clark of Paisley, and was to provide his son with a buoyancy throughout his career that was an essential factor in the maintenance of his independence. The Clarks lived at Sudbourne Hall, Suffolk, and Ardnamurchan, Argyllshire.

Kenneth Clark was educated at Winchester and won a scholarship to Trinity College, Oxford, where he gained a second class in modern history in 1925. A first class had been expected, but his interests had already turned conclusively to the study of art. An inborn sensitivity of response (sometimes described in almost mystical terms) to works of art, together with an insatiable appetite for them, allied to an acute, fastidious, and articulate intelligence, had developed from childhood on. He was a workmanlike draughtsman himself, and read widely and voraciously.

His introduction to the severer disciplines of close scholarship and analytic connoisseurship were provided at Oxford by Charles F. Bell amongst the superb drawings and prints of the Ashmolean Museum. In autumn 1925 Bell introduced Clark to the renowned art historian, Bernhard Berenson, at the latter's house, I Tatti, on the hills above Florence. Berenson gave him lunch and afterwards forthwith asked Clark to come to assist him in the revision of his great corpus of Florentine drawings.

There followed two years of concentrated practice and refinement of judgement of works of art together with Berenson in the library at I Tatti and in the great collections of Italy. He was however also discovering, in the course of completing his first full-length book, *The Gothic Revival* (the rather unexpected subject suggested by Bell, published in 1928), the controlled exhilaration of prose composition, an art which (he was to observe later) was to yield him as much pleasure in life as anything he ever did.

In 1929 he was offered the enviable task of cataloguing the rich hoard of Leonardo da Vinci drawings at Windsor Castle, and he also acted as joint organizer of an exhibition, which later became legendary, of Italian painting at the Royal Academy, involving masterpieces never seen before (or since) out of Italy. In 1931 his career took a decisive turn when he accepted Bell's former post, as keeper of the department of fine art at the Ashmolean, and became a full-time professional museum man. Less than three years later, in 1933, at the unprecedentedly early age of thirty-one, he was appointed director of the National Gallery in London (1934-45), and shortly afterwards became also surveyor of the King's Pictures (1934-44).

In 1927 he had married Elizabeth Winifred ('Jane'), daughter of Robert Macgregor Martin, a businessman, of Dublin, and his wife, Emily Dickson, a medical doctor. Between 1934 and 1939, lodged in an almost palatial house in Portland Place, with his wife as president of the Incorporated Society of London Fashion Designers and a leading hostess, the Clarks in joint alliance became stars of London high society, intelligentsia, and fashion, from Mayfair to Windsor. This was a period he later dubbed 'the Great Clark Boom'.

In the National Gallery he made a considerable contribution. Acquisitions during his directorship included such masterpieces as Rubens's 'Watering Place', Constable's 'Hadleigh Castle', Rembrandt's 'Saskia as Flora', Ingres's 'Mme Moitessier', and Poussin's 'Golden Calf'. He established a scientific department, and a carefully supervised programme of picture cleaning. In the art of administration however his touch failed with his senior staff, whom he alienated. Major crises occurred in 1937, culminating when, against the united advice of his professional staff, he persuaded the trustees to acquire four small minor Venetian School paintings and labelled them as Giorgiones. The virulent subsequent controversy became public scandal, and contributed to a lingering mistrust of his integrity, especially amongst some fellow art historians, that was perhaps never quite dispelled. On the other hand, he fought off the reappointment of Lord Duveen [q.v.] as trustee, believing that a dealer's interests were irreconcilable with those of a trustee and that, although Duveen's natural charm and indeed generosity were compelling, they were about matched by his natural duplicity.

In 1939, at the outbreak of war, Clark was responsible for the evacuation of virtually the entire collection from London, ultimately into a cavern in the slate quarries of north Wales. In the emptied Gallery, he organized with Dame J. Myra Hess [q.v.] the very popular and morale-raising lunchtime music recitals that continued despite the blitz, and then a scheme by which he brought back one masterpiece each month for display. He also found time to serve in the Ministry of Information, first as director of the film division and then as controller, home publicity. Perpetually frustrated however by Ministry bureaucracy, he resigned in 1941.

He was meanwhile establishing himself in the activity in which his greatest talents were to find their most congenial and successful employment. His catalogue of the Windsor Leonardo drawings (1935) still holds an honourable place in the Leonardo literature, but was to be his last exercise in what he came to term 'plodding scholarship'. His monograph, *Leonardo da Vinci . . . His Development as an Artist* (1939) established itself at once as the best general introduction in English, and was widely acclaimed. In 1946, following the successful return of the collections to the National Gallery, he resigned as director, to devote himself to his now paramount interest in writing. Between 1946 and 1950 he was Slade professor of fine art at Oxford, developing an already highly accomplished lecturing technique. For a more popular audience, he became known as a broadcaster in such programmes as 'The Brains' Trust'. In the drab post-war years of austerity, a new public avid for the arts was emerging.

He had bought works of art for himself since his schooldays, but his very generous and imaginative activity as collector-patron was especially fruitful. Several artists, struggling in the 1930s and kept afloat by Clark, later became national and international figures, most notably Henry Moore, who like Graham Sutherland [q.v.] and John Piper were to become lifelong friends. In 1939-40 Clark had been involved with the setting up of the Council for the Encouragement of Music and the Arts (CEMA). This, the first step towards state support for the living arts, subsequently became the Arts Council. As the Council's third chairman (1953-60), Clark underwent a frustrating ex-

perience. He felt little more than a figurehead, and his own commitment to the validity of state support of individual creative artists was ambivalent. His last appointment as a public administrator, as the first chairman of the Independent Television Authority (1954-7), was more rewarding. The concept of commercial television, in rivalry with the BBC, was controversial, and so too was Clark's appointment, though his claim that he was subsequently booed at the Athenaeum is disputed. The prime task was to ensure the quality, balance, and political impartiality achieved by the licensed programme contractors. Almost at once Clark had to fight off an attempt by the government to use the new channels for its advantage. Clark was not averse to an injection of 'vulgar vitality', but managed to preserve a news service under the direct control of the Authority rather than of the contractors. Though not reappointed in 1957, he had exercised a constructive and beneficent influence on the development of the British version of the most powerful instrument of mass communication yet known to mankind.

His finest books, mostly allied to lecture series, were now behind him: *Piero della Francesca* (1951); *The Nude* (1956, from the Mellon lectures in Washington, 1953); and *Rembrandt and the Italian Renaissance* (1966, from the Wrightsman lectures in New York). He was now to emerge as an outstanding writer and performer for television. In 1966 the programme series *Civilisation* was mooted with the BBC and finally broadcast in 1969. Its success in English-speaking countries, and beyond, was spectacular and unparalleled; he became as celebrated as a film star, known to a wider audience personally than even John Ruskin [q.v.] had been a century before. Ruskin had been the greatest single influence on his mind since his schooldays, and he aspired to being Ruskin's heir in spreading the gospel of art, though with little of Ruskin's social, political, and moral involvement. The style too was very different; the prose in which the message was delivered was coolly lucid and elegantly lapidary but never rhapsodic. *Civilisation* was avowedly a very personal, selective interpretation and illustration of the title's grandiose theme. Though in it he ranged widely, with great erudition, in time and space, his conceptions generally were conditioned by Mediterranean values. The areas in the study of art—psychological and philosophical; sociological, iconographical, scientific—that challenged some of his most remarkable contemporaries, were generally not for him, though he acknowledged, for example, the fruitful influence of Felix ('Aby') Warburg. Though he recognized the achievement of Mondrian, and even of Jackson Pollock, he did not respond generally to abstract art, and his primary concern was to arouse response in individual human beings to individual works of art, essentially by accounting for his own response. If he was in a sense old-fashioned, with a sensibility prone (though not in his prose) to tears and visionary flashes, he made art accessible to a whole generation as no other English-speaking writer was able to do. To this Dictionary he contributed the notice of Roger Fry.

Despite his proven qualities as a lecturer, that he was able to succeed so well, in the demandingly personal medium of television appearances, was to many unexpected. He proved able to achieve that most difficult balance on screen between the presenter's own presence and the object that he is presenting, never obscuring the object by his own brilliance. Though (in the interests of 'style') there might be 'an occasional sacrifice of the whole truth in the interests of economy', he did not talk down. Yet in more private life, while often charming and entertaining, and imaginatively generous to younger students, for many he remained aloof and arrogant, and finally, even to some amongst his closest friends, elusive behind an impenetrable urbanity. Probably he was most relaxed in self-revelation only with a few gifted women: in the earlier days of his marriage, with his first wife (his loyalty to whom in the later distress of her prolonged decline from health due to alcoholism caused his friends such admiration and such embarrassment), and then with others.

His two autobiographical volumes are elegantly and subtly polished, at times very moving, often very funny, ironical, entertaining and perceptive, if somewhat distanced as if about someone else. There is modesty but little searching self-examination.

He was of medium height, slender, and elegantly tailored, then thickening somewhat in later life and becoming more relaxed in dress. His head was handsomely browed and aquiline, though its carriage was such that in repose his features could too easily (and not necessarily accurately) convey the impression that he was looking down his nose. Portraits include a profile painting by Graham Sutherland, of which two versions (1963-4) are in the National Portrait Gallery.

He was much honoured: KCB, 1938; FBA, 1949; CH, 1959; a life peerage, 1969; OM, 1976. He was chancellor of York University (1969-79), and a trustee of the British Museum. Honorary degrees, fellowships, and distinctions were conferred on him by universities and academies in Britain, America, France, Spain, Italy, Sweden, Austria, and Finland, but the honour he cherished most was his appointment to the Conseil Artistique des Musées Nationaux, Paris.

His first wife died in 1976; they had a son and a twin son and daughter. The elder son, Alan, became a military historian and Conservative politician. In 1977 Clark married Nolwen, former wife of Edward Rice, cattle and sheep farmer, and daughter of Frederic, Comte de Janzé. Clark died in a nursing home in Hythe, Kent, 21 May 1983. At the close of a crowded memorial service in St James's, Piccadilly, most of the congregation were startled to hear that in the last days of his life he had been received into the Roman Catholic Church.

[*The Times*, 23 and 27 May 1983; Kenneth Clark, *Another Part of the Wood*, 1974, and *The Other Half*, 1977 (autobiographies); Meryle Secrest, *Kenneth Clark*, 1984; private information; personal knowledge.]

DAVID PIPER

CLARK, WILLIAM DONALDSON (1916-1985), journalist, international civil servant, and author, was born at Bellister Castle, Haltwhistle, Northumberland, 28 July 1916, the fifth son and sixth and youngest child of John McClare Clark, land agent, and his wife, Marion, daughter of Daniel Noel Jackson, medical doctor, of Hexham. His father was sixty-two when he was born and his mother died when he was seven months old. Though his father wrote to him every week and his family was close-knit, his perception of his father as remote undoubtedly left some emotional scars. Clark attributed to it his competitiveness with his brothers and a degree of insecurity. He was educated at Oundle where he was head of school and won a major scholarship to Oriel College, Oxford. His university career encompassed a half blue for athletics and was crowned with a first class honours degree in modern history (1938) and the prestigious Gibbs prize. A Commonwealth scholarship took him in 1938, as a fellow and lecturer in humanities, to the University of Chicago, where his lifelong affection for the United States was nurtured and some enduring transatlantic friendships formed. Bertrand Russell [q.v.], who was an early influence on him, arrived in Chicago on the same day.

Knee trouble, which saddled him with a serious limp in later life, ruled him out of military service and in 1941 Clark began three years with the British Information Services in Chicago followed by an appointment as press attaché in the embassy in Washington (1945-6). After the war he returned to England as London editor of the *Encyclopaedia Britannica* (1946-9), but the job did not allow full rein to his talents, which were better displayed as diplomatic correspondent of the *Observer* (1950-5) and as the first British television interviewer to put a cabinet minister on television.

At the *Observer* F. David L. Astor, under whose aegis he joined the paper in 1950, became another major influence and Clark's concern for developing countries took root when, at the 1950 Colombo conference, Ernest Bevin [q.v.] outlined to him a vision of a British Commonwealth devoted to economic development. It was to become Clark's mission in life, but not before an unhappy interlude as press secretary to Sir R. Anthony Eden (later the Earl of Avon) at No. 10 Downing Street, which culminated in his resignation because of the Suez crisis in 1956.

Always attracted by power, Clark was flattered by Eden's offer and persuaded himself that their political incompatibility would prove no bar to a successful relationship. In retrospect it was a bad mistake, and it was a bitter blow, when after little more than a year, appalled by the invasion of Egypt and the attempts to browbeat the BBC and other media into supporting it, he decided that he must resign. He kept his letter to the prime minister brief, but to Sir Norman Brook (later Lord Normanbrook, q.v.), the secretary to the cabinet, he undertook to remain silent and make no demonstration on leaving. In 1966 he published a novel, *Number 10*, which became a successful play but his diary of the Suez period was first published in his posthumous memoirs, *From Three Worlds* (1986), thirty years after the event.

He could not return to his old job but Astor asked him to go to India and Africa for the *Observer* and he decided to devote himself to the north–south divide. After two years editing 'The Week' for the paper, he left in 1960 to set up and run, as its director, the Overseas Development Institute, which quickly established a reputation for benevolent realism in a field where both attributes were rare. It was a good time, but he accepted readily when, in 1968, he was invited to return to Washington to become director of information and public affairs at the World Bank. Arriving on the same day as the Bank's new president, Robert S. McNamara, he struck up an immediate rapport with him.

Clark believed the cataclysm of world poverty to be just as threatening as the nuclear holocaust and his twelve years at the Bank, during which he rose, in 1974, to be vice-president, external relations, were the most fulfilling of his life. With McNamara he travelled to over eighty countries, arguing that third world poverty was an obscene tragedy and a threat to the future of all mankind, and urging the richer nations, in their own interest, to help developing nations help themselves. His admiration for his chief was reciprocated. McNamara said of him: 'He came to the World Bank with more experience in the politics of aid than almost anyone else in the international community. He was invaluable to a newly arrived chief executive much in need of the guidance of someone who already knew so well so many of the personalities and places with whom we were to deal in the years ahead . . . He was an indispensable colleague.'

When he left the Bank in 1980, Clark succeeded, as president of the International Institute for Environment and Development, another friend, Barbara Ward (Baroness Jackson of Lodsworth, q.v.) with whom he had played an important part in the setting up of the Pearson and Brandt commissions. He also became an independent director of the *Observer*.

Clark's personality, sometimes exotic dress sense, willingness to laugh at himself, and unpatronizing generosity made him a memorable friend. He was told he had incurable cancer of the liver on All Hallows' eve in 1984, not long after the publication of his book, *Cataclysm*, which forecast the collapse of the world economy due to the third world refusing to pay its debts. He faced death stoically, saying his illness enabled him to prove he was as brave as his brothers who had fought in World War II. Sustained by his many friends, the need to complete his memoirs, the wish to design his own tombstone, and the news of an honorary fellowship at Oriel, he was determined to die at home. He had shared The Mill in the Oxfordshire village of Cuxham for twenty-five years with his friend and associate, David Harvey, and he died there 27 June 1985, and was buried in the churchyard nearby. He was unmarried.

[William Clark, *From Three Worlds*, 1986; *The Times*, 29 June 1985; *Observer*, 30 June 1985; *New York Times*, 29 June 1985; private information; personal knowledge.]

NICHOLAS HERBERT

CLEGG, HUGH ANTHONY (1900–1983), medical editor, was born 19 June 1900 at St Ives, Huntingdonshire, the third in the family of four sons and three daughters of the Revd John Clegg, headmaster of the local grammar school, and his wife, Gertrude Wilson, of Hull. He won a King's scholarship in classics to Westminster and thence an exhibition to Trinity College, Cambridge, where he obtained a senior scholarship in natural sciences. Taking first class honours in part i of the natural sciences tripos in 1922, he qualified MRCS (Eng.) and LRCP (Lond.) from St Bartholomew's Hospital Medical School in 1925. His MB, B.Ch. followed in 1928 and his MRCP in 1929.

After junior hospital appointments Clegg joined the *British Medical Journal* in April 1931 as sub-editor, becoming deputy editor in 1934 and editor in 1947 (until 1965). Clegg was also active in international medico-politics, helping establish the precursor of the British Council's medical department and the World Medical Association. After retirement in 1963 he was the first director of the international relations office of the Royal Society of Medicine and the first editor (1971-2) of its journal, *Tropical Doctor*.

Clegg's career spanned immense changes in medicine—from few useful therapeutic drugs to antibiotics in profusion, and from the disorganized health care before the National Health Service to the later co-ordinated network of district general hospitals and health centres. As the editor of one of Britain's two principal medical journals he was well placed to influence events. For some years before he became titular editor, moreover, he had had this role owing to the inertia and ill health of his predecessor. His first task was to recover editorial freedom from the BMA officers, to whom it had latterly been so lightly relinquished. Next he had to create policy (for both the *BMJ* and the BMA) in response to the novel proposals for a national health service—given that, until the mid-1930s at any rate, the BMA was more a gentleman's club than a streamlined trade union.

Ultimately policy arose out of the mixture of reports of debates, correspondence, and particularly editorials in the *BMJ* in the run-up to the NHS Act of 1948. But editorials did not arise in a vacuum and even before the war Clegg was playing a prominent part in the discussions about the future organization of medicine. He was careful to allow free debate within the journal; nobody knew all the answers and under Clegg the correspondence columns assumed a leading place in the *BMJ*, where anybody could express his opinions provided these were well argued. BMA medico-politicians might not like these views or those in the editorials but their plaints were brushed aside, as were the calls (both then and subsequently) to discipline or even dismiss him. Some have portrayed the *BMJ* of this time as reactionary, implacably opposed to any health service, and poles apart from its sister weekly, the liberal *Lancet*. Little could be further from the truth: the BMA, its journal, and Clegg had favoured a health service all along, but they foresaw that the details were as important as the principle. And with his sixty editorials, almost all of them written by himself, Clegg did affect the details—today's independence of the family doctor, for example, is one direct result. Clegg contributed several articles to this Dictionary.

Clegg was not an easy man to get to know or to retain as a friend. Outside politics and literature he had few interests, though he returned to reading the classic Greek and Latin authors during his retirement. Stocky and neat

with a military moustache, he hated pomposity, and had an endearing tendency to wear brown shoes with blue suits. His greatest regret was being born too late to fight in 1918, his greatest pride to have organized the first world conference on medical education. *Sub specie aeternitatis* neither matters, but three other achievements do: his recreation of an independent international journal, designed by Stanley Morison and A. Eric R. Gill [qq.v.]; his creation of specialty journals for complex research topics; and his co-authorship with the secretary of the Finnish Medical Association of the Declaration of Helsinki, an epoch-making event in medical ethics.

Elected FRCP in 1944, Clegg received an honorary MD from Trinity College, Dublin, in 1952, an honorary D.Lit. from the Queen's University, Belfast, in 1962, and the Gold medal of the British Medical Association in 1966, the year he was created CBE.

In 1932 Clegg married Baroness Kyra Engelhardt, of Smolensk, Russia, daughter of Baron Arthur Engelhardt; they had a daughter and son. Clegg died in London 6 July 1983.

[Richard Smith in *British Medical Journal*, 5 July 1982; *The Times*, 7 July 1983; Stephen Lock in *British Medical Journal*, 16 July 1983; *Lancet*, 16 July 1983; Stephen Lock in *Munk's Roll*, vol. vii, 1985; private information; personal knowledge.]

STEPHEN LOCK

CLIFTON-TAYLOR, ALEC (1907-1985), lecturer, broadcaster, and architectural historian, was born 2 August 1907 at Sutton, Surrey, the eldest of three children and the only son of Stanley Edgar Taylor, corn merchant, and his wife, Ethel Elizabeth Hills. He adopted the hyphenated form of his surname by deed poll in the 1930s. He was educated at Bishop's Stortford College, Queen's College, Oxford, where he read modern history (he obtained third class honours in 1928), and the Sorbonne. His father was a good photographer, but his mother provided his main cultural and intellectual stimulus. It was his father's expectation that one day he would run the family business. He therefore joined Lloyd's, but the commercial world proved so uncongenial that when the Courtauld Institute of Art opened in 1931 he persuaded his father to allow him to give up insurance and pursue aesthetics.

He graduated from the Courtauld with first class honours in 1934 and went to live in South Kensington, London, where he remained until the time of his death. He began lecturing for the Institute of Education at London University, and learned to paint. In the war years he served in the Admiralty, first in Bath and in

1943-6 in London as private secretary to the parliamentary secretary.

He resumed lecturing extramurally for London University in 1946. Much of the conventional art-historical teaching he considered arid and unexciting, and he persuaded the authorities to institute a new diploma course of his devising, 'The Aesthetic Approach to the Visual Arts'. In this he pioneered a fresh approach, looking at works of art less in terms of their provenance than in the ways in which they were composed in colour, texture, and material, considering above all whether or not they were great works and why. He was reputedly among the first in England to use colour slides in his lectures, which caused some stir at the time within the art establishment.

In 1956 he went free lance, writing articles and reviews for the *Connoisseur* and *Listener*, with an occasional broadcast. Over the next twenty years he was much in demand as a lecturer, notably for the National Trust, British Council, and English-Speaking Union. He lectured in every continent and thirty-two American states. He now concentrated his visual approach towards architecture and the traditional building materials. In 1962 he published his master-work *The Pattern of English Building*, revised twice by himself and published again with his posthumous amendments edited by Jack Simmons, in 1987. This is the fullest work in the English language on the history of building materials, considered comprehensively in terms of their history, use, and aesthetic qualities. Six books followed, notably *The Cathedrals of England* (1967), and *English Parish Churches as Works of Art* (1974, reissued 1986). Despite Clifton-Taylor's love of church architecture he was not religious. He contributed essays to eighteen volumes of *The Buildings of England* edited by Sir Nikolaus Pevsner [q.v.] whose close friend he was for forty years. Much of the material for all this output was readily available; everything he saw he noted in tiny, neat handwriting in diaries and jottings carefully indexed. To this Dictionary he contributed the notice of W. G. Constable.

At the age of sixty-eight he made his television début. His immediate success led to three series of programmes about English small towns for BBC television. He was a brilliant presenter; when all eighteen programmes were repeated shortly after his death, the audience rose on two occasions to over five millions. He was president of the Kensington Society from 1979, a vice-president of the Men of the Stones, and a trustee of the Historic Churches Preservation Trust. He became FSA in 1963, honorary FRIBA in 1979, and OBE in 1982.

He expressed his likes and dislikes forcefully, yet his opinions were always reasoned and informed by a lifetime's keen observation. Glazing bars should always be in place in a Georgian façade, and creepers never allowed to obscure good craftsmanship. Not much Victorian building in machine-made materials pleased him, and he was unsympathetic with the modern movement, particularly with high rise and naked concrete. His aim was to convey learning by looking; even his most cogent criticism was tempered with humour and sincerity. He was an incessant worker and indefatigable traveller. Late in his fifties he walked the formidable Strada degli Alpini and he returned from a strenuous tour of southern India only six weeks before his death. His lifestyle was modest, yet he was generous in his friendships, in many public causes, and with help to private individuals. The residue of his substantial estate he left to the National Trust. His life was happy and fulfilled; he died, unmarried, 1 April 1985, in St Stephen's Hospital, London.

[Private information; personal knowledge.]

DENIS MORIARTY

CLITHEROE, first BARON (1901-1984), politician and businessman. [See ASSHETON, RALPH.]

COCKBURN, (FRANCIS) CLAUD (1904-1981), author and journalist, was born at the British embassy in Peking 12 April 1904, the younger child and only son of Henry Cockburn, CB, Chinese secretary in the Diplomatic Service in Peking and later consul-general in Korea, and his wife, Elizabeth Gordon, daughter of Colonel Stevenson. He was the great-grandson of Henry (Lord) Cockburn [q.v.], the Scottish lawyer. At the age of four he was sent to Scotland with his Chinese nanny to be cared for by his grandmother. His father retired from the Diplomatic Service in 1909 and after renting a number of houses eventually settled near Tring, Hertfordshire. Cockburn was sent to Berkhamsted School where Charles Greene was headmaster. He became a close friend of Greene's son Graham with whom he shared a liking for mischief-making and adventure stories, especially the yarns of John Buchan (later Lord Tweedsmuir, q.v.) in which brilliant but corrupted villains seek to overthrow the established order from within. Graham Greene's younger brother, (Sir) Hugh Carleton Greene, was a pupil of Cockburn's when he briefly took over the classical sixth form during an Oxford vacation and remembered him as the most brilliant teacher he ever encountered.

Cockburn entered Keble College, Oxford, where he obtained second classes in classical honour moderations (1924) and *literae hu-*

maniores (1926). At Oxford he joined the 'smart set' which included Robert Byron, Evelyn Waugh (a cousin) [qq.v.], and (Sir) Harold Acton. With Graham Greene he also joined the Communist Party, as a joke, in the vain hope of travelling to Russia. In 1926 he won a travelling scholarship from The Queen's College, Oxford. He went to France and then Germany, where he attached himself to *The Times*'s correspondent Norman Ebbutt [q.v.]. His experiences in Germany kindled an interest in politics and after reading the communist anthology *Against the Stream*, he first felt attracted to Marxism. In 1929 he accepted a full-time post on *The Times*, the setting of many of his best stories. They included a sub-editor who spent a whole day researching the correct spelling of Kuala Lumpur and his own victory in a competition for the most boring headline with 'Small Earthquake in Chile. Not Many Dead' (although this became part of Fleet Street folklore, it has to be said that extensive research failed to locate it in *The Times*'s back numbers).

In 1929 Cockburn went to New York as *The Times* correspondent, from time to time reporting from Washington where he stood in for the well-known reporter (Sir) Willmott Lewis [q.v.] who gave him what he always regarded as an essential piece of advice: 'I think it well to remember that when writing for the newspapers, we are writing for an elderly lady in Hastings who has two cats of which she is passionately fond. Unless our stuff can successfully compete for her interest with those cats, it is no good.' Meanwhile, influenced by the Wall Street crash and subsequent depression he became more and more drawn towards Marxism. In 1932 he returned to England, gave up his employment on *The Times* and joined the Communist Party—this time in earnest.

Cockburn now embarked on his most successful venture, *The Week*, a cyclostyled newssheet inspired by the French satirical paper *Le Canard Enchaîné*. Started on a capital of £50 provided by his Oxford friend Benvenuto Sheard, the paper, which was all his own work, was produced at 34 Victoria Street (now part of Scotland Yard) and was obtainable only by subscription. Although he relied on information supplied by a number of foreign correspondents including Negley Farson (*Chicago Daily News*) and Paul Scheffer (*Berliner Tageblatt*), it was his own journalistic flair which gave the paper its unique influence. Cockburn was not an orthodox journalist. He pooh-poohed the notion of facts as if they were nuggets of gold waiting to be unearthed. It was, he believed, the inspiration of the journalist which supplied the story. Speculation, rumour, even guesswork, were all part of the process and

an inspired phrase was worth reams of cautious analysis. (It was Cockburn who coined the expression 'the Cliveden Set' to describe the pro-appeasement lobby.) In other hands it might have been a fatal approach but Cockburn had great flair and although many stories in *The Week* were fanciful there was enough important information to win it an influence out of all proportion to its circulation. Cockburn boasted eventually among his subscribers the foreign ministers of eleven nations, all the embassies in London, King Edward VIII, (Sir) Charles Chaplin [q.v.], and the Nizam of Hyderabad.

At the same time as producing *The Week* Cockburn joined the staff of the *Daily Worker* in 1935 as diplomatic correspondent, reporting the Spanish civil war under the pseudonym of Frank Pitcairn. Following the declaration of war in 1939 the government suppressed the *Daily Worker* and *The Week* though they both were later allowed to resume publication once the USSR became one of the Allies. The new situation, which conferred respectability on the communists, was not to Cockburn's liking and his Marxist fervour began to wane. He was further influenced by an interview with Charles de Gaulle in Algeria in 1943 in which the general suggested that his loyalty to the communist movement might perhaps be 'somewhat romantic'. Following the Labour victory in 1945 he became convinced of the ineffectiveness of the communists as a political force.

The following year he decided to burn his boats, giving up his job on the *Daily Worker* and retiring with his second wife Patricia to her home town of Youghal in county Cork. The move suited him well as, having spent so much of his life abroad, he had never felt part of the English scene. But despite resigning from the *Daily Worker* he never formally renounced communism. He wrote a number of novels including (as James Helvick) *Beat the Devil* (1953) which John Huston made into a film starring Humphrey Bogart. In 1953 Anthony Powell, an Oxford contemporary, introduced him to the then editor of *Punch*, Malcolm Muggeridge, who became a close friend. Cockburn contributed humorous articles for a number of years and later became a regular columnist on the *Sunday Telegraph*. In 1963 he was guest editor of *Private Eye* at the height of the scandal involving John Profumo and continued to write for the magazine until his death.

Cockburn was a man of great charm, modest, unassuming, and possessed of a schoolboyish zest for life. His appearance was donnish and with his deep bass voice he spoke in staccato bursts in the manner of Mr Jingle in *Pickwick Papers*. Both in conversation and in print he was an anecdotalist. His highly diverting memoirs published originally in three volumes are full of very entertaining stories (many of them embellished over the years) as well as valuable and profound reflections on politics and journalism. During the final decade of his life he suffered from increasingly bad health. But his constitution was remarkably tough and he survived attacks of tuberculosis, cancer, duodenal ulcers, and emphysema before he finally died. For one whose life had been so full of ironies, it was fitting that five priests celebrated a requiem mass for him in Youghal, although he had been a committed atheist.

In 1932 Cockburn married the left-wing American journalist Hope Hale, daughter of Hal Hale, high school principal and superintendent of schools, and Frances, née MacFarland. They had a daughter Claudia, who later married the humorous song-writer Michael Flanders [q.v.]. Cockburn also had a daughter Sarah by Jean Ross, who inspired the character of Sally Bowles in *Goodbye to Berlin* (1939) by Christopher Isherwood. Cockburn's first marriage ended in divorce in 1935 and his second wife, whom he married in 1940, was Patricia, daughter of Major John Bernard Arbuthnot, of the Scots Guards, and his wife Olive Blake. A highly resourceful and energetic woman, who had been an explorer in her youth, Patricia helped to support her husband, who was invariably short of money, first by selling ponies and then by making shell pictures. They had three sons, all of whom became journalists. Cockburn died 15 December 1981 in St Finbarr's Hospital, Cork.

[Claud Cockburn, *In Time of Trouble*, 1956, *Crossing the Line*, 1958, *View from the West*, 1961, *I Claud*, 1967, and *Cockburn Sums Up*, 1981; Patricia Cockburn, *The Years of The Week*, 1968, and *Figure of Eight*, 1985; personal knowledge.]   RICHARD INGRAMS

COIA, JACK ANTONIO (1898–1981), architect, was born 17 July 1898 in Wolverhampton, the eldest of the nine children of Giovanni Coia, originally from Naples, who was a craftsman, hurdy-gurdy player, and latterly successful café owner, and his wife, Ernestina, who was born in Florence, started as assistant in her step-father's circus dog act, and also trained as a dancer. Later Coia revelled in telling of his parents' trek north, financed by his father's playing on the organ and his mother's dancing. Once settled in Glasgow, Giovanni Coia opened a shop and soon became proprietor of an East End café. Jack (previously called Iacomo) Coia studied at St Aloysius College, Garnethill, but left without formal qualifications.

While at school Coia started work in his father's business. After leaving St Aloysius the tedium of the café soon persuaded him to seek employment elsewhere. In 1915 he started work for J. Gaff Gillespie as an architectural apprentice. He enrolled in the Glasgow School of Art and soon was combining work and study in an all-consuming passion.

As poor eyesight exempted him from military service his work and studies continued uninterrupted throughout World War I. On qualifying ARIBA in 1924, he headed for London but in 1927 Gillespie's death prompted a return to Glasgow to join the old firm as William Kidd's partner. On Kidd's own death soon afterwards, he became owner of the practice of Gillespie, Kidd, and Coia. With commissions few and far between Coia began teaching at the School of Architecture. In 1931 an inspired visit to the Roman Catholic archbishop of Glasgow resulted in the first of a series of commissions which would prove the staple of the practice for the rest of his life. St Anne's, Whitevale Street, Glasgow, in red brick, is the first of five churches which Coia completed before World War II. These churches served as unemphatic foils to the art which adorned them. Coia's skill in selecting artists later became legendary. In 1938 Coia was selected to work on the Glasgow Empire Exhibition. The Roman Catholic Pavilion was unspectacular but his Post Office and Industry North Pavilion, the fastest built of the exhibition's major structures, was hailed for its modernity. In 1941 Coia became FRIBA.

In 1939 Coia married Eden Bernard. They had two daughters. Eden was a considerable support in the war years when Coia was shunned as an enemy alien. Work did not pick up again until the early 1950s. For the next three decades, however, Gillespie, Kidd, and Coia flourished. The prosperity of the practice was partly a result of a total break with historical styles in favour of an uncompromising modernism. Coia's talkative persuasiveness ensured a steady flow of commissions but the greatest factor in the firm's success was his skill or good fortune in selecting his assistants. Indeed two would later become professors of architecture of international repute.

During the 1950s Coia passed much of the responsibility for individual jobs to his senior assistants. Work of this time includes St Paul's, Glenrothes, with its dominating façade window. St Charles', Kelvinside, Glasgow, of 1959 looks to the Continent for inspiration. The 1950s also saw the firm's first forays into house and school design.

In the 1960s the practice achieved four RIBA bronze medals for architecture including one for its most controversial church design, St Bride's, East Kilbride, a tight collection of solid brick masses. Sadly the campanile which balanced the complex has been demolished, robbing the building of much of its drama. St Peter's College at Cardross was another notable medal winner. The 1960s were also a decade of personal honours. Coia was appointed CBE in 1967. In 1969, while president of the Royal Incorporation of Architects in Scotland, he received the Royal gold medal for architecture. In 1970 he obtained honorary doctorates from the universities of Glasgow and Strathclyde.

In the last decade of his life Coia relinquished control of the practice but retained a lively interest in architecture. On regular study trips to his favourite Tuscan haunt of St Gimignano he indulged his passion for sketching. All who knew Jack Coia were impressed by his forceful character though many were bemused by his cantankerousness. One of his closest friends, the architect Patrick Nuttgens, summed up this paradox: 'He was small, intense, unkempt, angry, and bloody minded . . . and I loved him.' Coia died 14 August 1981 in Glasgow.

[Jack Coia, unpublished notes for his Royal gold medal acceptance speech; Robert W. K. C. Rogerson, *Jack Coia: His Life and Work*, privately published, 1986, with bibliography.]　　　　　　　　NEIL BAXTER

COLLINS, (LEWIS) JOHN (1905–1982), canon of St Paul's Cathedral and social reformer, was born 23 March 1905 at Hawkhurst, Kent, the youngest of the four children (two daughters and two sons) of Arthur Collins, master builder, and his wife, Hannah Priscilla Edwards. He was brought up in a conservative, church-going home and at the age of six he felt called to the church's ordained ministry. This was reinforced during his time at Cranbrook School, and also at Sidney Sussex College, Cambridge, where he went as a scholar. He obtained a third class in part i of the mathematical tripos (1925), and a second in part i (1927) and a first in part ii (1928) of the theological tripos. He was ordained in Canterbury Cathedral in 1928 and became curate of Whitstable (1928–9).

Within a year, however, he was invited to return to Cambridge as chaplain of his old college, where he remained until 1931. During this time he became interested in the work of Albert Loisy, a French Roman Catholic scholar who had been excommunicated because of his liberal interpretation of the Bible. The two men became friends and Collins began to question various elements in his own faith, as well as his conservative approach to politics and the ordering of society.

In 1931 Collins became an assistant lecturer in theology at King's College, London, and minor canon of St Paul's, but three years later returned to Cambridge to spend four years as vice-principal of Westcott House (1934–7). Soon after his appointment in 1938 as dean of Oriel College, Oxford, he joined the Labour Party, having noted the effects of the 1930s economic depression on the working class.

In 1940 Collins left Oxford in order to become a chaplain in the Royal Air Force. For most of the war he was at a training station in Wiltshire where he conceived the idea of forming a fellowship of Christian airmen and airwomen who would meet regularly to study their faith and its practical implications. This experiment aroused considerable interest, though his choice of socialist speakers, and his frequent challenges to authority brought him into serious conflict with his senior officers.

When the war ended he resumed his post at Oxford and in December 1946 convened a meeting in Oxford town hall which was addressed by several prominent speakers, all of whom urged a large audience to take their religious convictions into the social and political life of the nation. As a result of this meeting Christian Action came into being, and in 1948 the prime minister, C. R. (later Earl) Attlee, appointed Collins to a canonry at St Paul's so that he might devote more time to the new movement and provide it with a London headquarters.

Before long Collins had become a national figure and for the next three decades was one of the world's leading Christian protagonists of action in the causes of justice, freedom, and peace. Christian Action itself never had a very large membership but it provided an organization to support his own highly controversial personal work and its influence was quite out of proportion to its size.

In home affairs Christian Action gave strong support to a successful campaign for the abolition of capital punishment and undertook pioneering work among the homeless and persons displaced by war. But the emphasis was soon to change. In 1956, after Collins had visited South Africa, over £20,000 was raised on behalf of some 156 opponents of apartheid who had been arrested and imprisoned in Johannesburg. This was intended to pay for their legal defence and to provide support for their families. Two years later a separate organization known as the International Defence and Aid Fund was set up, with Collins as president and director, and this soon became an important instrument of British, and later international, opposition to apartheid. Collins's work in this field was recognized in 1978 by the award of the gold medal of the United Nations special committee against apartheid.

The explosion of the first nuclear bomb at Hiroshima in 1945 disturbed Collins greatly and during the early part of 1958 he was one of the sponsors of a national Campaign for Nuclear Disarmament (CND), which had the philosopher Bertrand Russell [q.v.] as its president and Collins as its chairman. The main public manifestation of the campaign's activities was a series of Easter marches to and from the nuclear research establishment at Aldermaston. The numbers taking part ranged from 7,000 to 20,000. Soon however there were serious disagreements. A breakaway Committee of 100 was formed in 1960 to organize civil disobedience. This caused dissension, indiscipline, and violence in CND and Collins resigned from the chairmanship, which he had held since 1958.

Collins continued to serve as a canon of St Paul's where he held, successively, the offices of chancellor (from 1948), precentor (from 1953), and treasurer (from 1970). The cathedral pulpit was used by him to promote Christian Action causes, and although Collins did not himself attract large congregations, his controversial sermons received wide publicity. Collins would have been a disturbing member of any cathedral chapter, but he had a great affection for St Paul's and never lost his vision of this national cathedral as a centre of culture and of Christian faith and witness. It was therefore a great blow to him when, on the retirement of W. R. Matthews [q.v.] from the deanery in 1967, he was not chosen as his successor.

He remained at St Paul's until his seventy-sixth birthday and after a brief retirement died in a London hospital 31 December 1982. At the end of his life the British churches were more deeply aware than ever before of their social and political responsibilities, and the man who symbolized this change and helped to bring it about was John Collins of St Paul's. There is a memorial stone to him in the crypt of the cathedral he served for a total of thirty-six years.

In 1939 he married Diana Clavering, a gifted and dynamic woman who shared fully in every aspect of his work. She was the daughter of Iion Elliot, company director. They had four sons.

[Canon L. John Collins, *Faith under Fire* (autobiography), 1966; Ian Henderson (ed.), *Man of Christian Action*, 1976; private information.]   TREVOR BEESON

COLLINS, NORMAN RICHARD (1907–1982), television pioneer and author, was born in Beaconsfield 3 October 1907, the only son and youngest of the three children of Oliver Norman Collins, a publisher's clerk and illustrator, and his wife, Lizzie Ethel Nicholls. One of his sisters died in childhood and the family was left hard up after the father's death when his son was ten. Being attracted towards the written word from an early age, Collins joined the publicity department of the Oxford

University Press on leaving the William Ellis School in Hampstead. In 1929 he moved to the *News Chronicle* as assistant to Robert Lynd [q.v.], the paper's literary editor. In January 1933, when he was twenty-five, he became assistant managing director in the publishing house run by (Sir) Victor Gollancz [q.v.].

Gollancz and Collins made an effective if incompatible pair, each needing the other but without the bonds of mutual affection or unquestioning trust. The firm prospered on Gollancz's flair and drive, but it was as much Collins's managerial competence as deputy chairman from 1934 which kept the venture moving forward during the preoccupation of its mercurial governing director with the Left Book Club and socialist politics in the later 1930s. Delegation did not come easily to Gollancz, however, and in 1941 after a business association lasting for eight years their partnership was terminated when Collins departed to join the BBC in the relatively lowly position of talks assistant (Empire Talks) at a starting salary less than the amount he had been paying in income tax in his previous occupation.

At the wartime BBC Collins was soon marked out as a coming man. He had a talent for administration and a feel for corporate life; his interests and contacts were wider than most; he had already made a name for himself as a popular novelist (published, although without enthusiasm, by Gollancz); while in a larger organization his urbanity and witticisms were more readily appreciated. Energetic and ambitious, Collins rose fast in the General Overseas Service and by the time the war ended he was its director. After nearly two years in charge of the Light Programme (1946–7), he was selected by the Corporation's director-general, Sir William Haley, in November 1947 to head the BBC's television service.

Collins felt cramped by what he regarded as an unadventurous and hesitant attitude on the part of the hierarchy at Broadcasting House. In peace, as in war, radio was the BBC's *raison d'être*. When Collins took over, television licences stood at 27,850; three years later the total had risen to 440,550 and a new transmitting station at Sutton Coldfield had been opened to extend the service from the London area to the midlands. To reflect the growing significance of television Collins pressed for its interests to be represented on the BBC's board of management. In this contention, far-sightedness and personal advancement ran hand in hand, and in October 1950 he experienced the mortification of seeing his proposal accepted but an existing member of the board of management, (Sir) George Barnes [q.v.], appointed over his head as director of television. Faced with this rebuff, Collins resigned immediately. Insisting that the clash was one of principles rather than of personalities, he condemned the apathy and open hostility towards television to be found in some parts of the Corporation, protesting that television was being merged into the colossus of sound broadcasting. Prophetically, he added that the future of television did not rest solely with the BBC.

For the next three years Collins was at the heart of an organized campaign which coordinated a variety of political and commercial interests held together by a common dislike of monopoly. He eloquently and persistently maintained that competition need not result in a diminution of standards and urged the hybrid concept of a public agency regulating private enterprise companies which would be licenced to produce the programmes. The dignified title he devised, Independent Television Authority, ultimately found its way into the Television Act of 1954. The BBC monopoly was ended when Parliament responded to demands from the Conservative back-benches. It was ironic that shortly after the great prize had been secured, Collins's personal influence went into decline. Although a company he formed with his backers in 1952 was successful in obtaining one of the earliest programme contracts to be awarded, it was compelled to merge with a rival group owing to lack of adequate financial resources before going on the air in September 1955. Collins became (and remained) deputy chairman of Associated Television (ATV) but lacked management control.

Over the next quarter century Collins was the elder statesman of Independent Television, serving his own company and the industry in numerous capacities, notably by a long and satisfying connection with Independent Television News (ITN) of which he was a director for many years, periodically acting as chairman. In the non-profit making news company Collins may have seen a miniature BBC thriving in the more open and competitive structure of the commercial system. A wealthy man from his original investment in ATV, Collins retained the instincts of a public service broadcaster throughout his career. He retired from the board of Associated Communications Corporation (as ATV had become) only five months before his death.

An unmistakable mark of Collins's power of application and creative energy was that he continued to write fiction throughout such a busy life. Although never a full-time writer he was a fluent and prolific author with sixteen titles to his credit between 1934 and 1981. An autograph edition of twelve of his novels was published during the 1960s. His best known book, *London Belongs to Me*, was an instant success in 1945. 884,000 copies were sold and the novel was adapted both for a film and a television series.

A sociable man, Collins enjoyed conversation and clubs: the political committee of the Carlton being a particular favourite in his later life. He relished political as well as literary friendships. On 26 December 1930 Collins married Sarah Helen, daughter of Arthur Francis Martin, mining engineer. They had a son and two daughters. Collins died in London 6 September 1982.

[H. Hubert Wilson, *Pressure Group: the Campaign for Commercial Television*, 1961; Asa Briggs, *The History of Broadcasting in the United Kingdom*, vol. iv, 1979; Bernard Sendall, *Independent Television in Britain*, vol. i, 1982; private information; personal knowledge.]   WINDLESHAM

COOPER, (ARTHUR WILLIAM) DOUGLAS (1911-1984), art historian, critic, collector, and champion of modern art, was born at 49 Sloane Gardens, Chelsea, London, 20 February 1911, the first of three sons (there were no daughters) of Arthur Hamilton Cooper, captain in the Essex Regiment and of independent means, and his wife, Mabel Alice Smith-Marriott. Cooper's parents were British, but the family fortune had been made, generations back, in Australia. Cooper's estrangement from his national background constituted a lifelong theme. He chose to live in France for much of his adult life and was always severely critical of the British for what he saw as their ignorance about and failure to patronize the great art of the twentieth century. This criticism was the basis of his much publicized role in the 1950s when he attacked the Tate Gallery and its director Sir John Rothenstein.

Cooper was educated at Repton and at Trinity College, Cambridge, where in 1930 he obtained a third class in the French section and a second (division ii) in the German section of the modern and mediaeval languages tripos. He then went briefly to the Sorbonne and the University of Freiburg, concentrating on the study of art history. In 1932 he decided to devote one-third of his inheritance to the creation of an art collection which would represent the development of the four major Cubist artists— Pablo Picasso, Georges Braque, Juan Gris, and Fernand Léger—from 1906 until 1914. Cooper proceeded to build his collection with such alacrity and concentration that it was essentially formed by 1945.

His early experiences in World War II, serving with a French ambulance unit, are recorded in *The Road to Bordeaux* (1940), written with Denis Freeman. Subsequently he obtained a commission with the Royal Air Force, working in intelligence and (drawing upon his linguistic skills) interrogation. From 1944 to 1946 he was deputy director of the monuments and fine arts branch of the Control Commission for Germany, helping to identify, protect, and repatriate works of art.

Cooper's residences in London included 8 Groom Place (decorated with furniture designed by the painter Francis Bacon), and 18 Edgerton Terrace (shared with Basil Amulree, a noted gerontologist and a friend since Cambridge days). It was in 1949, while on holiday with Lord Amulree and an art historian, John Richardson, that Cooper discovered and subsequently purchased the dilapidated Château de Castille in Argilliers, Gard. That grand eighteenth-century house, filled with Cooper's impressive collection, and animated by Cooper's own colourful and controversial personality, became a popular end to a pilgrimage for members of the art world. Léger was among the first house guests; Picasso was a neighbour in the south of France and a frequent visitor.

Though the exact configuration of the collection changed frequently, the focus always remained the Cubists. Cooper also collected major groups of works by Paul Klee and Joan Miró and works by other modern artists including César, Cocteau, Giacometti, Guttuso, Hockney, Hugo, Marini, Masson, Matisse, Modigliani, de Staël, Graham Sutherland [q.v.], and Ubac (many of whom he knew personally). Later works of Braque ('Atelier VIII', for example), Léger, and Picasso also formed part of the collection. In 1963 five drawings by Picasso, including three studies relating to Manet's 'Déjeuner sur l'Herbe', were executed in enlarged format, in concrete and stone, on a loggia wall in the garden of Castille.

Cooper's was by no means an easy personality and even his friendships often had bitter endings. His lifelong friendship with Picasso concluded in alienation, capped rather tragically (even after Picasso's death) with a vicious commentary on the artist's late work exhibited in Avignon in 1973. This contentiousness characterizes many of Cooper's numerous reviews and letters which frequently appeared in the *Burlington Magazine* and the *Times Literary Supplement*.

Cooper was, indeed, a prolific and formidable scholar and critic. In his early years he wrote under the pseudonym Douglas Lord. His contributions to the study of Cubism remain landmarks in that field. His bibliography includes notably: *Fernand Léger et le Nouvel Espace* (1949); *Pablo Picasso: Les Déjeuners* (1962); *Picasso, Théatre* (1967); and *Juan Gris: Catalogue Raisonné de l'Oeuvre Peint* (1977). Cooper also organized numerous important exhibitions culminating in two extraordinary presentations:

'The Cubist Epoch' at the Los Angeles County Museum and the Metropolitan Museum in 1970, and 'The Essential Cubism' at the Tate Gallery in 1983. Cooper wrote extensively about other twentieth-century artists and major figures of the nineteenth century.

Cooper was a lecturer at the Courtauld Institute, Slade professor of fine art at Oxford (1957-8), and Flexner lecturer at Bryn Mawr (1961). His honours include membership in the Real Patronato of the Prado Museum and chevalier of the Legion of Honour.

Cooper died 1 April 1984 in the Royal Free Hospital, Camden, London. He was unmarried but in 1972 he adopted a son, William Augustine McCarty-Cooper, aged thirty-five.

[Dorothy Kosinski, *Douglas Cooper und die Meister des Kubismus*, Basel, Kunstmuseum, 1987, and London, the Tate Gallery, 1988; John Richardson, 'Remembering Douglas Cooper', *New York Review of Books*, 25 April 1985, pp. 24-6.]   DOROTHY KOSINSKI

COOPER, JOSHUA EDWARD SYNGE (1901-1981), cryptanalyst, linguist, and intelligence officer, was born 3 April 1901 in Fulham, the son of Richard Edward Synge Cooper, a chartered engineer, and his wife, Mary Eleanor, youngest daughter of William Burke. Joshua was the eldest of five children, four sons and a daughter, all brought up in England, but both parents came from Ireland, the Coopers being a well-known family associated with Castle Markree in county Sligo. Joshua's great-grandfather was the astronomer Edward Joshua Cooper, FRS [q.v.].

Cooper was educated at Shrewsbury and took a scholarship in classics to Brasenose College, Oxford. After taking a third class in classical honour moderations in 1921 he entered King's College, London, to read Russian and Serbian in which he took a first in 1924. In 1925 he was one of a small number of graduates recruited into the Foreign Office to work in the Government Code and Cypher School by its director, A. G. Denniston [q.v.]. It is said that these recruits were not told at their interviews the kind of work in which they would be engaged.

In 1934 the Air Ministry decided that, like the Royal Navy and the army, the RAF also needed to have its own stations to intercept the signals of potential opponents, and in 1936 Cooper, who was already a distinguished cryptanalyst in GC and CS, was seconded to the Air Ministry as head of AI 1(e) to analyse the intercepted material, and remained head of what became known as the Air Section until 1943, when he was transferred back to the Foreign Office. At the beginning of this period the role of his section (like GC and CS) was seen simply as that of cipher-breaking, and the section only received enciphered messages, the plain language being analysed elsewhere. Cooper changed this curious arrangement and thus prepared for the wartime work against the Luftwaffe. With the outbreak of war, and the great increase in Luftwaffe traffic, a number of bright young men and women, service and civilian, were assigned to its analysis at Bletchley or at the interception stations. Cooper was their mentor and inspiration.

The Air Section and its associated stations provided immediate and longer-term information to the Air Ministry and RAF commands on every aspect of the operations of the Luftwaffe. For example, in 1939-40 there were serious questions about the size of the Luftwaffe resulting from differing views in the Air Ministry and the Ministry of Economic Warfare. (Sir) Winston Churchill appointed Sir J. E. Singleton [q.v.] to conduct an inquiry and this was followed by studies by F. A. Lindemann (later Viscount Cherwell, q.v.). It was finally agreed that there had been a substantial overestimate of German strength, because the basic unit of the Luftwaffe, the Staffel, had been wrongly assessed by the Air Ministry as consisting of twelve aircraft. It was the Air Section which demonstrated conclusively that the true size was nine, and Cooper himself presented the evidence in these inquiries.

In his subsequent career at GCHQ, ending up as an assistant director in charge of GCHQ's research work, Cooper demonstrated the great range of his mind and his ability to comprehend in fields such as mathematics and physics which were outside those in which he had been educated. He realized very early the potential significance of the post-war development of computers and ensured that his colleagues understood too.

Cooper was not an administrator but was always admired and beloved by those who worked for him and with him; his mannerisms were endearing, his eccentricities much embellished in the telling. The latter arose from his concentration on the subject occupying his mind. With his extraordinary memory and instant recall, he would resume a conversation without preamble or reference after a lapse of weeks. He retired in 1961 and published *Russian Companion* in 1967 and *Four Russian Plays* in 1972. He was always close with his brother Arthur R. V. Cooper, the distinguished Chinese scholar, and shared ideas with him based on their complementary knowledge of languages.

Cooper, who was appointed CMG in 1943 and CB in 1958, took it as axiomatic that the safety of Britain and its citizens in peace and war depended on the effectiveness of its intelligence services; and that, to be effective, this work must remain secret—he deplored the spate of wartime reminiscences—and must be with malice toward none. He died at Amersham, Buckinghamshire, 14 June 1981.

In 1934 he married Winifred Blanche de Vere, daughter of Thomas Frederick Parkinson, a civil engineer in India. They had two sons, the elder of whom died in 1956.

[Ronald Lewin, *Ultra Goes to War*, 1978, pp. 135-6; *The Times*, 18 June 1981; Nigel West, *GCHQ*, 1986, p. 140; F. H. Hinsley and others, *British Intelligence in the Second World War*, vol. i, 1979, pp. 299-302; private information; personal knowledge.]

D. R. Nicoll

COOPER, THOMAS FREDERICK (1921-1984), comedian, was born 19 March 1921 at Caerphilly, Glamorgan, the elder son (there were no daughters) of Thomas Cooper, a coalminer hewer, and his wife, Catherine Gertrude Wright. Although born in Wales, Cooper was mainly brought up in Exeter, where he attended Radcliff College, and Southampton, where he began his working life as an apprentice shipwright. There then followed seven years in the Horse Guards, which included service in the Middle East during World War II. Like many of his generation of comedians, he became interested in show business during this period. And, like many of these other ex-Service entertainers, he was then blooded at the Windmill Theatre in the late 1940s.

He was soon successful, making his first appearance at the London Palladium in 1952; and the first of his regular royal command performances in 1953, early accolades for a rising star. He continued to be in enormous demand in theatre and cabaret, abroad as well as in Britain, but it was television that proved to be his most natural *métier*. Either in his own series, *It's Magic*, or as an ebullient guest in other shows, he remained extremely popular for thirty years, a most consistent achievement in the sometimes ephemeral world of entertainment. The hysteria wrought by his butter-fingered magic, and the response, horrified and gibbering, it drew from him, spread in epidemic proportion to his audiences, who often became locked in a genuine and convulsive mirth, no mean result for a comedian in the supposedly sophisticated post-war decades.

A reliable story tells of how, at an early nervous audition, his fumbled conjuring fooled an amused agent, who mistook it for the purposeful spoof it soon became. Only a most inventive and gifted conjuror—he was a member of the inner six of the Magic Circle—could have essayed his joyous romps through accident-strewn leger-demain, as, with ferocious grin and guttural chortle, Tommy Cooper surveyed the mounting anarchy of his magic. The 'cod' conjuror has a longish lineage in British comedy, but Cooper more or less obliterated the post-war competition. He underscored his bungled conjuring with an excruciating collation of hilarious puns, and his gabbled catch-phrase, 'just like that', normally coincided with his most egregious blunder.

His second chief routine was another burlesque, this time of the narrative monologist, who like the conjuror, was another one-time prototype of the music-hall boards. With the suspect assistance of a choice collection of hats, Cooper assumed the characters of the tale he recounted, the headgear and verses colliding in the most devastating of muddles. Eventually, he forgot the lines completely, and, returning to the start, attempted a whirlwind whisper and swift exchange of hats through the entire ballad, in the hope of reminding himself of his cue.

In a real sense, his was the tradition of the clown, rather than the comedian. A huge, heavy, shambling figure, with popping, staring eyes, surmounted by expressive brows, and with mouth lollingly agape, his very presence threatened dislocation. Like the best of clowns, he conveyed that innocent spirit of optimistic anticipation, assailed by physical misfortune. Grotesquely mobile features, swirling limbs, and the inevitable red fez completed this picture of a lord of misrule. It was this supremely visual factor which enabled him to make so telling a contribution to British television comedy.

In 1947 he married Gwen, daughter of Thomas William Henty, farmer. They had a daughter and a son. In 1977 Cooper had a heart attack while in Rome, and after lung trouble he had to forgo his affection for cigars. On 15 April 1984 he collapsed on the stage of Her Majesty's Theatre, London, while appearing in a 'live' television show: indeed, as the curtains closed on him, many viewers imagined his tumble to be part of his clowning. He was taken to Westminster Hospital, where he was adjudged dead.

[*The Times*, 16 and 17 April 1984; John Fisher, *Funny Way to be a Hero*, 1973; Mary Fieldhouse, *For the Love of Tommy*, 1986; information from Miff Ferrie; personal knowledge.]　　　　　ERIC MIDWINTER

CORBETT ASHBY, DAME MARGERY IRENE (1882–1981), feminist and internationalist, was born 19 April 1882 in London, the eldest of three children (two girls and a boy) of Charles Henry Joseph Corbett, buisnessman and landowner, of Woodgate, Danehill, Sussex, and his wife, Marie, daughter of George and Eliza Gray, of Tunbridge Wells. Both parents were keen feminists, involved in many public and local affairs, with liberal views of a radical tinge, her father being Liberal MP for East Grinstead from 1906 to 1910. She was educated at home by her parents and governesses, and at Newnham College, Cambridge, where she obtained a third class (third division) in part i of the classical tripos in 1904. She had deliberately chosen classics because it was a tough subject. She then took a teachers' training course, but did not go into teaching. In 1907 she became secretary of the National Union of Women's Suffrage Societies, a full-time and responsible job which she held for a year.

The event which probably had the most lasting influence on her life had already occurred in 1904 when she went with her mother to an international congress in Berlin, where the International Woman Suffrage Alliance was founded. She became closely identified with this body, working for it as secretary, member of the board, president from 1923 to 1946, and attending its congresses until 1976. As after World War I women obtained the vote in many countries, the Alliance, renamed the International Alliance of Women in 1926, widened its interests and concerned itself with other issues of importance to women—for example, equality of opportunity in employment, adequate representation on public bodies, nationality of married women, equal moral standards for both sexes, and family allowances. It also took up the cause of peace.

Margery Corbett Ashby was a keen supporter of the League of Nations, whose Assembly she attended regularly, and she wanted to widen women's sphere of interest and activity from the home, to the city, to the nation, to the world. But on this issue she had to overcome opposition from some members of the Alliance who argued that it should not be diverted into causes, such as peace, which were common to men and women. She herself was a substitute delegate for the United Kingdom at the disarmament conference of 1932 until she resigned from this position in 1935, discontented with her government's attitude. She continued throughout the 1930s to criticize the British government's foreign policy because it gave so little support to League of Nations principles and machinery.

When younger women got the vote in England in 1928, Margery Corbett Ashby and others in the National Union of Societies for Equal Citizenship, in particular Eva Hubback, were anxious to train the new voters for political responsibility. Their plan was to bring women together to share the interests they had in common, such as domestic affairs and arts and crafts, and then to broaden their minds and so foster citizenship. The result was the creation of Townswomen's Guilds, of whose National Union Margery Corbett Ashby was president until 1935.

In politics she was a staunch Liberal, holding a number of posts in the party, including that of president of the Women's Liberal Federation. She stood for Parliament eight times unsuccessfully in five different constituencies—at the general elections of 1918, 1922, 1923, 1924, 1929, and 1935, and at by-elections in 1937 and 1944. Her aim was not so much to become an MP as to get public opinion to accept women candidates and to publicize the causes she cared about.

She packed many other activities into her life. In 1919 she went on a deputation to the Versailles peace conference and made successful representations on several issues—for example, the opening of all posts at the League of Nations equally to women and men, the right of women to vote in plebiscites, and various social problems which were to be tackled by the International Labour Office. The same year she visited Germany to investigate and advise the War Office on some of the problems caused by occupying troops. She lectured from Liberal platforms, and all over the world—in Europe, India, Pakistan, the Near East, the USA, and Canada. In 1942 she went on a government propaganda mission to Sweden. After the war she worked in association with the United Nations. She was a Poor Law guardian for a time, and she also held posts in the Association for Moral and Social Hygiene, the County Federation of Women's Institutes, the British Commonwealth League, and the Women's Freedom League. In later life she became increasingly involved in local affairs in Sussex. She contributed to suffrage and liberal journals and was editor of *International Women's News* in 1952–61.

It has been said with some justification that she remained a Liberal of the Campbell-Bannerman school throughout her life and that there were problems she did not see, questions that for her did not arise. But about the issues she did see, she felt keenly, and for them she worked incessantly without personal ambition. She was a gifted linguist, a fluent, rousing speaker, but tactful and unaggressive. There can be little doubt that her activities assisted the cause of sex equality in many parts of the world, though the prevention of war was beyond her power and that of the many women she inspired.

In 1910 she married Arthur Brian Ashby (died 1970), barrister, and was thereafter known as Mrs Corbett Ashby. She had one son, born in 1914. Her home combined middle-class affluence with personal austerity. In old age she

considered her life had been remarkably happy, though in fact she was in her later years inevitably saddened by the decline in liberal values. Nor did she like the permissive society or political militancy. She was however helped by strongly-held religious beliefs. She received an honorary LLD at Mount Holyoke College, USA, in 1937 in recognition of her international work, and was appointed DBE in 1967. She died 15 May 1981 at Wickens, Horsted Keynes, Sussex, where she had lived since 1936.

[*The Times*, 16 May 1981; Newnham College records; Mary A. Hamilton, *Remembering my Good Friends*, 1944; Adele Schreiber and Margaret Mathieson, *Journey Towards Freedom*, 1955; Mary Stott, *Organization Woman*, 1978; Arnold Whittick, *Woman into Citizen*, 1979; Brian Harrison, *Prudent Revolutionaries*, 1987.]   JENIFER HART

CORNWALL, Sir JAMES HANDYSIDE MARSHALL- (1887–1985), soldier, linguist, and author. [See MARSHALL-CORNWALL.]

COX, Sir CHRISTOPHER WILLIAM MACHELL (1899–1982), a pioneer of university education in the Commonwealth, was born 17 November 1899 at Hastings, the eldest of three sons (there were no daughters) of Arthur Henry Machell Cox, a schoolmaster and later headmaster of Mount House School, Plymouth, and his wife, Dorothy Alice Wimbush. He was educated at Clifton (of which he later became an energetic and devoted governor). He was commissioned as a second lieutenant in the Royal Engineers (Signals) in 1918. When the war ended he went to Balliol College, Oxford, where he obtained firsts in classical honour moderations in 1920 and *literae humaniores* in 1923. In 1923 he was elected to a War Memorial studentship at Balliol, and in the following year to a university Craven fellowship and to a senior demyship at Magdalen. During these years he was engaged in archaeological exploration in Turkey, which resulted eventually, in 1937, in his only publication, *Monumenta Asiae Minoris Antiqua* vol. v (with Archibald Cameron). Some further documents from this period were found among his papers and were published posthumously.

In 1926 he was elected a fellow and tutor of New College, Oxford, to teach the Greek section of ancient history, and thus began an association with New College which lasted until his death. He was a vigorous, enthusiastic, and inspiring teacher. He held a succession of college offices, including two years as dean (1934–6), when his handling of disciplinary matters was held to be effective and also sympathetic. But it never seemed likely that he would stay for ever as a college tutor, and in 1937 he took two years' leave to become director of education in the Sudan and principal of Gordon College, Khartoum. This was a period of expansion, when Gordon College changed from a secondary school to a university college. Cox is remembered as an excellent chairman of committees, able to grasp what each person was trying to say and to express it more clearly for them.

In 1939 Cox returned to England and resumed teaching at New College. But with the outbreak of war normal university life went into eclipse, and in 1940 Cox became educational adviser to the colonial secretary, and thus entered upon what was to be the main activity of his career. It is true that when the war ended he was tempted to return to teaching at New College, but in the end he decided to remain with the Colonial Office; the college elected him to a supernumerary fellowship, so that his connection with it remained unbroken. He retained his rooms there and his membership of the governing body, and spent most of his weekends, when he was in England, in Oxford. He continued to work as educational adviser first at the Colonial Office, till 1961, then at the Department of Technical Co-operation, and finally at the Ministry of Overseas Development from 1964 until his retirement in 1970.

During this period he travelled widely throughout the Commonwealth, particularly in Asia and Africa and the Caribbean, in the colonies that were moving towards independence. He became the best known and best informed person in educational matters in the Commonwealth, and his wide contacts in the university world at home enabled him to involve in Commonwealth university affairs academics of the standing of Sir Alexander Carr-Saunders, Sir W. Ivor Jennings [qq.v.], Sir Douglas Logan, and Lord Fulton. Most of the universities in the 'New Commonwealth' regarded him as a father figure, and most of them gave him honorary degrees. But perhaps the academic distinction that pleased him most was the DCL that his own university conferred on him in 1965. He was appointed CMG in 1944, KCMG in 1950, and GCMG in 1970.

His supernumerary fellowship at New College was tied to his position as educational adviser and so lapsed on his retirement. But the college then elected him to an honorary fellowship and, uniquely, allowed him to retain his rooms in college (he was unmarried). Though he ceased to be a member of the governing body, he remained a member, and a very lively and entertaining member, of the senior common room. He also made many friends among the undergraduates, and one of the main interests of his life after he left government service was the reading parties for New College undergraduates that he organized annually at the chalet in the French Alps that once belonged to F. F. Urquhart, the dean of Balliol, where he presided as the genial, authoritative 'patron'.

Cox, a stocky, broad-shouldered, physically active figure, was a remarkable combination of order and disorder. His handwriting was illegible. He never threw anything away, and his rooms at New College were a scene of chaos and apparent confusion. Yet he in fact mastered his material completely. He was a compulsive talker. He would say 'I want to pick your brains for five minutes' and would then talk uninterruptedly himself for an hour, at the end of which his interlocutor would find that his brains had, in fact, somehow been efficiently picked. He enjoyed jokes, teases, reminiscences. He took enormous numbers of photographs. He was inclined to worry obsessively over details, to a point where symptoms of mental breakdown occasionally appeared, but in general he showed an impressive grasp of an immense range of knowledge. The vast accumulation of papers in his rooms was sorted out after his death and constitutes a massive archive, held at the Public Record Office for his official life and at New College for his personal affairs. He died in an Oxford hospital 6 July 1982.

[*The Times*, 7 and 13 July 1982; private information; personal knowledge.]

WILLIAM HAYTER

CRANKSHAW, EDWARD (1909–1984), writer, and commentator on Soviet affairs, was born 3 January 1909 in London, the elder son (there were no daughters) of Arthur Crankshaw, later chief clerk to Old Street magistrates' court, and of his wife, Amy Bishop. He was small in stature and suffered from a weak chest, and the family soon moved out to Letchworth for the sake of his health. He was educated at Bishop's Stortford College.

His parents wished him to go to university, but in a first display of his romantic and headstrong side he went alone to Vienna, aged eighteen, and taught English in the Berlitz School. Here began the deepest intellectual attachment of his life. He learned excellent German, and immersed himself in German and Austrian history and culture, developing not only his talent as a writer but his particular sense of politics as expressions of inherited 'national character'. On his return, he worked in the advertising department of *The Times* and began to write reviews—mostly musical—for the *Spectator*, *Bookman*, and other periodicals.

In 1931 Crankshaw married Clare, daughter of Ernest Carr, a civil servant, and they went to live in Hampstead. Here, after 1933, they gathered around them many of the German and Austrian intellectual refugees from Hitlerism. Crankshaw began to translate German plays and his versions were staged; when the revolutionary Ernst Toller arrived in England, he 'adopted' Crankshaw and his wife, the former of whom put five of his plays into English in the next few years. Crankshaw was by now writing his own books: *Joseph Conrad: Some Aspects of the Art of the Novel* appeared in 1936, and in 1938 he published *Vienna: a Culture in Decline* and *Nina Lessing*, the first of his three novels. In spite of these successes, the Crankshaws were existing on erratic scraps of income, and in the mid-1930s they left London for a cottage in Kent. Most of the rest of Crankshaw's life was spent in the village of Sandhurst, in west Kent; he worked alone and found the research and contacts which he needed on regular forays to London.

A man who was a friend both of Toller and of the novelist Ford Madox Ford [q.v.] was hard to catalogue politically. Crankshaw has been called a 'romantic Conservative', but his gesture of joining the Territorials in 1936, which dismayed some left-wing friends, turned out only too practical. 'If there is going to be a war', he observed, 'I might as well learn how to do it.'

When war came, his knowledge of German brought him into military Intelligence. Then, to his surprise, he was posted to the British military mission in Moscow. He had to learn Russian, and was in the country for less than three years. But, while deploring the regime, he came to love the Russian people as they fought for their survival, and to understand the historic roots of their political system in the tsarist autocratic tradition.

When Crankshaw joined the *Observer* in 1947 he was at first reluctant to write about Soviet affairs. He felt his strength lay not in political analysis but in literary and especially musical criticism (Artur Schnabel was a close friend). However, he let himself be tempted, and within a few years had become Britain's most authoritative and persuasive commentator on the USSR and its 'sphere of influence'. He wrote almost weekly until his retirement in 1968, and not much less frequently after that.

Crankshaw was not a true 'Kremlinologist', remarking once that 'the difference between Brezhnev and his colleagues seemed of no more interest than the difference between a number of stale buns'. Instead, he treated Soviet affairs much in the manner of the theatre reviewer he had once been. Neither was he an ideological 'cold warrior', but—for all his loathing of the Soviet system and his pessimism about its capacity for reform—a firm advocate of peaceful co-existence. His real testament is the preface,

written when he was a dying man, to *Putting Up With the Russians* (1984), his aptly named collection of shorter writings. In this preface, he furiously attacked the contemporary American ambition to evict the Soviet Union from all influence outside its own borders.

Crankshaw published numerous books on the changing Soviet scene, and his book on Nazi terror, *Gestapo* (1956) was widely read. But in 1963 he began to produce the ambitious, deeply researched books which are his main literary work. *The Fall of the House of Habsburg* appeared in that year; *Maria Theresa* in 1969; and his masterly *The Shadow of the Winter Palace*, a study of nineteenth-century Russia, in 1976.

Slight and gentlemanly in appearance, Crankshaw controlled a wild and independent nature. But even at the height of his fame, his modesty was phenomenal. He justified his own retirement by saying that he hated the Brezhnev leadership too much to be able to be fair to it. The Austrian government awarded him the Ehrenkreuz für Wissenschaft und Kunst in 1964, and the British followed with a series of prizes for his books including the Heinemann award in 1977 and the Whitbread prize in 1982 (for his *Bismarck*, 1981). He died at his home, Church House, Sandhurst, Hawkhurst, Kent, after a long illness, 30 November 1984, continuing to write even when he was too sick to leave his bed. He had no children.

[*The Times*, 3 December 1984; *Observer*, 2 December 1984; personal knowledge.]

NEAL ASCHERSON

CRAWLEY, LEONARD GEORGE (1903-1981), sportsman and golf correspondent, was born at Nacton, Suffolk, 26 July 1903, the eldest in the family of three sons and a daughter of John Kenneth Crawley, land agent, and his second wife, Cecily Frances Booker. He was educated at Harrow and Pembroke College, Cambridge, where he showed himself a celebrated player of games. After making a hundred for Harrow against Eton at Lord's in 1921, he missed by two runs equalling his Uncle Eustace's record as the only player to have scored a hundred in the Eton and Harrow cricket match and in the University match. At Cambridge he studied for the ordinary BA degree (not the tripos) but he did not complete this.

After Cambridge Crawley became a schoolmaster at Farnborough and in 1932 he started his own preparatory school, Warriston, at Moffat, in the Scottish borders. He played county cricket for Worcestershire (1922-3) and Essex (1924-37) and went on the MCC tour of the West Indies in 1925-6 but his working life as a schoolmaster meant that his appearances were limited to the summer holidays. Nevertheless, he made eight first-class hundreds, twice hitting Maurice Tate on to the pavilion roof in one innings at Leyton (1927). Less enviable was the task of opening the innings for Essex in 1932 after Yorkshire had declared, having made a world record of 555 for the first wicket. He played cricket with (Sir) John (Jack) Hobbs, Walter Hammond, and Frank Woolley [qq.v.]. He was a great admirer of Jack Hobbs whom he invited to coach the boys at his preparatory school, for Crawley was a strong advocate of the importance of good teaching at games. He was also a fierce critic of how poor the standard of teaching was for both cricket and golf.

Crawley was a natural stylist, someone to whom the art of hitting any ball was automatic. He had an elegance and power that people appreciated. One Glamorgan fast bowler of his day said that the only way to bowl at him was to deliver from 27 yards and hide behind the umpire. This was probably the result of Crawley's 222 against Glamorgan in 1928. Crawley abhorred the bodyline tour of Australia for which he was nearly picked in 1932, a year in which he averaged 51.87. He had a letter printed in *The Times* on the subject.

Crawley was the last of the great all-rounders and perhaps the best. He played golf with Ouimet, Sarazen, Snead, and Cotton. In addition, he was an outstanding rackets player, won the Northern lawn tennis doubles championship with an uncle, and captured a gold medal for skating. He was also a good and keen shot. Yet, like many a fine player of games, he was totally unaware of how good he was or how much pleasure he gave to others. Unless asked, he never talked about his achievements, but he presented an imperious figure at the crease or on the golf course immediately recognizable in his younger days by his ginger moustache. In later years an occasional irascible streak contrasted with a tendency to mislay things but he was also enormously kind.

In 1932 Crawley went with the British Walker Cup golf team to the United States where his singles victory over George Voigt gained the team's only point. By that time he was established more as a golfer than a cricketer, having won the English championship at Hunstanton in 1931 although it was, he always maintained, 'before I was anything like any good'. He played in the first victorious British team in 1938, winning his foursome with Frank Pennink—a feat he repeated in 1947 with Percy Belgrave ('Laddie') Lucas as a partner. It was Lucas who, in his book *Five Up* (1978), suggested that Crawley was the best all-round sportsman of all, adding 'I would put C. B. Fry a loser in the final against him.' Although Crawley was runner-up in the English cham-

pionships of 1934 and 1937 and, when he retired, had made more appearances for England than anyone, he may not quite have done his huge talent justice. Some ascribed this to the fact that his lovely, rhythmic swing rubbed off on others, inspiring them to play above themselves.

For a short spell before World War II, he abandoned schoolmastering for stockbroking. During the war he served in the RAF. In 1946 he was appointed golf correspondent of the *Daily Telegraph*, thus maintaining his wide-ranging interest in sport. He wrote felicitously for the paper until 1971 and also contributed regular articles to the *Field*. He loved helping the young with their golf and had a lot to do with improving coaching standards and urging the adoption of the American-sized ball in Europe.

In 1929 he married Elspeth, daughter of Rear Admiral John Ewen Cameron. They had two sons. Crawley died at his home in Worlington, Suffolk, 9 July 1981.

[*The Times*, 10 July 1981.]   DONALD STEEL

CRONIN, ARCHIBALD JOSEPH (1896–1981), novelist, was born 19 July 1896 at Cardross, Dunbartonshire, the only child of Patrick Cronin, a clerk and commercial traveller, and his wife, Jessie Montgomerie. When he was seven his father died and Cronin and his mother went to live with her family. His mother became a travelling saleswoman and the first woman public health inspector with Glasgow Corporation.

Cronin was educated at Dumbarton Academy and Glasgow University, where he studied medicine. He graduated MB, Ch.B. in 1919. His years at Glasgow were interrupted by service in 1916 as surgeon sub-lieutenant in the Royal Naval Volunteer Reserve and by three months at the Rotunda Hospital in Dublin, where he took his midwifery course. His first practice was in a mining district in Wales. During this period he obtained a diploma in public health (Lond. 1923), an MRCP (Lond. 1924) and an MD (Glasgow, 1925)—a considerable achievement which involved unremitting work. In 1924 he was appointed medical inspector of mines for Great Britain. His work at this time led to two reports on dust inhalation and first aid in mines.

Between 1926 and 1930 he practised in London, but ill health took him to the West Highlands, and there he wrote *Hatter's Castle* (1931). It made him famous overnight; he was able to give up his medical practice and become a full-time writer, as he had always wished to do. The second novel, *Three Loves* (1932), was 'torture to write', as he expressed it, and did not do well. However, *The Stars Look Down* (1935) was an instant favourite with his public. His next book, *The Citadel* (1937), which fiercely attacked the greed in Harley Street, caused a sensation. Launched with a brilliant publicity campaign by his publisher (Sir) Victor Gollancz [q.v.], it probably played some part in creating the climate of opinion which led to the National Health Service.

In July 1939 Cronin went with his family to the United States. Two of his novels were filmed in Hollywood at about this time (several of his books were made into successful films). Between 1941 and 1945 he worked in Washington for the British Ministry of Information and wrote *The Keys of the Kingdom* (1942) and *The Green Years* (1945). After the war he lived permanently in Switzerland, writing novels at roughly two-yearly intervals, notably *The Spanish Gardener* (1950). He was an honorary D.Litt. of Bowdoin and Lafayette universities.

Cronin's strength as a novelist lay in his narrative skill, his acute observation, and his graphic powers of description. His plots were often over-dramatic and his characters were in general unremarkable—he needed, as he himself remarked, to have real people to base them on (the tyrannical James Brodie, in *Hatter's Castle*, is said to be a portrait of his maternal grandfather, which caused consternation in the family). But as a craftsman he was meticulous and highly professional, and there is some refreshing humour in his books. He was not an intellectual, and enjoyed simple pleasures such as watching cricket matches and talking to the people round him, the kind of people who might have been his patients. A very hard worker, he greatly enjoyed writing. He also loved travelling, and this gave him material for his books which he used to good effect. He was a Catholic, and several of his novels are concerned with religion and matters of conscience. Though extremely tough in business dealings, in private life he was a happy, good-humoured person to whom each day was an adventure. His last years, however, after his wife became ill, were lonely, for he had always been a solitary individual, and his wealth cut him off from other people.

Cronin's experience in Dublin and Wales made him keenly aware of the evils of extreme poverty, and his skill in combining romantic, compelling narrative and vivid, realistic portrayal of life among the poorer members of society is one of the most striking facets of his novels. Nearly all his books have a strong autobiographical element, returning again and again to his childhood in Dumbarton, which is thinly veiled under the fictitious name of Levenford. Episodes in his autobiography *Adventures in Two Worlds* (1952) are repeated in his novels, making it difficult to disentangle fact from fiction. The immensely popular television and ra-

dio series which he wrote, *Dr Finlay's Casebook*, is also based on his experiences as a doctor.

Although Cronin's powers flagged in his later years, the best of his novels are extremely readable and accomplished, and they had deservedly large sales. He was a middlebrow writer *par excellence*, and above all a masterly story-teller.

He married Agnes Mary Gibson, MB, Ch.B., in 1921. She was the daughter of Robert Gibson, a master baker, of Hamilton, Lanarkshire. They had three sons, the eldest of whom, Vincent, became a writer. Cronin died 6 January 1981 at Glion, near Montreux, Switzerland.

[*The Times*, 10 January 1981; private information; personal knowledge.]

SHEILA HODGES

CUNNINGHAM, SIR ALAN GORDON (1887-1983), general, was born in Edinburgh 1 May 1887, the youngest child in the family of three sons and two daughters of Daniel John Cunningham [q.v.], professor of anatomy, of Dublin and Edinburgh universities, and his wife, Elizabeth Cumming Browne. His elder brother, Andrew Browne Cunningham [q.v.], became an admiral and Viscount Cunningham of Hyndhope. Alan Cunningham was educated at Cheltenham College and the Royal Military Academy, Woolwich.

Commissioned into the Royal Artillery in 1906 he served throughout World War I on the western front with the Royal Horse Artillery and on the staff, was decorated with the MC (1915) and DSO (1918) and was five times mentioned in dispatches. In 1919-21 Cunningham served in the Straits Settlements, then passed Naval Staff College, instructed at the Small Arms School, Netheravon, and in 1937 as lieutenant-colonel attended the Imperial Defence College. In the same year he became commander, Royal Artillery, 1st division, and in 1938 was promoted major-general to command 5th Anti-Aircraft division. Early in World War II he commanded several infantry divisions, and then in 1940 was selected as GOC East Africa for the campaign to reconquer Abyssinia led by Sir A. P. (later Earl) Wavell [q.v.].

Here Cunningham—a slight, fine looking, charming but sometimes choleric man—showed himself a brilliant, daring leader, moving with astonishing speed and achieving startling results. The campaign started in late January 1941 with the forces of General Sir William Platt [q.v.] advancing from Sudan and Cunningham's, consisting of four brigades mainly of South, East, and West African troops, from Kenya. Thrusting into Somaliland, he captured Mogadishu on 25 February, and by using seaports as supply bases, reached Harar a month later, having advanced 1,000 miles. He then turned on Addis Ababa which fell on 5 April. In two months he and his men had covered 1,700 miles, liberated nearly 400,000 square miles of country, and taken 50,000 prisoners, all at the cost of 500 casualties. The last stages of the campaign saw Platt and Cunningham converging on Amba Alagi where the Duke of Aosta surrendered on 16 May. Italian East Africa had been conquered in four months. Cunningham was appointed both CB and KCB in the same year (1941).

In June 1941 Wavell, C-in-C Middle East, was replaced by Sir Claude Auchinleck [q.v.], who chose Cunningham to command what was now called the Eighth Army. He therefore found himself running a mobile, fluctuating battle against Rommel in Auchinleck's winter offensive, Operation Crusader in the Western Desert. Cunningham planned two separate actions—one, an armoured thrust with XXX Corps which would outflank the frontier defences and concentrate at Sidi Rezegh, so drawing Rommel's armour to its destruction; second, a mainly infantry operation with XIII Corps and the Tobruk garrison to overcome frontier defences. It turned out otherwise. The British armoured brigades were dispersed, outfought by Rommel's superior tactics and tanks, and after a week's fighting, Rommel led the Afrika Korps on a raid to the rear of Cunningham's forces. This move, together with his tank losses, greatly disconcerted Cunningham and he recommended to Auchinleck that the battle be broken off. Auchinleck correctly insisted that the offensive must continue, forcing Rommel back for essential replenishment, and, because he believed that Cunningham was now 'thinking defensively', he decided to relieve him of his command, replacing him with (Sir) Neil Ritchie [q.v.]. The fact was that Cunningham, with no experience of the pell-mell style of desert fighting practised by Rommel, was unable to control such fast-moving operations. Nor was his health up to the strain of command under such conditions. He accepted his dismissal with staunch dignity, and after hospital treatment, returned to England.

He then held a series of appointments at home—commandant Staff College (rare for a non-graduate) and GOC Northern Ireland and Eastern Command. In November 1945 he was appointed high commissioner and C-in-C Palestine and high commissioner Transjordan. He brought his customary courage and shrewdness to the difficult problem of attempting to mediate between Arabs and Jews, holding the appointment until May l948 when the British left Palestine.

In 1951 Cunningham, who had been appointed GCMG in 1948, married Margery Agnes, widow of Sir Harold Edward Snagge KBE, and daughter of Henry Slater, of the Indian Civil Service. After leaving Palestine he lived in Hampshire, becoming deputy lieutenant, and was able to enjoy his favourite pastimes of gardening and fishing. He was also colonel commandant, Royal Artillery, from 1944 to 1954, and president of the council of Cheltenham College from 1951 to 1963. He lived until he was ninety-five, dying at the Clarence Nursing Home in Tunbridge Wells 30 January 1983.

[*The Times*, 1 February 1983; I. S. O. Playfair, *The Mediterranean and the Middle East* (official history), vols. i (1954) and ii (1956); *International Affairs*, October 1948.]

JOHN STRAWSON

CURTIS, DUNSTAN MICHAEL CARR (1910-1983), lawyer and European civil servant, was born 26 August 1910 at 6 Cheyne Gardens, Chelsea, the only child of Arthur Cecil Curtis, a civil servant, and his wife, Elizabeth, a teacher and painter, the daughter of Austin Cooper Carr, of Broxton Lower Hall, Cheshire. In 1923 he went to Eton, where he achieved the sixth form, was in 'Pop', and excelled at rugby football, abandoned in favour of sailing, his lifelong passion after he went up to Trinity College, Oxford, in 1929. He gained a third class honours degree in philosophy, politics, and economics in 1933. In 1932 he spent the long vacation learning French with Professor Martin at Wimereux with a group of friends, one of whom was (Sir) Terence Rattigan [q.v.], who wrote his play *French Without Tears* based on this occasion.

Qualifying as a solicitor in 1937, Curtis became legal adviser and business manager to Michel Saint-Denis, the French theatrical director at the Old Vic Drama School. When Saint-Denis was broadcasting to occupied France as Jacques Duchesne, Curtis broadcast for him in French after the St Nazaire raid. War found Curtis with an RNVR commission, commanding a motor torpedo boat with coastal forces engaged in landing Allied agents in France. In the St Nazaire raid of March 1942 he commanded the motor gunboat from which Commander R. E. D. Ryder, leading the naval forces, directed operations. The purpose was to destroy the dry dock and deny its use to the battleship *Tirpitz*. Once the ex-American destroyer *Campbeltown* had successfully rammed the all-important lock gate with her cargo of explosives, Curtis put his motor gunboat alongside the old mole and landed the ground forces. He remained until ordered by Ryder to withdraw with survivors. He was awarded the DSC (1942).

Later he commanded the naval wing of No.

30 Assault Unit in North Africa, Sicily, and north-west Europe where his knowledge of German helped him to receive the surrender of Kiel, after a spirited telephone conversation with Admiral Dönitz, the German naval commander-in-chief. Curtis was promoted commander and received a bar to his DSC and the croix de guerre.

When the war was over he was a tireless advocate for peace. In 1947 he became deputy secretary-general of the European Movement. When the consultative assembly of the Council of Europe met in August 1949 at Strasbourg University, he helped in drafting their proposals, including the European Convention on Human Rights, the Council's outstanding achievement. In spring 1950 he joined the Council secretariat as counsellor in charge of assembly committees. In this key post, efficiency and a delightful personality made him the trusted adviser of all parties.

In 1954 the assembly elected him deputy secretary-general. The deaths of both the secretary-general and the clerk of the assembly left Curtis in sole charge of all Council branches pending new appointments to those posts. Everything worked admirably. Curtis should have been appointed secretary-general in 1960 but the assembly decided in favour of a politician. Two of Curtis's projects stand out. From 1955 a chance occurred to merge the Council of Europe and the OEEC. Although Curtis won support for this from both organizations he failed because of the abortive British plan to create an enlarged European Free Trade Area, which was opposed by the six countries of the OEEC. When Britain first sought membership of the European Community in 1960, Curtis undertook two secretariat studies on future links between the Community and Commonwealth and between the six countries of the OEEC and the seven-power EFTA. When Britain joined the EEC in the 1970s both of Curtis's reports were put into practice.

Curtis left the Council of Europe in 1962. From 1964 to 1973 he was a senior partner in the Paris office of the law firm Herbert Smith & Co. When Britain entered the EEC in 1972 Curtis enjoyed an Indian summer as secretary-general to the Conservative (later Democrat) group in the European Parliament, a post from which he retired in 1976. He was appointed CBE in 1963.

In 1939 he married Monica, daughter of James Grant Forbes, lawyer, of Boston, Massachusetts. They had a son and a daughter. After a divorce he married in 1950 Patricia ('Tony') Elton, sociologist and daughter of George Elton Mayo, an industrial sociologist at Harvard University. Curtis was good looking— his fair hair was an oriflamme. His personality radiated fun, his courage, especially in final ill health, was profound, and he had a great sense of humour. He died 9 September 1983 in Montgomery, Powys.

[*The Times*, 13 September 1983; private information; personal knowledge.]

COSMO RUSSELL

CURZON, SIR CLIFFORD MICHAEL (1907-1982), pianist, was born 18 May 1907 in London, the younger son and second of three children of Michael Curzon, antiques dealer, and his wife, Constance Mary Young, an accomplished amateur singer. His uncle, the composer Albert Ketelbey [q.v.], tried out his latest compositions on the family piano and gave the boy his first abiding musical memories. Curzon's first studies were on the violin. At the unusually early age of fourteen he was admitted to the senior school of the Royal Academy of Music where his professor was Charles Reddie through whose own teacher, Bernhard Stavenhagen, Curzon could claim to be a great-grand-pupil of Liszt. Curzon's pianistic ability to learn new repertory at speed impressed Sir Henry J. Wood [q.v.], then conductor of the RAM first orchestra. Wood gave Curzon his first promenade concert appearance in 1922 as one of the soloists in a Bach triple-keyboard concerto and took him as his concerto pianist on concert tours of Britain. Curzon left the RAM with the McFarren gold medal and other prizes. At this time his repertory centred on Romantic and post-Romantic virtuoso piano works which better-known pianists did not play—for example, pieces by D'Indy and Frederick Delius [q.v.]. He also gave the first performance of Germaine Tailleferre's *Ballade*. Although later Curzon regretted his 'neglect of music of the first quality' this was a suitable repertory for an ambitious pianist whose seniors might well have found him too immature for great classic masterpieces. Nevertheless the young Curzon was specially praised for his account of Schubert's *Wanderer* Fantasia in Liszt's then more popular transcription for piano and orchestra.

It was through his familiarity with Delius's Piano Concerto that Curzon gained his entrée to the repertory which was to become his speciality. The pianist Katharine Goodson wished to rehearse this work with another pianist taking the orchestral part; Reddie recommended her to Curzon who subsequently accompanied her at home in numerous of the great piano concertos which he had hitherto neglected, an experience of value when he came to learn the solo parts himself. In 1926 his father had to abandon his business through illness: the son took a sub-professorship at the RAM to support his family by teaching the piano, while still undertaking concert engagements. An unexpected legacy enabled him to spend two years in Berlin as a pupil of Artur Schnabel. It was from him that Curzon inherited the intellectual seriousness and perfectionism of technique and style which subsequently established his international reputation as an interpreter par excellence of the Viennese classics and German romantics. Among these were Liszt, whose B Minor Sonata Curzon included in the Berlin recital which he gave before leaving Schnabel's tutelage, together with Beethoven's 'Les Adieux' Sonata, Schubert's 'Moments Musicaux', and a recent work by Ernst Lothar von Knorr, a Berlin pedagogue—Curzon always preferred his recitals to include a contemporary work.

Curzon then went to Paris where he studied the harpsichord with Wanda Landowska and attended the classes of Nadia Boulanger. These two great musicians undoubtedly supplemented Schnabel's Teutonic practical and intellectual tuition. In Paris he also met and married in 1931 the American harpsichordist Lucille Wallace (died 1977), daughter of Edward Wallace, a Chicago businessman. Her acute sense of style in performance came to match his own. They adopted the two sons of the soprano Maria Cebotari after her and her husband's untimely death in 1949.

Curzon returned to England in 1932 to build a new international career in the classic repertory, though his programmes still included more recent music. He was the best exponent of the Piano Concerto of John Ireland [q.v.] and was a witty and poetic first soloist in the Second Piano Concerto of Alan Rawsthorne [q.v.] during the Festival of Britain in 1951. In 1946 he introduced (Sir) Lennox Berkeley's Piano Sonata which is dedicated to him. A wartime friendship with Benjamin (later Lord) Britten [q.v.] found them giving concerts as a two-piano team for which Britten composed the *Scottish Ballad*, premièred by them at the Proms in 1944. Later, at Britten's Aldeburgh Festival, Curzon was often a visiting soloist.

In America, which he visited for the first time in 1938, Curzon continued regularly to play a large repertory. His concert schedule was calculated to allow for lengthy preparation with frequent intervals for sabbatical study. In Britain in 1945 he concentrated increasingly on that 'music of the first quality' which he had ignored in his youth—Mozart, Beethoven, Schubert, Brahms. It had been Schnabel's repertory; Curzon played, not in Schnabel's way, which was sometimes uncommunicative, but frankly, generously, yet with the utmost attention to every note and its relative weight in context. The virtues which he applied to Mozart's piano concertos—he regarded them as the most perfect music ever composed—included line-drawing that colours itself and a control of structure through harmony and feeling for ensemble, which was overwhelming when the conductor was sympathetic. He achieved them with Britten often, and also with Daniel Barenboim and Sir Colin Davis. In chamber music he gave unforgettable readings of Schubert's Trout Quintet, Dvorak's and Elgar's Piano Quintets, and the Mozart and Schumann concerted works with piano. Curzon seldom played chamber music at public concerts, but it was evident that chamber music was a necessary element of his art. He was an ideal host, a lively raconteur, a keen connoisseur of painting and literature, and appreciative of other countries and their cultures, food, drink, and language. On the concert platform he appeared nervous in his middle years (he always played from score) but latterly had learned to calculate every note for perfect effect and when he was clearly no longer physically powerful, his mastery of the piano seemed even more magical and potent.

Curzon was awarded many honours, notably honorary doctorates in music at Leeds (1970) and Sussex (1973) and the gold medal of the Royal Philharmonic Society in 1980. He was appointed CBE in 1958 and knighted (a rare honour for a pianist) in 1977. He died in London 1 September 1982.

[*The Times*, 3 September 1982; Max Loppert in *The New Grove Dictionary of Music and Musicians*, 1980 (ed. Stanley Sadie); A. Blyth, 'Clifford Curzon', *Gramophone*, vol. xlviii, 1971, p. 1794; information from Kenneth Loveland.]   WILLIAM MANN

D

DANCKWERTS, PETER VICTOR (1916–1984), chemical engineer, was born 14 October 1916 at Emsworth, Hampshire, the eldest of five children (three sons and two daughters) of (Rear-Admiral) Victor Hilary Danckwerts CMG, of the Royal Navy, and his wife, Joyce Middleton. He came from a family with distinguished naval and legal experience. One of his grandfathers was a highly successful QC and an uncle became a lord justice of appeal. Nevertheless his own inclination, as a boy, was towards neither the navy nor the law, but rather to chemistry. He constructed his own laboratory in an attic at his home. He was educated at Winchester College and at Balliol College, Oxford, where he obtained first class honours in chemistry (1939). He held a post in a small chemical company in 1939–40.

With the coming of World War II in 1939 Danckwerts became a sub-lieutenant in the Royal Naval Volunteer Reserve for training in bomb disposal. He was subsequently posted as bomb disposal officer to the Port of London Authority in time for the beginning of the blitz in September 1940. Although he had learned about the defusing of bombs his training had not included anything on magnetic mines whose use against cities had evidently not been anticipated by the Civil Defence authorities. When some of these bombs were dropped on a south London suburb Danckwerts was informed of it by telephone, and he there and then volunteered to attempt defusing them. This he did at great personal risk. On one occasion he worked for two days almost without rest and dealt with sixteen mines during the period. For his outstanding bravery he was awarded the George Cross (1940). Later in the war he was transferred to bomb disposal work abroad and was wounded in a minefield in Sicily. Following this episode he joined the Combined Operations headquarters in Whitehall and was subsequently appointed MBE (1943).

At the end of the war Danckwerts's interest in chemistry reasserted itself, but he decided to take up the closely related subject of chemical engineering as offering better prospects. Since chemical engineering was much more firmly established in the USA than in Britain, he decided to apply for a Commonwealth Fund fellowship for the purpose of obtaining a qualification in the subject at the Massachusetts Institute of Technology. Having been awarded the fellowship he stayed at MIT for two years and obtained the Master's degree.

Danckwerts's return to Britain in 1948 was at a fortunate moment because, due to a generous benefaction to the university by the Shell group of companies, a new department of chemical engineering was in the process of being formed at Cambridge, under the leadership of T. R. C. Fox [q.v.]. Danckwerts did much excellent research at Cambridge, but he was conscious that, on the teaching side, he had insufficient industrial experience. For this reason he accepted, in 1954, an invitation to join the Industrial Group of the UK Atomic Energy Authority at Risley, as deputy to Leonard Rotherham. However his stay at Risley was short-lived for in 1956 he was appointed to a professorship of chemical engineering science at Imperial College, a newly created chair within the department of which D. M. Newitt [q.v.] was the head. In this post Danckwerts continued with research and teaching and also played an active part in the affairs of the college.

In 1959 T. R. C. Fox resigned from the Shell chair at Cambridge and Danckwerts was elected in his place. He thus reached his final appointment at the early age of forty-two. He made many innovations in the Cambridge teaching course and also did research of great originality, especially in the fields of mixing phenomena and gas absorption. He visited the USA several times and gave lectures in France, Holland, India, Italy, Japan, and the USSR. Indeed he was much sought after as an international leader, and it was thus natural that he was elected FRS in 1969. Several other honours came to him, notably honorary degrees of the universities of Bradford (1978), Loughborough (1981), and Bath (1983), and the foreign associateship of the National Academy of Engineering, USA (1978). In 1959 he was elected to a professorial fellowship at Pembroke College, Cambridge.

Danckwerts was the president of the Institution of Chemical Engineers during 1965 and 1966. Another great service to his profession was his acting as executive editor of the journal *Chemical Engineering Science*, and he continued with this work until 1983, six years after he had already retired (1977) from the Shell chair due to a prolonged illness.

Danckwerts had a complex personality and was not an easy person to know. He was very reserved, even aloof, and yet had a strong sense of humour. His outstanding talents were combined with personal charm and considerable forcefulness of character. In 1960 he married Lavinia, daughter of Brigadier-General Duncan Alwyn Macfarlane. They had no children. Danckwerts died in Cambridge 25 October 1984.

[K. G. Denbigh in *Biographical Memoirs of*

Fellows of the Royal Society, vol. xxxii, 1986; private information; personal knowledge.]
K. G. DENBIGH

DANIELLI, JAMES FREDERIC (1911–1984), biologist, was born at 36 Swinderby Road, Wembley, London, 13 November 1911, the only son and the elder child of James Frederic Danielli, civil servant, of Alperton near Wembley, and his wife, Helena Mary Hollins. He was educated at Wembley County School and University College, London, and held a Commonwealth Fund fellowship at Princeton University in 1933–5, where he began his well-known work on the structure of the living cell membrane, returning in 1935 to University College, London, where he extended this research to the problems of membrane permeability and function.

In 1938 he moved to Cambridge University to work on oedema, becoming a fellow of St John's College in 1942. With the outbreak of World War II he became involved in defence research, initially on the problems of wound healing, and later in finding an antidote to the chemical warfare poison Lewisite.

In 1946 he became reader in cell physiology at the Royal Cancer Hospital, London, and in 1949 he became professor of zoology at King's College, London. There Danielli rapidly collected round himself a team of active young research workers, and enormously widened the scope of the department. Amongst many other things, he extended his work on the cell membrane, developed methods for the quantitative study of cellular chemistry, and jointly with the Chester Beatty Cancer Research Institute became involved in the development of anti-cancer drugs. He was elected FRS in 1957.

In 1962 he became chairman of the department of biological pharmacology in the rapidly expanding University of Buffalo, shortly to become the State University of New York. In 1965 he was appointed director of the university's new interdisciplinary Centre for Theoretical Biology, and became involved, amongst much else, in exploring the possibility of life on other planets. But the venture began to falter as state funding diminished, and in 1974 he became chairman of life sciences at the Worcester Polytechnic Institute (Massachusetts). He finally retired in 1980, but as late as 1982, when his health was already failing, he was actively campaigning for a cause that later became fashionable, namely the preservation of the DNA of endangered species, so that posterity could recreate them if they became extinct.

Danielli's place in twentieth-century biology is unusual. He was not one of the very few who made dramatic discoveries that stood the test of time unaltered. Nor was he simply one of the many who only made lesser discoveries that became minor elements in the growing structure of knowledge. His life's work can perhaps be best described as providing a constant stimulus to research over a remarkably wide area. His best known work on the structure and behaviour of the cell membrane was regarded as seminal, but it was to be extensively amended and refined by others as the years went by. His work on quantitative methods in cell chemistry was a notable stimulus to others as was his work on anti-cancer agents. His research on the transplantation of nuclei from one amoeba to another likewise prompted much research that ultimately threw new light on many aspects of the living cell, most notably on the function of the nucleus and the mechanisms of developmental biology.

It was however not only his research that provided these stimuli. In all his jobs he collected round himself a great many young workers who went on to make their mark world-wide. His immense activity in founding and editing scientific journals meant that his ideas spread out to the whole biological community. These included the *International Review of Cytology*, *Progress in Surface and Membrane Science*, the *Journal of Molecular Biology*, and the *Journal of Theoretical Biology*. The recognition of cell biology as one of the most important new areas of research owed much to his immense energy and constant flow of ideas, invariably stimulating, if often controversial. Danielli was a kind, generous, imaginative, and quixotic man to whom young research workers flocked.

In 1937 he married Mary, a poet and anthropologist, the daughter of Herbert Spencer Guy, accountant. They had a son and a daughter. Danielli died, after a long illness, at Houston, Texas, 22 April 1984.

[W. D. Stein in *Biographical Memoirs of Fellows of the Royal Society*, vol. xxxii, 1986; private information; personal knowledge.]
MICHAEL SWANN

DARLINGTON, CYRIL DEAN (1903–1981), cytologist and geneticist, was born in Chorley, Lancashire, 19 December 1903, the younger son (there were no daughters) of Henry Robertson Darlington and his wife, Ellen Frankland. His father was a schoolmaster at Chorley until 1911 when he was appointed secretary to the chief chemist of Crosswells Soap Ltd. and moved to Ealing, London. He attended Heathside Elementary School and Boteler Grammar School, both at Warrington, Lancashire. In London he attended Mercer's School, Holborn, and was a foundation scholar at St Paul's School. In 1920 he was enrolled at South Eastern Agricultural College, Wye,

Kent, obtaining a B.Sc. (Agric., London) in 1923. Darlington's future career and life's work was inspired at college when he read *The Physical basis of Heredity* by T. H. Morgan, A. H. Sturtevant, and C. B. Bridges (1921).

He was accepted in 1923 as a volunteer unpaid worker at the John Innes Horticultural Institution, Merton Park, London, by its director, William Bateson [q.v.]. He rose to become the head of the cytology department when it was first founded in 1939, and from 1939 to 1953 was director of the institution. Given every encouragement and a free hand to work on the physical basis of heredity, the chromosomes Darlington demolished most of the contemporary theories of their structure and behaviour, replacing them with his own, particularly the theory of chromosome pairing and crossing-over. His work culminated in a comprehensive book, *Recent Advances in Cytology* (1932), which had great influence and attracted many talented students to work in Darlington's laboratory. In consequence, in the 1930s the John Innes Horticultural Institution became the nursery of genetics in Britain because it supplied eight professors to establish genetics in the universities where, with two exceptions, it had been neglected.

Darlington's influence was universally recognized: a *Dictionary of Genetic Terms* (ed. R. Rieger, A. Michaelis, and M. N. Green, 1968) contained sixty terms including seven theories attributed to him. The study of the cell, *cytology*, had become the study of only a part of the cell, the chromosomes, and the term *cell biology* had to be adopted. A study of sex chromosomes led to a broader enquiry into breeding systems and the novel idea that the chromosomes were themselves objects of evolution and selection. These ideas were synthesized into Darlington's masterpiece, *The Evolution of Genetic Systems* (1939). Among the many books he wrote was a compilation with E. K. Janaki Ammal, *Chromosome Atlas of Cultivated Plants* (1948) and a book on technique with L. F. La Cour, *The Handling of Chromosomes* (1942), both of which were factual and uncontroversial unlike his other books.

Resigning from the directorship and appointed to the Sherardian chair of botany at Oxford University in 1953, Darlington encouraged students and staff to study chromosomes. He also organized chromosome symposia which evolved into international chromosome conferences. His personal work turned to the genetics of man and society with *The Facts of Life* (1953) and *The Evolution of Man and Society* (1969). His final conclusion was that innovation (genius) arises from outbreeding and the environment, and to preserve man's future we must preserve diverse environments

and diverse people. Outbreeding preserves the necessary uncertainty principle generated by the randomness of chromosome segregation and crossing-over, whereas the environment provides the interacting challenge. Darlington held his chair from 1953 to 1971, at the same time being a fellow of Magdalen College and keeper of the Oxford Botanic Garden.

A prolific writer himself, he encouraged others with the remark 'work not published is work not done'. To fulfil a need after the only British *Journal of Genetics* became irregular and finally extinct when J. B. S. Haldane [q.v.], its owner, left to become an Indian citizen, Darlington with (Sir) R. A. Fisher [q.v.] in 1947 started a new journal *Heredity* which was later presented as a gift to the Genetical Society. Darlington was a fearless critic of the establishment and derided the work of committees. He led the exposure of the extermination of genetics and the murder of geneticists in the USSR by Stalin. He detested organized games, had a fine appreciation of the arts and literature, and loved gardens and gardening, particularly at his last home at south Hinksey, Oxford.

He became FRS in 1941 and was awarded the Royal medal in 1946. In 1951 he was given the Trail award of the Linnean Society and in 1956 an Oxford D.Sc. He became an honorary fellow of Magdalen College, Oxford, in 1971. He was president of the Genetical Society (1943-6) and of the Rationalist Press Association (1948).

In 1939 he married his cytological collaborator, Margaret Blanche, daughter of Sir Gilbert Charles Upcott, civil servant, comptroller and auditor general to the Exchequer. They had two sons, the younger of whom died in 1970, and three daughters. The marriage was dissolved in 1949 and in 1950 he married Gwendolen Mabel Harvey, the daughter of Edward Davies Adshead, bank clerk, and former wife of John Dean Monroe Harvey, architect and perspective artist. Darlington died 26 March 1981 in Oxford.

[D. Lewis in *Biographical Memoirs of Fellows of the Royal Society*, vol. xxix, 1983; private information; personal knowledge.]

DAN LEWIS

DAVIDSON, (FRANCES) JOAN, BARONESS NORTHCHURCH and VISCOUNTESS DAVIDSON (1894-1985), politician, was born in Kensington, London, 29 May 1894, the younger daughter and second of three children of Willoughby Hyett Dickinson, first Baron Dickinson, barrister and MP, and his wife, Minnie Elizabeth Gordon Cumming, daughter of General Sir Richard John Meade, who served in India for forty-six years and raised Meade's Horse in the

Indian mutiny of 1857. She was educated at Kensington High School, at Northfields, and in Germany.

In World War I she served in the wounded and missing department of the Red Cross. She was appointed OBE in 1920. Stanley Baldwin (later Earl Baldwin of Bewdley) introduced Joan Dickinson to John Colin Campbell Davidson [q.v.], the son of (Sir) James Mackenzie Davidson, surgeon. He was then parliamentary private secretary to Andrew Bonar Law. A few months later, on New Year's eve 1918, they became engaged in the Baldwins' house, and they married in April 1919. Baldwin remained friendly with them until his death, and he wrote on average three times a week for twenty-five years to Joan Davidson ('Mimi' to many of her friends and family).

In 1920 John Davidson was elected MP for Hemel Hempstead, and at once his wife took on his constituency, releasing her husband for his work in Parliament and government, as well as giving him time for special work with Baldwin, whose parliamentary private secretary he became.

When Joan Davidson's husband became first Viscount Davidson in 1937 the Hemel Hempstead constituency unanimously asked her to stand as their MP. Duly elected, after canvassing on horseback, she became the only Conservative woman to hold her seat in the general election of 1945. She also retained it in February 1950, October 1951, and May 1955. Possessing much warmth, enthusiasm, and personal charm, she was greatly loved by her Labour women colleagues and respected by her fellow members and the House of Commons staff.

She was a member of the kitchen committee of the House of Commons during the war years of 1939-45. She started parties in Westminster, to which the leader of the Conservative Party always came, for the wives of Conservative candidates in order to interest them in their husbands' work. She was the only woman MP to be a member of the national expenditure committee throughout World War II. After the war she served continuously on the estimates committee of the House of Commons until the beginning of 1957, when she was obliged, through pressure of work, to resign. She was the first woman to be a member of the executive of the 1922 Committee of the House of Commons, and was re-elected to it in June 1955. She was also an elected member of the Inter-Parliamentary Union executive committee. Senior officials of her party frequently sought her advice.

In 1955 she became a member of the council of the National Union of the Conservative Party, a position she held until her death. She started the Young Britons, the junior branch of the Young Conservatives, and she also served on the policy committee of the Conservative Party. In 1964-5 she was president of the National Union of the Conservative and Unionist Association, and chairman of the party conference. She piloted the Anaesthetics for Animals Bill through the House of Commons in 1955, the first Act of Parliament making it compulsory for anaesthetics to be used when research and experiments were carried out on animals.

Lady Davidson decided to stand down in the 1959 election, after twenty-two years in the Commons (and a period of forty years when she and her husband had represented Hemel Hempstead). She had been appointed DBE in 1952 and in 1963 she was created a life peer, as Baroness Northchurch. The Davidsons were the first husband and wife both to be made peers. John Davidson died in 1970.

All Joan Davidson's life in London was spent in Westminster. She lived first in an eighteenth-century house in Great College Street for some forty years, and then in a smaller but equally attractive house in Lord North Street. This meant she was a short walking distance from the Houses of Parliament, and also almost within the precincts of Westminster Abbey, where she attended services every Sunday when she was in London. The Davidsons were an exceptional partnership, in public and in private life. They had two sons and two daughters. Baroness Northchurch died 25 November 1985 at Great Leighs, near Chelmsford, Essex.

[Personal knowledge.] ELLIOT OF HARWOOD

DAVIDSON, SIR (LEYBOURNE) STANLEY (PATRICK) (1894-1981), physician and professor of medicine, was born in Ceylon 3 March 1894, the second son in the family of three sons and one daughter of (Sir) Leybourne Francis Watson Davidson, later of Huntly Lodge, Aberdeenshire, coffee, tea, and rubber planter, and his wife Jane Rosalind Dudgeon Brown. He was educated at Cheltenham College and went to Trinity College, Cambridge, and to Edinburgh to read medicine. His studies were interrupted on the outbreak of World War I and he served for three years with the Gordon Highlanders in France and Belgium, where he was seriously wounded. Returning to Edinburgh in 1917 he then completed his medical studies graduating MB, Ch.B. with honours in 1919.

He held house physician appointments in Edinburgh hospitals and in 1923 became lecturer in bacteriology in Edinburgh University. He obtained a gold medal for his MD thesis (1925) on 'Immunization and Antibody Reactions' and became a fellow of the Royal College of Physicians of Edinburgh in 1926 (MRCP 1921) and of the Royal Society of Edinburgh in 1932. Although athletic by nature he was advised to seek a warm climate in the winter months. He made several voyages to New Zealand on cargo boats where he enjoyed shark and sword fishing.

He probably decided to take medicine seriously at this time and he was appointed as assistant physician to the Royal Infirmary in Edinburgh and pursued research in haematology. Intuitively thinking that pernicious anaemia was related to a disturbance in the gastro-intestinal tract he wrote a monograph on the subject with G. L. Gulland in 1930 summarizing the knowledge of the time. To the surprise of some of his close friends and possessing limited clinical experience he was appointed regius professor of medicine at Aberdeen University in 1930. There he collected around him a number of talented young men; he applied vital stains to blood films, drew attention to the reticulocyte response in the correction of anaemia, established the role of iron deficiency in the condition then referred to as 'chlorosis', and showed the importance of Weil's disease among fish workers in Aberdeen. In 1942 he wrote *A Textbook of Dietetics* with I. A. Anderson. Its successor, *Human Nutrition and Dietetics* (1959) ran into many editions.

With an established reputation in clinical research, and rapidly acquired clinical and teaching skills, he was appointed to the chair of medicine in Edinburgh in 1938. Initially he conducted almost all the undergraduate lectures himself. At the request of the secretary of state he organized the emergency medical services (in six new hutted hospitals of over 1,000 beds each) throughout Scotland during World War II. He also served on many national committees on food production with John (later Lord) Boyd Orr [q.v.].

In 1946 he turned his considerable administrative abilities to the upgrading of several local authority hospitals in Edinburgh. Anticipating the shape of modern medicine, he encouraged the development of major medical specialities of cardiology, gastro-enterology, nutrition, rheumatology, respiratory medicine, and neurology, as well as general medicine. He became a fellow of the Royal College of Physicians of London in 1940, was awarded an honorary MD of Oslo University in 1946, was a distinguished president of the Royal College of Physicians of Edinburgh from 1953 to 1957, received a knighthood in 1955, and was president of the Association of Physicians of Great Britain and Ireland in 1957. He was physician to King George VI (1947-52) and to the Queen in Scotland (1952-61). He had an honorary LLD from both Edinburgh (1962) and Aberdeen (1971) universities.

Davidson had a remarkable personality which all who met him were quick to appreciate. Not an intellectual, he depended largely on his younger colleagues to keep abreast of scientific advances. He possessed a penetrating mind which brushed jargon aside and quickly grasped the essentials of diverse problems. He loved to talk with specialists who were astonished at his capacity to simplify and uncover contradictions in their accounts of their own specialities. He was an excellent editor of undergraduate texts, the *Principles and Practice of Medicine* (1952) running into nine editions before he retired. His success was largely due to his tireless efforts to put into uncomplicated English the tortured drafts submitted by the members of his staff. He was unaffected and sincere with all who met him; he had a remarkable capacity to select young men of talent, many of whom later occupied chairs of medicine in Britain and overseas. Firm and scrupulously fair in his decisions, he was occasionally ruthless but decisions were always accompanied with a charming smile. He was a keen fisherman, golfer, and shot. He retired in 1959.

In 1927 he married Isobel Margaret ('Peggy') (died 1979), daughter of Andrew Macbeth Anderson (Lord Anderson), senator of the College of Justice and solicitor general for Scotland (1911-13). They had no children. Davidson died in Edinburgh 22 September 1981.

[R. Passmore in Royal Society of Medicine *Year Book*, 1983, pp. 142-8; obituaries in *British Medical Journal* and *Lancet*, October 1981; appendix to minutes of the senate meeting of Edinburgh University, 18 November 1959, pp. 87-9; private information; personal knowledge.]   J. S. ROBSON

DAVIES, RICHARD LLEWELYN, BARON LLEWELYN-DAVIES (1912-1981), architect. [See LLEWELYN DAVIES.]

DEACON, SIR GEORGE EDWARD RAVEN (1906-1984), oceanographer, was born 21 March 1906 at Leicester, the younger child and only son of George Raven Deacon, a boot and shoe factory worker, of Leicester, and his wife, Emma, daughter of David Drinkwater, stone quarrier, from Enderby, near Leicester. He was educated at Newarke Secondary School, Leicester; the City of Leicester Boys' School; and, as a King's scholar, at King's College, London, where he was awarded a first class honours degree in chemistry in 1926 and a diploma of education in 1927. He was awarded a D.Sc. degree by London University in 1937.

After a brief spell of teaching chemistry and mathematics at Rochdale Technical School Deacon was appointed as a hydrographer to the *Discovery* committee, a post that involved making accurate determination of the temperature and salinity of the sea from small research vessels in the stormy Southern Ocean. He worked in the ships *William Scoresby* in 1927–9 and *Discovery II* in 1929–31, 1931–3, and 1935–7 when he served as principal scientist. His observations provided the basis for two important *Discovery* reports on the hydrology of the Southern Ocean.

In 1939 Deacon was seconded by the *Discovery* committee to work for the Admiralty at HMS *Osprey*, a shore establishment at Portland concerned with Asdic and other anti-submarine equipment. His office was destroyed by a bomb during an air raid and he and his work on underwater sound transferred to Fairlie, Ayrshire, in 1940. From there he was sent in 1944 to the Admiralty research laboratory in Teddington, Middlesex, to lead a small group (Group W) that had been set up to study ocean waves and swell.

He encouraged his younger colleagues to combine theory, observations, and analysis (using a specially developed frequency analyser) and rapid progress was made in understanding the propagation of the energy in storm waves over distances as large as thousands of kilometres. Deacon left the employment of the *Discovery* committee in 1947 to take up his first permanent post, in the Royal Naval Scientific Service.

By then his group had branched out into other aspects of physical oceanography and it formed a major component of the National Institute of Oceanography when it was set up in 1949 with Deacon as director. His institute at Wormley, Surrey, rapidly acquired an international reputation in all branches of marine science. His contribution was to encourage his staff and to shield them from administrative distraction while he struggled to obtain funds for their long-term research.

In 1966 the institute became a component body of the Natural Environment Research Council, which did not please him: he formally retired in 1971 but continued to work in the institute, publishing papers and a book about the Southern Ocean and remaining active until his final short illness.

Deacon was a kind man whose modest manner belied his active mind, his determination to succeed, and a proper pride in his achievement. He was a major figure in the modern development of marine science.

Deacon was awarded a Polar medal in 1942, was appointed CBE in 1954, and was knighted in 1971. He was elected FRS in 1944, FRSE in 1957, and a foreign member of the Swedish Royal Academy in 1958. His elections to honorary membership included those of the Royal

Society of New Zealand (1964), the Challenger Society (1971), the Marine Biological Association (1973), the Royal Institute of Navigation (1975), the Scottish Marine Biological Association (1976), and the Royal Meteorological Society (1982). He was made a fellow of King's College in 1948 and given honorary D.Scs. by the universities of Liverpool in 1961 and Leicester in 1970. He was awarded the Alexander Agassiz medal of the US National Academy of Sciences (1962), a Royal medal of the Royal Society (1969), the Albert memorial medal of the Institut Océanographique, Monaco (1970), a Founder's medal of the Royal Geographical Society (1971), and a Scottish Geographical Society medal (1972). The American Miscellaneous Society presented him with their Albatross award in 1982.

In May 1940 Deacon married Margaret Elsa (died 1966), daughter of Charles David Jeffries, a company secretary. They had one daughter, Margaret Brenda Deacon, who became an expert in the history of oceanography. Deacon died in hospital at Guildford 16 November 1984.

[M. B. Deacon, 'Sir George Deacon, British Oceanographer', *Oceanus*, vol. xx, pp. 30–4; H. Charnock in *Biographical Memoirs of Fellows of the Royal Society*, vol. xxxi, 1985; private information; personal knowledge.]

HENRY CHARNOCK

DEE, PHILIP IVOR (1904–1983), physicist, was born 8 April 1904 in Stroud, the second of three sons (there were no daughters) of Albert John Dee, schoolmaster in Cainscross, and his wife, Maria Kitchen Tiley, whose father, William Tiley, was a butcher of Ebley. He was educated at Marling School, Stroud, and, as a scholar, at Sidney Sussex College, Cambridge, where he obtained a first class in both parts of the natural sciences tripos (1925 and 1926). He then became Stokes student at Pembroke College, Cambridge (1930–3). In this period he strongly impressed Lord Rutherford of Nelson [q.v.], head of the Cavendish Laboratory. He had started research under C. T. R. Wilson [q.v.] and by 1931 could use cloud chambers with wonderful skill. He provided new insight into their mode of operation and extended their potential uses. His mastery of technique meant he could prove with certainty the lack of significant interaction of neutrons with electrons. The great discoveries in 1932 by (Sir) James Chadwick and by (Sir) John Cockcroft [qq.v.] and E. T. S. Walton persuaded him to switch to study transmutation of atoms by bombardment with accelerated ions. His own early exciting and vivid research concerned the two modes of reaction between deuterons to yield tritium and protons or helium 3 and neutrons. The results of bombardment of lithium 6 with deuterons and of lithium 7 with protons followed and he proved elegantly that fast protons with boron 11 gave 3 helium nuclei. He became a fellow of Sidney Sussex (honorary fellow, 1948) and university lecturer in 1934 and later a deputy of Lord Rutherford, assuming responsibility for the construction of 1.2 MeV and 2.0 MeV accelerators. Notable researches with these ensued between 1937 and 1939. Dee was a lecturer in physics at the Clarendon Laboratory from 1934 to 1943.

The wartime phase of his work was of great national importance. As a superintendent of the Ministry of Aircraft Production TRE (telecommunications research establishment) from 1939 to 1945 he took charge of most of the work on centimetric devices. Leading a team of brilliant young scientists he proved new radar equipments for AI (aircraft interception), ASV (anti-surface vessel), and H2S (blind bombing) gave excellent performance. He was eminently practical in this phase and contacts with politicians, Service officers of all ranks, and fellow scientists proved equally easy. He was a great catalyst whose unflagging energy inspired all connected with radar. Honours came to him, among them fellowship of the Royal Society (1941), OBE (1943), CBE (1946), and the Hughes medal (1952). Informed outsiders see the welding by Dee of a disparate body of scientists into a creative team as his outstanding achievement.

In 1943 he was appointed professor of natural philosophy at Glasgow University and from 1945 he began to modernize physics there. Cambridge University soon approached him but he decided to remain in Glasgow to fulfil what he saw as an obligation. In fact he did much to ensure CERN at Geneva was created in its university-orientated form and he served important national organizations. He constructed a modest DC accelerator, and then a 30 MeV synchrotron followed by a 300 MeV machine. Later at the Kelvin laboratory he built a large linear accelerator yielding electrons of 160 MeV energy. His Glasgow department of physics, from which he retired in 1972, became one of the leaders in nuclear science and thus Dee achieved another major goal. Both the work done and quality of its graduates testified to this success. Dee was awarded an honorary D.Sc. by the University of Strathclyde (1980).

Dee was very dark, over six feet tall, slimly built, and with his commanding presence and direct manner of approach he was held in awe by most of his contemporaries and certainly by his students. Those who came to know him well realized his sincere interest in their well-being.

In 1929 Dee married Phyllis Elsie, daughter of George Williams Tyte, clockmaker. They had two daughters. Dee died in Glasgow 17 April 1983.

[Sir Samuel Curran in *Biographical Memoirs of Fellows of the Royal Society*, vol. xxx, 1984; P. I. Dee, Rutherford memorial lecture, 1965, *Proceedings* of the Royal Society, A298, p. 103; R. V. Jones, *Most Secret War*, 1978; private information from Mrs Anne Buckler, daughter.]   SAMUEL C. CURRAN

DEMANT, VIGO AUGUSTE (1893-1983), theologian and social critic, was born 8 November 1893 in Newcastle upon Tyne, the eldest of three children and only son of Thorvald Conrad Frederick Demant, a professional linguist and translator, who started a language school in Newcastle, and his wife, Emilie Thora Wildemann. On the paternal side he came from a line of Huguenot organ builders and piano makers who had worked their way northwards to settle in Denmark. T. C. F. Demant was a Unitarian and a follower of Auguste Comte.

After school in Newcastle Demant was sent on a six-month exchange to Tournan, to finish his schooling in France. He spent some time at the Sorbonne then returned to the north-east to study engineering at Armstrong College (later to become Durham University). Having obtained his B.Sc. and wishing to enter the Unitarian ministry Demant moved to Oxford in 1916 as a member of Exeter and Manchester colleges, taking the diploma in anthropology. He briefly ministered to a Unitarian congregation in Newbury, but meanwhile had met Charles Gore [q.v.]. Gore was a major influence, and through him Demant was converted to the Church of England and Anglo-Catholicism. After a short time at Ely Theological College he was ordained deacon in 1919 and priest in 1920.

He became a curate at St Thomas's in Oxford (1919-23), St Michael's in Summertown, Oxford (1923-4), and the mission of St Nicholas, Plumstead (1924-6). In 1926 Demant met M. B. Reckitt [q.v.], a prominent Anglican layman and author. Demant and Reckitt formed a close association which continued for the rest of their lives. Through Reckitt's patronage Demant was introduced into membership of the exclusive Chandos circle, which met regularly from 1926, surviving to the 1970s. Reckitt also encouraged Demant to use his abilities within a wider sphere than the parochial ministry, and in 1929, while based at St Silas's, Kentish Town, Demant became director of research for the Christian Social Council, which had emerged from COPEC (the Conference on Politics, Economics, and Citizenship). Demant produced as reports *The Miners' Distress and the Coal Problem* (1929),

The *Just Price* (1930), and *This Unemployment: Disaster or Opportunity?* (1931). In 1933 he became vicar of St John's, Richmond (Surrey), moving to St Paul's as canon residentiary in 1942. In 1949 he returned to Oxford as canon of Christ Church and regius professor of moral and pastoral theology, retiring to Headington in 1971.

Demant was the major theoretician in the Christendom Group of Anglican Catholic thinkers whose concern was to establish the centrality of what they termed 'Christian sociology', an analysis of society fundamentally rooted in a Catholic and incarnational theology. The group included Reckitt, W. G. Peck, and P. E. T. Widdrington, and in addition to numerous books their ideas were propounded in a quarterly journal, *Christendom*, which ran from 1931 to 1950. Annual summer schools sought to extend their influence further. William Temple [q.v.] was often an ally, but the Christendom Group placed greater emphasis on the need to derive Christian principles about society from Christian doctrine. Demant was one of the speakers at Temple's 1941 Malvern conference, and Malvern probably marked the high point of the Christendom Group's influence on the Church of England.

Demant was at home across a range of subjects and languages unusual in the Church of England of his time. His upbringing had been Darwinian. He had received an early training in Henri Bergson's philosophy. Much of this he rejected, but it gave him an access to Continental thought. He helped introduce the work of N. Berdyaev and J. Maritain to Anglican circles, and he also brought the thought of S. Kierkegaard before an English audience. His reading embraced anthropology, economics, sociology, philosophy, and theology, in French, German, and Danish. His clarity of mind fed on this varied diet and produced a coherent theology which carried forward an Anglican Catholic tradition of social thought.

Demant's conviction that an incarnational theology cannot be separated from the condition of human beings in society, and that the problems of society are structural rather than individual, can be seen in some of the titles of books written in the London years: *This Unemployment* (1931), *God, Man and Society* (1933), and *Theology of Society* (1947). One of the most influential was *Religion and the Decline of Capitalism*, the Scott Holland lectures for 1949, broadcast in 1950 and published in 1952. Taking up the theme earlier expounded by R. H. Tawney [q.v.], Demant examined contemporary religion and society to argue that capitalism was inherently self-destructive, being inimical to organic association, and thus contrary to human nature and needs. In 1957-8 he

delivered the Gifford lectures at St Andrews, surveying Christian ethics, but he chose not to publish them.

Demant was a sensitive and kindly priest, and an approachable and helpful figure to students and other scholars. He much valued Christ Church life, and became sub-dean of the Cathedral. He obtained an Oxford D.Litt. (1940) and an honorary DD from Durham University.

In 1925 he married Marjorie, daughter of George Tickner, zoologist; they had one son and two daughters. Demant's programme of reading and writing, in tandem with his priestly duties, required a rigid daily timetable but within this ordered framework he was much involved in family life. This setting also provided an outlet for his considerable practical talents, such as carpentry and kite making. A persistent interest was the authorship of the works of Shakespeare: Demant urged the claims of Edward De Vere, Earl of Oxford [q.v.], even travelling to the USA to lecture on the theme. He died 3 March 1983 in the John Radcliffe Hospital, Oxford.

[*The Times*, 17 March 1983; private information.]   ANGELA CUNNINGHAM

DIPLOCK, (WILLIAM JOHN) KENNETH, BARON DIPLOCK (1907-1985), judge, was born 8 December 1907 at 8 Barclay Road, South Croydon, the only child of William John Hubert Diplock, solicitor, and his wife, Christine Joan Brooke. He was educated at Whitgift School and University College, Oxford, of which in 1958 he became an honorary fellow and to which he remained devoted throughout his life. He passed chemistry (part i) in 1928 and took a second class in chemistry in 1929. He was called to the bar by the Middle Temple in 1932 where he swiftly made his mark. But in 1939 he left his practice for war service, joining the Royal Air Force two years later and reaching the rank of squadron leader. He returned to the bar in 1945. He took silk in 1948 at the early age of forty-one, and acquired an extensive practice in substantial commercial work and as adviser to Commonwealth governments. When he appeared for the exiled kabaka of Buganda in 1955, the British government had to allow the chief to return home even though the case was lost on a technicality. He was recorder of Oxford from 1951 to 1956. In 1956 he began his judicial career when he was appointed a judge in the Queen's Bench division, with the customary knighthood. Promotion to the Court of Appeal and the Privy Council followed in 1961 and to the House of Lords in 1968.

To his generation Diplock was the quintessential man of the law, serving as a judge with great distinction and complete dedication for nearly half a century. During the two decades in which he served in the House of Lords as a judge, in the opinion of his brethren and the advocates who appeared before the judicial committee he was second only to the Scottish jurist Lord Reid [q.v.] as the most distinguished judge of his day.

His was a supremely logical mind. With due respect for precedent, he would remorselessly pursue this path of logic confident that all legal issues were capable of solution. Many were the reversals in the House of Lords which he initiated on appeals from adventurous experiments by the Court of Appeal. Few thought that he was wrong, although some felt that he was so deft in demonstrating the misuse of logic in the Court of Appeal that he sometimes failed to appreciate that the lower court might have a point which required attention. But if there were, he would have said that that was a matter for Parliament and not for the judiciary. He was immensely industrious (he read Sir W. S. Holdsworth, q.v., for his light literary entertainment), and he was so well versed in the law that in discussion with his brethren some of them claimed that he had an almost hypnotic effect in swaying others to his point of view. At the same time, he always fiercely rejected any advantage gained by a legal trick. He had a clear idea of what the law should do to protect the citizen in dispute with modern authority and a clear concept of the consequences for municipal law arising from the new place of the United Kingdom within Europe. He believed that the progress towards a comprehensive system of administrative law, in which he played such an influential part, was the greatest achievement of the English courts in his judicial lifetime. From 1971 to 1982 he was chairman of the Permanent Security Commission and led inquiries into a number of security scandals, among them the Sir Roger Hollis [q.v.] affair.

In October 1972 the government of Edward Heath, which was anxious over the intimidation of juries in the courts of Northern Ireland, appointed Diplock to chair a commission of three to consider legal proceedings to deal with terrorist activities in the province. By December of that year the Diplock commission was ready with its report which recommended that terrorist offences as defined in a schedule should henceforth be tried by judge alone. This was at once implemented and the courts became known as Diplock courts. The speed and certainty of the recommendations were principally due to the guidance and clarity of the commission's chairman. He gained honorary degrees from Alberta (1972), London (1979), and Oxford (1977).

While his career was the law and his home was in the Temple, Diplock's other enthusiasms were horses and the sport of fox-hunting which he pursued until late in life, indeed too late for the comfort of some lord chancellors. At the time of the Diplock commission he attended one meeting adorned with a resplendent black eye sustained by a fall from his horse. In 1974, when treasurer of the Middle Temple, on one Grand Day he headed his guest list with the Duke of Beaufort [q.v.], master of the horse, and packed it with seven masters of hounds.

He married in 1938 Margaret Sarah, a nurse, the daughter of George Atcheson, who started and owned a shirt factory in Londonderry. The couple had no children. During their long life together they were rarely separated until 1984 when illness and particularly Sarah's loss of memory and ability to look after herself required her admission to hospital. There Diplock would visit her daily but she gradually became incapable of even remembering his visits. Her condition was a source of much grief to him in his last years at a time when he was presiding over the judicial committee of the House of Lords. In the end she survived him.

He died in King Edward VII's Hospital for Officers, London, 14 October 1985, only a few weeks after he had brought to a close his outstanding judicial career.

[Personal knowledge.]   Rawlinson

DIRAC, PAUL ADRIEN MAURICE (1902–1984), theoretical physicist, a founder of quantum mechanics, and Nobel prize-winner, was born at Bishopston, Bristol, 8 August 1902, the younger son of the family of two sons and one daughter of Charles Adrien Ladislas Dirac, a native of Monthey in the canton of Valais, Switzerland, and a teacher of French in the Merchant Venturers' Technical College at Bristol, and his wife, Florence Hannah, daughter of Richard Holten, master mariner in a Bristol ship. Dirac's secondary education was in the Technical College, from which he went on to the University of Bristol, graduating as an electrical engineer in 1921. As he found no job, he accepted the Bristol mathematics department's proposal that he stay on and take their course, and graduated in 1923. During this period he attended lectures on relativity by C. D. Broad [q.v.] which profoundly influenced his later thinking about problems in physics. In the autumn of 1923 he went up to Cambridge, to become a research student in mathematics at St John's College under the supervision of (Sir) R. H. Fowler [q.v.].

In August 1925 Fowler received a proof copy of Werner Heisenberg's seminal paper which was the first spark in the development of modern quantum mechanics, and passed it on to Dirac for study. It contained a mysterious equation, not fully understood by Heisenberg at the time, according to which two quantities P and Q describing a dynamical system, were to be multiplied in such a way that the products P.Q and Q.P are different. After several weeks Dirac suddenly realized the mathematical meaning of their commutator, (P.Q-Q.P), as a generalization of the Poisson Bracket in classical dynamics. This understanding led him to an independent approach to quantum mechanics, apparently different from those developed by Heisenberg (matrix mechanics) and by Erwin Schrödinger (wave mechanics). This was based on an operator formalism, dynamical quantities such as P and Q becoming operators whose results depend on the sequence in which they operate, their commutator being derivable from the classical theory, apart from a factor h, the quantum unit. When h is taken to zero, Dirac's formalism reduces to classical dynamics. Dirac next developed his 'transformation theory', which showed that the Heisenberg and Schrödinger formalisms were special cases of his more general theory. He took his Ph.D. degree at Cambridge in the spring of 1926 on the basis of the earlier parts of this work. His international reputation rose meteorically as these developments proceeded and he was invited to visit Göttingen, Copenhagen, and Leiden, centres for this work, for discussions in the following academic year. In 1927 he was invited to speak at the Solvay conference at Brussels. He was elected a fellow of his college in 1927 and appointed university lecturer in 1929.

While at Copenhagen he began to develop 'second quantization' for boson fields, which he applied to calculate from first principles Albert Einstein's coefficients for stimulated emission and absorption of photons by atoms. This work laid the foundation for quantum field theory, the formalism which later underlay all theoretical work on 'elementary particles'. Dirac's outstandingly significant achievement was his relativistic wave equation for the electron, published early in 1928. His deep concern for relativity drove him to make his quantum mechanics compatible with it. His transformation theory had impressed him with the necessity for such an equation to be of first order in the time derivative, and hence also of first order in space derivatives, if space and time were to appear on the same footing. The equation he obtained satisfied these requirements and implied that the electron should have intrinsic spin $1/2$, as was already known empirically. Sir Nevill Mott has described it as 'the most beautiful and exciting piece of theoretical physics that I have seen in my lifetime—comparable with Maxwell's deduction that the displacement current, and therefore electromagnetism, must

exist'. An even more remarkable prediction from Dirac's equation was the existence of an 'anti-electron', as Dirac termed it in 1931, with the same mass value as the electron but opposite charge. It was first observed in the cosmic radiation in 1932 and was later named the 'positron'. It was the first of the many antiparticles which later became well established.

Dirac's name is also well known for work on many other topics, such as magnetic monopoles, the Large Numbers Hypothesis, the separation of isotopes by diffusion and centrifuge processes (work done during World War II), and the quantization of constrained dynamical systems. These all pale into insignificance in comparison with his equation for the electron.

Dirac was a tall, slender man, with little regard for comfort, a great walker and solitary thinker. He was a legend in his lifetime, and difficult to know. He had no casual conversation and in his responses to others his thoughts followed a path all of their own. He applied strict logic to all aspects of life, often with surprising outcome. In lectures he often spoke about the importance of beauty in the fundamental equations of physics, and there is no doubt among scientists about the beauty of the Dirac equation. He lectured clearly but sparingly in words and with compelling logic; many physicists chose to attend the same lecture by Dirac more than once.

Dirac was elected FRS in 1930 and Lucasian professor of mathematics at Cambridge University in 1932. In 1933 he was awarded the Nobel prize, jointly with Schrödinger, for their discoveries in quantum mechanics. In 1930 Dirac published *The Principles of Quantum Mechanics*, an authoritative text.

From Cambridge Dirac travelled widely to scientific meetings. His early friendship with Piotr Kapitza [q.v.] at Cambridge led him to visit the USSR in 1928 and he returned there almost every year up to 1936. He spent many sabbatical periods in the USA. After retirement in 1969 he took up in 1971 a research professorship at Florida State University, Tallahassee. During his life he received many honours, among them the Copley medal of the Royal Society (1952), membership of the Pontifical Academy of Sciences (1958), and admission to the Order of Merit (1973).

Early in 1937, in London, Dirac married Margit Balazs, daughter of Antal Wigner, manager of a leather factory at Budapest, and sister of the nuclear physicist Eugene Paul Wigner. Her son and daughter from her previous marriage at Budapest also adopted the surname Dirac. The Diracs had two daughters. Dirac died at Tallahassee 20 October 1984.

[R. H. Dalitz and R. E. Peierls in *Bio-*

graphical *Memoirs of Fellows of the Royal Society*, vol. xxxii, 1986; J. Mehra and H. Rechenberg, *The Historical Development of Quantum Theory*, vol. iv, 1982; J. G. Taylor (ed.), *Tributes to Paul Dirac*, 1985; B. Kursunoglu and E. P. Wigner (eds.), *Paul Adrien Maurice Dirac, Reminiscences About a Great Physicist*, 1987; private information.]

R. H. DALITZ

DIXEY, SIR FRANK (1892-1982), geologist, geomorphologist, and hydrogeologist, was born 7 April 1892 at Bedminster, Bristol, the third of three children and second of two sons of George Dixey, journeyman boilermaker and riveter in the ship repair yards of Barry, Glamorgan, and his wife, Mary Nippress. The family moved to Barry when Frank was two, and he grew up, with a father whose favourite recreation was walking, within easy reach of the rocky coast. His elder brother died when Frank was fourteen. From Barry Grammar School he entered University College, Cardiff, to read chemistry and physics with the intention of becoming a teacher, but he changed to geology and gained a first-class degree in 1914.

After a short academic spell he became a gunner in the Royal Garrison Artillery, serving on the western front from 1915 to 1918. He was gazetted out, to go to Sierra Leone and make a reconnaissance survey of the territory. By foot traverse, using compass and barometer, he produced single-handed a geological and topographic map, and he recognized the importance of erosion surfaces in the physiography of Africa.

Having been appointed government geologist to Nyasaland (later Malawi) in 1921 (and later director), the remarkable landscape seen from the headquarters at Zomba stimulated his interest in landscape evolution which became such an important element in his scientific work. He investigated the coal deposits and described the dinosaur beds of Lake Nyasa. He prospected the bauxite deposits of Mlanje mountain, and described the carbonatites, making the first record of these remarkable rocks in Africa, with W. Campbell Smith.

Owing to the economic depression the Geological Survey, in the early 1930s, turned its attention to matters immediately productive, and especially to the provision of groundwater supplies from boreholes. This initiated one of Dixey's major and continuing interests. The influence of his work spread throughout Africa, with hand-pumped boreholes, often drilled in crystalline rocks, contributing greatly to the supply of pure water for rural populations. Dixey set down the knowledge and experience he gained in his *Practical Handbook of Water Supply* (1931, 2nd edn. 1950).

Dixey served in Northern Rhodesia (later Zambia) as director of water development (1939-44), and travelled widely in north-east Africa on advisory services, which further enlarged his experience of African landscape. He then became director of geological survey in Nigeria (1944-7).

The interplay of tectonics, erosion cycles, and sedimentation was the theme explored by Dixey in a series of papers. He recognized, classified, and dated the major planation surfaces and showed that they were developed throughout Africa; also that they were deformed and disrupted by the Rift Valley movements, and that they could be used to elucidate the history of rifting and the geomorphic development of the continent.

After World War II, to meet the demands for increased mineral exploration and mapping, geological surveys were initiated or expanded in many British colonies with funding from the colonial development and welfare funds. A headquarters was set up in London with Dixey as director of colonial (later overseas) geological surveys. This provided (both before and after the territories reached independence) co-ordinated recruiting, and specialist services such as geophysics, geochemistry, and notably photogeology in which training was provided for the many young geologists on their way from the universities to overseas posts. Dixey's practical experience and scientific prestige, as well as his diplomatic personality, contributed to the strength of the organization and to its *ésprit de corps*.

Dixey officially retired in 1959 but thirteen more years of activity lay ahead, much of it serving the United Nations in a consultative capacity, especially in hydrology. He was joint founder of the *Journal of Hydrogeology* and was one of its editors almost until his death. As consultant geologist to the Cyprus government (1967-73) he undertook active field-work in his seventies.

In appearance Dixey was not commanding in stature, and was slightly stooping, with heavy shoulders. He was reserved, almost diffident in manner, and invariably courteous. He was not an easy man to know, had few interests apart from his work, and was kind and thoughtful to his staff. His steady determination, meticulous observation, interest in his subject, and ability to write at length, established him as an important figure in African geology. He was honoured by many learned societies and he served on their councils. He received the Murchison medal (1953) and was elected FRS (1958). He was appointed OBE in 1929, CMG in 1949, and KCMG in 1972.

In 1919 Dixey married (Henrietta Fredrika Alexandra) Helen, daughter of Henry Golding, engineer, of Cardiff. She accompanied him on his overseas work from the early days in Sierra Leone. Dixey's first wife died in 1961 and in 1962 he married Cicely Mary Hepworth, daughter of Herbert Milner, bank branch manager. Dixey died at his home in Bramber, Steyning, Sussex, 1 November 1982.

[K. C. Dunham in *Biographical Memoirs of Fellows of the Royal Society*, vol. xxix, 1983; *The Times*, 8 November 1982; J. W. Pallister in *Annual Report* of the Geological Society, 1983; personal knowledge.]

JOHN V. HEPWORTH

DOLIN, SIR ANTON (1904–1983), ballet dancer, was born (Sydney Francis) Patrick (Chippindall Healey) Kay in Slinfold, Sussex, 27 July 1904, the second of three sons (there were no daughters) of Henry George Kay, amateur cricketer, and later master and owner of the South Coast Harriers and Staghounds, and his wife, Helen Maude Chippindall Healey, from Dublin. Patrick Kay took his first dancing lessons at the age of ten, in Hove, after much pleading with his reluctant father. He went to a Miss Clarice James, then to the two Cone sisters Lily and Grace. His parents moved to London to further his stage training and he auditioned successfully for the Black Cat in *Bluebell in Fairyland* at Christmas 1915. This role was followed by that of John in *Peter Pan*. Dolin then trained as an actor and dancer at the Italia Conti School, which arranged engagements and tours for him.

In August 1917 he began serious ballet training with Seraphine Astafieva. He studied with her for five years during which he was engaged, under the name 'Patrikéeff', by Diaghilev to appear in his 1921 production of *The Sleeping Princess*. More commercial engagements followed and on 26 June 1923 he appeared for the first time as Anton Dolin at the Royal Albert Hall in the 'Anglo-Russian Ballet'. He was by now also studying with Nicolas Legat, who had recently escaped from Russia. In November 1923 he joined Diaghilev's Ballets Russes.

He immediately established himself as an outstanding dancer and was particularly successful as Beau Gosse in *Le Train Bleu* for his Paris début in the summer season of 1924. He made his London début that autumn in the same role.

Less than two years later he left Diaghilev, whose favourite had become Serge Lifar, during the Paris summer season of 1925. He undertook many engagements in revues and musicals in England. In 1927 he danced with Tamara Karsavina at the Coliseum, and with (Dame) Ninette de Valois, in *Whitebirds* at His Majesty's. He also undertook a continental tour with Vera Nemchinova, with whom in 1927–8 he founded the first English Ballet Company, which included (Sir) Frederick Ashton, Harold Turner [q.v.], Mary Skeaping, and Margaret Craske. Dolin created several ballets for his company, the best being *The Nightingale and the Rose*, Chopin's *Revolutionary Étude*, and George Gershwin's *Rhapsody in Blue*. By the end of 1928 he had rejoined Diaghilev in Monte Carlo and danced with more great ballerinas—Olga Spessivtseva in *Lac*, Karsavina in *Petrushka*, and Alexandra Danilova in *Le Bal*. It was while filming in *Dark Red Roses* with Lydia Lopokova [q.v.] in August 1929 that news of Diaghilev's death reached him and he was forced to become free lance again. That autumn he performed at the London Coliseum with Anna Ludmilla, to whom he became engaged. Later she terminated the engagement.

During the 1930s Dolin appeared in various revues and as a soloist in the new Vic-Wells Ballet from 1931 to 1935. He helped to launch the Camargo Society in 1930, for which he created Satan in *Job* (1931). In 1935 he and (Dame) Alicia Markova founded their own ballet company with Bronislava Nijinska as the ballet mistress and Dolin as a director and first soloist. The company, which was financed by Laura Henderson who ran the Windmill Theatre, lasted until December 1937. During the next two years Dolin danced in Paris, Blackpool, Australia, New Zealand, Honolulu, and London.

He spent the war years in America and Australia. He helped build up the American Ballet Theatre for whom he restaged several of the classical ballets and danced the lead in Michel Fokine's *Bluebeard* in Mexico City in 1941. He returned to London in 1948 as a guest star with the Sadler's Wells Ballet. In August 1950 Julian Braunsweg founded the London Festival Ballet, of which Dolin was artistic director and first soloist until 1961. The company travelled extensively, with the young John Gilpin as one of its dancers. Dolin and Gilpin formed a lasting friendship, Dolin outliving Gilpin by only three months.

After leaving the Festival Ballet Dolin directed the ballet at the Rome Opera for two seasons and then became a freelance producer all over the world. He was hospitalized in New York in 1966 for a hernia, but later in 1967 performed as the Devil in Stravinsky's *The Soldier's Tale*. He taught at summer schools, adjudicated at festivals, and acted in his own one-man show *Conversations with Diaghilev*. In 1979 he played the part of the ballet teacher Enrico Cecchetti in Herbert Ross's film *Nijinsky*. He remounted *Giselle* in Iceland in 1982 and taught in Hong Kong and China in 1983. His last engagement was in Houston before he returned to London and Paris where he had a medical check-up in a Paris hospital which pronounced him in good health before he collapsed and died immediately afterwards in Paris 25 November 1983.

Knighted in 1981, Dolin was the recipient of the Royal Academy of Dancing's Queen Elizabeth II award in 1954 and the Order of the Sun from Peru in 1959. He was the author of several books, including autobiographies. He spotted and sponsored many talented young dancers, to whom his generosity was remarkable. He partnered many great ballerinas who knew him as a most courteous partner. He was one of the best *danseurs nobles* in classical ballet and was the first British male dancer to be acclaimed internationally. He was also anxious to take his art to the widest possible public and managed to combine strict classicism with the instinct of a showman. He was unmarried.

[Anton Dolin, *Autobiography*, 1960; Andrew Wheatcroft (compiler), *Dolin, Friends and Memories*, 1982; *The Times*, 27 February 1980; personal knowledge.]   BERYL GREY

DORS, DIANA (1931–1984), actress, was born Diana Mary Fluck 23 October 1931 in Swindon, Wiltshire, the only child of Albert Edward Sidney Fluck, of Swindon, a railway clerk and former army captain, and his wife, Winifred Maud Mary Payne. She was educated at local schools and when she was nine she wrote in a school essay, 'I am going to be a film star, with a swimming pool and a cream telephone.' At thirteen, pretending to be seventeen, she entered a beauty contest and came third, and during World War II she entertained troops at camp concerts.

At fifteen she enrolled at the London Academy of Dramatic Art, where she was spotted in a production and put into films, making her début in a thriller, *The Shop at Sly Corner*, in 1946. After other parts she was offered a ten-year contract by the Rank Organization and she joined the Rank Charm School, which had been set up to discover and groom British stars. She changed her surname to Dors, after her maternal grandmother.

Though as the cousin in the popular Huggett films she hinted at a flair for comedy, her screen career failed to develop and the Rank contract lapsed in 1950. But the publicity machine was already starting to take over. With her long, platinum blonde hair, sensational figure, and colourful private life, she was projected as the British answer to Marilyn Monroe; and for the rest of her life her professional achievements came a very poor second to her status as a celebrity.

The early publicity stunts were masterminded by her first husband, Dennis Hamilton, whom she married in 1951. Born Dennis Hamilton Gittings, he was the son of Stanley Gittings, manager of a public house in Luton. A Svengali figure, ruthless and domineering, he fed the gossip columns with a stream of Dors stories, many of them fabricated. The couple took off for Hollywood. Diana Dors continued to appear in films, most of them forgettable. An exception was *Yield to the Night* (1956), loosely based on the Ruth Ellis case, in which Dors eschewed her usual glamour roles to play a condemned murderess.

It showed her potential as a serious actress, though the public found the switch from blonde bombshell difficult to take. Her marriage foundered, and ended, in the now customary blaze of publicity, in 1957. Dennis Hamilton died in 1959 and in the same year she married in New York an American comedian, Dickie Dawson. They had two sons, Mark and Gary. In 1960 she was paid £35,000 for her memoirs by the *News of the World*. Lurid by the standards of the time, the series ran for twelve weeks. The archbishop of Canterbury denounced her as a wayward hussy.

By now the film parts were getting smaller. She put on weight and the erstwhile sex symbol gave way to a middle-aged mother figure. She had to return to England in 1966 to support her family, for she was sole breadwinner. She played Prince Charming in pantomime and did a cabaret act in the northern clubs. Her private life continued to make the headlines. Her marriage to Dawson ended after eight years and she lost custody of her two sons. In 1968 she married an actor, Alan Lake, son of Cyril Foster Lake, glaze maker. They had a son, Jason. In 1967 she was declared bankrupt, owing the Inland Revenue £48,413 in tax. She admitted to being hopeless with money. In October 1970 Lake was sent to prison for eighteen months for his part in a public house brawl.

Her acting career enjoyed a brief revival when she played a brassy widow in *Three Months Gone* at the Royal Court Theatre in London (1970) and there was a strong part in Jerzy Skolimowski's film, *Deep End* (1970). But a television series written for her, *Queenie's Castle* (1970), proved disappointing. Her Jocasta in Sophocles' *Oedipus* at the Chichester Festival in 1974 was a brave, but isolated, stab at the classics.

In 1974 she came close to death from meningitis and she underwent operations for cancer in 1982 and 1983. Resilient and cheerful in the face of such adversity, she produced further instalments of her memoirs (1978 and 1979), ran an agony column in a daily newspaper, and, by now well over fourteen stone, did a slimming series for breakfast television. A celebrity to the end, she died of cancer in hospital at Windsor, Berkshire, 4 May 1984. Her death was widely and genuinely mourned. Vulgar she may have been but there was admiration for her courage and tenacity. Alan Lake never got over his grief and he killed himself on 10 October 1984, the sixteenth anniversaty of their first meeting.

[*The Times*, 7 May 1984; Joan Flory and Damien Walne, *Diana Dors: Only a Whisper Away*, 1987; personal knowledge.]

PETER WAYMARK

DOUGLAS-HOME, CHARLES COS-PATRICK (1937–1985), journalist, was born in London 1 September 1937, the second son and youngest of three children of the Hon. Henry Montagu Douglas-Home, ornithologist, known as the BBC Bird Man, the second son of the thirteenth Earl of Home. His mother, Lady Margaret Spencer, a talented pianist, was the daughter of Charles Robert Spencer, the sixth Earl Spencer. Douglas-Home won a foundation scholarship to Eton College. This was followed by two years' compulsory National Service when he obtained a commission in the Royal Scots Greys, becoming a somewhat unruly subaltern (1956–7).

After leaving the army he travelled in Canada for nine months before taking up the post of aide-de-camp to Sir C. Evelyn Baring (later Lord Howick of Glendale, q.v.), the governor of Kenya. During this year (1958–9) he became fully involved for the first time with people who were shaping world events. He was very much influenced by Lord Howick, whose biography he wrote (*Evelyn Baring, the Last Proconsul*, 1978). He also wrote Howick's entry for this Dictionary, as he did that of Sir Seretse Khama.

Douglas-Home then opted for a career in journalism. Pulled towards Scotland by his family roots, he started as a general reporter on the *Scottish Daily Express*, a tough training ground. However, he soon developed a reputation as a 'hard news man' with a sure nose for a good story. In 1962 he moved to London as deputy to H. Chapman Pincher, defence correspondent to the *Daily Express*. This sharpened his interest in political intrigue and espionage. After eighteen months he was appointed political and diplomatic correspondent, at that time a powerful post. He used this opportunity to develop many lasting friendships on both sides of the House of Commons. In 1965 the breakthrough to the higher ground of journalism came when, at the age of twenty-eight, he was appointed defence correspondent to *The Times*. He adjusted quickly, bringing a new emphasis to news stories from the Services, and travelling extensively abroad. He covered the Arab–Israeli war in 1967 and the Soviet invasion of Czechoslovakia in 1968, being arrested and expelled by the Russians when found examining a Russian tank.

From 1970 to 1982 Douglas-Home spanned the full field of office on *The Times*. In 1970 he became features editor, bringing many new ideas to the newspaper. In 1973 he became home editor; in 1978 foreign editor; and in May 1981, after K. Rupert Murdoch had become owner, deputy editor under Harold Evans. When, after a year in the editorial chair, Evans left in 1982, Douglas-Home succeeded him and took over the paper whilst it was facing one of its worst crises. However, his true vocation as a journalist and his combined qualities of courage, integrity, and judgement were quickly to revitalize the paper, enabling it to achieve its highest circulation ever.

Douglas-Home gave *The Times* a consistent philosophy. The paper moved to the right, supporting many (but by no means all) Thatcherite policies. Always well informed, Douglas-Home regularly contributed leaders of forthright decisiveness on a broad range of important topics, whilst giving full hospitality in his columns to the views of opposition parties. He brought radical changes to the paper, even though the last three years of his editorship were dogged by illness.

A committed Christian, Douglas-Home was a man of many interests with a great diversity of friends. Amongst his pastimes were writing (he wrote four books), travelling, music, reading, speculating on the metaphysical, wildlife, sailing, and hunting. He was a highly competent pianist, as well as being a fearless rider to hounds who broke many bones in the hunting field. He had a bubbling sense of humour and although he brought a military approach to the order of his professional life, his personal style was remarkably informal. Indeed the more he grew in eminence the more striking was his total lack of pretension. His dress was unconventional, with a firm attachment to spectacularly shabby clothes.

Douglas-Home was a fellow of the Royal College of Music. His awareness of the visual arts was greatly stimulated by Jessica Violet Gwynne, an accomplished artist and stage designer, whom he married in 1966. She was the daughter of Major John Nevile Wake Gwynne, of Knightsmill, Quenington, Gloucestershire. They had two sons and their family life was exceptionally happy. Douglas-Home demonstrated great courage in continuing to edit *The Times* to the last despite the advance of cancer of the bone marrow, an illness which was eventually to overwhelm him. Indeed the rearguard action which he conducted against it was an epic of human valour which his friends will never forget. He died, aged forty-eight, in the Royal Free Hospital, London, 29 October 1985.

[*The Times*, 30 October 1985; Philip Howard, 'An Appreciation, Charles Douglas-Home', *UK Press Gazette*, 4 November 1985; personal knowledge.]   EDWARD CAZALET

D'OYLY CARTE, DAME BRIDGET (1908–1985), theatrical manager and proprietor of the D'Oyly Carte Opera Company, was born in London in Suffolk Street, Pall Mall, 25 March 1908, the only daughter and elder child of Rupert D'Oyly Carte, proprietor of the D'Oyly Carte Opera Company, and his wife, Dorothy Milner, third daughter of John Stewart Gathorne-Hardy, second Earl of Cranbrook. Her second forename was Cicely but she later abandoned this name by deed poll. Educated privately in England and abroad, she later went to Dartington Hall, Totnes, where her artistic talents and abiding interest in the arts were much encouraged. At the age of eighteen, in 1926, she married her first cousin, John David Gathorne-Hardy, fourth Earl of Cranbrook, but the marriage was dissolved in 1931. There were no children. In 1932 she resumed her maiden name by deed poll and helped her father in the Savoy Hotel. From 1939 to 1947 she worked in the poorer districts of London in child welfare.

However, a great change in her life became inevitable when, as the granddaughter of Richard D'Oyly Carte, she unexpectedly became the inheritor of a great theatrical tradition, as a result of the death of her brother Michael in a motor accident in Switzerland in 1932. Becoming, as a consequence, her father's sole heir when he died in 1948, there passed to her the proprietorship of the D'Oyly Carte Opera Company, and all the family rights in the Gilbert and Sullivan operas. This was a challenge she accepted, and with marked accomplishment she continued without a break the presentation of the operas, both in Britain and in the United States and Canada, until thirteen years later the copyright expired, fifty years after the death of Sir W. S. Gilbert [q.v.], and, with that, the performing rights owned by her family.

When the lapse of the copyright drew near, she at first thought it would be right that the long reign of her family, begun by her grandfather and faithfully continued and developed by her father, should be discontinued, and she gave no support to a petition to Parliament seeking a privileged position for the operas by a perpetuation of the copyright under some public body. However, those near to her suggested that she should continue to present the operas, for which there was still a great following, wherever the English language prevailed, and to this she consented.

The result was the formation of the D'Oyly Carte Opera Trust, a charitable organization, which she endowed, giving it her company's scenery, costumes, band parts, and other assets, worth then at least £150,000, to which she added £30,000 in cash. Later, she gave the trustees the original score of *Iolanthe*, which had been presented to her grandfather by Sir Arthur Sullivan [q.v.].

The new trust, which took over in 1961, included representatives of the three families, D'Oyly Carte, Gilbert, and Sullivan, and, guided by A. W. Tuke, then chairman of Barclays Bank, who became chairman, it assigned to Bridget D'Oyly Carte the theatrical presentation of the operas by a company formed for this purpose, Bridget D'Oyly Carte Ltd., of which she became chairman and managing director. In this capacity, she continued to present the operas until their last season in London in 1982, and, having succeeded Tuke as chairman of the trustees, she held this office until her death. In 1975 she was appointed DBE.

By nature shy and retiring, characteristics she inherited from her father, but always entertaining in her methods of expression, she invariably gave evidence of her shrewd observation and artistic judgement and ability. It was these qualities which made her of exceptional value to the Savoy group of hotels and restaurants, founded by her grandfather, of which her father was the chairman, and of which she became a director when he died. In this additional capacity, she actively controlled with considerable talent the furnishing and decoration departments.

At the time of her death she was president of the Savoy Company, in which she was a large shareholder, and at the same time she maintained her family connection with the Savoy Theatre, built by her grandfather, of which she herself was a director. She was also chairman of Edward Goodyear Ltd., the royal florists.

She had a great love of gardening and when her father died, leaving her their beautiful family home in Devon (now the property of the National Trust) between Brixham and Kingswear, she decided to part with it and acquired instead the estate of Shrubs Wood in Buckinghamshire, which remained her home for more than thirty years. There, as her parents had done in Devon, she transformed the grounds with rare trees and shrubs of exceptional beauty. She also spent some time in Scotland, where she lived in the seventeenth-century castle of Barscobe in the stewartry of Kirkcudbright.

She died at her Buckinghamshire home 2 May 1985. When her will was published, her benefactions were seen to be munificent. After leaving her shares in the Savoy Company, worth several million pounds, to the D'Oyly Carte Charitable Trust, which she had established in her lifetime, she bequeathed to the D'Oyly

Carte Opera Trust the residue of her estate, which amounted to a gift of £1½ million, for the future presentation of the operas, with which her family had been identified for over one hundred years.

[Personal knowledge.]   HUGH WONTNER

DUFFY, TERENCE (1922–1985), president of the Amalgamated Union of Engineering Workers, was born 3 May 1922 in Wolverhampton, in a typical industrial back-to-back house, the second child in the family of five sons and six daughters of John Duffy, engineering worker at Thomson's Motor Pressings, Wolverhampton, and later at the Goodyear tyre factory, and his wife, Ann Lockrey. He was educated at two local Roman Catholic schools — St Patrick's and St Joseph's. He left school at the age of fourteen and took a job as a sheet-metal worker in the local Standard Motor Company's factory. His main hobby as a youth was amateur boxing for which he gained local notoriety. In 1939 soon after the outbreak of war he enlisted in the Leicester Regiment as an infantryman, seeing service in the ensuing six years in Greece, Burma, North Africa, and Italy. He attained the rank of regimental sergeant-major, adorning himself with that grade's typical pencil moustache, which he kept until his death.

After his demobilization in 1945 he returned to his native Wolverhampton where he worked in a number of local engineering factories as a sheet-metal worker. In 1951 he was elected a shop-steward in the Norton motor-cycle factory. He then became involved in the broader trade-union and Labour movement as an executive member of the Wolverhampton Trades Council (1955–75), and of the Wolverhampton Constituency Labour Party over the same period. He was chairman of the latter from 1965 to 1970. Since such activities caused him to become reasonably well known locally, he moved in 1969 from being a senior shop-steward in the Lucas Aerospace plant to his first full-time office as the union's assistant divisional organizer in division no. 16. He was re-elected in 1973.

The democratic structure, power, and influence of the AUEW made it a most attractive union not just for those whose only ambition was to earn their living in the engineering industry, but also to the ambitious and politically motivated who recognized it as an instrument to change society to their liking. Consequently two schools existed side by side within the union—the democratic socialists and the broad communist (Trotskyist) Militant Tendency group. Duffy's religious faith and upbringing caused him to be a member of the former. Hence his union officership was stormy, for very little brotherly love existed between the two schools.

In 1975 Duffy was nominated to stand for the executive position which at that time was held by an official supported by the communists. Against all predictions he won, and when he was selected by the union's democratic forces to contest the presidency three years' later he repeated his success. Thereafter he was elected or appointed to many additional positions. He was president of the International Metalworkers Federation (British section), president of the European Metal Workers Federation, chairman of the Confederation of Ship-building and Engineering Unions' engineering committee, and a member of the general council of the Trades Union Congress and of its finance, economic, and international committees. He was also a member of the National Economic Development Council, the British Overseas Trade Board, and the Think British Campaign.

His educational limitations were by far outweighed by his open uncomplicated nature, sagacity, and common sense. He will be remembered for his leadership in the engineering workers' struggle to reduce the forty-hour working week, which caused him to be catapulted into a bitter national strike which he successfully concluded. He thus provided the inspiration for other European metal-working unions to seek the same success. His second triumph was to persuade his union's policy conference to accept government funding for postal balloting. Threatened with suspension, and expulsion from the TUC, he tenaciously insisted that no trade-union principle was being violated in accepting such money to pay for the purely mechanical process of postage. Little by little the opposition withered away, the TUC amended its policy, and many other unions subsequently followed the AEU's lead.

In 1946 he married Eileen Stokes, from Wolverhampton. They had two sons and a daughter. She died in 1954 and in 1957 he married Joyce, daughter of Alexander Sturgess, metal machinist at Guys' Motors, Wolverhampton. They had a son and a daughter. After a lingering lung illness Duffy died in the Brompton Hospital, London, 1 October 1985.

[AEU records; *The Times*, *Guardian*, and *Daily Telegraph*, 2 October 1985; information from Joyce Duffy (wife) and Denis Duffy (brother); personal knowledge.] JOHN BOYD

DYKES BOWER, SIR JOHN (1905–1981), cathedral organist, was born at Gloucester 13 August 1905, the third of four sons (there were no daughters) of Ernest Dykes Bower MD, surgeon, and his wife, Margaret Dora Constance Sheringham. Two brothers (Michael and

Wilfrid) became well-known doctors; Stephen, architect, designed the baldaquin at St Paul's. All four sons inherited from their parents a powerful interest in music, and as children were daily set to practise the piano during the hour before breakfast. The family worshipped regularly at Gloucester Cathedral.

Dykes Bower was educated at Cheltenham College and at the same time was a pupil of (Sir) A. Herbert Brewer [q.v.], organist of Gloucester Cathedral. From Cheltenham he went in 1922 to Corpus Christi College, Cambridge, having won the John Stewart of Rannoch scholarship in music. He was again awarded a Rannoch scholarship, together with his brother Wilfrid, in 1925. At Corpus he was organ scholar in succession to Boris Ord [q.v.]; his brother Wilfrid succeeded him. Ord and Dykes Bower were lifelong friends, both dedicated to the pursuit of perfection in the performance of church music and very austere in the demands they made on choirs. Both hated any element of 'show-biz' about the conductor's role.

From Cambridge Dykes Bower went to be organist at Truro (1926-9), where Bishop Walter Frere [q.v.] as musician and liturgist made his stay congenial. At Truro he succeeded in expelling from his choir a tone-deaf lay clerk who was mayor and a potentate in the city. This difficult achievement commended him to H. A. L. Fisher [q.v.], warden of New College, Oxford, and Sir Hugh Allen [q.v.], when New College needed an organist in 1929. In 1933 he was invited to be cathedral organist at Durham, with a university lectureship. His Cambridge college simultaneously elected him a (non-resident) fellow, 1934-7. The incomparable acropolis of Durham was congenial to him, but there were also difficulties to contend with (he did not get on well with those who wanted no changes and resented his perfectionism); and in 1936, aged only thirty-one, he was appointed by W. R. Matthews [q.v.], the dean, to St Paul's to succeed (Sir) Stanley Marchant. Matthews and Dykes Bower became instinctively drawn together in friendship as well as by their common responsibility for cathedral services. They perfectly understood their respective spheres. Moreover, Dykes Bower was a punctilious administrator and letter-writer. He enjoyed to the full the great occasions that come to St Paul's, like the thanksgiving service after World War II or Sir Winston Churchill's state funeral in 1965, when the huge congregation singing the 'Battle Hymn of the Republic' was totally controlled by masterly rhythmic playing on the part of Dykes Bower.

St Paul's had resources making it possible for him to include music of a complexity that other cathedrals could hardly attempt. Characteristically, unless an anthem were unaccompanied, he would always direct from the organ loft; his intense sense of pulse and rhythm was conveyed with the minimum of external sign. He disliked anything flamboyant or histrionic. In part this reflected the quiet reticence of his personality. But it was more an expression of his deep feeling that the sublimity of church music is diminished or even destroyed if the performance and the performers are perceived to be somehow distinct from the act of worship to which they help to give expression.

During World War II the cathedral was under frequent threat from the air. In the destruction of the city of London by fire-bombs in December 1940 Dykes Bower lost everything, including his exquisite grand piano to which he was devoted; he was at least as fine a performer on the piano as on the organ. In 1940 he joined the RAFVR and, with the rank of squadron leader, worked in the Air Ministry with a group which included the viola player Bernard Shore, with whom he used to give occasional wartime recitals when life made such relaxation possible. After the war he combined his continued work at St Paul's with the post of associate director of the Royal School of Church Music (1945-52). He held the professorship of organ at the Royal College of Music (1936-69), and sent out a series of distinguished pupils to many of the major cathedral posts in England. Only Boris Ord at King's College, Cambridge, had a comparable influence on the standard of musicianship in English cathedrals.

In 1967, aged sixty-two, his eyesight was threatened by cataract and once, playing some difficult Bach at the end of a service, he suddenly found himself unable to see the printed page. Immediately he decided to retire from the great position he had held so long. He also had such an attachment to W. R. Matthews, with whom he had collaborated for thirty-one years, that he did not want to continue after Matthews's retirement from the deanery. He took a flat near Westminster Abbey which he attended regularly. Weak sight robbed him of the earlier pleasure of reading Victorian novels, and railway timetables, on which he was remarkably expert. (He loved to plan imaginary cross-country journeys with *Bradshaw*.) But he continued to do much for the Royal College of Organists, of which he was president (1960-2).

He was appointed CVO in 1953 and knighted in 1968. Oxford made him an honorary D.Mus. (1944) and Corpus Christi, Cambridge, made him an honorary fellow (1980). He was master of the Worshipful Company of Musicians (1967-8). He did much for the council of *Hymns Ancient & Modern*, of which he was chairman. The hymn writer J. B. Dykes (1823-76, q.v.) was his forebear.

His fastidiousness and relentless quest for

flawless performance made him hard to please, and could combine with his quiet reticence to make him silent where a word of encouragement could have been beneficial. A very private man with a horror of the limelight, he asked only to be allowed to offer perfection through music in the worship of the Church of England. He inspired awe but also deep affection in everyone who worked alongside him. He was unmarried. He died 29 May 1981 in hospital at Orpington, Kent.

[Personal knowledge.]   HENRY CHADWICK

E

EASTWOOD, SIR ERIC (1910-1981), physicist and engineer, was born 12 March 1910 at Rochdale, the youngest in the family of five sons and one daughter of George Eastwood, who worked in cotton mills, and his wife, Eda Brooks. He was educated at Oldham Municipal Secondary School and Manchester University where, in 1931, he came top of the class list for B.Sc. honours in physics. He was awarded a two-year grant for research but, after working for one year on spectroscopy for his M.Sc. degree, he moved to the department of education and gained its diploma in 1933. In that year he was awarded a further grant for research at Cambridge. He was admitted to Christ's College and worked in the department of physical chemistry to gain his Ph.D. in 1935.

At this stage Eastwood intended to become a schoolmaster and taught with success at King's School in Taunton, Runcorn Grammar School, and Liverpool Collegiate. In 1941 he volunteered for service with the Royal Air Force. Commissioned in the education branch, he was soon transferred to 60 Group, which was responsible for the operation and maintenance of radar stations. Initially he worked on the calibration of a station to take account of the topographical features of the surrounding country. Submission of a detailed theoretical report on the influence of atmospheric refraction on the performance of the station led to his promotion and later to transfer to 60 Group headquarters as squadron leader in charge of calibration of all RAF radar stations in the United Kingdom. For this work he was mentioned in dispatches.

After release from the RAF in 1946 Eastwood joined the research laboratories of the English Electric Company. He was placed in charge of the physics department, with a major commitment to the development of wartime radar for civilian purposes. When the English Electric Company acquired the Marconi Wireless Telegraph Company in 1947, the radar work was moved to the Marconi laboratories at Great Baddow, to which Eastwood went as deputy director of research in 1948 and, in 1954, as director. During this period, in his own time and with the help of some colleagues, he carried out extensive scientific research on the migration of birds, using the high-power Marconi radar transmitter, to which he had access. This work was published in a book *Radar Ornithology* (1967). From 1952 to 1968 he was director of research for the whole of the English Electric Company and, when the company merged with the General Electric Company in 1968, he became director of research for the combined group. However, he was not happy in this post and in 1972 he retired and went back to the Marconi laboratories as chief scientist, a post which he held until shortly before his death.

Eastwood was a good committee man and his services were in great demand. Amongst many bodies which profited from his advice, often as committee chairman, were the Electronics Research Council, the Appleton Laboratory, and the Defence Scientific Advisory Council. He devoted much time to the Institution of Electrical Engineers and became its president in 1972-3. He was appointed CBE in 1962 and knighted in 1973. He was awarded honorary degrees by Exeter, Cranfield, Aston, City, Heriot-Watt, and Loughborough universities and medals by the Royal Aeronautical Society, the Institute of Physics, and the Institution of Civil Engineers. He was elected FRS in 1968 and honorary fellow of the Institution of Electrical Engineers in 1979.

Eastwood was a kind and gentle person who would nevertheless defend strongly any cause in which he believed. His early years had been spent in a happy and close-knit family, though his father was sometimes out of work and moves to different towns in search of employment were not unknown. A love of music was handed down from the father, a skilled violinist, to the children, several of whom played musical instruments. Eric Eastwood himself joined the family band as a cellist and later learned to play the flute. The love of music remained with him throughout his life; he had a good voice and often sang in choirs. In his early years he had been a regular attender at Sunday morning church and afternoon Bible class. To this may be ascribed his absolute integrity in later life— and his habit of producing apt biblical quotations at unexpected moments.

In 1937 he married Edith, daughter of Harry Butterworth, an engineer. They had two sons, who both became engineers. Eastwood died 6 October 1981 at Little Baddow near Chelmsford.

[F. E. Jones in *Biographical Memoirs of Fellows of the Royal Society*, vol. xxix, 1983; personal knowledge.]   CHARLES OATLEY

ELLERMAN, (ANNIE) WINIFRED (1894-1983), writer under the name of BRYHER and private philanthropist, was born, fifteen years in advance of her parents' marriage, 2 September 1894 at Margate, the elder child (by fifteen years) and only daughter of (Sir) John Reeves Ellerman [q.v.], later first baronet, shipping magnate and newspaper owner, of London W1, formerly of Brough near Hull, and Hannah, daughter of George Glover. This circumstance diminished her rights of inheritance when her father died in 1933 leaving one of the largest private fortunes in Britain. She used the name Bryher from 1918 and confirmed this by deed poll some years later.

Virtually self-educated by voracious reading especially of history and G. A. Henty [q.v.], and extensive foreign travel, her world was shattered when she was sent, at fifteen, to Queenswood School, Eastbourne. Her analyst later asked of her: 'What do you expect me to do for you? As a child you have been in Paradise!'

The experience precipitated her first novel, *Development* (1920). Childhood had bred in her 'a desire of expression, love of freedom'. School presaged future constraint: 'To possess the intellect, the hopes, the ambitions of a man, unsoftened by any feminine attribute, to have these sheathed in convention, impossible to break without hurt to those she had no wish to hurt, to feel so thoroughly unlike a girl—this was the tragedy.' After school, she planned a career in journalism and during the war tried to sign up as a land worker.

In 1918 she sought out in Cornwall the Imagist poet HD (Hilda Doolittle), later writing: 'There will always be one book among all others that makes us aware of ourselves; for me, it is *Sea Garden* by H. D.' Their close personal association lasted until HD's death in 1961.

On 14 February 1921 in the City Hall of New York City, Bryher married a writer, Robert Menzies McAlmon. There was some mutual regard but principally he gained funds and access to travel whilst she acquired the freedom reserved for a married woman. McAlmon then founded in Paris with Bryher's money the Contact Publishing Company which, over eight years, published Ernest Hemingway, James Joyce [q.v.], Ezra Pound, William Carlos Williams, Gertrude Stein, Nathanael West, HD, and Djuna Barnes. Bryher also supported Joyce and his family with a monthly allowance, helped George Antheil, and financially assisted Sylvia Beach in her running of the influential Shakespeare & Co. bookshop.

Divorcing McAlmon in 1927, Bryher then married a lover of HD's, Kenneth Macpherson, son of John Macpherson, an artist, on 1 September 1927 at Chelsea Register Office. The web of loyalties apparently worked. Nurturing Macpherson's interest in cinema, Bryher started a film company, POOL Productions. Its works included *Foothills*, which when later shown in Berlin attracted the approbation of G. W. Pabst, and *Borderline* with Paul Robeson. She founded *Close-Up*, a film magazine that ran suc-

cessfully until 1933 and which introduced Sergei Eisenstein to a wider public.

Her building in 1930-1 of a late Bauhaus home, Kenwin, above Lake Geneva, coincided with the formal adoption of HD's daughter Perdita by Bryher and Macpherson. Bryher had no children of her own. At this time she also supported the psychoanalytic movement in Vienna by providing the money for the publication of their *Psychoanalytic Review*, and, as well as revelling herself in analysis, she paid for HD to be analysed by Sigmund Freud.

The overseeing of the literary magazine *Life and Letters Today* which she formed in 1935 exercised her noteworthy business acumen as well as literary inclinations. Her involvement persisted into the war years which she spent in London, learning Persian and getting to know the Lowndes Square Group, especially the Sitwells (she bought a house for Edith Sitwell, q.v.). The period also saw Bryher properly embarked upon her own literary career. Between 1948 and 1972 she published twelve books including *The Fourteenth of October* (1954), *The Player's Boy* (1957), *Roman Wall* (1955), *Gate to the Sea* (1959), *The Coin of Carthage* (1964), *This January Tale* (1968), and two moving autobiographical accounts, *The Heart to Artemis: a Writer's Memoirs* (1962) and *The Days of Mars: a Memoir 1940-1946* (1972). The novels are distinctive historical imaginings, intense but passionless, frequently using the narrative perspective of a young man, and with settings ranging from Paestum to the battle of Hastings. They are minutely researched but the detail is awkwardly assimilated. The rest of her output, besides reviews, included poetry, literary criticism, an account of her responses to the USA, and *Film Problems of Soviet Russia* (1929).

Bryher's wealth gave her something of the freedom and independence she so envied as the especial province of men. Valuing autonomy and work so highly, she helped others to achieve it by thoughtful and responsible financial sponsorship. She said of herself: 'I have rushed to the penniless young not with bowls of soup but with typewriters' and, even after their divorce in February 1947, she supported Macpherson. Loyalty and friendship were a passion, as were activity and adventure and she embraced the new ('I was completely a child of my age')— psychoanalysis, air travel, inoculation against flu, modernist fiction, experimental cinema. She died, rather lonely, in Vaud, Switzerland, 28 January 1983.

[Barbara Guest, *Herself Defined: The Poet HD and her World*, 1984; Bryher, *The Heart to Artemis*, 1962, and *The Days of Mars*, 1972 (autobiographies); private papers at Magdalene College, Cambridge; private information.]　　　　　　　KIM SCOTT WALWYN

ELLIS, CLIFFORD WILSON (1907–1985), artist, designer, teacher, and promoter of education through art, was born at Bognor, 1 March 1907, the eldest in the family of two sons and two daughters of John Wilson Ellis, a designer who was trained in cabinet making, and his wife, Annie Harriet Westley. His grandfather, William Blackman Ellis, artist, naturalist, and taxidermist, who took him as a child for walks in Arundel Park, taught him much about the flora and fauna of the area, and this background of a love of nature and of skill in craftsmanship doubtless sowed the seed in him of a passion to perfect such love and skills in himself and, through teaching, to develop them in others. At a slightly later stage when the family moved to Highbury he fostered his knowledge of natural history by keeping stick insects, lizards, toads, and frogs in his bedroom and by frequent sketching visits to the London Zoo.

Ellis was educated first in small rural schools, and from 1918 to 1923 at Dame Alice Owen's Boys' School to which he gained the scholarship. He then won a place at the Regent Street Polytechnic School of Art, qualified, and through the University of London took the Board of Education's art teachers' diploma, studying under Marion Richardson. He was invited to join the Polytechnic staff in 1928 and became head of department for teaching first-year students.

In 1931 he married (Dorothy) Rosemary Collinson, herself an artist, daughter of the designer Frank Graham Collinson, who shared his work and ideals throughout his life. Their joint signature, C & RE, on book-jackets, posters, and other designs indicated work of a highly original and imaginative order based on a close knowledge of form and structure. The series of jackets for Collins's *The New Naturalist* was begun in 1943. Over ninety were produced. In the 1930s they collaborated in poster designs for London Transport (in 1938 London buses carried their 'Summer is Flying' poster), the Empire Marketing Board, the General Post Office, and Shell-Mex ('Antiquaries Prefer Shell' was one of the most admired in the Shell 'Professions' series which Ellis initiated), and they designed a mosaic floor for the entrance to the British pavilion at the Paris Exhibition of 1937. In 1947 they were among the group of distinguished artists invited by J. Lyons & Co. to cheer up the post-war public with lithographs for the Corner Houses.

In 1936 Ellis was appointed assistant master at the Bath School of Art; he became headmaster in 1938. At this time the School was attached to the Technical College but after the bombing of the city in 1942 when the rooms of the Art School were destroyed it moved to new premises in Sydney Place. This was the beginning of what proved to be a long-term and almost visionary scheme which culminated in 1946 in a change of status from School to Academy (Teachers' Training College as well as Art School) and the removal from Sydney Place to Corsham Court, the home of Lord Methuen, RA, some eight miles away in Wiltshire. In this year Ellis was offered, but refused, the chair of fine art in the University of Durham.

He was a man of great enthusiasm, determination, and vigour, and he conveyed a sense of urgency, importance, and delight to all who worked with him. Early on in the Bath days he attracted the friendship of Walter Sickert [q.v.], who came voluntarily once a week to talk to the students. Similarly he found a firm ally in Lord Methuen who was glad to see a large part of his house and grounds used by the Academy, and put the fine Capability Brown picture gallery and the collection at its disposal. Among the staff were Isabelle Symons, Kenneth Armitage, and William Scott who moved out from Bath and, later, Peter Potworowski and Peter Lanyon. Sir Kenneth (later Lord) Clark [q.v.] and John Piper were on the board of governors as co-opted members, the overall responsibility remaining with the City of Bath Education Authority. Corsham was a great experiment and was unique. Students were taught to understand the making of everything from the elaborate plaster-work ceilings and the finely-chased door locks in the state rooms at the Court to the glazing bars of terraced houses in Bath, from the culture of orchids to drawing from wild nature, from music and dance on the lawns to music and dance in painting and sculpture, from history to art history. The production of an idiosyncratic but enchanting adaptation of Gluck's *Orfeo ed Euridice* was memorable. All this was due to the inspiration of Ellis and his conviction that the arts are the staple of the fulfilled life. His approach to all these matters was essentially realist. Ellis retired from the Bath Academy in 1972.

Ellis's activities were ceaseless and multifarious. He saved fine railings in Bath from being sent for scrap metal. He advised the authorities on camouflage and devised apparatus for the visual training of tank commanders and gunners. He served on advisory committees of the Arts Council and Unesco, and formed a link between Corsham and the new University of Bath by establishing a research centre in art and education.

Clifford and Rosemary Ellis had two daughters. Ellis died at Urchfont, Wiltshire, 19 March 1985.

[*The Times*, 29 March 1985; Clifford Ellis,

'Preparing Art Educators' in *Education and Art*, a symposium by Edwin Ziegfeld for Unesco, 1955; private information; personal knowledge.]   KENNETH GARLICK

ELMHIRST, SIR THOMAS WALKER (1895–1982), air marshal, was born 15 December 1895 in Laxton, Yorkshire, the fourth of eight sons and nine children of the Revd William Heaton Elmhirst and his wife, Mary, daughter of the Revd William Knight, of Hemsworth, Yorkshire. He was a younger brother of Leonard Knight Elmhirst [q.v.]. In 1899 his parson father became head of a small estate which had been in the family since 1340. From the age of six he intended to become a sailor and in 1908 at the age of twelve he entered the Royal Naval College at Osborne. He then went to Dartmouth. He saw naval action at the Dogger Bank under Admiral Sir David (later Earl) Beatty [q.v.] in January 1915. In March 1915 he was one of twenty young officers selected by Admiral Lord Fisher [q.v.] to start a new Naval Airship Service to combat the German submarine threat. After qualifying as a balloon pilot he captained airships on anti-submarine patrols over the North and Irish Seas. From April 1918 until the end of the war of 1914–18 he commanded, at the age of twenty-two, the airship stations on Anglesey and at Malahide in Ireland. He was awarded the Air Force Cross (1918).

In 1919 he transferred to the RAF and flew seaplanes and flying boats. On graduating from the Staff College in 1925 he served for three years in Intelligence on the staff of Sir Hugh (later Viscount) Trenchard [q.v.] in the Air Ministry before returning to flying duties at RAF Leuchars. After further staff appointments he was sent to form and command No. XV Bomber Squadron at Abingdon (1935–7). He then went to Ankara as air attaché to Turkey (1937–9).

Elmhirst was a good all-round sportsman, excelling at rugby football and playing at scrum half for the RAF against the Royal Navy and the army in 1921. He later became honorary secretary of the RAF Rugby Union.

In 1939 he returned to England to command RAF Leconfield in the rank of group captain and on 3 September prepared to dispatch the first bombers detailed to attack Germany though at the last moment the bombs were off-loaded and replaced with leaflets. On promotion to air commodore in January 1940 he was posted to the Air Ministry as director of Intelligence before going to Fighter Command headquarters as duty air commodore during the crucial months of August and September in the Battle of Britain. In December his experience in Intelligence and his service in Turkey led to his

selection, together with an admiral and a general, for a mission for conversations with the Turkish general staff. He was retained in Cairo after the talks and was appointed in April 1941 to command No. 202 Group which was responsible for the air defence of Egypt. His ability for senior command was demonstrated in the build-up of an efficient organization modelled on that in Britain.

In February 1942 he was asked by his commander-in-chief, Sir Arthur (later Lord) Tedder [q.v.] to go to the Desert Air Force as chief administrative officer to its recently arrived commander, Sir Arthur Coningham (whose notice Elmhirst later wrote for this Dictionary). Though reluctant to give up his command in Egypt he recognized the importance of the desert campaign and went cheerfully to begin a partnership with Coningham that continued through Libya, Tunisia, Sicily, Italy, Normandy, and Brussels to final victory in Germany. In this operational administrative role he planned and organized mobile air forces which enabled the development of Coningham's concept of tactical air forces for operations with armies, an idea which was subsequently adopted in other theatres of war. He regarded this as his greatest single achievement in his service career and it was marked by the award of the US Legion of Merit and the CB (1945).

Returning to England early in 1944 to prepare (for the invasion of Europe) the 2nd Tactical Air Force, which Coningham was to command, he was promoted air vice-marshal. The achievements of 2nd TAF from the invasion of Normandy onwards owed much to Elmhirst's preparatory work. In 1945 his service was recognized when he was appointed KBE (he was made CBE in 1943).

His first peacetime post was as assistant chief of the air staff (Intelligence) in the Air Ministry but within two years he was pressed to go to India as chief of administration on the staff of Field-Marshal Sir Claude Auchinleck [q.v.] and was promoted air marshal. When India achieved independence in 1947 he was asked by Jawaharlal Nehru and Earl Mountbatten of Burma [qq.v.] to remain and become the first commander-in-chief of the new Indian Air Force. In 1950 ill health forced his resignation but as a tribute he was made an honorary air marshal for life in the Indian Air Force.

On retirement to Scotland he became deputy lieutenant (1960–70) and county councillor for Fife and controller of Civil Defence for east Scotland. In 1953 he was called to join the Ministry of Supply responsible to Sir William (later Lord) Penney for the organization of his atom bomb team and its transportation to Australia for the first British atom bomb to be exploded on land. On completion of this task in the same year he was appointed lieutenant-governor of Guernsey where he served until 1958. He then retired for the second time to Fife.

Elmhirst was a small man, always impeccably dressed in his uniform. His outstanding physical feature was his large bushy eyebrows which rose perceptibly when he saw anything that displeased him. He married in 1930 Katharine Gordon, daughter of William Black, chartered accountant, of Chapel, Fife. They had a son and a daughter. She died in 1965 and in 1968 he married Marian Louisa, widow of Colonel Andrew Ferguson and daughter of Lieutenant-Colonel Lord Herbert Montagu-Douglas-Scott. Elmhirst died 6 November 1982 at Basingstoke.

[Private information; personal knowledge.]
KENNETH CROSS

EMERY, RICHARD GILBERT (DICK) (1915–1983), stage, radio, and TV comedian, was born 19 February 1915 at University College Hospital in St Pancras, London, the only son of Laurence Cuthbert Emery, an actor, and his wife, Bertha Gilbert Callen, a former Gaiety girl. During most of Dick Emery's childhood his mother and father performed as a music-hall act, Callen and Emery, and young Emery had little formal education as his parents were in a different town each week appearing at the local theatre while Emery attended the local school. Later a half sister and half brother were born.

The marriage broke up in 1926 and Emery, offered the choice, opted to stay with his mother. She exerted great influence on him and he is quoted as saying of her, 'I adored her but I was also frightened of her.' After her divorce Dick Emery's mother gave up show business but was determined that her son should be a star. She was not disappointed for by the time she died, aged ninety-five in 1977, her son was one of the biggest stars on British television and one of the BBC's most reliable crowd pullers, featuring frequently in the Top Ten audience ratings.

For Dick Emery performing was a nightmare and he was always apprehensive and often physically sick before a performance. He possessed, from childhood, a good singing voice—when he was small his mother always insisted he sang to visitors, an early taste of what became for Emery in later life the torture of performing in public. He is quoted as saying that he would have preferred a career in opera, but family background and natural accomplishments directed him towards comedy.

His first big chance came during the war when serving in the RAF. In 1942 he joined one of the RAF Gang Shows organized by

W. H. Ralph Reader where he made his first appearance as a female impersonator, the forerunner of roles which were subsequently to make him famous. After demobilization he spent some time fruitlessly auditioning for various agents and managers but at last, like so many of his contemporaries—for example, Tony Hancock, Peter Sellers [qq.v.], Jimmy Edwards, and (Sir) Harry Secombe—he became a mainstay of the Windmill Theatre in London, working there for nine months in 1948.

From those small beginnings Dick Emery began to emerge as a first-class character actor on radio in such series as *Pertwee's Progress* (1955), which starred Jon Pertwee, *We're in Business* (1960), with Peter Jones, Harry Worth, and Irene Handl, and *Educating Archie* (1958), where he created the character of the elderly, grumbling odd-job man, Lampwick.

On television he appeared with Michael Bentine in *It's a Square World* (1962), and in Granada TV's *The Army Game* (1960). In 1963 came *The Dick Emery Show* in which, for the first time, Dick Emery was the eponymous star. Its success with the public if not always with the critics lasted until his death in 1983. For nearly twenty years his loyal public laughed at Emery's characterizations which included, among others, a vicar, a homosexual, a traffic warden, a frustrated spinster called Hettie, a leather-clad motor cyclist, the aged Lampwick, the bovver boy (an early version of the skinhead), 'College' the tramp, and, best remembered and certainly the most popular with the audience, the ageing sex-kitten Mandy with her inevitable catchphrase, 'Ooh you are awful—but I like you.'

Dick Emery's work in the cinema was mainly limited to minor roles. In 1968 he appeared in the X-certificate *Baby Love* with Diana Dors [q.v.] and Keith Barron, and in 1970 he appeared as Bateman in the film of *Loot*, the play by Joe Orton [q.v.]. In 1972 he starred in *Ooh, You Are Awful*.

But if Emery's professional life prospered his private life appears to have been a series of crises. In 1952 his self confidence was at such a low ebb he resorted to hypnosis, and subsequently psychoanalysis to pull him out of a deep depression which lasted for five years.

He was married five times. His first wife, Zelda, left him after five years taking with her their son, Gilbert. His second marriage, to Irene in 1945, only lasted a few months and there were no children. Emery's third wife, Iris, who bore him a son, Nicholas, died from a brain tumour. Her father was William Paulk Tully, butcher. His fourth wife, Victoria, was eighteen when they married and Emery was then forty-three. She was the daughter of Harold Booth Chambers, a musical director. They had two children, Michael and Eliza. His fifth wife was a singer and actress, Josephine Blake. They married in 1969 and had no children. When Emery died he had by then left Josephine and was living with a dancer, Fay Hillier. Dick Emery died 2 January 1983 at King's College Hospital, London.

It would seem that Emery's childhood insecurity inhibited him throughout his adult life and prevented his undoubted comic genius from flowering completely. He will be remembered for his flashes of inspiration of cartoon brevity rather than for the depth and insight of the truly great character comedian. His was a talent sadly unfulfilled.

[Personal knowledge.]   BARRY TOOK

EMPSON, SIR WILLIAM (1906-1984), poet, literary critic, and teacher, was born at Yokefleet Hall, Howden, Yorkshire, 27 September 1906, the youngest of four sons and five children of Arthur Reginald Empson, of Yokefleet Hall, later captain in the Royal Field Artillery, and his wife, Laura, daughter of Richard Micklethwait, of Ardsley House, Yorkshire. His father died when he was ten. He was educated at Winchester College (1920-5) and Magdalene College, Cambridge, where his results indicated his unusual range: in the mathematics tripos he gained a first in part i (1926) and was a senior optime in part ii (1928) and in the English tripos he obtained a first in part i (1929). He impressed his English supervisor, I. A. Richards [q.v.], then at the height of his reputation as a critic with a particular interest in linguistic theory and the workings of the mind. One of the insights that Richards threw off almost casually was that ambiguity, traditionally seen as a blemish in poetry, could actually be one of the ways in which the poetic language gained its effects. Empson, acting on this hint, asked if Richards would permit him, instead of turning in the usual weekly essay, to work on his own for a while, and after a few weeks came back with a 30,000 word screed which turned out to be the central core of the book that established his critical reputation (*Seven Types of Ambiguity*, 1930). At the same time he was finding his highly individual voice as a poet; he contributed poems to the *Cambridge Review* and *Cambridge Poetry 1929* which became famous far beyond the undergraduate literary world, and were subsequently much reprinted.

Immediately on graduating, Empson was elected to a research fellowship at Magdalene. Unfortunately, the servant who was moving his effects from his undergraduate rooms to those he was to occupy as a fellow found three contraceptives in the pocket of his jacket, and, on this slender foundation—coupled with the un-

disputed fact that he had received lady visitors in his rooms—Empson was deprived of his fellowship. Since offences against chastity were regarded as breaches of university, not merely of college, discipline, his Cambridge career was, at least for the foreseeable future, at an end.

The decision of the fellows of Magdalene seems in retrospect a regrettable one, designed to uphold a morality blemished by a streak of bigotry; the chief personal effect on Empson, a generous-minded man, was to instil his lifelong conviction that people who talk about 'religion and morality' should be aware that there is more than one religion and more than one morality. The next move in his life would in any case have implanted such notions, for it was to the Far East.

From 1931 to 1934 he was professor of English literature at Tokyo University of Literature and Science (Bunrika Daigaku), and while in Japan he managed to ignore the ugly militarism then gathering pace and turn his gaze on the beautiful and permanent and especially on the religion of the Buddha. *Some Versions of Pastoral* (1935), his second critical book, written during these years, is as brilliant as his first but more widely speculative; not only was his mind growing at a great rate, but the move to the East had brought it into contact with pastures new. Three years in England, living in Marchmont Street, Bloomsbury, on a small private income and consolidating his work as critic and scholar, were followed by the offer, which he accepted, of a professorship of English at Peking National University.

When, in the autumn of 1937, Empson arrived in Peking, he found that the university had fled before the Japanese invasion. He caught up with them at Changsha, whence they journeyed overland to Kunming and set up long-term though makeshift headquarters. 'Imagine camping out with a set of dons', Empson wrote to a friend, ' . . . You must think of Oxford and Cambridge contriving to sink their differences and combine in the Highlands.' Books were scarce; thrown back on his memory, Empson found that he could write out the complete text of *Othello* and also recite long passages from *Paradise Lost* and Swift's *Modest Proposal*. Difficult as these two years were, he was on his mettle and wonderfully productive; during them he roughed out most of the leading ideas that went into the first major work of criticism he produced in the post-war period, *The Structure of Complex Words* (1951).

Empson identified deeply with the Chinese people in their resistance to invasion (and, to glance ahead once more, the moving 'Chinese ballad', adapted from a twelfth-century original but with emotional reference to this period, was in 1952 to be his last published poem), but the

European struggle against Fascism seemed to him the primary duty, and in 1939 he returned home to give his services to Allied propaganda at the BBC. During these years, so total was his commitment to his war work (he was editor in the BBC monitoring department in 1940-1 and Chinese editor in the Far Eastern Section, 1941-6), his literary life ceased, though there was always his friendship with writers such as George Orwell and Dylan Thomas [qq.v.], and he met and married (1941) Hester Henrietta Crouse, a South African sculptor settled in London. She was the daughter of Johannes Jacobus Crouse, cattle dealer. They had two sons. His one gesture towards imaginative work during the war years was the sketch of a ballet, 'The Elephant and the Birds' (1942) which blended a Buddhist with an Ancient Greek legend, the kind of amalgam that always attracted him.

In 1947 Empson, now with his wife and children, returned to Peking and resumed his professorship at the National University. Their return to England in 1952 was not the result of any disharmony with either the regime or their circle of friends and colleagues, who seem to have been genuinely sorry to see them go; but the Korean war was dragging on and the international future was uncertain. As the only foreigners left in Peking, they had to reckon with the possibility of not being able to get back at all, and wanted their sons to have the later stages of their upbringing in England. Finding that during his absence a new generation of English writers had become deeply interested in his work, Empson took a chair of English at Sheffield University from 1953 until his retirement in 1971.

Empson wrote a small though highly potent and much discussed body of verse (incidentally his recorded reading of his own poems is one of the finest of such performances by any twentieth-century poet). As a critic his output was much larger (five books and much material still uncollected at his death). The prose and the verse share many characteristics, coming as they do from a mind highly singular and also remarkably consistent with itself. Empson's chief mark as a thinker is his superb buccaneering assurance, his refusal to halt at frontiers. Though he works through the medium of literary criticism, he never gives the impression of trotting round a paddock labelled 'Literary Studies'. Since literature takes the whole of human experience as its province, an experience that includes metaphysical speculation, anyone who deals with it will ideally be prepared to discuss virtually anything. If Empson's mind had been less than first-rate, and his training less than rigorous, this willingness might have resulted in jungles of irrelevance; but he had

studied mathematics to a level that gave him the key to much scientific theory, and his acquaintance with literature was wide and deep; of all modern writers, he comes closest to the ideal of a universally competent intelligence. Add to this the strong individual flavour of his work, at once humorous and peppery, so that books like *Milton's God* (1961) and *Essays on Shakespeare* (ed. David B. Pirie, 1986) are read not only for instruction but with the fullness of response with which we read literature itself.

Empson's poetry is strongly rhythmical (to the point, sometimes, of being hypnotic) and permeated everywhere by a tragic beauty and a dignity in confronting the suffering of life. He became a fellow of the British Academy in 1976 and was knighted in 1979. He had honorary D.Litts. from the universities of East Anglia (1968), Bristol (1971), Sheffield (1974), and Cambridge (1977). He died in London 15 April 1984.

[Roma Gill (ed.), *William Empson: the Man and His Work*, 1974; John Haffenden (ed.), *The Royal Beasts and Other Works by William Empson*, 1986; Christopher Ricks in *Proceedings* of the British Academy, vol. lxii, 1985; personal knowledge.]   JOHN WAIN

ENGLEDOW, SIR FRANK LEONARD (1890-1985), plant scientist and agriculturalist, was born 20 August 1890 at Dartford, Kent, the fifth and youngest child of Henry Engledow, of Norfolk, and Elizabeth Prentice, from Essex. Henry Engledow was police station sergeant at Bexleyheath, Kent, and on retirement became agent to a brewery. Engledow attended Dartford Grammar School from 1904 to 1909, and then he entered University College, London, with a one-year scholarship to study pure and applied mathematics with physics. He obtained a London B.Sc. and proceeded to St John's College, Cambridge, in 1910 on an exhibition to read for the mathematical tripos. At half term he transferred to the natural sciences tripos and he obtained a first in part i (1912). St John's made him a scholar and awarded him a Slater studentship. He also received a Ministry of Agriculture research scholarship and entered the school of agriculture for the two-year diploma course under the care of (Sir) Rowland Biffen, whose notice Engledow later wrote for this Dictionary.

The first two years of Engledow's research career resulted in three papers in 1914, two written in collaboration. They dealt with the statistical analysis of acquired data, the genetics of wheat, and the quality of wool. But then came the outbreak of World War I, anticipated by Engledow's voluntary enlistment in the 5th battalion of the Queen's Own Royal West Kent

Regiment. He had a distinguished war career, rising to the rank of lieutenant-colonel in the Territorial Army (1921) and being mentioned in dispatches and awarded the croix de guerre (1918) while serving in the Middle East and India. He had a short spell as assistant director of agriculture in Mesopotamia.

On his return to Cambridge in 1919 Engledow continued his researches on the genetics and yield of wheat, and the use of statistical methods in data analysis and field experimentation. He published this work in a series of important papers over ten years, while he later bred new wheat varieties, the most important of which was Holdfast, a top-class bread wheat. Engledow was a very effective lecturer, contributing most importantly at the school of agriculture at graduate level for the diploma courses, designed for the training of officers for the Colonial Service and for the Empire Cotton Growing Corporation. The combination of the rapidly developing relevant sciences and the receptive state of agriculture were a great encouragement to research and teaching, and Engledow seized the opportunity with great effect. He thus fulfilled the expectations of his college which had elected him to a Founders fellowship in 1919.

In 1930 Engledow was elected Drapers' professor of agriculture and head of the department of agriculture at Cambridge, and an expansive period followed as his reputation as an academic agriculturalist grew, and the school of agriculture's reputation became international. Engledow was associated with much change and advancement including the development of the agricultural services and education. He became an adviser to the minister, and with the threat and advent of the war of 1939-45 became increasingly involved with policy in food production, not only in this country but on a world scale. He served on many official bodies, including the Agricultural Research Council and the Agricultural Improvement Council, and produced an official memorandum on food and agriculture.

These involvements were complemented by his close association with scientific and agricultural problems of the tropical empire. For thirty years from 1927, he had travelled extensively in the tropics reporting to some fifteen royal commissions and inquiries. He became much involved with major tropical crops, particularly tea, cotton, and rubber and their associated industries. His concern with the significance and importance of agricultural policy was widely based, and his last publication in 1980 was *Britain's Future in Farming*, written with Leonard Amey and other collaborators. Engledow retired from his chair in 1957. He was appointed CMG in 1935 and knighted in

1944. In 1946 he was elected a fellow of the Royal Society, on whose council he served in 1948-9.

Engledow was a deeply religious practising Christian and a considerable student of the Bible. He and his family were regular churchgoers and he was a churchwarden. He believed in self-discipline and practised it physically as well as mentally by exercise, games, horseriding, and carpentry. In 1921 he married Mildred Emmeline (died 1956), daughter of Frederick Edward Roper, merchant, of Cape Town. There were four daughters. Engledow died 3 July 1985 in the Hope Nursing Home, Cambridge.

[G. D. H. Bell in *Biographical Memoirs of Fellows of the Royal Society*, vol. xxxii, 1986; personal knowledge.]   G. D. H. Bell

ESCOTT, BICKHAM ALDRED COWAN SWEET- (1907-1981), SOE officer, banker, and businessman. [See Sweet-Escott.]

EVANS, (BENJAMIN) IFOR, Baron Evans of Hungershall (1899-1982), scholar and administrator, was born in Soho, London, 19 August 1899, the younger son and younger child of Benjamin Evans, journeyman carpenter, and his wife, Ann Powell, both of whom originated in Wales. He was educated at the Stationers' Company's School and at University College, London, obtaining first class honours in English (1920), the teaching diploma with honours (1921), and MA with distinction (1922). He won several college prizes and medals and the Early English Text Society's prize (1920), and became president of the Union. His first post was as lecturer in English at Manchester University (1922-4). He rose rapidly, spurred on by ability and the pursuit of worthy objectives rather than ambition, starting and finishing in universities with a brief but significant excursion into the outside world and numerous subsequent contacts with it.

He was professor of English at Southampton (1925-6), Sheffield (1926-33), and Queen Mary College, London (1933-44); the years 1940-4 being spent as educational director of the British Council. He was principal of Queen Mary College in 1944-51 and provost of University College, London, 1951-66. Part-time commitments included service as vice-chairman of the Arts Council (1946-51); chairman of Thames TV's educational advisory council, the Linguaphone Institute, the Observer Trust (1957-66), the Royal Society of Literature (1975-7)—and its vice-president (1974); governor of the Old Vic and Sadler's Wells; and trustee of the British Museum.

The springboard for this diverse career was

a deep love of English literature. Evans always spoke dismissively of his scholarly work but, though administrative talents took him along different paths, his list of publications is nevertheless long and varied. A very precise kind of textual scholarship is firmly evinced in his edition (with (Sir) W. W. Greg, q.v.) of *The Most Virtuous and Godly Susanna* and *Jack Juggler* (both 1937, for the Malone Society). His main gift, however, lay in exposition. A witty and pithy lecturer, lucid and sensible, he also wrote works which generations of students have found valuable and admirably attuned to their particular needs, encapsulating the quality, character, and place of an author in the pattern of literary history.

Evans's work in the University of London, hard hit by the war, deserves special mention. In 1945 he brought Queen Mary College back, after evacuation, to bomb-scarred buildings in a devastated area. Seeing its potential, he set about acquiring land which doubled the college's site and made possible the expansion necessary for its survival as bearer of light in a dark corner of London. His subsequent appointment as provost of University College crowned his career; it brought his varied abilities and his intense loyalty back, when they were most needed, to an institution he loved. He owed much to UCL and amply repaid that debt. Reconstruction, after catastrophic bombing, had barely begun and the college was just settling down after wartime dispersal. Evans's energy and leadership brought all the threads together and released a creative spirit of unity and co-operation. Although public money was becoming less scarce, the college's needs, particularly for building purposes, far outstripped support available from the University Grants Committee, despite generous help from the University Court, and he realized that he must go out and get funds if UCL was to prosper in teaching and research and especially to provide adequate student living accommodation. He was remarkably successful in attracting benefactions, and enjoyed the careful preparation, discreet matching of potential donor with appropriate cause, and the interplay of foresight, patience, timing, and presentation involved. He managed these operations with a rare combination of drive and charm. Charm, founded on true human warmth, not on flattery or superficial glitter, indeed played a part in all he did. Many acts of kindness stand to his credit: most of them unprompted, springing from his interest in people and understanding of their needs.

He maintained that the provost of UCL had no powers and few duties, the latter consisting mainly in chairing professorial board meetings, getting good heads of departments, and then backing them. As to powers, he needed no formal statement; his persuasiveness and evident disinterestedness and dedication combined to melt opposition. As to his style as a chairman, his obituarist in *The Times* wrote, with some justice, of the 'affectionate disdain' with which he treated his colleagues; though 'affectionate leg-pulling' might have been more accurate. During a discussion on tree planting in the college quadrangle, one professor opined that 'trees should be chosen that will flower at times when we can enjoy them'; to which Evans replied: 'Then, my dear professor, we must have trees that will flower between 11 a.m. on Tuesdays and 4 p.m. on Thursdays.' His irony could also be self-directed; he once said 'Bill Coldstream painted my portrait, and I've been trying to look like it ever since.'

He had a love-hate relationship with the Federal University of London. He endorsed the dictum of Lord Beveridge [q.v.] that it could achieve more than could its component colleges separately; but it irked him when UCL's desire to introduce new material into syllabuses was balked by boards of studies because not every college could teach it. He welcomed the Saunders reforms of 1966, which gave more freedom to colleges within the federal framework. He had, earlier, helped to found the University Students' Union and later worked for the erection of its fine new building. He discharged the duties of public orator with supreme elegance. After retirement he became director of the Wates Foundation—an interesting example of enterprising poacher turned generous gamekeeper.

Evans was knighted in 1955 and became a life peer in 1967. He received honorary degrees from Paris and Manchester, and became an officer of the Legion of Honour and commander of the Orders of Orange Nassau and Dannebrog.

He married, in 1923, Marjorie Ruth, daughter of John Measures, of Ifield, Sussex. She was a fellow student at UCL. They had one daughter. Evans died at Tunbridge Wells 28 August 1982.

[Records of University College, London; *The Times*, 1 September 1982; private information; personal knowledge.]

ARTHUR TATTERSALL

EVANS, SIR DAVID GWYNNE (1909–1984), microbiologist, was born in Atherton, Lancashire, 6 September 1909, the third in the family of four of Frederick George Evans from Pembroke, an Atherton headmaster, and his wife, Margaretta Eleonora Williams, a school teacher from Bangor. His elder brother, Meredith Evans [q.v.], was professor of physical chemistry at Leeds and Manchester and a fellow of the Royal Society. His younger brother, Alwyn, was such a professor at Cardiff, and his sister, Lynette, a high school teacher with a language degree from Manchester.

Evans left Leigh Grammar School in 1928 and after two years with the British Cotton Growers Association went to Manchester University with Alwyn; both graduated in physics and chemistry in 1933, gaining their M.Sc. degrees a year later. David Evans then joined the Manchester University department of bacteriology under Professor H. B. Maitland and so started his career in microbiology. Their research on *H. pertussis* led to his lifetime interest in whooping cough. He took his Ph.D. in 1938.

In 1940 Evans went to the department of biological standards at the National Institute for Medical Research, Hampstead, under (Sir) Percival Hartley, FRS. He worked on clostridial antitoxin standardization and tetanus vaccination schedules. In 1947 he returned as reader in the bacteriology department at Manchester to restart the diploma in bacteriology course. In spite of heavy teaching commitments he and his associates continued their research into whooping cough and clostridial toxins. He became secretary of the Medical Research Council whooping cough vaccination committee which organized trials of *H. pertussis* vaccines leading to the establishment of a British Standard vaccine.

In 1955 Evans went back to Hampstead as director of the new Biological Standards Control Laboratory which was formed initially to monitor poliovaccines and formulate tests for their safety and potency. In 1958 he also became director of the parent Department of Biological Standards. In 1957 he was appointed chairman of the MRC committee on the standardization of freeze-dried BCG vaccine which was successfully introduced as a replacement for the current liquid one: in consequence he was invited to be president of the international symposium of BCG vaccine at Frankfurt-on-Main in 1970.

From 1961 to 1971 Evans was professor of bacteriology and immunology at the London School of Hygiene and Tropical Medicine. He became secretary and then chairman of the Medical Research Council's measles vaccine committee and organized trials of this product in children. He served on the World Health Organization expert panel on biological standardization, the Central Health Services council, the British Council medical advisory committee, and the Medical Research Council.

From 1964 Evans became involved with veterinary problems, being on the governing body of the Animal Virus Research Institute, at Pirbright, and the Northumberland committee investigating the foot-and-mouth disease outbreaks of 1967. In 1971 the Joint Racing Board asked him to undertake a study of influenza in racehorses and in 1972 he became a member and then chairman (1973-9) of the board's veterinary advisory committee. He organized an investigation into contagious equine metritis which identified the agent responsible and recommended methods of prevention. In 1971-2 Evans was director of the Lister Institute and struggled in vain to save its Chelsea laboratory from financial failure.

In 1972 he was appointed director of the new National Institute for Biological Standards and Control in order to prepare for its transfer in 1976 from the MRC to the National Biological Standards Board and its move to South Mimms. From 1973 he was on the committees of safety of medicines and the British and European pharmacopoeia commissions. In 1976 Evans went to Oxford and taught medical students at the Sir William Dunn School of Pathology until 1979 when he retired to north Wales. To this Dictionary he contributed the notice of A. T. Glenny.

Enthusiasm and commitment were Evans's main attributes both in his scientific and social life. These made him an outstanding student and graduate teacher; he had the knack of presenting subjects lucidly and simply. He always had time to listen to others, however junior, and bring out the best in them. He was always welcome socially as well as scientifically and liked to take colleagues on trips to the Lake District and Snowdonia, places he loved very much. Opera was his special interest and later in life he enjoyed gardening.

Evans gained his D.Sc. in 1948, and was elected FRS in 1960 and FRCPath. in 1965. In 1968 he was awarded the Stewart prize by the British Medical Association for his studies on epidemiology and in 1977 the Buchanan medal of the Royal Society because his work had 're-volutionized the picture of childhood disease'. In 1969 he was appointed CBE and in 1977 received a knighthood. He was president of the Society of General Microbiology (1972-5) and received an honorary doctorate from Surrey University in 1982.

In 1937 he married Mary, a fabric designer and artist, the daughter of Ben Darby, district electrical engineer, of Atherton. They had two children, John, an archaeologist, and Mary, a schoolmistress and lacrosse international. Evans died at Rhos-on-Sea 13 June 1984.

[A. W. Downie, E. C. Gordon Smith, and J. O'H. Tobin in *Biographical Memoirs of Fellows of the Royal Society*, vol. xxxi, 1985; personal knowledge.]　　　J. O'H. TOBIN

EVANS, SIR TREVOR MALDWYN (1902-

1981), journalist, was born in Abertridwr, Glamorgan, 21 February 1902, the only son and elder child of Samuel Evans, a police sergeant, and his wife, Margaret Jones. He was educated at Pontypridd Grammar School. When he was fifteen his father was killed in a railway-crossing accident and in order to bolster the family finances Evans put aside his ambition to go to the University College of Wales, Aberystwyth, to train as a teacher. Instead he went down the local pit and, for four years, worked as a trainee electrician until the miners' strike of 1921 threw him out of work. Evans went to night school, obtained a London matriculation certificate, became a pupil teacher, and added to his earnings by doing free-lance reporting for the Glamorgan Free Press, for a payment of 25p a column. One week the paper left out his column, so Evans tramped over the hills to Pontypridd and confronted the editor, who at once appointed him as his assistant. His life's work in newspapers had begun.

In 1926 he moved to the *Daily Dispatch* in Manchester; in 1928 to the *Daily Mail*, and in 1930 to the *Daily Express* as news editor in the Manchester office. In 1933 he moved to the *Daily Express* London office and became one of a renowned editorial team—to be known as the 1933 Club—that Arthur Christiansen [q.v.], editor at the age of twenty-nine, was building up around him. Increasingly, during these nomadic years, Evans had become a writer deeply involved in labour and industrial affairs. These had been years of depression and unemployment; the trade unions were emerging as a powerful influence in the nation and Evans, with his mining background and a steadfast devotion to the miners' cause, saw a supreme opportunity to pioneer a new era in industrial reporting.

He was not alone in this ambition. Two other distinguished Fleet Street industrial correspondents, Ian Mackay (of whom he later wrote a biography, 1953) of the *News Chronicle* and Hugh Chevins of the *Daily Telegraph*, joined him in establishing a standard of reporting that did justice to the growing importance of industry and labour in national life. They cultivated close and friendly relations with employers, unions, and civil servants. Evans in particular practised a rare political detachment; within the National Union of Mineworkers, for example, he was on intimate terms with right-wing leaders such as (Sir) Will Lawther [q.v.] and Sam Watson, as well as with communists such as Arthur Horner [q.v.]. Yet this detachment never robbed him of a fearless invective when he considered that either employers or unions were guilty of conduct that fell short of the highest standards.

The trust and confidence he established soon made him a widely known and respected writer at all important industrial conferences, especially the Trades Union Congress. He acquired an easy sociability that often masked a steely determination to get every detail of a story right and in the correct perspective. He was a superb after-dinner speaker and raconteur, fluent, engaging, with a touch of Welsh exuberance. He put his diplomatic skills to good use, especially during the war years when on one famed occasion he tried to bring together those two irreconcilables, Ernest Bevin [q.v.], then minister of labour, and Lord Beaverbrook, then minister of production. He arranged a dinner party at Beaverbrook's home. 'A nice place you have here', said Bevin on arrival. 'Yes', replied Beaverbrook, 'and I expect you fellows will take it away from me some day.' Bevin departed at once without his dinner. Temperamental incompatibility had, for once, defeated Evans's diplomacy. He later wrote a biography of Bevin (1946).

In the course of his career, Evans was offered a high appointment with the Federation of British Industries and invited to stand for Parliament by the Labour Party. His devotion to journalism overrode all temptation. He was also later to become a director of Express Newspapers, but the executive role never greatly appealed to him; he remained always the good reporter.

He became chairman of the London Press Club, later vice-president, and finally honorary life member emeritus. He was appointed to the Press Council in 1964 and served as a member until 1975. Throughout all this full life, he found time to teach and inspire a new generation of industrial reporters. Many of his pupils achieved fame in both newspapers and in public life. He was appointed CBE in 1963 and was knighted in 1967.

In 1930 he married Margaret ('Madge') Speers, daughter of John Best Gribbin, journalist, of Heaton Moor, Cheshire. They had a son and a daughter: Richard Trevor, a senior journalist on the *Financial Times*, and Marilyn Speers, King Edward VII professor of English literature at Cambridge University. Evans died suddenly 10 June 1981, while travelling from his home in Kingston upon Thames, Surrey, to a meeting at the London Press Club.

[*The Times* and *Daily Express*, 11 June 1981; private information; personal knowledge.]

EDWARD PICKERING

EXETER, sixth MARQUESS OF, and BARON BURGHLEY, (1905-1981), athlete and parliamentarian. [See CECIL, DAVID GEORGE BROWNLOW.]

F

FAIRFIELD, CICILY ISABEL (1892–1983), author, reporter, and literary critic. [See WEST, DAME REBECCA.]

FARMER, HERBERT HENRY (1892–1981), theologian, was born in Highbury, London, 27 November 1892, the youngest of four sons (there were no daughters) of William Charles Farmer, journeyman cabinet-maker, and his wife, Mary Ann Buck. He attended Owen's School in Islington, going from there as a scholar to Peterhouse, Cambridge, where he read for the moral sciences tripos. Awarded a first class in part i in 1914, he was elected Burney student in the philosophy of religion, proceeding to Westminster College in 1916 to prepare for the ministry of the Presbyterian Church of England. His first pastorate was at Stafford, where he went in 1919, moving in 1922 to New Barnet, where he remained till 1931.

Farmer was already a talented preacher, who eschewed flamboyant rhetoric, but who succeeded in a rare way in combining reasoned argument with a most searching appeal to his hearers, as a first volume of sermons, *Things Not Seen* (1927), shows.

In 1931 he joined the staff of Hartford Seminary Foundation in Connecticut, returning to England four years later to the Barbour chair of systematic theology at Westminster College in succession to John Oman (whose notice Farmer later wrote for this Dictionary). Farmer acknowledged Oman as his master, dedicating to him his first major work: *The World and God* (1935). They were both Reformed theologians who had been emancipated through study of Kant's ethics, from a narrowly predestinarian understanding of the relation of grace to human freedom. But as his sermons, and also his pupils attest, Farmer combined with Oman's austerity a sensitivity and delicacy very much his own. He was a rigorous and perceptive teacher, as the achievement of those whom he supervised at Cambridge indicates; they include J. A. T. Robinson [q.v.], Professor John Hick, and Professor Allan Galloway.

His teaching in Cambridge (including three years as Stanton lecturer) made him a natural choice in 1949 for the Norris-Hulse chair of divinity. The first two incumbents of that chair had been New Testament scholars. But the situation in Cambridge in 1949 made it desirable to appoint a philosopher of religion, and Farmer was duly elected, becoming a fellow of Peterhouse in 1950.

The Cambridge in which Farmer had read philosophy was intellectually very different from the one in which he began his work as professor. James Ward and W. R. Sorley [qq.v.] were the professors of his student days: since then the influence of Bertrand Russell, G. E. Moore, Ludwig Wittgenstein, C. D. Broad [qq.v.], and Arthur Wisdom had introduced very different styles of philosophizing. Farmer's *œuvre* seems to come from another world. Yet such a judgement is superficial, as familiarity with his book on the preacher's task—*The Servant of the Word* (1941)—makes plain. The influence of Martin Buber's *Ich und Du* (1923) is strongly marked in this work; but it is primarily reflection by a master of the pulpit on his task. Shortly before, Farmer had published his second collection of sermons, *The Healing Cross* (1938), and to read these two books together provides the critical student of religious language with an indispensable introduction to evaluation of the preacher's address as a fundamental dimension of that language.

Although Farmer's understanding of religious experience invites comparison with the strikingly similar emphases of his older Oxford contemporary, C. C. J. Webb [q.v.], it sprang out of his interpretation of the preacher's *ministerium verbi divini*. His conception of God as 'unconditional demand' and 'absolute succour' might be thought to presuppose acceptance of the ontological argument, in the form in which Webb had recast it. But it was first and foremost the God whom Farmer proclaimed, one who addressed men and women in terms of an unrelenting demand on their obedience, but who in the same moment revealed himself as infinitely compassionate. One of Farmer's finest sermons treated of the loneliness of Christ in his Passion; the theme elicited that rare combination of tenderness and austerity which was one of the pervasive marks at once of his doctrine and his whole ministry.

Farmer was a pacifist. His pacifism was to him a very costly part of his faith, and its texture reveals the way in which such a commitment may be born of an understanding of the ways of God with men that is of an undaunted austerity of conception. He rejected every form of facile optimism; if he remained curiously untouched by the work of Karl Barth, his theology was a *theologia crucis*. It is significant that if his Gifford lectures (1950), were somewhat disappointing, his last book published seventeen years later—*The Word of Reconciliation* (1966)—is a profound essay in the theology of the Atonement, haunted by the memory of issues dividing James Denney from John McLeod Campbell [qq.v.], and both alike from Oman, but suffused with Farmer's own reverent, yet always restlessly interrogative spirit.

He was a warm and affectionate man, rooted in the traditions of the Presbyterian Church of England, and convener of its Assembly's committee on doctrine, which, during the years 1943–8, drafted and finally brought before the Assembly the document now known as 'Statement of the Christian Faith'. His influence on the ministers of his own Church, whom he helped train at Westminster, was considerable, and if his role in the university faculty of divinity in promoting his subject was significant, his more pastoral work at Westminster in preparing ministers for their work should not be forgotten. Indeed Farmer's strength lay in the fact that he was all of a piece. It is characteristic that after appointment as Norris-Hulse professor, he continued to serve Westminster without payment.

Farmer retired in 1960, living first in Hove and later in Lancashire. For a long time he continued to preach, and his book on the Atonement (originally lectures given in America) belongs to those years.

In 1923 he married Gladys Sylvie, daughter of Thomas Offord, motor engineer, of Brighton. They had a son and two daughters. Farmer died 13 January 1981 in Birkenhead.

[*Yearbook* of the United Reformed Church in the UK, 1982; *Reform* (URC monthly), March 1981; *Presbyterian Messenger*, May 1960; private information; personal knowledge.]   DONALD M. MACKINNON

FENDER, PERCY GEORGE HERBERT (1892–1985), cricketer, was born 22 August 1892 at Balham, London, the elder son (there were no daughters) of Percy Robert Fender, head of a wholesale stationery firm, and his wife, Lily Herbert. He was educated at St George's College, Weybridge, and St Paul's School.

Fender played first for Sussex in 1910 when a schoolboy of seventeen, and showed much youthful all-round promise. In 1914 he transferred his allegiance to the county of his birth and played a useful part in the championship won by them in a season cut short by World War I. He enlisted in the Royal Fusiliers, then moved in 1915 to the Royal Flying Corps which shortly became the RAF. He served also with the RAF in World War II, becoming a wing commander.

A leg injury on the football field denied him cricket in 1919. The following summer he caught the limelight as an all-rounder and won a place in the MCC team which toured Australia under J. W. H. T. Douglas. Fender emerged well enough from the tour to play also against Warwick Armstrong's formidable Australians in two of the 1921 tests. In 1922/3 he toured South Africa with the MCC as vice-captain to F. T. Mann. In thirteen tests over his career his all-round contributions were respectable, no more.

His true *métier* was county cricket and as an innovative, frequently daring captain of Surrey. For them his feats with the bat, as a leg-spin and googly bowler and as a tip-top slip fielder, had a uniquely spectacular quality. With his horn-rimmed spectacles, crinkly hair, short moustache, and sweaters almost down to his knees, his appearance was as unusual as his cricket. He was the cartoonists' delight.

Fender is remembered for the fastest ever hundred (since equalled) in 35 minutes at Northampton in 1920, immediately prior to Surrey's declaration at 619 for five. But there were many more meritorious feats of hitting against stronger opposition. He once took 24 off an over by Maurice Tate. He made a hit of 132 yards out of the Oval over long-off against Kent, and also cut another ball square out of the ground. The cricket historian Gerald Brodribb reckoned his big innings were scored at 62 runs an hour. (G. L. Jessop is rated by the same authority at 80 runs an hour and Fender is among his ten fastest scorers.)

Fender could play a waiting game, but Surrey's batting strength was such that he rarely needed to. His problem on the flawless Oval pitches of his time was usually to get the other side out twice, and here he ingeniously enjoyed full scope both as bowler and tactician. He was always prepared to buy wickets with his wrist-spin, sometimes by means of the dropping full pitched delivery.

In 1923 he made 1,427 runs and took 178 wickets. Five other times he did the double and three times more scored 1,000 runs. In his career he made 19,034 runs including 21 hundreds, average 26, and took 1,894 wickets at 25 a time. He took 598 catches, almost all in the slips. His ploys and love of a gamble meant that Surrey's cricket during his eleven years' reign was never dull. True, he never had the bowling quite to win a championship. Yet there may have been substance in the comment that he was the best captain England never had. There were undoubtedly times when he must have been a contender to lead England. He was however too outspoken—and even tactless—to suit the hierarchy at Lord's, especially as embodied in the fourth Baron Harris [q.v.]. While always highly popular with the teams he led his relations with the Surrey authorities were also apt to be difficult. He was superseded as captain by D. R. Jardine [q.v.] rather than retiring from office in 1932, and the termination of his cricket for Surrey after 1935 was accompanied by some friction.

Fender was in business a wine merchant,

founder and for fifty-four years until his retirement at the age of eighty-four managing director of Herbert Fender & Co. He wrote on the
game of cricket with discernment and a technical
insight uncommon in his day. There were four
books on test tours, two on the visits of the
MCC to Australia in 1920/1 (in which he took
part) and 1928/9, and two about the Australians
in England in 1930 and 1934. He wrote regularly for the *Evening News*, and was the first
man to use a typewriter in the press box, an
innovation not at first well received.

Fender married first in 1924 Ruth Marion
(died 1935), daughter of William Clapham, jew-
eller and silversmith, of Manchester. They had
a son and a daughter. In 1962 he married Susan
Victoria Gordon (died 1968), the daughter of
Captain J. T. Kyffin, of the Royal Flying Corps
in World War I and of independent means.
Having been inflicted for several years with
almost total blindness, Fender died at Exeter 15
June 1985.

[*Daily Telegraph* and *The Times*, 17 June
1985; Wisden, *Cricketers' Almanack*; Richard
Streeton, *P. G. H. Fender*, 1981; personal
knowledge.]   E. W. SWANTON

FFORDE, SIR ARTHUR FREDERIC
BROWNLOW (1900–1985), lawyer, headmaster, and public servant, was born 23 August
1900 in Watford, Hertfordshire, the only son
and eldest of four children of Arthur Brownlow
fforde, of the Indian Civil Service, and his
wife, Mary Alice Storer Branson. Educated at
Rugby, where he was a scholar and head of the
school, he came under the liberal influence of
his housemaster and headmaster, A. A. David
[q.v.], who later became bishop of Liverpool. A
scholar of Trinity College, Oxford, from 1919,
he disappointed himself and his tutors by
achieving only a third class in *literae humaniores*
(1922).

He had been drawn to both teaching and law
as a career, but this result settled the issue. He
joined the well-known firm of London solicitors, Linklaters & Paines, qualifying in 1925
and being admitted to partnership in 1928. As
a leading partner during the 1930s in what became the largest firm of solicitors in the City he
gained an outstanding reputation as a company
lawyer, not least for the concision and electric
speed of his draftsmanship, and was involved in
several important developments such as establishing under English law the basis of unit
trusts. In 1937 he became a member of the
council of the Law Society, and was much concerned with institutional and educational aspects of his profession.

During World War II he dealt with supply
contracts and was successively deputy

director-general and under-secretary in the
Ministry of Supply and in January 1944 under-
secretary in the Treasury. Returning in 1945
to Linklaters & Paines, he and another senior
colleague, (Sir) Samuel Brown, together had the
main responsibility for re-establishing the firm
after the destruction of its premises and records
in the blitz.

His appointment therefore in 1948 to the
headmastership of Rugby School surprised
everyone, and dismayed some. The governing
body's explanation that, since the post-war public schools' problems were likely to be economic
and political, they needed a man of proved
ability in these fields, convinced few critics.
However, the school flourished under him, the
personal qualities for which he had been selected more than compensating for his lack of
school experience. He was at his best in his
pastoral capacity, not least as housemaster of
School House, when boys could appreciate his
discernment of their character and difficulties,
his humanity and understanding, and his respect for their individuality.

On retirement in 1957 he was appointed
chairman of the British Broadcasting Corporation, accepted several City directorships,
and joined the Church of England's central
board of finance (1957–70), which he chaired
from 1960 to 1965.

His chairmanship of the BBC was widely
regarded as most successful. Under his sensitive, gentle, yet firm leadership the board of
governors fully supported the policy of (Sir)
Hugh Greene, the director-general, of opening
the windows of Broadcasting House to meet the
challenge of commercial television. He handled
external relations so skilfully that the BBC's
independence was never seriously endangered.

Forced by illness to retire in 1964, he was
appointed GBE in recognition of his distinguished chairmanship. He had been knighted
in 1946 for his wartime service, and honoured
in 1960 by the University of Wales with an
LLD.

Though slight in build, with a puckish air
about him in his more quizzical moods, he never
seemed slight in stature. His qualities were so
unusual that even his closest friends did not
find him an easy man to know fully; yet there
was a simplicity about him which revealed the
essence of his character, as being deeply religious, sensitive, perceptive, tolerant, and immensely kind. His acute and original sense of
humour, expressed orally, in verse, and in drawings, was always a delight, if sometimes too
subtle for immediate understanding. Children
in particular found him the most companionable
and congenial of men, and appreciated his serenity and wisdom. Though somewhat handicapped by ill health in his later years he was an

outstanding public figure, but of a particularly private kind. He never sought or welcomed the limelight and remained dedicated to a spiritual philosophy of thought, as shown in the Christmas poems that he sent to his friends.

In 1926 he married (Mary) Alison, younger daughter of James MacLehose, printer to the University of Glasgow, who was a steadfast support to him at all times, especially during his headmastership of Rugby where, keeping open house for all comers, they enabled boys, particularly those in School House, to feel part of an extended family. They had two sons and a daughter. fforde died 26 June 1985 at Bramley near Guildford, Surrey, and his ashes were laid on the island in the Close at Rugby School, as was his wish. There is a portrait in the Temple speech room at Rugby by Sir William Hutchison.

[*The Times*, 29 June 1985; Rugby *Meteor*, 1986; Old Rugbeian Society *Newsletter*, no. 214, September 1985; J. B. Hope-Simpson, *Rugby since Arnold*, pp. 234-42; private information; personal knowledge.]

MICHAEL MCCRUM

FITTON, JAMES (1899-1982), painter and designer, was born 11 February 1899 in Herbert Street, Oldham, the second son and youngest of three children of James Fitton, iron plater and union leader, and his wife, Janet Chadwick, mill weaver. He was educated at the Watersheddings Board School, leaving in 1913. Recuperation from a bungled mastoid operation had left him partially deaf and made a gap in his schooling during which he discovered the joy of drawing. His father, sacked after a lockout, could only find night work at a distance under an assumed name; but his Fabian meetings brought James Keir Hardie and Emmeline Pankhurst [qq.v.] to the house.

Fitton's first solid employment was as a sorter of textile samples in Manchester, but he was accepted without fee as an evening student at Manchester Art School. Here he attended for six years under Adolfe Vallette, one-time assistant to Degas, with Sam Rabin and L. S. Lowry [q.v.] as fellow students. With Lowry he drew the countryside and the music-hall and educated himself at concerts, opera, ballet, and the public library.

In 1921 he went to London. Hawking his work, he found employment at J. S. Riddell, the printer, stayed for eighteen months lowly, despised, but avidly learning, and gave up when he was paid fifty pounds for two twelve-foot murals for the British Gas Association at the Ideal Home Exhibition. On the embankment he bumped into J. R. Clynes [q.v.] from Oldham who insisted on taking him back to stay at 11

Downing Street while he looked for work. He encountered H. G. Wells, the Webbs, J. Ramsay MacDonald [qq.v.], spotted a letter bomb, and came under police observation while designing a poster for the Russian trade delegation just before the Arcos raid. When his father's promotion to national organizer of the Amalgamated Engineering Union brought the family to London, Fitton rejoined them.

He was relieved from the drudgery of making oil paintings from film stills for Lasky's Film Corporation by a two-year spell as illustrator to a monthly adventure magazine edited by Eric Maschwitz which gave him time to visit the Victoria and Albert Museum and the British Museum—and to paint. By 1925 he was able to go to evening classes in lithography at the Central School of Art and Crafts with A. S. Hartrick who had shared a studio with Van Gogh. He met Royal College and Slade students at the class, notably Barnett Freedman [q.v.], and he was absorbed into the world of young London artists. In 1928 he married a fellow student Margaret Mary Elizabeth Cook, the painter and illustrator, daughter of Walter Cook, civil servant (and also a painter). They had one son and one daughter. James and Peggy Fitton moved to Pond Cottages close to his family in Dulwich, and he took a job with Vernon's, a small advertising agency. In 1930 he became a member of the New English Art Club, the London Group, and the Senefelder Club; he was appointed art director of Vernon's agency. His talent as designer and his gift for friendship, translated into public relations, built up the firm to one of the most successful of pre-war days. He remained in the agency for over fifty years while engaging in multifarious other activities.

In 1937 he was commissioned to design two posters for London Transport by Frank Pick [q.v.], and in 1938 murals for the United Kingdom government pavilion at the Empire Exhibition in Glasgow. In the 1940s he undertook free-lance illustration in a great number of newspapers and magazines ranging from *Housewife* to *Lilliput*, posters for Ealing Studios, and the art work for the film *Kind Hearts and Coronets* (1949).

In 1933 he had inherited Hartrick's class at the Central and it filled up with his friends. They were appalled by mass unemployment, Nazi aggression, and the threat of war, and inspired to action by the graphics of *Krokodil*, *Simplicissimus*, and the drawings of Grosz. In the studio of (Sir) Misha Black [q.v.] James Fitton, James Boswell, and James Holland together with Pearl Binder and Clifford Rowe, fresh from a trip to Russia, founded the Artists' International Association. They plunged into the production of cartoons, posters, banners, pamphlets, and exhibitions. By 1935 virtually all respected London artists had become members. Cartoons by 'the three Jameses' in the *Left Review* (from 1934) began the revival of satirical drawing which, persisting through the war, led to a new age of British caricature exemplified by Gerald Scarfe and Ralph Steadman.

Apart from a single one-man exhibition at Arthur Tooth (1933) Fitton's career as an exhibiting painter was within the Royal Academy. He was elected ARA in 1944, and RA in 1954. From 1929 when he first showed there he appeared as the leader of a new popular realism, in which both visual description and social comment could be easily appreciated, and which might be seen now as the precursor of both Pop Art and (as he noted) post-war Kitchen Sink. Sir Alfred Munnings [q.v.] singled him out as a dangerous modernist. Modernist he was not, but he was sympathetic to advanced European art. He was one of those few artists who revolutionized commercial graphics by an infusion of modernism, and brought the art of the poster to a peak in the 1930s. He also entered the Academy with the zeal of a reformer. As chief assessor for the Ministry of Education diploma in design (1940-65) his thoughts on the future of art or the role of the Royal Academy were influential. He was runner-up for the presidency in the elections of 1954 and 1956 and a strong contender in 1966. He was a trustee of the British Museum (1968-75).

His speech was rapid, his eyes twinkled, and a red handkerchief flopped from his breast pocket. He had great gusto for life, humorous enthusiasm, and common sense. He died at Pond Cottages in his sleep 2 May 1982.

[John Sheeran, 'James Fitton, an Appreciation' in catalogue *James Fitton RA 1899-1982* (Dulwich Picture Gallery, 1986); introduction by Sidney Hutchison in catalogue *James Fitton RA 1899-1982* (Oldham Art Gallery, 1983); Fitton's unpublished autobiography; private information.]

FREDERICK GORE

FITZGIBBON, (ROBERT LOUIS) CONSTANTINE (LEE-DILLON) (1919-1983), writer, was born 8 June 1919 at Lenox, Massachusetts, the only son and youngest of four children of Francis Lee-Dillon FitzGibbon, who had retired from the Royal Navy as commander and later served in the French Foreign Legion, and who was a direct descendant of John FitzGibbon [q.v.], first Earl of Clare, lord chancellor of Ireland, whose granddaughter and heiress married a Dillon but made him change his name to FitzGibbon. Constantine Fitz-Gibbon's mother was Georgette May Folsom, from New England, and as his parents were parting at that time, he was born in the home

of her father. He was brought up at Lenox, and later in France, with his three elder sisters. Both his father and his paternal grandfather had been born abroad, so the only citizenship that he could claim was American.

FitzGibbon went to school first at the New Beacon, Sevenoaks, and then, with a scholarship, to Wellington College, which he left before he was sixteen. After a period in Germany, studying painting and immersing himself in European literature, he went in October 1937 with an open scholarship in modern languages to Exeter College, Oxford. He left in the spring of 1939 without taking a degree, and then married in the same year, briefly and unsuccessfully, Margaret Aye Moung, a beautiful half-Burmese, half-Irish fellow student from Somerville College, the daughter of M. R. Aye Moung, superintending engineer in the Public Works Department, Rangoon.

An attempt to join the Fleet Air Arm was frustrated by defective eyesight. In 1940 Fitz-Gibbon joined the Irish Guards, but later transferred to the Oxford and Buckinghamshire Light Infantry and in 1942 to the US army. He then served as an intelligence officer on the staff of General Omar Bradley at 12th Army Group, throughout the preparations for the invasion and the campaign in France, and then in Germany. There he saw the workings of power at first hand, and it left him forever fascinated by authority, and profoundly suspicious of it.

On the dissolution of his marriage in 1944, he married Joan ('Theodora') in Chelsea. She was the daughter of Adam Rosling. As soon as he was out of the army in 1946 they went together to Bermuda, where an aunt had lent him a cottage, so that he could settle down to write. In 1949 FitzGibbon's first novel, *The Arabian Bird*, a wartime drama of considerable psychological depth, was published by Cassell to enthusiastic reviews. He returned to England with Theodora, and they settled at Sacombs Ash in Hertfordshire. There ensued a productive decade in which FitzGibbon confirmed his reputation as a writer. He always worked very fast, and now reckoned to spend three-quarters of the year on reviews, articles, and translations, leaving three months to complete a novel or other book of his own. The Fitz-Gibbons became close friends of their neighbour Henry Moore, the sculptor, and Theodora started her notable career as a writer of cookery books.

Two of FitzGibbon's most successful books soon appeared. *When the Kissing Had to Stop* (1960) was a futuristic political morality about the Soviet take-over of a Britain in decline—a threat by which the author had become increasingly concerned, particularly since his extensive research in Germany for *The Shirt of*

Nessus (1956). He had been impelled to write this, a history of the tragic failure of the Stauffenberg conspiracy to kill Hitler on 20 July 1944, by his feeling that the true story of that courageous exploit had been suppressed by the wartime propaganda of both sides.

FitzGibbon was an industrious and urbane writer, all of whose work was characterized by a lucid and elegant use of English. His true love was the novel, and he published fourteen works of fiction, but he also wrote books of autobiography, historical works about Ireland and Germany, and a distinguished biography. He translated a number of books from French and German (of which the most notable were the novels of Manés Sperber and *Der Fragebogen von Ernst von Salomon*, 1954) and there was a constant output of political and literary journalism, short stories, radio features, and occasional verse.

His marriage to Theodora broke down in 1959, and in the following year he married Marion McFadyean, née Gutmann. They moved to Dorset, where he took a happy pride in his brief ownership of Waterston, with its splendid Elizabethan façades and lovely setting. His only son Francis was born 18 July 1961. In 1965 he published *The Life of Dylan Thomas*, an objective yet affectionate biography of his old friend, and a fascinating evocation of the atmosphere of the time. It was perhaps Fitz-Gibbon's most considerable work.

The strain of dealing with that tragic story left him emotionally exhausted, with his marriage coming to grief. He decided to live in Ireland, and went there with Marjorie Sutton, who had previously been married to Huntington Hartford. As Marjorie Steele (daughter of Harold Wright Steele of Minnesota) she had made a career as actress and painter. They were married in 1967 and moved to Killiney near Dublin, where their daughter Oonagh was born the following year.

Marjorie began to work successfully as a sculptor, and FitzGibbon, who had grown a grizzled beard which, together with his thick spectacles, gave him a surprisingly professorial aspect, continued to write novels and histories, becoming a member of the Irish Academy of Letters and an Irish citizen in 1970.

FitzGibbon's passionate nature lived uneasily with his sceptical, probing intelligence, and he sometimes drank heavily if his work was not going well. Quarrels and aggression would suddenly erupt, although normally he was a man of great charm and courtesy, a delightful raconteur, and much loved by his numerous friends, to whom he was warmly loyal. Eventually he realized that his drinking had taken on the nature of a progressive disease, as he candidly described in *Drink* (1980). In 1978 he

took a successful cure in the Hospital of St John of God, Dublin, and then espoused the cause of the conquest of alcoholism, but the publication of his book did not achieve the important public results he had hoped for.

Illness clouded his last two years, and he died in hospital in Dublin 23 March 1983. He maintained his professionalism to the last and at the time of his death was writing a biography of Charlemagne.

[Constantine FitzGibbon, *Through the Minefield*, 1967, *Man in Aspic*, 1977, and *Drink*, 1980; Theodora FitzGibbon, *With Love: An Autobiography 1938-46*, 1982; personal knowledge.]   Roger Lubbock

FITZMAURICE, Sir GERALD GRAY (1901-1982), judge of the International Court of Justice, was born 24 October 1901 in Storrington, Sussex, the elder son of Vice-Admiral (Sir) Maurice Swynfen Fitz Maurice, and his wife, Mabel Gertrude Gray. Fitzmaurice was educated at Malvern and Gonville and Caius College, Cambridge, where he was a pupil of Arnold (later Lord) McNair [q.v.] and of which he became an honorary fellow in 1961. He obtained a first class (division II) in both parts of the law tripos (1923 and 1924). He was called to the bar (Gray's Inn) in 1925.

After five years of practice, in 1929 he was appointed assistant legal adviser to the Foreign Office. He remained in government service until 1960. He was principal legal adviser to the Ministry of Economic Warfare from 1939 to 1943, during the early war years, dealing with the important subject of contraband control, and returned to the Foreign Office as deputy legal adviser in 1943. In 1945 he was appointed second legal adviser, becoming first legal adviser in 1953, a demanding office which he retained until his election as a judge of the International Court in 1960. After his retirement from the Court in 1973 he was a judge of the European Court of Human Rights from 1974 to 1980. At the same period he took up practice at 2 Hare Court in the Temple and spent most of the last day of his life at a meeting of a team involved in the Libya/Malta continental shelf case.

Fitzmaurice was one of the greatest international lawyers of his generation. Above all, he was the practitioner *par excellence*, and this no less because his role was, for most of his career, as a lawyer working for his own government within the Civil Service. However, and this was a combination peculiar to him, he was also an excellent scholar whose formidable intellect found expression in a massive flow of lucid and authoritative writing, in which he combined excellent analysis with a shrewd and informed perspective of the diplomatic and political realities. The principal writings, which are of enduring importance, were the series of major articles in the *British Year Book of International Law* on 'The Law and Procedure of the International Court of Justice' (1950–60), the 'general course' at the Hague Academy (1957), and the valuable and candid essay on the future of international law in the *Livre Centenaire* of the Institute of International Law (1973). In addition, in his role as special rapporteur of the International Law Commission, he prepared five elaborate reports on the law of treaties. There are few writers able to present such an impressive alloy of scholarship and experience. This combination, it is important to note, was not a mere happy coincidence but a matter of essence, since for Fitzmaurice public international law was an intellectual system with its own internal logic and integrity and not a servant of political expediency.

The background to the vigorous intellectual output was a career which encompassed major episodes of diplomatic and professional life. Of his life at the Foreign Office in the post-war years a colleague of that period has written: 'Within the Foreign Office, he had an authority which made his advice widely sought and accepted. He approached his task with an understanding of foreign relations and a clinical detachment which made his advice both pertinent and precise. He was also very easy and companionable to work with: never ruffled or excited, always patient and charged with what seemed to be a quiet and slightly suppressed good humour.' From 1945 onward, Fitzmaurice played a very active role in a number of international conferences, which included San Francisco, at which the United Nations charter was drafted, and the two conferences on the Italian and the Japanese peace treaties. He also acted as counsel for the United Kingdom in several cases before the International Court of Justice.

In addition, Fitzmaurice was legal adviser to the United Kingdom delegation in the United Nations General Assembly in 1946 and 1948–59, and played a significant role in the United Nations conference on the law of the sea at Geneva in 1958. He was attuned to practice, and thus after his retirement from the International Court in 1973 he took up free-lance work (as a member of the English bar), becoming involved in two sets of proceedings before the International Court. He was also the very capable president of the Beagle Channel court of arbitration, and drafted the preponderance of the report of the tribunal.

As a judge of the International Court, Fitzmaurice displayed toughness of mind and independence, and his style of reasoning placed him in the less popular camp in the controversial South-West Africa cases. He did not become president of the Court, but left an imposing series of separate and dissenting opinions. A similar pattern developed when he became a judge of the European Court of Human Rights (1974–80). Whilst Fitzmaurice was in some sense judicially conservative, his carefully reasoned opinions did not derive from a simple aversion to judicial innovation, and he was very conscious of the interaction of rule and policy. He was part of the majority in the markedly innovative judgment in the North Sea continental shelf cases. Moreover, his approach to the interpretation of treaties derives precisely from a concept of workability and his own experience in negotiating treaties. Similarly, his views on the application of the European convention of human rights were based upon notions of how the system should function and its attractiveness to states generally. As he saw the matter, 'a court situated as the International Court is, must steer a course between being over-conservative and ultra-progressive'.

It would not have troubled Fitzmaurice one bit that he was in no sense a public figure, and that much of his important work was in a real sense arcane. He had his own rigorous standards (he was as fastidious in his dress as in his language), his professionalism was much appreciated by his colleagues, and he was able to make a highly practical contribution to the rule of law in international affairs. His honours included the CMG (1946), KCMG (1954), and GCMG (1960). He was appointed QC (1957), was elected a bencher of Gray's Inn (1961), and became president of the Institute of International Law (1967–9). He had honorary degrees from Edinburgh (1970), Cambridge (1972), and Utrecht (1976).

In 1933 Fitzmaurice married Alice Evelina Alexandra, daughter of Christer Peter Sandberg, CBE, consulting engineer to the Chinese government and other railways. They had two sons, the elder of whom died in 1979. Fitzmaurice died in London 7 September 1982.

[*The Times*, 9 September 1982; Sir Francis Vallat in *Graya*, 1982, p. 30; Sir Robert Jennings in *British Year Book of International Law*, vol. lv, 1984, p. 1; personal knowledge.]

IAN BROWNLIE

FLUCK, DIANA MARY (1931–1984), actress. [See DORS, DIANA.]

FORTES, MEYER (1906–1983), anthropologist, was born in Britstown, South Africa, 25 April 1906, the eldest child in the family of four sons and two daughters of Nathan Fortes, who left Russia as an adolescent to escape being drafted into the army. He went to the English town of Leeds, lived for a while in Memphis, Tennessee, then returned again to Leeds where he worked in the clothing industry with his two sons, before leaving for South Africa. There he met and married Bertha Karbel, of Yamshik in Lithuania, and settled down as an innkeeper. Meyer Fortes attended the South African Collegiate High School at Cape Town, dominated by Scottish teachers, and then with the aid of various scholarships took a BA in English and psychology at the University of Cape Town (MA, 1926).

On the basis of two scholarships and strong recommendations from the University of Cape Town, Fortes was accepted as a postgraduate student in psychology at University College, London, where he carried out research for a Ph.D. (1930) on non-verbal intelligence tests for inter-racial use under C. E. Spearman [q.v.]. From 1930 he held the Ratan Tate research studentship at LSE, working with Emanuel Miller at the first child guidance clinic in the East End of London on the effects of sibling order on adolescent behaviour. His association with the clinic and his contact with J. C. Flugel led to a permanent interest in psychoanalytic theory, especially as it affected interpersonal interaction within the family.

It was through Flugel that Fortes met the influential anthropologist, Bronislaw Malinowski, in 1931 and was invited to join his seminar at the London School of Economics. As a result he got to know (Sir) E. E. Evans-Pritchard [q.v.] and through him the other professor of anthropology at the School, C. G. Seligman [q.v.], both somewhat critical of Malinowski's functional theories as well as his style of functioning. But his seminars were the prevailing intellectual feature in the field and he was the dominant patron. He it was who backed Fortes for a fellowship (1934-8) with the newly founded International African Institute, financed by the Laura Spellman Rockefeller Foundation, which enabled a body of talented scholars from many countries to undertake field research in Africa. He was Rockefeller fellow in 1933-4, and a fellow of the Institute in 1934-8.

Advised by the administrator and anthropologist R. S. Rattray, Fortes left to carry out field-work among the Tallensi of the northern territories of the Gold Coast in December 1933. The initial direction of his research was towards 'the psychological approach to the study of African societies', with particular reference to the family. Under the influence of Malinowski's seminars the project became more sociological and in his major accounts of his research, *The Dynamics of Clanship among the Tallensi* (1945, but largely written in 1938) and *The Web of*

Kinship among the Tallensi (1949), he explored the nature of tribal social organization which was largely based upon kin groups and kin relationships, to whose comparative study he made major contributions. At the same time he was deeply interested in the link between the family, morality, and religion; in these studies, notably the Frazer lecture of 1957, entitled *Oedipus and Job in West African Religion* (1959), and in his essays *Religion, Morality and the Person*, published posthumously in 1987, his psychological and psychoanalytic interests came to the fore.

When his fellowship came to an end, Fortes took a temporary lectureship at the LSE (1938-9), then moved to Oxford to join Evans-Pritchard and A. R. Radcliffe-Brown who had taken up the chair of social anthropology in 1937. There he held a research lectureship for the first two years of the war. In 1941-2 Fortes carried out research in Nigeria under a project organized by (Dame) Margery Perham [q.v.], remaining in West Africa to carry out intelligence work. In 1944 he became head of the sociological department in the West African Institute, Accra, a forerunner of the new University of the Gold Coast. There he directed the Ashanti Social Survey (1945-6), one of the first major socio-economic enquiries in an oral culture carried out in conjunction with a geographer and economist, and making use of modern data processing methods. Returning to England in 1946, he joined Evans-Pritchard and became reader in social anthropology. The two of them had long planned for a 'new' anthropology, less functionalist, more structuralist, than the Malinowskian variety and they built up a strong department. In 1950 Fortes moved to the William Wyse chair at Cambridge, a position which he held until his retirement in 1973.

Fortes was one of the leading members of that outstanding generation of British scholars who followed Malinowski and Radcliffe-Brown at a time when social anthropology in this country was at the peak of its reputation. He made important advances in the field, merging empirical enquiry with wider theoretical concerns in a profitable if unobtrusive manner. He had wide interests in social theory, using demographic techniques and automatic data processing, as well as being an excellent linguist. He will be particularly remembered for his work in the field of interpersonal relations, that is, of kinship and the family in relation to ritual and political concerns, not only among the patrilineal Tallensi and the matrilineal Ashanti of Ghana, but also on a broader, comparative, canvas. He had honorary degrees from Chicago and Belfast, was an honorary fellow of LSE (1975) and was elected FBA (1967).

In 1928 he married Sonia (died 1956), daughter of N. Donen, of Worcester, South Africa; they had one daughter. In 1960 he married Doris Yankauer Mayer, MD, a psychiatrist, daughter of David Sigmund Yankauer, wholesale textile dealer, of New York. Fortes died 27 January 1983 in Cambridge.

[Personal knowledge.]   JACK GOODY

FOSTER, SIR IDRIS LLEWELYN (1911–1984), Welsh and Celtic scholar, was born 23 July 1911 at Bethesda, Caernarvonshire (later part of Gwynedd), the elder child and elder son of Harold Llewelyn Foster, shopkeeper, of Bethesda, and his wife, Anne Jane Roberts. He was educated at Bethesda County School and the University College of North Wales, Bangor, from which he graduated BA (Wales) with first class honours in Welsh (with accessory Latin) in 1932 and MA (Wales) with distinction in 1935. From 1932 onwards he pursued research on the complex but fascinating and important Middle Welsh prose tale *Culhwch ac Olwen* under the guidance of (Sir) Ifor Williams. The award of a University of Wales fellowship in 1935 enabled him to spend periods of study also at the National University of Ireland with Professor Osborn Bergin and the University of Bonn with Professor Rudolf Thurneysen: he could thus claim to have sat at the feet of possibly the three greatest Celtic scholars of the earlier twentieth century.

In 1936 he was appointed head of the department of Celtic at the University of Liverpool where he remained for eleven years, except for three and a half years (1942–5) spent at Cambridge during World War II in the Intelligence division of the naval staff, where he worked mainly on material in Serbo-Croat; during his last year at Liverpool he was also warden of the men's hostel, Derby Hall. In 1947 he was elected to the Jesus chair of Celtic in the University of Oxford and a fellowship of Jesus College; on his retirement in 1978 he was made emeritus professor of the university and honorary fellow of the college. He brought to the chair not only extraordinary erudition in the major Celtic languages and their literature but also an avid interest in a wide range of more or less kindred disciplines: history, archaeology, anthropology, art, music, and theology. Thus equipped he conferred a new lustre on Celtic studies at Oxford, which was early recognized by the elevation of his chair to schedule A status in 1950.

His publications were sparse but uniformly learned, penetrating, and judicious: he wrote not only on Middle Welsh narrative and religious prose (for example, his chapter in *Arthurian Literature in the Middle Ages*, ed. R. S.

Loomis, 1959, and his British Academy Rhys lecture of 1950, *The Book of the Anchorite*) but also on the earliest Welsh poetry (as in his chapter in *Prehistoric and Early Wales*, edited by himself and Glyn Daniel, 1965). He edited with Leslie Alcock the volume *Culture and Environment* in 1963. His interest in the early poetry led to the formation of the Hengerdd colloquium which met two or three times a year under his chairmanship at Jesus College between 1972 and 1978 and which bore fruit in the volume *Astudiaethau ar yr Hengerdd/Studies in Old Welsh Poetry*, eds. R. S. Bromwich and R. Brinley Jones, published in 1978 and dedicated to Foster.

Regrettably his *magnum opus*, the definitive edition of *Culhwch ac Olwen*, remained unfinished at the time of his death. This was partly due to the clear priority he gave during his Oxford years to teaching. He not only lectured on a wide variety of topics within the Celtic field but also supervised the research of a long sequence of able graduate students. At one time four of the five professors of Welsh language and literature in the constituent colleges of the University of Wales were former pupils of his. His penetrating intelligence and vast learning were matched by a sustaining interest in the lives and doings of his pupils and an impish sense of humour. Short and rotund in appearance, he in fact commanded general affection among a wide circle of friends. Although unmarried, he was particularly fond of children and they of him.

Many institutions and societies both in England and in his native Wales benefited from his steadfast service. He was a member of the royal commission on ancient monuments in Wales and Monmouthshire 1949–83, the standing commission on museums and galleries 1964–82, and the council for the Welsh Language 1973–8. In Oxford he was successively chairman of both the modern languages board and the anthropology and geography board and was a select preacher to the university 1973–4. For many years he acted as external examiner in both Welsh and Welsh history for the University of Wales. Among other bodies he served were the Society of Medieval Languages and Literature (president, 1953–8), the Honourable Society of Cymmrodorion (editor, 1953–78), the Cambrian Archaeological Society (president, 1963–9), the Irish Texts Society (president, 1973–84), the Gwynedd Archaeological Trust (chairman, 1974–9), the Ancient Monuments Board for Wales (chairman, 1979–83), and especially the National Library of Wales (treasurer, 1964–77; vice-president, 1977–84) and the National Eisteddfod of Wales (chairman of council, 1970–3, president of court, 1973–7). He was also a member of the governing body of the

Church in Wales: a convert to Anglicanism from the Nonconformity in which he had been brought up, his strong and vital attachment to his new faith did not preclude a continuing appreciation of the virtues of the old.

He was appointed Sir John Rhys memorial lecturer of the British Academy in 1950; O'Donnell lecturer in the University of Edinburgh in 1960 and the University of Wales in 1971-2; G. J. Williams lecturer at University College, Cardiff, 1973; and James Ford special lecturer in the University of Oxford, 1979. He was elected a fellow of the Society of Antiquaries in 1954 and was knighted in 1977.

Foster died 18 June 1984 at the Caernarfon and Anglesey General Hospital, Bangor. He had retired to his native Bethesda for which his affection had never diminished throughout his forty years' exile in England.

[*The Times*, 25 June 1984; D. Ellis Evans, 'Sir Idris Foster', *Transactions of the Honourable Society of Cymmrodorion*, 1984, pp. 331-6; Meic Stephens (ed.), *The Oxford Companion to the Literature of Wales*, 1985; personal knowledge.]

R. Geraint Gruffydd

FOSTER, Sir JOHN GALWAY (1903/4-1982), lawyer, was the only child (he also had a half brother and two half sisters) of Lieutenant-Colonel Hubert John Foster, Royal Engineers, military attaché at the British embassy in Washington, and a lady whose surname was Galway, an Irish Canadian. No birth certificate can be found. He was educated, first in France and then at Hildesheim in Germany. At the age of eight he was trilingual and he always spoke French and German fluently. After attending A. J. Richardson's private school in Broadstairs, in 1917 he won a scholarship to Eton where he immediately gave evidence of his cleverness. When he left at the age of sixteen he had won thirteen school prizes. He continued his scholastic success at New College, Oxford. He achieved first class honours in modern history (1924) and was elected a fellow of All Souls College in 1925. He remained a fellow until his death, having been advanced to a distinguished fellowship in 1980. The popular press tipped him as a future prime minister. Foster, however, was not a man with strong allegiance to any particular party. He admitted that his loyalties were to justice and the welfare of mankind in general, and felt that a conventional political career was well nigh impossible for him.

Foster was called to the bar in 1927 (Inner Temple) and took silk in 1950. Before the outbreak of World War II he had a number of appointments such as secretary to the Thetis enquiry, the budget leak enquiry, and the law reform committee. He was a lecturer in private international law in Oxford (1934-9) and at the Hague, and recorder of Dudley in 1936-8 and of Oxford (1938-51 and 1956-64). On his ceasing to be recorder he was appointed KBE (1964).

In 1939 Foster became first secretary and legal adviser to the British embassy in Washington and later, in SHAEF, legal adviser to General Dwight Eisenhower. He was awarded the Legion of Honour, American Legion of Merit, and croix de guerre (with palms), and played a part in the Nuremberg trials. In 1944 he was accorded the rank of brigadier.

Foster was elected a Conservative member of Parliament for the Northwich division of Cheshire in 1945, a seat he held without interruption until his retirement in 1974. Between 1951 and 1954 he served as under-secretary of state for commonwealth relations but greatly perplexed the establishment by his rapid and informal methods of administration. Do what he might he found that he invariably finished the week's work in under twelve hours; furthermore he was reluctant to keep minutes, considering they were an unnecessary waste of time.

After Gerald (later Lord) Gardiner (subsequently lord chancellor) retired from the chairmanship of the Council of Justice, Foster succeeded him in the post, which he held until his death. He was also active in bringing about the European Convention of Human Rights. Throughout his life he fought tirelessly on behalf of the victims of persecution, be they Jews or Tamils or some unfortunate stranger who had experienced any degree of injustice.

Foster possessed a powerful brain. His mind was nimble and quick, and he used his intellect with virtuosity, perceiving connections between ideas, propositions, theories, and hypotheses with lightning speed and consummate skill. In pleading a case he showed himself both innovative and imaginative. Not altogether surprisingly some judges (and, indeed, some of his fellow members in the House of Commons) found him difficult to follow. His style was much more appreciated in the appellate courts, especially the House of Lords; a search of post-World War II reported cases shows that Foster had been counsel in more than 160. His greatest expertise was in private international and domicile law. He was an outspoken critic of the accusatorial system of justice and would have liked to remake it, based rather on the continental mould. He greatly enjoyed his arbitration cases at the International Court of Justice at The Hague.

Foster was a most unusual and remarkable character, for his public achievement was negligible compared with his private and personal influence which was considerable in England but especially in North America. First and foremost he was a true egalitarian. He possessed a rare combination of ebullience, cleverness, good humour, delight in the kaleidoscope of human affairs, an unselfish and practical interest in the successes and tribulations of friends and acquaintances, and the gift of combining a crystal-clear objectivity with great kindness and an infinite capacity for helping those in trouble. He possessed a fund of amusing anecdotes, which were enriched by the fact that he was a gifted extrovert. He was unmarried. He died in London 1 February 1982.

[Personal knowledge.] MIRIAM ROTHSCHILD

FRASER, BRUCE AUSTIN, first BARON FRASER OF NORTH CAPE (1888-1981), admiral of the fleet, was born in Acton, London, 5 February 1888, the younger son (there were no daughters) of General Alexander Fraser, CB, of the Royal Engineers, and his wife, Monica Stores Smith. He was educated at Bradfield College before passing into HMS *Britannia* in September 1902. He completed his cadetship with distinction, and was appointed a midshipman in HMS *Hannibal*, a battleship with the Channel Fleet, in January 1904.

In the years that followed, Fraser served in a succession of battleships and destroyers in home waters, being promoted sub-lieutenant in September 1907. In 1911, having determined to become a gunnery specialist, he was posted to HMS *Excellent* at Whale Island. Fraser passed out top of the course. He acted as gunnery officer in the cruiser *Minerva* (1914-16), and saw action in the Dardanelles. He spent some months of 1916 on the senior staff of HMS *Excellent*, thus missing Jutland. At the end of the year, he was posted to the new battleship *Resolution*, in which he became commander in 1919. Ironically, for an officer who had shown exceptional leadership and technical capabilities, Fraser was obliged to end the war without experiencing a major action in a modern warship.

In April 1920 he suffered a bizarre misfortune. Because he was on poor terms with his captain in *Resolution*, he sought escape by responding to a call for Mediterranean Fleet volunteers to travel to Baku and assist the White Russian fleet against the Bolsheviks. He arrived in command of his detachment of thirty-one men, just in time to be caught up in a local Bolshevik coup. The British party was imprisoned in wretched conditions until freed in November, when Fraser came home to spend a further two years on the staff of *Excellent*.

Despite favourable reports and widespread acceptance as a popular and able officer, Fraser's career thus far had been sluggish. But

from 1922 onwards he was plainly marked for high rank, earning the commendation of the Admiralty Board in 1924 for his work on a new fire control installation. As fleet gunnery officer in the Mediterranean (1925-6) and as a captain in the Admiralty tactical division (1926-9), he worked close to the heart of the navy's gunnery development. From 1929 to 1932 he held his first seagoing command in the cruiser *Effingham* in the East Indies. As director of naval ordnance (1933-5), he devised the armament for Britain's last generation of battleships, the 14-inch King George V class.

In 1936-7 Fraser commanded the aircraft carrier *Glorious*. In January 1938, just short of his fiftieth birthday, he was appointed rear-admiral, and chief of staff to Sir A. Dudley Pound [q.v.], C-in-C Mediterranean. It was in this role that he forged the close relationship with Pound that continued in World War II, when Pound was first sea lord.

In March 1939 Fraser became controller of the navy and third sea lord. In this role, for three testing years he bore responsibility for the navy's construction and repair programme, perhaps above all for the creation of the corvette, the mainstay of the Atlantic convoy escort system. He also played an important part in the development of warship radar. In May 1940 he became vice-admiral.

Fraser won the confidence of (Sir) Winston Churchill in this period at the Admiralty, and never lost it for the remainder of the war, despite periodic differences of opinion, for instance when the controller opposed Churchill's enthusiasm to build a new battleship. In June 1942 Fraser was sent to sea once more, as second-in-command of the Home Fleet under Sir John (later Lord) Tovey [q.v.]. He arrived just before the tragedy of convoy PQ17, one of the darkest naval episodes of the war.

By 8 May 1943, when Fraser was appointed to succeed Tovey as C-in-C Home Fleet, he could claim wide experience of both naval operations at sea, and their strategic direction ashore. A bluff, cheerful, straightforward officer with much shrewdness and technical knowledge but no pretensions to intellectualism, he was committed throughout his career to making inter-service co-operation a reality. He had shown remarkable gifts for winning the confidence of his peers and political masters at home, while commanding the affection and loyalty of subordinates afloat. Essentially a simple man who used to declare without embarrassment that he had never read a novel in his life, a bachelor who had made the Royal Navy his life-work, he was acknowledged as one of the outstanding naval professionals of his generation. His elevation was widely welcomed.

Yet in the strategic situation in the summer

of 1943, it seemed unlikely that Fraser would be granted the opportunity to conduct a major fleet action. The Russian victory at Stalingrad, and the consequent shift in the balance of advantage against the Germans, diminished the importance of the western Allies' Arctic convoys. These now offered a lure to the three German capital ships based in Norwegian waters—*Tirpitz*, *Scharnhorst*, and *Lutzow*—but it seemed unlikely that the Germans would consider the bait worth the hazard to their remaining fleet. Correspondingly, the British Home Fleet had been weakened by the transfer of ships to the Mediterranean. *Anson* and *Duke of York*—in which Fraser flew his flag—were now the only British battleships at Scapa.

Yet in September, to the surprise of the British, *Tirpitz* and *Scharnhorst* sortied for two days to bombard Spitsbergen. This was a negligible feat, yet sharply reminded the Royal Navy of the difficulties of keeping effective watch on the German ships. A fortnight later, a substantial British success was gained, when midget submarines successfully crippled *Tirpitz* in Kaafiord. She was rendered unfit for active operations for six months. Four days afterwards, *Lutzow* escaped into the Baltic. *Scharnhorst* was now alone.

As a succession of Allied convoys sailed to Russia that autumn, almost unmolested but shadowed by units of the Home Fleet, Fraser, having declined an offer by Churchill to become first sea lord, continued to believe that *Scharnhorst* would sooner or later come out. Earlier in the war, the German battle cruiser had inflicted major damage upon British shipping. On 19 December British Ultra decrypts revealed that *Scharnhorst* had been brought to three hours' readiness for sea. Fraser's Force 2, led by the cruiser *Jamaica* and the *Duke of York*—the only British ship with the armament to match that of *Scharnhorst*—sailed from Icelandic waters at 23.00 hours on 23 December. Fraser had carefully briefed his captains, and carried out repeated exercises in identifying and engaging hostile ships by radar, given the almost permanent Arctic darkness.

On the afternoon of 24 December 1943, in atrocious weather, the British convoy JW55B was ordered to slow to eight knots because Force 2 was 400 miles behind, too distant for comfort when the convoy was only the same distance from *Scharnhorst* in Altenfiord. *Scharnhorst* sailed to attack JW55B at 19.00 on 25 December, commanded by Rear-Admiral Erich Bey.

Eight hours later, this news was passed to Fraser, whose ships were still struggling through mountainous seas to close the gap with the convoy. Fraser accepted the risk that *Scharnhorst* would turn away if he broke wireless silence, and ordered JW55B to turn northwards, away from the Germans. Bey was still searching in vain for the convoy at 07.30 on 26 December, when the 8-inch cruiser *Norfolk*, a unit of Force 1, led by Vice-Admiral (Sir) R. L. Burnett [q.v.], located *Scharnhorst* on radar at 33,000 yards. The British ship, with her 6-inch consorts *Belfast* and *Sheffield*, opened fire at 09.29. They strove to close the range speedily, and avoid the sort of mauling *Graf Spee* had inflicted upon a British cruiser squadron four years earlier. At this early stage, *Norfolk* damaged *Scharnhorst*'s radar. Bey, as the British had expected, at once withdrew at 30 knots.

The British now suffered almost three hours of acute apprehension, having lost touch with *Scharnhorst*, and being fearful that she might break away into the Atlantic. Only at 12.20 did *Belfast* triumphantly report the battle cruiser once more in sight. Bey had turned north, still searching for JW55B. Burnett's cruisers and destroyers lay between the Germans and the convoy.

In the second twenty-one minute cruiser action, the British ships suffered significant damage before *Scharnhorst* broke away unscathed. Yet Bey now hesitated fatally. He knew that a British battle group was at sea. But he believed it was too distant to harm him. Only at 14.18 did he abandon the attempt to engage the convoy and turn for home, independent of his destroyer escort.

Fraser was now racing to cut across *Scharnhorst*'s southward track. At 16.17, *Duke of York*'s 14-inch guns opened fire at 12,000 yards, and straddled their target—clearly illuminated by starshell—with the first broadside. Critics subsequently suggested that *Scharnhorst* might have been destroyed at this stage of the battle, had Fraser not delayed ordering in his destroyers. He was fearful that a premature torpedo attack would drive the German ship away north-eastwards, beyond his grasp.

Under fire from *Duke of York*, *Scharnhorst* turned away at full speed first north, then east. By 17.13, when at last Fraser loosed his destroyers, he had left it too late. The enemy was outrunning both the British cruisers and destroyers. Only *Duke of York*'s guns were still within range. At 18.20, an 11-inch shell from *Scharnhorst* temporarily severed the flagship's radar cables, blinding her gunners. For a few terrible minutes, Fraser believed that victory had been snatched from him.

Yet just before the British ship was hit, although *Scharnhorst* had succeeded in opening the range to 20,000 yards, a shell from *Duke of York* burst in her number 1 boiler room, abruptly cutting the ship's speed to eight knots. Power was restored soon afterwards. But the brief crisis allowed three of Fraser's destroyers to close. Four of the twenty-eight torpedoes which they launched hit *Scharnhorst*, drastically reducing her power and ensuring her destruction. At 19.45, after enduring concentrated gunfire and torpedo attacks for almost three hours more, *Scharnhorst* sank. Her guns continued firing almost to the last. Out of her complement of 1,803, thirty-six survivors were plucked from the Arctic darkness.

Duke of York's gunnery had plainly been the decisive factor in the victory. For all the undoubted British advantage of strength, the entrapment and destruction of *Scharnhorst* had been a considerable achievement, ensuring Fraser's perpetual celebrity in the annals of the Royal Navy, alongside that of the Norwegian North Cape beyond which the battle was fought.

Fraser's remaining service with the Home Fleet was dominated by the conduct of further Russian convoys. But with the shift of strategic attention from the Mediterranean to north-west Europe, force was now available to provide massive escorts, and Allied losses declined steeply. On 16 June 1944 Fraser relinquished command. He was now assigned to become commander-in-chief, Eastern Fleet. In November he became C-in-C Pacific Fleet.

Fraser's task in the Pacific was delicate. The US navy dominated the theatre, and the Royal Navy's contribution seemed not merely modest, but even unwelcome. The Americans were deeply suspicious of British imperial motives in the eastern hemisphere. It is a tribute to Fraser's competence and transparent good nature that he achieved an amicable working relationship with the Americans. He believed passionately in the need to develop—and to perpetuate post-war—Anglo-American co-operation. Such gestures as volunteering to adopt US navy signalling procedures went far to encourage trust.

His command continued to suffer from lack of resources, and it was only in the last weeks of the war that its forces achieved real weight. The British were hampered by the acute discomfort of their ships in tropical conditions. But the fleet made a useful contribution to the last stages of the Pacific war. It was Fraser who signed the Japanese surrender document for Britain, aboard the battleship *Missouri* in Tokyo Bay on 2 September 1945.

On his return home in 1946, Fraser was saddened to be denied the succession as first sea lord. He was appointed C-in-C Portsmouth in May 1947, and at last gained the first sea lordship in September 1948, together with promotion to admiral of the fleet a month later.

Fraser's tenure at the Admiralty embraced a series of cold war crises, and finally responsibility for British naval participation in the Korean war. For all the affection and respect

that he commanded as an old 'sea-dog', Fraser was considered by some critics to possess too limited an intellect to distinguish himself at the summits of power. He retired from the Royal Navy in April 1952, and passed the next twenty-eight years in almost uneventful retirement.

A barony was conferred upon him in 1946. He was appointed GCB in 1944 (KCB 1943, CB 1939); KBE in 1941 (OBE 1919). He was first and principal naval ADC to the King (1946-8), and held honorary degrees from the universities of Oxford (DCL 1947), Edinburgh (LLD 1953) and Wales (LLD 1955). He held the American DSM. He was a member of the Russian Order of Suvarov and of the Grand Order of Orange Nassau (Netherlands); was a chevalier of the Legion of Honour and holder of the croix de guerre with palm (France); and held the grand cross, Order of St Olav (Norway).

A lifelong bachelor, he died in London 12 February 1981. The barony became extinct. A portrait of him by Sir Oswald Birley hangs in the Royal Naval College, Greenwich.

[Richard Humble, *Fraser of North Cape*, 1983; Stephen Roskill, *The War at Sea 1939-1945* (official history), 3 vols., 1954-61; F. H. Hinsley and others, *British Intelligence in the Second World War*, vol. III, pt. i, 1984; John Winton (John Pratt), *The Forgotten Fleet*, 1969; Stephen Roskill, *Churchill and the Admirals*, 1977.]   MAX HASTINGS

FRASER, SIR HUGH CHARLES PATRICK JOSEPH (1918-1984), soldier and politician, was born 23 January 1918 in Westminster, the third of five children and younger son of Simon Joseph Fraser [q.v.], soldier, the fourteenth (sometimes reckoned sixteenth) Baron Lovat, of Beaufort Castle, Beauly, Inverness-shire, and his wife, Laura, second daughter of Thomas Lister, the fourth and last Baron Ribblesdale. He was educated at Ampleforth College, the Sorbonne, and Balliol College, Oxford. He led the school debating society. As head of St Oswald's House, Fraser declined his headmaster's invitation to become head boy, preferring to study in Paris before going up to Balliol as an exhibitioner. As chairman of the Oxford University Conservative Association, he was less hostile to Neville Chamberlain's foreign policy than was his college contemporary Edward Heath, whom he succeeded as president of the Oxford Union in 1939. Fraser was proud to preside over the reversal in debate of the notorious 'King and Country' resolution of 1933. His patrician manner and grandiloquent debating style, both in the Union and later in the House of Commons, were lightened by wit and self-mockery. He gained a second class in modern history in 1939.

Starting the war with the family Yeomanry regiment, the Lovat Scouts, Fraser soon sought more adventurous service with Phantom (GHQ Liaison Regiment) and the Special Air Service Regiment. After the battle of Arnhem (1944) he took part in the 'Pegasus' operation for the rescue of stragglers. In one night, as joint beach-master with Airey Neave [q.v.], then also serving with Intelligence School No. 9, he took part in saving 138 evaders from across the Rhine. Nicknamed 'Fearless Fraser', he was appointed MBE (military division, 1945) and became a chevalier (with palm) of the Belgian Order of Leopold II. He was also awarded the Belgian and the Netherlands Order of Orange Nassau.

Fraser was twenty-seven when he won Stone in the general election of 1945. He was the sole successful Conservative candidate in Staffordshire and was one of only four Roman Catholic Tories elected. He compared the 'nightmare' of his maiden speech to 'a descent by parachute behind the enemy line'. Early on, he displayed his independent judgement by going against the opposition whips and voting against the Washington loan agreement. In 1950 the seats of Stafford and Stone were united. So, until his return in the general election of that year, Fraser shared the constituency with a Labour MP, Stephen Swingler.

In 1951–4 Fraser was parliamentary private secretary to the colonial secretary, Oliver Lyttelton (later Viscount Chandos, whose notice Fraser wrote for this Dictionary). Together, they toured Malaya during the communist insurgency, once crash-landing at Penang, and a Kenya afflicted by Mau Mau rebellion. In 1956, after the seizure of the Suez Canal Company, Fraser was vehement for military action against Nasser's Egypt; but he did not belong to the Suez group of Tory critics of the government's failure to complete the Suez operation.

In 1958 Fraser joined the succeeding ministry headed by Harold Macmillan (later the Earl of Stockton) as parliamentary undersecretary of state and financial secretary to the War Office. He held those posts until 1960. In 1962, he became, successively, parliamentary under-secretary of state for the colonies and the last secretary of state for air and a privy councillor. In April–October 1964 he was minister of defence for the Royal Air Force in the new ministry that had absorbed the three Service departments. After the Conservative electoral defeat that year Fraser was dropped from the shadow cabinet. Temperamentally different from Edward Heath, he opposed entry into the European common market which was so dear a

cause to his leader. In 1967 Fraser's fire was directed against another Balliol contemporary, Denis Healey, for the cancellation of the TSR 2.

He never held office again, although he kept his seat until his death. To some contemporaries he seemed at times to have lost a sense of political purpose. But he espoused causes out of conviction, irrespective of party policy or establishment opinion. In 1968 he was one of those who, like Enoch Powell and Michael Foot, opposed the reform of the House of Lords on lines drawn, in part, by the Labour peer, the seventh Earl of Longford, Francis Aungier Pakenham, whose daughter Antonia he had married in London in 1956. They had three sons and three daughters. The marriage was dissolved in 1977.

Fraser called for the return of conscription. The romantic appeal of Zionism and his own assessment of British interests in the Middle East made him a consistent and eloquent partisan of the state of Israel. This involved some financial loss because of disagreement on the Arab–Israel question with business associates. Fraser also championed, and visited, the largely Roman Catholic people of eastern Nigeria during their struggle for the independent state of Biafra, at a time when the integrity of the Nigerian federation was the common policy of Labour government and Conservative opposition. Ever loyal to friends, Fraser backed Jonathan Aitken MP, then a young journalist, when he was prosecuted under the Official Secrets Act in connection with a *Sunday Times* article he wrote on Biafra. Fraser demanded that he too be prosecuted. But the attorney-general did not oblige; and Aitken was acquitted.

At the party leadership election of 1975 Fraser stood in the first round. Margaret Thatcher was the successful candidate. That same year Fraser was shocked and distressed when an Irish terrorist bomb intended for him killed a neighbour instead. But he spoke out all the more against the IRA. Knighted in 1980, he was a gallant cavalier descended from Jacobite Frasers who turned from rebellion to the wars of empire. He was a charming and entertaining companion with a high sense of the ludicrous. He could jest about the sacred; but faith and family were very dear. A traditional Roman Catholic, he sympathized with Anglicans resentful of liturgical innovation; and he spoke in the Commons against the Church of England measure of 1974 as ousting parliamentary 'power and responsibility' for the established church.

Having undergone surgery, Fraser died 6 March 1984 in St Thomas's Hospital, London. A portrait painted in 1966 by Olwyne Bowey is the property of his eldest son, Benjamin.

[Personal knowledge.] JOHN BIGGS-DAVISON

FRASER, SIR ROBERT BROWN (1904–1985), the first director-general of the Independent Television Authority, was born 26 September 1904 in Adelaide, South Australia, the elder son (there were no daughters) of Reginald Fraser, an Adelaide businessman, and his wife, Thusnelda Homberg. He was educated at St Peter's School, Adelaide; Trinity College, Melbourne (1924–6), where he obtained a third class honours degree in classics; and London School of Economics (1927–30), where he obtained a B.Sc.(Econ.) with second class honours.

At LSE his political leanings as an intellectual socialist caught the attention of E. Hugh J. N. (later Lord) Dalton, whose literary executor he became, and of Harold Laski [qq.v.]. In 1930, on the advice of Laski, he became leader writer for the *Daily Herald*, where he stayed until 1939. He stood unsuccessfully as Labour candidate at York in 1935.

On the outbreak of war, he joined the empire division of the Ministry of Information. In 1941 he became director of the publications division and was responsible for the enormously successful series of MOI booklets about major aspects of the war. Despite Dalton's urging that he stand for Parliament in 1945 he elected to continue with the MOI where he was appointed that year as controller of production. In the following year he became director-general of the newly formed Central Office of Information. This proved a demanding post, calling for the creation of a new organization largely out of the remnants of the MOI, many of whose senior staff had departed with the ending of the war. When he left it eight years later, the COI was established as a permanent part of the government's publicity machinery and the ground rules which distinguished between publicity for government activity and that of a politically partisan kind had been clearly defined.

In August 1954 the Independent Television Authority was set up by statute to run a new advertising-supported television service. In September Fraser was appointed its first director-general. Within weeks he was able to work out with Sir Kenneth (later Lord) Clark [q.v.], the ITA chairman, and other members of the authority a framework for the operations of the private companies which would provide the programmes to be broadcast by the ITA. (In Fraser's time, there were to be fifteen companies, in addition to a central news company.)

In 1963 Lord Hill of Luton became ITA chairman, determined that the authority should play, and be publicly seen to play, a more commanding role in the supervision of the programmes. Fraser's relationship to Hill's two predecessors had been that of a chief executive with influential, but non-executive, chairmen. Now, as he himself said, it became more analogous to that of a Civil Service permanent secretary with his minister. The new relationship worked well and Fraser must share the credit for the fact that when Hill went to the BBC as chairman in 1967 the public standing of ITV had been greatly improved. (Its standing with the viewing audience had never been in doubt, and remained consistently high.)

Fraser's leadership of the authority's staff and his management of its operations were notable. The authority's responsibilities included the planning and provision of the network of stations transmitting the Independent Television programmes. When Fraser retired in 1970, the authority had completed virtual national coverage in black and white signals from a network of nearly fifty transmitters and was already, from a further dozen stations, providing two-thirds of the population with higher standard and colour transmissions which had begun the previous year. After retirement from the ITA, Fraser was independent chairman of Independent Television News for three years from 1971 to 1974.

Fraser was invariably elegant in writing, in public speaking, and in dress. He had great mental and physical stamina and the absorptive capacity of the first-rate civil servant. He was decisive and not prepared to fudge on important issues. Only his close associates knew the immense amount of careful thought and sheer hard work which went into his mastery of any subject and into his decisiveness. In spite of his own high standards, he was tolerant—sometimes perhaps over-tolerant—of the shortcomings of others. He had an intense dislike, surprising in so prominent a figure in the communications world, for the personal publicity which his position occasioned. Although his appointment as ITA director-general in 1954 gave rise to criticism in some newspapers because of his socialist past, his earlier political sympathies were by then no longer evident. He was always punctilious, to the point of being over-cautious, about questions of political impartiality arising in ITV programmes.

Fraser was appointed OBE (1944) and was knighted in 1949. He was an honorary fellow of LSE (1965) and life member of the Royal Institute of Public Administration (1975).

He married in 1931 Betty, daughter of George Harris, a government scientist. They had one daughter. Fraser died in London 20 January 1985. His wife died the following day.

[Bernard Sendall, *Independent Television in Britain*, vol. i, 1982; Ben Pimlott (ed.), *The*

Second World War Diary of Hugh Dalton, 1940-45, 1986; private information; personal knowledge.]   ANTHONY PRAGNELL

FRERE, ALEXANDER STUART (1892-1984), publisher, was born 23 November 1892 in Emsworth, Hampshire, the only child of Arthur Reeves and his wife, Mary Frere. As a young man he adopted the surname Frere-Reeves and in middle age he dropped the Reeves; Frere *tout simple* became among friends his Christian as well as his surname. He was educated at a local school in Southampton. In August 1914 he enlisted in the Royal East Kent Yeomanry and in 1915 he fought at Gallipoli. In 1916 he was seconded to the Royal Flying Corps but was invalided out in 1917 after his plane crashed. His education was completed at Christ's College, Cambridge, where he read economics and in 1920-1 edited *Granta*. He obtained a third class in part i of the economics tripos in 1921.

After a short spell with the *Evening News* he accepted a job offered by the American publisher F. N. Doubleday who on the sudden death of its founder had bought William Heinemann Ltd. In 1926 Frere was made a director of Heinemann's and when after the Wall Street crash Doubleday sold his shares he became managing director in 1932 under the chairmanship of the legendary C. S. Evans. Their relationship was often strained but together they presented a remarkable list, which included John Galsworthy, Richard Aldington, J. B. Priestley, John Masefield, W. Somerset Maugham [qq.v.], Thomas Wolfe, and Graham Greene.

Frere's and Evans's tastes were happily complementary, Frere being more in tune with younger, more 'advanced' writers—early on he was in serious trouble for telling people at a cocktail party that Galsworthy, then the firm's star author, went about erecting stiles in order to help lame dogs over them. Frere's personal and publishing lives were intertwined, several of his authors becoming intimate friends. He was as equally at home at soirées given by Lady Cunard as he was drinking Pernods at the Dome or Coupole or demonstrating his remarkable skill as a tap dancer.

For most of World War II he was director of public relations at the newly formed Ministry of Labour and National Service, where he worked closely with Ernest Bevin [q.v.]. On Evans's death in 1944 he returned to Heinemann to become, for the next seventeen years, its chairman.

Under his guidance Heinemann's opened four Commonwealth offices. A major acquisition was a controlling interest in Secker & Warburg.

Among authors published during the post-war years were Nevil Shute, Olivia Manning [qq.v.], Anthony Powell, and Pamela Frankau; non-fiction titles included books by Clement (later Earl) Attlee, Aneurin Bevan [qq.v.], and General Dwight D. Eisenhower. In Frere's day the Heinemann list was eclectic, allowing room for frankly commercial best sellers, such as the historical romances of *Angelique*, as well as the journals of James Boswell [q.v.].

Frere was very opposed to what he called 'publishing by committee'. Trusting his own, often idiosyncratic, judgement, he encouraged his editors to do likewise. His belief in his authors was confirmed when he courageously stood trial on a criminal charge of publishing an obscene libel. By the standards of the early 1950s *The Image and the Search*, a novel by Walter Baxter (1953), was undoubtedly sexually frank, but Frere told the court that he felt 'publishers owed a duty to such gifted writers and to the public to ensure that their creative work was not stillborn'. After two successive jury disagreements, at a third trial the Crown offered no evidence and the jury was directed to return a verdict of not guilty. It was a sensational victory and Frere was its hero. The verdict helped pave the way to the more tolerant Obscene Publications Act of 1959.

Despite the excellence of the list, shortage of working capital led Frere to approach his friend Lionel Fraser, head of the powerful conglomerate, Thomas Tilling Ltd.; eventually Heinemann became its wholly owned subsidiary. After an abortive attempt to sell Heinemann to McGraw-Hill in New York relations between Frere and Tillings soured and he was 'kicked upstairs' to be president. Soon after this he felt obliged to resign—a sad end to a brilliant career. Even so, he became an adviser to the Bodley Head whence some of his most established and faithful authors followed him. They included Graham Greene, Georgette Heyer [q.v.], and Eric Ambler.

Frere was short of stature and had a debonair, aristocratic presence aided by a fine Wellingtonian nose. While much loved by his authors and most of his staff, his talent for dry, humorous sarcasm also made enemies. He was an active Freemason, reaching the exalted rank of grand superintendent. He was appointed CBE in 1946 and a chevalier of the Legion of Honour in 1953.

In 1930 he married Jessica, daughter of Henry Rayne, shoemaker. He married in 1933 Patricia Marion, daughter of (Richard Horatio) Edgar Wallace [q.v.], novelist, playwright, and journalist. An experienced journalist herself, she shared in much of Frere's life as a publisher. They had two sons and a daughter. Frere died 3 October 1984 at Hythe, Kent, near his home.

[*The Times*, 6 October 1984; personal knowledge.]   JOHN ST JOHN

FREUD, ANNA (1895–1982), psychoanalyst, was born 3 December 1895 in Vienna, Austria, the third daughter and youngest of six children of Sigmund Freud, founder of psychoanalysis, and his wife, Martha Bernays. She was educated at the Cottage Lyceum in Vienna and qualified as a teacher. From her father, however, she received the educational experience that was to shape her entire life. As a shy young girl she used to sit quietly in the background in Freud's study when he discussed psychoanalysis with visitors. In her early twenties she was in analysis with her father, an extraordinary arrangement, demonstrating that Freud was not an orthodox Freudian. Neither, as it was to emerge, was she. While she remained Freud's most loyal representative, she went beyond his work in method and theory.

Her innovations were the result of continuous practical involvement with normal and disturbed children. In the 1920s she worked in a nursery for deprived Viennese children while conducting individual child and adult analyses, and taking courses at Wagner Jauregg's psychiatric clinic. At the same time she lectured to psychoanalysts, parents, and teachers, began to publish, and represented her cancer-ridden father publicly and acted as his secretary and nurse privately.

Her first paper appeared in 1922, her first book *The Ego and Mechanisms of Defence* in 1937 (in German in 1936). This book became a classic of psychoanalysis and was translated into many languages. In it she added to Freud's identification of nine ways in which the ego protects itself against inner conflict five more, of which 'identification with the aggressor' has become the best known. Her own life, however, was an outstanding example of an already established mechanism: sublimation. This period saw the beginning of her lifelong friendship with Dorothy T. Burlingham and of her lasting public controversy with Melanie Klein's approach to child analysis.

Hitler's invasion of Austria ended her Viennese career. For five hours she was interrogated by the Gestapo. Released with the help of powerful international friends, in 1938 she, her father, and his closest circle escaped to London, where they were warmly received by British psychoanalysts and officials. Four of Freud's sisters, all aged over eighty, perished, however, in concentration camps.

Anna Freud began the second half of her life under the shadow of her father's deteriorating health; he died in 1939. In 1940 she and Burlingham established the Hampstead Wartime Nurseries for children suffering from family disruption. At the end of the war (she was naturalized in 1946) she founded the Hampstead Child Therapy Course and Clinic, devoted to training, therapy, and research and offering educational and medical services. The Hampstead seminars in the late 1950s showed Anna Freud at the height of her powers. Sitting regally under a portrait of her father she dominated the proceedings not only intellectually but also by her unparalleled knowledge of every detail of the clinic's work. She knew every child, had read every clinical report, and was actively involved in research. With unflagging energy she continued individual analyses, supervised trainees, lectured in many countries, edited, and wrote profusely. In 1966 she published another important book, *Normality and Pathology in Childhood*, in which she summarized her innovations, demonstrated her openness to ideas and methods outside psychoanalysis, and recognized the importance of genetic equipment, medical conditions, and above all environmental influences.

Taking the child as her focus she branched out relatively late in life into relevant matters of law. The Yale Law School appointed her visiting lecturer. A number of papers on psychoanalysis and law concerning children followed, culminating in two widely read books co-authored with Yale faculty members.

Her appeal to diverse audiences owed much to the clarity of her written and spoken word, a clarity that could diverge into sharpness. Some people experienced her as authoritarian, particularly in guarding her father's legacy. She arranged for some of his papers to be withheld well into the twenty-first century. Those who knew her privately spoke of her charm, humour, modesty, and enjoyment of the good things in life, particularly her cottage in Ireland, where swimming and riding alternated with work. Her other hobby, knitting, she engaged in during analytic hours.

She received honorary degrees and fellowships from many universities and medical institutions in England, the USA, Germany, and Austria, of which a medical degree from the University of Vienna (1975) and a Ph.D. from the Goethe University in Frankfurt (1981) meant most to her. She was appointed CBE in 1967. She died 9 October 1982 in London. She was unmarried.

[Elizabeth Young-Bruehl, *Anna Freud: A Biography*, 1989; *The Writings of Anna Freud*, 8 vols., 1975; obituary by Clifford Yorke in *Journal of Child Psychology and Psychiatry*, vol. xxiv, 1983; obituaries by Joseph Goldstein, Leo Rangell, and Robert S. Wallerstein in *The Psychoanalytic Study of the Child*, vol. xxxix, 1984;

personal knowledge.]   Marie Jahoda

FRYE, LESLIE LEGGE SARONY (1897–1985), song writer and entertainer. [See Sarony, Leslie.]

FULFORD, Sir ROGER THOMAS BALDWIN (1902–1983), author and journalist, was born 24 November 1902 in the vicarage at Flaxley in Gloucestershire, the younger son and second of the three children of Frederick John Fulford who was vicar (he was later an honorary canon of St Edmundsbury and Ipswich), and his wife, Emily Constance, daughter of W. H. Ellis of Ottermouth, Budleigh Salterton. Roger's brother died young and his sister spent a happy life as a nun. He was educated at Lancing (where he was a contemporary of Evelyn Waugh, q.v.) and Worcester College, Oxford, where he took a second in modern history (1927) and was Liberal president of the Union in 1927. He was called to the bar in 1931, though he never practised. He joined the editorial department of *The Times* in 1933 and remained a contributor for many years. A fervent and lifelong Liberal, he was thrice defeated in parliamentary elections (1929, 1945, and 1950), and during the 1930s his third Christian name produced many jokes at his expense. In 1964–5 his constancy was rewarded by the presidency of the Liberal Party.

In 1932 G. Lytton Strachey [q.v.] died, leaving unfinished the editing of the first unexpurgated edition of *The Greville Memoirs*, on which he had been working for several years. His brother James asked Fulford to complete the editing, which he did, with the assistance of Ralph and Frances Partridge who had already done much of the donkey work for Strachey. The first of the ten volumes appeared in 1938 with a short preface by Fulford and his name as co-editor on the title page.

But all the time his heart was in history, particularly that of the English royal family in the late eighteenth century and all the nineteenth. His first book, *Royal Dukes* (1933), brought him great praise. In it he gave detailed accounts of the lives of the six younger sons of George III. Lampooned and often execrated in their lifetime, to Queen Victoria they were, except for her father, unmentionable. Never before had they been treated on their merits and demerits, and Fulford did this with wit and, wherever possible, sympathy. A slightly revised edition of the book, published in 1973, seemed as fresh and absorbing as it had forty years before.

George the Fourth (1935), following the same pattern of sorting good from bad, true from false, was as successful as its predecessor.

During the war Fulford was an assistant censor, a civil assistant in the War Office (1940–2), and assistant private secretary to the secretary of state for air (1942–5), and after the war he returned to the royal family with *The Prince Consort* (1949) and a short life, *Queen Victoria* (1951).

Next he turned aside to write a history of Glyn's Bank (*Glyn's 1753–1953*, 1953), *Votes for Women* (1957), a history of the suffragette movement, which received a prize of £5,000 from the *Evening Standard*, and *The Liberal Case* (for the general election of 1959).

Then he returned once more to the royal family with *Hanover to Windsor* (1960), a study of the monarchy from William IV to George V, and his final task was the editing of five volumes of the correspondence between Queen Victoria and her eldest daughter the Empress Frederick of Germany; *Dearest Child* (1964), *Dearest Mama* (1968), *Your Dear Letter* (1971), *Darling Child* (1976), and *Beloved Mama* (1981). He contributed many notices to this Dictionary.

In 1937 he married Sibell Eleanor Maud, daughter of Charles Robert Whorwood Adeane, of Babraham Hall, Cambridge, and widow of the Hon. Edward James Kay-Shuttleworth (died 1917) and of the Revd. Hon. Charles Frederick Lyttelton (died 1931). There were no children of the marriage but Sibell had a son and a daughter by her first marriage and two sons by her second, one of whom died in infancy and the other of wounds in 1944. The Fulfords' married life, which was happy and lasting, was mostly spent at Barbon Manor, on a hilltop near Kirkby Lonsdale, which was then in Westmorland. They filled the house with engaging Victoriana and established a woodland garden in which numerous Himalayan rhododendrons, including some of the large-leaved and frost-tender species, still flourish at 700 feet above sea-level.

Fulford became a great favourite with his wife's numerous relations, who often called on his aid, and they repaid his kindness by helping to look after him in his last few years.

He was a small man, always courteous and serene. His brother-in-law George Lyttelton described him as 'demure and impish', and indeed his wry humour was ever present. He was a regular attender at committee meetings of the London Library, but he never came all the way from Westmorland for an uneventful meeting. Often when some matter had apparently been dealt with, a quiet voice would say 'But, Mr Chairman . . .' and the whole discussion would be reopened. He was a loyal and affectionate friend, the most stimulating and delightful of companions.

The death of his wife in 1980 was a terrible blow, his health deteriorated, and he died at

Barbon 18 May 1983. He was appointed CVO in 1970 and knighted in 1980.

[*The Times*, 19 and 26 May 1983; personal knowledge.]   RUPERT HART-DAVIS

FURY, BILLY (1941–1983), singer of popular music, was born Ronald Wycherley 17 April 1941 in Garston, Liverpool, the elder son (there were no daughters) of Albert Wycherley, shoe repairer, and his wife, Jean Homer. As a child he suffered rheumatic fever which left him with a damaged heart. He was educated at Wellington Road School, Liverpool. On leaving he worked as a tugboat hand on the River Mersey. His interest in music was stimulated as he listened to records brought back from New York by seamen, particularly the rock and roll records of Elvis Presley. He bought a guitar, dressed as a Teddy Boy, began to write songs, and started emulating Marty Wilde, then one of Britain's most popular performers.

In 1958 he auditioned for Larry Parnes who immediately engaged him for a series of package shows touring Britain. It was Parnes who conceived the name Billy Fury and negotiated a recording contract with Decca Records. His first single, 'Maybe Tomorrow', was released in March 1959 and was an instant hit. It climbed to no. 18 and remained for a total of nine weeks on the British list of bestsellers.

Fury was hailed by the media as 'The Blond Presley' and 'The British Teen Dream' and was certainly Britain's first major rock singer. Within the first six months of his professional career, he appeared in a television play, *Strictly for Sparrows*, by Ted (later Lord) Willis. Fury had a large number of fans but he also had his critics. In October 1959 the curtain was dropped on his act at Dublin's Theatre Royal when part of his performance was considered too suggestive. With his hair combed slickly back at the sides, ruffled at the front, his shoulders hunched, eyes hooded, dressed in bomber jacket, open-necked plaid shirt, and cowboy boots, Fury conveyed a look which with his hip swivelling and hand movements impressed his female fans, but brought so much criticism that the act was 'cleaned up'. The jackets, plaid shirts, and boots were replaced by a golden suit, shiny silk shirts, and shoes. On stage Fury continued to include beat numbers in his act as part of his rock image, but more ballads were included to project a wider image.

In the two years following 'Maybe Tomorrow' Fury enjoyed moderate success with his next six records. Then, in May 1961 his recording of the rock ballad, 'Halfway to Paradise', reached no. 3, stayed for twenty-three weeks in the bestseller lists, and brought him his first silver disc. Fury followed this with 'Jealousy' which went to no. 2. However, in the years between 1959 and 1966 when all his twenty-six records appeared on the top forty, he never had a no. 1 hit in Britain. It was one of his greatest disappointments. On record he concentrated more on ballads. On stage he developed a 'little boy lost' look which also helped stimulate his female following. He also appeared frequently on television in such programmes as *Oh! Boy*, *Boy Meets Girls*, *Wham*, *Thank Your Lucky Stars*, and *Saturday Spectacular*. In 1962 Fury appeared in the first of three films, *Play it Cool*, directed by Michael Winner. It was essentially a low budget picture and did little for his career. Three years later Fury appeared in *I've Gotta Horse* which told of a singer who put his horse before his career. It was a biographical reflection of his love for animals and also featured his own racehorse, Anselmo, which finished fourth in the 1964 Derby. His only other movie appearance was as a rock singer in *That'll be the Day* in 1973.

With the coming of the Beatles, solo singers began to struggle and Fury was feeling the winds of change. He tried to enlarge his career by attracting a wider audience through Christmas pantomimes such as *Aladdin*, in which he played the title role, and the night club circuit.

In 1966 Billy Fury parted company with Decca Records and signed with Parlophone and later Red Bus. Neither company, however, brought him any new recording success.

Shortly after the release of the film *That'll be the Day* in 1973 Fury underwent open heart surgery and retired to his farm near Brecon, Wales. Only occasionally did he appear on stage, preferring to spend his time breeding horses and bird-watching. In 1978 Fury was declared bankrupt and compelled to pay a tax bill of £16,000.

He returned to the recording studio in July 1981 and recorded a new single, 'Be Mine Tonight', for Polydor but was unsuccessful. On 8 March 1982 Fury collapsed, suffering partial paralysis and temporary blindness and was rushed to hospital. He survived and was planning to return to the concert stage when he collapsed and died 28 January 1983, in St. John's Wood, London.

In 1969 Billy Fury married Judith Hall. They divorced in 1973. There were no children.

[Personal knowledge.]   DAVID JACOBS

G

GALE, SIR RICHARD NELSON (1896–1982), general, was born in London 25 July 1896, the only child of Wilfred Gale, a merchant from Hull, and Helen Webber Ann, daughter of Joseph Nelson, of Townsville, Queensland, Australia. After early years in Australia and New Zealand, the family returned to England when Richard was ten. He entered Merchant Taylors' School, by his own account a 'day-dreamer'. He then went to Aldenham School and began work in the City in 1913. Keen to become a regular army officer, he attended a crammers and passed the entrance examination for the Royal Military College, Sandhurst, in 1915. From a shortened war course, he was gazetted to the Worcestershire Regiment in the same year.

A particular interest in the Vickers machine-gun carried him into a unit of what was to become the Machine Gun Corps. He took part in the major battles in France and Flanders from the Somme, through Passchendaele to the defeat of the final German offensive in 1918, winning an MC as a company commander in that year. He remained in the Corps until, to his vexation, it was disbanded in 1922, when he reverted to his peacetime seniority. He served in India from 1919 to January 1936, attending the Staff College at Quetta in 1930–2.

Although he was graded above average in his annual reports and selected for the staff, he was a subaltern for fifteen years until accelerated to captain in the Duke of Cornwall's Light Infantry in 1930. These reports remarked, 'a strong personality and leader with plenty of ambition . . . a readiness for hard work, practicality, and great keenness on the polo field'. Promoted to major in 1938 in the Royal Inniskilling Fusiliers, he never served with this regiment; for he was then a general staff officer grade 2 in the War Office and was advanced on the outbreak of war to grade 1. In 1940 he was given command of a Territorial Army battalion, the 5th Leicesters.

General Sir Alan Brooke (later Viscount Alanbrooke, q.v.), commander-in-chief Home Forces, was much impressed by the high morale and standards of Gale's battalion. In September 1941 he promoted him to form and command the 1st Parachute brigade, raised by direction of the prime minister. His height and weight were not well suited to parachuting, but his earthy humour and direct methods appealed to the men who volunteered for this duty. With a threefold expansion of airborne forces in view, a specialist director was required in the War Office. Gale was ordered to take up this post in 1942. He was thus well fitted in practical experience, including acquaintance with the Royal Air Force and the development of glider aircraft, for command of the 6th Airborne division on its formation in April 1943.

Gale's task was to make his division ready for the invasion of Europe in June 1944. No British airborne division had ever taken to the battlefield as an entity by air; the technical and concomitant tactical options had to be tried out. In the subsequent training and planning, he drew shrewdly on the ideas of able subordinates, but his was the binding influence throughout the division. He made friends with the American air transport forces. The result was an outstanding operational success in the airborne assault, transcending that of the two United States airborne divisions involved.

Because of its spirit and operational skill, his division was chosen to help counter the German Ardennes offensive at the end of 1944. It repeated its earlier airborne success in the Rhine crossing in March 1945. In the remaining months of the war Gale was given command of the British airborne corps.

After the war many of Gale's contemporaries were retired or reverted to lower permanent ranks. Valued by Field-Marshal Viscount Montgomery of Alamein [q.v.], he was selected for a series of senior appointments: command of 1st division in the Palestine operations, 1946–7, of British troops in Egypt, 1948–9, director-general of military training, 1949–52, and commander-in-chief of the British Army of the Rhine and the NATO Northern Army Group, 1952–6, a unique extension of that post. Almost two years after retirement, he returned to succeed Montgomery as deputy to the NATO supreme Allied commander in Europe. There was no command function, which fell to the American principal alone. Gale wisely made no attempt to follow his predecessor's ways; rather, he made his own reputation among the alliance, relying on his experience in training, and international co-operation, playing up to the continental image of a British general. He took pains to manifest loyalty to the supreme Allied commander. These accomplishments overcame international resentment of continuous British tenure of the post, the final achievement of a career in which a simple intelligence combined fruitfully with an instinctive understanding of those with whom he worked.

He was appointed GCB in 1954 (CB 1945, KCB 1953), KBE in 1950 (OBE 1940), DSO in 1944, commander of the Legion of Merit (United States), and grand officier de la Couronne (Belgium). He was awarded the croix de guerre with palm (France). He was

ADC (general) to the Queen, 1954–7; colonel, the Worcestershire Regiment, 1950–61; and colonel-commandant, the Parachute Regiment, 1956–67. A good portrait of him is owned by the Worcestershire Regiment trustees.

In 1924 he married Ethel Maude Larnack (died 1952), daughter of Mrs Jessie Keene, of Hove, Sussex. In 1953 he married Daphne Mabelle Eveline, daughter of Francis Blick, of Stroud, Gloucestershire. She died after a candle overturned in her flat in Hampton Court Palace in March 1986, burning down part of the palace. Gale died in Kingston upon Thames General Hospital 29 July 1982. He had no children.

[Sir Richard Gale, *Call to Arms* (autobiography), 1968.]

ANTHONY FARRAR-HOCKLEY

GANDHI, INDIRA PRIYADARSHANI (1917–1984), prime minister of India, was born 19 November 1917 at Allahabad, the only child of Jawaharlal Nehru [q.v.], politician and prime minister of India, and his wife, Kamala, daughter of Jawaharmal Kaul, businessman, of Delhi. She was educated at the International School in Geneva, L'École Nouvelle Bex, St Mary's Convent in Allahabad, the Pupil's Own School in Poona and Bombay, Vishwa Bharati University at Shantiniketan, Badminton School near Bristol, and Somerville College at Oxford. She left Oxford before completing her degree course.

Indira Gandhi was born into one of the leading families of the Indian independence movement. For her the loneliness of being an only child was heightened by her father's absences in jail for his political activities, her mother's ill health and death when Indira was eighteen, and her own frequent changes of school. The dominant influence on her was her father Jawaharlal Nehru. In 1947 he became the first prime minister of independent India and remained in office until his death in 1964. Indira Gandhi lived in the prime minister's house and acted as his official hostess.

In 1955 she was nominated a member of the Congress Working Committee, the national executive of her father's party. This was her first senior political post. Then in 1959 she was elected to the prestigious office of president of the Congress Party. She strongly supported the Congress Party's participation in the agitation against the communist government in the state of Kerala, and called for its dismissal. Her father reluctantly agreed to the first dismissal of a democratically elected state government.

Indira Gandhi remained influential in the Congress Party up to her father's death. He

was succeeded by one of his closest ministerial colleagues, Lal Bahadur Shastri. Indira Gandhi's relations with Shastri were not cordial, but she agreed to be his minister of information and broadcasting. When Shastri died unexpectedly in 1965 she was chosen to succeed by a powerful group of Congress leaders who wanted to keep out the conservative Morarji Desai. They believed that she would be their creature. Indira Gandhi did indeed make a faltering start to her premiership. Her public appearances and her performances in Parliament were unconvincing. In the general election of 1967 the Congress Party only just scraped home. The results in elections to the state assemblies held at the same time were even more dismal. The power-brokers within the party moved against Indira Gandhi and she for the first time showed the political courage and skill which was to make her such a formidable leader. After open challenges to her authority she decided to allow the party to split in 1969, leaving her dependent on the support of the communists in Parliament.

In 1971 Indira Gandhi called an early general election, routed the old guard of the Congress Party, and emerged as the undisputed leader of India. The same year her stature was further enhanced by the defeat of the Pakistan army in the then East wing of that country and the emergence of Bangladesh as an independent nation. In the state assembly elections next year Indira Gandhi's party swept the polls. She now set about gathering all power into her own hands to prevent the re-emergence of any rivals.

By 1975 Indira Gandhi's triumph had turned sour. She was under attack for giving her younger son Sanjay a much sought after licence to manufacture cars, a veteran leader of the opposition closely associated in the public mind with M. K. Gandhi [q.v.] had launched a crusade against corruption, election results were going against her, and there was raging inflation. To cap it all a judge found Indira Gandhi guilty of corrupt electoral practices. Although the charge against her was nothing more than a technicality, the sentence, if upheld, would have obliged her to resign. A group of Indira Gandhi's colleagues persuaded her not to wait for her appeal to be heard but to declare a state of emergency. Under the emergency the powers of the judiciary were curtailed, the press was strictly censored, and Indira Gandhi's leading opponents were arrested. Her son Sanjay emerged as the second most powerful person in the government although he held no official position. He pioneered compulsory sterilization and slum clearance programmes which were deeply resented. It was largely because of these programmes that Indira Gandhi's party was routed in the general election she called in 1977.

She had hoped the electorate would endorse the emergency.

The unstable coalition of Indira Gandhi's opponents which came to power did not last long. In 1980 she was re-elected prime minister. Later that year Sanjay, who was by now acknowledged as her political heir, died piloting a light aircraft. That was a tragedy from which Indira Gandhi never seemed to recover. Problems were left unsolved, in particular the problem of Sikh militancy. This became so acute that in June 1984 Indira Gandhi ordered the Indian army to enter the golden temple complex in Amritsar to capture the Sikh militants who had turned it into an armed fortress. A bloody battle ensued during which tanks battered one of the most sacred buildings in the complex. Operation Blue Star, as it was known, caused widespread outrage in the Sikh community, and led directly to Indira Gandhi's assassination by two Sikh members of her bodyguard on 31 October 1984. She was succeeded by her elder son Rajiv.

Indira Gandhi was a small but imposing woman, who dressed simply but with impeccable taste. Her lifestyle was simple too. She could be awe-inspiringly remote, but she could also be utterly charming. On many occasions she showed great physical courage. She had interests in art, literature, and nature but her all-consuming passion was politics. Her memory for the names, faces, and life histories of men and women active in state and national politics was remarkable. After dominating the political scene for eighteen years she bequeathed to India a democratic system still intact but the Congress Party was bereft of leadership at all levels, and the bureaucracy had been politicized. Her critics accuse her of undermining the institutions of India and of professing socialism but practising pragmatism. Internationally she was widely respected and played key roles in the Non Aligned Movement, the Commonwealth, and other organizations.

During her long career Indira Gandhi was awarded honorary doctorates by a large number of Indian and foreign universities including Agra, Oxford, the Sorbonne, Moscow State University, and the Soviet Academy of Sciences. In 1972 she was given India's highest award, the Bharat Ratna. Among the international awards she received were the Order of the Republic of Tunisia, 1982; the Order of the Gold Ark from the World Wild Life Fund, 1982; the Olympic Gold Order, 1983; and the United Nations Population award, 1983. She was awarded the Lenin prize posthumously in 1985.

In 1942 she married Feroze Gandhi, journalist and politician, the son of Jehangir Fardoonji Gandhi. There were two sons. Indira Gandhi was shot dead in the garden of the prime minister's residence in Delhi, 31 October 1984.

[Zareer Masani, *Indira Gandhi, a Biography*, 1975; Uma Vasudev, *Indira Gandhi, Revolution in Restraint*, 1974; Sarvepalli Gopal, *Jawaharlal Nehru: A Biography*, 3 vols., 1975, 1981, and 1984; personal knowledge.]

MARK TULLY

GARNER, (JOSEPH JOHN) SAVILLE, BARON GARNER (1908–1983), civil servant and diplomat, was born 14 February 1908 at Muswell Hill, north London, the second of three sons (there were no daughters) of Joseph Garner, general draper, of Highgate, London, and his wife, Helena Maria, daughter of John Culver, of London. He was educated at Highgate School and Jesus College, Cambridge. He obtained a second class (division I) in part i of the modern and medieval languages tripos (1927) and a first in part ii (1929). After a short time as a schoolmaster teaching modern languages at Haileybury, he entered the Dominions Office in 1930 as an assistant principal and began his lifelong career of service to the Commonwealth. He joined the private office of the dominions secretary in 1935 as an assistant private secretary and enjoyed a formative experience under three strikingly different personalities. He served two secretaries of state— J. H. (Jimmy) Thomas [q.v.], a former railwayman, and Malcolm MacDonald [q.v.], the prime minister's son and someone with whom his life was to be inextricably linked in later years. His immediate mentor as principal private secretary was (Sir) Edward Marsh [q.v.], one of the last of the great amateur tradition of literary dilettantes in Whitehall.

It was Thomas who refused to call the young Garner by his customary Christian name of Saville ('What sort of fancy name is that? Ain't you got a plainer name?') and insisted to the dismay of his parents in calling him Joe. Joe Garner he remained for the rest of his life and it was the name he preferred to use on his monumental and definitive history, *The Commonwealth Office 1925–1968* (1978), which he wrote in retirement.

Garner was thus early identified as a high flyer in the Dominions Office. He returned to the private office as principal private secretary to the secretary of state during the first three years of World War II when relations between the United Kingdom cabinet and the dominion governments were of vital importance. In 1943 he went to his first overseas post in Ottawa as a deputy to Malcolm MacDonald, who was then high commissioner to Canada. In 1948 Garner returned to the Dominions Office as assistant under-secretary of state and in 1951 he had his first experience of the new post-war Commonwealth as deputy high commissioner in Delhi. In 1953 he was brought back to what was now the Commonwealth Relations Office as deputy permanent under-secretary of state. In 1956 he went back to Canada as high commissioner and in 1962 he returned to London to take over the leadership of the Commonwealth Office as permanent under-secretary of state at a time of great change in Britain's international role generally.

Garner personified in his life the historic transition from Empire into Commonwealth. Within Whitehall he was the quintessential Commonwealth man. His entry into the old Dominions Office in 1930 coincided with the Statute of Westminster, and his departure in 1968 marked the end of the separate identities of the great departments of state for the Commonwealth and for the colonies.

The creation of the contemporary Commonwealth demanded diplomatic skills which Garner had in generous measure. It was a distinctive kind of diplomacy, different in character from the classical inter-state diplomacy of the Foreign Office. Bringing dependent territories forward to full sovereignty demanded an intimacy of relationship with the new leaders and human qualities of inspiring respect and trust.

These qualities were also required in Whitehall in the delicate operation of persuading the separate colonial, Commonwealth, trade, and Foreign Office organizations to pool their sovereignty in a single external representative system for Britain. It was a tribute to Garner's personal contribution in bringing this about, that when it was completed in 1965, and after Lord Caccia, his fellow permanent under-secretary at the Foreign Office, had served barely two months, Garner was chosen to succeed him for the first three formative years of a unified Diplomatic Service. And this was the foundation stone on which was subsequently built the integrated structure of a single Foreign and Commonwealth Office. Here again Garner played a sensitive and constructive role, and with this task accomplished in 1968, he retired from public service. But he did not retire from Commonwealth affairs and he continued to give voluntary service as chairman of the Commonwealth Institute (1968–74), of the Commonwealth scholarship commission for the UK (1968–77), and of the Institute of Commonwealth Studies (1971–9). He also served in retirement as a member of the Security Commission (1968–73).

Garner stood at six feet and, broad in proportion, was an impressive figure. Because of his glasses, and a gap in his front teeth, he never thought of himself as handsome; others are more likely to remember his robust and cheerful manner, his lively eyes, his ever youthful appearance.

Garner was an honorary LLD of the universities of British Columbia (1958) and Toronto (1959), and an honorary fellow of Jesus College, Cambridge (1967). He was also a governor of Highgate School and its treasurer and chairman from 1976. He was appointed CMG in 1948, KCMG in 1954, and GCMG in 1965, and was created a life peer in 1969.

He married in 1938 Margaret (Peggy), daughter of Herman Beckman, of Cedar Lake, Indiana, USA, an inventor with the DeLaval Separator Company, of Chicago. Her effervescent, outgoing mid-Western personality was a great support in the social and personal side of their diplomatic work together. There were two sons and one daughter. Garner died 10 December 1983 at Bournemouth, while visiting his family.

[Personal knowledge.]

THOMSON OF MONIFIETH

GARNETT, DAVID (1892–1981), writer, editor, and publisher, was born 9 March 1892 at Brighton, the only child of Edward Garnett, author and publishers' reader, of The Cearne, Edenbridge, and his wife, Constance [q.v.], distinguished translator of Russian classics. She was the daughter of David Black, coroner. His grandfather was Richard Garnett (1835–1906, q.v.), prolific man of letters and keeper of printed books at the British Museum. David was educated at University College School and the Royal College of Science, South Kensington, where he studied botany, and fungi in particular. At twelve years of age he paid a memorable visit to Russia with his mother, while through his father's gift for discovering literary talent he became friendly from boyhood with such writers as Joseph Conrad, D. H. Lawrence, and W. H. Hudson [qq.v.]. Later he got to know the Stracheys, J. Maynard (later Lord) Keynes, Duncan Grant [qq.v.], and the rest of the Bloomsbury Group, with whom he was associated all his adult life.

During the war of 1914–18 he went to France with the Friends' War Victims' Relief Mission, and afterwards worked on the land. After the war he opened a bookshop in the heart of Bloomsbury with his friend Francis Birrell, but writing had already become his chief interest, and his marriage in 1921 together with the outstanding success of his first serious novel *Lady into Fox* (1922; Hawthornden and James Tait Black memorial prizes for 1923) persuaded him to buy Hilton Hall, a Queen Anne house near Cambridge, and devote himself entirely to literature. His output of over a dozen novels included *The Sailor's Return* (1925), *Aspects of*

Love (1955), and *Up she Rises* (1977). Among his autobiographical works were *A Rabbit in the Air*, a delightful account of learning to fly (1932), *The Golden Echo* (1953), and *Great Friends* (1979). Garnett was also responsible for editing *The Letters of T. E. Lawrence* (1938), *The Novels of Thomas Love Peacock* (1948), and Dora Carrington's *Letters and Extracts from her Diaries* (1970), as well as being involved in two successful publishing ventures: the Nonesuch Press—with (Sir) Francis Meynell [q.v.]—in 1923, and Rupert Hart-Davis Ltd. in 1946.

In 1939 Garnett became literary editor of the *New Statesman*, and he continued contributing to it after the outbreak of World War II, when he joined the Air Ministry with the rank of flight-lieutenant RAFVR, and subsequently was an intelligence officer in the Political Warfare Executive, whose secret history he wrote. His was a large and vigorous output, based on a variety of interests and wide reading. On first starting as a novelist he had taken Daniel Defoe [q.v.] as his model, and the same combination of an imaginative or fantastic premiss with a sturdy, objective, and masculine style can be seen in both writers. Many of his plots were markedly original, and have attracted the interest of artists in other media. Ballets founded on *Lady into Fox* (1939, with music by Arthur Honegger) and *The Sailor's Return* (1947) were successfully staged by the Ballet Rambert with Sally Gilmour in the leading roles, while the latter novel and *A Man in the Zoo* (1928) were made into films and shown on television.

As a young man Garnett, known as 'Bunny' to his friends, was good-looking, fair-haired, and blue-eyed. His energy and enterprise—and perhaps a streak of recklessness—found an outlet in learning to fly, among other outdoor activities such as fishing, swimming (he could be seen diving into cold water when he was over eighty), and bee keeping. For a time he kept a small Jersey herd at Hilton, and he took a sympathetic interest in most agricultural processes and liked to describe them. As he said himself: 'I believe I can write about grass growing as well as anyone.' Indeed he had always been more at home in the country than in town, from his days as a rather solitary child exploring the woods around his parents' house and developing an enthusiasm for the wild flowers and creatures he saw there until his discovery in middle age of the attractions of the Yorkshire dales. Travel was another source of pleasure, particularly in France, but he also paid several visits to the United States, where he made many friends and sometimes gave lectures. A life-lover, gregarious, and a genial host, he gave and received much affection and was an excellent letter-writer. He spent the last ten years of his life in a modest stone cottage in Charry,

Moncuq, Lot, France, surrounded by his fine library and a small collection of Bloomsbury paintings. Here he cooked and bottled wine for his many visitors, and could be seen sitting out of doors under a large straw hat typing away at his latest book until he died there 17 February 1981.

David Garnett was appointed CBE in 1952, was elected a fellow of the Imperial College of Science and Technology in 1956, and became C.Lit. in 1977. In 1977 he received an honorary D.Litt. from Birmingham University.

In 1921 he married Rachel Alice ('Ray'), daughter of William Cecil Marshall, architect, and sister of the author of this notice. She illustrated her husband's books and some others, and died in 1940. In 1942 he married Angelica Vanessa, daughter of Duncan James Corrowr Grant and Vanessa Bell [qq.v.], both painters, of Charleston, Sussex. There were two sons of the first marriage and four daughters, of whom the final two were twins, of the second. His eldest daughter died in 1973.

[*The Times*, 19 February and 4 and 23 March 1981; Carolyn G. Heilbrun, *The Garnett Family*, 1961; David Garnett, *The Golden Echo*, 1953, *Flowers of the Forest*, 1955, and *The Familiar Faces*, 1962 (autobiographies); private information; personal knowledge.]

FRANCES PARTRIDGE

GEORGE-BROWN, BARON (1914-1985), politician. [See BROWN, GEORGE ALFRED.]

GÉRIN, WINIFRED EVELEEN (1901-1981), biographer, was born 7 October 1901 in Hamburg, the youngest of four children and second daughter of Frederick Charles Bourne, senior manager in Nobel's Chemical Industries, and his wife, Katharine Hill, who was in Hamburg studying music and German. The family returned to England and their London home in South Norwood, where Winifred was brought up. She was educated at Sydenham High School for Girls in Croydon, and at Newnham College, Cambridge, where she read English and history, and where she later became an associate. She obtained second classes in modern and medieval languages (1922) and English (1923).

Before World War II she lived in France, and it was there that she met Eugène Gérin, a Belgian poet, and a professional cellist performing regularly with the Monte Carlo Symphony Orchestra. They were married at the Croydon register office in September 1932. This period between the wars was for her a blissful one. She was of a romantic disposition. Her chief interests and inspirations were literature, music, and 'wild nature'; to the last she was to attach much importance in later years. For the moment she had to her credit a volume of verse, *The Invitation to Parnassus* (1930), but she had no thought of being anything but a writer. She accompanied her husband on concert tours, and according to the fashion of the times fell in love with Paris.

When the war came they were trapped in Vichy France. It took two years to negotiate their passage back to England, their nationalities by this time being in doubt. During those years they sheltered many refugees and helped them on their journeys across the Pyrenees. Winifred Gérin was fluent in German, French, and Spanish. For the rest of the war she and her husband worked together in the political intelligence department of the Foreign Office. Eugène Gérin was a broadcaster to the Resistance, putting out his daily bulletins to Belgium. In 1945 he died, at the age of forty-eight.

For the next ten years Winifred Gérin lived in Kensington with her sister Nell Bourne, a painter. Her devotion to the Brontës lured her eventually to the 'wild nature' of the west Yorkshire moors. There she at last found her vocation, but she was fifty-eight when her first biography was published. Before that she met in Haworth another seeker after the Brontës, John Lock, who was working on Patrick Brontë [q.v.], and equally susceptible to natural beauty. He was the son of William Lock, FCIS, a secretary of the Chartered Institute of Secretaries. They were married in 1955 and made their home on the edge of Haworth Moor. They lived and worked there until 1964.

Winifred Gérin's first Brontë biography was of Anne [q.v.], published in 1959; it was followed by her study of (Patrick) Branwell [q.v.] (1961). Her major work, *Charlotte Brontë, The Evolution of Genius*, was published by the Oxford University Press in 1967. It was awarded the James Tait Black memorial prize for the best biography of the year, the W. H. Heinemann award, and the British Academy Rose Mary Crawshay prize (all 1968). In 1971 she produced *Emily Brontë*.

Charlotte Brontë [q.v.] left an abundance of source material for the biographer, including some 700 letters: only three letters from Emily [q.v.] have been preserved, and little else. Winifred Gérin approached her biography of Emily with trepidation, fearing that she could not do justice to her. She had only the words of others as a guide, her intuition, her sense of place, and her ten years in Haworth—'at all seasons and in all weathers walking Emily's moors'. By then the Brontë industry was developing, painstakingly scholarly work was taking place, concentrating for the most part on textual problems and the juvenilia. Winifred Gérin regarded the labours of the scholars respectfully, and

with some wonder, but that was not her way. She herself became passionately fond of her subjects, and cherished them as friends or daughters. Warmth, compassion, and enthusiasm characterize her work. A more academic biographer might have thought it unnecessary to get quite so weather-beaten in the course of research, but Winifred Gérin needed 'atmosphere'. Of the biographies of Charlotte and Emily, C. P. (Lord) Snow [q.v.] wrote that the 'two together rank as one of the monuments, scholarly, literary, intuitive, of our time'.

In 1975 Winifred Gérin was appointed OBE for services to literature. She was a fellow (1968) and member of council of the Royal Society of Literature. In 1976 her *Elizabeth Gaskell* won the Whitbread literary award for biography. During these later years she devoted much of her time to her ailing sister, Nell Bourne, who died in March 1981. She herself died two months later, 27 June 1981, in Kensington, three days after the publication of her last book, *Anne Thackeray Ritchie*. This was a first introduction to the daughter of W. M. Thackeray [qq.v.] for many readers, and it gave much pleasure to readers as well as to the author. Winifred Gérin identified with the gregarious, high-spirited London hostess just as thoroughly as with the lonely sisters of Haworth. Her sympathy encompassed them all. She had no children.

[Winifred Gérin, 'The Years that Count' (unpublished autobiography); private information; personal knowledge.]

PETER SUTCLIFFE

GIBBERD, SIR FREDERICK ERNEST (1908-1984), architect, town planner, and landscape architect, was born in Coventry 7 January 1908, the eldest of five sons (there were no daughters) of Frederick Gibberd, a shopkeeper (gentleman's outfitter) of Earlsdon, Coventry, and his wife, Elsie Annie Clay, of Nuneaton. Gibberd was educated at the King Henry VIII School at Coventry. He was articled to an architect in Warwick and subsequently entered the Birmingham School of Architecture. Here he met, and was initially strongly influenced by, F. R. S. Yorke [q.v.], one of the pioneers of modernism in English architecture. He shared an office with Yorke when they both moved to London in 1930, but he soon set up on his own and built up a successful architectural practice in which the modern movement was reflected, notably in his first substantial building, Pullman Court, an elegantly designed block of flats at Streatham. His work afterwards conformed more closely to the idiom of the day although at its best it retained qualities of invention and sensibility that most commercial architecture lacked. Unlike many heads of large architectural offices Gibberd refused to become a mere administrator and insisted on himself spending time on the drawing-board.

His most important architectural works, executed between 1930 and 1970, include the first terminal buildings at Heathrow airport; civic buildings at Doncaster, St Albans, and Hull; power stations at Hinkley Point and Didcot; the Inter-Continental Hotel at Hyde Park Corner, London; the reconstruction of the Coutts Bank in the Strand (a building by John Nash, q.v., for which Gibberd's glass-fronted centre, replacing a Victorian block, caused much controversy in 1969); the London mosque in Regent's Park; and the Roman Catholic cathedral at Liverpool. The last two were the outcome of architectural competitions won by Gibberd in 1969 and 1960. The cathedral's circular plan and crown-like tower were original conceptions but their total effect was criticized as being somewhat brittle in view of their monumental purpose.

Gibberd's most considerable achievement however was Harlow New Town, one of the ring of new towns round London initiated by the government after World War II, for which he was appointed planner and chief architect in 1946 and of which he remained in charge until 1972. Throughout his career he had shown a special talent for landscape—he was responsible for the landscaping of several new reservoirs, notably at Llyn Celyn, Dyfed—and Harlow gave him an opportunity to which he devoted unremitting care and energy until the end of his life. With its green spaces penetrating the housing areas and its carefully considered planting it is probably the most successful of the post-war new towns.

Gibberd was also an expert and imaginative gardener. The garden he laid out round his own house near Harlow, which occupied much of his later years, was a beguiling mixture of the romantic and the monumental and revealed qualities as a designer that circumstances had not always allowed him to display in his buildings. His aesthetic sensibility was also shown in his discriminating collection of paintings and drawings.

He was a man of attractive presence as well as professional ability, but delicate health prevented him from playing the prominent part in affairs for which he was equipped. He was however twice vice-president of the Royal Institute of British Architects, was a hard-working president of the Building Centre, and served for twenty years on the Royal Fine Art Commission (1950-70). In 1942-4 he was principal of the Architectural Association School of Architecture in London. He was a fellow of the Royal Institute of British Architects (1939, having be-

come a licentiate the previous year), of the Royal Town Planning Institute, and of the Institute of Landscape Architects, was elected RA in 1969 (having become ARA in 1961), appointed CBE in 1954, and knighted in 1967. In 1969 Liverpool University awarded him an honorary LLD. He also won many medals and awards. He published several books, the most notable being *Town Design* (1953), a well-balanced and well-illustrated survey.

Gibberd was twice married: first in 1938 to Dorothy Pryse Phillips who was of Welsh descent and a Welsh speaker, the daughter of John Hugh Phillips, a London bank manager. They had a son and two daughters. She died in 1970. He then married, in 1972, Patricia Fox Edwards, the widowed daughter of Bernard Joseph Spielman, of Bristol, like Gibberd's father a gentleman's outfitter. With his second wife's collaboration, notably over the choice of sculpture to embellish it, Gibberd maintained his involvement with Harlow New Town although after 1972 he was no longer officially in charge and no longer took an active part in the architectural practice that continued to bear his name. His Harlow garden then provided the main outlet for his creative instincts. He died at home in Harlow 9 January 1984.

[Private information; personal knowledge.]

J. M. RICHARDS

GIELGUD, VAL HENRY (1900-1981), novelist, playwright, and head of radio drama at the BBC, was born 28 April 1900 in South Kensington, London, the second of the four children and second of three sons of Frank Gielgud, a stockbroker of Polish descent, and his wife, Kate Terry, daughter of Arthur Lewis. The family influences were strong. His Polish forbears, some of whom had been soldiers, had a romantic fascination for him and awakened his lasting passion for military history. From both his father and his mother, who was daughter of the actress Kate Terry and niece of Dame A. Ellen Terry [q.v.], he acquired a love of the theatre.

Though not such a scholar as his admired elder brother Lewis, Val Gielgud was always a voracious reader and began early to write himself. In 1918 he went from Rugby to the Household Brigade Officers' Cadet battalion. At the end of the war he took up a history scholarship at Trinity College, Oxford, where he 'read . . . talked and . . . began to learn how to think'. He also directed some drama, but left after three years without taking his degree.

From 1921 to 1928 he worked briefly in various jobs before following his younger brother, (Sir) John Gielgud into acting. It amused him to say that while his brother had become arguably the best actor in Britain, he himself had undoubtedly been the worst. The claim reflected both his natural modesty and his excellent judgement. Throughout this period, he continued to write, following a precept given to him by Michael Sadleir [q.v.]—that good fiction depended upon telling a good story. He wrote entertainingly for the rest of his life: over twenty novels, seven volumes of autobiography and essays, and eighteen plays and screen-plays. To this Dictionary he contributed the notice of Julia Neilson. Yet writing was never his primary occupation.

In May 1928 his friend Eric Maschwitz, then editor of the *Radio Times*, invited him into the BBC as his assistant. The job was not arduous, and left time for such activities as an amateur production of a play by Ian Hay [q.v.], *Tilly of Bloomsbury*, with Sir John (later Lord) Reith [q.v.], the BBC's director-general, playing a drunken broker's man. Less than a year later, at the beginning of 1929, Gielgud was appointed director of productions, in succession to R. E. Jeffrey, with responsibility for all drama and variety.

It was a bold and unexpected appointment but it was to prove inspired. Gielgud shared some of Reith's characteristics: vision, enthusiasm, and commitment to quality. He took over a small but diverse department with an annual output including only about twenty plays. Thirty-four years later, despite competition from television, he handed to his successor, Martin Esslin, the largest radio drama department in the world, with an output of over a thousand productions a year.

During those years Gielgud was widely involved in broadcasting and in the lively discussion which always surrounds it. His policy for drama was not only formative for radio but pioneered much that was later found on television. He fought, and won, the battle for drama of varying length, maintaining that plays should always be special events in the schedule. He made the serialization of classic novels one of the staple ingredients of broadcasting. He started series of popular plays, such as those heard on Saturday Night Theatre from 1943, which have run continuously to a cumulative audience of billions. In presenting the widest range of plays from all over the world, he could justly claim to have run a true National Theatre long before there was one for the stage.

Like most pioneers, he was often attacked for going too far or too fast but he was a valiant defender of what he believed to be good. In 1941 his production of *The Man Born to be King*, by Dorothy L. Sayers [q.v.], was almost prevented from reaching the air but became one of radio's most triumphant and popular successes. Ten years later, a parliamentary row

over his own play *Party Manners* helped to reinforce the BBC's independence from direct political pressure.

He had taken part in early experiments with television yet, sadly, his period running television drama from 1949 to 1951 was not a success. He returned to radio, heading a lively department and producing many plays, until his retirement in 1963. Like many shy people, Gielgud was sometimes mistaken for an autocrat. In fact, though firm in his beliefs, he often accepted the enthusiasm of others for work in which he saw little merit; honest to a fault, he would later take no credit for what he had allowed to happen. His achievements enriched the life of the country during a critical and formative period. He was appointed OBE in 1942 and CBE in 1958.

His retirement gave him more time to write and to travel, to watch polo, and to enjoy the company of his cats and many friends at the house near Lewes in East Sussex to which he and his wife moved from London in 1963.

In 1921 he married Natalie Mamontoff; the marriage was dissolved in 1925. In 1928 he married the actress Barbara Dillon, daughter of Herbert Druce, actor. They had one son, Gielgud's only child. This marriage was dissolved and in 1946 he married Rita Vale, daughter of Samuel Weill, lawyer. After this marriage was dissolved, in 1955 he married Monica Joyce, daughter of Arthur Hammett Greey, company director of a Birmingham store. The marriage was dissolved in 1960 and in the same year he married (Vivienne) June ('Judy'), daughter of John Bailey, farmer. Gielgud died 30 November 1981 at Eastbourne.

[Val Gielgud, *Years of the Locust*, 1947, *One Year of Grace*, 1950, and *Years in a Mirror*, 1965; BBC archives; personal knowledge.]

RICHARD IMISON

GILLIE, DAME ANNIS CALDER (1900–1985), general practitioner and medical politician, was born in London 3 August 1900, the eldest daughter and first of the four children of the Revd Dr Robert Calder Gillie, a minister in the Presbyterian Church of England, and his wife, Emily Japp. She was educated at Wycombe Abbey School; University College, London; and University College Hospital, graduating MB, BS in 1925 and taking the membership of the Royal College of Physicians two years later. She was proud to be a daughter of the manse and equally proud of her family which included a number of distinguished journalists. She entered general practice in London as assistant to a partnership of three women, one of whom, Christine Murrell, was herself a well-known figure in medical politics. When one senior partner died and the others retired Annis Gillie worked single-handed from her London home in Connaught Square. During World War II she moved out of London with her two young children to her country cottage at Pangbourne from whence she drove in and out of London and over a wide area of countryside visiting patients. She resumed in partnership in 1954 until her retirement in 1963.

Annis Gillie played a predominant part in the renaissance of general practice that began after World War II. She was a member of the General Medical Council in 1946–8 and was president of the Medical Women's Federation in 1954–5. She became a member of several statutory bodies concerned with general practice including the Medical Practices Committee, the Executive Council of London, the Standing Medical Advisory Committee, and the Central Health Services Advisory Council. She was a member (and for some years vice-chairman) of the British Medical Association central ethical committee and a member of the BMA council in 1950–64.

She was a founder member of the College of General Practitioners, an early member of its council, and its chairman in 1959–62. This was a period of intense depression within general practice. Recruitment was falling, general practitioners were leaving the country in large numbers, under-doctored areas were increasing, and all the advantages, particularly financial, lay with medical careers other than general practice. It was also a time of difficult relationships between the new college and the British Medical Association. In 1961 the Standing Medical Advisory Committee set up a sub-committee, chaired by Annis Gillie, to predict the course along which general practice should develop over and beyond the next decade. Its report, 'The Field of Work of the Family Doctor', was published in 1963. It was unusual in that the chapters were written by members of the committee rather than departmental officials but it provided a statement of confidence in the future drawn from a strong factual base. It made its mark on practice throughout the English-speaking world and profoundly influenced the 'charter' agreed between the BMA and minister of health in 1964 which restored the fortunes of general practice. It had also a major effect upon the report of the royal commission on medical education which, appearing in 1968, recognized general practice as a separate speciality and gave pride of place to its separate postgraduate training.

During the time that Annis Gillie held office in the college, first as chairman of council and then president in 1964–7, many important decisions were taken, such as the institution of an examination for membership and the purchase of 14 Princes Gate, the college's subsequent

home. This period was crowned with the award to the college of the royal charter. Annis Gillie had a strong personality and a very tough centre. She dealt with patients in a gentle, firm, authoritarian manner. In meetings she was able when necessary to silence unconstructive contributions but always did this with courtesy. On public occasions she never failed to find the right words. She was not an original thinker or innovator. Her skills were those of facilitating, encouraging and promoting others, and persuading committees to produce sensible and balanced conclusions. She was a medical politician and as one of her contemporaries said, was an 'Attlee' rather than a 'Bevan' and was always nearer to the political than the academic side of her discipline.

Many honours came her way. She was appointed OBE in 1961 and DBE in 1968. In 1964 she was elected FRCP and was awarded an honorary MD by Edinburgh University (1968). She was elected a fellow of the BMA in 1960 and in 1968 elected a vice-president of the association in recognition of her valuable service, the first woman to be so honoured.

In 1930 she married Peter Chandler Smith, an architect. His practice was destroyed by the war and he was rejected for military service, and so she was for many years the 'bread-winner'. They had a daughter, who died in 1974, and a son. Her husband became incapacitated with a chronic illness that made him very dependent upon her for the last fifteen years of their life together. He died in 1983. Annis Gillie died 10 April 1985 at her home in Bledington, Oxfordshire.

[Personal knowledge.]　　　　V. W. M. DRURY

GODWIN, SIR HARRY (1901–1985), botanist, was born 9 May 1901 at Holmes, Rotherham, Yorkshire, the only son and elder child of Charles William Thomas Godwin, grocer and licensed victualler, and his wife, Mary Jane Grainger. Soon after his birth the family moved to Long Eaton in Derbyshire. There Harry Godwin attended the High Street council school, and from the age of twelve the co-educational Long Eaton County Secondary School. He always spoke and wrote in highly appreciative terms of the quality of the education he received at this school, then under the headship of the outstandingly able Samuel Clegg. The war of 1914–18 unfortunately restricted the school's activities, to Godwin's lasting regret reducing the time available for teaching mathematics in the upper classes. He gained a distinction in botany in the Oxford senior local examination and attended laboratory classes in geology at University College, Nottingham. In 1918 he obtained a Derbyshire county major scholarship and an open scholarship of £60 a year at Clare College, Cambridge. In his last year at school he wrote a paper for the school magazine on the vegetation of an outlier of Liassic limestone near Gotham. He was now aware of a keen interest in ecology.

Godwin entered Clare College in 1919 and obtained first classes in both parts of the natural sciences tripos (1921 and 1922). He was much excited and influenced by the ecological work of (Sir) A. G. Tansley [q.v.], then a university lecturer in the Cambridge botany school, but was accepted by F. F. Blackman as a research student in plant physiology. He explained that he wished to apply the experimental methods of leading plant physiologists to the ecological research he hoped to undertake later. His work on the mechanism of starvation in leaves gained him his Ph.D. in 1926.

In 1923 he was appointed a junior university demonstrator in botany and ran the practical classes supporting the first-year lectures of (Sir) A. C. Seward [q.v.]. In 1927 he was made senior university demonstrator and in 1934 university lecturer, giving for many years a much appreciated elementary course for first-year botanists. In the preface to his highly successful book *Plant Biology* (1930, 4th edn. 1945) he rightly claims that the emphasis on physiology rather than on morphology and evolution secured a deeper understanding than previously of the basic role of the green plant and a general recognition of the scientific and social value of ecological studies. Meanwhile he was appointed a research fellow of Clare in 1925 and was an official fellow from 1934 until his retirement in 1968. He always took a keen interest in college affairs and was acting master in 1958-9 and senior fellow for many years. He continued to live in Cambridge until his death.

In 1923 Godwin began systematic and detailed studies of Wicken Fen, in particular of the mechanism of the vegetational succession. He and Tansley published in 1929 an interim account of their findings under the title *Vegetation of Wicken Fen* (part v of *The Natural History of Wicken Fen*). He continued research on fen vegetation, at Wicken and elsewhere, applying the methods of pollen analysis to the deep deposits of peat in order to work out the past history of the fens and their surroundings. Godwin's main contributions were undoubtedly in establishing the relationship between peat stratigraphy and the pollen zones based on the identification and estimate of relative abundance of pollen grains of different tree species in different strata. More accurate identification of non-tree pollen aided accounts of the correlation between climate, forest composition, prehistoric agriculture, and peat stratigraphy. This wealth of data and interpretation were brought together in Godwin's extremely readable *The History of the British Flora* (1956, 2nd edn. 1975), in addition to numerous papers in scientific journals.

In the following decades he gained a high reputation as a leader in ecological thought and practice, both in the university and further afield. He was president of the British Ecological Society in 1942-3, editor of the *Journal of Ecology* (1948-56), and joint editor of the *New Phytologist* (1931-61). In 1948 he was appointed the first director of the newly formed sub-department of quaternary research in the Cambridge botany school, a post which he held until 1966. Special mention must be made of his notable contributions to our knowledge of radiocarbon dating and of geologically recent changes in relative land/sea levels as well as to the archaeological implications of his findings. From 1960 until his retirement in 1968 he was professor of botany at Cambridge University.

Godwin was a big man, tall and robust-looking, with brown hair and pale grey eyes. He smiled freely and entered into lively and humorous conversation with both friends and strangers, who all saw him as a most kindly person. He was recognized and accepted as a natural leader and he exerted a very considerable influence over all his acquaintances.

Godwin was elected FRS (1945), a member of the Royal Society council (1957-9), and Croonian lecturer in 1960. In 1970 he was knighted. He received the Prestwich medal of the Geological Society of London in 1951 and the gold medal of the Linnean Society in 1966, with comparable awards from several non-British institutions. He was also made an honorary member of various academies including the Royal Irish Academy in 1955, the Botanical Society of Edinburgh in 1961, and the International Union for Quaternary Research in 1969. He was made Sc.D. (Cantab.) in 1942 and received honorary doctorates of science from Trinity College, Dublin (1960), and the universities of Lancaster (1968) and Durham (1974).

In 1927 Godwin married Margaret Elizabeth, daughter of James Daniels, butcher. She had attended the Long Eaton Secondary School with him and had then taken a London B.Sc. degree in botany at Nottingham University. She was calm and clear-headed, an ideal partner for Godwin. She collaborated with her husband in the early work on pollen analysis of fen peats, until the birth in 1934 of their only child, a son. He became a general practitioner but to the great sorrow of his parents died in 1974. Godwin died in Cambridge 12 August 1985.

[R. G. West in *Biographical Memoirs of Fellows of the Royal Society*, vol. xxxiv, 1988;

personal knowledge.]   A. R. CLAPHAM

GOODALL, NORMAN (1896-1985), missionary statesman and pioneer in ecumenism, was born in Birmingham 30 August 1896, the twelfth of the thirteen children, five of whom died before he was born, of Thomas Goodall and his wife, Amelia Ingram. The family lived in cramped conditions over their father's sweet shop in Handsworth and poverty was never far from their door. Goodall left school at fourteen to work as an office boy, but ambition soon led him to apply for a clerical post in the Birmingham city treasurer's department. In 1915 he enlisted in the Royal Army Medical Corps and was transferred to the Artists' Rifles, from which he was seconded a year later to the Ministry of Munitions. Soon afterwards he became the first member of the staff of the Department of National Service. Thus, as a very junior civil servant, Goodall embarked on a brief career which was to bring him into contact with ministers of the Crown and give him responsibilities far beyond anything he could have anticipated.

When the war ended, he was urged to enter the permanent Civil Service, in which he would doubtless have had a distinguished career. But he felt an irresistible call to the Congregational ministry. Apart from evening classes he had received no formal education since leaving school, though his father and mother had instilled in him a love of literature which was the foundation of the mastery of the English language of which he was to be such an elegant exponent.

Despite his lack of qualifications he was granted special admission to Mansfield College, Oxford, in 1919 where he took a third class honours degree in theology (1922). In 1950 he was awarded a D.Phil. in the same university. In 1922 he was ordained to the ministry at Trinity Congregational church, Walthamstow, and six years afterwards moved to the church in New Barnet.

In 1922 he and his wife had offered themselves for service abroad with the London Missionary Society, but the regulations in force prevented them being accepted. However, this was to be only a postponement; for in 1936 Goodall was invited to become a staff member of the Society with secretarial responsibility for India and the South Pacific. This led to extensive travel throughout these regions and a widening circle of personal contacts with missionaries and government officials. In 1944 he was appointed to succeed William Paton [q.v.] as London secretary of the International Missionary Council and this was to place him at the centre of the developing ecumenical movement. He had always believed that church and mission were indivisible and for seventeen years worked with great skill and patience to bring the fledgling World Council of Churches into integral relationship with the older organization. He was instrumental in persuading the two bodies to work in association with one another, and in 1954 became secretary of a joint committee to explore their full integration. He saw this consummated at the third assembly of the World Council at New Delhi in 1961 and could rightly be called the architect of that achievement.

After retirement in 1963 Goodall devoted himself to inter-church relations. He was moderator of the International Congregational Council in 1962-6, moderator of the Free Church Federal Council in the following year, and he played an influential part in the establishment of the United Reformed Church. He was the author of standard works on the history of the ecumenical movement and the London Missionary Society, and he edited the report of the fourth assembly of the World Council at Uppsala in 1968. To this Dictionary he contributed the notice of B. J. Mathews. He lectured extensively as visiting professor at the Selly Oak Colleges, Birmingham (1963-6), at the Irish School of Ecumenics in Dublin (1971-3), at Heythrop College (1970-1), and at the Pontifical Gregorian University in Rome (1975). His semester in Rome broke fresh ground in ecumenical understanding and fittingly crowned his contribution to the world church.

Goodall was a gracious man with an extraordinary capacity for making friends all over the world. He was excessively modest, evidenced in the title of his autobiography, *Second Fiddle* (1979). Although he was content to play a supporting role to the leaders of the ecumenical movement, he was the architect of some of its most important developments.

In 1920 he married a medical doctor, Doris Elsie Florence, daughter of William Thomas Stanton, barrister. They had two sons and one daughter. His wife died in 1984 and at the end of his life he was cared for by an old friend, Dr Elizabeth Welford, whom he engaged to marry. On 1 January 1985, two days before the wedding, he died from a heart attack at her house in Oxford.

[*The Times*, 3 January 1985; Norman Goodall, *Second Fiddle* (autobiography), 1979; private information; personal knowledge.]

     PAUL ROWNTREE CLIFFORD

GORE, (WILLIAM) DAVID ORMSBY, fifth BARON HARLECH (1918-1985), politician and ambassador. [See ORMSBY GORE.]

GORE-BOOTH, PAUL HENRY, BARON GORE-BOOTH (1909-1984), diplomat, was born

3 February 1909 at Doncaster, Yorkshire, the elder son of Mordaunt Gore-Booth, of 37 Hall Gate, Doncaster, manager of the Vickers Tyre Mill at Doncaster, and his wife, Evelyn Mary, daughter of Robert Stanley Scholfield, of Sandhall, Howden, east Yorkshire. The second of their three children, a daughter, died in infancy. He was educated as a King's scholar, at Eton, whence he won an open scholarship to Balliol College, Oxford, in 1928. He obtained second classes in *literae humaniores* (1931) and philosophy, politics, and economics (1932). An active Christian Scientist from boyhood and a lifelong total abstainer, he came from families known in public life in Ireland and in Yorkshire. His aunt, Constance, Countess Markiewicz, the Sinn Fein leader, was the first woman elected to Westminster. He inherited musical and literary talent, an engaging streak of eccentricity, and a keen sense of duty.

In 1933 he entered the Diplomatic Service. Having served in the Foreign Office and in Vienna he was moved in 1938 to Tokyo. After Japan's entry into the war he was evacuated from the Far East in 1942 and joined the highly talented team assembled in the British embassy in Washington. There he participated in several major conferences at which post-war reconstruction was being planned and which culminated in the San Francisco conference of 1945 which established the United Nations.

From 1945 to 1949 he was engaged in political and economic work in the Foreign Office as head successively of the UN (economic and social) and the European recovery departments. In 1949 he came for the first time into the public eye. His appointment as director of the British Information Services in the US was initially questioned by the British press. Despite the difficult background of declining British power he was successful and retained the respect of the press for the remainder of his career.

In 1953 he was sent as ambassador to Burma, where he and his wife made a particularly strong impression at a difficult time. In later years the Gore-Booths gave a home to the daughter of Aung San, the murdered Burmese leader, whilst she studied in Oxford. In 1956 he returned to London as deputy under-secretary in charge of economic affairs—a newly created post. His arrival coincided with the Suez crisis. Consistent with his character he was not silent in his opposition to government policy and came near to resignation over Suez. For the next four years he was prominently involved in the unsuccessful negotiations to reconcile the conflicting views of the United Kingdom and the Continental powers on the economic development of Europe. His own disappointment in the failure was diminished by his personal role in the establishment in 1960 of the Or-

ganization for Economic Co-operation and Development, which included the USA and Canada.

In 1960 he was chosen, unusually for a Diplomatic Service officer, for the Commonwealth post of high commissioner in India at Delhi. The United Kingdom was still able to influence Indian policy but relations were always sensitive. During his appointment Delhi was frequently the scene of great diplomatic activity involving Pakistan, Kashmir, China, and the Congo. Gore-Booth was characteristically active and was well liked by the Indians for his frankness.

In 1965 he was recalled to the Foreign Office to be permanent under-secretary. This was a surprise to him and others who thought his talents suited him better for another important post abroad. But political factors had intervened. Gore-Booth, who had never been closely associated with Conservative policies, was the choice of the new Labour foreign secretary, Patrick (later Lord) Gordon Walker [q.v.]. Gordon Walker failed to secure a parliamentary seat and was succeeded by Michael Stewart (later Lord Stewart of Fulham), who worked smoothly with Gore-Booth, describing him in his autobiography as 'wise and urbane'. In mid-1966 conflicts in the cabinet led to George Brown (later Lord George-Brown, q.v.) becoming foreign secretary. There was nothing in common between Brown and Gore-Booth who found it a hard task to adjust himself and his department to the foreign secretary's flamboyant, provocative, and often brutal style. In spite of this he was scrupulously—some said unduly—loyal to his chief.

In 1969 Gore-Booth retired. His four years of office had been particularly anxious. The economic situation of the country hampered an international role when ministers wished to show a high profile. The atmosphere was poisoned by the Vietnam war, fighting in the Middle East, the Nigerian civil war, and the beginning of trouble in Rhodesia. In addition, there were jealousies in the cabinet. The Diplomatic Service itself had been thrown off balance by the Duncan report proposing internal changes and by the abrupt amalgamation with the Commonwealth Office. Few contested the logic of the amalgamation but not everyone was prepared for its effect on their own careers. Gore-Booth oversaw the inevitable reductions of senior posts with fairness. His sympathy was sincere but it was not his nature to parade it. No one could impugn his integrity or his intense devotion to the public service.

He was appointed CMG in 1949, KCMG in 1957, KCVO in 1961, and GCMG in 1965. He was made a life peer in 1969 and became a regular attender at the House of Lords where he spoke elegantly from the cross-benches on a wide variety of subjects.

Although in later years he was much troubled by his eyesight he was an active chairman of Save the Children Fund (1970-6), chairman of the board of governors of the School of Oriental and African Studies (1975-80), and chairman of the Disasters Emergency Committee (1974-7). From 1967 to 1979 he was president of the Sherlock Holmes Society which in the words of an observer permitted him, 'suitably attired and giant eyebrows bristling, to play the lead in a reconstruction on the spot of his hero's dramatic end'. His autobiography *With Great Truth and Respect* appeared in 1974 and he edited the valuable fifth edition of *Satow's Guide to Diplomatic Practice* (1979). To this Dictionary he contributed the notices of Sir T. L. Rowan and Sir A. G. M. Cadogan.

In 1940 he married, in Tokyo, Patricia Mary, daughter of Montague Ellerton, company secretary, of Kobe, Japan. The marriage was outstandingly happy and his wife gave him great support in his public life. They had twin sons and two daughters. The Gore-Booths made a tall and handsome couple. He died at Westminster 29 June 1984.

[Paul Gore-Booth, *With Great Truth and Respect*, 1974 (autobiography); personal knowledge.]   GREENHILL OF HARROW

GOUDGE, ELIZABETH DE BEAUCHAMP (1900-1984), author, was born 24 April 1900 at Wells, Somerset, the only child of the Revd Henry Leighton Goudge DD, at that time vice-principal of Wells Theological College, and his wife, Ida de Beauchamp, daughter of Adolphus Collenette, of Guernsey, Channel Islands. She grew up in Wells and Ely and was educated at Grassendale School, Southbourne, Hampshire, and the Art School at Reading College. When her family moved to Oxford in 1923 on her father's appointment as regius professor of divinity, Elizabeth Goudge worked as a handicraft teacher. But in her spare time she began to write seriously, composing plays and poems before finding her *métier* as a novelist.

Her first novel, *Island Magic* (1934), was inspired by her childhood holidays with her grandparents in Guernsey, and won immediate acclaim in England and America. A succession of novels and short stories followed, including *A City of Bells* (1936) which depicted the Wells of her childhood and *Towers in the Mist* (1938), a novel of sixteenth-century Oxford. Both novels were based on cathedral cities in which her distinguished father had exercised his teaching ministry.

On her father's death in 1939 she moved with her mother to a small bungalow at Marldon in south Devon. Deeply committed to nursing her ailing mother, the author required exceptional determination to complete the prize-winning novel *Green Dolphin Country* (1944). This romantic novel achieved spectacular success and was made into the film *Green Dolphin Street* in 1947. Set in the Channel Islands and New Zealand of the nineteenth century, it gave ample proof that Elizabeth Goudge possessed the fertile imagination of the born story-teller as well as a special facility for describing the country places and people she had come to love.

After her mother died (1951), Elizabeth Goudge moved to Oxfordshire. In 1952 she settled in a seventeenth-century cottage at Peppard Common near Henley on Thames. She loved the unhurried country life and never moved house again, happy in the companionship of a friend, Jessie Monroe, and a succession of much-loved dogs.

With *The Heart of the Family* (1953) she completed the family trilogy which she had started with *The Bird in the Tree* (1940) and *The Herb of Grace* (1948). The three novels, collectively known as *The Eliots of Damerosehay*, won for the author a host of new readers, charmed by her portrayal of family life centred on a great country house. Damerosehay was situated near Lymington on land familiar to the author from her schooldays and from later residence in her father's 'vacation' bungalow at Barton-on-Sea, Hampshire. This love of familiar places was a notable feature of *The Dean's Watch* (1960) set in Ely which was her favourite home among her father's residences; and of *The Child from the Sea* (1970) which stemmed from a blissful holiday with Jessie Monroe in Pembrokeshire. Apart from anthologies and short stories, Elizabeth Goudge wrote sixteen novels, six children's books, a life of Christ for younger readers (*God so loved the World*, 1951), and a life of *St Francis of Assisi* (1959).

Among her later novels, *The Scent of Water* (1963) gave rise to a huge correspondence which afforded exceptional evidence of the rapport established between the author and her readers. It would seem that the men and women she created to inhabit the pleasant places of memory reflected the inherent charity and faith of an author who had come to believe in the love of God and possessed the ability to convey that love to many unknown readers. Her autobiography *The Joy of the Snow* (1974) reveals the strength of this Christian conviction.

To her delight, she received grateful letters from many men and women whose faith and confidence had been restored by one or other of her novels. Some had been helped to accept disabling disease, to face life with a spastic or crippled child, or to solve marriage problems.

Others had been inspired to paint and compose and regain the ability to write creatively. These letters meant much more to Elizabeth Goudge than fame or the financial rewards of a best-selling author.

Fragile in appearance, though her features reflected her father's good looks and her mother's early beauty, Elizabeth Goudge never married. But few novelists have so well conveyed to their readers the goodness of human nature and the beauty of childhood. In choosing to write of things that were lovely and of good report, she followed Jane Austen's advice 'to let other pens dwell on guilt and misery'. She was elected FRSL (1945). In addition to winning a Literary Guild award and the MGM film prize for *Green Dolphin Street*, she received the Carnegie medal for her children's book, *The Little White Horse* (1946). She died 1 April 1984 at her home, Rose Cottage, Peppard Common.

[*The Times*, 3 April 1984; private information; personal knowledge.]

JOHN ATTENBOROUGH

GOUZENKO, IGOR SERGEIEVICH (1919–1982), Soviet Russian defector and author, was born 26 January 1919 at Rogachovo near Moscow. He was the third son and the youngest of four children and was named after his deceased older brother. His father, Sergei Davidovich Gouzenko, a native of Rostov-on-Don, was a bank accountant who died while serving with the Red Army. His mother, Praskovia Vasilyevna Filkova, whose father was a building contractor, attended university which was unusual in Tsarist Russia. Gouzenko commenced his primary school education in Rostov-on-Don, continued his studies at the Maxim Gorky School in Moscow and graduated in June 1937 with high honours. However, before matriculating at the University of Moscow he studied drawing and drafting for one year in Moscow's prestigious Art Studio Chemko. In September 1938 he entered the Architectural Institute of the University of Moscow (the pre-revolutionary Strogonov School of Art). It is here that he met his future wife whom he married 20 November 1942, Svetlana Borisovna Gouseva. They had four boys and four girls, most of whom attended university specializing in the sciences. Mrs Gouzenko's father was a mining engineer and a member of the faculty of the Mining Institute of the University of Moscow and her mother, Anna Constantinovna Kosciuszko, was a lineal descendant of General Tadeusz Kosciuszko, the Polish revolutionary.

In June 1941 Gouzenko was nominated for a lucrative Stalin fellowship because of his outstanding academic record, but the German surprise attack on 22 June ended his formal education and his hopes of becoming an architect. He was posted to the Kuibishev Engineering Academy and then selected to study ciphering. In the spring of 1942 he was assigned to the Moscow headquarters of military intelligence, the GRU (Glavnoye Razvedyvatelnoye Upravleniye), where he worked as a cipher clerk until the spring of 1943 when he was transferred to the office of the military attaché at the legation in Ottawa.

From the moment of his arrival Gouzenko experienced the exhilaration and freedom offered by an open society. The material comforts available even in a wartime Canada seemed unimaginable considering the deprivations that he had experienced as an ordinary Russian citizen prior to the outbreak of the war. The ambience of the legation was oppressive and when his return to Moscow could no longer be delayed Gouzenko decided to defect and, as described in his 1948 memoirs, *This Was My Choice*, carefully prepared the documentation he would take with him to alert the Canadian authorities and the West that Moscow was involved in blatant spying activities directed against its allies. Initially no one in Ottawa believed him, but an attempt by legation personnel to kidnap him convinced the Canadian authorities of his sincerity.

Gouzenko's defection in September 1945 in line with Russian activities in other parts of the world contributed to the intensifying of the cold war. His testimony before the royal commission of inquiry established to investigate his charges of Russian espionage in Canada plus the documentation he brought with him led to the prosecution and conviction of eleven people, including a member of Parliament. In Britain Gouzenko's information led to the arrest of the atomic physicist Allan Nunn May. He maintained that a Soviet agent, 'Elli', was at work in MI5. Though Gouzenko's information dealt with only a narrow spectrum of the GRU's espionage, H. A. R. ('Kim') Philby subsequently admitted that his defection had been a counter-intelligence 'disaster' for Moscow. In particular, the telegrams and other materials that he removed from the legation proved of enormous assistance to western decrypters in unravelling the GRU codes. That no further arrests followed can probably be explained by the fact that it was impossible to fathom the cryptonyms assigned to those working for the GRU. Fearing Stalin's retribution Gouzenko was given Canadian citizenship and a new identity, and was closely guarded by the Royal Canadian Mounted Police. In 1954 his best-selling novel, *The Fall of a Titan*, which dealt with the inhumanity and brutality of the Stalinist regime, won the Governor-General's award. In the 1950s Gouzenko appeared frequently on television, but to disguise his identity wore a bag over his head.

Gouzenko was of medium height, broad shouldered, had a military posture, and dressed conservatively. He was partial to classical music, was an avid photographer, enjoyed doing drawings and paintings, many of which he sold, wrote several unpublished works including a novel *Ocean of Time*, and was interested in astronomy and gymnastics. His diabetes led to blindness, but he read and wrote Russian and English in braille. He died suddenly 25 June 1982 near Toronto, thirty-nine years to the day after his arrival in Canada.

[Igor Gouzenko, *This Was My Choice*, 1948; Svetlana Gouzenko, *Before Igor*, 1961; *Report* of the royal commission to investigate the circumstances surrounding the communication, by public officials and other persons in positions of trust, of secret and confidential information to agents of a foreign power, 1946; *Toronto Star*, 4 April 1988, p. A12; private information.]　　　JAMES BARROS

GRAVES, ROBERT RANKE (1895–1985), writer and poet, was born at Wimbledon 24 July 1895, the eldest of the three sons and the third of five children of Alfred Perceval Graves, an inspector of schools, and his second wife, Elizabeth Sophie ('Amy'), daughter of Heinrich von Ranke, a medical doctor. The family ancestry included German, Danish, Irish, Scottish, and English elements. Literature of many kinds was in the boy's blood; his maternal great-uncle was the historian Leopold von Ranke; an eighteenth-century ancestor, Richard Graves [q.v.], had written *The Spiritual Quixote* (1772), a novel remembered in the twentieth century; and his own father had composed the words to a popular Irish song, 'Father O'Flynn'.

Educated at various preparatory schools and Charterhouse (1907–14), Graves was about to proceed to St John's College, Oxford, when World War I broke out. Being in north Wales at the time (the family had a house at Harlech where the boy spent much, and formative, time) he reported to the nearest depot where he could be accepted for training as an officer, and since this was Wrexham he found himself commissioned, for the duration of the war, in the Royal Welch Fusiliers. After serving through some of the heaviest fighting, being twice mentioned in dispatches, and once having his death in action reported in *The Times*, he was demobilized in 1919.

In 1918 Graves married Annie Mary Pryde ('Nancy') (died 1977), daughter of (Sir) William Newzam Prior Nicholson [q.v.], artist, and sister of Ben Nicholson [q.v.]. They had two

daughters and two sons. The couple moved to Oxford in 1919, and Graves began to read English at St John's College, living in a cottage at the end of the garden of John Masefield [q.v.], the poet. They stayed near Oxford till 1926, Graves first supporting his growing family by running a small grocery business and then solely by writing. He did not take an undergraduate degree because of illness but gained a B.Litt. (1925). He also became acquainted with T. E. Lawrence [q.v.] and heard his version of wartime events in Palestine. After spending part of a year (1926) as a lecturer at the Royal Egyptian University, he returned to England and settled to the life of a professional writer.

In 1929, parting company with Nancy, he established a household in Deyá, Majorca, that included the American poet Laura Riding, with whom he had been associated in his work since 1926. She was the daughter of Nathaniel Reichenthal, a tailor, and the former wife of Louis Gottschalk, a history teacher at Cornell University. Until the Spanish civil war uprooted them in 1936, Graves and Riding both produced a steady flow of work, some of it published by the Seizin Press, a small company they founded in England and continued in Deyá.

In 1936 they moved to London. Two years later they took a flat in Rennes, France, and in 1939 went to America. There Laura left Graves to live with Schuyler Jackson. Graves returned to England and set up house with Beryl, daughter of Sir Harry Goring Pritchard, solicitor and parliamentary agent, and wife of Alan Hodge, poet and writer; they married in 1950. They had three sons and a daughter.

The years of World War II, paradoxically, were a good period for Graves; he was productive in his country fastness at Galmpton, Devon, and happy in his second marriage (with a rich harvest of love poems to show for it); though doubtless deeply stricken by the death of his first son David, killed in Burma while serving with his father's old regiment.

Graves and his family returned to Majorca in 1946. Graves spent the rest of his life there, with excursions, notably to give the Clark lectures in Cambridge (1954) and to be professor of poetry (1961–6) at Oxford, where St John's College elected him an honorary fellow in 1971.

In the preface to his *Poems 1938–45* (1948) Graves declared: 'I write poems for poets, and satires or grotesques for wits. For people in general I write prose, and am content that they should be unaware that I do anything else.' 'People in general' did indeed know him mainly as a writer of historical novels, notably *I, Claudius* (1934, which won the Hawthornden and James Tait Black memorial prizes), *Claudius the God* (1934), *Count Belisarius* (1938), *Sergeant Lamb of the Ninth* (1940), *King Jesus* (1946), and *Homer's Daughter* (1955). His wider fame was first established, however, by the autobiography *Goodbye to All That* in 1929. This admirable book (perhaps best read in its slightly tidied-up version of 1957, which corrects the hasty syntax without blurring the honest sharp edges) stands in a perceptible relationship to the Claudius novels. The 'I' in each case is very much the product of the surrounding civilization, even while evaluating and indeed rejecting it.

As a poet Graves employs cool, restrained rhythms, often showing a fine lyrical sense of the musical possibilities of pause and stress, and a diction very close to the central core of English poetic tradition. Without actually making a count, one has the impression that his vocabulary contains very few words that do not occur in classical English poetry from Dryden to Wordsworth. His metrical and stanzaic forms, likewise, draw on a traditional, rather than an experimental vitality. The tmesic element in Graves's poetry, the element that represents a break with tradition, is sited in the subject-matter rather than the form. A large proportion of his verse, especially in the second half of his working life from about 1950 onwards, is devoted to exploring and celebrating a certain kind of relationship between the sexes, a relationship which, like the religion of matriarchal societies, makes the female the dominant principle. Graves evidently regarded this view of the sexes as determining the nature of much in his personal life and almost everything in his poetry, and he expounded it at length in *The White Goddess* (1948, revised and expanded edition 1952), a book that stands in the same relationship to his mature poetry as *A Vision* does to Yeats's.

Graves was also active as a literary critic, especially of poetry. His criticism, though always interesting, is often vitiated by a curious combativeness; at times he seems almost comically aggressive, a Mr Punch thwacking at anything in reach with a clumsily held truncheon. An early collaboration with Laura Riding, *A Survey of Modernist Poetry* (1927), seems at first an exception, since it is positively in favour of something—the 'modernist' movement. The analysis of Shakespeare's Sonnet 129 in juxtaposition with a poem by E. E. Cummings, demonstrating that 'modernist' techniques are just about equally present in both, is a superb *tour de force* of analytical criticism, to which his fellow poet doubtless contributed her share. Yet it has to be borne in mind that to advocate anything as alarming as modernism was, in the 1920s, a stick to beat the establishment. Graves particularly enjoyed an academic platform from which he could compel the learned and august to listen submissively while he assaulted their

sacred cows, as can be seen from *The Crowning Privilege* (1955)—the text of his Clark lectures in Cambridge—and *Oxford Addresses on Poetry* (1961).

Graves had his share of insights, but it would seem fair comment to say that he was less the kind of critic who patiently illuminates work he admires than the kind who finds emotional release in attacking what he dislikes. He appears to have had a particular hostility to Milton, and, though never mounting a full-scale attack on that poet's art, devoted much energy to blackening his character in a novel, *Wife to Mr Milton* (1943).

Turning from Graves the writer to Graves the man (strictly speaking an impossibility, since *le style* was *l'homme même* and even the most transient of his writings always concerned something of importance in his life), one's most abiding impression is of singularity. The gigantic stature, the clipped speech (to the end his diction was that of a subaltern of 1916), the erect officer-like bearing that never seemed in conflict with the bardic majesty, the wide-brimmed hats (of black felt on formal occasions, often of ragged straw in his leisure on the beach), all made a unique presence. As much as any Irish or Welsh medieval bard, Graves held that being a poet set one apart totally from the rest of mankind; and, while this did not prevent his sharing many of the best qualities of common humanity—courage, humour, cheerful hospitality, philoprogenitiveness, to name no more—it is true that one often felt in his presence a certain otherness, a wind blowing from an undefined quarter of the spirit: whether caused by the cloud of abstraction that sometimes came down over him, or by the colossal size that made him seem like a giant out of a folk tale, or the knowledge of the terrible ordeals he had endured and suffered so long ago, or most likely, a combination of all these, he was *sui generis*, and unrepeatable. He died at his home in Deyá, Majorca, 7 December 1985.

[Robert Graves, *Goodbye to All That*, 1929; Alfred Perceval Graves, *To Return to All That*, 1930; *The Times*, 9 December 1985; Richard P. Graves, *Robert Graves: the Assault Heroic*, 1986; personal knowledge.]

JOHN WAIN

GREENWOOD, ARTHUR WILLIAM JAMES ('ANTHONY'), BARON GREENWOOD OF ROSSENDALE (1911–1982), politician, was born at Leeds, 14 September 1911, the only son and younger child of Arthur Greenwood [q.v.], MP, and his wife, Catherine Ainsworth, daughter of John James Brown, clerk, of Leeds. He was educated at Merchant Taylors' School and

Balliol College, Oxford, where he became president of the Union (1933), and also gained third-class honours in politics, philosophy, and economics (1933). In 1938–9 he worked with the National Fitness Council. At the outbreak of World War II he joined the Ministry of Information and from 1942 served in the RAF as an intelligence officer. He worked with the Allied reparations commission and attended the conference at Potsdam in 1945.

Greenwood had developed strong political interests early on: he joined the Labour Party at the age of fourteen. Before the war he became prospective candidate for Colchester, and he led the Labour group on Hampstead borough council from 1945 to 1949. In a by-election in February 1946 he was elected for Heywood and Radcliffe. After redistribution, he moved to become member for Rossendale in 1950 and held the seat for the next twenty years. He served as parliamentary private secretary to the postmaster-general in 1949 and became vice-chairman of the Parliamentary Labour Party in 1950–1.

Greenwood was a strong supporter of Aneurin Bevan [q.v.] in the internal Labour Party disputes of the 1950s. In 1954 he was elected to the Labour Party national executive committee as part of the left-wing tide in the constituency section at that period. He served on the NEC until 1970. He shortly entered the shadow cabinet where he served successively as spokesman on Home Office affairs and education. He also became a very popular platform speaker and television performer. When the Campaign for Nuclear Disarmament was founded Greenwood was a staunch supporter and took part in the first Aldermaston march in 1958. He challenged Hugh Gaitskell [q.v.] vigorously on defence policy and on 13 October 1960 resigned from the shadow cabinet in protest. He backed Harold Wilson (later Lord Wilson of Rievaulx) against Gaitskell for the party leadership that month. In November 1961 he stood himself for the leadership but was defeated by Gaitskell (by 171 votes to 59).

Greenwood's career gained a lift in February 1963 when Wilson succeeded Gaitskell as leader, following the latter's unexpected death. Also Greenwood served as party chairman in 1963–4. He was assured of cabinet office in a future Labour government, and duly became colonial secretary under Wilson in October 1964.

There was much for him to do at the Colonial Office. He visited the Caribbean, Aden, and Mauritius. He took a close interest in the affairs of Belize, and helped create a unitary state in southern Arabia. In December 1965 a cabinet reshuffle saw him moved to the Ministry of Overseas Development. Following Labour's

handsome election victory, he became minister of housing and local government in August 1966. His private ambitions were for the Home Office, but housing was an area in which he had long had a specialist interest. In 1966–7 house construction went ahead with great rapidity, with over 400,000 houses—a record—completed in 1967. However, Greenwood's reputation suffered from the Draconian cuts in the housing budget imposed in 1968 in the aftermath of the devaluation of the pound. He was attacked by the left for not resisting cuts in public expenditure more vigorously. In October 1969 his housing department was placed under the expanded department of local government and regional planning headed by C. A. R. Crosland [q.v.]. Greenwood's being dropped from the cabinet was another sign of waning fortunes.

He had suffered another blow in July 1968 when, after being persuaded to stand for the general secretaryship of the Labour Party, he was unexpectedly opposed on the NEC by right-wing trade unionists. Greenwood was defeated 14–12 by Harry Nicholas of the Transport Workers. A final setback came in 1970 when he was offered the chairmanship of the Commonwealth Development Corporation by Wilson. However, Labour lost the election and the new Conservative prime minister, Edward Heath, rescinded the appointment. Greenwood, who had resigned his seat, was out of Parliament and in serious financial straits as well. A life peerage (1970) was scant consolation.

However, after 1970 he showed much resilience in developing a new career. His finances were repaired by business directorships. He was very active in a host of social and charitable causes, including the Greenwood Development Housing Company to build 'Greenwood Homes', and the Piccadilly Advice Centre for the homeless. He was deputy-lieutenant of Essex in 1974 and also pro-chancellor of the University of Lancaster (1972–8). He became a privy councillor in 1964, an honorary LLD of the University of Lancaster in 1979, and was also a JP (1950).

'Tony' Greenwood was an attractive politician, tall, handsome, and elegantly dressed. He cut a fine figure on the platform and was a most effective public speaker. It may have been a lack of ruthlessness on his part that prevented his rising to the highest offices. His later years were marked by numerous set-backs. But they never soured him, while he played a considerable part in the revival of the British left down to 1964. Unlike many political sons of famous men, he carved out a notable career of his own.

In 1940 he married an artist, Gillian, daughter of Leslie Crawshay Williams, an engineer and inventor, of Bridgend. They had two daughters. Greenwood died suddenly, of a heart attack, 12 April 1982 at home in Essex. There is a portrait of him by his wife in the Greenwood Homes head office at Northampton.

[Private papers in the possession of Lady Greenwood; Labour Party archives, Walworth Road; *The Times*, 14 April 1982; Janet Morgan (ed.), *The Backbench Diaries of Richard Crossman*, 1981; Richard Crossman, *The Diaries of a Cabinet Minister*, 3 vols., 1975-7; Barbara Castle, *The Castle Diaries, 1964-70*, 1984; private information.]

KENNETH O. MORGAN

GRIGSON, GEOFFREY EDWARD HARVEY (1905-1985), poet, critic, anthologist, and man of letters, was born 2 March 1905 at Pelynt, Cornwall, the seventh and last son (there were no daughters) of Canon William Shuckforth Grigson, vicar of Pelynt, and his wife, Mary Beatrice, daughter of John Simon Boldero, vicar of Amblecote, near Stourbridge, Staffordshire. At the time his father was fifty-nine years old and his mother forty-two. His childhood and adolescence, as described in his vivid, impressionistic autobiography *The Crest on the Silver* (1950) were deeply unhappy. He was sent as boarder to a preparatory school at the age of five, then to a minor public school in Leatherhead which he detested. This was succeeded by what he called a profitless sojourn at St Edmund Hall, Oxford, where he gained a third class in English in 1927. These years were marked also by grief at the death of three of his brothers during World War I, including one whom he particularly loved. Three more were to die during World War II, so that by the end of it this seventh son was the only survivor of his father's large family.

He came down from Oxford an awkward, reticent young man, loving Cornwall and the countryside, knowing little of London; yet there his talent flowered. He worked on the *Yorkshire Post*, then the *Morning Post* where he became literary editor. He founded *New Verse* (1933-9), a 'malignant egg' as he later called it, the most influential British poetry magazine of the 1930s. In *New Verse* he printed the finest poets of the W. H. Auden [q.v.] generation, but the magazine was almost equally known for his own criticism of his contemporaries, always unsparing and at times ferocious. He later regretted what he called such savage use of the billhook, but still employed it often, in hundreds of reviews written for the *Observer*, *Guardian*, *New Statesman*, and other papers. Some of the best and sharpest of them were gathered together in *The Contrary View* (1974) and *Blessings, Kicks and Curses* (1982).

New Verse was the beginning of a literary career which embraced art criticism, anthologies of verse and prose, guides to the countryside and its flora, and the writing of poems. He spent the war years in the BBC at Evesham and Bristol, but thereafter was a free lance, making his home at the farmhouse he had found in Wiltshire, living mostly by literary journalism, but never relaxing his standards. The results were remarkable, both in quality and variety. The finest of his essays on art are to be found in *The Harp of Aeolus* (1947), which includes appreciations of artists as diverse as George Stubbs, Francis Danby [qq.v.] and De Chirico, and along with this book should be put *Samuel Palmer, the Visionary Years* (1947), which prompted a revised view of Palmer's genius. *The Romantics* (1943) and *Before the Romantics* (1946), two of the first among many anthologies, mark the immense scope of his reading, his endless curiosity about the relationship of man and nature, his concern with the shape and sound of language. He had a quite separate fame as author of *The Englishman's Flora* (1955), *The Shell Country Book* (1962), and other works about the English countryside.

And, last and to him by far the most important, he was a poet. His first book, *Several Observations* (1939), fulfilled his own requirements of 'taking notice, for ends not purely individual, of the universe of objects and events'. All his poems do this, whether they are lyrical, satirical, or views of scenes and people. A genuineness of feeling and observation, and a refusal of rhetorical gestures, are the hallmarks of his poetry. *Collected Poems* appeared in 1963, and a further volume covering the years up to 1980 in 1982. In the same year he published *The Private Art*, a 'Poetry Notebook' of comments and quotations that emphasized again the generosity and tough delicacy of his mind. *Grigson at Eighty*, edited by R. M. Healey, a collection of tributes from friends and admirers, appeared in the year of his death.

Both the breadth of his knowledge and interests, and the integrity with which he pursued them, made Grigson a uniquely valuable figure in twentieth-century literary Britain. In person he was tall, handsome, enthusiastic, with an attractive blend of sophistication and innocence. The fierceness of his writing was belied by a gentle, sometimes elaborately polite manner. He distrusted all official bodies dealing with the arts, and served on no committees. When, in 1972, he received the Duff Cooper memorial prize for a volume of poems, his short speech made clear his uneasiness on such large formal occasions.

Like his father Grigson was married three times. In 1929 he married Frances, daughter of Thomas Trauldin Galt, attorney, of St Louis, Missouri. They had one daughter. Frances died

in 1937 and in 1938 he married Berta Emma
Beatrix, daughter of Otto Kunert, an Austrian
major from Salzburg. They had a son and
daughter. This marriage was dissolved and he
married thirdly Jane, daughter of George Shipley McIntire, CBE, town clerk of Sunderland.
Jane Grigson became a celebrated cookery expert; they had one daughter. Grigson died 28
November 1985 at Broad Town Farm, his Wiltshire home, and is buried in Christ Church,
Broad Town, churchyard.

[R. M. Healey (ed.), *Grigson at Eighty*, 1985;
Geoffrey Grigson, *The Crest on the Silver*,
1950 (autobiography); private information;
personal knowledge.]   JULIAN SYMONS

GROSS, (IMRE) ANTHONY (SANDOR)
(1905-1984), painter and etcher, was born in
Dulwich, London, 19 March 1905, the elder
child and only son of Alexander Gross, a naturalized British subject of Hungarian origin,
who had settled in the London area as a map
publisher, and his wife, Isabella Crowley, who
was a playwright and suffragette.

In 1923, after general education at Repton
School where his first attempts at etching were
encouraged, Gross began full-time art training
in London, studying painting at the Slade
School of Fine Art and etching, in the evenings,
at the Central School of Arts and Crafts. Later
the same year he moved to Paris, continuing
with these subjects respectively at the Académie
Julian and the École des Beaux Arts, and in
1925, the year of his first one-man exhibition in
London, he went on to Madrid, enrolling at the
Escuela de San Fernando. Thence, with the aid
of a donkey, he spent the summer roaming the
countryside, his sketch-book always at the
ready, and in subsequent years he made similar
study tours in Andalusia, Spanish Morocco,
and Algeria, followed by visits to Italy, Belgium,
and France. Thus he not only experienced considerable variety in his formal training but
gained first-hand knowledge of these places and
their peoples, even taking part in some bullfighting, before returning to Paris in 1926.

Gross made friends with many artists in
France, not least the engravers S. W. Hayter
and Joseph Hecht, and he was at first best
known for his topographical works but his series
of paintings and prints entitled *Sortie d'Usine*
(completed 1930-1) had as their subject-matter
the workers from the local factories and these
pictures, so vital and subtle in line, together
with those called *La Zone*, a Parisian shanty
suburb, reveal his understanding of their life.
In contrasting mood and in collaboration with
the American Hector Hoppin, he then set about
thousands of drawings for the cartoon film *La
Joie de Vivre* (1933), followed by *Fox Hunt*
(1936), meanwhile illustrating Jean Cocteau's
Les Enfants Terribles for an edition in 1936.

He stayed in London from about 1934 to
1937 and returned to England again at the outbreak of World War II, first of all recording the
routine activities at various training depots and
then, in 1941, being appointed an official war
artist. These duties took him with the troops to
Egypt and the Middle East, North Africa,
India, and Burma, then back to Europe. He in
fact crossed to Normandy on D-Day, wading
ashore with his water-colour paper held high
above his head. The Imperial War Museum
possesses over 300 examples of the resulting
work.

Soon after the war he produced illustrations
for editions of Emily Brontë's *Wuthering Heights*
(1947) and John Galsworthy's *The Forsyte Saga*
(1950), and colour lithographs for J. Lyons'
restaurants. He also taught at the Central
School from 1948 to 1954. In 1955 Gross
bought a house, Le Château du Boulvé, near
his wife's home town of Villeneuve-sur-Lot
(hence the large etchings *Le Boulvé Suite*, 1956)
and thereafter spent the summer months there
each year and the winters working in London.
He was head of the etching and engraving department of the Slade School from 1955 to 1971
and his authoritative book *Etching, Engraving
and Intaglio Printing* was published in 1970.
Writing of his work he said 'I have always held
an unfashionable view that drawing is the prime
factor in pictorial art—the understanding of
how things are constructed, the shape of everything, be it flower or factory girl, landscape or
insect' and there can be no better example of
this than in the text and illustrations for *The
Very Rich Hours of Le Boulvé* (1980).

Gross was a prolific artist (among other
things producing over 360 copperplate etchings
and engravings), inventive in his methods, full
of energy and working seven days a week, an
optimist who enjoyed good company, walking,
fishing, and cooking. He was a member of the
London Group from 1948 to 1971, was made
honorary RE and ARA in 1979 and RA in 1980
(his oil painting *Les Causses* won the Wollaston
award that year), and was appointed CBE in
1982.

He married in 1930 the painter Marcelle
Marguerite ('Daisy'), whose father, Henri Flo-
renty, owned an artificial flower factory and
shop, and they had one son, Jean-Pierre Anthony ('Pete'), a historian, and one daughter,
Mary (Mrs West) who was her father's printer
from 1973. Gross died at Le Boulvé 8 September 1984.

[*The Times*, 12 September 1984; private information; personal knowledge.]
                                    S. C. HUTCHISON

GUILLUM SCOTT, SIR JOHN ARTHUR (1910–1983), secretary of the Church Assembly and first secretary-general of the General Synod of the Church of England. [See SCOTT.]

GUTHRIE, (WILLIAM) KEITH (CHAMBERS) (1906–1981), classical scholar, was born in London 1 August 1906, the younger child and only son of Charles Jameson Guthrie, who had moved from Scotland to London to work for the Westminster Bank, and his wife, Katharine Chambers, who was also of Scottish descent, her father William being an elder of the Presbyterian Church. He was educated at Dulwich College whence he proceeded with an entrance scholarship to Trinity College, Cambridge, in 1925. He was placed in the first class in both parts of the classical tripos (1926 and 1928), with distinction in Greek philosophy in part ii. He won the Browne university scholarship (1927) and the university Craven studentship (1928) and Chancellor's medal (1929).

He took part between 1929 and 1932 in expeditions to Asia Minor with the American Society for Archaeological Research, though his interest was epigraphy, not archaeology. His work in this field was included in *Monumenta Asiae Minoris Antiqua*, vol. iv (ed. W. H. Buckler . . . W. K. C. Guthrie). But his abiding interests were Greek religion and philosophy, and his first lectures were given in this field; he became a full university lecturer in 1935. In 1930 he was elected bye-fellow and in 1932 a full fellow of Peterhouse. In the course of his college teaching he supervised a Newnham undergraduate from Melbourne, Adele Marion, the daughter of Adam Loftus Ogilvie, civil servant. She obtained a first class in the tripos, and they married in 1933. It was a marriage of true minds, of shared interests (he pays tribute to her in *The Greeks and their Gods*) and of temperaments which both disliked fuss and falsity. They had two children, Robin who was later director of the Joseph Rowntree Memorial Trust, and Anne whose death in her twenties left Guthrie under the permanent shadow of a great sorrow.

During the war of 1939–45 he served in the Intelligence Corps, looking with his habitual neatness more like a soldier than his rather miscellaneous wartime colleagues. He reached the rank of temporary major. A posting to Istanbul made use of his knowledge of Turkish. Resuming his academic career after the war he became successively Laurence reader in 1947 and then Laurence professor of ancient philosophy (1952–73). Always ready to take his share in administrative affairs, he was one of the university proctors in 1936–7, served on the council of the senate and the library syndicate, and was for eighteen years public orator (1939–57),

presenting a succession of candidates for honorary degrees in felicitous Latin and meticulous detail.

In 1957 he was elected master of Downing College. His early years there with his wife as hostess were active and happy, but the end of his time was overshadowed by student unrest, and when the rewritten statutes of the college set fifteen years as the length of tenure of the mastership he took the opportunity to resign in 1972 and to devote himself to his main interest, Greek philosophy. He had already, in addition to his earlier *Orpheus and Greek Religion* (1935) and *The Greeks and their Gods* (1950), published work in this field: *The Greek Philosophers* (1950), two seminal articles on the development of Aristotle's thought (*Classical Quarterly*, 1933 and 1934) and his Messenger lectures at Cornell (*In the Beginning*, 1957).

In 1956 the officers of the Cambridge University Press invited him to undertake a comprehensive *History of Greek Philosophy*. The original plan was to cover in five volumes the whole field down to the Hellenistic philosophers, but to stop short at the Neo-Platonists. When he died after the completion of the volume on Aristotle in 1981 the number was already six. The loss of the remainder was irreparable, but the achievement was great. The *History* is meticulously comprehensive; anything omitted is of no importance, yet for all its thoroughness it never becomes tedious. There is a beautiful clarity in the treatment, and when the evidence has been reviewed the reader feels that however difficult the process has been the answer arrived at is right. The *History* is in fact one of the great works of scholarship of this century— of scholarship for though it was a history of philosophy he was a historian and not a philosopher, as he was himself the first to emphasize; though he could nod towards modern philosophy, fear of contamination made him keep it at a distance.

A man of slight physique, he was an elder of his church and a man of steady but unostentatious Christian faith, in whom a certain surface austerity would mellow into the warmth of friendship. He was elected a fellow of the British Academy in 1952 and became an honorary fellow of both Downing and Peterhouse. President of the Classical Association in 1967–8, he received a Cambridge Litt.D. (1959) and honorary D.Litts. from Melbourne (1957) and Sheffield (1967). In June 1980 he suffered a stroke which incapacitated him for further work and he died 17 May 1981, in Cambridge.

[G. E. R. Lloyd in *Proceedings* of the British Academy, vol. lxviii, 1982; *The Times*, 18 May 1981; address given by the writer at a memorial service, October 1981; personal knowledge.]   DESMOND LEE

GWYNNE-JONES, ALLAN (1892–1982), artist, was born 27 March 1892 in Richmond, Surrey, the only son and younger child of Llewellyn Gwynne-Jones, solicitor, and his wife, Evelyn Hooper, who translated French and German verse. He was educated at Bedales, where he developed a precocious talent for calligraphy and illumination, and formed a deep and lasting admiration for the work and poetry of William Morris [q.v.]. In obedience to his father's wishes he studied law, and from 1911 to 1914 he was articled to William Joynson-Hicks (later Viscount Brentford, q.v.). He was awarded a senior scholarship by the Law Society and qualified as a solicitor, but never practised. During these years he became acquainted with Albert Rutherston. It was with his encouragement that he painted a series of exquisite water-colours on silk. They were mainly of East Anglian subjects and included one of Southwold Fair. This was the first of several paintings of fairs, including the large oil in the Tate Gallery (1937–8).

Gwynne-Jones became a student at the Slade School of Fine Art in May 1914, but his career there was cut short by the outbreak of war. He joined the army in August 1914, and served throughout World War I. A second lieutenant in the 1st Cheshire Regiment, he was wounded twice, mentioned in dispatches twice, and appointed to the DSO during the battle of the Somme (1916); the award was made on the spot in special recognition of his bravery, and the ribbon was pinned on his tunic as he lay wounded on a stretcher. He was later transferred to the Welsh Guards and demobilized in 1919.

Gwynne-Jones then rejoined the Slade where he received much encouragement from Henry Tonks [q.v.] and won several prizes. In the years 1920–2, towards the end of his time at the Slade, he painted a number of haunting pictures of farmyard scenes. As compositions they are strikingly original, and their dramatic effect is largely due to their dark tonality.

In 1923 Gwynne-Jones was invited to join the staff of (Sir) William Rothenstein [q.v.] at the Royal College of Art, where he became professor of painting. While teaching there he studied etching under Malcolm Osborne. Although he produced only eleven plates, most of which were executed in 1926 and 1927, they are of great beauty, and bear a distant affinity with the work of Samuel Palmer [q.v.] for which he

professed considerable admiration. From 1930 until his retirement in 1959 he was a senior lecturer at the Slade.

Many artists have paid tribute to what they owed him as a teacher. His views about painting and his practical advice are embodied in two of the books that he wrote, *Portrait Painters* (1950), and *Introduction to Still-Life* (1954). In his preface to the first of these books he defines the qualities that are required to paint a fine portrait; these are based on his profound study of the old masters. In the course of his long life he painted the portraits of many distinguished sitters such as Lord Beveridge, Lord Ismay, Lord Butler of Saffron Walden, and Sir R. George Stapledon [qq.v.]. The best of these portraits have a sympathetic perception of character which few English painters have rivalled in the twentieth century. In his book *Still-Life* Gwynne-Jones reserves his fullest praise for the 'grandeur, calm and simplicity' of Francisco Zurbarán's still lifes. And, indeed, his own work has much in common with that of the Spanish painter, and of his other idol, J.-B.-S. Chardin. For, like them, he was a master of tonality, and could create on canvas as much beauty out of a loaf of brown bread as out of a basket of peaches. The same can be said of his countless flower paintings for he found the same inspiration in the commonest wayside plants as in the rarest rose.

Gwynne-Jones was elected ARA in 1955, and RA in 1965. He served as a trustee of the Tate Gallery from 1939 to 1946. His distinguished achievements as an artist were recognized belatedly in 1980 with his appointment as CBE, two years before his death, at the age of ninety, 5 August 1982 in the Northleach Hospital, Gloucestershire.

Gwynne-Jones was the most generous and modest of men. He loved people, conversation, food, and good wine. In 1937 he married Rosemary Elizabeth, daughter of Henry Perceval Allan, ship-broker. She had been a pupil of his at the Slade. They had one daughter Emily (Mrs Beanland) who, like her mother, became a gifted artist.

[Humphrey Brooke, introduction to the catalogue of the exhibition held at Agnew's in 1972 to mark Gwynne-Jones's eightieth birthday; *The Times*, 6 August 1982; private information; personal knowledge.]

BRINSLEY FORD

H

HADRILL, (JOHN) MICHAEL WALLACE- (1916-1985), historian. [See WALLACE-HADRILL.]

HAILWOOD, (STANLEY) MICHAEL (BAILEY) (1940-1981), world champion motor-cycle rider and car driver, was born in Great Milton, Oxfordshire, 2 April 1940, the only son and elder child of Stanley William Bailey Hailwood, millionaire businessman and managing director (motor-car sales), and his wife, Nellie Bryant. Hailwood was educated at Pangbourne Nautical College, Berkshire. He left after two and a half years. Despite a lack of academic achievements he was a champion boxer and a very accomplished musician on a variety of instruments. Even in his early days he displayed that will to win that made him such a talented competitor.

He spent a short time in the family business before his father sent him to work at Triumph Motorcycles near Coventry. His first race was at Oulton Park, Cheshire, on 22 April 1957. He finished eleventh in the 125cc race, riding an Italian MV Augusta machine. With encouragement and financial backing from his father his career blossomed. He won his first race on 10 June 1957 at Blandford Camp, Dorset. He repeated that success many times during his first season and gained more experience racing in South Africa in the winter.

In 1958 he competed in his first TT race in the Isle of Man and finished third in the 250cc event. At the end of an outstanding year he was British champion in the 125, 250, and 350cc classes and had won seventy-four races. He was fourth in the world 250cc championship and sixth in the 350. A year later Hailwood won the first of his seventy-six Grands Prix victories in the Ulster 125cc race.

In 1961 he captured the first of his nine world titles, winning the 250cc championship on a Japanese Honda machine. The same year he made history at the TT races by becoming the first rider to win three TT races in one week. At the end of the season he joined the world-famous MV Augusta team and dominated the 500cc world championship for the next four years. The combination of his ability and the superb Italian machinery proved unbeatable but typically Hailwood was bored at winning so easily. The large Japanese Honda factory had achieved great success in the smaller classes but had not won the blue riband, of the 500cc class. Hailwood joined Honda in 1966 and won both the 250 and 350cc world titles for the next two years. The unstable handling of the 500 Honda prevented him retaining his 500 crown. When Honda quit Grand Prix racing in 1968 Hailwood switched to car racing.

He had already tried to combine the two sports, without having much success on four wheels although he scored one Formula 1 point in 1964. He won the European Formula 2 championship in 1972 but never repeated his motorcycle racing successes. He competed in fifty Formula 1 Grand Prix races, scoring a second place in the 1972 Italian Grand Prix. He was nineteenth in the 1964 world championship, eighteenth in 1971, eighth in 1972, and tenth in 1974. His career came to a premature halt in the 1974 German Grand Prix at the Nurburgring. He crashed and broke his right leg in three places and could not race cars again.

He then retired to New Zealand. In 1978 he made an exceptional come-back when at the age of thirty-eight he decided to return to the Isle of Man to compete in the TT races after an absence of eleven years. In front of record crowds he won the Formula 1 race on the Italian Ducati machine. A year later he returned and won his fourteenth TT race, the Senior, riding a 500cc Suzuki. He then felt it was time to retire. He was appointed MBE in 1968 and was awarded the George medal when he rescued Clay Regazzoni from a blazing car in 1973. Following his TT come-back he was also awarded the prestigious Segrave trophy (1979).

Away from the track Hailwood was a modest, shy person who had a great sense of fun. He hated fuss and unwanted attention but was hero-worshipped by thousands of fans throughout the world.

With his wife Pauline, Hailwood had a daughter and a son. He died 23 March 1981 in Birmingham Accident Hospital two days after being involved in a traffic accident on the A435 road in Warwickshire. His daughter Michelle also died in the crash. Thousands of mourners attended his funeral to honour a man simply known to many as 'Mike the Bike'.

[Private information; personal knowledge.]

NICK HARRIS

HALL, PHILIP (1904-1982), mathematician, was born 11 April 1904 in Hampstead, the natural son of George Hall and Mary Laura Sayers, dressmaker, of Hampstead, daughter of Joseph Sayers, gardener, of Balcombe. George Hall disappeared from his life soon afterwards and he was brought up by his mother. He was educated at Christ's Hospital and, as a scholar (1922-5) and senior scholar (1925-6), at King's College, Cambridge. He was in the first class in part i (1923) and a wrangler in part ii (1925) of the mathematical tripos. In 1927 he was elected to a fellowship at King's, which he held for the rest of his life. He was appointed university lecturer in 1933. From 1941 to 1945 he worked at the Government Code and Cipher School at Bletchley Park, first on Italian ciphers and later on the Japanese diplomatic ciphers. He was reader in algebra at Cambridge from 1949 and Sadleirian professor of pure mathematics from 1953 until 1967, when he elected to retire.

While still an undergraduate he began to study the works on group theory of William Burnside [q.v.], whom he never met but who was the greatest influence on his ways of thinking. In 'A Note on Soluble Groups' (1928), he proved a result as important for the theory of finite soluble groups as Sylow's theorem of 1872 is for finite groups in general, that if G is a soluble group of order mn, where m and n are coprime, then every subgroup of G whose order divides m is contained in some subgroup of order m, and these subgroups of order m are all conjugate in G. Ten years later he characterized soluble groups by such arithmetic properties, and went on to develop a general theory of finite soluble groups. His most influential pre-war paper was 'A Contribution to the Theory of Groups of Prime-power Order' (1934), in which he discovered many features which underlie the structure of the most general p-group, initiated the theory of regular p-groups, laid down the basic laws of the commutator calculus, and revealed one of the links connecting the study of groups with that of Lie rings. In 1935 he proved a fundamental combinatorial result known as the marriage theorem which became part of mathematical folklore.

Between 1940 and 1954 he published no mathematics. His work in the mid-1950s on theorems like Sylow's and, in collaboration with Graham Higman, on the p-lengths of p-soluble groups was indispensable for the great achievements in finite group theory of the 1960s. He also studied combinatorial questions arising from the appearance of partitions in different parts of group theory and invented an important and elegant algebra of partitions. He made many contributions to the theory of infinite groups. Of seminal importance was his systematic investigation between 1954 and 1961 of certain finiteness conditions in soluble groups, which, for example, initiated the representation theory of polycyclic groups. In 1959 he constructed a universal locally finite simple group, in 1963 a non-strictly simple group, and in 1974 he proved very general theorems about embedding groups in simple groups.

Both through his own work and through that of his students, to whom he was always most generous with his ideas, Hall exercised a profound influence on English mathematics, and an influence which was felt throughout the mathematical world.

He had unusually wide interests in both the sciences and humanities, with an encyclopaedic knowledge and prodigious memory. For many years when doing mathematics he would smoke one cigarette after another, but, to test his will, used from time to time and for predetermined periods, to give them up. He was unmarried and for much of his life lived alone, always studious and always caring nothing for hot water or central heating. He had little liking for large gatherings or formal occasions, but behind his shyness lay a particularly friendly disposition, and when he was with friends he was the best company in the world. He had an extensive, but discriminating, love of poetry, which he spoke beautifully, not only in English. He enjoyed music and art, flowers and country walks. He was gentle, amused, kind, and the soul of integrity.

He was elected FRS in 1942 and received the Sylvester medal (1961). He was president of the London Mathematical Society (1955-7) and was awarded both the senior Berwick prize (1958) and the De Morgan medal and Larmor prize (1965). He had honorary doctorates from Tübingen (1963) and Warwick (1977). From 1976 he was an honorary fellow of Jesus College, Cambridge.

Hall died in Cambridge 30 December 1982. He left half his residuary estate to the National Trust.

[J. A. Green, J. E. Roseblade, and J. G. Thompson in *Biographical Memoirs of Fellows of the Royal Society*, vol. xxx, 1984; personal knowledge.]   J. E. ROSEBLADE

HALLIDAY, EDWARD IRVINE (1902–1984), artist, was born 7 October 1902 in Liverpool, the second of the three sons and the third of the four children of James Halliday, company director, of Garston, Liverpool, and his wife, Violet Irvine, only daughter of Edward Irvine, of Orkney. He was educated at the Liverpool Institute and Liverpool College and then at Liverpool City School of Art, the Académie Colarossi in Paris, and the Royal College of Art in London. He won the Rome scholarship in decorative painting in 1925 and spent three years in Rome at the British School. Subsequently he painted a number of murals for the Athenaeum Club in Liverpool, the ship *Hilary*, and several London restaurants. But from 1929 he was a regular exhibitor at the Royal Academy and gradually turned towards traditional portraiture and, his great favourite, conversation pieces.

While in Rome he met the classical scholar and archaeologist Dorothy Lucy Hatswell, the only daughter of Robert Hatswell MBE, senior staff officer in the secretary's office of the General Post Office. They married in 1928 and she was a constant and loyal support to him throughout his career. They had a son, Stephen, born in 1933, and a daughter, Charlotte, born in 1935.

Halliday's career was interrupted by World War II, when he joined the Royal Air Force (Bomber Command). He worked in Air Traffic Control but was shortly seconded for special duties with the Foreign Office—a euphemism which described the small and carefully selected 'black propaganda' team led by D. Sefton Delmer [q.v.].

After the war he returned to his painting and also to his rather unusual sideline—broadcasting. This had begun in Savoy Hill days with BBC radio programmes related to art, but soon expanded to include writing and reporting news items and commentaries on royal occasions. He also made similar contributions to early television. For over five years, until 1954, his was the voice which the rapidly expanding television audience heard each evening behind the BBC television newsreel.

His career as a royal portrait painter began with a portrait of Princess Elizabeth for the Draper's Company in 1948, and of her with Prince Philip the following year. After her accession to the throne he painted the pair many times for various cities, embassies, and regiments. One of the most appealing of the portraits is the family conversation piece at Clarence House in 1952. Other portraits of the royal family include those of the Queen Mother, Princess Alice, Countess of Athlone, the Earl and Countess Mountbatten, and the Prince of Wales. He also painted foreign heads of state, among them Jawaharlal Nehru, Nnamdi Azikiwe, and Kenneth Kaunda.

His sitters were from many walks of life and frequently became his close friends. But he also enjoyed the company of his fellow artists and devoted much time to their welfare in many ways. One such was in founding in 1958, with (Sir) James Gunn, Sir William Hutchison [qq.v.], and Cosmo Clark, the Federation of British Artists, and in establishing it, with Maurice Bradshaw as its first administrator, at the Mall Galleries. He was also president of the Royal Society of Portrait Painters from 1970 to 1975 and of the Royal Society of British Artists, whose fortunes he had done much to rebuild, from 1956 to 1973. He was involved with the Artists' League of Great Britain, but his greatest contribution was through the Artists' General Benevolent Institution. Possibly feeling that life had treated him more generously than others, he served on the latter's council from the early 1950s and as chairman from 1965 to 1981. This group of artists made themselves responsible for the less fortunate of their profession and their widows and families. Halliday presided over the monthly meetings, keeping every member of the council interested and participating, letting each have his say, and smoothing over controversy by some sensible compromise or piece of well-timed wit. He was a great raconteur and many a City and club diner will remember his very witty after-dinner speeches. He was, at various times, a member of the Savage Club, the Garrick, and the Athenaeum, as well as chairman of the Arts and the Chelsea Arts.

The central character of Halliday's work was the distinguished portraits which, for the last forty years, he painted in the studio at his home in St John's Wood. His gifts as a draughtsman, his liking for people, and his humour, were all essential, useful attributes to his success as a portrait painter. Excessive bravura, prettiness, and obvious flattery were not for him. He understood people and wanted to create a true and dignified image. He became ARCA (1925), RBA (1942), RP (1952), and FRSA (1970). He was appointed CBE in 1973. He died at his London home 2 February 1984.

[Private information; personal knowledge.]

GEORGE BUTLER

HALLILEY, JOHN ELTON (1912–1983), actor. [See LE MESURIER, JOHN.]

HAMPSON, FRANK (1918–1985), strip cartoonist, was born in Audenshaw, Manchester, 21 December 1918, the elder son and eldest of three children of Robert Hampson, a police constable, and his wife, Elsie Light. He was educated at the King George V Grammar School at Southport. He had no formal art training but liked to draw. He was fascinated by American newspaper supplements sent by relatives from Canada, and was particularly inspired by the strip cartoons of Hal Foster, Milton Caniff, and Alex Raymond (whose Rip Kirby had a 'cliff-hanger' every three frames). He learnt how to tell a story in pictures, using few words. He left school at fourteen and became a telegraph boy. His father got him admitted to life classes at the local art school. He enrolled at Victoria College of Arts and Science in 1938 and obtained his national diploma in design just before being called up in 1939. Rescued from Dunkirk as a private in the Royal Army Service Corps, he was landed in Normandy as a lieutenant: seeing the doodle-bugs he later wrote that 'on the quays of Antwerp you could watch the birth of space travel'.

In 1944 he married Dorothy Mabel, daughter of William Rosser Jackson, surveying engineer.

They had one son. After the war he set up home with her in Southport and with his friend Harold Johns enrolled in 1946 at the Southport School of Arts and Crafts. Their tutor, Raymond Geering, said Hampson was an outstanding draughtsman 'prepared to go to endless trouble to get a thing right'. This was to prove both his strength and his undoing.

He could not have dreamed that his big chance was to come through a local parson, the Revd Marcus Morris, vicar of the affluent parish of St James, Birkdale. Morris felt the Church of England was not publicizing its message effectively. He remodelled his parish magazine, *Anvil*, and in 1948 engaged Hampson to illustrate it. At a conference of diocesan editors, it was agreed to form a Society for Christian Publicity, 'Interim', to syndicate the best Christian writers, produce magazines to appeal to the man in the street, and explore the use of strip cartoon.

Britain was being saturated with American 'horror comics'. On 13 February 1949 Morris published an article, 'Comics that bring horror to the nursery', in the *Sunday Despatch*, saying he wanted to see a popular children's comic with clean and exciting adventures: for example, those of Grenville of Labrador, and 'the daily dangers St Paul met'.

Hampson, who had become Interim's full-time artist at £8 per week, began work on a dummy with a Life of St Paul and a story he himself created, Dan Dare, Space Pilot of the Future. Morris now had something to show to publishers; a magazine with a large page, superb drawing, bright colour, and every kind of adventure designed to appeal to girls and boys of the ages eight to twelve. After numerous refusals, Hulton Press wired: 'Definitely interested do not approach any other publisher.' Thus *Eagle* magazine was launched on 14 April 1950, with Marcus Morris as editor. Hampson worked more than full-time with a team chosen by himself, first in 'The Bakery' at Southport, and then in houses in Epsom (to which town he and Morris moved), the last one, Bayford Lodge, having a bedroom floor removed so that photographs could be taken of dressed-up posed models from above or below.

Hampson's perfectionism increased with his skill. He would think out his next episode of Dan Dare during the weekend and do a rough in pencil, and then in ink and colour. These were good enough for any other comic but only the start for Hampson. With the aid of his research and reference sheets, models of space ships, plaster heads of characters, mock-ups of space suits, and a complete deep space station, every frame would be posed and photographed on Monday and Tuesday and on Wednesday the photos and roughs together were given to the artists, Hampson taking the big opening frame which was dramatic, technologically ingenious, or hilarious. On Friday the artwork had to go to Bemrose of Liverpool's German rotogravure presses which could turn out $3\frac{1}{2}$ copies of *Eagle* a second to supply a million copies six weeks later (because of the time needed for artists to work on the photographic plates in three colours). The team then went home after a week of five thirteen-hour days, but Hampson would spend another weekend of creation, invention, and nodding off over his drawing boards until disturbed by 'them dratted birds'. He was a genius of the genre.

In 1959 Hultons was sold to Odhams Press, who wished to alter Dan Dare. Marcus Morris left as the editor of *Eagle* and its sister papers *Girl*, *Swift*, and *Robin*, and Hampson's studio was disbanded. He himself left in the summer of 1961. Thereafter he undertook small advertising commissions, illustrations for magazines, and the pictures for seven Ladybird books. In 1970 he developed cancer of the trachea. He then became a graphics technician at Ewell Technical College. In 1978 he sat for an Open University degree, which he received in January 1979.

In November 1975, at the biannual Lucca comics convention, the Mecca of strip cartoon enthusiasts, Hampson was awarded the Yellow Kid award and declared 'prestigioso maestro' as the best writer and illustrator of strip cartoons since the end of World War II. At the British comics convention, Comics 101, he was presented with the Ally Sloper award as the best British strip artist. Hampson died 8 July 1985 at the Cottage Hospital, Epsom.

[Alastair Crompton, *The Man Who Drew Tomorrow*, 1985; information from Marcus Morris; personal knowledge.] CHAD VARAH

HANSFORD JOHNSON, PAMELA, LADY SNOW (1912–1981), novelist, dramatist, and critic. [See JOHNSON.]

HARDY, SIR ALISTER CLAVERING (1896–1985), zoologist and investigator of religious experience, was born in Nottingham 10 February 1896, the youngest of three sons (there were no daughters) of Richard Hardy, architect, and his wife, Elizabeth Hannah Clavering. He was educated at Oundle and Exeter College, Oxford (1914–20) where, after war service (1915–19) as lieutenant and captain in the Northern Cyclist battalion and a camouflage officer on the staff of XIII Army Corps, he was awarded the Christopher Welsh scholarship and graduated with distinction in zoology (1920).

He was immediately appointed assistant naturalist at the Ministry of Agriculture and Fisheries laboratory in Lowestoft to work on herring and plankton; and in 1924 chief zoologist to the Colonial Office *Discovery* Expedition, to study the biology of the Antarctic whales. After returning in 1927 he was appointed (1928) professor of zoology (extended to include oceanography in 1931) in the newly founded University College of Hull. In 1942 he became regius professor of natural history in Aberdeen, and in 1945 Linacre professor of zoology in Oxford, until 1961 when he retired to become professor of field studies there until 1963. He immediately founded and became director of the religious experience research unit in Manchester College, Oxford (later renamed the Alister Hardy Research Centre, now established in Oxford and Princeton, USA).

Hardy's professional career ranged through zoology, with special emphasis on plankton and its distribution and behaviour, insects, and evolution. On the *Discovery* expedition he invented, and in Hull developed, the continuous plankton recorder, with which the spatial and seasonal distribution of the drifting life in the sea has been surveyed for many years: first in the North Sea and then over the whole North Atlantic Ocean, becoming the broadest and most regular ecological monitoring system in the world (*Bulletin of Marine Ecology*, 1935 onwards). In his teaching and research he stimulated innumerable students and colleagues, many of whom have paid tribute to his influence, especially his continuous plankton recorder team and at Oxford where he is said to have revitalized and integrated the work of the department.

He never retired, however. Having had a sense of the numinous from youth onwards, in 1963 he set out to lay the foundations of what he termed 'a future science of natural theology', by investigating religious experience. In this he aimed to remove the barriers between science and religion, and to reconcile current theories of evolution (he was a neo-Darwinian) with his view that religious experience is a biological fact which can be studied by scientific methods. Hardy sketched and painted in water-colour, and wrote easily and extensively. This is illustrated not only in many conventional scientific papers (e.g. his *Discovery Reports*, 1928–35), but in three fascinating books, *The Open Sea*: Pt. I, *The World of Plankton* (1956) and Pt. II, *Fish and Fisheries* (1959), and *Great Waters* (1967). The transition to religion is developed in his Gifford lectures, *The Living Stream* (1965) and *The Divine Flame* (1966), followed by *The Biology of God* (1975) and *Darwin and the Spirit of Man* (1984) among many other books and essays.

Hardy was a tall man of enormous energy, a great walker and cyclist, and fascinated by flying. He was extremely loyal and his courage and encouragement never failed, as demonstrated by his annual meetings with the pitmen members of his wartime cyclist company, and by the famous Xmas cards, illustrated by him with his wife's help, sent annually, to countless friends.

He received many distinctions and honours, from his Oxford D.Sc. in 1938 and Scientific medal of the Zoological Society in 1939, to his election as FRS in 1940 and his knighthood in 1957. He became emeritus professor on retirement in 1961 and fellow and then honorary fellow of Merton and Exeter Colleges, receiving the honorary degrees of D.Sc. of Southampton and Hull and the LLD of Aberdeen. Finally, just before being crippled by a stroke which shortly killed him, he was awarded the Templeton prize for progress in religion ($185,000), which he planned to spend on his new institute, and which in his illness was received by his daughter from the Duke of Edinburgh.

In 1927 he married Sylvia Lucy (died 1985), second daughter of Walter Garstang, professor of zoology at Leeds. He and his wife had been fellow students, and she was behind all he did. They had a son and a daughter. Hardy died in Oxford 23 May 1985.

[N. B. Marshall in *Biographical Memoirs of Fellows of the Royal Society*, vol. xxxii, 1986; *The Times*, 24 May 1985; private information; personal knowledge.]   CYRIL LUCAS

HARE, JOHN HUGH, first VISCOUNT BLAKENHAM (1911–1982), politician and farmer, was born 22 January 1911 in London, the third of the four sons and the third of the six children of Richard Granville Hare, the fourth Earl of Listowel, and his wife, Freda, daughter of Francis Vanden-Bempde-Johnstone, second Baron Derwent. He was educated at Eton. He became a member of the London county council and was from 1937 to 1952 an alderman of the LCC. He was chairman of the London Municipal Society from 1947 to 1952. During World War II he served in North Africa and Italy with the Suffolk Yeomanry and was mentioned in dispatches and appointed MBE (1943) and OBE (1945) and awarded the US Legion of Merit (1944). He left the army with the rank of lieutenant-colonel.

After the war he returned to politics and, in addition to his local government work, he was returned in the 1945 general election as Conservative MP for Woodbridge in Suffolk, later to become Sudbury and Woodbridge, which he represented until he was made a peer in 1963.

In December 1951 he was made vice-chairman of the Conservative Party in charge of the selection of candidates. Thereafter he was, in succession, minister of state for colonial affairs (1955-6); secretary of state for war (1956-8), when he was at the heart of the Suez operation; minister of agriculture, fisheries, and food (1958-60); and minister of labour (1960-3).

Throughout his ministerial and parliamentary career his great strength was his capacity for understanding human reactions and so for dealing successfully with a wide variety of people. This skill was shown to best advantage during his time as minister of labour. He quickly gained the complete confidence of the trade-union leaders through his obvious sincerity and his genuine concern for the interests of their members. He was therefore in a strong position when it came to promoting co-operation and understanding in industry. He was particularly interested in industrial training and some of the early efforts to establish industrial training boards with government assistance owed much to his initiative.

In 1963 he was created first Viscount Blakenham and became chancellor of the Duchy of Lancaster. He was appointed chairman of the Conservative Party and deputy leader of the House of Lords. He was admitted to the Privy Council in 1955. He took on as chairman of the Conservative Party when Sir Alec Douglas-Home (later Lord Home of the Hirsel) became prime minister in October 1963 the task of preparing the organization for the imminent general election. He did not naturally enjoy the electioneering side of party politics but nevertheless threw himself wholeheartedly into a job which he certainly did not seek. In the event, he very nearly succeeded in helping to achieve a Conservative victory which few considered possible. Sadly he did not appreciate himself the full measure of his achievement and so was deeply disappointed. This disappointment, coupled with the great strain he placed on himself during the period leading up to the election, affected his health and led to his retirement from active politics and to his taking no further part in government. This decision was much regretted by his many political friends in all parties.

Thereafter he gave much of his time to charities. He was on the council of Toynbee Hall from 1966 and on the governing body of the Peabody Trust from 1967. He was also treasurer from 1971 of the Royal Horticultural Society—a position he greatly enjoyed through his love of gardening. For this work he obtained the Victoria Medal of Honour in 1974. Otherwise he developed his farming activities in Suffolk (he had 600 acres) and made a beautiful woodland garden at his home, Cottage Farm, Little Blakenham. In 1968 he was appointed a deputy lieutenant for Suffolk.

He was not an orator, nor ever claimed to be, but his ability to get on with people enabled him to gain the trust of members on all sides of the House of Commons and so ease the passage of any legislation for which he was responsible. He was therefore a most successful minister and a very deceptive politician. Behind an apparently diffident manner and a halting delivery when speaking, he hid a fund of common sense, considerable understanding, and unshakeable determination.

In 1934 he married (Beryl) Nancy, daughter of Weetman Harold Miller Pearson, second Viscount Cowdray. They had one son and two daughters. Blakenham died in London 7 March 1982. He was succeeded in the viscountcy by his son, Michael John (born 1938).

[Personal knowledge.] WILLIAM WHITELAW

HARGRAVE, JOHN GORDON (1894-1982), artist and writer, was born 6 June 1894 at Midhurst, the son of Gordon Hargrave, landscape painter, and his wife, Babette Bing. He attended Hawkshead Grammar School for a short time, and when he was fifteen illustrated *Gulliver's Travels* and *The Rose and the Ring*. Two years later he was appointed chief cartoonist of the *London Evening Times* and in 1914 joined the staff of C. Arthur Pearson Ltd.

On the outbreak of the 1914-18 war he enlisted in the Royal Army Medical Corps and later served with the 10th (Irish) division at Gallipoli and in Salonika. Having been invalided from the army at the end of 1916 he became art manager of C. Arthur Pearson Ltd. from 1917 to 1920. He continued his work as a cartoonist over the next thirty years, and in 1952 created the animal character 'Bushy' for the *Sketch*.

At the age of fourteen he had joined the Boy Scout movement, recently formed by Robert (later Lord) Baden-Powell [q.v.], and by 1921 he was commissioner for woodcraft and camping. In that year, however, he left the movement, having inaugurated in 1920 the Kindred of the Kibbo Kift (an old Scottish term meaning a feat of strength). The purpose of this organization was to encourage youth 'to seek health of body, mind, and spirit', and to become expert at woodcraft and other open-air pursuits whilst believing in disarmament and the brotherhood of man. In its early years the Kibbo Kift, whose members wore a woodcraft costume of cloak, hooded jerkin, and shorts, numbered only a few hundred, but it was supported by such eminent people as H. Havelock Ellis, (Sir) Julian Huxley, H. G. Wells [qq.v.], and Emmeline Pethick-Lawrence. After 1931, when the movement developed into the Green Shirts, a paramilitary organization, the membership increased to thousands.

In the meantime Hargrave had met and come under the influence of Clifford Hugh Douglas [q.v.], the originator of the theory of Social Credit. In brief, the ideas of Douglas were based on the premiss that the economic problems of his time were caused by a shortage of purchasing power which stemmed from a flaw in the system of price fixing. To remedy these defects he advocated the issue of additional money to consumers or of subsidies to producers to enable them to set prices below the costs of production.

Hargrave had sufficient influence with his followers to persuade the movement of which he was the leader to adopt these theories in opposition to the prevailing banking and financial structure of the capitalist system. Although Hargrave and the Green Shirts were opposed to the activities of Sir Oswald Mosley [q.v.] and his Black Shirt Fascists, both movements were equally affected by the Public Order Act of 1937 which banned the wearing of political uniforms; and two years later the outbreak of the 1939–45 war effectively brought to an end the activities of both organizations. After the war Hargrave attempted to revive the Green Shirts, but his efforts ended in failure, and in 1951 the organization was dissolved.

In 1936–7 Hargrave had been honorary adviser to the Alberta government planning committee which was seriously concerned with the introduction of the Social Credit system into that province of Canada. The Alberta government, led by William Aberhart, accepted the theories set out in the report in the preparation of which Hargrave had played an influential part, but never succeeded in putting its plans into effect as, in the last resort, the governor-general declared them to be unconstitutional.

Hargrave was indefatigable in setting out his ideas in writing. By 1913 he had written Lonecraft and five other books on camping and similar outdoor activities. His Social Credit Clearly Explained (1934) gave chapter and verse for his theories on that subject, and in other publications, such as Professor Skinner, alias Montagu Norman (1939), he castigated mercilessly the exponents of monetarism and other policies of contemporary financiers. The Confession of the Kibbo Kift (1927) explained the purpose of that movement; he also wrote Words Win Wars (1940).

He demonstrated his versatility in many other books including Summer Time Ends (1935) and five other novels. He was also a lexicographer and published his The 'Paragon' Dictionary (1953); he wrote The Life and Soul of Paracelsus (1951), the German–Swiss physician and alchemist, on whom he was a recognized authority; and his wartime experiences led to The Suvla Bay Landing (1964).

Another aspect of the career of this extraordinary man, with his intense personal enthusiasms and keen, if unorthodox opinions, was his invention of an automatic navigator for aircraft which he claimed to be the basis for equipment used in Concorde and other supersonic aeroplanes. These claims were argued in The Facts of the Case concerning the Hargrave Automatic Navigator for Aircraft (1969) but he failed to win his case in a public inquiry held in 1976.

Hargrave was twice married. In 1919 he married Ruth, daughter of William Clark, engineer; they had one son. This marriage ended in divorce in 1952, and in 1968 he married Gwendoline Florence Gray, an actress. He died 21 November 1982 at Branch Hill Lodge, Camden.

[The Times, 25 November 1982; Hargrave's own books.]   H. F. Oxbury

HARLAND, SYDNEY CROSS (1891–1982), agricultural botanist and geneticist, was born 19 June 1891 at Snainton, near Scarborough, the son of Erasmus Harland and his wife, Eliza. He was educated at the Municipal Secondary School in Scarborough and at King's College, London, where he graduated in 1912 with honours in geology. Seized by the wanderlust which was to be a feature of his life, he took a job as a schoolmaster in St Croix, then a Danish colony but now part of the US Virgin Islands. Within a year he accepted a post as assistant to the director of an agricultural experiment station on the island, where work centred on the cultivation of sugar-cane and cotton. He was given responsibility for investigating hybrids between local and North American cottons, and his enthusiasm for the work was immediate. So began a distinguished career during which he greatly advanced understanding of the genus Gossypium, to which the cottons belong, and successfully applied that knowledge to the breeding of improved varieties.

In 1915 he moved to St Vincent as assistant agricultural superintendent. His studies on the genetics of cotton continued, and his published work from this period was recognized by the award of the London D.Sc. degree. He came back to England in 1920 as head of botany at the British Cotton Industry Research Association (the Shirley Institute) in Manchester, but he preferred to be where cotton was grown rather than where it was spun, and in 1923 he returned to the West Indies as professor of botany and genetics at the Imperial College of Tropical Agriculture, Trinidad. As well as resuming field studies on cotton, he became an inspiring teacher, bringing out the best from his students by imaginative projects, penetrating discussion, and an infectious liking for tropical crops.

His growing reputation attracted the attention of the Empire Cotton Growing Corporation which invited him to take charge of cotton genetics at their new research station, also in Trinidad, and this he did from 1926 to 1935. The outstanding success of his work during these years, though largely due to his own vision and originality, was greatly helped by the freedom and independence allowed him by the Corporation. He gained an encyclopaedic knowledge of the world's cottons, established a classification of *Gossypium* based on sound genetic criteria, and thus provided cotton breeders with a new basis for breeding strategies. His achievements were academic as well as agricultural. He made notable contributions to the understanding of evolutionary processes and of speciation; his paper on 'The Genetical Conception of the Species' (1936) was of crucial importance. It was followed in 1939 by his only book, *The Genetics of Cotton*, in which the results of his own research rightly formed a prominent part.

Though generally affable and greatly respected by close colleagues, Harland could also be outspoken, even obstinate, and he did not suffer gladly the curbs of authority. His contract with the ECGC came to an end during a time of disagreement in 1935. He moved first to Brazil as general adviser to the state cotton industry, and then in 1940 to Peru as director of the Institute of Genetics at Lima. In 1949 he was appointed reader in genetics at Manchester University, and he became professor of botany in 1950. By this time his most productive research had been done, but he continued to travel widely to cotton-growing regions in an advisory capacity. Within the university he was tireless in his insistence that genetics should feature largely in the curriculum of any student of biology in its broadest sense, and the introduction of genetical teaching for students of medicine and psychology was due to his missionary zeal.

Among the honours he received were FRS (1943), FRSE (1951), honorary fellowship of the Textile Institute (1954), and an honorary D.Sc. from the University of the West Indies (1973). He was also president of the Genetical Society from 1952 to 1955.

In 1915 he married Emily Wilson Cameron, and they had two daughters. After his first marriage ended in divorce he married, in 1934, Olive Sylvia Atteck, daughter of a prominent Trinidad family. There was one son of the second marriage who became professor of child health in the University of the West Indies.

Harland retired in 1958 and lived for some years on a small private estate in Peru where he indulged his abiding interest in breeding

tropical crops. But he never ceased to be a Yorkshireman at heart, and he finally returned to Snainton, his birthplace, where he died 8 November 1982.

[Sir Joseph Hutchinson in *Biographical Memoirs of Fellows of the Royal Society*, vol. xxx, 1984; personal knowledge.]

J. N. Hartshorne

HARLECH, fifth Baron (1918-1985), politician and ambassador. [See Ormsby Gore, (William) David.]

HARRIS, Sir ARTHUR TRAVERS, first baronet (1892-1984), marshal of the Royal Air Force, was born 13 April 1892 at Cheltenham, the second youngest in the family of four sons and two daughters of George Steele Travers Harris, an engineer-architect in the Indian Civil Service, and his wife, Caroline Maria, daughter of William Charles Elliot, a surgeon in the Madras Cavalry. The second son died in 1917 and the youngest son died as an infant. Educated at Gore Court, Sittingbourne, and All Hallows, Honiton, while his parents were in India, he soon became exceptionally self-reliant. At seventeen, against his father's wishes, he went out to farm in Rhodesia.

In October 1914 Harris joined the new 1st Rhodesian Regiment as a bugler—almost the only vacancy. He took part in the successful operations against German South-West Africa, and on the disbanding of the regiment returned to England to enlist. Failing to find a place in the cavalry or artillery, he applied to join the Royal Flying Corps. Half an hour's tuition at Brooklands qualified him as a civilian pilot, and in November 1915 he was commissioned as a second lieutenant.

In the RFC Harris became an 'anti-zeppelin night pilot' and soon commanded No. 38 Squadron. Posted to No. 70 Squadron in France, he was invalided home after a crash, but returned in May 1917 for fighter work with No. 45 Squadron. During 1918 he commanded a night-training and night-fighting squadron (No. 44) in England, and was awarded the AFC.

In August 1919 Harris was offered a permanent commission in the RAF as a squadron leader. Between then and 1925 he commanded No. 2 Flying Training School at home, No. 31 Squadron in India (with operations on the North-West Frontier), and No. 45 Squadron in Iraq, where he successfully adapted his troop-carrying Vernons for day and night bombing. He also invented an electric truck by which these cumbersome aircraft could be moved on the ground by two men instead of sixteen. His spell in Iraq completed, Harris then commanded (1925-7) No. 58 Squadron of Virginia bombers at Worthy Down and made it one of the most efficient in the Service. Again his work involved intensive training by night. In 1927 he was appointed OBE.

Having passed successfully through the Army Staff College (1928-9) and declined an invitation to stay on as an instructor, Harris then served for two years on the air staff in the Middle East. He was next posted home, in 1933, to take a flying-boat pilot's course and briefly command No. 210 (Flying-boat) Squadron. Six months as deputy director of operations and intelligence at the Air Ministry then followed, after which he became deputy director of plans (1934-7). Here, as a member of the chiefs of staff joint planning sub-committee, he was much concerned with national defence policy as a whole. On the RAF side he urged the development of mines for aircraft to lay at sea, and he gave ardent support to the conception of new ultra-heavy (including four-engined) bombers. These were specified in 1936, and used by Harris with devastating effect in 1942-5.

A year's command of the newly formed No. 4 (Bomber) Group, during which Harris became an air commodore (1937), then followed. The squadrons of this group, on Whitleys, were specially trained for night bombing. Before taking up his next post in 1938 Harris was then sent to the United States to study the aircraft and aircraft equipment position there, and make purchases. He and his small staff hit on two winners: the Hudson for general reconnaissance and the Harvard advanced trainer.

Harris took up his next appointment as air officer commanding Palestine and Transjordan (1938-9). In his task of helping to restrain the growing hostility between Jews and Arabs Harris co-operated smoothly with the army and devised what he called the 'air pin'—pinning down villagers by the threat of bombing from patrolling aircraft until the army could get to the scene and root out the trouble-makers. During the course of this posting Harris was promoted air vice-marshal (1939) and mentioned in dispatches. He returned home with a duodenal ulcer which was to give him trouble later.

Shortly after the outbreak of war in 1939 Harris was given command of No. 5 (Bomber) Group. His Hampden aircraft soon proved unsuitable for daylight raids against the German fleet and became largely confined to operating by night. They did useful work against German barge concentrations and airfields during the invasion threat, and began the task which Harris was to develop so greatly later, of mining German waters. In the course of his year's command Harris secured many improvements in these aircraft, initiated operational training under simulated battle conditions, and received recognition in appointment as CB (1940).

In November 1940 Sir Charles Portal (later Viscount Portal of Hungerford, q.v.) summoned Harris to serve under him at the Air Ministry as deputy CAS. Six months later he sent him out to Washington as head of the RAF delegation there, to speed up the purchase of aircraft and air supplies. Harris brought his characteristic drive to the task, got on well with the Americans, and had secured a marked improvement when in 1942 Portal called him home to become AOC-in-C Bomber Command.

The success of Harris's work at Bomber Command, though it seemed obvious at the time, has been questioned subsequently. Critics have denounced area bombing, his most favoured form of attack, as not only immoral but ineffective compared with precision bombing. Harris, however, did not determine bombing policy, which was a matter for Whitehall. Area bombing had come about because the British bombers of 1940-1 had proved incapable of penetrating Germany by day and of finding and hitting precision targets by night. It was a policy already in operation when Harris took over Bomber Command. Behind it lay the view that Germany had set ample precedents; that the destruction of industrial cities and the homes of the workers would seriously weaken the enemy; and that unless Bomber Command were given a practicable task no offensive action at all could be taken against Germany for years to come.

It was under Harris, who was selflessly supported by an outstanding senior air staff officer, Air Marshal (Sir) Robert Saundby, that Bomber Command developed from a force of modest size and doubtful efficiency into a truly formidable instrument of war. Well versed in technical matters, Harris eagerly grasped at developments which transformed the capabilities of his force: the Lancaster bomber, new bombs (up to 22,000 lb.), the succession of navigational or bombing aids (Gee, H2S, Oboe, GH), new techniques of target-marking and saturation raids under the direction of a master bomber—these are but a few of the best-known examples. By the latter part of 1944 his force had in fact become capable of precision bombing by night over Germany in cloudy weather, and was so employed with decisive results against German oil plants and communications. Area bombing, however, also remained in his directive from the Air Ministry, though in lower priority, until April 1945; and critics have suggested that Harris's own preference for this led him to continue it after precision bombing became regularly possible. But to Harris the retention of area bombing in his directive provided a flexibility he considered essential—the freedom to choose within a diversity of large targets in the light of meteorological conditions and the need to achieve tactical surprise. He was promoted air chief marshal in 1943.

On many occasions Harris strenuously resisted pressure from the Air Ministry; but when his objections were overruled, he prided himself on obeying his orders wholeheartedly. Perhaps the most striking instance was his initial opposition to the idea that Bomber Command should give direct assistance in the invasion of Normandy: he maintained that the best help he could give would be to continue bombing German cities. When overruled, he abundantly fulfilled his orders by attacks on railway and coastal targets in France which greatly facilitated the Allied landings and helped to prevent reinforcement of the forward German troops.

The contribution of Bomber Command under Harris to the Allied victory was in fact outstanding. It took many forms: sea mining, attacking German warships and U-boat bases, co-operating with the Allied armies, disrupting German V-weapon activity, and waging a strategic offensive against German industry and urban industrial life. It cannot, however, be properly considered in isolation from the work of the Americans, without which it would not have been so effective. It was the combination of the two forces in 'round the clock' bombing from 1943 onwards that pinned down over a million civil workers, hundreds of thousands of German soldiers, and nearly 20,000 guns to the air defence of the Reich. According to Albert Speer, the Anglo-American air offensive 'opened a second front long before the invasion of Europe' and constituted 'the greatest lost battle on the German side'. Without this joint offensive, which among other things crippled the Luftwaffe and its sources of replenishment, it is inconceivable that the Allied armies could have successfully invaded Normandy in 1944.

For his achievements Harris was mentioned in dispatches (1939 and 1941) and appointed KCB (1942). An abundance of foreign honours descended on him: the Russian Order of Suvorov in 1944 and then in 1945 the Order of Polonia Instituta, the Brazilian National Order of the Southern Cross (grand cross), the French Legion of Honour (grand officer) and the croix de guerre with palm. In 1946 came the US DSM.

In Britain, the story was different. In June 1945 Harris accepted the GCB reluctantly—he was pressing at the time for proper recognition for his crews, and had said that if this were denied he wanted nothing for himself. He was omitted from the New Year's honours of 1946, though other commanders received peerages; but in his own Service he was promoted to the highest rank—in 1946 he became a marshal of the RAF. Also in 1946 Liverpool University

awarded him an honorary LLD. Later, after (Sir) Winston Churchill's return to power, he was asked if he wished to be recommended for a peerage, but he preferred to retain his familiar style, and accepted a baronetcy (1953).

After the war Harris, in poor health and aggrieved at what he felt was lack of understanding of the magnitude of his Command's efforts, retired and departed for South Africa. Through a friend he started up, as a subsidiary of a larger concern, a shipping firm, the South African Marine Corporation. He ran this until 1953 when he returned to England and settled down amiably at Goring-on-Thames, where his favourite diversions included entertaining, cooking, and carriage-driving. Surprisingly, perhaps, for the commander whose forces destroyed most of the Ruhr, Hamburg, Berlin, and Dresden, he enjoyed the company of children and excelled at making up stories for them.

Until his old age Harris was of heavy build and reddish hair and complexion. A man of combative temperament who could be gruff and dismissive as well as affable, he was nevertheless popular with most of his colleagues, including those from other Services and nations. He was on the best of terms with Dwight D. Eisenhower and Viscount Montgomery of Alamein [q.v.], and his collaboration with the American air forces based in Britain was virtually perfect. His own crews in Bomber Command rarely, if ever, saw him, but somehow the power of his personality and his utter commitment to his task radiated out from High Wycombe and provided a grim inspiration: his airmen knew he might be sending them to their death, but they also knew that their lives would not be lightly gambled, and that he would fight to his last gasp on their behalf. His nicknames—'Bomber' to most of his crews and the world at large, 'Butcher' to some, 'Bert' to his colleagues (from naval associates, for in the navy all Harrises are Berts), and 'Bud' or 'Buddy' to his second wife and family friends—testify to the varying elements within him. But his main characteristics were throughout clear: drive, efficiency, courage, determination, loyalty, common sense, practicality, humour, and a habit of exaggeration, particularly on paper, which got him into trouble but also resulted in perhaps the most readable official letters ever written by an air marshal.

There is a crayon portrait of Harris by Eric Kennington in the Imperial War Museum; a portrait by Anna Zinkeisen at Strike Command HQ; a picture by Herbert Olivier entitled 'Ops. Room Conference' showing Harris, Saundby, and others, in the RAF Museum; and a portrait with General Fred Anderson, USAAF, by Herbert Olivier in the possession of the family.

In 1916 Harris married Barbara Daisy Kyrle,

daughter of Lieutenant-Colonel Ernle William Kyrle Money, of the 85th King's Shropshire Light Infantry. A son and two daughters were born of this marriage, which was dissolved in 1935. In 1938 Harris married Thérèse ('Jill'), daughter of Major Edward Patrick Hearne, of Carlow, Eire. There was one daughter of this marriage, born in 1939. Harris died 5 April 1981 at his home in Goring. He was succeeded in the baronetcy by his only son, Anthony Kyrle Travers (born 1918).

[A. T. Harris, *Bomber Offensive*, 1947; Dudley Saward, *'Bomber' Harris*, 1984; Sir C. Webster and N. Frankland, *The Strategic Air Offensive Against Germany, 1939–1945* (official history), 4 vols., 1961; private information; personal knowledge.]

DENIS RICHARDS

HASTINGS, ANTHEA ESTHER (1924–1981), publisher (known professionally as the Hon Mrs Michael Joseph), was born 6 March 1924 in London, the only daughter of (Francis Lord) Charlton Hodson (Lord Justice Hodson, PC, MC) [q.v.], a lord of appeal in ordinary, and his wife, Susan Mary, daughter of Major William Greaves Blake. Educated at Queen's Gate School, London, she worked at the American embassy in London during World War II. In 1946 she became secretary to Michael Joseph, the publisher, who had founded the company bearing his name ten years earlier. He was the son of Moss Joseph, diamond merchant. They married in 1950 following the death of Michael Joseph's second wife.

Eight years of marriage (Michael Joseph died in 1958) saw her do a great deal to establish the family home at Brown's Farm, Old Basing, Hampshire. She had two children, a daughter Charlotte and a son Hugh, and was a sympathetic stepmother to Michael Joseph's children (three sons and a daughter) from his earlier marriages. During this period she lacked time to work closely with her husband in his business but accompanied him on visits to the United States (in those days a restful voyage on the *Queen Elizabeth*) and South Africa, events which broadened her circle of friends and professional acquaintances to her benefit in the years following Michael Joseph's early death.

His death was the start of Anthea Joseph's real career in publishing. The years 1958–61 were difficult for the company. They culminated in the resignation of the majority of the board, leaving her and one other director in charge. It was largely due to Anthea Joseph's energy and determination that the company survived intact to settle down as a part of the Thomson Organization and to enjoy two decades of very successful publishing.

Anthea Joseph was not a publisher in the broadest sense of the word. The mechanics of the business, the printing and binding, the marketing and other commercial aspects of publishing, interested her only to the extent that they were the means of presenting the work of an author she believed in to the public. That her judgement often proved to be correct and commercial criteria were met, was secondary: the act of publication meant everything to her. This somewhat romantic attitude towards her profession could be a trial to her colleagues at times, but her lively sense of humour, her subtle powers of persuasion and, above all, her enthusiasm invariably disarmed her keenest critics. She was, however, far more than an editor: rather, she believed that the true task of the publisher is to inspire and support creators of literature, a position that was becoming difficult to sustain towards the end of her life.

She was a sensitive judge of fiction: Stan Barstow, Dick Francis, James Baldwin, James Herriot, H. E. Bates [q.v.], Barry Hines, Alun Richards, and Julian Rathbone were among the writers she worked with and published. Many others, lesser known, enjoyed equal dedication, understanding, and sympathy for their work, for she believed that a writer ought to be given time to develop and the publisher should, if necessary, have the patience to wait. She was a critic of the 'hyped' book that in the 1970s became a prominent feature of publishing, both in the UK and the USA. While accepting that some were necessary for the income they produced she argued consistently that a part of their profits should be set aside to support creative literature.

She had an outstanding flair for friendship which stemmed from her deep interest in people. She was an excellent listener. Her loyalty to friends, authors, colleagues, and her family was widely acknowledged. A committed Christian who was modest about her faith, she was determined that her religious belief should shape every aspect of daily life.

In 1962 she became deputy chairman of Michael Joseph Ltd., and chairman in 1978. In 1963 she married a close friend of her first husband, (Douglas) Macdonald Hastings (died 1982), the son of Basil Macdonald Hastings, playwright. They had one daughter, Harriet. Anthea Hastings died after a long illness 23 January 1981 at her home in Hampshire.

[*The Times*, 26 January 1981; *Bookseller*, 31 January 1981; *At the Sign of the Mermaid: Fifty Years of Michael Joseph*, 1986; private information; personal knowledge.]

VICTOR MORRISON

HEAD, ANTONY HENRY, first VISCOUNT

HEAD (1906-1983), soldier and politician, was born in London 19 December 1906, the only son and younger child of Geoffrey Head, of 51 South Street, London, and his wife, Ethel Daisy, daughter of Arthur Flower, of Prince's Gate, London. His father, a Lloyd's broker, was one of eleven children. One uncle went down with the *Titanic* and another, Sir Henry Head [q.v.], was an eminent neurologist and FRS.

Educated at Eton and the Royal Military College, Sandhurst, Head worked for a summer as a stable lad in the well-known racing establishment of Atty Persse. He retained an interest in racing all his life. Tall, slim, and athletic, he was commissioned in the 15th/19th Hussars, but an illness which rendered him temporarily unsuited for service abroad led him in 1924 to transfer to the Life Guards; he became an adjutant ten years later. In the intervening years he rode in the Grand National and in many other races as an amateur jockey. In 1932 he obtained leave to sail in the great Finnish four-masted barque, *Herzogin Cecilie*, in the annual grain-race to Australia. The voyage round the Cape lasted three months and he worked his passage before the mast, receiving at the end of the voyage the able bodied seaman's certificate.

In his regiment and elsewhere he was highly regarded for his energy, enterprise, and love of adventure. A few days before the outbreak of war in 1939 Head was engaged in manœuvres with a Polish cavalry regiment. He contrived to return home before war was finally declared and wrote a report on the Polish cavalry's tactics and equipment. He was sent to the Staff College and by the spring of 1940 he was brigade-major of the 20th Guards brigade. He took part in the fighting at Boulogne and was awarded the MC. He was then appointed an assistant secretary to the Committee of Imperial Defence; but by 1942 he was GSO 2 in the Guards Armoured division. His clear brain and notable endowment of common sense were recognized in high military quarters. This led to his being appointed chief military planner at combined operations headquarters. From 1943 to 1945 he worked with the directors of plans on amphibious operations, made frequent visits to North Africa, and was much concerned with the planning of the landings in Sicily and Italy. He worked closely with the chiefs of staff and the joint planning staff, with the rank of brigadier. He attended the Casablanca, Tehran, Quebec, and Yalta conferences; and after Yalta he went to Moscow with his close friend, Major-General (Sir) Robert Laycock [q.v.], chief of combined operations, to discover how best to elicit information about Soviet operational intentions from the Russian general staff who until then had been uncooperatively reticent.

He was a member of the British delegation at the Potsdam conference in July 1945, but he was already a parliamentary candidate for Carshalton in the general election of that year. He won the seat for the Conservatives with a majority of little more than 1,000. His success and popularity in his constituency can be measured by the uninterrupted rise in his majority at subsequent elections until in 1959 it reached 13,244.

While his party was in opposition, in 1945-51, he made a name for himself in the House of Commons as a fluent and persuasive speaker on defence. It was not therefore surprising that on 31 October 1951 (Sir) Winston Churchill, who liked him personally and admired his drive and sharpness of intellect, made him a privy councillor and secretary of state for war, a position he filled to Churchill's satisfaction and continued to hold when Sir Anthony Eden (later the Earl of Avon) became prime minister in April 1955.

One of his major achievements at the War Office was the establishment of Welbeck College, which provided a broad education for young men, mostly the sons of artisans, showing an aptitude for technology and mechanization, as a preliminary to entering Sandhurst and obtaining a commission in the technical corps of the army.

On 18 October 1956 Head was appointed minister of defence in succession to Sir Walter Monckton (later Viscount Monckton of Brenchley, q.v.) who resigned because of his objection to the projected Suez operation. Head was thus placed in a position of vital and immediate significance only a fortnight before the Suez landings, a venture of which the three Service ministers had not even been informed and with which the chiefs of staff were not wholly in accord. He was distressed that the operation once begun was not completed. However he endorsed the policy and was unshaken in his loyalty to the prime minister, Eden. In January 1957 Harold Macmillan (later the Earl of Stockton) succeeded Eden. Against the advice of Head and the chiefs of staff, Macmillan offered what they considered inadequate financial backing for the armed forces. Head therefore declined an offer to remain minister of defence and retired to the back-benches.

In 1960 he was created first Viscount Head. On Nigerian independence in the same year he became the first British high commissioner in Lagos. Relations with the Nigerian government were at that time excellent and it presented the high commission with the best site in Lagos. The Heads established a close friendship with the prime minister, Sir Abu Bakar Tafawa Balewa [q.v.]. They travelled assiduously throughout the country, associated with Nigerian politicians, businessmen, and artists, and their three years' tenure of the post was an undiluted success.

Head, who had been appointed CBE in 1946, and KCMG in 1961, was made GCMG on leaving Lagos in 1963. He was then, until 1966, the first British high commissioner in Malaysia, a duty he performed with his usual enthusiasm, although the political and racial animosity of Malays and Chinese made his task exceedingly difficult and resulted in Singapore leaving the Malaysian Federation in 1965. The British army was engaged in military operations to oppose the invasion of North Borneo by the Indonesians. Head played an important part in smoothing relations between the British expeditionary force and the Malaysian government which was technically sovereign in the former British North Borneo.

He was all his life a keen and expert ornithologist and entomologist, and he loved butterflies. When high commissioner in Lagos he would rise early and set forth with a butterfly net in pursuit of rare varieties. His charm never failed, and he had a way of expressing himself which delighted his friends and amused the staff of the offices he held, all of whom were consistently devoted to him. Sometimes outrageous, especially in his youth, he was endowed with an original and decisive mind, and was never anything but honest in speech and action, patriotic, generous, and considerate. He adamantly refused to have his portrait painted, though he himself had skill as an artist.

In July 1935 he married Lady Dorothea Louise, daughter of Anthony Ashley-Cooper, ninth Earl of Shaftesbury. They had two sons and two daughters, one of whom died in infancy. A portrait painter of outstanding ability and, like her husband, witty and entertaining, Lady Head was a constant help to her husband throughout his varied career, and in both Nigeria and Malaysia was a tower of strength to the Red Cross and the Order of St John of Jerusalem. On retirement Head and his wife settled at Throope Manor near Salisbury and he devoted his still abounding energy to helping others. A trustee of the Thompson Foundation from 1967, he was appointed colonel commandant (1968–76) of the SAS regiment in 1968, and he became chairman and subsequently president of the Royal National Institute for the Blind and also chairman of the Wessex region of the National Trust. He died at Throope Manor 29 March 1983 and was succeeded in the viscountcy by his elder son, Richard Antony (born 1937).

[Private information; personal knowledge.]

JOHN COLVILLE

HEATHCOAT AMORY, DERICK, fourth baronet, and first VISCOUNT AMORY (1899–1981), industrialist and statesman. [See AMORY.]

HEFFER, REUBEN GEORGE (1908–1985), bookseller, was born 29 January 1908 in Cambridge, the younger son and third of three children of Ernest William Heffer, bookseller, of Cambridge, and his wife, Louisa Marion Beak, and grandson of the founder, William Heffer, of W. Heffer and Sons Ltd., booksellers, publishers, stationers, and printers. He was educated at the Perse School where he acquired a lifelong interest in modern languages, especially French; that was the subject he read at Corpus Christi College, Cambridge, gaining a second class (division 1) in part ii of the modern and medieval languages tripos (1928) before completing his degree with a third in part ii of the economics tripos (1929). He trained for a period at the London School of Printing.

He then joined the family firm, with the intention that he should be involved with the printing side. But the untimely death of his elder brother, Arthur, in 1932 necessitated his transferring his attention to the bookshop in Petty Cury where he worked with his father, taking charge in 1948. He became chairman of the firm in 1959 and continued thus until 1975, remaining a director after his retirement. A strong supporter of his trade, he was a member of the council of the Booksellers Association, of the 1948 Book Trade Committee, of the Society of Bookmen, and of the Sette of Odde Volumes. Under his direction the family firm expanded to become international booksellers of high repute and Heffer printed, published, and sold a wide range of books, particularly in medicine, oriental studies, and phonetics. Not by nature a committee man, his shyness and his distrust of his natural speaking voice dictated his eschewing public office; he was one of the shrewdest and most balanced of booksellers never to be president of the Booksellers Association.

In his youth a keen Morris dancer and a lifelong Francophile, his favourite occupation was people, an interest he could indulge fully in his shop, particularly with eccentric dons or itinerant visitors. After the war he found a haven on the north Norfolk coast where he could talk and entertain simply, and where he became a keen and accomplished sailor. He kept his connections with his school and his college, presenting and maintaining there the splendid gown for schoolmaster fellow-commoners. He was also responsible in many ways for the continued existence of the *Cambridge Review*.

A man of considerable charm, he was unfailingly generous of his time and quiet advice. He was not at home with detail or administration or planning, but he was always meticulous in valuing and thanking those of his colleagues who undertook those boring chores on his behalf; he was a good master. A convivial man, with a wry linguist's sense of fun, he enjoyed good company as much as he contributed to it. He and his wife made many friends among booksellers and publishers. When his wife died in 1981 he was bereft of a lively and positive companion, and sadly he remained in many ways lost without her vivacity. Honest and caring, he was above all a liberal man. Holding firm views, he never inflicted them on anyone. His great talent and his abiding pleasure was to encourage those younger than himself to succeed.

In 1942 Heffer joined the RAF, serving in flight control, Fighter Command, and then as a squadron leader at the Air Ministry, being demobilized in 1945. Outside his career in the book trade he served as a JP for twenty-seven years, and with the Marriage Guidance Council, the Trustee Savings Bank, and the Cambridge Preservation Society. He helped to found the Bell School of Languages. All these activities were, typically for him, centred in Cambridge, for he had a strong sense of family and community. In recognition of his support for the Open University, he was awarded an honorary MA by that university in 1979.

In 1935 he married Nesta May, daughter of Owen Thomas Jones [q.v.], Woodwardian professor of geology at Cambridge. They had two sons, Nicholas and William. Nicholas became chairman of Heffers in 1984, third in direct line from the founder, William. Heffer died 17 July 1985 at his home in Cambridge. He suffered a heart attack while swimming, as he loved to do, in the lake at the bottom of his garden.

[Private information; personal knowledge.]
JOHN WELCH

HEITLER, WALTER HEINRICH (1904-1981), physicist, was born 2 January 1904 in Karlsruhe, the second of three children and elder son of Adolf Heitler, a professor of engineering in the Technical High School of that city, and his wife, Ortilie Rudolf. His father described his family as being of Jewish peasant stock. His interest in science awoke early, between the ages of ten and eleven, without particular stimulus from home or school. His school education was mainly in classical languages, and after leaving he was uncertain what to do next. He spent three semesters at the Technische Hochschule in Karlsruhe attending lectures in chemistry and mathematics. Then a friend told him that he ought to be a theoretical physicist and to work with A. J. W. Sommerfeld in Munich.

After rather unsatisfactory sojourns in Berlin, Munich, and Copenhagen (but not with Niels Bohr), feeling strongly attracted by the new wave mechanics, he prevailed on its originator, Erwin Schrödinger, to let him join him in Zurich. Even from Schrödinger there was little help to be had apart from excursions in the country and plenty of wine, but he managed to study his very recent papers and to understand them. Meanwhile he gained his doctor's degree from Munich University in 1926. While in Zurich he met a contemporary, Fritz London, and between them they applied wave mechanics to the hydrogen molecule and published their famous paper about it in 1927 when Heitler was only twenty-three, a paper which made known the Heitler–London theory. Shortly after its publication he met Max Born, professor of theoretical physics at Göttingen and another of the originators of wave mechanics, who offered him an assistantship there in 1929. Here he worked happily till 1933, when, because of his Jewish ancestry, he had to leave, and a modest position was created for him at the University of Bristol. Here he changed his subject, working on the theory of cosmic radiation and collaborating with Cecil Powell [q.v.], who later gained a Nobel prize, and with other British and refugee scientists.

In 1940, however, along with other 'enemy aliens' who had not yet received British nationality, he was interned in the Isle of Man and elsewhere and released three months later. In 1941 he was offered the position of assistant professor at the Dublin Institute of Advanced Studies, of which Schrödinger was now the director; he succeeded him as director in 1946. He was impressive for the clarity of his lectures and his courtesy and helpfulness towards students. He received Irish citizenship in 1946. In 1949 he was appointed Ordinarius for theoretical physics and director of the institute for that subject in the University of Zurich—a chair held previously by many men of the very highest distinction, including Albert Einstein and Schrödinger. He continued to work on the more abstract aspects of atomic physics. But from 1960 onwards, perhaps feeling that leadership in his subject had passed elsewhere, he turned to philosophy and religion, publishing books such as *Man and Science* (1963), writing newspaper articles, and giving lectures. He sought to close the gulf between religion and science. His expositions had great success with the general public; some wrote to him that he had given the strongest proof of the existence of God that they had seen. In his last years, he joined the Swiss reformed church.

Heitler, a man of small stature, had great charm and a natural courtesy. He was glad in 1949 to return to a German-speaking country for he acclimatized himself to England and perhaps even Ireland less easily than many other immigrants of that period. He received many honours: he was elected into the Royal Irish Academy in 1943, became FRS in 1948, and was an honorary doctor of the universities of Dublin, Göttingen, and Uppsala. From Germany he received the Max Planck medal in 1968 and the gold medal of the Humboldt Gesellschaft in 1979.

He married in Dublin in 1942 Kathleen Winifred, a research worker in biological science at Bristol, daughter of John Winifred Nicholson, merchant, and sister of John Henry Nicholson, later vice-chancellor of Hull University. They had one son. Heitler died in Zurich after a long illness 15 November 1981.

[Sir Nevill Mott in *Biographical Memoirs of Fellows of the Royal Society*, vol. xxviii, 1982; information from Mrs Kathleen Heitler; personal knowledge.]   NEVILL MOTT

HERMES, GERTRUDE ANNA BERTHA (1901–1983), engraver and sculptor, was born 18 August 1901 in Bickley, Kent, the second of five children, four girls and a boy, of German-born parents, Louis Auguste Hermes, of Krefeld, and his wife, Helene Augusta Gerdes, of Altena. Her father, domiciled in England since the age of three, had been artistic director of a firm manufacturing silk wear with hand-blocked patterns. Her mother had studied at an art school in Munich.

Gertrude Hermes was educated at Belmont School for Girls in Bickley. She excelled at sports and played hockey for Kent and Eastern Counties. On leaving school she spent a year on a farm in Essex but her interest in art asserted itself and in 1919 she became a pupil at the Beckenham School of Art and from 1922 to 1925 studied at the Leon Underwood School of Painting and Sculpture in Hammersmith. She always emphasized her debt to G. C. L. Underwood [q.v.] both as teacher and friend. Her talent was for sculpture rather than painting, and for engraving, a subject not on the school syllabus. With other students she experimented with wood engraving and within a few years was producing prints of exceptional originality and technical mastery. In 1925 she qualified as a finalist in the prix de Rome (engraving).

In 1926 she married a fellow student and wood engraver, Blair Rowlands Hughes-Stanton, son of Sir Herbert Edwin Pelham Hughes-Stanton, president of the Royal Society of Painters in Water-Colours (1920–36), and his wife, Elizabeth Cobden Rowlands. Gertrude and Blair Hughes-Stanton had two children, a girl and a boy. In 1928 they were living in Hammersmith when the Thames flooded their home, destroying wood-blocks and many drawings. They moved to Hacheston, Suffolk, in the same year and in 1931 to central Wales where they were employed to engrave for the Gregynog Press. For a time each appears to have influenced the work of the other. She illustrated books for private presses and later for the new and popular Penguin Books which brought her wood engravings before a wide public for the first time.

In 1932 her marriage was dissolved and she returned with her young children to live in London. She became a member of the Society of Wood Engravers in 1933 when she exhibited at the Royal Academy for the first time. The following year she was elected to the London Group. In 1938 her name was entered in the register of Industrial Art Designers and in 1939 she was chosen as one of the wood engravers to represent Great Britain at the Venice International Exhibition.

But she had not ceased to carve and model, working in wood, stone, and metal to produce a variety of sculptures, some large scale, as well as architectural and decorative pieces of which her mosaic pool floor and stone fountain (1933) for the Shakespeare Memorial Theatre at Stratford is the best known. Her greatest contribution to the art of sculpture, her portraits in bronze, gathered momentum in the later 1930s when she concentrated particularly on portraying her own children and those of friends. These activities were interrupted, though not completely stemmed, by the war.

From 1940 for the next five years she was in New York and Montreal with her children. In Canada she undertook to make working drawings in tank factories and shipyards. She recalled that she became 'heartily sick of black and white' and turned after the war to broader, bolder designs and to the use of cool colours in her prints.

Returning to London in 1945 she devoted herself mainly to print-making and to sculpture, portrait sculpture in particular. For her prints she chose to work mainly on lino, occasionally on wood plank, to gain a larger format. She was a pioneer of the large linocut and woodcut in which she frequently used colour. Through the Giles Bequest her linocuts passed into the collections of the national museums. At the same time she was producing bronze portrait heads, sometimes half-lengths, of contemporary artists, architects, poets, and musicians, work of acknowledged distinction. She became a fellow of the Royal Society of Painters, Etchers, and Engravers in 1951, was awarded the Jean Masson Davidson prize for portrait sculpture in 1967, was elected associate of the Royal Academy in 1963 and academician in 1971.

She was also in demand as a teacher in the

London art schools and taught at Camberwell, St Martin's, the Central School, and finally the Royal Academy Schools. Former pupils gratefully remember how this unpretentious artist infected them with her passion and pleasure in drawing and engraving.

In 1981 to honour her eightieth birthday the Royal Academy held a major retrospective exhibition of her work and in the following year she was appointed OBE for 'services to art'. As a person she was open, generous, and mild-mannered which sometimes belied her sense of principle. Though 'not a feminist', as she said, she was the first to insist that women members of the Royal Academy be invited to attend the annual and presidential dinners on equal terms with men. The rules were in consequence changed to allow this in 1967.

Her work is to be found in most major collections in Europe and North America. Her art is highly concentrated and her drawings are models of economy, the line sinewy and vital. The sculptures and prints made from them simplify the theme but emphasize its very essence with a tendency to abstraction. It is an art full of surprises and these together with its strong individuality make it difficult to fit it neatly into any school or tradition. She died in Bristol 9 May 1983.

[*Royal Academy Exhibitors*, vol. iv (1905 70), 1979, p. 40; *Gertrude Hermes*, catalogue of an exhibition at the Whitechapel Gallery with notes on the artist by Bryan Robertson and Naomi Mitchison, 1967; *Gertrude Hermes R.A.*, catalogue of an exhibition at the Royal Academy of Arts, introduction by David Brown, 1981; *The Times*, 11 May 1983; information from Mrs Judy Russell, daughter of the artist; personal knowledge.]

PAUL HULTON

HIBBERD, (ANDREW) STUART (1893–1983), BBC announcer, was born at Village Canford, Dorset, 5 September 1893, the youngest child of William Henry Hibberd, farmer and corn dealer of Village Canford, and his second wife, Mary Catherine Edney. He had one brother and seven half-brothers. From Weymouth College he won a science exhibition to St John's College, Cambridge. He had a fine tenor voice and became one of four choral scholars. He enlisted in 1914, after only a year and a half at Cambridge, and was awarded a wartime degree. He served in the Dorset Regiment at Gallipoli, and later in Mesopotamia, then with the Punjabi Regiment of the Indian Army until it was disbanded in 1922.

In November 1924 Hibberd joined the two-year-old British Broadcasting Company Ltd. at Savoy Hill as an announcer. His cultivated voice and clear enunciation typified what came to be known between the wars as the BBC accent. The words he spoke were usually written by others, and announcers then were anonymous. Nevertheless he soon acquired a special position in the affection of listeners, and was generally known as the BBC's chief announcer, in recognition of his seniority and the respect he enjoyed, although during his twenty-six years of the same work he was never officially appointed to such a post.

Hibberd's diary recorded the eminent speakers and musicians invited to broadcast in the early days, and the national events in which he was involved. It formed the basis of his book '*This—is London . . .*' (1950) which gives a vivid picture of the BBC's first quarter-century. He noted that on 4 January 1926 the evening announcer first wore a dinner jacket as a courtesy to the people broadcasting, many of whom were similarly clad. Hibberd himself found evening dress unsatisfactory for reading the news; he wanted nothing tight round his neck, and the microphone sometimes transmitted the creak of his starched shirt-front.

After he had closed down the Savoy Hill studio at 10.30 p.m. Hibberd would walk up to the Savoy Hotel to announce, from backstage, the dance-band tunes played at the end of the broadcasting day. Microphones suspended from the ceiling, well above the band of Debroy Somers or Carroll Gibbons, would pick up the music and the swish of dancing feet, giving listeners at home a vicarious evening out. Deliberately there was no visible stand microphone which might tempt dancing couples to send greetings to friends, or—Heaven forfend—to put over an advertisement. Meanwhile at nearby Charing Cross station friendly railway staff would sometimes delay the departure of a train for Hibberd's Kent home when the much loved familiar figure was seen racing to the platform, his white scarf trailing.

In May 1926 the general strike began, and with it the first broadcast of news before 6 p.m. An irksome restriction, imposed by the news agencies in the interest of the newspaper proprietors, permitted the BBC access to agency news only on condition that it neither gathered news itself nor transmitted it before the evening papers were on sale. During the strike, when no newspapers appeared save the *British Worker* and the *British Gazette*, edited by (Sir) Winston Churchill, this restriction was temporarily lifted; it took World War II to remove it altogether. On each day of the strike the BBC broadcast five news bulletins, mostly read by Hibberd. Their calm tone, markedly different from the *British Gazette*, helped to steady a divided nation. Broadcasting became responsibly established, and the way was paved for the private company to be transformed, at the end of that year, into a public corporation operating under royal charter, still with the initials BBC.

Apart from news-reading, Hibberd's main work in the 1920s was to present musical programmes. He was a kindly, courteous man whose best-remembered announcement was his rendering of the bulletin issued by the doctors of King George V on the evening of 20 January 1936: 'The King's life is moving peacefully towards its close.' With all other programmes cancelled, Hibberd poignantly repeated the words each quarter-hour, following the chimes of Big Ben, until Sir John (later Lord) Reith [q.v.] announced the death of the monarch at midnight.

During World War II it was clear that Hibberd had passed his peak. A critic in the *Listener* noted in 1943 'He frequently misses his footing as he delivers a bulletin.' However he continued to retain the affection of many listeners, especially when he began to present, in 1949, a Thursday afternoon programme, *The Silver Lining*, which brought comfort to people handicapped or housebound, and reflected his own strong Christian faith. Hibberd retired from the BBC to Budleigh Salterton in 1951, suffering from fatigue, but he continued to introduce *The Silver Lining* each week from a small self-operated studio in Exeter until the programme ended in 1964. He also gave much time to recording talking books for the blind.

Hibberd was appointed MBE in 1935. He married in 1923 Alice Mary (died 1977), daughter of Lieutenant-Colonel Gerard Chichester, of the North Staffordshire Regiment. They had no children. Hibberd died in a nursing home at Budleigh Salterton, 1 November 1983.

[*The Times*, 2 and 7 November 1983; *Daily Telegraph*, 2 November 1983; *Listener*, 29 July 1943; Stuart Hibberd, *This—is London . . .*, 1950; radio broadcast from BBC Plymouth, *Stuart Hibberd talking to Guy Slatter*, 13 November 1972; private information; personal knowledge.]

LEONARD MIALL

HILL, SIR (JOHN) DENIS (NELSON) (1913-1982), psychiatrist, was born 5 October 1913 in the village of Orleton, Herefordshire, the only child of Lieutenant-Colonel John Arthur Hill, farmer, and his wife, Doris Nelson. He was educated at Shrewsbury School and studied medicine at St Thomas's Hospital, where he became interested in psychiatry. He won prizes in the subject, and as a twenty-first birthday gift chose the collected works of Sigmund Freud.

Soon after qualifying MB, BS (Lond. 1936),

Hill went to Maida Vale Hospital to study neurology with W. Russell (later Lord) Brain [q.v.]. There he met Grey Walter who was beginning to use the new technique of electroencephalography (EEG) to investigate neurological disorders. Hill was impressed by Walter's early successes in locating cerebral tumours; and fascinated by the technical aspects of the EEG because, as a schoolboy, he had been an amateur radio enthusiast and had studied electronics.

In 1938 Hill returned to St Thomas's as an assistant in the department of psychiatry, but this work was soon interrupted by the outbreak of war. Because of severe asthma, Hill was unfit for military service and was sent instead to the Emergency Hospital at Belmont in Surrey where he joined a talented group of psychiatrists, including Eliot Slater and William Sargant, who were to be influential in the development of psychiatry after the war. An EEG machine was obtained from America, and Hill set up a laboratory in which he worked with such energy and enthusiasm that by the end of the war more than 20,000 recordings had been made. When the war ended Hill was the obvious person to set up an EEG laboratory as part of the new Institute of Psychiatry at the Maudsley Hospital (where he was a senior lecturer from 1948 to 1960), and carry out similar work at the National Hospital for Nervous Diseases in London. Before long he was an acknowledged expert in his field and he edited, with Geoffrey Parr, the first comprehensive textbook of clinical EEG, published in 1950. He became FRCP in 1949 (MRCP, 1940, and DPM, 1940) and FRCPsych. in 1971.

Hill was no narrow specialist. He used the opportunities provided by his special expertise in EEG to widen his clinical interests, and his influence grew correspondingly. A study of an important case of murder, in which it was shown that the accused was likely to have been in a state of altered consciousness at the time of the act, led to an invitation to carry out EEGs on all cases of murder awaiting trial in the home counties. This and other experience with forensic problems led to invitations to serve on several important government committees: in 1959 the working party on special hospitals; in 1962 the working party on the organization of prison medical services; and in 1972 the Aarvold committee, and the Butler committee on mentally abnormal offenders (1972-5).

Hill's wide experience and balanced judgement led to many other invitations to take part in important committees. In 1956-60 he was a member of the Medical Research Council; in 1961-7 of the central health services council of the Ministry of Health; and from 1961 until his death of the General Medical Council. In the GMC he held the important offices of treasurer, and chairman of the penal cases committee and the special committee on mental health.

Despite these demanding administrative duties, Hill remained above all an outstanding clinician and teacher. First as lecturer in psychological medicine at King's College Hospital (1947-60) he worked to give psychiatry an established place in the medical curriculum. Moreover, because of his achievements, he was influential in bringing about the foundation of chairs of psychiatry in other medical schools. In 1966 he was appointed professor of psychiatry at the Institute of Psychiatry, a postgraduate teaching school of the University of London, a post he held until 1979. In this new role, Hill soon impressed his personality on the workings of the Institute and the associated Bethlem Royal and Maudsley Hospitals. Under his leadership, the department of psychiatry grew in strength and numbers, with endowments for new chairs and a growing and wide-ranging programme of research. In all these activities he was, above all, a bridge builder—between neurology and psychiatry, between psychoanalysis and biological psychiatry, and between medicine and philosophy.

Hill's achievements led to many honours. He was knighted in 1966. He gave the most prestigious named lectures in his subject: the Maudsley lecture of the Royal College of Psychiatrists in 1972, the Adolf Meyer lecture of the American Psychiatric Association in 1968, the Bickers lecture of the Mental Health Research Fund in 1972, the Ernest Jones lecture of the Institute of Psychoanalysis in 1970, and in 1969 the Rock Carling lecture.

Hill was married twice: in 1938 to Phoebe Elizabeth Herschel, daughter of Lieutenant-Colonel H. H. Wade, and, after this marriage was dissolved in 1960, in 1962 to Lorna, daughter of John Fleming Wheelan, tax inspector. His second wife was a consultant child psychiatrist. There were a son and a daughter of each marriage. Hill was devoted to his family, and generous and loyal to his friends. In his leisure, he retained an abiding interest in the countryside and especially in the Herefordshire village in which he had grown up. In his later years he spent much time and money on the restoration of Orleton Manor, the half-timbered manor house in which his family had lived for generations; and he enjoyed weekends there spent tending the garden, managing the farmland, and showing the surrounding countryside to his friends. He died suddenly in London 5 May 1982.

[A. H. Crisp in *Munk's Roll*, vol. vii, 1984; personal knowledge.]   M. G. GELDER

HILLIER, SIR HAROLD GEORGE

KNIGHT (1905-1985), horticulturalist, was born in Winchester 2 January 1905, the second son in a family of three sons and two daughters of Edwin Lawrence Hillier, nurseryman and head of the family firm Hillier Nurseries, and his wife, Ethel Marian Gifford. He was educated at Peter Symonds School in Winchester and King Edward VI Grammar School in Southampton.

In 1921 he entered the family firm Hillier Nurseries. He spent his formative years helping his father rebuild the firm's collection of plants and its stocks which had been sadly depleted by World War I. Accompanying his father he visited estates and gardens throughout the country collecting plants. New plant material also flowed in from abroad. This was the foundation of his deep appreciation and love of plants and he initiated correspondence with horticulturalists and estate owners which he carried on for the rest of his life. He was made a partner in 1930 and became head of the firm in 1944 on the death of his father. Then began a period of expansion, with a large increase in the staff and nursery land. He assembled a vast collection of trees and shrubs from the northern temperate region, larger than that of any other nursery.

In the 1950s he made his first visit to the United States. This was followed by visits to many countries in Europe, Asia, Australasia, and the Americas where he gathered seeds and new plants for his collection. Hillier always said he put plants before money and he distributed a great many rare and endangered species. His dedication to plants was equalled by his generosity. He gave a collection of tender plants to Ventnor Botanic Gardens in the Isle of Wight where they could grow more satisfactorily than in Hampshire and in recognition of this he was the first recipient of the Hillier trophy presented by the South Wight borough council. He donated many plants to the Royal Horticultural Society for its collection at its Wisley Gardens. Westonbirt Arboretum was the recipient of a collection of ornamental cherries (Prunus species).

Hillier's greatest contribution to the field of conservation was the creation as a national and international asset of the Hillier Arboretum at Ampfield near Romsey. To secure the future of this unique collection of many thousands of different species and varieties of plants he presented the arboretum as a gift to Hampshire county council in 1977. Unfortunately he rarely put pen to paper. Apart from his correspondence he left little record of his work in the family firm or the people he met, although he had a story for every plant. His main achievement as an author was the preparation and publication of *Hillier's Manual of Trees and Shrubs* which was first published in 1972.

He showed a dedicated commitment to the

Royal Horticultural Society, with which he had a close relationship and of which he was made an honorary fellow in 1972. He served with distinction on its council for twenty-five years and was elected a vice-president in 1974. He continued to attend the society's shows and to contribute to its committees. He was awarded the Victoria medal of honour in 1957 and the Veitch memorial medal in 1962. The family firm received a gold medal at every Chelsea Flower Show since 1922. Hillier also served for many years on the Westonbirt Arboretum advisory committee.

In 1954 he was elected a fellow of the Linnean Society of London and in 1965 he was awarded the Thomas Roland medal of the Massachusetts Horticultural Society. In 1976 he was made an honorary member of the Garden Society and he was also made an honorary fellow of the Japanese Horticultural Society. In 1971 he was appointed CBE and he was knighted in 1983 for services to horticulture.

He was a dedicated Christian being a deacon (later a life deacon) of the Congregational Church to which he was most beneficent. When this Church became part of the United Reformed Church he was appointed an elder. He was a man of infinite courtesy and modesty, with a delightful sense of humour and complete integrity. He was warm and sincere in his affections.

In 1934 he married Barbara Mary, daughter of Arthur Phillip Trant, flour miller. They had two sons, who succeeded their father in the family firm, and two daughters. It was a close and united family. Hillier died 8 January 1985 at Romsey in Hampshire.

[Personal knowledge.]   J. G. S. HARRIS

HILLIER, TRISTRAM PAUL (1905-1983), painter, was born in Beijing 11 April 1905, the youngest in the family of two daughters and two sons of Edward Guy Hillier, banker and diplomat, and his wife, Ada Everett. His brother was killed in World War I. Hillier came to England at an early age but, after schooling at Downside, returned to China in 1922-3 to study its language before continuing his education at Christ's College, Cambridge. In due course his father's death released him from any obligation to study there.

He began his training as an artist, first of all in London, at the Slade School, in 1926, where he studied under Henry Tonks [q.v.], and then for two further years as a pupil of André Lhôte in Paris. Here he came to know most of the Surrealist painters, and it is this strand of influence which thereafter substantially conditions the nature of his art. He was particularly influenced by Max Ernst and the early de Chirico, although his brand of Surrealism had none of the histrionics of Salvador Dali, or the ferocity of André Masson. He brought to the exposition of his personal Surrealist vision a quality of pure, dreamlike serenity which was wholly English in spirit. Very much a part of the British Surrealist avant-garde in the 1930s, Hillier was a member of the Unit One Group led by Paul Nash [q.v.]. To all ends and purposes he always remained a Surrealist, a fact clearly stressed in the exhibition, 'A Timeless Journey', presented at the Bradford Art Gallery immediately following the artist's death in 1983. As Nicholas Usherwood wrote in the introduction to the exhibition catalogue: 'The gradual drift of his style in the late 1930s towards an apparently more traditional figurative manner, based on Flemish fifteenth-century art, was taken by its protagonists as "a fall from grace". It was seen as a sell-out to academic conservatism for which he has to this day not been fully forgiven.'

His stature as a painter rests on the magical way in which he invests inanimate objects, and his subjects in general—harnesses, still lifes, boats, harbours, and especially anchors—with an ambiguous, dream-like presence, creating a unique world of images, symbols, and metaphors peculiar to himself.

In 1945 after a spell of living in France and Spain he settled permanently in Somerset, near Castle Carey. He was not by nature a countryman, although he enjoyed riding. The depths of the country provided him with the peace and isolation that were so important to the realization of his art. During World War II he served with the Royal Navy, and with the Free French naval forces in the rank of lieutenant RNVR. He was elected ARA in 1957 and RA in 1967. He exhibited mainly at Arthur Tooth & Sons, but also at the Lefevre Gallery. He is represented in the Tate Gallery, and in many private and public collections.

As a man he was handsome, elegant, and something of a dandy. He was fastidious to a fault. His family tell how he insisted on the immaculate, virtually mathematical placing of the cutlery on the dining-table, this insistence growing more intense and demanding if his mood blackened, as it frequently did. This sense of precision relates perfectly to the character of his aesthetic vision and technical style. As a father he was strict and aloof, forbidding his daughters to enter his studio, except by his express invitation. Born into the Catholic faith, he lapsed, returning to the Roman Church in 1945. After the conversion of the Tridentine mass from Latin into English and other vernaculars in 1964, he reacted passionately to the change, and in later years increasingly set himself against the Vatican and the papacy.

Hillier was twice married, first in 1931 to Irene Rose, daughter of Horace Hodgkins, an off-course bookmaker. There were twin sons, Jonathan and Benjamin. The marriage was dissolved in 1935. In 1937 he married Leda Millicent, daughter of Sydney Hardcastle, captain in the Royal Navy, who invented the Hardcastle torpedo used in World War I. There were two daughters, Mary and Anna-Clare. Hillier died in hospital at Bristol 18 January 1983.

[*A Timeless Journey*, catalogue of a retrospective exhibition, with an introduction by Nicholas Usherwood, 1983; Tristram Hillier, *Leda and the Goose* (autobiography), 1954; private information.]   MERVYN LEVY

HINTON, CHRISTOPHER, BARON HINTON OF BANKSIDE. (1901-1983), engineer, was born 12 May 1901 at Tisbury, Wiltshire, the third of the four children (the eldest of whom died at the age of two) and elder son of Frederick Henry Hinton, the village schoolmaster, and his wife, Kate, formerly a children's nurse and daughter of Samuel Charles Christopher, of Ware, Hertfordshire. All four grandparents had been teachers. Hinton attended his father's next, larger, school in Chippenham before entering Chippenham Secondary (later Grammar) School; he performed precociously at elementary school but not at secondary school until his last year.

In 1917 he became a premium engineering apprentice with, first, a small firm and then the Great Western Railway at Swindon. His foreman's tribute—'you're the best craft apprentice I've ever had'—gave him as much pleasure as his later first class degree. After evening study at Swindon Technical College—on top of a 54- (later 47-) hour week—he won in 1923 an Institution of Mechanical Engineers scholarship at Trinity College, Cambridge, gained a first class degree in the mechanical sciences tripos (1925) in two years, and spent his third year in research, winning university and college awards. Trinity made him an honorary fellow in 1957.

With this perfect blend of practical and theoretical training he joined Brunner Mond, soon to become the Alkali Group of the new Imperial Chemical Industries (ICI), and at the age of twenty-nine became chief engineer. Engineers were then second-class citizens but he achieved equal status with management and research, and great authority. Here he met and married in 1931 Lillian (died 1973), head of the tracing office and daughter of a powerhouse operator, Thomas Boyer, of Winnington, Northwich, Cheshire. They had one daughter who married the son of Sir Charles Mole, director-general of the Ministry of Works.

In 1940 Hinton became director of ordnance factory construction at the Ministry of Supply and in 1941 of the explosive filling factories where he replaced chaos, and fear of an ammunition shortage scandal, with great efficiency; in 1942 he became their deputy director-general. Exhaustion, enhanced by sleeping in Ministry air-raid shelters, led him to the verge of breakdown.

When the war ended he returned briefly to ICI but thereafter worked exclusively in the public sector. The government had decided to establish a native atomic project and ICI, refusing to be a main contractor for the factories, urged that they should be a government undertaking under Hinton. He accepted and arrived with six of his former colleagues at his Risley (Lancashire) headquarters in February 1946.

From this nucleus grew the Industrial Group of the British atomic project, collaborating with Harwell's Research Establishment and the Establishment for Weapons Research (later at Aldermaston). The government regarded production of fissile material as supremely urgent for atomic bombs and industrial power. This meant four very different types of plant: nuclear reactors—two experimental reactors at Harwell and two plutonium producers at Windscale in Cumberland; a plant near Preston to produce fuel rods from uranium ore; a chemical plant to separate plutonium, and associated plants at Windscale; and a gaseous diffusion plant at Capenhurst, Cheshire, to enrich uranium. United States law permitted no transfer of information to Britain. Hinton relied on teamwork but played a crucial part in all phases and parts of the enterprise. All plants were built to programmed times and cost and fulfilled their task, although in 1957 the two Windscale reactors were closed after a fire.

After the first British bomb test in 1952, it was decided to meet increased demands for plutonium from reactors that would also produce power. In 1956 Calder Hall's nuclear reactors were the first in the world to feed power into a national grid. Hinton's organization built them faultlessly to time and cost and they had an excellent operating record. Even before Calder 'went critical' Britain had announced a modest civil nuclear power programme based upon its reactors but with an open mind about future types to be built by consortia of private firms. Hinton's staff were also designing and building an experimental fast breeder reactor at Dounreay.

In 1954 atomic energy had been transferred from the Ministry of Supply to the quasi-independent Atomic Energy Authority with Hinton as member for engineering and production. In 1956, during his absence (on a triumphant visit to Japan which ordered a British reactor) a new greatly enlarged nuclear power programme was produced but Hinton feared the effects of this general nuclear euphoria. Nervous strain affected relations with some of his staff.

In 1956 the government appointed Hinton chairman of the new Central Electricity Generating Board, established to supply electricity in bulk to the retailing area boards. He regretted that as chairman (rather than board member for engineering) he could not build the strong engineering design and construction department necessary to ensure prompt commissioning and high plant availability. Transmission engineering was however excellent and Hinton developed both CEGB's research and its concern for the environment. He questioned the size of the enlarged nuclear power programme and insisted on basing atomic, like other, judgements on engineering and economic criteria rather than prestige.

Knighted in 1951, appointed KBE in 1957, he retired with a life peerage in 1965. He called himself the 'Odd Job Man'—chairing the world energy conference, advising the World Bank, serving vigorously as president of the Institution of Mechanical Engineers, the Council of Engineering Institutions, and the new Fellowship of Engineering. He was active in the House of Lords and as the first chancellor of Bath University. An excellent lecturer, he also wrote five short lucid books and contributed the notice of Sir Claude Gibb to this Dictionary. He especially enjoyed his deputy chairmanship, from retirement to his death, of the generating industry's research council. A fall shortly after their party for his eighty-second birthday led to his death in a London hospital 22 June 1983.

Hinton's deeply probing mind encompassed country churches, history, and Jane Austen as well as engineering. He was proud, uncompromising in his standards, incapable of dissimulation, and neither gave, nor expected, soft answers. With his commanding presence, intellect, creativity, and managerial skills he was considered by his colleagues at the time and in retrospect to be one of Britain's relatively few truly great engineers. In 1976 he was admitted to the Order of Merit, crowning his many fellowships (FRS in 1954) and honorary degrees, which included an Oxford D.Sc. (1957) and a Cambridge Sc.D. (1960). He also received medals from five British societies and honours from Austria, Japan, Sweden, and the United States.

[Margaret Gowing, *Independence and Deterrence, Britain and Atomic Energy, 1945–1952*, vol. i Policy Making, vol. ii Policy Execution, 1974, and article on Hinton in *Dictionary of Business Biography*, vol. iii, 1985; Leslie Hannah, *Engineers, Managers and Politicians*, 1982; papers in the archives of the Institution of Mechanical Engineers; personal knowledge.]   MARGARET GOWING

HODSON, (FRANCIS LORD) CHARLTON, BARON HODSON (1895–1984), lord of appeal in ordinary, was born 17 September 1895 at Charlton Kings' vicarage, Cheltenham, the seventh son and seventh of eight children of the Revd Thomas Hodson, rector of Oddington in Gloucestershire, and his wife, Catherine Anne, daughter of Thomas Maskew, headmaster of Dorchester Grammar School. He was educated at Cheltenham College, where he won a classical scholarship to Wadham, his father's college at Oxford. Always known as 'Charles', he was a keen oarsman and stroked the college boat in 1913.

The outbreak of World War I (in which his sixth brother was killed in 1915) interrupted his studies and settled his immediate future. He joined the 7th battalion of the Gloucestershire Regiment and fought in Gallipoli and Mesopotamia where he was wounded several times. In the fighting for the relief of Kut he won the MC. He was demobilized in 1919 with the rank of captain. He decided to read for the bar, but beforehand took the shortened course in jurisprudence at Oxford (1920). By then he was a married man. In 1918 he had married Susan Mary (died 1965), daughter of Major William Greaves Blake, DL, of Eccleshall, Sheffield. They had a daughter and two sons, the elder of whom was killed in 1943 during World War II. The daughter, Anthea Hastings [q.v.], died in 1981.

Reading for the bar was shortened for those who had served in the forces and Hodson was called by the Inner Temple in 1921. Earning a livelihood could not be so swift. He began to practise in common law chambers and would have stayed on there if he could have afforded to wait five years and upwards for an income. As a married man he could not. Progress was much faster at the divorce bar. The consequence of an easy start is proverbially a dead end. So it was thought to be in divorce, a jurisdiction which in the great amalgamation of 1873 had been bundled into the new High Court with probate and admiralty as its companions. The few at the divorce bar who took silk earned less than busy juniors. They could not even dream of the bench since on the reasonable assumption that ignorance of adultery could be more swiftly dispelled than ignorance of navigation, the judges were promoted from the admiralty bar.

Until 1936 the basis of divorce law was that the bond could be dissolved only by adultery. In that year (Sir) A. P. Herbert [q.v.] persuaded Parliament to add cruelty and desertion as grounds for divorce. It was rightly anticipated that the number of petitions would greatly increase; indeed, ten years later they had multiplied tenfold. The appointment of a judge from the divorce bar could not be evaded. None of the silks was found worthy. Hodson was by then recognized as the most capable of the juniors; in 1935 he had been made Treasury counsel in probate matters. So when Herbert's bill was about to become law the lord chancellor sent for Hodson and advised him to apply for silk. He was of course given it (1937) and a few months later he was appointed a justice of the High Court, at forty-two the youngest ever to hold that office, and duly knighted (1937). Thus was the proverb dishonoured. In 1938 he became a bencher of the Inner Temple.

In 1951 the increase in the number of lords justices of appeal from six to nine made it imperative that one should come from divorce. Hodson was obviously the man and at the same time he was admitted to the Privy Council. Nine years later seniority made him the natural choice for the succession as a lord of appeal in ordinary, with a life peerage (1960). He retired in 1971 and lived quietly at Rotherfield Greys. Though he was a reserved man, he had the gift of making friends and made many among his colleagues.

His thirty-four years of judicial service left little or no mark on the law. He took the law as he found it, whether he liked it or not. He did not like the treatment of unhappiness in married life as a reason for the dissolution of marriage. His memorandum to the royal commission on marriage and divorce (1951–5) summarized his opposition to any extension of the grounds for divorce; he returned to the theme in his presidential address to the Holdsworth Club in 1962. As a churchman he was an active member of the laity.

In appellate work he was at ease with the common law. But it was at first instance that he won his high reputation. He was a man of strong moral principle and with a clear perception of what was just and fair. These virtues flowed evenly within the banks of the law into his judgments. His judicial demeanour was near to perfection. He talked little and listened attentively. He was quiet, courteous, and firm. In his court justice proceeded *de die in diem* with the strength and serenity of a Bach mass. His quality was much admired in his one excursion outside the field of English law. This was in the British branch of the International Law Association where he presided in 1955. To this Dictionary he contributed the notices of Sir John Singleton and Lord Merriman. He became an honorary fellow of Wadham College. Hodson died 11 March 1984.

[Personal knowledge.]   DEVLIN

HOLLOWAY, STANLEY AUGUSTUS (1890–1982), actor and singer, was born 1 October 1890 in Manor Park, London, the younger child and only son of George Augustus Holloway, a law clerk, and his wife, Florence Bell. His family was fairly prosperous, and he attended the Worshipful School of Carpenters, where local engagements as a boy soprano encouraged him to contemplate a career in singing. As soon as his voice matured after breaking, he took a teenage plunge into the world of entertainment. He studied singing in Milan and then served with the Connaught Rangers during World War I.

With other stars, such as Leslie Henson [q.v.], he was a successful concert party entertainer after the end of the war. In 1921 he became a utility-man-cum-baritone for a West End success—the Co-Optimists pierrot show. The show waned in 1927; two years later it revived.

In 1927 he first introduced the monologue into his work, with the story of 'Sam "pick oop tha musket" ' Small. He delivered comic narratives such as these in a flat, unemotional Lancastrian fashion. The droll accounts, both in variety and on twelve-inch records, of Albert Ramsbottom, who was swallowed by the lion at Blackpool zoo, and other such tales, ensured Holloway a deep and lasting popularity.

Holloway also appeared in light theatre. His first West End showing was in 1919 as Captain Wentworth in *Kissing Time*, and his first pantomime was in Birmingham in 1934, where he was Abanazar in *Aladdin*. He starred on variety bills as well as in musical comedy and revue, and his personable and dominating presence was much in demand.

Next, there were his legitimate theatre successes, notably in Shakespeare, as First Gravedigger in *Hamlet* (with (Sir) Alec Guinness, 1951) and Bottom in the Old Vic's Edinburgh Festival and American coast-to-coast tour of *A Midsummer Night's Dream* (1954). In the Shaw Festival at Niagara-on-the-Lake, Canada, he appeared as Burgess in *Candida* (1970) and William in *You Never Can Tell* (1973). He also played Pooh Bah in an impressive USA television production of *The Mikado*. As late as 1977 he toured Australia and Hong Kong with Douglas Fairbanks jun.

A generation of cinema audiences also learned to appreciate his delightful comedy acting, and he appeared in over thirty films, among them *This Happy Breed* (1944), *Brief Encounter* (1945), *The Way to the Stars* (1945), *The Lavender Hill Mob* (1951), and *The Titfield Thunderbolt* (1952). On American television he appeared with his son, Julian, in the series *Our Man Higgins* (1962–3).

Ultimately, he crowned a satisfying career by

his legendary creation of Alfred Doolittle, the philosophic dustman, in *My Fair Lady*, Lerner and Loewe's musical version of *Pygmalion* by G. B. Shaw [q.v.]. In a jubilant hat-trick, he starred in the Broadway première (1956-8), the London production at Drury Lane (1958-9), and the film version of 1964. With his cockney authenticity, his splendid baritone voice, and his wealth of comedy experience, he seemed, in this one superb role, to encapsulate all the exuberance and gusto that had marked his long years of well-merited achievement. Few actors so bestrode the world of entertainment and mastered so many of its facets. Above all, Holloway's expansive personality relaxed and pleased audiences of all kinds. He evoked for them what the critic (Sir) Harold Hobson called 'a maelstrom of uncomplicated happiness'.

He was appointed OBE in 1960, and was awarded the Variety Club of Great Britain special award in 1978. He published his autobiography, *Wiv a Little Bit o' Luck* in 1967 and three anthologies of monologues in 1979, 1980, and 1981.

In 1913 he married Alice Mary-Laure ('Queenie') Foran (died 1937), daughter of John Thomas Foran, who lived on the income from inherited property. They had three daughters and a son. In 1939 he married Violet Marion Lane, actress, daughter of Alfred Lane, civil engineer. They had one son. Holloway died at Littlehampton, Sussex, 30 January 1982.

[*The Times*, 1 February 1982; Stanley Holloway, *Wiv a Little Bit o' Luck*, 1967; private information; personal knowledge.]

ERIC MIDWINTER

HOLST, IMOGEN CLARE (1907-1984), musician, was born in Richmond, Surrey, 12 April 1907, the only child of the composer Gustav Theodore Holst [q.v.] and his wife, Isobel, daughter of an artist, Augustus Ralph Harrison. From the first music was her environment and her nourishment. She was formally educated at St Paul's Girls' School where the curriculum was wide, interesting, and lively. There she developed her lifelong habit of disciplined study and took part in much music-making. Her father was the school's director of music and a major influence on her life and that of many of his other pupils.

In 1926 she entered the Royal College of Music to study piano with Kathleen Long, composition with (Sir) George Dyson and Gordon Jacob [qq.v.], and paperwork with her father's great friend Ralph Vaughan Williams [q.v.]. She also studied dancing and took part in many college stage performances. In 1927 she won an open scholarship for composition. In 1930 a scholarship for study abroad allowed her to travel widely in Europe. Although she lived frugally she was able to enjoy concerts and picture galleries and to experience the civilization of Europe between the wars.

When phlebitis in her left arm prevented her from becoming a concert pianist she turned to teaching and to arranging music for plays, for the English Folk Dance and Song Society as well as for amateur choirs and orchestras which she conducted and for whom she also composed original works. After her father's death in 1934 she wrote his biography (*Gustav Holst*) which was published in 1938.

In April 1939 she went to Switzerland to study the music of fifteenth- and sixteenth-century composers. Returning before the outbreak of war she served on the Bloomsbury House refugee committee concerned with musicians from Germany and Austria. In January 1940 she became one of the six organizers appointed by the Pilgrim Trust (taken over by the Council for the Encouragement of Music and the Arts, the forerunner of the Arts Council) to organize amateur music in wartime Britain. She had responsibility for seven counties in the west. With little help, means of transport, or money she managed to create choirs and instrumental groups among local residents, evacuees, land girls, and service men and women whom the chances of war had temporarily brought together. To them she offered days of music-making which enriched their leisure. In 1942 ill health forced her to resign, and after a recuperative year at Dartington Hall in Devon she became director of music at the arts department there. She trained her students to become music teachers and organizers, teaching them as her father had taught at St Paul's and at Morley College, always drawing in others to share the pleasures of making and of listening to music. She found time to study medieval European music and wrote a severely critical book about her father's works (*The Music of Gustav Holst*, 1951). In 1951 she spent two months at Santiniketan, the university of Sir Rabindranath Tagore [q.v.] in West Bengal, learning about the folk music of India. That summer she left Dartington to travel and study in Europe.

In 1952 she moved to Suffolk where Benjamin (later Lord) Britten [q.v.] had invited her to be his amanuensis. She also helped with the Aldeburgh Festival, of which she was an artistic director (1956-7). In 1953 she founded the Purcell Singers, a group she conducted until 1967. She wrote books for children on Henry Purcell, William Byrd [qq.v.], Bach, and Britten.

In 1964 she ceased to work for Britten in order to concentrate on Holst's work, editing scores for performance and publication and conducting and supervising recordings. In 1974 she produced an informative, definitive *Thematic Catalogue of Gustav Holst's Music* and just before her death a reconsidered edition of her 1951 book.

Though an intensely private person she enjoyed good company and friendship. She had a charming speaking voice which added to her success as a lecturer. Her books are written with the same clarity and persuasive talent that marked her work as teacher and as lecturer. Her own compositions she described as 'useful'. Though written to serve individual needs or special occasions they bear her stamp of professionalism and originality.

Imogen Holst was appointed CBE in 1975. She became a fellow of the Royal College of Music in 1966, and an honorary member of the Royal Academy of Music in 1970. She received honorary degrees from the universities of Essex (1968), Exeter (1969), and Leeds (1983). She died 9 March 1984 at Aldeburgh, where she is buried. She was unmarried.

[Personal knowledge.]
URSULA VAUGHAN WILLIAMS

HOME, CHARLES COSPATRICK DOUGLAS- (1937-1985), journalist. [See DOUGLAS-HOME.]

HOOKER, SIR STANLEY GEORGE (1907-1984), mathematician, aerodynamicist, and engineer, was born 30 September 1907 in the Isle of Sheppey, Kent, the fifth son and ninth and youngest child of William Harry Hooker, innkeeper and later corn miller, and his wife, Ellen Mary Russell, of Ruckinge, Ashford, Kent. From village schools and Borden Grammar School a Royal scholarship took him in 1926 to Imperial College, London, to read mathematics. He won the BUSK studentship in aeronautics (1928) and the Armourers and Braziers research fellowship (1930), and gained his D.Phil. at Brasenose College, Oxford, in 1935.

From Oxford Hooker went to the Admiralty Laboratories, Teddington, to Woolwich Arsenal, and then in 1938 to Rolls-Royce at Derby. After suggesting improvements which raised the efficiency of Merlin superchargers from 68 to 76 per cent, Hooker found himself in charge of supercharger development. Under his leadership progress was rapid, and by 1941 a two-speed, two-stage intercooled supercharger for the Merlin 60 engine was provided. This increased the speed of the Spitfire aircraft from 300 to 400 m.p.h. at 30,000 feet, and also increased the rate of climb from 650 to 2,300 feet per minute. This development was prominent in enabling the RAF to win the air war against the German air force.

By August 1940 the Rover Company had Ministry contracts to produce (Sir) Frank Whittle's jet engine at Barnoldswick and Clitheroe, but progress was slow. In November 1942 S. B. Wilks of Rover agreed with E. W. (later Lord) Hives [q.v.] of Rolls-Royce to exchange the work on the Whittle jet engine for the Rolls-Royce tank work at Nottingham, and in January 1943 a Rolls-Royce team, with Hooker as chief engineer and Leslie Buckler as works manager, took over the Rover workshops. Buckler quickly overcame turbine blade shortages which had delayed progress and Hives turned production resources over to development, thus increasing the effort tenfold, with the result that a squadron of RAF Meteors was delivered in July 1944. The B37 turbo-jet, the Nene, and the Derwent V, all centrifugals, were rapidly developed at Barnoldswick under Hooker, but the Avon axial flow turbo-jet which followed gave trouble for years. Believing that turbo-jets would be unsuitable for transport aircraft, Hooker started turbo-prop work in 1943. The Trent became the world's first turbo-prop to fly in 1944. The 2-shaft Clyde turbo-prop with axial flow LP and centrifugal HP compressors became the first to do the 150-hour military/civil type test.

By 1946 Hooker was suffering from overwork. The war over, Hives sent him to Argentina to recuperate, and moved Barnoldswick Engineering to Derby. On his return Hooker struggled with stubborn Avon troubles but failed to solve them quickly. Hives, whose first priority had to be the health of Rolls-Royce, found himself in an awkward situation. He put A. Cyril Lovesey in charge of the Avon and made Hooker chief research engineer. As a result, the relationship between Hives and Hooker was permanently soured. In September 1948 Hooker left Rolls-Royce for Bristol Siddeley.

January 1949 found Hooker tackling the Bristol Proteus turbo-prop, power plant of the Bristol Britannia. Proteus pioneered the use of gas turbines in the navy and in the Central Electricity Generating Board. The 2-shaft Olympus turbo-jet went to test in 1950, as Hooker became chief engineer. It powered the Vulcan Bomber, the cancelled TSR2, and Concorde. In 1953 Bristol started the Orpheus turbo-jet for the 'Gnat' fighter designed by W. E. W. Petter [q.v.]. Rolls-Royce pursued vertical take-off with small multiple jet engines but Hooker at Bristol Siddeley preferred the 'Gyroptère' concept of Michel Wibault. Using the Orpheus compressor, Bristol created the Pegasus vectored thrust vertical take-off turbofan, which emerged as the world's only operational VTOL (vertical take-off and landing) fighter engine, and was used in the Harrier in the Falklands war.

In 1966 Rolls-Royce acquired Bristol but did

not appoint Hooker to the main board and in September 1970 he retired. A month later he returned as technical director to help the Derby works with RB211 troubles, which bankrupted Rolls-Royce soon afterwards. The government acquired the aero assets of Rolls-Royce and continued to fund RB211 development.

During the 1960s the supply of Viper turbo-jets to Romania enabled Hooker to develop a relationship with Romanian communism, which gave him an entrée to China. An honorary professorship of Peking Institute of Aeronautical Sciences followed in 1973, and substantial business later. Of his many distinctions and medals only his FRS (1962) was nearer to his heart than this Chinese professorship. Hooker was appointed OBE (1946) and CBE (1964), and was knighted in 1974.

Mechanical engineering was not Hooker's forte. He excelled in the physics of aviation, in procuring funding, in salesmanship, and in exposition to laymen. His charisma, and his humour, ensured an easy relationship with people in all walks of life.

He was twice married. In 1937 he married Margaret, daughter of John Swanwick Bradbury, first Baron Bradbury, civil servant. They had one daughter. The marriage was dissolved in 1950 and in the same year Hooker married Kate Maria, daughter of Herbert George Pope, licensed victualler, and former wife of Gordon Garth. They also had one daughter. After retirement his interest in Rolls-Royce only yielded to terminal illness. Hooker died 24 May 1984 in the Chesterfield Hospital, Clifton, Bristol.

[P. H. J. Young, L. Haworth, H. Pearson, G. L. Wilde, and J. E. Ffowcs-Williams in *Biographical Memoirs of Fellows of the Royal Society*, vol. xxxii, 1986; Sir Stanley Hooker, *Not Much of an Engineer*, 1984; personal knowledge.]   LIONEL HAWORTH

HORROCKS, SIR BRIAN GWYNNE (1895–1985), soldier, was born at Ranniken in India 7 September 1895, the elder child and only son of a doctor of medicine, Colonel Sir William Heaton Horrocks, and his wife, Minna, daughter of the Revd J. C. Moore, of Connor, Antrim, Ireland. Educated at Uppingham and the Royal Military College, Sandhurst, he was commissioned into the Middlesex Regiment on the outbreak of war in 1914. Captured during the battle of Ypres in October 1914, he passed the war as a prisoner of war in Germany, and was a Soviet prisoner for another eighteen months after going to Vladivostok in 1919 as a staff officer in a mission under Major-General Sir Alfred Knox to the White Russian Admiral Alexander Kolchak. He was awarded the MC (1919).

Horrocks spent fifteen years as an infantry captain, but such was his love of the army that he was not dismayed by the poor career prospects. In 1924 he won the British modern pentathlon championship, and took part in the Olympic Games. Having passed out of the Staff College in 1933 he was well poised to achieve high rank when World War II started in 1939.

Horrocks was something of an actor, tall and good looking with charisma and charm, and a capacity to make friends although he could sometimes quarrel. He became friendly with A. P. (later Earl) Wavell and B. L. Montgomery (later Viscount Montgomery of Alamein) [qq.v.] during peacetime exercises when he shared their enthusiasm for mechanized mobile warfare at a time when some senior officers still hankered after horse cavalry. (Sir) Frank Simpson had become a close friend at the Staff College, and during the war as director of military operations (War Office) he helped Horrocks to achieve promotion.

Horrocks went to France in 1939 commanding the 2nd battalion, Middlesex Regiment, and was promoted brigadier during the evacuation of Dunkirk. Soon he was a divisional commander, and in 1942 Montgomery summoned him to the Western Desert to command XIII Corps and later IX and X Corps. At El Alamein and the Mareth line his troops performed magnificently, and Sir C. Denis Hamilton claimed that a combination of Montgomery as army commander and Horrocks as corps commander made 'an unbeatable team'.

Horrocks was seriously wounded at Bizerta in an air raid in August 1943 as he was preparing X Corps for the Italian campaign. Five operations and eighteen months of illness and convalescence at Aldershot followed. However in July 1944, when Montgomery sacked the XXX Corps commander after a lack-lustre tank battle in Normandy, he sent for Horrocks to replace him. Everyone connected with XXX Corps at the time testified to the spectacular way in which he revived the morale of his war-weary troops. Still suffering from recurrent fever, Horrocks led XXX Corps in the dramatic dash from the Seine to Antwerp. Then he made a mistake by ordering his corps to rest for three days at a moment when it was imperative to rush a crossing of the wide Albert canal while the Germans were still in confusion. Hitler used the delay to send up fresh formations and the British drive into Holland was halted.

Both the British and Polish airborne commanders in their autobiographies criticize Horrocks for his failure to link up with them after the Arnhem drop in September 1944. According to some of his colleagues Horrocks was unwell, but he crossed the Nijmegen bridge within a few hours of its capture by the Grenadiers to congratulate them. The truth is Montgomery had set XXX Corps an impossible task.

When Karl von Rundstedt's 1944 Christmas offensive threatened Brussels and Antwerp, Horrocks told Montgomery the German armour should be allowed to cross the Meuse and then his corps would annihilate them on the battlefield of Waterloo. This did not appeal to Montgomery who sent him home on compulsory sick leave. He was quickly back to lead his corps over the Rhine and in the final triumphal drive to the Elbe. Friends of General (Sir) Miles Dempsey [q.v.], commander of the Second Army, have criticized Horrocks during this period for bypassing Dempsey and taking his instructions direct from Montgomery. The fault lay with Montgomery who was nostalgic for his previous relationship with Horrocks in Africa. Horrocks was appointed CB (1943), to the DSO (1943), KBE (1945), and KCB (1949).

After the war Horrocks became GOC Western Command in February 1946 and then GOC-in-C British Army of the Rhine early in 1948. The problems of defeated Germany put too much of a strain on his health and he was invalided out of the army in 1949, with the rank of lieutenant-general.

Immediately he was offered the post of gentleman usher of the Black Rod in the House of Lords, which he discharged with distinction from 1949 to 1963. He also became a television star, making around forty programmes with (Sir) Huw Weldon. Horrocks loved the bustle and immediacy of television and also the floods of fan mail which in no way affected his character. After fourteen years as Black Rod Horrocks resigned to become a director of Bovis, and to leave himself more time for television, sailing, and charities. He had an honorary LLD from Belfast.

He married in 1928 Nancy, daughter of Brook Taylor Kitchin, architect, of the Local Government Board. They had one daughter who was drowned when swimming in the Thames in 1979. Horrocks died 4 January 1985 in Chichester.

[Brian Horrocks, *A Full Life*, 1960 (2nd edn. 1974); Philip Warner, *Horrocks*, 1984; Richard Lamb, *Montgomery in Europe*, 1983; Sir Frank Simpson's papers in the Imperial War Museum.]   RICHARD LAMB

HOWELLS, HERBERT NORMAN (1892–1983), composer, teacher, and writer on music, was born at Lydney, Gloucestershire, 17 October 1892, the youngest in the family of six sons and two daughters of Oliver Howells, painter and decorator, and his wife, Elizabeth Burgham. Both parents were of Gloucestershire stock. While he was still at Lydney Grammar School his talent came to the notice of (Sir) A. Herbert Brewer [q.v.], organist of Gloucester Cathedral, who taught him until in 1905 he was ready to be formally articled. Howells was therefore one of the last English musicians to be brought up in the old apprentice system under which the aspirant was bound to a master, generally the nearest cathedral organist, who accepted full responsibility for his training and entry into the profession.

Howells accordingly spent the next three years in the daily routine of cathedral music, playing and singing in the services, and open to the influences of the Anglican liturgy, the psalms, the Book of Common Prayer, and the Authorized Version of the Bible, read by educated men in a noble building. He admitted, in later life, that this discipline had largely moulded him, and that the main lines of his development had already been laid down when in 1912 he won an open scholarship at the Royal College of Music and became a pupil of Sir Charles Stanford and Sir C. Hubert H. Parry [qq.v.], under whom, during the next five years, he wrote a great deal of music and acquired a formidable technique. Even more significant than Stanford's discipline and Parry's generous friendship were the social and literary influences that flooded over him in the new world that welcomed his romantic good looks, charm, and lively intelligence.

Recognition and a kind of maturity came early. 'Lady Audrey's Suite' (1916) and 'A Spotless Rose' (1918) were widely performed, but it was the Piano Quartet (1916) and the Phantasy String Quartet (1918) that gave fuller insight into the composer's powers. In 1916, whilst acting as assistant organist at Salisbury Cathedral, Howells suffered a complete breakdown of health, and it was not until late in 1918 that he was able to return to work and to London, where he lived for the rest of his life except for a sojourn in Cambridge during World War II.

At all times he had his living to earn, and after 1920 a family to support. He did this by teaching theory and composition at the Royal College of Music for over fifty years, and by lecturing and writing, for which he had a great gift. He was also much in demand as an adjudicator at music festivals. If he had not been a musician he could have excelled as a poet and essayist like his friend Walter de la Mare [q.v.]. From 1936 to 1962 he was director of music at St Paul's Girls' School, Brook Green. From 1955 to 1962 he was King Edward professor of music at London University. It was his powers of withdrawal and concentration that enabled Howells in a busy life to produce so much music, including, besides early orchestral works, three concertos, six quartets, five instrumental sonatas, a number of large choral works, among them the masterly 'Hymnus Paradisi' (1938), forty songs, and much church and organ music. He had an unfailing insight into what is possible with a small choir and organ. Howells also wrote many short choral pieces for school and festival use, some of which are among his most distinguished works, imaginative, beautifully composed, and fastidious in choice and treatment of poetry. In this and other respects he may be compared with his Tudor predecessors, John Dowland and Thomas Campion [qq.v.], with whom he clearly felt an affinity. It is significant that he chose to write for the clavichord, most intimate of instruments, for his music often conveys a sense of privacy, of direct address to one friend, or of solitary meditation.

Yet he was in no sense a miniaturist. Although he did not choose, in maturity, to work in the major symphonic forms, there are movements in the chamber music and the choral works that are designed on a massive scale; and he had, moreover, the power to create a sense of space even in short pieces. The Nunc Dimittis of the St Paul's service, for instance, in the building for which it was composed, sounds as vast as the cathedral itself. Howells's harmonic idiom was based on the tonal tradition, derived through Parry, Stanford, and Sir Edward Elgar [q.v.], but extended by bold use of chords superimposed or used as appoggiaturas. No composer of his generation could remain aloof from the recent revivals of Tudor music, folk-music, and modal harmony, and Howells took from these sources all that he needed: but he was never dominated by them, or by the powerful proximity of Ralph Vaughan Williams [q.v.], as were some of his contemporaries. There is, in his melodic style, as much of S. S. Wesley [q.v.] and Parry as of folk-song.

One finds in Howells's music depth of feeling, tenderness, humour, and a love of the English landscape, but seldom great exuberance. Especially after the death of his son in 1935, there is often a persistent note of nostalgia and regret that has led critics to compare him with Frederick Delius [q.v.]. But in Howells's music there are gleams of visionary hope that are not to be found in Delius and it is noticeable that in the last years of his life he wrote little or nothing except church music, as if, for the finale, he was returning home to Gloucester. Although as a young man he had not expected to live very long, he remained active, in spite of increasing deafness, until his death in Putney at the age of ninety, 23 February 1983.

In 1920 he married Dorothy (died 1975), daughter of William Goozee. There were two children, Michael, who died of poliomyelitis in 1935, and Ursula, the distinguished actress.

He received many honours. He was appointed CBE in 1953 and CH in 1972. Honorary degrees were conferred on him by Cambridge (1961) and the RAM. He was a D.Mus. of Oxford. He was elected an honorary fellow of St John's College, Cambridge, in 1962 and of The Queen's College, Oxford, in 1977. He served as master of the Worshipful Company of Musicians in 1959. In 1952 he was president of the Incorporated Society of Musicians, and in 1958–9 of the Royal College of Organists. After his death a memorial tablet was unveiled in Westminster Abbey, where his music is so often to be heard. There are portraits by Sir William Rothenstein and Brenda Moore in the RCM, and a fine photograph of Howells as a young man by Herbert Lambert of Bath.

[Christopher Palmer, *Herbert Howells*, 1978; Hugh Ottaway in *The New Grove Dictionary of Music and Musicians*, 1980 (ed. Stanley Sadie); private information; personal knowledge.]   THOMAS ARMSTRONG

HUMPHREYS, (TRAVERS) CHRISTMAS (1901–1983), judge, was born in Ealing 11 February 1901, the younger son (there were no daughters) of (Sir) (Richard Somers) Travers (Christmas) Humphreys [q.v.], a barrister, and his wife, Zöe Marguerite, the daughter of Henri Philippe Neumans, an Antwerp artist. The elder son was killed in France in 1917. His father was to become senior Treasury counsel at the Central Criminal Court, a High Court judge, and a member of the Privy Council. He was educated at Malvern and Trinity Hall, Cambridge, where he obtained a second class in part i of the law tripos (1922) and also a second (division II) in part ii (1923).

He chose the bar as his profession, probably more out of a sense of filial duty than inclination. When he made his decision he knew that his father wanted his son to carry the practice of the law into a third generation. His paternal grandfather, Charles Humphreys, had been the solicitor for Oscar Wilde [q.v.]. He was called to the bar by the Inner Temple in 1924. In 1927 he married Aileen Maude, the daughter of Charles Irvine, a Yorkshire doctor of medicine. They had no children.

Like his father and paternal grandfather he chose the criminal courts for his practice. He made rapid progress. He was probably helped at first by his father's influence and connections but he had much natural ability. He had a fine presence, being tall and slim, and a pleasant, distinctive voice. In 1934 he was appointed junior Treasury counsel at the Central Criminal Court; in 1942, recorder of Deal (until 1956); in 1947, deputy chairman of east Kent quarter-sessions (until 1971); in 1950, senior Treasury counsel at the Central Criminal Court (until 1959); and in 1956, recorder of Guildford.

In 1946 he was appointed junior counsel, together with (Sir) A. Comyns Carr KC as his leader, to the war crimes tribunals trying Japanese war criminals. He did much work sifting and evaluating the evidence and drafting the charges but took little part in presenting the cases. His presence in the Far East enabled him to travel widely there and to increase his knowledge of eastern religions which was already extensive.

In 1955 he joined his father as a bencher of the Inner Temple. All seemed set for him to have a similarly distinguished career in the law but that was not to be. Unlike his father who lived for the law and amidst lawyers, he had many interests outside. As the years went by these interests seemed, in the opinion of the director of public prosecutions, to interfere with the performance of his duties as senior Treasury counsel. The resulting loss of confidence, one with the other, led him to decide in 1959 to apply for silk, which by custom necessitated his giving up his appointment. It was given it. By this time he was fifty-eight, somewhat old for starting practice as a QC. In 1962 he was appointed a judge at the Central Criminal Court and he sat there until he retired in 1976. As a judge he was competent and kindly, too kindly for many who considered that his sentences were over lenient. His last years on the bench cannot have been satisfying for him. Those in charge of administration at the Central Criminal Court seldom gave him the more interesting cases to try. For about four years before her death in 1975 his wife had been suffering from a terminal illness which had the distressing symptom of loss of memory.

Some lawyers might say that Humphreys did not make the most of his considerable legal talents. He did not consider legal fame worth seeking. From his student days onwards his interests had ranged far beyond the law. As a youth he had become interested in Buddhism. In 1924 he became the founding president of the Buddhist Lodge, now the Buddhist Society. In 1928 he wrote a book, *What is Buddhism?* From that time onwards he led his life according to his Buddhist beliefs. He studied eastern philosophy and learned to appreciate Asian culture, particularly Chinese art. Between 1928 and 1962 he wrote twenty books, including four books of poetry, and numerous pamphlets and articles. To this Dictionary he contributed the articles on Sir Bernard Spilsbury, Hilda Leyel, and Sir Archibald Bodkin. He pursued his interests outside the law with enthusiasm and sometimes with more vigour than sound judgement. He was, for example, convinced that the Earl of Oxford had written the plays usually attributed to Shakespeare.

In his dealings with people 'Toby' Humphreys was always friendly, courteous, and considerate; but he was not gregarious and in his later years he did not seem to enjoy the company of the lawyers with whom he had worked. He died at his London home 13 April 1983. He left his house in St John's Wood, which for many years had been a meeting-place for Buddhists, to the Zen Society.

[Christmas Humphreys, *Both Sides of the Circle* (autobiography), 1978; private information; personal knowledge.]

FREDERICK LAWTON

HUNTER, LESLIE STANNARD (1890–1983), bishop of Sheffield, was born in Glasgow 2 May 1890, the younger son and younger child of John William Hunter, Congregational minister at Trinity Congregational church in Glasgow, and his wife, Marian Martin, formerly of Hull. J. W. Hunter was a liberal preacher, concerned for women's rights, and of considerable civic influence. Educated at Kelvinside Academy, and from 1909 at New College, Oxford, Leslie Hunter obtained a second class honours degree in theology in 1912. He spent time in France with the YMCA in 1916, and also became a friend of Baron von Hügel, growing to admire his liberal Catholicism and mysticism. His elder brother was killed in World War I.

Charles Gore [q.v.] confirmed him into the Anglican Church in 1913. His first post as study secretary of the Student Christian Movement (1913–20) was a decisive influence. After ordination (1916), he also served part-time curacies in Brockley and St Martin-in-the-Fields (hero-worshipping H. R. L. Sheppard, q.v., the vicar), and as a hospital chaplain. He published the first of many books, *The Artist and Religion* (1915), arising from work with students in colleges of art.

A Newcastle upon Tyne canonry (1922–6) preceded four creative years as vicar of Barking (1926–30). During the depression he returned to the north as archdeacon of Northumberland (1931–9), where he became the effective force in the diocese. *A Parson's Job* (1931) argued for a strongly led team ministry approach in growingly secularized communities. In Newcastle he founded the Tyneside council of social service, recruiting the sociologist Henry Mess, and committed himself to bringing the needs of the north east—unemployment relief and more generous and imaginative policies in health, housing, and education—to the attention of the government. He defended the 1936 Jarrow march as a necessary expression of frustration at indifference by government and City. He preached a Sandringham sermon and wrote letters to *The Times* querying establishment attitudes. He forged links with the German church resistance, joined those working to create the World Council of Churches, and led the 'Men, Money, and the Ministry' movement within the Church of England for the sake of a more militant church with new methods of paying and deploying the clergy. Reforms he suggested were later adopted.

As bishop of Sheffield (1939-62) he transformed the diocese into the most forward looking and strategy conscious within the established church. He was haunted by the gulf between workers and the church. He created a team, including Oliver Tomkins, Alan Ecclestone, and others who respected his convictions and shared his aims. Trade-union leaders were invited to his home, though at first they were prepared to come only at night. He brought in Edward R. Wickham to build a new type of industrial mission, enabling the church to meet workers on their own ground in factories and mines. Unfortunately, timorous ecclesiastical authority later modified this industrial mission, making it pietistic rather than pioneering.

Sociologically, he understood the alienation of working-class people and worked tirelessly to alter the restricted attitudes of church-goers. He helped to found William Temple College, Whirlow Grange conference centre, and Hollowford youth centre. In the House of Lords he pleaded the cause of German prisoners of war, the need for clean air, and justice in industrial relations. He was deeply disappointed when introverted central church leadership at Lambeth concentrated on canon law. Distrusting narrow ecclesiastical attitudes, he looked forward to a ministry which would include ordained women; in this as in the range of his concerns, this unassuming man became the heir of Archbishop William Temple [q.v.].

Short of stature, with piercing eyes, often difficult to hear, Hunter was no orator, but as a listener with an imaginative and critical approach, he valued and resolutely used episcopal office to relate Christian faith and activity to twentieth-century life. Of his fourteen books, the most perceptive is *A Mission of the People of God* (1961). His italic handwriting, often inviting the reader to accept some difficult task, was known as the 'snare of the hunter'. Critics detected Machiavellian skill, but the young appreciated being taken seriously. Honorary degrees (DCL, Durham, 1940; DD, Lambeth, 1949; LLD, Sheffield, 1953; DD, Trinity College, Toronto, 1954) indicated the respect in which he was held by universities and policy makers. The European Bursary Fund set up in his memory witnessed to his lifelong interest in Scandinavian and continental Christianity, and his admiration of the Taizé community.

In 1919 Hunter married Grace Marion (died 1975), a Cambridge graduate and SCM staff member, the daughter of Mary and Samuel McAulay, farmer, of Aylesby, Lincolnshire. Her hospitable home complemented Hunter's taciturnity, which could alarm visitors. They had no children. Hunter died 15 July 1983 at York.

[Gordon Hewitt (ed.), *Strategist for the Spirit*, 1985; *The Times*, 19 July 1983; private information; personal knowledge.]

ALAN WEBSTER

HUSBAND, SIR (HENRY) CHARLES (1908-1983), engineer, was born 30 October 1908 at Sheffield, the youngest of the three sons (there were no daughters) of Joseph Husband and his wife, Ellen Walton Harby. His father (1871-1961) was the founder (1892) of the civil engineering department of the Sheffield Technical School. This eventually merged with the University of Sheffield where he was the first professor of civil engineering (1921-36). His youngest son was educated at King Edward VII School, Sheffield, and read engineering in the University of Sheffield where he graduated B.Eng. in 1929.

After gaining early experience with the Barnsley Water Board, Husband served in 1931-3 as an assistant to the civil engineer and architect Sir (Evan) Owen Williams. He was then appointed engineer and surveyor to the First National Housing Trust Ltd. In 1936 he founded the firm of Husband & Clark, consulting engineers. As a partner in this firm with his father he was involved in a variety of public works—these included major road and railway bridges, and drainage and water supply schemes. During World War II he was a principal technical officer in the central register of the Ministry of Labour and National Service (1939-40); and then assistant director on the directorate of aircraft production factories of the Ministry of Works.

After the war, Antony Clark having retired from their partnership, Husband, who had become the senior partner in 1937, was the major force in the development of the partnership, now Husband & Co., into one of the foremost firms of consulting engineers in the United Kingdom. Under Husband's guidance the firm quickly became involved in a wide range of construction projects. In 1946 Husband designed the first high altitude testing plant for the continuous running of jet engines and from 1947 he received extensive commissions from the National Coal Board. He was also responsible for the construction of new buildings for various industrial organizations, and for the research establishments of the British Iron and Steel Research Association and the Production Engineering Research Association.

In 1949 Husband was introduced to (Sir) Bernard Lovell at Jodrell Bank who had been rebuffed by a number of major engineering firms in his attempt to interest them in the construction of a large steerable radio telescope. Characteristically Husband did not regard the task as impossible and in the autumn of 1957 the 250-foot aperture telescope came into operation. The success of this instrument is a tribute to Husband's engineering skill and courage. Designed before the age of digital computers many difficulties were surmounted and at the time of Husband's death the telescope had withstood the forces of nature for more than a quarter of a century and continued to be a major facility in the international research field. After this experience Husband designed a number of smaller telescopes including the first satellite/earth stations for the GPO at Goonhilly Downs, Cornwall.

When the Britannia rail bridge over the Menai Strait was destroyed by fire in May 1970 Husband was presented with the task of rebuilding. Officially opened ten years later this combined rail and road bridge, like the Jodrell telescope, stands as an example of Husband's many fine engineering accomplishments. For his work on the bridge Husband was awarded the James Watt medal of the Institution of Civil Engineers in 1976. In addition to many engineering projects in the United Kingdom, Husband was involved with a variety of overseas activities, particularly in Sri Lanka where he was responsible for a number of public works.

Husband brought to his work much of the persistence and width of interest shown by the great engineers of the nineteenth century. Especially in his work on the radio telescopes it was his skill in co-ordinating the efforts of engineers and scientists in the solution of unusual structural difficulties which led to their successful completion. He was appointed CBE in 1964 and knighted in 1975. In 1965 he was the first recipient of the Royal Society Royal medal for distinguished contributions to the applied sciences. He was made honorary D.Sc. of the University of Manchester (1964) and honorary D.Eng. of the University of Sheffield (1967). He was awarded the Sir Benjamin Baker gold medal of the Institution of Civil Engineers (1959), the Wilhelm Exner medal for science and technology of the University of Vienna (1966), and the gold medal of the Institution of Structural Engineers (1974). He was a founder fellow of the Fellowship of Engineering, and a fellow of the Institution of Civil Engineers, Mechanical Engineers, Structural Engineers (of which he was president 1964–5), and the American Society of Civil Engineers. In 1967 he was chairman of the Association of Consulting Engineers.

Husband was an accomplished sailor and undertook a number of voyages under hazardous conditions. On 19 March 1932 he married Eileen Margaret, daughter of Henry Nowill, an architect, of Sheffield. She survived him with their two daughters and two sons, the elder of whom (Richard William) succeeded Husband as head of the firm of Husband & Co. Husband died 7 October 1983 at Sheffield.

[*The Times*, 8 and 15 October 1983; private information; personal knowledge.]

BERNARD LOVELL

J

JACKSON OF LODSWORTH, Baroness (1914–1981), journalist and broadcaster. [See Ward, Barbara Mary.]

JACKSON, BRIAN ANTHONY (1932–1983), educational reformer, writer, and teacher, was born 28 December 1932 in Huddersfield, the son of (John) Henry Jackson, labourer at ICI, and his wife, Ellen O'Brien. Though he never wrote his autobiography he gave many glimpses of himself as a child in his writing, particularly in his first and most influential book, *Education and the Working Class* (1962), written jointly with Dennis Marsden, a fellow pupil at Huddersfield College. The book is a brilliant study of the effect of grammar-school education on working-class children and their families in which the authors transform their old school into 'Marburton College'. Jackson was a scholarship boy who went on to win an open exhibition to St Catharine's College, Cambridge, where he managed to hold on to working-class values without decrying the middle-class values which so often supersede them.

Some of the scholarship boys and girls became conformists. Not Jackson. He showed his independent spirit by the way he gained admission to university. Travelling up to Cambridge bearing the all-important headmaster's letter of reference with him on the train, he wondered whether he might be the messenger for his own death sentence. So he opened the letter, found his fears were right, and threw it out of the window. Asked at the interview where the letter was, he said he had read it, decided it would do him no good, and destroyed it. To the credit of the college he was accepted and justified his choice by getting a first class honours degree in English (1955).

Orthodox careers for him might have been in Labour politics or in teaching. Instead, after four years in which he combined teaching in a primary school and supervising undergraduates in a Cambridge college, he took the path of research and reform which dominated his life. After the Marburton book, for which he did the research while on the staff of the Institute of Community Studies in Bethnal Green, he was from 1962 to 1965 a visiting fellow at the department of applied economics at Cambridge. His next book again showed his empathy with children. *Streaming: an Education System in Miniature* (1964) was an attack on the manner in which children were winnowed and divided, enjoined to become what their teachers expected them to become. Later, the problems of culture

and class were still under his microscope in *Working Class Community* (1968).

In his other career of reformer, he achieved results not by haranguing his audiences (though he was a persuasive writer and speaker) so much as by running organizations where actions spoke louder even than his words. With his ever-buoyant energy, his flair, and his talent for enthusing people to work hard for little money, he managed to be simultaneously director of the Advisory Centre for Education (1962–74) and the National Extension College. ACE was an information co-operative for parents and teachers, taken together rather than separately, which played a part in every educational reform of the period. NEC was expressly designed by Jackson and Lord Young of Dartington as a distance teaching initiative which would be a pilot for the idea of an Open University. Until then correspondence education had been held in almost as low repute as television teaching. NEC helped to raise the standard and standing of both by teaching home students effectively and making a convincing case for an Open University. Jackson managed to build up both institutions without substantial funding and with no money from the state. NEC continued as a pre-university college for many thousands of students.

From there Jackson went on to start in 1976 the National Research and Development Trust. From that sprang the National Children's Centre and Childminding Research Unit. He brought the work of the small armies of women who look after other people's children before they go to school into focus for the first time. *Childminder* (1979), which he wrote jointly with Sonia Jackson, was the result of that research. While remaining chairman of the trust he became a research fellow at the department of child health at Bristol University where he opened up yet another new subject in *Fatherhood* (1984). This book portrayed the 'invisible man', or rather the shadowy man who is feeling his way towards a new role in which releasing the full force of fatherhood would mean breaking the masculine taboo on tenderness.

Jackson was married first in 1956 to Sheila Mannion, a childhood friend from Huddersfield. She was the daughter of John Henry Mannion, unskilled employee of the Gas Board. They had a son and a daughter. After they were divorced in 1977 he married in 1978 Sonia Abrams who already had a son and a daughter and together they had another son and daughter. Sonia, who was the daughter of (Israel) Maurice Edelman, MP [q.v.], and formerly the wife of Philip Abrams, sociologist, became a lecturer at Bristol University and their house in Bristol was the centre of an abundant life with the six children and numerous friends. Intellectual and family life were infused with the same zest and energy and hatred of humbug. Jackson died suddenly 3 July 1983 in Huddersfield after taking part in a charity run there.

[*The Times*, 5 and 13 July 1983; Sonia Jackson, preface to *Fatherhood*, 1984; Harry Rée, foreword to reprint of *Education and the Working Class*, 1986; private information; personal knowledge.]   MICHAEL YOUNG

JACKSON, DEREK AINSLIE (1906-1982), physicist, was born in Hampstead 23 June 1906, identical twin brother of Vivian and son of Sir Charles James Jackson, barrister, art collector, and author of books on the history of art, and his wife, Ada Elisabeth, daughter of Samuel Owen Williams. The only other child, Daphne, ten years older than the twins, died during World War I. The father had a great influence on the early development of his sons and on their interests but he died in 1923 and their mother died a year later. They were then left largely to themselves, with an ample supply of money, restricted by a guardian in a rather remote way. They were lively, very fond of one another, and also very like one another, in their appearance, movements, and gestures; they also shared a strong interest in science. Vivian's death in 1936 in a sleigh accident came as a great shock to Derek. Jackson was educated at Rugby and, as a scholar, at Trinity College, Cambridge. He obtained a first class in part i of the natural sciences tripos (1926) and a second in part ii (1927). His supervisor at Cambridge was H. W. B. Skinner, known for his work on Zeeman effects in very strong fields.

For his doctorate work, Jackson accepted an offer by F. A. Lindemann (later Viscount Cherwell, q.v.) of spectroscopic work at the Clarendon Laboratory in Oxford where he had his own spectrograph. For buying other optical equipment Jackson established a university fund. The results of this work on the hyperfine structure in the arc spectrum of caesium were published in the *Proceedings* of the Royal Society in 1928 and led to Jackson's Oxford D.Sc. (1935). It was followed by similar work for the elements In, Rb, Tl, and Ga, leading to some information on their nuclear spins and magnetic moments.

In 1934 Jackson became a lecturer in spectroscopy at Oxford and in 1947 professor of spectroscopy. Both these positions were unpaid but he regularly gave lectures at the Clarendon Laboratory. For several years he worked with H. G. Kuhn and their collaboration resulted in publications on hyperfine structures and iso-tope shifts in the optical spectra of the elements K, Ag, Al, Na, and Li. A Fabry-Pérot étalon was still used for the high resolution required, but its power was greatly increased in several ways. Now hyperfine structures and isotope shifts could be measured more accurately.

The outbreak of World War II brought this joint work to a sudden end. Jackson wanted to take an active part in the war and joined the Royal Air Force Volunteer Reserve. Cherwell and others realized that Jackson's contribution to the war could be more valuable if his experience and knowledge as a scientist were fully used. Thus, he was not only trained as a navigator, but also played an important part in the air war, in a very direct way, by working on the development of 'Window' and the interference with German night defences.

After the war Jackson tried to find his way back into civil life and scientific activity, but found it hard to come to terms with an altered world. Changes in the tax laws had now substantially reduced his income, and emigration to Ireland or to France had to be considered. His personal life had also become unsettled. In 1952 he accepted an invitation from Professor P. Jacquinot to work in his laboratory at Bellevue (this was later transferred to Orsay). At this 'Laboratoire Aimé Cotton' Jackson's large spectroscope and other equipment were installed, and he continued, together with French colleagues, to work on hyperfine structure. Much interesting work, especially on isotope shifts, was carried out and published jointly with Duong Hong Tuan and M. C. Coulombe. Jackson often took an active part at international conferences such as EGAS (the European group of atomic spectroscopy).

Jackson gained several honours. In 1947 his contributions to spectroscopy were recognized by a fellowship of the Royal Society, his contribution to the air war led to the awards of DFC (1941), AFC (1944), and OBE (1945), and the USA made him an officer of the Legion of Merit. For his research in France he was made a chevalier of the Legion of Honour. Jackson was a lively person with an interest in art, literature, and politics, about which he had strongly held views. He was excellent company, a stimulating conversationalist about his many occupations, which included fox-hunting, weekend parties, horse-racing, and steeple-chasing. He was himself a steeplechase rider of some standing; he rode twice in the Grand National. He was a part-owner of the *News of the World*.

Jackson was married six times. In 1931 he married Poppet, daughter of Augustus John [q.v.], painter. This marriage was dissolved in 1936 and in the same year he married Pamela, daughter of David Bertram Ogilvy Mitford, second Baron Redesdale. This marriage was dissolved in 1951 and in that year Jackson married Janetta, daughter of the Revd Geoffrey Harold Woolley. They had one daughter. The marriage was dissolved in 1956 and the following year Jackson married Consuelo Regina Maria, widow of Prince Ernst Ratibor zu Hohenlohe Schillingsfürst and daughter of William S. Eyre. This marriage was dissolved in 1959. In 1966 Jackson married Barbara, daughter of Eric Georg Skelton; the marriage was dissolved after a year. In 1968 Jackson married Marie-Christine, daughter of Baron Georges Reille. She was a young widow with two daughters. Jackson died 20 February 1982 at Lausanne.

[H. G. Kuhn and Sir Christopher Hartley in *Biographical Memoirs of Fellows of the Royal Society*, vol. xxix, 1983; private information; personal knowledge.]   H. G. KUHN

JACOB, GORDON PERCIVAL SEPTIMUS (1895-1984), composer, was born in London 5 July 1895, the seventh son and youngest of the ten children of Stephen Jacob, comptroller-general of Indian finance in the Indian Civil Service, and his wife, Clara Laura Forlong. He was educated at Dulwich College where, though specializing in science, he shone at music and wrote pieces for the school orchestra and band. In 1914 he joined the army and, after three years at the front in which he was badly gassed, was taken prisoner near Croiselle. There followed twenty months in a German prison camp where he organized, composed for, and conducted a small heteroclite orchestra formed from his fellow prisoners.

On demobilization in 1919, in his own words he 'wasted a year at a so-called School of Journalism', a view he was to temper later when he admitted that he 'did learn a bit about writing in a way that would interest people'. Certainly his books *Orchestral Technique* (1931), *How to Read a Score* (1944), and *The Composer and his Art* (1955) are models of clear unstuffy exposition.

In 1920 he entered the Royal College of Music to study composition with Sir Charles Stanford and Ralph Vaughan Williams, conducting with (Sir) Adrian Boult [qq.v.], and organ with George Thalben-Ball. Of Stanford he did not wholly approve, though he learnt from him much about the technique of composition. On one occasion the story goes that, when Stanford remarked of one of Jacob's more up-to-date harmonic progressions that 'any fool could write that', Jacob went to the piano, played an Amen cadence and said 'And that too!'

After winning the Foli scholarship for his

ballet *The Jew in the Bush* and the Sullivan prize for some William Blake settings, Jacob remained at the college as a member of the teaching staff. This did not stem the flow of works: a Viola Concerto (1925) and a Piano Concerto (1927), a Symphony in 1929, and many orchestrations of ballets for the Sadler's Wells company.

During World War II he became well known to a wider public for his almost surrealist arrangements of popular tunes for the top-rating radio show ITMA. He did not neglect serious composition, however, being helped by the award of the Musicians' Company's Collard fellowship in 1943, which gave him an annual stipend of £300 for three years. In 1944 his epic Symphony no. 2 emerged and in 1949 the brilliantly delightful Suite no. 3 which is full of good tunes, rhythmic vitality, and the clear textures which characterized his best music.

Jacob retired from the college in 1966 to devote the rest of his long life to composition. At his death he had over 200 works to his credit: many film scores, two ballets, two symphonies, no fewer than twelve concertos for various instruments (of which that for flute is perhaps the most widely performed), over twenty chamber works for various combinations, many cantatas and shorter choral works, and some music for military and brass bands. He also edited from 1947 to 1957 the now defunct series of Penguin Scores in which, at his far-sighted behest, all transposed lines were printed at sounding pitch: an innovation regrettably not yet taken up in other editions.

Jacob's place in English musical history is as an eclectic, versatile member of the school of Vaughan Williams and Gustav Holst [q.v.] rather than of the more cosmopolitan group which includes Sir Edward Elgar, Sir William Walton, and Constant Lambert [qq.v.]. Pungent and closely argued though much of his music is, its diatonic and often modal lyricism sets it strongly apart from the prevalent style of the next generation, many of the outstanding representatives of which were his pupils, including such diverse composers as John Addison, Malcolm Arnold, Philip Cannon, Humphrey Searle [q.v.], and Bernard Stevens. He had the knack as teacher of quickly spotting a faultily contrived texture in a score and at the same time being open-minded and undidactic about it.

Jacob suffered from a cleft palate: a handicap which often coloured his speech with a dry irony that was the delight of his pupils and friends. Though the kindest and gentlest of men, he had a gift for sardonic yet wise comment that could bring the members of the many committees on which he sat sharply to their senses.

In 1949 he was awarded the Cobbett medal for chamber-music composition. He was a D.Mus. of London University and was appointed CBE in 1968. In 1924 he married Sidney Wilmot, daughter of the Revd Arthur Wollaston Gray, of Ipswich. She died in 1958 and in the following year he married her niece, Margaret Sidney Hannah, daughter of Cuthbert Arthur Gray, insurance agent, of Helions Bumpstead. They had a daughter and a son. He died at his home in Saffron Walden 8 June 1984.

[Frank Howes in *Grove's Dictionary of Music and Musicians*, 5th edn., 1954 (ed. Eric Blom); R. S. Pusey, 'Gordon Jacob: A Study of the Solo Works for Oboe and English Horn and their Ensemble Literature', thesis for Peabody Conservatoire of Music, Johns Hopkins University, Baltimore, 1979; *The Times*, 11 June 1984; private information; personal knowledge.]   JOHN GARDNER

JANNER, BARNETT, BARON JANNER (1892–1982), parliamentarian, solicitor, and Jewish leader, was born 20 June 1892 at Lucknick, Lithuania, the only son and the second of three children of Ofse Vitum-Janner (known as Joseph Janner), a small shopkeeper, and his wife, Gertrude (Gittel) Janner. He was taken at the age of nine months to Barry, Glamorgan. He was educated at Barry County School, which he entered by scholarship, graduating BA from the University of Wales. Shortly after taking his law exams, Janner enlisted in the Royal Garrison Artillery as a gunner, serving in France in 1917–18 and being gassed in the trenches.

He established a modest solicitor's practice in Cardiff where his defence of people with rent and lease problems established his reputation in these fields which were to be a primary preoccupation throughout his subsequent public career. It was in Cardiff, too, that he embraced another lifetime passion, the Zionist cause. A large, bluff man with a ready smile and conciliatory nature, he nevertheless enjoyed a good fight and he was to find many a one in politics and Jewish community affairs. Adopted on 8 April 1929 as prospective Liberal Party candidate for Cardiff Central, he doubled the Liberal vote but came third. In 1930, by which time he had moved his law practice to London, Lloyd George urged him to fight Whitechapel and St George's, which then embraced the heartland of metropolitan Jewry, and he was elected there at the second attempt in 1931. Janner's maiden speech on 23 November 1931 was on leasehold reform. One of his last speeches in the House of Lords, in 1980, was on the same topic.

In 1935 Janner lost Whitechapel to Labour and a year later, to no one's surprise, he left the Liberals for the Labour Party with which his views on social and international issues had increasingly coincided. He was not to return to the Commons until 1945 when he won Leicester for Labour (later Leicester North West), a seat he held until he was made a peer. The years out of Parliament he devoted to extending his practice, fighting the growing menace of Fascism and, when war came, to pressing the case for the establishment of a specifically Jewish fighting formation within the Allied forces, in which Jews from Palestine could participate. In 1944 the government agreed to set up the Jewish Brigade. His parliamentary activities after 1945 reflected his unceasing concern with the underprivileged at home—he was vice-president of the Leasehold Reform Association, a member of the committee on Wage Earning Children, and honorary rents adviser to the Labour Party—and the plight of post-Holocaust Jewry abroad, especially in relation to the new-born State of Israel.

In 1955, with his election as president of the board of deputies of British Jews, he became the lay leader of the Anglo-Jewish community and its premier spokesman, a position in which he gloried and for which he was knighted, the first president of the board to be so honoured specifically for that role. His defeat in 1964 in that post, in which he had spearheaded the campaign on behalf of the Jews of the Soviet Union and opposed the application of a statute of limitations for war criminals in the German Federal Republic, distressed him deeply but, with his characteristic largeness of heart, he continued to give unstintingly of his time and advice to the Jewish community. In Parliament, his persistent campaign for British ratification of the UN genocide convention (sometimes spoken of by colleagues as the 'Jannercide convention') was rewarded in 1969.

Only once did he come under strong criticism from some sections of the Jewish community and that was when, in 1956, he voted with his party against Britain's involvement with France and Israel in the Suez affair. His explanation to the Jewish community was that while Israel had a right to act in her self-defence against Egypt and could not therefore be regarded as an aggressor, the British government had failed to establish any legal grounds for its use of force. His argument swayed the European council of the World Confederation of General Zionists, of which he was president, and the board of deputies, both of which adopted votes of confidence in him.

In 1957 he received an honorary LLD at Leeds University. In 1963 he was made commander of the Order of Leopold II, in recognition of his contribution to Anglo-Belgian relations, particularly his service as honorary secretary of the Anglo-Belgian parliamentary group. In 1970 Queen Juliana of the Netherlands made him a commander of the Order of

Orange Nassau. He was knighted in 1961 and made a life peer in 1970.

He married in 1927 Elsie Sybil, daughter of Joseph Cohen, furniture retailer. She later became a well-known social worker and JP and was appointed CBE in 1968. They had two children, Greville Janner, QC, MP, who held his father's old seat of Leicester North West for Labour, and Ruth, who became Lady Morris of Kenwood. Janner died in London 4 May 1982.

[*The Times*, 6 May 1982; *Jewish Chronicle*, 7 May 1982; private information; personal knowledge.]   GEOFFREY D. PAUL

JENKINS, RICHARD WALTER (1925–1984), actor. [See BURTON, RICHARD.]

JOHN, SIR CASPAR (1903–1984), admiral of the fleet, was born at his parents' home, 18 Fitzroy Street, London W1, 22 March 1903, the second of the five sons (there were no daughters) of Augustus Edwin John [q.v.], artist, and his wife, Ida, daughter of John Trivett Nettleship [q.v.], animal painter. Augustus John also had two other sons and four daughters. Caspar John's mother died when he was four. Due to his father's love of gypsies and itinerant life-style during the years preceding World War I, John was not subjected to any form of systematic schooling until the age of nine, when he was sent with his brothers to Dane Court Preparatory School in Parkstone, Dorset. There he won the prize for the best gentleman in the school and a copy of *Jane's Fighting Ships*, and it was this, together with a wish to seek a more orderly existence, that inspired him to join the Royal Navy. In 1916 he entered the Royal Naval College, Osborne, in the Isle of Wight, at the age of thirteen. He transferred to the RNC, Dartmouth, in 1917 and passed out eighty-third of a hundred in 1920. He became a good long-distance runner and rackets champion: physical fitness being of paramount importance in naval training, he developed a lifelong love of athletics.

His midshipman years were spent aboard the flagship of the Mediterranean Fleet, the *Iron Duke*, against a background of Graeco-Turkish disturbances and the problem of Russian refugees caused by the revolution of 1917. It was at this time (1922–3) that the future of naval aviation was being debated. The issue caught his imagination, and, heeding advice he had previously had from Admiral of the Fleet Lord Fisher [q.v.] to 'Look forward, not backward', decided that this was to be his future. 'I was the angry young man of the day', he later wrote, and questioned the need to clutter up the navy with outdated battleships. He envisaged the role

of the aeroplane as broadening the naval horizon, and during his qualifying exams for lieutenant in 1925 (he gained first class certificates in gunnery and torpedo), he applied to train as a pilot in the Fleet Air Arm, then under the dual administration of the navy and Royal Air Force. His request was not welcomed and was considered a grave risk to his promotion prospects. However, after gaining his wings in 1926, he became passionate about flying, and thenceforth devoted his naval career to building up the strength of the Fleet Air Arm, of which he was one of the founding fathers.

In the aircraft carrier *Hermes* he spent the years 1927–9 in the China station during the warring between the communists and Chiang Kai-Shek's nationalist armies. On returning from China he bought his own aeroplane, an open cockpit Avro Avian, and had many flying adventures in France and England.

A loner by nature, he became known for his zeal, clear-headedness, loyalty, and resilience. He had few interests other than the affairs of the navy: his work was his life. Throughout the 1930s he devoted his time to all aspects of naval flying—deck landing, demonstrating, surveying, testing new designs, and night flying—and gradually became involved in the design and production of naval aircraft.

Promotion to lieutenant-commander and commander came in 1933 and 1936 respectively, and, as a result of Italy's war with Abyssinia, he spent 1936 based in the Western Desert outside Alexandria, attached to the carrier *Courageous*. He spent much time practising carrier night flying which was then an innovation. In 1937 he was appointed to the Admiralty's naval air division, where he worked ceaselessly to free the Fleet Air Arm from what he described as 'the folly' of dual control between the navy and RAF. The RAF's hold on the FAA was ended with the 'Inskip award' in July 1937.

During the war of 1939–45 he spent eighteen months as second-in-command of the cruiser *York*, patrolling the North Sea, participating in the Norwegian campaign, and transporting arms around the coast of Africa to Egypt for the campaign in the Western Desert. Subsequently he had eighteen months at the Ministry of Aircraft Production (he was promoted to captain in 1941), and in 1943–4 he was in the USA as naval air representative in the British Admiralty delegation in Washington, and naval air attaché at the British embassy. His main task, and one which he considered of supreme importance, was to procure US naval aircraft for the under-equipped FAA and to set up the organization and training of British pilots in Canada and the USA. His meeting with the Russian aircraft designer Igor Sikorski, was in large part responsible for the introduction of the helicopter into its first practical military use by the navy after the war. He spent the last year of the war in home waters in command of two aircraft carriers, *Pretoria Castle*, and, until 1946, *Ocean*, a brand-new light carrier. As captain of *Ocean*, his main concerns were to boost the morale of his men (with the war ended, many longed to return home), and to maintain strict discipline in all flying activities.

In 1947 he attended the Imperial Defence College, London, for a course in world affairs and in 1948 he was given the command of the large and complex naval air station, Lossiemouth. He then returned to the Admiralty, first as deputy chief of naval air equipment and then as director of air organization and training. Although it was not his personal ambition, he excelled at administrative work, and his promotion to rear-admiral in 1951 ushered in his last year at sea, in command of the heavy squadron. In 1952 he was appointed CB. Two years (1952-4) at the Ministry of Supply updating naval aircraft preceded the important administrative post of flag officer (air) Home at Lee-on-Solent, the 'Clapham Junction' of all naval air stations. He was promoted viceadmiral in 1954, was appointed KCB in 1956, and was made a full admiral in 1957, the year he became vice-chief of naval staff to Earl Mountbatten of Burma [q.v.]. This was a period of great uncertainty for the navy, and ranked as one of the most demanding periods of concentrated activity experienced by a naval officer in modern times.

He crowned his career by becoming first sea lord (1960-3), the first ever naval aviator to have done so, and was appointed GCB in 1960. The major issues he dealt with while in office were characteristic of those facing any first sea lord as professional head of the navy but particularly those concerned with plans for the building of a new generation of large aircraft carriers. In 1962 he was promoted admiral of the fleet but later declined a peerage offered to him by Sir Alec Douglas-Home (later Lord Home of the Hirsel).

He embraced a number of widely differing jobs throughout the 1960s and 1970s: member of the government security commission (1964-73), chairman of the Housing Corporation (1964-6), member of the Plowden committee and of the Templer committee (1965), chairman of the Star and Garter Home for disabled servicemen, chairman of the Back Pain Association, and chairman of the tri-service Milocarian Club (athletics). He was made an honorary liveryman of the Fruiterers' Company.

John inherited his parents' good looks. He was tall and slim, with brown eyes and dark bushy eyebrows. His penetrating look and brusque speech, alarming to some, belied a sens-

itive, moody nature. He loved children, the works of Thomas Hardy [q.v.], and the music of Claudio Monteverdi. He had the ability to go straight to the point in argument, had a quick wit, and excelled at speech-making. His father painted several oil portraits of him as a child and young man, and there are a number of superb line drawings of him as a child.

He married in 1944 Mary, daughter of Stuart Vanderpump, of New Zealand. They had two daughters and one son. In 1978 he had both his legs amputated because of vascular trouble. His extraordinary courage and his determination to regain some degree of independence gave him the spirit to face life in a wheelchair for six years. His wife was a great support during this difficult time, and together they made their home in the Cornish village of Mousehole, where John became a much loved and familiar figure on the quayside and in the Ship Inn. He died 11 July 1984 at Hayle, Cornwall.

[Rebecca John (daughter), *Caspar John*, 1987; private information; personal knowledge.]   DAVID WILLIAMS

JOHNSON, ALAN WOODWORTH (1917-1982), organic chemist and university teacher, was born in South Shields 29 September 1917, the eldest in the family of three sons and a daughter of James William Johnson and his wife, Jean Woodworth. His father, trained at Cammell Laird at Birkenhead, worked as a nautical optician with John Lillie & Gillie at North Shields. A lifelong connection with the National Adult School Union culminated in his presidency in 1966. Johnson won a scholarship to King Edward VI's School, Morpeth, but his family could not afford to foster his interest in chemistry with a university education and so he went initially as a laboratory assistant to Swan Hunter and then as an analyst to Thomas Hedley, while studying part-time at Rutherford Technical College. An exhibition and some additional help from a local benefactor got him to Imperial College, London, where, his talent soon recognized, he became a Royal scholar. A first class B.Sc. (1938) led to a university research studentship and a Ph.D. (1940).

Employed now by ICI, he enlisted as a research assistant with (Sir) Ian Heilbron [q.v.] and (Sir) Ewart R. H. Jones on acetylene chemistry and the vitamin A synthesis. In spite of wartime hardships this was a very fruitful time for the development of Johnson's scientific and cultural interests, especially music, in close association with the neighbouring Royal College. At ICI Dyestuffs Division (1942) at Manchester he benefited from contact with a host of talented chemists and he was fortunate to work with J. D. Rose [q.v.] on acetylene chemistry in the new exploratory research section. He burned much midnight oil writing his two-volume *The Chemistry of the Acetylenic Compounds* (1946 and 1950), timely works in a newly developing area of organic chemistry.

He went to Cambridge (1946) as one of the first of the ICI fellows and later became an assistant director of research (1948) and a fellow of Christ's College (1951). There A. R. (later Lord) Todd introduced him to natural products with exciting studies on insect pigments. In that favourable environment he could not fail to prosper, with able collaborators from all over the world, while organic chemistry derived unprecedented benefit from the discovery and application of many novel techniques, which Johnson applied skilfully to a wide range of problems. Microbial metabolites, from *Penicillium* species, led to studies on modern aspects of aromatic character, while the actinomycins introduced him to vital polypeptide studies. His great opportunity came with vitamin B12, at that time a terrifyingly large and complex molecule, and the degradation fragments isolated at Cambridge were vital in elucidating the complete structure. Most importantly these researches furnished fundamental problems of corrin and porphyrin synthesis which occupied Johnson for the rest of his career in chairs at Nottingham (1955–68) and at Sussex (1968–82). At Nottingham he was head of the chemistry department, deeply involved in the design of a splendid new building, and at Brighton he was also honorary director of the Agricultural Research Council Institute of Invertebrate Chemistry and Physiology.

His loyalty to the Chemical Society (later the Royal Society of Chemistry) was great indeed and his service on its behalf outstanding. More than two-thirds of his publications are in its journal, and he was its honorary secretary (1958–65), vice-president (1965–8), and finally president (1977–8). A clear and stimulating lecturer, much in demand world-wide, he gave four of the society's named lectures and received the Meldola medal (1946) and its award for synthetic organic chemistry (1972). Elected FRS in 1965, he served thrice on the Royal Society's council, and was a vice-president (1981–2) and Davy medallist (1980). His 1978 presidential address was on 'The Interface of Academic and Industrial Chemistry', his industrial connections were continuous, and he became a member of the National Research Development Corporation in 1976 and the National Enterprise Board from 1981.

Notable as were Johnson's scientific contributions, his influence on the many associated with him was at least as important. Remarkably and perennially youthful in appearance and spirit, all who came under his spell were affected by the enthusiasm he radiated and his deep love for chemistry. No gathering could for long be unaware of his presence, for he was a grand social ice-breaker. But behind his ever cheerful and informal approach there was a dedicated and extraordinarily hard-working scientist.

Johnson married in 1941 Lucy Ida Celia, daughter of William Lionel Gaulton Bennett, a gentleman of leisure; they had a daughter and a son. Johnson died at his home in Lewes 5 December 1982, two months after his retirement.

[Sir Ewart Jones and R. Bonnett in *Biographical Memoirs of Fellows of the Royal Society*, vol. xxx, 1984; private information; personal knowledge.]　　　E. R. H. JONES

JOHNSON, DAME CELIA (1908–1982), actress, was born 18 December 1908 at Richmond, Surrey, the younger daughter and second of three children of John Robert Johnson, physician, of Richmond, Surrey, and his wife, Ethel Griffiths. She was educated at Miss Richmond's private school and then at St Paul's Girls' School, Hammersmith. Having obtained a GCE first-year award to the Royal Academy of Dramatic Art, she won, during her training there, a special prize as well as a French prize.

In 1928 she played her first professional part as Sarah in *Major Barbara* at the Theatre Royal, Huddersfield. The following year she went south, to the Lyric Theatre, Hammersmith, to take over from Angela Baddeley as Currita in *A Hundred Years Old*. In *Cynara* (1930) she stole the notices from the two stars, Sir Gerald du Maurier and (Dame) Gladys Cooper [qq.v.]. In 1931 she made her first trip to America to play Ophelia in *Hamlet*. Back in London the following year, she acted in *The Wind and the Rain*, a play set in a Scottish University. From that date she became a star, admired by all and unscarred by adverse criticism.

In 1935 she married (Robert) Peter Fleming [q.v.] (died 1971), traveller and author, the son of Major Valentine Fleming, a merchant banker. They had one son and two daughters. With a growing family in wartime, plus the prospect of her husband being overseas on active service, Celia Johnson faced difficulties. Typically, she solved them neatly. After playing Mrs de Winter in *Rebecca* in 1940 and then taking over from Vivien Leigh [q.v.] in *The Doctor's Dilemma* in 1942, she retired from the stage for five years, returning in 1947 to play the title role in *St Joan* with the Old Vic Company at the New Theatre. During the war she was an auxiliary policewoman. She starred in every kind of play, from Shakespeare to *Ten Minute Alibi* (1933), from Chekhov to *Flowering Cherry* (1957), by

Robert Bolt. She also starred in four successful plays written by William Douglas-Home: *The Reluctant Debutante* (1955), *Lloyd George Knew My Father* (1972), *The Dame of Sark* (1974), and *The Kingfisher* (1977). She was, to put it in a single phrase, a playwright's dream.

As a film star, she was equally successful as the captain's wife in *In Which We Serve* (1942) and as the housewife in *Brief Encounter* (1945). Her other films included *A Kid for Two Farthings* (1955), *The Good Companions* (1956), and *The Prime of Miss Jean Brodie* (1969). She was rarely seen on television but gave excellent performances in *Mrs Palfrey at the Claremont* (1973), *The Dame of Sark* (1976), and *Staying On* (by Paul Scott, q.v., 1980), in which she appeared with Trevor Howard, her co-star in *Brief Encounter*.

Though her dedication to her calling was immediately apparent to her public in the theatre, it rested on her shoulders lightly in her private life, which was devoted to her family, who lived in Nettlebed, Oxfordshire. She would often come out of a play before the end of its run in order to be with them. She seldom talked about the theatre and, when she did, approached the subject with a gay irreverence. She was a person of outstanding charm, with wide eyes, a retroussé nose, and a remarkable voice.

To her amused delight she was appointed CBE in 1958 and, to her great surprise, DBE in 1981. She died 25 April 1982, at Nettlebed, while playing bridge one weekend, during a pre-London run, in harness to the end.

[Personal knowledge.]

WILLIAM DOUGLAS-HOME

JOHNSON, PAMELA HANSFORD, LADY SNOW (1912–1981), novelist, dramatist, and critic, was born 29 May 1912 in London, the elder child (the younger daughter died as a baby) of Reginald Kenneth Johnson, a colonial administrator in the Gold Coast, and his wife, Amy Clotilda, daughter of C. E. Howson, actor and treasurer to Sir Henry Irving [q.v.] for twenty-five years. She was educated at Clapham County Secondary School, leaving at the age of sixteen to learn shorthand and typing. Her father had died five years earlier leaving only debts and there was no money for further education. She never regretted this deprivation, claiming that 'a course in Eng. Lit. has rotted many a promising writer' and finding in *Texts and Pretexts* (1933) by Aldous Huxley [q.v.] the key to her own 'higher education' in both English and French literature.

For five years she worked in a bank, writing (and occasionally publishing) verse, essays, and short stories in her spare time. In 1934 she won the *Sunday Referee*'s annual poetry prize—a subsidy for a book of her poems. She met and fell in love with Dylan Thomas [q.v.] who won the prize the following year. They contemplated marriage but decided amicably to go their separate ways. Her diaries for the two years of their friendship and his letters to her are now in the library of the University of New York at Buffalo.

She had an immediate success, both critical and commercial, with her first novel, *This Bed Thy Centre* (1935), although it was widely considered to be shocking. For a while she was disowned by her late father's family and was subjected to a shoal of anonymous letters. But, reassured and encouraged by Cyril Connolly [q.v.] and others, she abandoned the bank in favour of full-time writing. She went on to publish thirty-one novels; seven plays; a book of memoirs, *Important To Me* (1974); an appraisal of contemporary 'permissiveness' in the light of the infamous Moors murder case, *On Iniquity* (1967); and six Proust reconstructions (1958), the radio plays in which she placed Proust's characters in different settings and situations. These were commissioned by the BBC Third Programme and were repeated frequently during the 1950s. The tapes have now been erased.

After the great promise of her first novel it seemed that she had lost both her way and her touch. She wrote five undistinguished novels in four years and it was not until 1940 that she regained her true form with *Too Dear for my Possessing*. She was invited to review regularly for the *Sunday Times* and became well-known as a demanding but generous-minded and constructive critic.

She married in 1936 Gordon Neil (Stewart), an Australian journalist, and they had a son, Andrew, in 1941, and a daughter, Lindsay, in 1944. The marriage came to an end in 1949.

She was married again in 1950 to the writer Charles Percy (later Lord) Snow [q.v.]. Their considerable talents were complementary, their tastes and interests largely coincided, and they made a formidable and influential literary partnership. They travelled widely in the United States and in the Soviet Union and her books were published with success in both countries, leading to translations in many European languages.

She wrote ten of her best and best-known novels between 1950 and 1978 including *Catherine Carter* (1952), her only novel with a theatrical setting; *The Unspeakable Skipton* (1959), a satirical comedy based on the character of Frederic Rolfe, Baron Corvo, and generally accepted as her masterpiece; *The Humbler Creation* (1959) about an unfashionable London parish, and *The Honours Board* (1970) set in the enclosed world of the teaching staff of a boys' preparatory school. Her last novel *A Bonfire* (1981) was published two months before her death.

Her work was popular in the best sense: she had a large 'library' following as well as attracting the respect and admiration of the literary pundits of the day. She was generous with help to young writers and was a warm-hearted and entertaining friend and companion. She was a fellow of the Royal Society of Literature and was appointed CBE in 1975. She had four American honorary degrees. A sufferer for thirty years from migraine she helped to found the Migraine Trust in 1969.

She had a second son, Philip, in 1952 by her second husband, C. P. Snow. She died 18 June 1981 in the London Clinic after a series of strokes.

[Pamela Hansford Johnson, *Important To Me*, 1974; Isabel Quigly, *Pamela Hansford Johnson* in Writers and their Work, no. 203, 1968; personal knowledge.] ALAN MACLEAN

JOHNSON-MARSHALL, SIR STIRRAT ANDREW WILLIAM (1912–1981), architect, was born 19 February 1912 in India, the elder son (a daughter died in early infancy) of Felix William Norman Johnson-Marshall, a civil servant of Scottish descent who administered the salt trade, and his wife, Kate Jane Little. Stirrat's childhood was overshadowed by the fact that his father worked abroad, at first in India and later in Baghdad. This meant that while he and his brother (Professor) Percy Johnson-Marshall had stirring adventures, there were attendant miseries connected with being at boarding-school in England, and sometimes with being farmed out to reluctant hosts during the holidays.

He left the Queen Elizabeth School at Kirkby Lonsdale in 1930 as head boy and captain of cricket and rugby. He then went to the Liverpool School of Architecture which at the time was in transition from the classical mode of architecture to the modern movement proselytized in European schools like the Dessau Bauhaus. Gropius, its founder, made a profound impression on Johnson-Marshall during a visit to Liverpool. This may have been the point where the latter's devotion to the ideal of service to the community fused with his love of science and engineering, and his strong aesthetic feelings for a kind and inspiring environment for people.

He left Liverpool in 1935 with a first-class honours degree and in 1936 became ARIBA. In 1937 he married Joan Mary Brighouse, a fellow architect from the Liverpool School whose father Christopher Brighouse was an architect in Ormskirk and had given Stirrat holiday jobs.

The Johnson-Marshalls had three sons and a daughter.

Local authority posts first in Willesden and then in the Isle of Ely confirmed his conviction that public service was the right vehicle for the pursuit of social architecture and incidentally introduced him to people like (Sir) Donald Gibson and (Sir) Alan Harris who were to be influential in his career after the war.

His war in the Royal Engineers found him in Singapore in 1942 with the Japanese about to inflict an ignominious defeat on the British. When almost everybody had given up the ghost he valiantly organized a desperately dangerous escape with two colleagues island-hopping by small boat and running the Japanese gauntlet to Colombo. He hardly ever spoke about his ordeal and when he did it was dismissively. He was fascinated, however, by the nature of courage for the rest of his life.

The death of his second son in tragic circumstances brought him back to England in time for D-day preparations. He found himself in the Camouflage Development and Training Centre at Farnham Castle working in a team of many skills to design and build inflatable tanks, guns, and other imitation war machines whose disposition as decoys successfully fooled the Germans into defending non-existent landing sites. This was a seminal experience in the comprehensive organization of designing and manufacturing skills to make a new and unprecedented product. It remained the essence of his dream of a revitalized building industry for the rest of his life.

From 1945 to 1948 he was deputy architect of Hertfordshire County Council and from 1948 to 1956 chief architect of the Ministry of Education. If anyone could be called the architect of this country's achievement in building for education since the war, it must be Johnson-Marshall. His work won international recognition as a unique example of the marriage of innovative building technology to the satisfaction of user needs and the fulfilment of a desperately urgent need to build quickly and within stringent budgetary limits.

He was extraordinarily generous to his colleagues, being particular to make sure that the younger architects who were working with him got full credit for work which may well have originated in his own mind. At the same time his retiring modesty disguised a ruthless determination to get done what he believed it was right to do. This astonished those who mistook his shyness for timidity. He loved being with the young and had the gift of making them more confident and competent because of his interest. He was passionate about the basic elements of interaction between buildings and people — such as colour, light, sound, the sensitive control of temperature and air movement, and the more tactile experiences from the ergonomics of furniture design to the feel of a handrail. These passions united the design teams which he led and heightened their awareness of such issues. This in turn raised their creativity to high levels.

During the 1950s Johnson-Marshall was one of the leaders of the revolt of public architect members of the RIBA, in a majority for the first time in its history, who turned the Institute into a socially concerned body dedicated to improving the managerial and technical competence and political influence of the profession. His role in this revolution was to lead from behind. He hated writing and public speaking, although he was a first-class editor and literary critic, and refused many pressing invitations to become president. He was vice-president of RIBA in 1964-5. He was a clever devil's advocate in the testing of theories and rehearsal of campaigns but did his most creative work over a pink gin in pipe-smoke filled rooms late at night, or along the corridors of power. This made him enemies and earned him the reputation of a Machiavellian conspirator among those who failed to understand the profoundly moral and selfless nature of his motives.

His partnership with (Sir) Robert Matthew [q.v.], whom he joined in private practice in 1956, was not an entirely happy one, for they responded differently to the pressures of survival in the market place. The Edinburgh and London offices of the firm went on to work virtually as separate practices for a number of years.

The business flourished however and the two senior partners shared the high ground of strategy while agreeing to differ on the short term. In London Johnson-Marshall pursued his ideal of a better method of building through prefabrication mainly in the University of York whose earlier buildings were constructed in a heavily modified form of CLASP—the user-driven building system promoted by a group of local authorities. This project, preceded by the Commonwealth Institute and followed by the Central Lancashire New Town, embodied his major contribution to architecture and planning after he left the public service.

Johnson-Marshall retired in 1978. Towards the end of his life he knew that he was swimming against the tide of a more opportunistic architectural and political environment and this saddened him. However the record shows that he had his chance in his time to make history and he seized it with distinction. He became FRIBA in 1964, was appointed CBE in 1954 and was knighted in 1971. He died in his Bristol office 16 December 1981.

[Andrew Saint, *Towards a Social Architecture—The Role of School Building in Post-war England*, 1988; Stirrat Johnson-Marshall, *Five Commemorative Speeches given at the Royal Society of Arts*, 24 March 1982, privately published; personal knowledge.]　　　　　　　　ANDREW DERBYSHIRE

JONES, LADY (1889-1981), novelist and playwright. [See BAGNOLD, ENID ALGERINE.]

JONES, ALLAN GWYNNE- (1892-1982), artist. [See GWYNNE-JONES.]

JOSEPH, SIR MAXWELL (1910-1982), businessman, was born 31 May 1910, at 21 Whitechapel Road, London, the second of the three sons (there were no daughters) of Jack Joseph, a journeyman tailor, and his wife, Sarah Orler, who had been a schoolteacher in Stroud, Gloucestershire. Joseph seems to have skipped most of his education, and by his own account learned little from a private tutor. He learned shorthand and bookkeeping at Pitman's College. He was influenced in his choice of job by his father, who dealt in property in a modest way, and started work with a Hampstead estate agent putting up signs at thirty shillings a week. After several similar and not very successful jobs he set up his own business at the age of nineteen, with £500 of capital given by his father. He was quickly successful, and by the outbreak of World War II in 1939 was already prosperous, although trading on wide margins. He served in the Royal Engineers as a lance-corporal.

After the war Joseph turned his attention to dealing in small hotels (some rather more like boarding-houses than hotels), and these formed the basis of his Grand Metropolitan Hotels group. At about this time he started to call himself 'Maxwell', perhaps encouraged by embarrassed friends. Previously he had been Max. From this time onwards Joseph was able to give full play to an extraordinary instinct for property; he foresaw the enormous growth of leisure and travel earlier than others, and put a feeling for the value of properties ahead of the logic which controlled them. He would walk around a hotel, make a few notes, and put a bid price on it. He was sometimes beaten, but more often successful. Although it was in hotels that he made his fortune, he deluded himself into thinking he could run them as well as trade in them; others managed the daily business for him. In 1957 he diversified into a 'shell' company called Lintang, whose subsequent unhappy history in the 'Jasper affair' brought about important reforms in company law. Although Joseph had no liability for the losses incurred, he made a point of repaying them over the following years.

At this time, as well as carrying on the business of Grand Metropolitan, Joseph was trading on his own account in a number of associated deals. It must be borne in mind that in these times there was no capital gains tax, and the rules controlling business and finance either did not exist or were loosely enforced. By the end of the 1960s, despite many rebuffs and disappointments, Joseph had established himself as a brilliant and intensely ambitious financier, spurred by a strong resentment of the City establishment and what he saw as its clannish exclusiveness.

In 1971 he made a dramatic, fiercely opposed, but successful bid for the London brewery group, Truman Hanbury Buxton. This was the first invasion by outside interests of the British brewery business, and was quickly followed by the even bigger and more dramatic take-over of the Watney Mann business, at £400m the biggest such deal in British business history. A period of consolidation followed. The costs of the acquisitions had to be absorbed; much restructuring was needed; the group's shares fell. It was a difficult and testing time for Joseph. But he had by now built a team of lieutenants who restrained his wilder impulses while helping him to run the enormous conglomerate which Grand Metropolitan had become (by this time the group's interests included apart from hotels in the United Kingdom and abroad, beers, wines, and spirits, dairies, industrial catering, and gambling). By 1980 the group's affairs had been sufficiently stabilized for Joseph, aided and advised by a small group of close colleagues, to get back on the take-over trail; seeing the USA as the best place for diversification, he fought and won a bitterly contested bid for the Liggett drinks and tobacco group, which set the style for many British businesses to acquire their agents in this and other parts of the world.

A bid for the Coral bookmaking company was frustrated by the Monopolies Commission in 1980. But in August 1981 there was a final dramatic take-over when Grand Metropolitan bought Intercontinental Hotels from the troubled Pan American Airlines. This deal, like most of Joseph's, was marked by a speed of decision and action which took competitors by surprise. Joseph handed over direction of the group to Stanley Grinstead in July 1982, after an operation for cancer. He was knighted in the New Year honours of 1981.

Joseph was a quiet, almost shy man, who always felt himself not accepted by the world he had so successfully dominated. His deals, which seemed rash to many business classicists, were always founded on a sound, if instinctive, understanding of the capital values of the 'bricks and mortar' of any acquisition. He was sometimes a heavy gambler. His stamp collection of Cape of Good Hope triangulars was famous. He had many friends among his competitors, but they were men like himself, arrivals on the business scene, not old inhabitants.

In 1932 Joseph married Sybil Nedas. They had a son and a daughter. They were separated in 1953 and divorced in 1981. In 1981 he married Eileen Olive Simpson. Joseph died in London 22 September 1982, leaving an estate of over £17,000,000.

[*The Times*, 24 September 1982; private information; personal knowledge.]

THOMAS JAGO

K

KAPITZA, PIOTR LEONIDOVICH (1894–1984), physicist and Nobel prize-winner, was born in Kronstadt (near St Petersburg) 9 July 1894, the younger son of General Leonid Petrovich Kapitza, a military engineer, and his wife, Olga Ieronimovna, daughter of General Ieronim Ivanovich Stebnitski. His mother was a specialist in children's literature and folklore. He was educated at the local 'Realschule' from which he graduated with honours in 1912. He then entered the electrotechnical faculty of the St Petersburg Polytechnical Institute where he was appointed to a staff position after graduating in 1918. In spite of the difficult conditions of World War I, the revolution, and the civil war, he took an active part in the research activity initiated by his teacher A. F. Joffé and soon acquired a reputation for his skill and originality. In December 1919 he lost his father and his young son and in January 1920 his wife and newly born daughter in the prevailing epidemics and, partly in order to take his mind off this devastating blow, Joffé arranged for him to take part in a Soviet mission to renew scientific contacts with western Europe. He and Joffé went to Cambridge in 1921 and Sir Ernest Rutherford (later Lord Rutherford of Nelson, q.v.) agreed to have Kapitza stay in the Cavendish Laboratory for a winter to gain research experience. Rutherford was much impressed by Kapitza's success in tackling his first problem and a very cordial relationship was soon established between them. As a result Kapitza stayed in Cambridge for thirteen years rather than just for a winter as originally intended.

Some of his work on alpha particles required the use of high magnetic fields and Kapitza developed special techniques for producing such fields only momentarily, lasting long enough to measure what was needed, but not long enough to overheat the magnet coil. But once the equipment for producing these fields had been developed, Kapitza's interest shifted from nuclear physics to the study of magnetic phenomena in the new range of high fields available. Such properties become more interesting at very low temperatures and Kapitza developed ingenious methods of liquefying hydrogen and helium to cool specimens to temperatures close to the absolute zero. To house all this new activity more space was needed and Rutherford persuaded the Royal Society to provide £15,000 from the Ludwig Mond bequest for building what came to be known as the Royal Society Mond laboratory in the courtyard of the Cavendish Laboratory. The new laboratory was opened in 1933.

Kapitza rose rapidly in the scientific establishment. He gained his Ph.D. in 1923; in 1925 he was appointed assistant director of magnetic research and was elected to a fellowship at Trinity College (honorary fellow, 1966). In 1929 he was elected FRS and also a corresponding member of the Soviet Academy of Sciences (he became a full member in 1939) and in 1931 he was appointed Royal Society Messel professor. However, he had little opportunity to exploit the facilities of his new laboratory, for in the summer of 1934, following a routine holiday visit to the Soviet Union, he was refused permission to return to Cambridge. In spite of many appeals, the Soviet authorities would not reconsider their refusal and eventually he agreed to become director of a new Institute for Physical Problems in Moscow, built specially to his requirements. During his thirteen years in Cambridge Kapitza became somewhat of a legend, not only for the originality of his scientific achievements, in which he brought his early engineering training to bear on problems of physics, but also for his ebullient and sometimes eccentric personality and for the stimulating scientific discussions at his personal seminar, which was known as the Kapitza Club. He can be said to have laid the foundations of a new school of low temperature and solid state physics in Cambridge, which has continued ever since.

In his new Moscow Institute, Kapitza changed the direction of his personal activity from magnetism to study of the strange behaviour of liquid helium at very low temperatures. In the course of a series of ingenious experiments he discovered superfluidity and provided the basis for a fundamental theory of quantum liquids developed by Lev Landau, the Institute's house theoretician. He also made important contributions to the techniques of gas liquefaction and his turbine method of liquefying air became important during World War II because of its relevance to the bulk production of oxygen for the steel industry. Towards the end of the war Kapitza was briefly in charge of the Soviet oxygen industry. In 1946, however, his fortunes again suddenly changed. Probably mainly because of his criticisms of L. P. Beria, the notorious chief of secret police who headed the initial stages of the Soviet atom bomb project, Kapitza was suddenly dismissed from his Institute, allegedly for shortcomings in his oxygen work for which only a few months earlier he had received a high government decoration.

Though dismissed he was not arrested, and for the next seven years lived at his dacha (country house) outside Moscow where he managed to continue his scientific work, but yet again in a new direction. He set up a laboratory in some outhouses of the dacha and his main effort went into developing powerful new microwave sources for the intense heating of plasma. Effectively, he switched from very low to very high temperature physics. After Stalin's death in 1953, Kapitza was reappointed to his Institute and continued this plasma work to the end of his life. His aim was to reach temperatures high enough to produce thermonuclear fusion, but although he made interesting contributions to plasma physics he did not achieve his aim. Thermonuclear fusion has indeed not yet been achieved anywhere, but powerful machines such as the Tokamak seem to offer greater promise. In the course of his fifty years in the USSR, Kapitza became an important figure in the Soviet scientific and cultural establishment and often courageously promoted liberal causes. His work was recognized by many awards and academic honours both in the USSR and in the West, culminating in the Nobel prize in 1978.

In 1916 he married Nadezhda Kyrillovna, daughter of General Kyrill Kyrillovich Chernosvitov but, as already mentioned, she died in 1920. His second marriage in 1927 was to Anna Alekseevna, daughter of Admiral Alexei Nikolaievich Krylov, mathematician and naval architect, and she and their two sons survived him. Kapitza died in Moscow 8 April 1984.

[D. Shoenberg, *Biographical Memoirs of Fellows of the Royal Society*, vol. xxxi, 1985; private information; personal knowledge.]

DAVID SHOENBERG

KAY, (SYDNEY FRANCIS) PATRICK (CHIPPINDALL HEALEY) (1904–1983), ballet dancer. [See DOLIN, SIR ANTON.]

KEATING, THOMAS PATRICK (1917–1984), artist, restorer, and faker, was born 1 March 1917 in Forest Hill, London, the fourth child in the family of four sons and three daughters of Herbert Josiah Patrick O'Brian Keating, a house painter, and his wife, Louisa DeLieu, a charwoman. He attended the local infants' school in Dalmain Road, where he learned to draw, and at the age of seven ran away to stay with his maternal grandmother in Eltham, Kent. There he attended Roper Street School. Three years later he returned to Forest Hill and Dalmain Road School where he won a paintbox for swimming a width of the local baths underwater. Painting and drawing became his obsession.

Leaving school at fourteen Tom Keating took a variety of jobs, including working as a latherboy and as a lift boy at the Capitol cinema in the Haymarket, before joining his father as a decorator. It was there he learned decorative skills and how to mix paint. In the evenings he attended art school in Croydon and Camberwell.

During World War II Keating served as a stoker in the Royal Navy and saw service in the Far East and on Russian and Atlantic convoys. After his ship was torpedoed he was invalided out of the navy and, at the age of thirty, he became a full-time art student at Goldsmiths' College, south London, on an ex-serviceman's grant. He failed his exams twice. He had wanted to teach and without a diploma that career was closed to him. It was the start of his bitterness towards an establishment he always viewed as hostile.

He joined a restoration studio in London and while there was asked to make copies of a number of paintings. He was later horrified to discover them being sold as genuine. It was then that he decided to flood the market with fakes (or Sexton Blakes as he called them in his own variant of cockney rhyming slang) as a way of striking a blow for impoverished artists against rich dealers and collectors and of getting back at a world which he felt was both shunning and using him.

During the next twenty-five years he worked as a free-lance restorer, his most important commission being the two years he spent restoring the Laguerre murals at Marlborough House. But all the time he was painting both in his own style and in that of other artists including Rembrandt, Constable, Kreighoff, Degas, Renoir, and Turner. He later admitted to putting more than 2,000 fakes in the style of more than 130 artists into circulation. He released them on to the market gradually either by giving them away, selling them to recover the cost of his materials, or putting them into small auctions where they would not arouse suspicion.

Keating's faking became public knowledge in 1976 when it was revealed that thirteen watercolours attributed to Samuel Palmer [q.v.] were not by Palmer. Keating wrote to *The Times* and admitted he had done them. The newspaper hunt to find him (he was touring the West Country on his motor cycle) and subsequent revelations turned him into a folk hero. His trial at the Old Bailey in 1979 was stopped because of his ill health, and he returned home to Dedham to continue painting. It was important, he said, that his faking should be discovered in order that the 'joke' should become public knowledge. 'If I had wanted to be a real faker', he later said, 'you would never have heard of me.' He also said that fooling the experts was his greatest joy in life; the thought of it made him helpless with laughter. Keating's object was never to make money. He was generous to a fault and remained poor throughout his life.

In 1982 he found new fame as presenter of Channel 4's *Tom Keating on Painters*, a series in which he talked about his favourite artists and demonstrated their style. This won him the Broadcasting Press Guild award for the best on-screen performance in a non-acting role. He followed this with a further series on the Impressionists. In 1983, 135 of his paintings were sold at Christie's for £72,000 and, for the first time in his life, he had real money. However, his health, which had never been good since the war, was declining rapidly. In 1943 Keating married Ellen, daughter of James Graveney, printer. They had a son, Douglas, and a daughter, Linda. Keating died in Essex County Hospital, Colchester, 12 February 1984.

[*The Times*, 13 February 1984; Tom Keating, Geraldine Norman, and Frank Norman, *The Fake's Progress*, 1977; personal knowledge.]
RICHARD FAWKES

KENDALL, SIR MAURICE GEORGE (1907-1983), statistician, was born 6 September 1907 in Kettering, Northamptonshire, the only child of John Roughton Kendall, engineering worker, of Kettering, and his wife, Georgina, of Standon in Hertfordshire. At matriculation stage in the Derby Central School, he was primarily interested in languages, but he developed late in mathematics and won a scholarship to St John's College, Cambridge, where he was a keen cricketer and chess player. He obtained a first class in part i of the mathematical tripos (1927) and was a wrangler in part ii. He passed into the administrative class of the Civil Service in 1930.

At the Ministry of Agriculture he was soon made responsible for statistical work, and this led to his becoming co-author with George Udny Yule [q.v.] of *An Introduction to the Theory of Statistics* (11th edn. 1937). In 1941 he became statistician to the British Chamber of Shipping, and in the following years published many papers in the theory of statistics, primarily in the theory of rank correlation coefficients (one of the two most important of which is named after Kendall's discovery of it in 1938), of paired-comparison experiments and in the analysis of time-series, in the construction of a theory of randomness, and most originally and most intricately in the theory of symmetric functions of the observations in a sample. His two-volume treatise *The Advanced Theory of Statistics* (1943, 1946), was followed by the award of a Cambridge Sc.D. in 1949.

In 1949 he was appointed professor of statistics in the University of London, holding this chair at the London School of Economics, where he founded a research techniques division and published the first major dictionary of statistical terms (1957) and the first comprehensive bibliography of statistical literature (3 vols., 1962, 1965, 1968).

In 1961 he again changed the direction of his career joining the computer consultancy now called SCICON, where he was successively scientific director, managing director, and chairman. His interests spilled over from statistics into many adjoining fields: he was at various times president not only of the Royal Statistical Society and the Institute of Statisticians, but also of the Operational Research Society and the Market Research Society. When he retired from SCICON in 1972, he began yet another, and perhaps his most testing, career as the first director of the World Fertility Survey, a very large multinational sample survey project, from which illness forced his retirement in 1980.

Kendall had an orderly mind and an enormous capacity for hard work. He was at his best in the organization of research, where he had an almost infallible capacity to delegate responsibility. His own research was extensive, including seventeen books and about seventy-five papers in statistics alone; a memorial volume *Statistics: Theory and Practice* (ed. Alan Stuart, 1984) reprints seventeen of his papers and contains a select bibliography. But perhaps the greatest influence that he exerted was through *The Advanced Theory of Statistics*, which was revised into a three-volume work (1958, 1961, 1966) and later retitled *Kendall's Advanced Theory of Statistics*, and became the leading treatise on the subject.

Kendall's literary style, lucid, balanced, and sometimes ironical, reflects his early interest in language. He constantly played with words, writing sometimes under the anagrammatic names Lamia Gurdleneck and K. A. C. Manderville many pastiches, of which the most well known is 'Hiawatha Designs an Experiment'. His inaugural lecture 'The Statistical Approach' and his presidential address to the Royal Statistical Society 'Natural Law in the Social Sciences' will convince doubters that it is possible to write interestingly of statistical matters—they are both (as also are 'Hiawatha' and the well-known paper on the cube law in general election results) reproduced in the memorial volume. Kendall was a tall, good-looking, and friendly man, not the least of whose virtues were his avoidance of controversy and his consequent ability to get on with all those with whom he worked.

Many honours came to him. In 1968 the Royal Statistical Society, of which he had been president in 1960-2, awarded him its highest distinction, the Guy medal in gold. He was made D. Univ. of Essex (1968) and Lancaster (1975). In 1970 he became a fellow of the British Academy, and in 1974 a knighthood was conferred on him for services to the theory of statistics. When, finally, he retired from the World Fertility Survey, his exceptional work there was recognized by the award of a United Nations peace medal.

Kendall was twice married. In 1933 he married Sheila Frances Holland, daughter of Percy Holland Lester, Church of England vicar. They had two sons and a daughter. After the dissolution of this marriage in 1947, he married in the same year Mrs Kathleen Ruth Audrey Whitfield, daughter of Roland Abel Phillipson, dentist. They had one son. Kendall died in Redhill, Surrey, 29 March 1983.

[Private information; personal knowledge.]

ALAN STUART

KENNEDY, JAMES (JIMMY) (1902–1984), popular songwriter, was born 20 July 1902 at Omagh, county Tyrone, Ireland, the elder son of Joseph Hamilton Kennedy, of the Royal Irish Constabulary, and his wife, Annie Baskin. He was educated at Trinity College, Dublin, and spent some time as a graduate teacher before joining the Colonial Service as a political officer. Bored, he turned to songwriting, and in 1929 his first song was published, 'Hear the Ukuleles'. 'The Barmaid's Song', a summertime success with Blackpool holidaymakers in 1930, resulted in the London music publisher Bert Feldman offering him a contract as 'lyric editor'. He supplied words for 'Oh Donna Clara' and 'Play to me Gipsy', but it was the twee, appealing 'Teddy Bears' Picnic' (1932) that made Kennedy a force in popular music, with an astonishing sale of over four million records.

'Isle of Capri', to music by an Austrian refugee, Wilhelm Grosz, was the first of a string of anecdotal songs which told, with admirable compression, a simple, sad love story:

Summertime was nearly over,
Blue Italian sky above.
I said 'Lady, I'm a rover,
Can you spare a sweet word of love?
She whisper'd softly 'It's best not to
    linger',
And then as I kiss'd her hand I could see
She wore a plain golden ring on her finger—
'Twas goodbye on the Isle of Capri.

An international success, this was followed in 1935 by the equally affecting 'Red Sails in the Sunset', and 'Harbour Lights'.

It was at this stage in his career that Jimmy Kennedy met the somewhat larger-than-life Tin Pan Alley tunesmith Michael Carr, son of a featherweight boxer known as 'Cockney' Cohen. The partnership between the brash Carr and the charming, shy (if money-conscious) Kennedy worked well, and together they wrote hit after hit: first 'The Chestnut Tree', then 'South of the Border' (another tragic tale, ending with the hero returning to find that his girl—'a picture in old Spanish lace'—had taken the veil), and the ill-fated 1939 song which proclaimed that the British troops were going to hang out their washing on 'The Siegfried Line'. As Kennedy recalled, it enraged Hitler, but not for long.

Following Territorial Army service, he became a captain in the Royal Artillery. A friend of those wartime years recalls his charm, capacity for friendship, and (on occasions) violent Irish temper. Kennedy's unerring gift for sensing the public taste led to his 1942 action dance called 'The Cokey Cokey'. Though primarily a lyricist he composed the music as well as the words for 'Roll Along Covered Wagon', in which he liked to accompany himself on a halting, self-taught piano. Regrettably, 'The Angels are Lighting God's Little Candle' proved too much even for the general public to stomach, but success returned with 'An Apple Blossom Wedding', unseasonably celebrating the November 1947 marriage of the Princess Elizabeth.

Asked once 'Where do you get your inspiration?' Kennedy replied, with characteristic self-effacement, 'I don't think I've ever been inspired in my life. I regard songwriting as a skill and a craft.' At that skill and craft he was a master, and, in British popular music, something of a pioneer.

After the war Kennedy lived in America for some years, less happily in politically-torn Ireland, then settled in Switzerland as a tax exile when the teenage record-based rock and roll revolution shattered the dominance of Tin Pan Alley and its sheet music industry. ('Better to be rich and miserable in Switzerland than poor and happy in England', Kennedy told a colleague, without conviction.) Royalties poured in and honours came his way at last: a first Ivor Novello award (1971); an American ASCAP award (1976); an honorary D.Litt. from the University of Ulster (1978), of which, being 'merely a wordsmith', he was justly proud; and appointment as OBE in 1983. He became a fellow of the Royal Geographical Society and of the Royal Society of Arts.

In his later years Kennedy returned to Dublin under an Irish government tax concession, commuting at his own expense to London for meetings of the Songwriters Guild (later renamed BASCA), of which he was the much-loved conscientious chairman.

A lifelong charmer of women, he was three times married. In 1932 he married Margaret Elizabeth Winifred, daughter of Simon Galpin, hotelier. Their marriage was dissolved in 1946 and in the same year he married the actress and singer Constance, daughter of Harold Carpenter and former wife of Paul Ord Hamilton and John Lucas-Scudamore (both earlier marriages having been dissolved). After Kennedy's second marriage was dissolved in 1972 he married in 1976 Elaine Pobjoy, widow, daughter of Henry Holloway, solicitor. He died 6 April 1984 in a Cheltenham hospital, and he was buried at Staplegrove, near Taunton, Somerset. The eulogy at his memorial service on 2 October 1984 at St Giles-in-the-Fields, London, was given by his wartime colleague Denis Thatcher, husband of the prime minister.

[*The Times, Daily Telegraph*, and *Irish Times*, 9/10 April 1984; *Tyrone Constitution*, 25 July 1902; *Who's Who in Music*, 1972; *Performing Right Society Journal*, autumn 1972; BASCA *News*, summer 1984; Mark White, *You Must Remember This*, 1983; BBC sound archives; information from Elaine Kennedy, Derek J. Kennedy, Lesley Bray, and Denis Thatcher; personal knowledge. Excerpt from 'Isle of Capri' © 1934 Peter Maurice Music Co. Ltd., printed with kind permission of EMI Music Publishing Ltd., London WC2H OLD.]   STEVE RACE

KER, NEIL RIPLEY (1908-1982), palaeographer, was born 28 May 1908 in Brompton, London, the only child of Robert Macneil Ker, of Dougalston, Milngavie, later a captain in the 3rd battalion of the Argyll and Sutherland Highlanders, and his second wife, Lucy Winifred, daughter of Henry Strickland-Constable, of Wassand Hall, Yorkshire. Ker was educated at preparatory school in Reigate and at Eton. He went to Magdalen College, Oxford, in 1927, intending to read philosophy, politics, and economics, but on the advice of C. S. Lewis [q.v.] turned to English language and literature, in which he obtained a second class honours degree in 1931.

Even as an undergraduate, his antiquarian interests and enquiring mind led him to the Bodleian Library to examine the manuscript copies of the texts he was studying. His conviction that first-hand knowledge of the palaeographical and linguistic features of the manuscripts themselves was essential for the study of the texts they contained, influenced his choice of B.Litt. thesis: a study of the manuscripts of Aelfric's Homilies (1933). At this time classics, not English, was the traditional breeding ground for palaeographers, but Ker's palaeographical interests were encouraged by another Anglo-Saxon scholar, Kenneth Sisam, whose commendation of the benefits of 'foraging among manuscripts' exercised a strong influence on him.

Ker first began to give classes in palaeo-graphy in 1936, and in 1941 was appointed lecturer in palaeography. During the war, as a conscientious objector, he worked as a porter in the Radcliffe Infirmary, Oxford. In 1945 he was elected to a fellowship at Magdalen, and in 1946 was appointed reader in palaeography, holding both posts until his early retirement in 1968. At Magdalen he succeeded C. T. Onions [q.v.] as librarian (1955-68), and was vice-president (1962-3). In 1975 he was elected to an honorary fellowship.

For Ker, palaeography entailed the systematic examination of all aspects of the manuscript book, from the methods of its production, to its owners, past and present. He brought to the subject new standards of observation and description. He was concerned to advance learning by making information available, and to let the information speak for itself, often presenting it in list form. Although he had few pupils in the formal sense, numerous scholars benefited from the information and advice he was always ready to give: his name is ubiquitous in footnotes and acknowledgements.

Ker's acute powers of observation, visual memory, knowledge of manuscripts (especially those of English origin and provenance), and skills of description, are best represented by his many catalogues, among them: *Medieval Libraries of Great Britain* (1941, 2nd edn. 1964); the *Catalogue of Manuscripts containing Anglo-Saxon* (1957), which was awarded the British Academy's Sir Israel Gollancz memorial prize (1959) for its outstanding contribution to Anglo-Saxon studies; and his largest undertaking, *Medieval Manuscripts of Great Britain* (1969 onwards), in which he intended to catalogue all manuscripts in British institutions which had not yet been described; he was correcting the proofs of the third volume at the time of his death.

Ker's many other publications demonstrate the extent of his contribution to the study of scribal and binding practices, and library history. *English Manuscripts in the Century after the Norman Conquest* (1960), the published version of the lectures given as the first Lyell reader in bibliography, 1952-3, is perhaps his finest achievement: a remarkable study of English manuscript production in the twelfth century. A bibliography of his published writings is included in the Festschrift presented to him in 1978, and supplemented in his posthumously collected papers (1985).

Ker won national and international recognition. The honours awarded him include honorary doctorates from the universities of Reading (1964), Leiden (1972), and Cambridge (1975); and fellowships of the British Academy (1958), the Medieval Academy of America (1971), and the Bayerische Akademie der Wissenschaften (1977). He was a gold medallist of the Bibliographical Society (1975), and in 1979 was appointed CBE.

Ker showed astonishing powers of concentration and tireless energy—he was a keen hill walker. To readers in Duke Humfrey's library, his spare figure was a familiar sight, 'proceeding with quick step, all intent on the common pursuit, with a sharp eye . . . (missing) nothing, the very nose seeming to be on the scent for some recondite fact'. He was always as generous in sharing the fruits of his research with others as he was in giving hospitality at his homes in Kirtlington, near Oxford, and, after his retirement, at his lochside house at Foss near Pitlochry, and in Edinburgh.

He married his second cousin, Jean Frances, daughter of Brigadier Charles Bannatyne Findlay, soldier, in 1938. They had four children, a son and three daughters. He died 23 August 1982, after a fall on the hillside near his house at Foss.

[The Times, 25 August 1982; C. R. Cheney in Medieval Scribes, Manuscripts and Libraries: Essays Presented to N. R. Ker, 1978; T. J. Brown, obituary in Magdalen College Record, 1983; private information.] TERESA WEBBER

KERENSKY, OLEG ALEXANDER (1905–1984), consulting bridge engineer, was born in St Petersburg 16 April 1905, the elder son (there were no daughters) of Alexander Fedorovitch Kerensky, barrister, and later briefly the prime minister of the provisional government of Russia prior to the October revolution in 1917, and his wife, Olga Lvovna Baranovskaya. His parents, although not divorced until 1939, did not live together after 1917 and the two sons were brought up solely by their mother. After the Russian revolution they suffered much danger and many privations, including a short term of imprisonment, before escaping via Estonia to London in 1920. Thereafter the sons had regular meetings with their father in London and Paris. Their mother did not marry again.

Oleg and Gleb Kerensky attended a private school (Oakfield) in north London, both by 1923 obtaining the necessary examination qualifications to gain entry to London University. They studied engineering at the civil and mechanical engineering department of the Engineering College, St John Street (later to become first the Northampton Engineering College and finally developing into the City University). The brothers passed with honours, Gleb subsequently following a successful career as a mechanical engineer with English Electric, and surviving Oleg.

Following graduation in 1927 and a short period with Oxford city council, Oleg Kerensky joined the bridge department of Dorman Long & Co. Ltd., working first on the tender documents and then on the detailed design and erection scheme for the Sydney harbour bridge. Before leaving Dorman Long in 1937, he also worked on other schemes, including the Bangkok memorial bridge, Lambeth bridge in London, and the Kazr-el-Nil bridge in Egypt.

From 1937 to 1946 Kerensky worked with the steel construction firm of Holloway Brothers. His work during World War II included the construction of units for Mulberry harbours, which were used in the Normandy landings. Subsequently, becoming senior designer, he was in charge of designs for the Baghdad Railway bridge and the Lesser Zab and Euphrates bridges. He was naturalized in 1947.

Kerensky's final move was in 1946 to the London consulting firm of Freeman Fox & Partners, where he first worked on the design for a suspension bridge over the River Severn, not at that time proceeded with, although it later formed the basis for the Forth road bridge opened in 1964. Promoted to partnership in 1955, Kerensky assumed responsibility for many major bridges at home and abroad, including Auckland harbour bridge in New Zealand, the Ganga and Brahmaputra bridges in India, the Grosvenor railway bridge in London, the Medway bridge in Kent, and the Erskine and Friarton bridges in Scotland. There was notable use, among these works, of pioneering principles which Kerensky applied with courage and thorough attention to detail, including the use of high tensile rivets as well as bolts and forms of composite action between steel and concrete. In the case of the Medway bridge he overcame many problems in the construction of the then largest clear-span pre-stressed concrete bridge in the world.

Kerensky also had notable achievements in many highway schemes, such as on the two major motorways M2 and M5, the latter including the technically interesting Almondsbury interchange with the M4.

Throughout his life Kerensky was devoted to the service of his profession. He served as president of, successively, the Institutions of Structural Engineers (1970–1) and Highway Engineers (1971–2), and was an active member of the Institution of Civil Engineers and of the International Association for Bridge and Structural Engineering. He received many awards from these and other institutions, including the Telford and George Stephenson gold medals of the ICE, the gold medal of the IStructE, and the gold medal of the Worshipful Company of Carmen, London. He was elected FRS (1970) and was a founder member of the Fellowship of Engineering (1976). He took a lively interest in the education and training of engineers, and had a supportive association with Northampton Engineering College and then with the City University, from which he received an honorary D.Sc. in 1967. He was appointed CBE in 1964.

Kerensky was also very interested in engineering research, and worked in an advisory capacity on various committees, frequently as chairman, for the Welding Institute and the Construction Industry Research and Information Association (CIRIA). After many years serving as chairman of council of CIRIA, he was elected its president in 1979. He also gave advice on investigations into bridge topics carried out at the Transport and Road Research Laboratory. His most continuing, devoted service was probably on the British Standards bridge committee, (chairman from 1958), in particular overseeing for sixteen years the long preparation of the comprehensive and innovatory ten-part code, which was finally completed in 1984.

Kerensky was a man with a dominating presence, sometimes overwhelming to his juniors, but he showed great kindness and understanding. In 1928 he married Nathalie, daughter of James (Yaxa Moiyitch) Bely, from Russia, partner in Willis, Faber, Dumas, & partners, a firm of insurance brokers. They had one son, also called Oleg, a journalist and art critic. After his first wife's death in 1969, in 1971 Kerensky married his second wife Mrs Dorothy Harvey. She became crippled following a fall soon after the marriage and was shown the greatest of care and attention by her husband. He was a keen player of bridge, an interest he shared with both his wives. Throughout his life he liked to swim. In early days he was fond of playing rugby and latterly he was a proficient player of croquet. He died in London 25 June 1984, and his wife died the following year.

[M. R. Horne in *Biographical Memoirs of Fellows of the Royal Society*, vol. xxxii, 1986; personal knowledge.]   MICHAEL R. HORNE

KESWICK, SIR JOHN HENRY (1906-1982), merchant and chairman of Jardine Matheson & Co. Ltd. of Hong Kong and Mathesons of London, was born at Cowhill, Dumfries, 13 July 1906, the third and youngest son (there were no daughters) of Henry Keswick, of Cowhill, merchant, and his wife, Wynefride Johnstone, also of Dumfriesshire. He was educated at Eton and Trinity College, Cambridge, where he was remembered by his contemporaries for the ease with which he made and retained friends, a gift which remained with him all his life. He obtained third classes in both parts i and ii of the historical tripos (1927 and 1928).

One of his forebears was Andrew Jardine (1740-1793), father of the founder of Jardine Matheson and, in 1929, he followed his own father and grandfather and joined Jardines, initially in New York, moving on to Shanghai in 1931. During the 1930s he travelled widely in China and learnt to speak both the Shanghai dialect and serviceable Mandarin.

On the outbreak of war in Europe in 1939 he returned to Britain and was recruited by the Ministry of Economic Warfare. When Lord Louis Mountbatten (later Earl Mountbatten of Burma, q.v.) set up his South East Asia Command in 1943, he joined his personal staff as a political liaison officer in Chungking. In 1946 he returned to Shanghai to lead Jardine Mathesons' recovery in China and, at that time, he also became chairman of the British (and International) chambers of commerce, remaining there after the communists took over in 1948.

He was with Jardine Matheson as chairman in Hong Kong from 1952 to 1956, serving on the legislative and executive councils of Hong Kong until, in 1956, he returned to Mathesons in England where, apart from frequent visits to the Far East, he remained for the rest of his life. However, his involvement with, and influence on, Sino-British relations continued to grow and, as chairman (and later president) of the China Association for twenty-five years, president of the Sino-British Trade Council from 1963 to 1973, and one of the founder members of the Great Britain-China Centre, he played the leading part in British commercial relations with China during these difficult years.

Keswick's achievements were largely due to his genius for friendship and spontaneous hospitality, doubly enriched and shared by his wife. He was a great setter of scenes and creator of confidence and, during his time in Chungking, this flair for personal relationships was effective alike with the senior members of the British embassy, the Nationalist hierarchy, and the Communist Liaison Office, including Zhou Enlai. This entrée to the communist leadership was maintained for the rest of his life, and was of benefit to the British government as well as to British trade, including of course, that of his firm.

Thus it was Keswick who was largely instrumental in mounting the exhibition of British technology in Beijing in 1964 and leading so effectively numerous future trade missions to China thereafter. Even in the darkest days, his confidence in China was not extinguished and, when his brother, (Sir) William Keswick, and J. K. Swire of Butterfield & Swire called on the foreign secretary, Ernest Bevin [q.v.], in 1951 to announce that British firms had come to the end of their tether and must withdraw from China, Bevin replied, 'End of your tether is it?

Oh well, make sure some strands will 'old.' That they did so was, in great measure, due to John Keswick. For these services, he was appointed CMG in 1950 and KCMG in 1972.

Keswick had a remarkably pronounced back to his head, which so resembled the statues of the Chinese god of happiness, Fu Shen, that in the country Chinese often touched him in the belief that some of his happiness would rub off. It usually did with all who met him and this, and his exuberant self-mockery, lack of pomposity, and contemporaneous rapport with the young, were perhaps his strongest characteristics, his care for the young being marked not only in the numerous friendships with young people of many nationalities—his house was always open to Chinese and Japanese—but also by the work that he did for boys' clubs and other charities. The self-mockery is typified by his response to his mother's telegram when he was subjected to bombing during the Shanghai crisis of 1932: 'I am worried about you.' John Keswick: 'I am worried about myself.' Sometimes this lightheartedness was misunderstood. When he and his brother William were among the chief witnesses at the tribunal enquiring into the allegations about bank-rate disclosures in 1957, he was rebuked for the apparent frivolity of his replies although, at the end of the inquiry, the chairman, Lord Justice Parker (Baron Parker of Waddington, q.v.), exonerated Mathesons and their directors completely.

In 1940 he married Clare, youngest child of Gervase Henry Elwes [q.v.], singer, and his wife, Lady Winefride Mary Elizabeth Feilding, daughter of the eighth Earl of Denbigh. They had one daughter Maggie who, as the author of *The Chinese Garden, History, Art and Architecture* (1978), has written the authoritative modern book on this subject. Keswick died 5 July 1982 while fishing at Portrack, within half a mile of his birthplace, on the banks of his beloved River Nith.

[*The Times*, 8, 16, and 28 July 1982; family and private information; personal knowledge.]   JOHN SWIRE

KEYNES, LADY (1892–1981), ballerina. [See LOPOKOVA, LYDIA VASILIEVNA.]

KEYNES, SIR GEOFFREY LANGDON (1887–1982), surgeon, bibliographer, and literary scholar, was born in Cambridge 25 March 1887, the younger son and third and youngest child of (John) Neville Keynes, lecturer in moral science and later university registrary, and his wife, Florence Ada, daughter of the Revd John Brown, minister of Bunyan's chapel

at Bedford. His brother was John Maynard (later Lord) Keynes [q.v.] and his sister Margaret married A. V. Hill [q.v.], the physiologist. He was educated from 1901 at Rugby, before gaining in 1906 an exhibition to Pembroke College, Cambridge (of which he was made an honorary fellow in 1965), to study natural sciences, in which he received a first class (part i, 1909). He graduated MA (1913), B.Chir. (1914), and MD (1918). He also became FRCS (Eng. 1920), FRCP (Lond. 1953), FRCOG (1950), and FRCS (Canada 1956).

Keynes pursued several careers with equal intensity, and achieved prominence in each. After Cambridge, he trained at St Bartholomew's Hospital, where in 1913 he won a Brackenbury scholarship in surgery. At the outbreak of World War I he joined the Royal Army Medical Corps, and spent the following years in France, principally with the Royal Field Artillery and, after July 1916, with casualty clearing stations. This brought him practice as a surgeon on a vast scale; it also engendered a hatred of unnecessarily spilled blood and a lasting dislike of mutilation. He was mentioned in dispatches in 1918. He was a pioneer in blood transfusion, and after observing an American medical team went on to develop his own mechanism. The Keynes flask, which incorporated a device to regulate flow, became standard equipment after the war. In 1921 he joined P. L. Oliver in founding the London Blood Transfusion Service, and in 1922 he issued *Blood Transfusion*, the first textbook on the subject to be published in Great Britain.

After the war he joined the surgical team at Bart's headed by George Gask, and so encountered (Sir) Thomas Dunhill who with Sir B. G. A. (later Lord) Moynihan [qq.v.] remained among his most revered teachers and colleagues. Though he stayed in general surgery, Keynes gradually specialized in thyroid, gastrectomy, hernias, and breast cancer, supporting his work at Bart's with private practice. In 1928 he was appointed assistant surgeon at the hospital. He flouted widely held beliefs by his advocacy, from 1929 onwards, of conservative treatment for cancer of the breast, though he based his position on arguments of characteristically straightforward logic. The wisdom of his advocacy of radium, with moderate surgery when unavoidable, was acknowledged more slowly than he hoped, and he derived much satisfaction from living to see his work rediscovered fifty years later. Much of his research on the topic was however ended by his appointment as consulting surgeon to the RAF in World War II, a position from which he drew little satisfaction. He reached the rank of acting air vice-marshal in 1944. Work on the thyroid meanwhile inspired his demonstration in 1942 of the link between the thymus gland and myasthenia gravis, though here too he had to face considerable hostility. He retired from Bart's in 1952, and was knighted in 1955.

At Rugby Keynes was a contemporary and close friend of Rupert Brooke [q.v.], whose letters he edited in 1968. Brooke fostered a taste for seventeenth-century poetry, and in 1914 Keynes's first author bibliography, of John Donne [q.v.] (revised 1932, 1958, 1972), was published by the Baskerville Club, a group of Cambridge bibliophiles. Like all his subsequent bibliographies, it was compiled not least from his own library, which was to become one of the finest private collections assembled in the twentieth century. For John Evelyn [q.v.], in which he was much helped by A. T. Bartholomew of Cambridge University Library, he issued a preliminary handlist in 1916, only developing it into a full bio-bibliography in 1937: by then he could incorporate his Sandars lectures given at Cambridge in 1934. A revised edition followed in 1968. Another early enthusiasm, for Sir Thomas Browne, was encouraged by Sir William Osler [qq.v.]; Keynes's bibliography appeared in 1924 (2nd edn. 1968), and his edition of the works in 1928-31 (new edn. 1964).

His work on William Blake [q.v.], inspired originally by seeing plates from *Job* in a Cambridge shop window, was instrumental in establishing Blake as a central figure in the history of English art and literature. His bibliography, begun in about 1909, and much revised under the eye of John Sampson [q.v.] of Liverpool University, was published by the Grolier Club of New York in 1921. Keynes's own collection was to become formidable, rich particularly in drawings, separate prints, books from Blake's own library and the work of the Blake circle more generally. His edition of the *Writings* (3 vols., Nonesuch Press 1925) was succeeded by a series of single-volume editions culminating in that for the Oxford Standard Authors in 1966. These in turn were supplemented by editions of the drawings, studies of plates, a new census of the illuminated books (with Edwin Wolf II, 1953), an edition of the letters (1956, 3rd edn. 1980), and an iconography of Blake and his wife. Many of Keynes's shorter essays were gathered in *Blake Studies* (1949, 2nd edn. 1971). In 1949, with the help principally of Kerrison Preston, W. Graham Robertson's executor, Keynes established the Blake Trust, with the purpose of printing hand-coloured collotype facsimiles of all Blake's major illuminated books. On his death his collection was divided between a family trust, the Fitzwilliam Museum, and Cambridge University Library.

With the foundation by (Sir) Francis Meynell [q.v.] of the Nonesuch Press in 1923, Keynes found a congenial publisher for much of his literary work. Apart from Blake, he also edited volumes by John Donne, John Evelyn, William Harvey, William Hazlitt, and Izaak Walton [qq.v.] besides writing bibliographies of Jane Austen [q.v.] (1929) and Hazlitt (1931, revised edn. 1981): he became by far the most prolific of the press's authors. After World War II, the foundation of the firm of Rupert Hart-Davis brought another sympathetic outlet, and for the series of Soho Bibliographies begun in 1951 he wrote volumes on Rupert Brooke (1954, revised 1959) and Siegfried Sassoon [q.v.] (1962): he had first met Sassoon in 1933, and oversaw the production of several volumes of his poetry.

In his library Keynes created not only the foundations for a long series of literary bibliographies (work on Blake led naturally to the eighteenth-century philosophers, and a bibliography of George Berkeley, q.v., appeared in 1976), but also the means for an intensive study of seventeenth-century medicine and science. His biography of William Harvey (1966) gave him the opportunity, in its combination of medical, political, social, and literary history, to exploit his interests to the full: it gained him the James Tait Black memorial prize. He had already published a bibliography of Harvey (1928, revised 1953), while his several studies of the iconography helped his appointment in 1942 as trustee of the National Portrait Gallery, where he was chairman in 1958-66. These interests led also to bibliographies of John Ray (1951, revised 1956), Robert Hooke (1960), Sir William Petty (1971), and Martin Lister (1980) [qq.v.]. His work on Robert Boyle [q.v.], of whom he had an outstanding collection, was taken up by John Fulton of Yale, and work on Thomas Willis [q.v.] remained uncompleted at his death. As a bibliographer, Keynes's influence was magisterial; as a collector, he usually kept shrewdly ahead of others, and so avoided many of the pitfalls associated with collecting the merely fashionable. He was a member of both the Roxburghe Club and the Grolier Club. Most of his library passed by bequest and purchase to Cambridge University Library.

For much of his life Keynes's principal relaxation was the ballet, which he followed avidly, beginning with Serge Diaghilev's productions in the 1920s. Balletomania and admiration for Blake's art coalesced in his own ballet *Job*, with music by Ralph Vaughan Williams [q.v.], first performed by the Camargo Society in 1931 with sets designed by his sister-in-law Gwendolen Raverat [q.v.]: its success helped to ensure the survival of professional ballet in Britain.

Erect, vigorous, and constantly alert, Keynes could be a formidable figure, though his early shyness never entirely deserted him. Of all his vocations, surgery gave him the most satisfaction, and here his courage and decisiveness were ideal characteristics. In the eyes of many he lacked the patience to be a great hospital teacher; but he more than compensated for this by the lucidity of his writing, and the willingness with which he published his discoveries whether in medicine, bibliography, literature, or art history. As the owner of an important library and art collection he sometimes astonished scholars by his generosity, and here he was an inspiring guide.

He gained honorary degrees from Cambridge (Litt.D. 1965), Oxford (D.Litt. 1965), Edinburgh, Birmingham, Sheffield, and Reading. He was elected honorary FBA in 1980. He won many medals and prizes.

Keynes married 12 May 1917 Margaret Elizabeth (1890-1974), daughter of Sir George Howard Darwin [q.v.]; they had four sons. He died at Brinkley, near Newmarket, 5 July 1982.

A bust by Nigel Boonham is in the National Portrait Gallery, and there are drawings by Gilbert Spencer and T. L. Poulton in Cambridge University Library.

[*The Times*, 6 July 1982; *Annals of the Royal College of Surgeons*, 1983; *Geoffrey Keynes; Tributes on the Occasion of his Seventieth Birthday* (Osler Club), 1961; Sir Geoffrey Keynes, *The Gates of Memory*, 1981 (autobiography); personal knowledge.]

DAVID MCKITTERICK

KIRKMAN, SIR SIDNEY CHEVALIER (1895-1982), general, was born 29 July 1895 at Bedford, the elder son (there were no daughters) of John Parke Kirkman, schoolmaster, of Bedford, and his wife, Eva Anchoretta, daughter of Captain Henry Whyte, of the 14th Madras Native Infantry. He was educated at Bedford School, where he was a boarder in his father's house (Glanyrafon). After attending the Royal Military Academy, Woolwich, he was commissioned in the Royal Artillery in February 1915, serving on the western front and in Italy during World War I. He was wounded twice, at Loos and when commanding his battery near Cambrai. He was twice mentioned in dispatches and awarded the MC (1918).

Between the wars he served, at staff and regimental duty, in Egypt, Palestine, Malta, and India and attended the Staff College at Camberley in 1931-2. A skilled horseman, he excelled at polo, but as a serious professional soldier he chafed at the virtual standstill in promotion of those years. World War II introduced a dramatic change of tempo and from being a major with twenty-four years' service he was to reach full general's rank in eight years.

He returned home from India in 1940 to command the 65th Medium Regiment RA (TA) and within six months was appointed commander Royal Artillery (brigadier, 1941) of the 56th division. Soon after this, at an artillery exercise he was asked by his corps commander, General B. L. Montgomery (later Viscount Montgomery of Alamein, q.v.) what he thought of the latter's summing-up speech. His forthright reply that the general had omitted two most important lessons was to prove the start of a long and close association between the two men and to result in Montgomery describing Kirkman in his memoirs as 'the best artilleryman in the British army'. Within eighteen months he was successively appointed Montgomery's artillery commander in XII Corps, South Eastern Command, and Eighth Army in the Western Desert.

At Alamein and throughout the subsequent advance he displayed the techniques of artillery employment and fire planning in which he so fervently believed. He continued these methods as brigadier (1942) Royal Artillery 18th Army Group in Tunisia, but with the ending of the North African campaign he was promoted to command the 50th Northumbrian division in time to prepare it for the invasion of Sicily. His insistence on thorough training, as much as his own calm but forceful leadership, was reflected in the division's impressive performance throughout the island campaign, particularly at the hard-fought battle of Primasole Bridge. After Sicily the 50th division returned to England to prepare for the Normandy landings, but in January 1944 Kirkman returned to Italy to command XIII Corps.

In the ensuing year he played a leading part in the offensive led by Sir H. R. L. G. Alexander (later Earl Alexander of Tunis, q.v.) which finally broke the Cassino line and carried the Allied armies into the mountains south of Bologna. It was in these rocky heights that during the last winter of the war he spent the most demanding months of his whole career, serving under the American General Mark Clark, to whom high-minded, courteous, and imperturbable British officers like Kirkman were an enigma. In February 1945 he was 'invalided' home, but soon appointed in swift succession to Southern Command and I Corps in Germany, before moving in September to the War Office as deputy chief of the imperial general staff and two years later, on promotion, to quartermaster-general. In both appointments he proved a wise and energetic member of the Army Council, as Emanuel (later Lord) Shinwell, the secretary of state, was later to testify.

On retirement in 1950 he was appointed special government representative to investigate British military expenditure in Germany and subsequently to look for economies in the Home Forces. In 1954 he was made director general of Civil Defence, charged with creating a tenable basic doctrine and a functional organization to implement it. For six years he toured Britain, lecturing, and striving to animate national and local government planning—a difficult task in the prevailing political climate.

He was appointed OBE (1941), CBE (1943), CB (1944), KBE (1945), KCB (1949), and GCB (1951). He became commander of the American Legion of Merit, was an officer of the Legion of Honour, and was awarded the croix de guerre. From 1947 to 1957 he was colonel commandant of the Royal Artillery.

'Kirkie', as he was known, was a tall, handsome man, charming and amusing to his friends and considered a hard but fair taskmaster by those who worked with him. He had a clear, incisive mind, and as a practising Christian regarded integrity and honesty as paramount virtues. In his later years, despite worsening arthritis, he enjoyed the company of his grandchildren, his garden, and winter visits to southern France.

In 1932 he married Amy Caroline, second daughter of the Revd Charles Erskine Clarke, vicar of Reigate; they had two sons. Kirkman died at Southampton 5 November 1982, shortly after his fiftieth wedding anniversary.

[Kirkman's diaries and papers (Liddell Hart archives); tape-recorded interviews of Kirkman by Nigel Hamilton (Imperial War Museum) and by Mr Liddell of Sunderland; *The Times*, 6 November 1982; *Journal of the Royal Artillery*, March 1983; *Gunner Magazine*, January 1983; *Ousel* (Bedford School magazine), March 1983; Dominick Graham and Shelford Bidwell, *Tug of War*, 1986; private information; personal knowledge.]

TOM DAVIS

KNIGHT, (GEORGE) WILSON (1897–1985), literary commentator, was born in Sutton, Surrey, 19 September 1897, the younger son and younger child—the elder being W. F. Jackson Knight, the Virgilian scholar—of George Knight, of the Northern Insurance Company, and his wife, Caroline Louisa, daughter of Captain John Barclay Jackson, of the West Indian Regiment. Educated at Dulwich College (1909–14), he was employed in insurance offices until 1916 when he volunteered for the army. As a motor-cyclist dispatch rider in the Royal Engineers he served in Mesopotamia (Iraq), India, and Persia (Iran) where chess became a favourite off-duty recreation. Demobilized in 1920, he taught mathematics in preparatory schools until matriculating at St Edmund Hall, Oxford, in

1921. Graduating with second class honours in English language and literature in 1923, a year in which he represented Oxford against Cambridge at chess, he taught at Hawtreys, Westgate on Sea, until 1925 when he was appointed English master at Dean Close School, Cheltenham.

Now started a period of unceasing literary productivity. After three unpublished novels—though the first, *Klinton Top* (1927), was to be printed in 1984—he began the long series of Shakespeare studies that brought him renown. Rejecting the name of critic, Knight styled himself an 'interpreter'. Characters in the plays, a previous focus of attention, though highly important were but components in a 'spatio-lineal' design to which symbols, metaphors, images, and poetic evocations contributed. A play of Shakespeare was an 'expanded metaphor', the matrix of its characters.

In *Myth and Miracle: an Essay on the Mystic Symbolism of Shakespeare* (1929), Knight offered an original reading of Shakespeare's intuition of immortality in the last plays. This was followed by *The Wheel of Fire: Essays in Interpretation of Shakespeare's Sombre Tragedies* (1930), with an introduction by T. S. Eliot [q.v.], which led to Knight's election to the Chancellor's chair of English in Trinity College, University of Toronto, in 1931.

In Toronto, besides writing *The Imperial Theme* (1931), which focused on Shakespeare's Roman plays, and *The Shakespearian Tempest* (1932), Wilson Knight addressed himself to producing and acting Shakespeare at the Hart House Theatre. He was to say that in acting Shakespeare he found his supreme satisfaction. The results of theory and practice found expression in *Principles of Shakespearian Production* (1936). But he was also to extend his method of 'interpretation' to other poets with *The Starlit Dome* (1941) on the Romantics, and to religion with *The Christian Renaissance* (1933), and, in time, to international politics with *The Olive and the Sword* (1944) and *Hiroshima* (1946).

He resigned his chair in 1940 so as not to be distant from his mother and brother in wartime England. He was too a fervent patriot and produced 'This Sceptered Isle' at the Westminster Theatre, London, in 1941, with Henry Ainley [q.v.] as the narrator. After teaching geography at Stowe (1941–6), he was appointed reader in English literature at the University of Leeds in 1946 where he remained—much loved and admired by colleagues and students—until his retirement in 1962. He was elevated to a personal chair in 1956. Further books on Shakespeare appeared: *The Crown of Life* (1947) and *The Sovereign Flower* (1958) are especially notable. A lecture course at Leeds on British drama was published under the title of *The Golden Labyrinth* (1962). He also acted and produced at Leeds, revealing a particular interest in Shakespeare's *Timon of Athens*. He wrote a series of books on Lord Byron, and a notable work on Alexander Pope [qq.v.]. Indeed no poet, from Spenser to the present day, escaped his invigorating comment. *Neglected Powers* (1971) on recent writers, reveals a special interest in John Cowper Powys [q.v.].

After his mother's death in 1950 Knight became interested in spiritualism though he never proselytized.

On retirement he shared a home in Exeter with his brother until the latter's death. He published *Jackson Knight: a Biography* (1975) where he relates the story of his subject's family and his own part in that story. In 1974 he assumed by statutory declaration Richard as a second forename, for he had always been called Dick by his closest associates, from childhood onwards. Even in his eighties he made tours of North America to perform his one-man show 'Shakespeare's Dramatic Challenge', and he continued to publish.

Knight was appointed CBE in 1968, was elected an honorary fellow of St Edmund Hall, Oxford, in 1965, and received an honorary Litt.D. from Sheffield and D.Litt. from Exeter in 1966 and 1968 respectively.

He relished sympathetic humour, disliked cynicism, and was innocent of malice. His conversation was vitalizing. He liked the colourful whether in nature or art. He never married. He died at his home in Exeter 20 March 1985.

A sculpture of Wilson Knight as Timon by Robert Russin (1975) is privately owned.

[*The Times*, 23 March 1985; G. Wilson Knight, *Jackson Knight: a Biography*, 1975; personal knowledge.]   FRANCIS BERRY

KOESTLER, ARTHUR (1905–1983), writer, was born 5 September 1905 in Budapest, the son of Henrik Koestler, talented eccentric and usually prosperous Jewish businessman, and his wife, Adela Jeiteles Hitzig, of an old Prague-Viennese Jewish family. She considered Hungarians barbarians, refusing to speak their language. Koestler admired his self-taught father and disliked his snobbish mother who treated him unkindly, was shamed by his sensational political adventures, belittled his literary successes, and whom he hated meeting even when she was old. Educated originally in Budapest and then in the Vienna Polytechnic High School and University, where the family finally settled after the Hungarian communist revolution in 1919, he read engineering which stimulated his lifelong passion for science. When twenty-one he went to Palestine and became Middle East

correspondent of Ullstein, the great newspaper chain, soon moving to Paris and then to Berlin where in 1930 he became science editor of their *Vossische Zeitung* and foreign editor of their evening paper thus developing his knowledge both of science and politics.

Alert to the Hitler menace he joined the Communist Party as the best prospect of defeating Nazism and hence was courteously dismissed by the alarmed Ullstein in 1931. Financed by the Communist International to write a book on the first five-year plan he travelled widely in Russia, but the book, completed in 1934, was rejected by Soviet censorship as insufficiently adulatory. Back in Paris among anti-Nazi exiles he began his first novel *The Gladiators* (1939) showing that Spartacus was doomed because he would not apply 'the law of detours' requiring leaders to be 'pitiless for the sake of pity' in executing dissidents. He returned to the theme in *Darkness at Noon* (1940) in which the old Bolshevik Rubashov followed the 'law of detours' but was also doomed.

In 1936 he went to Franco's Seville headquarters to report for the Liberal *News Chronicle*. Denounced as a communist he escaped and returned in 1937 to report from the Republican side. Captured by Franco's troops he was imprisoned in Malaga and Seville from February to June, daily awaiting execution (*Spanish Testament*, 1937), an experience which made him disapprove of capital punishment against which he campaigned prominently during the post-war years in England. The vigorous activities of his first wife, Dorothy Asher, induced the British government to intervene to save him. He had married her in 1935; they had no children and were divorced in 1950. She came from Zurich.

In 1938 he left the Communist Party, in revulsion from Stalin's mass arrests and show trials and from the communist duplicity and incompetence he had seen in Russia and Republican Spain. His internment as a political suspect in a French concentration camp, from which he was released in 1940, gave him further writing material (*Scum of the Earth*, 1941). After joining the French Foreign Legion under a false name to avoid the Gestapo he escaped to England. Although his first language was Hungarian, he wrote in German until 1940 and thereafter in English. Put in the Pioneer Corps (1941–2), his latent love for England, despite his anglophile father forcing him to wear an Eton suit in Budapest when he was thirteen, blossomed. He became a British citizen in 1948. Such curiosities as compulsory tea breaks in the midst of war and the distinction between saloon and public bars fascinated him. Sexually hyperactive, which he thought normal for a Continental male, he was delighted by his successes with a series of pretty, upper-class Englishwomen, not least because they were upper-class. One liaison led to marriage in 1950 to Mamaine (died 1954), daughter of Eric Morton Paget, country gentleman. They had no children and were divorced in 1953. He treated her abominably, believing women's role to be that of a willing servant with the ability to please of a delicately nurtured, skilful courtesan. Possibly because of his mother's indifference he needed the unstinting love of women, 'but to write about them bores me'.

As well as the unattached, friends' wives or girl friends were fair game. He thought Bertrand (Earl) Russell [q.v.] petty to refuse to address a meeting organized by Koestler to protest against Russian suppression of the 1956 Hungarian revolution because Koestler was a co-respondent in Russell's divorce from Peter Spence.

Intolerant of disagreement, arrogantly certain of his genius, frequently and irrationally angry with friends, Koestler's questing charm and boyish curiosity made him deeply loved and liked by the many men and women he chose to please. A whole-hearted evening drinker with a lively humour he made parties rollick till the small hours. To be with him was to be aware of a considerable intellect brilliantly deployed in proselytizing the causes he believed mankind must adopt. The most important and seminal of these was his exposure by fiction and fact of the aridity and inevitably concomitant horrors of communism and its hopelessness as a better alternative to free societies based on capitalism. He shocked many gullible idealists into reality.

That is likely to be his greatest memorial though he would have preferred it to have been his scientific writings, particularly those relating to parapsychology, or telepathy, which he regarded as his most significant work, insisting that as nuclear particles can communicate with each other faster than light there must be a scientific basis for the belief that human minds can do the same. Among such books were *The Sleepwalkers* (1959), *The Act of Creation* (1964), and *The Case of the Midwife Toad* (1971). In his will he left over £500,000 to endow a chair of parapsychology, which was later founded at Edinburgh University. An atheist, his hope that there might be some vague continuation after death possibly lay behind his parapsychological enquiries. He was generous to persecuted writers and provided annual cultural prizes to prisoners in British jails for whose barren lives he had sympathy and understanding. He was appointed CBE in 1972 and C.Lit. in 1974. He had honorary degrees from Queen's University, Kingston, Ontario (LLD, 1968), and Leeds (D.Litt., 1977).

Koestler's third wife was Cynthia Jefferies, née Paterson, the daughter of a South African surgeon. She was twenty-two years younger than himself. He employed her in 1948 first intermittently and then permanently as secretary. There was the customary seduction and the starstruck, naïve South African girl became a willing slave in a harem which changed frequently even while Koestler was married to Mamaine. When her breakaway attempt ended in a failed marriage she was adoringly grateful to be put condescendingly on the strength again and eventually to marry him in 1965, not complaining at his unrestrained unfaithfulness. As his Parkinson's disease advanced over seven years culminating in terminal leukaemia, he became so dependent on her doglike devotion that there was a near reversal of roles. Their suicide with an overdose of barbiturates and alcohol on the night of 1 March 1983, in their London flat, was joint because she could not live without him. Both were adherents of Exit, the voluntary euthanasia group. Although his wives bore him no children he had a daughter.

[Arthur Koestler, *Spanish Testament*, 1937, *Arrow in the Blue*, 1952, and *The Invisible Writing*, 1954 (autobiographies); Iain Hamilton, *Koestler*, 1982; George Mikes, *Arthur Koestler*, 1983; Brian Inglis, *Arthur Koestler and Parapsychology*, 1984; Arthur and Cynthia Koestler, *Stranger on the Square*, 1984; Mamaine Koestler, *Living with Koestler*, 1985; Arthur Koestler's contribution in *The God That Failed* (ed. Richard H. Crossman), 1949; personal knowledge.]

WOODROW WYATT

KREBS, SIR HANS ADOLF (1900-1981), biochemist and Nobel prize-winner, was born 25 August 1900 at Hildesheim, Germany, the elder son and second of the three children of Dr Georg Krebs, an otolaryngologist, and his wife, Alma Davidson, the daughter of a banker. After being in the army for a few months in 1918 he was educated in a manner not unusual for those intending to qualify in medicine: pre-clinical studies at the universities of Göttingen and Freiburg-im-Breisgau, clinical work in Munich (from which university he graduated in 1923 with first-class marks overall), and two years of hospital experience in Berlin. What was unusual was that, whenever possible, Krebs combined these traditional pursuits with direct experience of work in research laboratories: this confirmed him in his resolve not simply to practise medicine but to concentrate on basic medical research. But his proper education in research techniques began only in 1925, when he was appointed research assistant to Professor Otto Warburg, who was undoubtedly the leading biochemist of that time. Krebs remained there for five years. What was perhaps also unusual was that he adhered steadfastly to his determination to undertake research, despite positive discouragement from his mentor.

After various vicissitudes, Krebs in 1931 secured a post at Freiburg in which, despite heavy clinical responsibilities, he could at last test his own abilities to advance learning through empirical enquiry. To his (and possibly, others') surprise, his work was immediately and spectacularly successful. Assisted by a medical student, Kurt Henseleit, Krebs elucidated in less than a year the chemical steps that enable the liver to form urea (the main nitrogenous material excreted by animals in their urine) from ammonia and carbon dioxide. Most remarkably, this sequence was revealed to be cyclic in form: the starting materials entered, and urea left, a set of reactions in which the initial reactant was ultimately regenerated.

The publication in 1932 of this 'urea cycle' was widely recognized as a milestone in biochemistry. Of particular importance to Krebs's subsequent fate was that Sir Frederick Gowland Hopkins [q.v.], was so impressed by this work that he not only referred to it in his presidential address to the Royal Society on 30 November 1932 but also invited Krebs to visit him in Cambridge. Two months later, Hitler came to power; four months after that, Krebs, as a Jew, found himself ignominiously dismissed from Freiburg. Hopkins immediately repeated his invitation and arranged temporary financial assistance from the Rockefeller Foundation; in July 1933 Krebs was at work in Cambridge.

After two years there, Krebs accepted a lectureship at Sheffield. This enabled him to build up his own small research team, and to apply the techniques and ideas that had proved so successful in the study of urea formation, to unravelling the chemical steps by which muscle and other cells oxidize sugars to carbon dioxide and water. Again working with a student, W. A. Johnson, Krebs in 1937 elucidated this process and found it also to be cyclic in form; however, universal acceptance of the validity of this 'tricarboxylic acid cycle' was not achieved until it had been fully vindicated some years later. The widest recognition of its fundamental importance, as the terminal route for oxidation of virtually all food materials, also had to wait until the end of World War II, when novel procedures (particularly the use of radioactive tracers and of spectroscopes for studying chemical reactions in living matter) became available to biochemists.

His discoveries led to an avalanche of honours for Krebs. He had not been granted the title of professor until 1945, when he was also given honorary direction of a research unit of the Medical Research Council. Thereafter, he was elected FRS (1947), awarded a half-share in the Nobel prize for physiology or medicine (1953), and appointed to the Whitley chair of biochemistry at Oxford (1954), where he became a fellow of Trinity College (1954-67). Krebs was naturalized as a British citizen in 1939 and was knighted in 1958; by the end of his life, he had also been awarded honorary degrees from twenty-one universities, honorary or foreign membership of at least thirty-five learned societies and academies, many medals (including a Royal (1954) and the Copley (1961) medal of the Royal Society) and—in view of his past, most pleasing to him—the German Order 'Pour le Mérite'. He became FRCP in 1958.

Despite the near-veneration that was increasingly accorded him, particularly by the young, Krebs remained all his life simple, approachable, unaffected, and totally devoid of malice or of arrogance. He also continued to be active in the laboratory. Krebs published more than 100 papers between 1954 and his retirement from the Oxford chair in 1967 and a further 100 from the research laboratory in the Radcliffe Infirmary in which he, his faithful colleagues, and a constant stream of academic visitors, worked for the subsequent fourteen years.

Krebs married, in 1938, Margaret Cicely, daughter of Joseph Leo Fieldhouse, owner of some shoe shops in Rotherham, of Wickersley, Yorkshire. They had two sons and one daughter; the younger son, John, was elected FRS in 1984 and holds a research professorship of the Royal Society at the University of Oxford. Krebs died in Oxford 22 November 1981.

[Hans Krebs, *Reminiscences and Reflections*, 1981; H. L. Kornberg, 'H. A. Krebs: a Pathway in Metabolism', *Biochemical Society Symposium*, vol. xxvii, 1968; Sir Hans Kornberg and D. H. Williamson in *Biographical Memoirs of Fellows of the Royal Society*, vol. xxx, 1984; personal knowledge.]

HANS KORNBERG

L

LANG, SIR JOHN GERALD (1896-1984), secretary of the Admiralty, was born in Woolwich 20 December 1896, the elder son (there were no daughters) of the second marriage of George Thompson Lang, an engineering tool maker, and his wife, Rebecca Davies. There were four sons and one daughter of George Lang's first marriage. Lang was educated at Haberdashers' Aske's School, Hatcham, and entered the Admiralty as a second division clerk in 1914. He saw service in World War I as a lieutenant in the Royal Marine Artillery (1917-18). After his return to the department, it was not long before he began to make a name for himself and his abilities were eventually recognized in 1930, by promotion to the administrative class as assistant principal.

In his early days Lang was engaged mainly on personnel administration but his mastery in this field was so marked that by the outbreak of World War II it was clear that he would go far. He became principal in 1935 and assistant secretary in 1939. During the war he was selected to fill the important post of director of labour, concerned with the recruitment, organization, and deployment of dockyard and shipyard labour. The contacts which he made with shipbuilders and officials of the shipbuilding unions were to stand him in good stead later on. From these civilian personnel tasks he moved on in 1946 to become the under-secretary concerned with naval personnel and he made a notable contribution to the reorganization of naval manpower as it was returned to a peacetime level.

By then his capacity had become so widely recognized that, when Sir Henry Markham died prematurely in December 1946, Lang was the obvious choice to succeed him as secretary of the Admiralty, although this involved a double promotion. His time as secretary (1947-61) was, in general, one of retrenchment, and his was the guiding hand in the process of reducing the Admiralty to its peacetime size.

Lang was endowed with a photographic memory. He could recall, over several years, not merely the contents of a paper, but also its registered number. This memory enabled him to build up an unrivalled knowledge of past discussions and events and this in turn made him a formidable advocate of any policy which he decided to support. He was, by nature, very deliberate, and he liked to have plenty of time to consider a question from all its angles before coming to a conclusion. As a result, he developed a very mature judgement, which was rarely at fault.

His staff held him in high regard and affection. He led them by his example of selfless service to the department's affairs, rather than by any strict discipline. He was always accessible and invariably calm and courteous. His retirement in 1961 was widely regarded as something of an end of an era.

In retirement he pursued several interests with undiminished enthusiasm, in spite of the handicap of impaired sight, which struck him not long after he left the Admiralty. From 1964 to 1971 he was principal adviser to the government on sport and for most of that time he was deputy chairman of the Sports Council. In 1969 he produced a report on crowd control at football matches, which formed the basis of many of the early measures to combat soccer hooliganism. In 1969 he headed an investigation which secured for the taxpayer a large refund on a defence contract. He was a governor and an officer of the Royal Bethlem and Maudsley hospitals (1961-70).

He was a vice-president of the Royal Naval Association and chairman of its standing orders committee, responsible for the running of the association's annual conferences. He was active in the Royal Institute of Naval Architects, of which he also became a vice-president (1977). He was associated with the Navy Records Society, and a member of the Worshipful Company of Shipwrights and of the Pepys Club, of which he became president. In his eighties he offered his services to Help the Aged, and though he became their adviser on VAT, to begin with he was content, with typical modesty, to perform very humble tasks for them.

He was appointed CB in 1946, KCB in 1947, and GCB in 1954.

He was twice married. In 1922 he married Emilie Jane (died 1963), daughter of Henry Shelley Goddard, interior decorator, of Eastbourne. They had one daughter. In 1970 he married her sister, Kathleen Winifred, widow of C. G. E. Edmeades. She survived him by only a few days. He died 22 September 1984, at his daughter's house at Walton-on-the-Hill, Surrey.

There is a portrait of him, by Trevor Stubley, in Admiralty House.

[Private information; personal knowledge.]

CLIFFORD JARRETT

LARKIN, PHILIP ARTHUR (1922-1985), poet, was born in Coventry 9 August 1922, the only son and younger child of Sydney Larkin, treasurer of Coventry, who was originally from Lichfield, and his wife, Eva Emily Day, of Epping. He was educated at King Henry VIII

School, Coventry (1930-40), and St John's College, Oxford, where he obtained a first class degree in English language and literature in 1943. Bad eyesight caused him to be rejected for military service, and after leaving Oxford he took up library work, becoming in turn librarian of Wellington, Shropshire (December 1943-July 1946), assistant librarian, University of Leicester (September 1946-September 1950), sub-librarian of the Queen's University of Belfast (October 1950-March 1955), and finally taking charge of the Brynmor Jones Library, University of Hull, for the rest of his life.

Larkin, while always courteous and pleasant to meet, was solitary by nature; he never married and had no objection to his own company; it was said that the character in literature he most resembled was Badger in Kenneth Grahame's *The Wind in the Willows*. A bachelor, he found his substitute for family life in the devotion of a chosen circle of friends, who appreciated his dry wit and his capacity for deep though undemonstrative affection. His character was stable and his attitude to others considerate, so that having established a friendship he rarely abandoned it. Most of the friends he made in his twenties were still attached to him in his sixties, and his long-standing friend and confidante Monica Jones, to whom he dedicated his first major collection *The Less Deceived* (1956), was with him at the time of his death thirty years later.

Larkin was a highly professional librarian, notably conscientious in his work, and an active member of the Standing Conference of National and University Libraries. In the limited time this left him he did not undertake lecture tours, very rarely broadcast or gave interviews, and produced (compared with most authors) very little ancillary writing; though his lifelong interest in jazz led him to review jazz records for the *Daily Telegraph*, 1961-71. Some of the reviews were collected in *All What Jazz* (1970). In his forties he discovered a facility for book reviewing, of which he had previously done very little, and a collection of his reviews, *Required Writing* (1983) reveals him as an excellent critic; though perhaps 'reveal' is not the right word, for a decade earlier he had done much to influence contemporary attitudes to poetry with his majestic and in some quarters highly controversial *Oxford Book of Twentieth-Century English Verse* (1973), prepared with the utmost care during his tenure of a visiting fellowship at All Souls College in 1970-1. He spent much time working on behalf of his fellow writers, as a member of the literature panel of the Arts Council, helping to set up and then guide its National Manuscript Collection of Contemporary Writers in conjunction with the British Museum, and serving as chairman for several

years of the Poetry Book Society. He was chairman of the Booker prize judges in 1977. To this Dictionary he contributed the notice of Barbara Pym.

Larkin's early ambition was to contribute both to the novel and to poetry. His first novel *Jill* (1946), published by a small press (which paid him with only a cup of tea) and not widely reviewed, did little to establish him, though its merits were recognized when it was reprinted in 1964 and 1975; but the second, *A Girl in Winter* (1947), attracted the attention of discerning readers, and the only reason he did not write more novels was that he found he could not, though he tried for some five years before giving up and working entirely in poetry, an art he loved but did not regard as necessarily 'higher' than fiction. The poet, he said, made a memorable statement *about* a thing, the novelist demonstrated that thing as it was in actuality. 'The poet tells you that old age is horrible, the novelist shows you a lot of old people in a room'. Why the second became impossible to him, and the first remained strikingly possible, it is useless to speculate.

As a poet, Larkin's early work, written when he was about twenty, already shows a fine ear and an unmistakable gift; but the breakthrough to an individual, and perfectly achieved, manner came some ten years later, in the poems collected in *The Less Deceived*. From that point on, his work did not change much in style or subject matter throughout the thirty years still to come, in which he produced two volumes, *The Whitsun Weddings* (1964) and *High Windows* (1974), plus a few poems still uncollected at his death. There were surprises, but then there had been surprises from the start, for Larkin's range was much more varied than a brief description of his work could hope to convey. He was restlessly alive to the possibilities of form, and never seemed constricted by tightly organized forms like the sonnet, the couplet, or the closely rhymed stanza, nor flaccid when he moulded his statement into free verse. It is instructive to pick out any one individual poem of Larkin's and then look through his work for another that seems to be saying much the same thing in much the same manner. As a rule one finds that there is no such animal. Most poets repeat themselves; he did not, and this should qualify the frequently repeated judgement that his output was 'small'.

Both in prose and verse, Larkin's themes were those of quotidian life: work, relationships, the earth and its seasons, routines, holidays, illnesses. He worked directly from life and felt no need of historical or mythological references, any more than he needed the cryptic verbal compressions that were mandatory in the 'modern' poetry of his youth. Where 'modern' poetry put its subtleties and complexities on the surface as a kind of protective matting, to keep the reader from getting into the poem too quickly, Larkin always provides a clear surface—one feels confident of knowing what the poem is 'about' at the very first reading—and plants his subtleties deep down, so that the reader becomes gradually aware of them with longer acquaintance. The poems thus grow in the mind until they become treasured possessions; this would perhaps account for the sudden explosion of feeling in the country at large when Larkin unexpectedly died at the Nuffield Hospital, Hull, 2 December 1985 (he had been known to be ill but thought to be recovering), and the extraordinary number who crowded into Westminster Abbey for his memorial service on St Valentine's Day 1986.

Philip Larkin was an honorary D.Litt. of the universities of Belfast, 1969; Leicester, 1970; Warwick, 1973; St Andrews, 1974; Sussex, 1974; and Oxford, 1984. He won the Queen's gold medal for poetry (1965), the Loines award for poetry (1974), the A. C. Benson silver medal, RSL (1975), the Shakespeare prize, FVS Foundation of Hamburg (1976), and the Coventry award of merit (1978). In 1983 *Required Writing* won the W. H. Smith literary award. In 1975 he was appointed CBE and a foreign honorary member of the American Academy of Arts and Sciences. St John's College made him an honorary fellow in 1973, and in 1985 he was made a Companion of Honour.

[Personal knowledge. See also Anthony Thwaite (ed.), *Larkin at Sixty*, 1982.]

JOHN WAIN

LASCELLES, SIR ALAN FREDERICK (1887-1981), royal secretary, was the sixth and youngest child and only surviving son (the other died in infancy) of Commander Frederick Canning Lascelles, second son of the fourth Earl of Harewood and godson of the viceroy of India, Earl Canning [q.v.]. His mother, Frederica Maria, daughter of Sir Adolphus Frederic Octavius Liddell, son of the first Baron Ravensworth, died in March 1891. 'Tommy', as he was known from childhood, was born 11 April 1887, the birthday of his ancestor George Canning [q.v.], prime minister, at Sutton Waldron House, Dorset, where his father had retired after naval service. Lascelles was brought up by a not very competent governess and his older sisters until he went to preparatory school and on to Marlborough (1900-5) where he was not very happy.

In 1905 he went to Trinity College, Oxford, and at once found himself at home. For him Oxford became 'the Beloved City'. Lascelles was one of a brilliant generation, centred on

Balliol, including Julian Grenfell [q.v.], Charles Lister, Edward Horner, and others. Most of this glittering circle were killed in World War I. In the vacations he was able to move from country house to country house and pursue his favourite sports of hunting and fishing; he rode well in the field and point-to-point. In 1909 he was placed in the second class in *literae humaniores*, though hopes had been entertained of a first. During his time at Oxford he had read widely and cultivated a growing taste in music, particularly that of Wagner and Brahms (he eventually became an honorary fellow of the Royal Academy of Music). He could hardly write a dull letter.

For some years Lascelles had difficulty in finding his vocation. He sat in vain twice for the Foreign Office, jettisoned the idea of the Indian Civil Service, and tried a brief spell in the City. In 1913 he joined the Bedfordshire Yeomanry, and with them he was mobilized on the outbreak of war. During its course he was wounded, won the MC (1919), and was mentioned in dispatches. At its close he noted: 'Even when you win a war, you cannot forget that you have lost your generation.'

Early in 1919 he sailed for India as aide-de-camp to his brother-in-law Sir George Ambrose (later Lord) Lloyd [q.v.], the designated governor of Bombay. During his stay he met and married the daughter of the viceroy; his cousin H. G. C. Lascelles (later the Earl of Harewood, q.v.) generously made the wedding possible from his own inheritance. On his return to England in 1920 he was appointed assistant private secretary to the Prince of Wales (later Edward VIII) and with one interval the rest of his active life was passed in royal service. In 1929 he left the Prince of Wales and from 1931 to 1935 he was private secretary to the governor-general of Canada (V. B. Ponsonby, Earl of Bessborough, q.v.); in 1933 he was appointed CMG for his work at the Ottawa conference. However in 1935 he returned to the palace as assistant private secretary to George V, and in succession to Edward VIII and George VI, to whom he became private secretary in 1943, retaining the post for the first year of Queen Elizabeth II's reign. By the time of Edward VIII's abdication he was out of sympathy with his master. He was keeper of the royal archives from 1943 to 1953. He was admitted to the Privy Council in 1943.

Lascelles was every inch a courtier, tall, slim, good-looking, and quick-moving. His discretion was impeccable, though he committed much to journals which have yet to be published. He instituted a press office at Buckingham Palace and encouraged a newer style of royal biography such as the life of George V by his old friend, John Gore. His memory was remarkable and his reading extensive, with a late devotion to Horace Walpole [q.v.]. He was promoted GCVO in 1947 (his KCVO was given in a royal train in America in 1939 and his MVO in 1926) and GCB in 1953. He was offered a peerage but refused it. He became a director of the Midland Bank; he was chairman of the Pilgrim Trust (1954–60) and chairman of the Historic Buildings Council for England (1953–63). His old college made him an honorary fellow in 1948, and Oxford University gave him a DCL in 1963. He also had honorary LLDs from Bristol and Durham.

In retirement he occupied a grace and favour house in Kensington Palace: in widowhood he grew a beard and remained the best of company. He married in 1920 Joan Frances Vere (1895–1971), eldest daughter of Frederic John Napier Thesiger, first Viscount Chelmsford [q.v.], viceroy of India. They had a son, who became ill and tragically died in 1951, just when the health of George VI was beginning to give anxiety, and two daughters, one of whom was married to the second Viscount Chandos. Lascelles died in Kensington Palace 10 August 1981. There is a drawing of 1922 by Oswald Birley.

[*The Times*, 11 August 1981; Duff Hart-Davis (ed.), *End of an Era, Letters and Journals of Sir Alan Lascelles from 1887 to 1920*, 1986, and *Letters and Journals . . . from 1920 to 1936*, 1989; private information; personal knowledge.]   MICHAEL MACLAGAN

LAUWERYS, JOSEPH ALBERT (1902–1981), educationist, was born in Brussels 7 November 1902, the eldest child in the family of two sons and two daughters of Henry Lauwerys, a tailor, and his wife, Louise Nagels. The family were mainly French speaking. In 1914 they fled to England just before the German occupation, establishing themselves there permanently. Lauwerys spent a year at Ratcliffe College, Leicester, and then completed his schooling at Bournemouth, where the family eventually settled. He left school early, and worked in a number of jobs, mainly as a shop assistant. Through the Co-operative movement he was encouraged to pursue his education part-time and in 1927 obtained a first class London University B.Sc. general degree, from King's College. This was followed in successive years by honours degrees in chemistry and physics. He taught for a while in Stirling House, a private school in Bournemouth, but in 1928, upon the recommendation of (Sir) T. Percy Nunn [q.v.], of London University Institute of Education, became physics master at Christ's Hospital, Horsham. He was naturalized in 1928. In 1932 he was appointed lecturer in the

methods of teaching science in schools at London University Institute of Education. During this period he wrote many science textbooks and works on scientific method and science teaching, as well as programmes for schools broadcasting. In 1946 Ghent University awarded him a doctorate for a thesis on film and radio as educational media. His interest in science teaching was lifelong.

However, World War II stimulated his different growing concern, education in other countries. In 1941 he had become reader in comparative education, and during the war led an enquiry initiated by the committee of Allied ministers of education that met in London under R. A. Butler (later Lord Butler of Saffron Walden, q.v.) to plan educational reconstruction after the conflict. Promoted in 1947 to be the first professor of comparative education at London University Institute of Education, his work as adviser for Unesco gave him further insights into how education might become an instrument for promoting peace and international understanding.

He also saw his professional role as one of establishing comparative studies in education more firmly as a legitimate field of enquiry. In this, through his writings, his indefatigable development of international contacts, and his own teaching he was eminently successful. Although he wrote no single major work, his contribution as senior editor of the *World Year Book of Education* from 1947 to 1970 was invaluable. The thematic and global approach adopted in these twenty-four volumes, which, for example, made more widely known such topics as economics and education (1956) and educational planning (1967), was highly innovative and acted as a catalyst for further research. His own numerous articles and monographs were equally influential. He was himself visiting professor at many universities ranging from the Americas to the Far East. He made the London Institute into the major graduate centre for comparative studies, particularly by his own charismatic lecturing. A polyglot, Lauwerys was a lively conversationalist and tenacious debater, communicating in his many varied activities an unbounded enthusiasm and zest for life.

Responsibilities and honours were showered upon him. Thus he was the British representative on the Unesco conciliation and good offices commission dealing with discrimination in education. With Lord Boyle of Handsworth [q.v.] he co-chaired the education committee of the parliamentary group for world government. He was active in such organizations as the Council for Christians and Jews, the International New Education Fellowship, the International Montessori Association, and the Basic English movement. He saw the possibilities of using Basic English as a lingua franca for education in the Third World. 'Professeur associé' at the Sorbonne in 1969-71, he had been made a commander of the Ordre des Palmes Académiques in 1961. His own university had conferred a D.Lit. upon him in 1958, and he was made a fellow of King's College, London.

He retired from London University in 1970, but until 1976 was director of the Atlantic Institute of Education, Halifax, Nova Scotia. In his latter years both his Catholic upbringing and his scientific background, as well as a natural philosophical bent, led him to interest himself in Japanese efforts to establish a 'universal moral science', or 'moralogy', and in 1976 he published *Science, Morals and Moralogy*. To this Dictionary he contributed the notice of Sir Fred Clarke.

In 1932 he married Waltraut Dorothy, from Germany, daughter of Hermann Bauermeister, publisher. They had three sons. Lauwerys died at Guildford 29 June 1981.

[M. McLean (ed.), *Joseph A. Lauwerys, a Festschrift*, London Institute of Education Library Bulletin, Supplement 22, 1981; V. Mallinson, 'Emeritus Professor J. A. Lauwerys (1902-1981)', *Comparative Education*, Oxford, vol. xvii, no. 3, 1981; H. van Daele, 'Joseph Lauwerys (1902-1981) en de vergelijkende pedagogiek', *Persoon en Gemeenschap*, Brussels, vol. xxxiv, no. 6 (1981-2); private information.]   W. D. HALLS

LAWSON, FREDERICK HENRY (1897-1983), academic lawyer, was born in Leeds, 14 July 1897, the only child of Frederick Henry Lawson, a wool merchant, and his wife, Mary Louisa Austerberry. Harry (as he was always called) was educated at Leeds Grammar School and The Queen's College, Oxford, where he won a Hastings exhibition in classics in 1915. He had always, however, wanted to be a historian and, after commissioned service in an anti-aircraft regiment from 1916 to 1918, switched to the school of modern history, obtaining first class honours in 1921. He then took the law school in one year, again with a first (1922), and went on to be called to the bar by Gray's Inn in 1923. He was determined, however, on an academic career and subsisted for some time as a lecturer at various colleges. Merton gave him a junior research fellowship in 1925 and he was encouraged to take up Byzantine law, the study of which took him for a year to Göttingen and later led to his appointment as university lecturer in the subject (1929-31). The lasting result of his year in Germany, however, was the turning of his interests towards foreign law and the comparative approach.

In 1930 he finally became established, as a tutorial fellow of Merton. He was a tutor in the old style, teaching almost every subject in the syllabus and concerned with the whole intellectual formation of his pupils. He retained to the end of his life a zest for new ideas and an eagerness to impart them which, coupled with a remarkable range of intellectual interests, made him, for those who could and would keep up with him, a fascinating teacher. The principal directions of his interests reflected his classical and historical backgrounds. While he was a lecturer at University College he had met (Sir) David Lindsay Keir [q.v.] and their *Cases in Constitutional Law* began its long and influential career in 1928. From 1931 to 1948 he was All Souls reader in Roman law and though, as he himself said, he was never a real Roman law specialist, the subject coloured all his work in private law. What mainly interested him in law was not the detail, but the conceptual structure of the systems of continental civil law and his gift for inspired generalization illuminated not only those systems, as in *A Common Lawyer Looks at the Civil Law* (1955), but also the English common law itself, as in *The Rational Strength of English Law* (1951). Perhaps his best book was *Introduction to the Law of Property* (1958), which, though not ostensibly a work of comparative law, could hardly have been written without his wide reading in other systems. In it he broke away from the traditional historical approach and treated the subject from the point of view of its function in modern life. He gained an Oxford DCL in 1947.

In 1948 he became the first occupant of the chair of comparative law and moved from Merton to Brasenose. He now became a figure on the international stage, his gregarious nature and his unusual gift for foreign languages helping him to establish contact with foreign lawyers. Their high opinion of him was reflected in honorary doctorates from Louvain, Paris, Ghent, and Frankfurt (as well as Glasgow and Lancaster) and membership of the Accademia dei Lincei. He became a fellow of the British Academy in 1956.

He was quite without false pride or any sense of distinctions of age or position. He excelled in the self-effacing task of editing other men's works (some of his most interesting *aperçus* are to be found in his edition of *Roman Law and Common Law* (1952) by W. W. Buckland and A. D. (later Lord) McNair [qq.v.]. To this Dictionary he contributed the notices of G. C. Cheshire, H. F. Jolowicz, M. Wolff, and F. de Zulueta.

He retired from his Oxford chair in 1964 and immediately began a fresh life in the new University of Lancaster where, before the creation of the law school, he taught a variety of legal subjects to a wide range of non-lawyers and once again created for himself a devoted following. He finally retired to his native Yorkshire in 1977 at the age of eighty.

In 1933 he married Elspeth, younger daughter of Alexander Webster, of Kilmarnock, a ship's captain with the Ben Line. They had a son and two daughters. Lawson died in Middlesbrough 15 May 1983.

[*The Times*, 17 and 28 May 1983; *Oxford Journal of Legal Studies*, 1984; P. Wallington and R. M. Merkin (eds.), *Essays in Memory of Professor F. H. Lawson*, 1986 (containing a list of his publications and a portrait); private information; personal knowledge.]

BARRY NICHOLAS

LE MESURIER, JOHN (1912–1983), actor, was born John Elton Halliley in Bedford 5 April 1912, the only son and younger child of John Halliley, a well-known and long-established family solicitor, and his wife, Mary Le Mesurier. He spent his early years at Bury St Edmunds before being sent away to school, firstly to Grenham House at Birchington in Kent and then to Sherborne in Dorset. Whilst at school (which he thoroughly disliked) he managed to visit a great many West End plays and in particular perhaps significantly the farces presented at the Aldwych Theatre by Tom Walls [q.v.] and Ralph Lynn. On leaving school he was already determined to be an actor but being diffident at expressing such an ambition to his family, he allowed himself to become an articled clerk to a firm of Bury St Edmunds solicitors, Greene & Greene. He then drifted into the theatre through the Fay Compton Studio of Dramatic Art. He adopted his mother's maiden name as his stage name.

He gained considerable experience in the theatre of the 1930s. His first job was at the Palladium Theatre, Edinburgh, at a salary of £3. 10s. per week. Then followed a season at Oldham and various touring shows until 1940 when he joined the Royal Armoured Corps. He was sent to India and enjoyed what he was to describe as a 'comfortable war with captaincy thrust upon me'. He was demobilized in 1946 and began the many roles which seemed to make few demands on his acting talents; barristers, doctors, vicars, naval commanders, family solicitors, courtiers, air force officers which he played to such perfection. He became an indispensable figure in the gallery of second-rank players which were the glory of the British film industry in its more prolific days.

It became almost impossible to sit through any home-grown comedy without expecting to encounter at some time that inimitable brand of bewildered persistence under fire which

Le Mesurier made so very much his own. The character he cumulatively created will be remembered when others more famous are forgotten, not just for the skill of his playing but because he somehow embodied a curiously British reaction to the madness of the modern world—endlessly perplexed as he was by the dizzying and incoherent pattern of events but doing his courteous best to ensure that resentment never showed. Like Woody Allen, he rarely smiled. His characteristic expressions were the twitch and the lopsided grin, the raised eyebrow, the grimace, and the world-weary sigh that passed over his face.

He first attracted critical attention as a supporting player in popular British comedies like *Private's Progress* (1955)—he was the psychiatrist with a tic—and *I'm All Right Jack* (1959)—as a time-and-motion expert. During this period too, he frequently appeared on television with his close friend A. J. ('Tony') Hancock [q.v.] and he had a leading part in the Hancock film *The Punch and Judy Man* (1962) giving a beautifully judged and extremely sad performance as a sand artist. Later films included *Carlton-Brown of the F.O.* (1958), *We Joined the Navy* (1962), *The Wrong Arm of the Law* (1962), *Mouse on the Moon* (1963), *The Pink Panther* (1963), *The Liquidator* (1965), *The Wrong Box* (1966), and *Casino Royale* (1967).

In 1966, on television, he was in the comedy series *George and the Dragon*, with Sidney James [q.v.] and Peggy Mount but it was for his unforgettable portrayal of 'Sergeant Wilson' in the Home Guard television series *Dad's Army* (from 1968) that he will be most remembered. This he wonderfully sustained over a period of nine years. The stroke of genius was the casting of Arthur Lowe as Captain Mainwaring, the bank manager, and Le Mesurier as Sergeant Wilson, his clerk, rather than the other way around. This made the series a much more acute commentary on class in World War II, with a lower-middle-class Mainwaring lording it over his upper-middle-class public school subordinate, to whom he constantly felt and was made to feel, inferior.

Though mainly in demand as a comedy actor, he could be equally effective in straight parts. One of his best was in Dennis Potter's television play *Traitor* in 1971. He played a character based on the spy Kim Philby and turned in a memorable performance of a drunken stammering wreck of a man, holding court to western journalists in a Moscow flat. It gained him the Best Television Actor award from the Society of Film and Television Arts and made one realize how much more he could have achieved if he had chosen to be other than what he called a 'jobbing actor'. His other straight work for television included the lead part in David Mercer's play *Flint* and Marley's Ghost in *A Christmas Carol*. He was last seen in an adaptation of the novel of Piers Paul Read, *A Married Man* (1983).

Tony Hancock, of whose unofficial repertory company he was long a member, loved his air of gloom and always called him 'Eeyore'. Behind this apparent gloom, however, the inner Le Mesurier was the very opposite; ringing up to propose a meal or a party or a night out, the sepulchral voice would first murmur 'Playtime?' He loved jazz clubs and was a talented jazz pianist himself. He liked late-night restaurants at midnight and was the one who would drive the others home if they had celebrated too extravagantly.

He married three times. In 1939 he married the actress June Melville, daughter of Frederick Melville, dramatic author and theatre manager. There were no children and after World War II this marriage was dissolved. In 1949 he married the comedienne Josephine Edwina ('Hattie') Jacques (died 1980), daughter of Robin Rochester Jacques, who had been a test pilot in the Royal Flying Corps. They had two sons, Robin and Kim. This marriage was dissolved in 1965 and in 1966 he married Joan Malin, former wife of Mark Eden and daughter of Frederick Daniel Long, manager of a Ramsgate funfair. Le Mesurier died 15 November 1983 at Ramsgate Hospital. His whimsical sense of humour and baleful view of life were typified by his death announcement in *The Times*: 'John Le Mesurier wishes it to be known that he conked out on November 15th. He sadly misses family and friends.'

[John Le Mesurier, *A Jobbing Actor*, 1984; *The Times*, 16 November 1983; private information; personal knowledge.]

DEREK NIMMO

LENNOX-BOYD, ALAN TINDAL, first VISCOUNT BOYD OF MERTON (1904–1983), politician, was born 18 November 1904 at Loddington, Boscombe, Bournemouth, the second of four sons (there were no daughters) of Alan Walter Boyd, a barrister, of Bournemouth, and his second wife, Florence Anne, daughter of James Warburton Begbie, MD, of Edinburgh. He had a half-sister born of his father's first marriage. His eldest and youngest brothers were killed on active service in World War II and his second brother died in mysterious circumstances on secret service in Germany in 1939. Alan Walter Boyd assumed the additional surname of Lennox by deed poll in 1925.

Lennox-Boyd was educated at Sherborne School and Christ Church, Oxford. He won the Beit prize in colonial history, was elected chairman of the Oxford University Conservative Association, and in 1926 became president of the Oxford Union. He obtained a second class honours degree in modern history (1927).

In 1929 Lennox-Boyd stood unsuccessfully for the Gower division of Glamorgan but in 1931 was elected Conservative MP for mid-Bedfordshire, a constituency he held until his retirement from politics in 1960. In December 1938 he married Patricia Florence Susan, daughter of Sir Rupert Edward Cecil Lee Guinness, second Earl of Iveagh [q.v.], chairman of Arthur Guinness, Son, & Company Ltd. They had three sons. She brought him a substantial fortune which spared him the anxieties that can beset a public man and enabled him to indulge his own generous sense of hospitality.

In the House of Commons Lennox-Boyd was a strong supporter of Neville Chamberlain's efforts to reach an accommodation with the Axis powers and was closely associated with R. A. Butler (later Lord Butler of Saffron Walden, q.v.), Sir Henry ('Chips') Channon, and Harold Balfour (later Lord Balfour of Inchrye). Foreign and colonial affairs were always his main interest but he gained his early ministerial experience between 1938 and 1940 as a junior minister in the Ministry of Labour, Ministry of Home Security, and Ministry of Food.

In May 1940 Lennox-Boyd left active politics to serve in the RNVR as a lieutenant commanding a motor-torpedo-boat on the east coast and in the Dover patrol. He was called to the bar (Inner Temple) in 1941. He was recalled to office by (Sir) Winston Churchill in 1943 as a junior minister at the Ministry of Aircraft Production where he served until the general election of 1945. For the next six years the Conservative Party was in opposition. In this period Lennox-Boyd established his reputation as an authority on colonial affairs. He travelled widely to the colonies and was often called upon to intervene in colonial debates from the front bench.

When the Conservatives returned to power in 1951 he was appointed minister of state at the Colonial Office, under Oliver Lyttelton (later Viscount Chandos, q.v.), and was admitted to the Privy Council. After only six months, however, he was promoted to become minister of transport, a post he held for two years. It was not until the summer of 1954 that he at last achieved his abiding ambition to be secretary of state for the colonies. He held this post until 1959 and was thus the longest serving secretary of state since Joseph Chamberlain [q.v.]. He never aspired to any other office.

Lennox-Boyd had a commanding presence (he was six feet five inches tall), a very quick mind, and an unusual capacity for work. As colonial secretary he would start work early, between five and six in the morning, arriving in the office having already mastered his briefs. Thereafter he would seldom read or write in the office but spend the day listening to advice or giving it. He was persuasive in council and dominated the House of Commons on colonial affairs, not so much by his oratory as by his knowledge of the facts and his grasp of the arguments involved.

He saw his task as colonial secretary as one of preparing the colonies for self-government within the Commonwealth. But he knew that political progress would be meaningless or worse unless underpinned by economic development. To this end he devised a series of federal systems for South East Asia, Aden and its associated protectorates, East Africa, Central Africa, and the West Indies. The essential feature of each of these was that it contained a substantial source of wealth and so could carry the less favoured territories involved. It was a grand and practical concept but could only be brought to fruition gradually and with British administrative and if necessary military support. But here Lennox-Boyd's classical concept of colonial evolution clashed with events outside Britain's control. The defeat suffered by Britain and France at Suez in 1956 undermined the French position in Algeria. With Algeria lost, General de Gaulle decided to decolonize the whole of French black Africa as soon as possible. Could Britain have continued with the gradual progress to which Lennox-Boyd aspired? This would not have been impossible but it might have involved a growing military commitment and would have clashed with the efforts of Harold Macmillan (later the Earl of Stockton) to convince de Gaulle that Britain was more interested in Europe than with hanging on to the colonial empire. Lennox-Boyd appreciated Macmillan's problem and could not resist his conclusions. But he knew that premature independence must mean the betrayal of friends, the installation of oppressive regimes, and the impoverishment of the mass of people—to say nothing of the sacrifice of British interests. As the last imperial statesman in a long and distinguished line he was not prepared to become the undertaker of imperial responsibility. He offered to resign but was persuaded, with some difficulty, to remain as colonial secretary until the 1959 election.

After the election Lennox-Boyd left the House of Commons and accepted a viscountcy (1960). He was also made a Companion of Honour (1960). He held certain directorships and found time for a wide range of charitable and social concerns. In particular he was a trustee of the British Museum and the Natural History Museum.

Lennox-Boyd returned briefly to colonial affairs in 1979 when Margaret Thatcher, then leader of the opposition, asked him to lead a mission sent by the Conservative Party to Rhodesia to observe the general election which was won by Bishop Muzorewa. The delegation found that the election was a valid test of opinion and recommended that a future Conservative government should recognize the Muzorewa–Smith government.

In private Lennox-Boyd had a wide circle of friends of every race, colour, and social origin. He also had a talent for taking trouble over seemingly unimportant things. Few colonial visitors came to London without being invited to his house in Chapel Street. Their particular interests or foibles were invariably carefully attended to. He always carried a small notebook in his pocket and would jot down any thought or suggestion as to how he might please visitors. On leaving politics he gave up his house in Chapel Street and retired to Ince Castle, a lovely seventeenth-century house in Cornwall overlooking the river Tamar. There at last he was able to enjoy his family, sail his yacht, and indulge his hobbies. He died on the evening of 8 March 1983 in Fulham Street, Chelsea, London, when run down by a car driven by an unaccompanied learner driver. He was succeeded in the viscountcy by his eldest son, Simon Donald Rupert (born 1939).

[Personal knowledge.]   JULIAN AMERY

LEWIN, (GEORGE) RONALD (1914–1984), military historian and biographer, was born in Halifax, Yorkshire, 11 October 1914, the eldest of four sons (there were no daughters) of Frank Lewin, patent agent, and his wife, Elizabeth Wingfield. He was educated at Heath Grammar School in Halifax and The Queen's College, Oxford, where he was a Hastings scholar and a Goldsmiths' exhibitioner. His first class in classical honour moderations in 1934 was the best of its year and was followed by a first in *literae humaniores* (1936).

From Oxford he went in 1937 as an editorial assistant to the publishers, Jonathan Cape. Having joined the Territorial Army as a gunner early in 1939, describing himself as 'probably the most inefficient civilian who ever put on uniform—and that includes the good soldier Schweik!', he was called up on the day war broke out and served in the Royal Artillery until 1945. He was in North Africa with the Eighth Army, was slightly wounded, and was mentioned in dispatches. In the winter of 1943 he was posted to England to train with a super-heavy regiment, and fought with 21st Army Group in north-west Europe from June 1944 until the end of the war, by which time he had become a captain.

He returned to Cape's briefly on being demobilized, but prospects seemed limited and in 1946 he joined the BBC as a talks producer in the Third Programme. In 1954 he became chief assistant, Home Service, its head from 1957, and its chief in 1963. He set up the Music programme and initiated the Today programme and other successful series. He was, however, not cut out for administration, and he retired ill in 1965. There followed some ten years of clinical depression, the strain of his work having brought on a delayed reaction to his wartime years.

He returned to publishing, joining Hutchinson's and specializing in works of military history and wartime experiences. But he also turned to writing his own books. Before the war he had written many book reviews and contributed poems and articles to various periodicals, and now he was commissioned to cover *Rommel* in the Great Commander series. This appeared in 1968 and *Montgomery as Military Commander* followed in 1971. In 1969 he also edited volume three of *Freedom's Battle, The War on Land, 1939–45*. In 1973 *Churchill as War Lord* was published, and in 1976 *Man of Armour*, about Major-General Vyvyan Pope. By now his reputation was well established and he was chosen to undertake the official biography of Field-Marshal Viscount Slim [q.v.]. It was a difficult task, for Slim's autobiography had been justly acclaimed, but *Slim the Standard-Bearer* (1976) was a triumphant success, lucid, intelligent, and exceptionally readable. It won the W. H. Smith literary award in 1977. *The Life and Death of the Afrika Korps* followed in 1977.

Lewin was now accepted as one of the leading military historians but *Ultra Goes to War*, published in 1978, presented a new challenge. Though he lacked scientific or mechanical training, his mastery of the techniques by which the allied cryptographers broke the German ciphers was so complete that he was able to explain them in words that the least qualified could understand, and assess their significance with magisterial authority. He followed this with an account of cipher-breaking achievements in the USA, published as *American Magic* there and *The Other Ultra* (1982) in Great Britain. For this he paid several visits to the United States and was the first to see many of the relevant papers in the National Archives. He moved on to *Hitler's Mistakes*, a study of Hitler's shortcomings as politician and general, but owing to a delay in publication this appeared only posthumously. To this Dictionary he contributed articles on Richard Dimbleby, Lord Ismay, Sir Basil Liddell Hart, Sir Desmond Morton, and Bernard Fergusson (Lord Ballantrae).

In 1982 he was struck down by a recurrence of the cancer for which he had had an operation eight years earlier. From then on until he died he was undergoing treatment and in constant pain. He was kept going by his determination to finish the one-volume history of World War II which he had been asked to write by the Oxford University Press. He achieved about a third of his objective, laying down his pen only on the day he went into hospital.

He was appointed CBE for services to military history in the New Year honours of 1983, but took even greater pleasure in receiving the Chesney gold medal of the Royal United Services Institute in the previous year for 'eminent work calculated to advance military science and knowledge'. His prodigious memory, clarity of mind, and immaculate prose style did indeed put him among the masters of his profession; his generosity to those less experienced who sought his help ensured that he was as well liked as he was respected. He was an FRSL and F.R.Hist.S.

In 1938 he married Sylvia Lloyd (died 1988), daughter of Philip Maximilian Sturge, industrial print manufacturer, of a Quaker family in Birmingham. They had three sons and a daughter, and the death in a road accident of the youngest son in 1963 was a terrible blow. He died 6 January 1984 at St Luke's Hospital, Guildford, Surrey.

[Private information; personal knowledge.]

PHILIP ZIEGLER

LEWIS, SIR ANTHONY CAREY (1915–1983), musician and founder of *Musica Britannica*, was born 2 March 1915 in Bermuda, the youngest of the three sons (there were no daughters) of Major (later Colonel) Leonard Carey Lewis, of the Lincolnshire Regiment and Royal Army Ordnance Corps, then chief ordnance officer in Bermuda, afterwards of Hampton, Middlesex, and his wife, Katherine Barbara, only daughter of Colonel Henry George Sutton, Indian Army, of Hartington, Derbyshire. At an early age Lewis revealed exceptional musical gifts. He went to Salisbury Cathedral choir school, and when he was eight was admitted a chorister at St George's chapel, Windsor, where he sang under Sir Walter Parratt, Edmund Fellowes, and Sir H. Walford Davies [qq.v.].

In 1928, after several months at the Royal Academy of Music, where his composition professor was William Alwyn, Lewis entered Wellington College as the first music scholar. He became proficient at the oboe, achieved concerto standard as a pianist, and in 1932 won the Bernard Hale organ scholarship at Peterhouse, Cambridge. He now studied composition and research with Professor Edward J. Dent [q.v.], whose teaching and example influenced him profoundly. He won the John Stewart of Rannoch scholarship in sacred music in 1933, and the award of the Leith studentship enabled him to study composition in Paris with Nadia Boulanger during the summer of 1934. A year later he graduated BA and Mus.B., winning the Barclay Squire prize for musical palaeography.

In September 1935 Lewis joined the music staff of the British Broadcasting Corporation, under (Sir) Adrian Boult [q.v.]. Inspired by Dent's view that the standard classics should not be allowed to obscure other music of importance, Lewis brought before the public many revivals of unfamiliar pre-nineteenth century works. He produced the long-running series 'Foundations of Music' and other similar programmes, and later became responsible for all broadcast chamber music. His composition *A Choral Overture*, in which an eight-part unaccompanied choir vocalizes to varying syllables, received its première in the 1938 season of Queen's Hall Promenade concerts.

On the outbreak of war in 1939 Lewis joined the Royal Army Ordnance Corps. He was posted as a major to the Middle East in 1942, became deputy assistant director of ordnance services in 1943, and for a short period in 1945 was assistant director, with the rank of lieutenant-colonel, displaying administrative abilities which were to benefit music and musicians greatly in the years ahead. Under the auspices of the British Council and ENSA he helped to organize the provision of music for the troops, himself conducting the Cairo Symphony and other orchestras.

He returned to the BBC in February 1946, undertaking the planning and supervision of all music for the new Third Programme. The introduction of the network on 29 September 1946, under (Sir) George Barnes [q.v.], was soon recognized as the most important development for music since the beginning of broadcasting, and Lewis's contribution to laying its foundations was not the least of his achievements.

At the end of the year he left the BBC, and in 1947 succeeded (Sir) Jack Westrup [q.v.] as professor of music in the University of Birmingham. It was from the Peyton and Barber chair that his greatest achievement as a musicologist was undertaken, the foundation of *Musica Britannica* as a national collection of the classics of British music. His proposals were submitted to the council of the Royal Musical Association in 1948. An editorial committee was appointed, with Lewis as general editor (which function he fulfilled for the rest of his life) and R. Thurston Dart [q.v.] as secretary; and the first three volumes were published in

1951, as part of the Festival of Britain celebrations.

It was Lewis's constant aim to see that the fruits of scholarly research should be enjoyed through practical performance; and during his time at Birmingham, where his compositions included concertos for trumpet, horn, and harpsichord, he conducted many revivals of baroque music, notably Handel operas, together with premier recordings of works by composers such as Lully, Rameau, Handel [q.v.], and especially Henry Purcell [q.v.]. He served as honorary secretary of the Purcell Society (1950-76) and artistic director of the Festival of Britain's Purcell series (1951), and was chairman of the Arts Council's music panel (1954-65), the Purcell-Handel Festival (1959), and the British Council's music committee (1967-73). He was also dean of the faculty of arts at Birmingham University (1961-4), and was president of the Royal Musical Association (1963-9).

In 1968 Lewis succeeded Sir Thomas Armstrong as principal of the Royal Academy of Music. The balance between scholarly and artistic work which characterized his life was now demonstrated by the fact that for the next fourteen years he presided over many important developments in an institution where the emphasis was on performance and composition. This phase saw the publication of his contributions to *The New Oxford History of Music*—'English Church Music' for volume v, *Opera and Church Music (1630-1750)* (1975), of which he was joint editor, and 'Church Music' for volume viii, *The Age of Beethoven (1790-1830)* (1982). During this period he was president of the Incorporated Society of Musicians (1968), a director of the English National Opera (1974-8), and chairman of the Purcell Society (1976-83).

Lewis was a musician of rare accomplishment, widely skilled in the science and practice of music. His knowledge as an editor, experience as a composer, insight as a conductor, and eloquence as a writer, were all combined in solving the manifold problems surrounding practical performance, particularly of the neglected treasures of the national heritage. His appearance was imposing, his manner reserved; yet behind this lay a vigour which was essentially creative. He liked to see things grow. His chief recreation was gardening, a pastime which gave him much pleasure. He was genial and kindly, and had a lively sense of humour. His sustained vision, patient advocacy, and practical wisdom enabled him to blaze fresh trails. The foundation of *Musica Britannica* was an act of high imagination and courage; and, with over fifty volumes completed by the time of his death, this growing collection had achieved the early aim of ranging from Dunstable to Parry, and stood four-square as 'a living tribute to British musical achievement through the centuries'.

He was appointed CBE in 1967 and knighted in 1972. The honorary degree of D.Mus. was conferred on him by Birmingham University in 1970. He was an honorary member of the Royal Academy of Music (1960) and the Guildhall School of Music and Drama (1969), and was also honorary FTCL (1948), FRCM (1971), FRNCM (1974), and FRSAMD (1980). He was a governor of Wellington College (1953-83).

In 1959 Lewis married Lesley Lisle, daughter of Frank Lisle Smith, bank manager, of Northland, New Zealand. There were no children. He died at his home in Haslemere, Surrey, 5 June 1983. A portrait by Pamela Thalben-Ball (1976) is at the Royal Academy of Music.

[Anthony Lewis, 'Musica Britannica', *Musical Times*, May 1951; *Royal Academy of Music Magazine*, summer 1982 and autumn 1983; David Scott in *The New Grove Dictionary of Music and Musicians*, 1980 (ed. Stanley Sadie); John L. Holmes, *Conductors on Record*, 1982; family papers; personal knowledge.]   MICHAEL POPE

LIDDELL, EDWARD GEORGE TANDY (1895-1981), physiologist, was born 25 March 1895 at Harrogate, the second in the family of two sons of John Liddell, physician, of Harrogate, and his wife, Annie Louisa Tandy. His first three winters were marred by life-threatening bouts of pneumonia which left him with a permanently damaged heart. He had to limit his way of life and was fortunate to escape further serious illnesses. He was educated at Harrow, where he spent two years on the classical side before going over to science. He lived for six months with families in Germany before entering Trinity College, Oxford, in October 1914 to read medicine. He took a first in physiology in 1918, was elected to a senior demyship at Magdalen, and went for his clinical training to St Thomas's Hospital, qualifying BM, B.Ch. (Oxon.) in 1921. A year before he qualified he had had a tentative offer of a research fellowship at Trinity, coupled with an assistantship to (Sir) Charles Sherrington [q.v.], then Waynflete professor of physiology. These appointments he took up in 1921. Already as an undergraduate he had come under Sherrington's spell and had learned, unlike the majority of undergraduates, how to derive inspired instruction from his notoriously difficult lectures.

Sherrington was president of the Royal Society from 1920 to 1925 and had to spend much time in London, sometimes even making two return rail journeys in a single day. Much,

therefore, of the responsibility for setting up their experiments necessarily devolved on Liddell. In those years he was Sherrington's sole collaborator in research. As a beginner, and self-effacing almost to a fault, his contribution was inevitably overshadowed by Sherrington's towering genius. It is clear however that his share in their classic researches went well beyond the level of consummate experimental skills. Their collaboration began at a turning-point in studies of reflex action, when Sherrington's interest was moving away from exteroceptive reflexes as items of animal behaviour and towards the cellular basis of synaptic excitation and inhibition, conceived as summing algebraically at the post-synaptic membrane. The exteroceptive reflexes were elicited by graded electrical stimulation of afferent nerves in spinal and decerebrate animals; the synaptic actions were detected by quantitative high-speed myography. The classical discovery of the proprioceptive stretch reflex was important because it could be elicited only by a pure physiological stimulus, minimal lengthening of a muscle; and because of its essential role in reflex posture. This work and its further development was summarized in the monograph *Reflex Activity of the Spinal Cord* (R. S. Creed, D. Denny-Brown, J. C. Eccles, E. G. T. Liddell, and C. S. Sherrington, 1932). The preface acknowledged Liddell's special editorial role. From 1930 onwards his experiments were concerned with the control of postural reflexes by impulses descending from different areas of the brain—work of special relevance to the understanding of 'spinal shock' in patients with lesions of the spinal cord.

Liddell was elected FRS in 1939 and Waynflete professor of physiology in 1940. In 1960, the year of his retirement, he published his monograph *The Discovery of Reflexes*. With the insights of an experienced neuroscientist combined with a scholarship equally at home in French, German, and eighteenth-century Latin, he traced the slow growth of knowledge and ideas about the nervous system that set in relief the veritably Harveian revolution which Sherrington had brought about. In 1975 he was awarded the Osler memorial medal for the science, art, or literature of medicine.

Liddell's college pupils held him in respectful awe. Those who penetrated his formidable reserve were rewarded by interesting tutorials and lifelong friendship. His university lectures lacked showmanship but those who stayed the course were apt to find, when re-reading their notes, that his reviews had been up-to-date and his laconic comments ahead of their time. As professor he had to run a shortened preclinical course that could be taught by the few demonstrators who remained in wartime Oxford.

After the war the old laboratory was overwhelmed by the resumption of honours work for unprecedented numbers of undergraduates. Liddell presided over the appointment of new demonstrators and the building of the new laboratory. Thanks to his close supervision of architects and contractors the move was accomplished with minimal interruption of teaching and research. As chairman of the organizing committee of the XVII International Congress of Physiologists which met in Oxford in July 1947 he coped with the many frustrations of the immediate post-war period. Until his retirement in 1960, and in spite of precarious health, he did his share, as he had done since the 1930s, of service on council and other university committees. Extramurally he served as external examiner, as member of the council of the Royal Society, and on the committee and the editorial board of the Physiological Society. He was chairman of the Oxford Eye Hospital.

He married in 1923 Constance Joan Mitford, daughter of Bertram Mitford Heron Rogers, physician, of Bristol. They had three sons, one of whom died in 1978, and one daughter. Liddell died in a nursing home in Witney 17 August 1981.

[C. G. Phillips in *Biographical Memoirs of Fellows of the Royal Society*, vol. xxix, 1983; personal knowledge.]   C. G. PHILLIPS

LIDELL, (TORD) ALVAR (QUAN) (1908-1981), BBC broadcaster, was born 11 September 1908 at Wimbledon Park, south London, the third of three children and younger son of Swedish parents, John Adrian Lidell, timber importer, and his wife Gertrud Lundstrom. He was educated at King's College School, Wimbledon, and Exeter College, Oxford, where he obtained a second class in classical honour moderations (1929). He studied piano, piccolo, and cello as a boy, and at seventeen began singing lessons. At Oxford he was an outstanding actor, notably in the production of *Comus* by (Sir) Arthur Bryant [q.v.].

After brief teaching jobs, and engagements as a singer with a puppet theatre company, Lidell joined the BBC as chief announcer at Birmingham; the following year he transferred to London, where he became deputy chief announcer in 1937. To him fell the task of reading the historic announcement of King Edward VIII's abdication, and on 3 September 1939 he read the ultimatum to Germany from a room at 10 Downing Street. He remained there to introduce the prime minister, A. Neville Chamberlain who, at 11 a.m., broke the news that Britain was from that moment at war. Lidell never forgot the experience of 'sitting there,

behind this figure of terrible grief'.

During World War II the BBC dispensed with the traditional anonymity of its newsreaders (to distinguish them from enemy propagandists) and Lidell was one of the named readers who brought the war news to the nation. The phrase ' ... and this is Alvar Lidell reading it' was a guarantee of clarity, intelligence, and cool objectivity. Only once did Lidell break with this principle. When going on the air with news of the victory at El Alamein, he allowed himself to say: 'Here is the news, and cracking good news it is too.' He was called up for war service in RAF Intelligence in 1943, but a year later returned to his nationally important work at the BBC.

In 1946 Lidell was appointed chief announcer on the new Third Programme. He remained for six years in this post, for which he was admirably qualified by his command of languages and knowledge of music. Artists of the calibre of (Sir) Clifford Curzon [q.v.] and members of the Amadeus Quartet insisted that Lidell should introduce their broadcasts, a state of affairs much appreciated by one who was always at his happiest with musical people. In his work for the Third Programme, he set and maintained the highest standards, taking infinite trouble over pronunciation and phrasing.

In 1952 Lidell returned to news-reading in a newly constituted team of specialists who, in 1954, added the presentation of television news to their other duties. Television work, however, did not appeal to Lidell, who devoted most of the remainder of his career to radio broadcasting. In 1964 he was appointed MBE and he retired in 1969. His influence, however, persisted. When, in 1979, the *Listener* published his article about deteriorating standards of speech at the BBC (it was headed 'Newsweeding'), the impact was considerable. As a result the BBC set up a panel of experts to report on the quality of spoken English on the air.

Lidell's talent as an exemplary stylist of the spoken word was not confined to announcing. He was in international demand as a narrator in such taxing works as Arnold Schoenberg's *Gurrelieder* and *A Survivor from Warsaw*, as well as *An Oxford Elegy* by Ralph Vaughan Williams [q.v.] and *Façade* by (Sir) William Walton [q.v.]. (Dame) Edith Sitwell [q.v.] admired his rhythmic perfection in *Façade*, and he performed it with a prologue he had written himself at a Downing Street party to mark the composer's seventieth birthday. He was a dedicated reader of 'Books for the Blind', recording a total of 237 volumes, including marathons such as *Anna Karenina*. As a baritone singer he also achieved distinction, giving *Lieder* recitals during and after the war, and memorably recording English ballads with Gerald Moore at the piano.

In appearance Lidell was tall (6 ft. $3\frac{1}{2}$ in.), aristocratic, and perhaps a touch reserved. In private life he was a very loving and devoted husband and father. In 1938 he married Nancy Margaret, daughter of Thomas Henry Corfield, lawyer. They had two daughters and a son. Lidell was compassionate and generous not only to his family and friends but to all those he considered to be in great need. By way of relaxation he enjoyed games. He played rugby, tennis, and cricket at school and later took up golf and darts; he could also solve *The Times* crossword in six minutes. After two years' illness he died of cancer 7 January 1981 at Michael Sobell House, Mount Vernon Hospital, Northwood, Middlesex.

[*The Times*, 9 January 1981; BBC records; family information; personal knowledge.]

RICHARD BAKER

LIMERICK, COUNTESS OF (1897-1981), leader of the British and International Red Cross movements. [See PERY, ANGELA OLIVIA.]

LINDSAY, SIR MARTIN ALEXANDER, first baronet, of Dowhill (1905-1981), soldier, explorer, politician, and author, was born in London 22 August 1905, the only son and elder child of Lieutenant-Colonel Alexander Bertram Lindsay, 2nd KEO Gurkhas, and his wife, Gladys, widow of Lieutenant Maurice Cay, RN, and daughter of William Hutton, of Beetham House, Milnthorpe, Westmorland. He came of a military family and traced his descent, as twenty-second feudal Baron of Dowhill, from Sir William Lindsay, born *c*.1350. His father died when he was fourteen and his only sister at the age of twenty-five. He was educated at Wellington (where he won English essay prizes) and at the Royal Military College, Sandhurst, which he left without distinction, a fact which he attributed to his idleness. In 1925 he was commissioned in the Royal Scots Fusiliers. He was a keen horseman and rode in steeplechases as an amateur.

In 1927 he was seconded to the 4th battalion of the Nigeria Regiment at Ibadan. In 1929 he travelled across Africa through the Ituri forest in the Belgian Congo, unaccompanied save by porters and enduring considerable hardship, collecting pigmy artefacts for the British Museum.

Back in England, Lindsay volunteered for the expedition to Greenland which H. G. ('Gino') Watkins [q.v.] was to lead and of which the objective was a survey of south-east Greenland with a view to the establishment of an Arctic air route to Canada. Watkins accepted Lindsay on condition that he learned surveying. Such was Lindsay's enthusiasm that he fulfilled the condition and overcame the objections of his commanding officer. In the course of this expedition (1930-1) Lindsay acquitted himself with distinction and gained invaluable experience. He was awarded the King's Polar medal and wrote an account of the expedition in *Those Greenland Days* (1932). He was now fired with ambition to lead his own expedition to the east coast mountains of Greenland and complement the work of Watkins.

Access from the east coast being difficult, he planned to approach his objective from the west, across the ice cap. He had difficulty in getting backing, but finally won the approval of the Foreign Office, the consent of the Danish government, and the patronage of the Prince of Wales. He was accompanied by (Colonel) Andrew Croft and Lieutenant A. S. T. Godfrey of the Royal Engineers (killed in action, 1942). Lindsay dispatched Croft in advance to train dogs for sledging, and, with Godfrey, joined him in May 1934. Thanks to careful planning and the courage and resource of all three, the expedition was a success. They sledged 1,050 miles in 103 days over the ice cap from Jakobshavn to Angmagssalik, carrying their equipment with them, and mapped 350 miles of mountains between Scoresby Sound and Mount Forel. With the approval of the Danish government Lindsay named two mountain ranges after Prince Frederick of Denmark and the Prince of Wales respectively. Lindsay received medals from the French, Belgian, and Swedish geographical societies. He described the expedition in *Sledge* (1935).

In 1936 Lindsay left the army and was adopted as prospective Conservative parliamentary candidate for the Brigg division of Lincolnshire, but war intervened. Lindsay saw active service in Norway and trained airborne troops in Britain and India. In 1944 he was appointed to command the 1st battalion, Gordon Highlanders, whom he led in sixteen operations in north-west Europe. He was wounded, twice mentioned in dispatches, and appointed to the DSO (1945). He described his experiences in *So Few Got Through* (1946).

In 1945 Lindsay was elected Conservative MP for Solihull and held the seat until 1964. He was a loyal Tory but did not fear to take an independent line. He conscientiously represented his constituency, greatly helped by his wife, and took a special interest in the motor industry. He was always quick to defend the interests of the armed forces and of ex-soldiers and their families. From 1949 to 1952 he was chairman of the West Midlands Conservative Party and his services were in demand to advise new parliamentary candidates. He was appointed CBE in 1952 and created a baronet in 1962. He was a Gold Staff officer at the coronation of Queen Elizabeth II and a member of the Queen's Body Guard for Scotland (Royal Company of Archers). From 1973 to 1979 he was vice-chairman of the standing council of the baronetage, of which he wrote a history in 1979.

A man of action, Lindsay has an assured place in the annals of Arctic exploration. He had considerable presence and charm as well as drive, energy, and organizing ability. Sometimes impulsive and frequently outspoken, he was not universally popular, but he enjoyed the affection and respect of many friends.

In 1932 Lindsay married Joyce Emily, daughter of Robert Hamilton Lindsay, of the Royal Scots Greys, son of the twenty-sixth Earl of Crawford [q.v.]. They had two sons and one daughter. This marriage was dissolved in 1967. In 1969 he married Loelia Mary, formerly wife of Hugh Richard Arthur Grosvenor, second Duke of Westminster, and daughter of Frederick Edward Grey Ponsonby, first Baron Sysonby, treasurer to King George V. There were no children of this marriage. Lindsay died at Send, near Guildford, 5 May 1981. He was succeeded in the baronetcy by his elder son, Ronald Alexander (born 1933).

[Martin Lindsay, *Sledge*, 1935, *Those Greenland Days*, 1932, and *So Few Got Through*, 1946; *The Times*, 7 May 1981; *Geographical Journal*, vol. liii, no. 10, July 1981; private information; personal knowledge.]

W. PERCY GRIEVE

LLEWELLYN, RICHARD (pseudonym), novelist and dramatist. [See LLOYD, RICHARD DAFYDD VIVIAN LLEWELLYN.]

LLEWELYN DAVIES, RICHARD, BARON LLEWELYN-DAVIES (1912–1981), architect, was born in London 24 December 1912, the elder child and only son of Crompton Llewelyn Davies, solicitor, and his wife, Moya, daughter of James O'Connor. He was educated privately in Ireland and at Trinity College, Cambridge. He obtained a third class in the mechanical sciences tripos in 1934 and also attended summer school sessions at the École des Beaux Arts in Paris. He obtained its diploma in architecture in the same year as his Cambridge degree.

At Cambridge he had friends from many disciplines, and he became convinced of the need for architects, no less than engineers, to understand social, as well as technical issues. In Britain, debate on the emerging 'modern' architecture was lively at only a few schools in 1933, notably Liverpool and the Architectural Association in London. He enrolled at the AA and was in the forefront of the group there striving to understand how new building materials and widening social awareness would affect the practice of architecture.

He gained his diploma, with honours, from the AA School in 1937 and completed his first building, a remarkable private house near Chichester, in partnership with his contemporary Peter Moro in 1939. He became ARIBA in 1939. During the war he worked with the engineering firm of Sir Alexander Gibb & Partners on the immense government programme of factory building.

In 1942 he joined the London, Midland, and Scottish Railway and developed there a system of prefabrication for railway stations. In the late 1940s he played a decisive role in the adoption of the metric system for post-war construction. In 1948 he joined the Nuffield Provincial Hospitals Trust and then the Nuffield Foundation directing multi-disciplinary study groups on hospital function and design. Pioneering programmes of architectural research were developed which influenced hospital design throughout the world.

He was impatient with a subjective basis alone for architecture, and with the narrow experience of most practitioners; he was convinced of the value of a scientific approach and of the necessity to consider buildings in relation to their social environment as well as aesthetic context. He became FRIBA in 1956.

In 1960 Llewelyn Davies was appointed to the chair of architecture at University College, London, and he started at once to develop a wholly new curriculum based on his beliefs, to replace the unconvincing version of the Beaux Arts system which still survived there. Architectural students had been selected hitherto from those with a clear interest in arts subjects, but for the new curriculum at least one science subject and mathematics were required. There were many new staff and research appointments, and for their postgraduate years students came from other disciplines, other schools, and other countries. The new curriculum at London University became a model for many other schools in the United Kingdom and elsewhere.

Llewelyn Davies was created a life peer in 1963. He held the chair of urban planning at University College, London, from 1969 to 1975. He was a member of the Royal Fine Art Commission from 1961 to 1972 and was largely responsible for the creation of the Centre for Environmental Studies in 1967, and was its first chairman. He was appointed an honorary fellow of the American Institute of Architects in 1970.

In 1960 he formed a partnership with an architect colleague with whom he had worked in the LMS development group and in the Nuffield teams. In a few years the new firm,

Llewelyn-Davies Weeks, became large and active internationally, and it had its own multidisciplinary research unit. Llewelyn-Davies took an active role in the practice, and in particular in the development of town planning as an integral part of the firm. This was paralleled at University College, where he combined the Schools of Planning and Architecture to form a School of Environmental Design. His major planning achievement was at Milton Keynes, where the development plan was his own concept, as a memorial to him at the town centre testifies.

Llewelyn-Davies was rather an enigmatic architect. His greatest influence was in the fields of architectural research rather than in individual buildings, and in education, planning, and the philosophy of design. In 1975 he gave the Gropius lecture at Harvard, an intense affirmation of his belief in the necessity for thought as the basis of successful design. Many of the major buildings bearing the firm's name, notably the London Stock Exchange, were designed according to his concept, but his most personal work was the creation of his own houses and houses for friends. For these he designed everything, including the central heating systems. They are handsome, comfortable, and quite free from stylistic tricks; they have wit and charm.

In 1943 he married (Annie) Patricia, daughter of Charles Percy Parry, engineer. She was created Baroness Llewelyn-Davies of Hastoe in 1967. They had three daughters. Llewelyn-Davies died in St Bartholomew's Hospital, London, 27 October 1981.

[Muriel Emmanuel, *Contemporary Architects*, 1980; Andrew Saint, *Toward a Social Architecture*, 1988; private information; personal knowledge.]   JOHN WEEKS

LLOYD, SIR HUGH PUGHE (1894–1981), air chief marshal, was born 12 December 1894 at Leigh, Worcestershire, the third of the four sons and the third of six children of Lewis Thomas Lloyd, schoolmaster, of Leigh, and his wife, Anne Pughe. From King's School, Worcester, he went up to Peterhouse, Cambridge, in 1913 and studied law before enlisting as a private in the Royal Engineers in February 1915.

On the western front Lloyd served as a dispatch rider and was wounded three times before being commissioned in the Royal Flying Corps in April 1917. From January 1918 he was a bomber reconnaissance pilot with No. 52 Squadron. He soon distinguished himself, and in quick succession won the MC, the croix de guerre, and the DFC.

In August 1919 Lloyd was granted a permanent commission in the Royal Air Force. The following month he married Kathleen (died 1976), daughter of Major Robert Thornton Meadows DSO, an army doctor. At the end of the year he was posted to India, where the only child of the marriage, a daughter, was born.

After four years mainly as a flight commander with No. 28 Squadron, Lloyd returned home in 1924. During the next five years he flew with No. 16 Squadron, passed through the RAF Staff College, and served for three years on the staff of a training group (No. 23). Promoted squadron leader, he was then for nearly a year chief flying instructor at No. 2 Flying Training School. Clearly marked out for senior posts, he was next sent out to take the Staff College course at Quetta (1931–2). Three years on the air staff at HQ No. 1 Indian Group at Peshawar followed. His two spells in India brought him operational experience on the North-West Frontier and mentions in dispatches.

Back in England, Lloyd became senior RAF instructor at the Staff College, Camberley (1936–8), and then commanded No. 9 (Bomber) Squadron. Soon after the outbreak of war he was given command, as a group captain, of the bomber station at Marham, but was quickly summoned to air staff duties at HQ Bomber Command. On 20 May 1940 he became senior air staff officer at No. 2 (Bomber) Group, where he organized many strikes against German shipping. This made him an ideal choice for his next post. On 1 June 1941, with the rank of air vice-marshal, he took over command of RAF Mediterranean at Malta. Its resources were pitifully small, and reinforcement appallingly difficult, but Lloyd's determination to maintain an offensive never faltered. Despite all the Italian and German bombing, which by June 1942 brought the island to the verge of starvation, Malta remained unsubdued. Throughout it all Lloyd kept the island in use as a staging post for bombers reinforcing RAF Middle East, and—except briefly during the very worst of the German air assault—he kept up attacks on the Axis shipping lanes across the Mediterranean. The result, in conjunction with the work of the submarines, was the denial of vital supplies to the Axis forces in North Africa and an easier task for the British Eighth Army. Of these experiences, Lloyd left a vivid account in his book *Briefed to Attack* (1949).

Lloyd left his Malta command on 15 July 1942 to become senior air staff officer at RAF Middle East and then, from March 1943, commanded the newly formed North-West African Coastal Air Forces. In this role he again waged a successful offensive against Axis shipping. For his work from 1941 onwards he received many honours. He was appointed CBE in 1941, and CB and KBE in 1942, officer of the French

Legion of Honour in 1943 (commander in 1945), and officer of the US Legion of Merit in 1944. He received an honorary degree from the University of Wales after the war.

At the end of 1944 Lloyd returned to England. In July 1945 he was appointed, as acting air marshal, to command Tiger Force, the long-range RAF bomber group intended to join the Americans in their air assault on Japan. The dropping of the two atomic bombs, however, obviated the need for this. For two years Lloyd was then RAF instructor at the Imperial Defence College before being sent out to command the RAF in the Far East (1947–9). Finally, in 1950, he was entrusted with Bomber Command, which benefited greatly from his drive. He was appointed air chief marshal and KCB in 1951, and retired in 1953.

After his retirement Lloyd reared pigs and poultry on a small farm in Buckinghamshire. His chief recreations were sailing and riding. Among voluntary activities he was for more than twenty-five years a very active president of the Polish Air Force Association and of the RAF Association, Wales.

In manner, Lloyd had something of the air of an attractive buccaneer. His intensely light blue eyes had a normally smiling look, but could fix an offender with a laser-like probe. He had the great gift of making people feel that they were the sole subject of his attention. He was popular—he was known throughout the Service as Hugh Pughe or Hughie Pughie—and extremely brave and tough. His humour, virility, and outgoing character to some extent masked his more intellectual qualities: in the 1920s he had gained high awards in the RAF's most prestigious essay competitions, and he was a respected instructor at the Staff Colleges. He died 13 July 1981 at Cheltenham.

[H. P. Lloyd, *Briefed to Attack*, 1949; private information; personal knowledge.]

DENIS RICHARDS

LLOYD, RICHARD DAFYDD VIVIAN LLEWELLYN (1906–1983), novelist and dramatist under the name of RICHARD LLEWELLYN, was born between 8 and 10 December 1906 at St David's, Pembrokeshire, the son of William Llewellyn Lloyd, hotelier, and Sarah Anne. His birth was not registered.

Llewellyn's scattered autobiographical sketches are not always consistent, but it is clear that his father's peripatetic career led to frequent changes of school in south Wales, London, and elsewhere. At sixteen he entered the hotel trade, briefly as a dishwasher in Claridge's and then in Italy. In 1926 he enlisted in the army and served in India and Hong Kong,

returning to Britain in 1931 to a period of near destitution before being taken on as a film reporter by *Cinema Express* and cobbling together a career as bit-player, assistant director, production manager, and script-writer. A play, 'Poison Pen', had a successful run in London in 1937 and this emboldened him to take time off to complete his first novel, of which a draft had been written during his army days in India. The publication in October 1939 of *How Green was my Valley* (originally entitled 'Slag') brought him instant celebrity and an assured income. The twenty-one translations include versions in Hindi, Japanese, and Turkish as well as two in German. The setting was his paternal grandfather's village of Gilfach Goch and, though Llewellyn later referred somewhat vaguely to a period underground and to an escape from a serious roof fall, it is safer to conclude that the novel was based on stories he had picked up there on holiday. The simple virtues of an idealized community, buttressed by the stereotypes of imagined Welshness, are portrayed against the stresses of industrialization, and the novel created an indelible mythology of the mining valleys, reinforced by the 1941 film directed by John Ford.

Llewellyn's subsequent novels are notable for their variety of backgrounds. *None but the Lonely Heart* (1943, expanded in 1970) grew from his difficult years in London in the early 1930s; *A Few Flowers for Shiner* (1950) was set in Italy, visited again as a captain in the Welsh Guards between 1941 and 1946. He soon established a pattern of going to live in a country in which he would then place a novel: Kenya and the culture of the Masai in *A Man in a Mirror* (1964); Israel, which he much admired, for *Bride of Israel, my Love* (1973) and *A Hill of Many Dreams* (1974). Other countries in which he lived included France, India, Switzerland, Brazil, Chile, Paraguay, Uruguay, Argentina, and Ireland.

Two novels of life among the Welsh of Patagonia, *Up, Into the Singing Mountain* (1963) and *Down where the Moon is Small* (1966) followed the fortunes of Huw Morgan from *How Green*. Then a visit to Wales in 1973 kindled a patriotic fervour that produced *Green, Green My Valley Now* (1975), involving Huw Morgan, somewhat implausibly, on the fringes of revolutionary activity. He wrote several other novels, bringing the total to twenty-three with four stage plays.

Llewellyn's novels are rich in character and incident, with high emotional impact and ingenious stylistic devices creating atmosphere and suggesting national diversity. There is much background detail, not all of it convincing, and he was not slow to supplement observation with copious injections from his imagination. If he did, indeed, speak eight languages the novels reveal a tenuous enough grasp—including his boyhood Welsh. The fictions are often garish and artificial and his apparent grasp of contemporary cultural clashes is frequently coloured by sentiment. Yet he created in *How Green was my Valley* a picture or a dream that appeared to bring the values and culture of a whole community to new life. In person he was military with a dash of Hollywood, his manner somewhat brusque, his opinions generally illiberal or naïve. He was profoundly indifferent to most contemporary literature.

He married in 1952 Nona Theresa Sonsteby, of Chicago. This marriage was dissolved in 1968, and in 1974 he married a publisher's editor from New York, Susan Frances, daughter of Heinrich Heimann, patent attorney and mechanical engineer. Llewellyn died of a heart attack in Dublin 30 November 1983.

[Stanley J. Kunitz (ed.), *Twentieth Century Authors*, 1950; Alan Road, *Observer* colour supplement, 20 April 1975; Mick Felton in *British Novelists 1930-59* (ed. Bernard Oldsey), 1983 = *Dictionary of Literary Biography*, vol. xv; *The Times*, 1 December 1983; private information.]   GLYN TEGAI HUGHES

LOBEL, EDGAR (1888-1982), Greek scholar, was the elder son (there were no daughters) of Arthur Lobel, a shipowner who underwent several marked changes of fortune, and his wife, Amelia. His place of birth, which was not in Britain, is unknown; he may have been born in Russia or Poland. No birth certificate can be found and his date of birth is 12 December 1888 in the Manchester Grammar School records, and 24 December 1888 by the time he attended Oxford. He himself celebrated 24 December as the date. He went to Kersal School and then Manchester Grammar School, where he was well taught by Harold Williamson and others, became head of school, and in 1906 won a scholarship to Balliol College, Oxford. At that time his father lost most of his money and took his family to America, but Lobel remained behind in order to go up to Balliol, boarding with his former teacher during the vacations. Matriculating in 1907, he was taught by A. D. Lindsay (later Lord Lindsay of Birker), Cyril Bailey, and J. A. Smith [qq.v.], but the tutor most congenial to him was (Sir) A. W. Pickard-Cambridge. Lobel flourished in the exhilarating atmosphere of the Balliol of that time, which contained so many undergraduates who were not only gifted scholars but witty and agreeable people, and he retained many of its characteristics, both superficial and serious, throughout his life. He had a highly successful undergraduate career, obtaining first classes in classical honour moderations (1909) and *literae humaniores* (1911), the Gaisford prize for Greek verse, and Craven and Derby scholarships.

After a year as assistant in humanity at Edinburgh University (1911-12), Lobel returned to Oxford as Craven fellow, and took up the study of papyri. He visited Paris, Lille, Bonn, and Dublin, and for some months worked in Berlin under Wilhelm Schubart. In 1914 he was elected to a research studentship at Queen's College, Oxford, the college of B. P. Grenfell and A. S. Hunt [qq.v.], the finders and for the first years the editors of the great hoard of papyri from Oxyrhynchus in Upper Egypt.

When war broke out in 1914, Lobel was rejected for military service on account of his short sight, a deficiency which is sometimes compatible, as in his case, with keen sight if an object is held close to the eyes. After teaching briefly at Repton and at Downside he worked in military intelligence, first in the Admiralty and later in the War Office. The loss of so many of his contemporaries caused him deep distress, and for the remainder of his life he never wore a tie that was not black.

After the war Lobel combined his research studentship at Queen's with a sub-librarianship in the Bodleian Library (1919-34); to this side of his activity belong his authoritative studies of the medieval Latin version of Aristotle's *Poetics* and of Cardinal Pole's manuscripts (both in *Proceedings* of the British Academy, vol. xvii, 1931) and of the Greek manuscripts of the *Poetics* (*Supplement to the Bibliographical Society's Transactions*, no. 9, 1933). But he continued to work on papyri, and produced two masterly editions of Sappho (1925) and Alcaeus (1927), whose introductions form a continuous essay on the textual history, language, and metre of the two poets, and whose texts were later subsumed in *Poetarum Lesbiorum Fragmenta* (1955), in which Lobel collaborated with (Sir) Denys Page [q.v.]. Important fragments of these poets which had been published from papyri had in many editions been defaced by ill-judged conjectural supplements, and by castigating these Lobel managed to arrest this tendency. In 1927 he assisted Hunt in the publication of important fragments of Callimachus in part xvii of the Oxyrhynchus papyri; in the same year he became a supernumerary fellow of Queen's College. At the same time, in 1927 he married Mary Doreen Rogers, later editor of the *Victoria History of the County of Oxford* (vols. iii-viii) and of maps and plans of historic towns in the British Isles. She was the daughter of Frederick William Rogers, director of a company dealing in asphalt, of Bristol. After a short stay in St Giles's they moved to 16 Merton Street. They had no children.

In 1931 Lobel became keeper of western manuscripts in the Bodleian Library, but in 1934 he succeeded Hunt as editor of the literary papyri in the Oxyrhynchus series, becoming reader in papyrology in the university from July 1936. In 1938 he resigned from the Bodleian, being at the same time elected a senior research fellow of Queen's. These appointments he retained until his retirement at the age of seventy in 1959, his tenure having been specially extended five years earlier. But retirement made little difference to his activities, since he did not cease to work at the papyri from Oxyrhynchus. After 1972 he was not again responsible for an entire volume in the series, but he continued work and intellectual activity until not long before his death.

Between part xviii of the series, published in 1941, and part xxxix, published in 1972, by far the greatest part of the literary papyri were published by Lobel, and in exemplary fashion. In the piecing together and reading of the papyri he showed unique palaeographical skill, and in the identification and interpretation of the texts contained in them he showed unique knowledge of and feeling for the Greek language and literature. Lobel had little use for speculation, his aim being, as he put it, 'to attain the measure of certainty possible in these studies'. But he made certain supplements in many places, including some where no other scholar would have thought of them, and his concise notes, restricted as they were by the limits which he set himself, show an astonishing familiarity with all the relevant material and the keenest critical intelligence.

Apart from Sappho and Alcaeus, it fell to Lobel to edit texts of Hesiod, Archilochus, Alcman, Stesichorus, Anacreon, Simonides, Pindar, and Bacchylides; of Aeschylus and Sophocles; of Callimachus, Antimachus, Rhianus, and Euphorion; and also many unidentified fragments. Euripides and Menander he left to others, but he handled all with equal mastery; Paul Maas, who had known all the leading Hellenists of his time and whose respect for his own teacher Wilamowitz was very great, held that Lobel knew Greek better than any of them.

Tall, erect, and distinguished in appearance, Lobel had a memorable presence. As a young man he acquired a reputation for being formidable that never quite left him: he did not suffer fools gladly, and had no use for teaching, academic gatherings, or anything else that might have distracted him from what he regarded as his proper work. He travelled widely in Britain and in Europe during vacations, and was remarkably well informed not only about European literature but about a whole range of topics, including wine, botany, and topography.

His wit, though not without a sardonic edge, was highly entertaining, but at the same time his old-fashioned courtesy and consideration for others preserved something of the old Oxford collegiate atmosphere. Those who met him in his own college and in the few places where he could be persuaded to dine out found him a most agreeable companion.

Lobel scorned academic honours (he refused fellowship of the British Academy), but accepted the honorary degree of Litt.D. from Cambridge (1954), and became an honorary fellow of Balliol College (1959) and of Queen's College (1959). He died in Oxford 7 July 1982.

[Eric G. Turner, *Gnomon*, vol. lv, 1983, pp. 175–80; B. F. McGuinness, *Balliol Record*, 1983, pp. 12–16; information from Mrs Lobel; personal knowledge.]

HUGH LLOYD-JONES

LOPOKOVA, LYDIA VASILIEVNA, LADY KEYNES (1892–1981), ballerina, was born 21 October 1892 at St Petersburg (Leningrad), the third of five children (three sons and two daughters) and second daughter of Vasili Lopukhov, impassioned theatre lover and an usher at the Imperial Alexandrinsky Theatre, and his wife, Rosalia Constanza Karlovna Douglas, daughter of the clerk to the municipality of Riga, of Scottish ancestry. Lopokova was educated at the Imperial Ballet School, St Petersburg, which she entered shortly before her ninth birthday in 1901. She graduated into the Imperial Ballet at the Maryinsky Theatre, St Petersburg, in 1909. Although a demi-charactère dancer she could also shine in the purely classical roles because of her strong technique, extreme lightness in jumping, and stylistic sensitivity.

In 1910 Serge Diaghilev included Lopokova in the company he had formed for the second summer running to tour European capitals. Although only seventeen when thrust among Diaghilev's exalted group of artists Lopokova quickly established herself and successfully danced Tamara Karsavina's roles, including those in *Firebird* and *Carnaval* (with Nijinsky) when Karsavina was away fulfilling other contracts.

Following the tour Lopokova, with her sister and elder brother, both dancers, sailed for America on eight-month contracts. She never returned to the Imperial Ballet nor danced again in Russia. When her brother and sister went home she chose to stay on to dance in assorted ballet groups, shows, and musicals, making a name for herself. She also ventured into straight acting.

In 1916 Diaghilev sent his Ballets Russes to America and Lopokova rejoined the company as the leading ballerina. During the tour she married in 1916 Randolfo Barocchi, Diaghilev's business manager and an older man of great charm. Thus Lopokova returned to Russian ballet and with it to Europe for the first time in six years. There followed seasons in Europe and North and South America but it was not until 1918 that Lopokova first danced in London. Her triumphs were now crowned by the roles created for her by Léonide Massine in *The Good-Humoured Ladies* (1917-18) and *Le Boutique Fantasque* (1919). Such was her fame that when she abruptly left her husband and the ballet company simultaneously in July 1919 her mysterious disappearance caused banner headlines in the London press. Once again Lopokova had abandoned the Russian Ballet as though it meant nothing to her. Yet two years later she returned to Diaghilev's company in Paris and then starred among five important ballerinas in Diaghilev's 1921 London production of *Sleeping Beauty*.

At this time the economist (John) Maynard (later Lord) Keynes [q.v.] became her ardent admirer. He was the brother of (Sir) Geoffrey Keynes, whose notice appears in this volume, and the son of (John) Neville Keynes, registrary of Cambridge University. Belonging to the Bloomsbury Group, he brought Lopokova into the circle—which was difficult at times for her, and for its members. When Keynes began to think of marriage some of his friends were filled with foreboding. They tended to find Lopokova bird-brained. In reality she was intelligent, wise, and witty, but not intellectual. E. M. Forster, T. S. Eliot [qq.v.], and Picasso were among her close friends. She artfully used, and intentionally misused, English to unexpectedly comic and often outrageous effect. Keynes was constantly amused and enchanted. Lopokova idolized him and they married in 1925, the year of her divorce from Barocchi. They had no children. When they were apart they wrote daily if only between King's College, Cambridge, of which Keynes was the bursar, and Bloomsbury.

Lopokova continued to dance and act intermittently, playing Ibsen, Molière, and Shakespeare, albeit with her charming Russian accent; and she helped the burgeoning British Ballet tremendously. But from when Keynes suffered his first serious illness in 1937 until his death in 1946 the total dedication she had never quite mustered for her career came to flower. She was a devoted wife, forsaking all interests save her husband's health and work while entertaining him and their friends with her unpredictable remarks. They now lived mostly at Tilton in Sussex, sharing their love of poetry, literature, and the countryside. Lopokova accompanied her husband on his economic missions abroad.

Lopokova was a diminutive figure with the natural air of an eager, enquiring child which caused her to hold her head tilted up towards anyone with whom she was conversing. This could give the impression that her nose, too, was up-tilted whereas, unusually, just the tip turned down as a Picasso drawing of 1919 clearly shows. Her face was round with alert eyes under perfectly curved eyebrows; her mouth was a well-defined feature. Both her face and hands were remarkably expressive. On or off the stage her vitality, originality, humour, and youthful enthusiasm were irresistible. Gaiety and good humour prevailed. In addition she was devoid of jealousy, malice, vanity, meanness, or pretension. After her husband's death she adopted a retired way of life but lost none of her originality and charm. She died in a home near Tilton 8 May 1981.

[Milo Keynes (ed.), *Lydia Lopokova*, 1983; Polly Hill and Richard Keynes (eds.), *Lydia and Maynard: Letters between Lydia Lopokova and John Maynard Keynes*, 1989; Lydia Yoffe, 'The Lopukhov Dynasty', *Dance Magazine*, New York, January 1967; Anatole Chujoy in *Dance Encyclopedia*, A. S. Barnes & Co., New York, 1949; personal knowledge.]　　　　　　MARGOT FONTEYN

LUTYENS, (AGNES) ELISABETH (1906-1983), composer, was born at 29 Bloomsbury Square, London, 9 July 1906, the fourth of five children and third of four daughters of (Sir) Edwin Landseer Lutyens [q.v.], architect, and his wife, Lady Emily, daughter of Edward Robert Bulwer Lytton, the first Earl of Lytton [q.v.], statesman. She was educated at Worcester Park School, Westgate-on-Sea. She turned to music at an early age, not because of any conspicuous talent for it, but 'as another form of my need for privacy'. The family's reaction to her musical aims was at first more of apathy than outright opposition, and after a period of private piano and violin lessons, in January 1922 she was permitted, at the age of sixteen, to go to France to study at the École Normale in Paris. In 1926 she entered the Royal College of Music, London, becoming a pupil of Harold Darke for composition and Ernest Tomlinson for viola. She had cause to be grateful to Darke, who managed to arrange for almost everything she composed at this time to be performed. This was most unusual for the RCM, where 'Brahms was the god of *new* music.' Lutyens was emphatically not sympathetic to Brahmsian ideals, yet working in this style for exercises enabled her to develop a powerful compositional technique. Among her friends from this time were Anne Macnaghton, the violinist, and Iris Lemare, and together they founded the Macnaghton-Lemare concerts which began in 1931 at the Mercury Theatre, Notting Hill Gate.

The main aim of the concerts was, and still is, to 'discover and encourage composers of British nationality by having their works performed'.

On 11 February 1933 Elisabeth Lutyens married Ian Herbert Campbell Glennie, who had also been an RCM student. He was the son of William Bourne Glennie, a minor canon of Hereford. They had a son and twin daughters. They were divorced in 1940, and in 1942 Elisabeth Lutyens married Edward Clark (died 1962), a tireless champion of new music, who had previously worked for the BBC in Newcastle and London. He was the son of James Bowness Clark, coal exporter, of Newcastle. They had one son.

For much of her life Elisabeth Lutyens endured relative poverty (with occasional help from her family over such things as housing and children's education), a considerable amount of ill health, physical and mental, and widespread lack of recognition of her originality and achievement. None the less, she never ceased to compose, her opus numbers extending at least to 135, in addition to which there are a hundred film scores from 1944 to 1969 and about the same number of musical commissions for radio. She admitted that the continual steady drinking involved during discussions of the radio projects turned her into an alcoholic. Most of her life, apart from moves to Northumberland and other areas during World War II, was spent in and around London.

She was a prolific and versatile composer, one for whom neglect may have been discouraging but made no essential difference to her development and productiveness. From the outset she veered away from what Constant Lambert [q.v.] termed 'the cowpat school of English music' and described integrity as 'not a virtue [but] a necessity for an artist'. She was the outstanding pioneer of serial music in England, and while she would have liked more performances of her works, her fulfilment was in the composition of them. Her first work to become known through performance was a setting for a ballet *The Birthday of the Infanta* (1932) after Oscar Wilde [q.v.]. This was given at an RCM Patrons' Fund concert while she was still a student (the score has since been withdrawn). The main work of the pre-war years was the Chamber Concerto op. 8 no. 1 for nine instruments (1939) which antedated by several years any knowledge of Webern's concerto for a similar ensemble. The first work to be performed abroad was the String Quartet no. 2 op. 5 no. 5 given at the International Society for Contemporary Music festival at Cracow in 1939. Possibly her best-known work is *O Saisons, O Châteaux!* op. 13 (1946) for soprano and strings, to a poem by Rimbaud. A very high proportion of her later works make use of words,

some written by herself, and others by an enormous range of writers, among them Stevie Smith and Dylan Thomas [qq.v.], with whom she was personally acquainted. At a relatively late stage she turned to dramatic music, and her opera 'Charade' *Time off? Not a Ghost of a Chance!* op. 68, 1967–8, to her own libretto, was staged at Sadler's Wells theatre in 1972.

She relished company and good talk, and could be provocative and outrageous, as when ringing up a Jewish pupil at 1 a.m., saying 'the PLO's all right'. Her sitting-room was very welcoming to a new pupil, with a large pot of steaming tea on the table, a standard lamp created from a French horn bell, and the work-desk with its stop-watches and slanting architect's board—the sense of excitement and joy in the act of composing this generated in a young composer can be imagined. She had various rather unsatisfactory arrangements with publishing firms, and eventually formed her own, the Olivan Press, which in the 1960s and 1970s published many more works than all the other publishers had managed over her entire career. In 1969 she was appointed CBE and awarded the City of London Midsummer prize. York University awarded her an honorary D.Mus. (1977). She died in Hampstead, London, 14 April 1983.

[Elisabeth Lutyens, *A Goldfish Bowl*, 1972 (autobiography); M. S. Harries, *A Pilgrim Soul: The Life and Work of Elisabeth Lutyens*, 1989; information from Robert Saxton.]   JAMES DALTON

LYONS, (FRANCIS STEWART) LELAND (1923–1983), historian, was born 11 November 1923 in Londonderry, the elder son (there were no daughters) of Stewart Lyons, bank manager, and his wife, (Florence) May Leland. He was educated at Dover College in Kent and Trinity College, Dublin, which he entered in 1941. Two years later he was elected a foundation scholar, and in 1945 took an outstanding first with a gold medal in modern history and political science. In 1947 he completed his doctoral thesis on the Irish Parliamentary Party, and was appointed lecturer in history at Hull University. He returned to Dublin in 1951, and held a fellowship at Trinity College until 1964.

His principal scholarly mentor then and afterwards was T. W. Moody, who pioneered a new objectivity and scientific precision in the writing of Irish history. The readiness to depart from comfortable orthodoxy led to what may justly be called a renaissance in Irish historical study. The embattled and embittered recent past of the Irish national movement rendered such 'revisionism' doubly difficult, yet by the same token doubly valuable. Well into the postwar years, Irish historiography remained bound up with politics, to their mutual impoverishment. Separation of the two was to be of incalculable importance not only in the establishment of a clearer-headed sense of Irish national identity, but also in the liberation of political activity from historical shackles. The work of Leland Lyons represents a decisive phase in this process. His view of revisionism, as he was to express it in his candid W. B. Rankin memorial lecture *The Burden of our History*, given in Belfast in 1978, was that it was 'proper' only in so far as it was 'a response to new evidence which, after being duly tested, brings us nearer to a truth independent of the wishes and aspirations of those for whom truth consists solely of what happens to coincide with those wishes and aspirations'.

Lyons, of Presbyterian and Church of Ireland stock, brought to the study of history the best of Anglo-Irish—though not Ascendancy—qualities. Sharp-sighted yet temperate in judgement, formidable in mastery of detail yet elegant in synthesis, he became by general consent the foremost Irish historian of his day. His early work on *The Irish Parliamentary Party 1890-1910* was published in 1951, and his striking treatment of a single critical year, 1890-1, in *The Fall of Parnell*, in 1960. On a very different canvas he wrote *Internationalism in Europe 1815-1914* (1963) for the Council of Europe.

He returned to England in 1964 as the first professor of modern history in the University of Kent at Canterbury. In so doing he went back to the county of his early schooling, which was now to benefit from his accomplishments in the establishment of its new university. He served as master of Eliot College at the University of Kent from 1969, and published two major works: *John Dillon* (1968), which won the Heinemann award of the Royal Society of Literature, and *Ireland since the Famine* (1971). The former, a relentlessly detailed account of the doomed parliamentarian nationalist, showed the capacity of biography to approach the complexity of total history. The latter was the first attempt at a true total history of modern Ireland, demonstrating the extent of the historical revision so far achieved. To this Dictionary he contributed the notice of Henry Harrison.

In 1974 he returned once more to Trinity, as provost. In an outward sense this was the pinnacle of a brilliant academic career. Inwardly, however, it may be thought that the most significant event of that year was his appointment as official biographer of W. B. Yeats [q.v.]. Yeats was the ultimate subject for the Anglo-Irish historian, and came to engross (to use his own word) his intellectual energies. In 1977 he published a second large-scale biography, *Charles Stewart Parnell*. What might for many a scholar have been his final masterpiece became a preparatory exercise for the daunting work on Yeats. By this time, Lyons's writing displayed such harmonious style, subtlety of argument, and understated judgement, that careless readers could miss much of his reinterpretation of Parnell. But the real earnestness of his future performance came with the Ford lectures which he gave in Oxford in 1977-8. Published as *Culture and Anarchy in Ireland 1890-1939* (1979), they won the Wolfson literary award in 1980. They challenged those who argued the 'essential unity' of Ireland to accept the 'essential diversity' of its cultures and to build on their strength.

In 1981 he resigned his office as provost of Trinity; his unstinting devotion of energy to the public role of a university head in times of stress and change had come into serious conflict with his overriding determination to pursue his work on Yeats. That work, so impressive even to those entrenched in the bastions of literary criticism, came to an end with awful suddenness when Lyons succumbed to acute pancreatitis in August 1983. He died in the Adelaide Hospital, Dublin, 21 September 1983.

In 1962 Lyons was elected a member of the Royal Irish Academy and in 1974 a fellow of the British Academy. He was F.R.Hist.S. and FRSL. He was elected an honorary fellow of Oriel College, Oxford, in 1975, and was awarded honorary doctorates by the universities of Pennsylvania in 1975, Hull, Kent, and Queen's, Belfast, in 1978, Ulster in 1980, and St Andrews in 1981.

Known as 'Lee', Lyons was a graceful and reserved man. He added to his academic achievements formidable gifts as a squash player. In 1954 he married Jennifer Ann Stuart, daughter of Lt.-Col. Archibald Donald Cameron McAlister, of the 11th Hussars and Cheshire Regiment; they had two sons.

[R. F. Foster in *Proceedings* of the British Academy, vol. lxx, 1984; *The Times*, 24 September 1983.]　　　　　CHARLES TOWNSHEND

LYONS, SIR WILLIAM (1901-1985), founder of Jaguar Cars Ltd., was born 4 September 1901 in Blackpool, the only son and younger child of William Lyons, dealer in musical instruments, and his wife, Minnie Bancroft. He was educated at Arnold House (later Arnold School), Blackpool, and left at seventeen to become a trainee at Crossley Motors Ltd. For a short time he worked with Jackson Brothers who held the franchise for Sunbeam motor cars, and then, in 1922, he went into partnership with one of his friends, William Walmsley, and they opened a small factory in Blackpool for the manufacture of Swallow motor-cycle side-cars. In spite of competition from cheap all-weather motor cars, particularly the Bullnose Morris, which threatened the market for side-cars, the Swallow business continued to flourish up to the outbreak of war in 1939.

Lyons greatly admired the machines being produced in the 1920s by William Morris (later Viscount Nuffield) and Herbert Austin (later Baron Austin) [qq.v.], but thinking that their bodywork could be improved, he and his partner extended their business into the Swallow Sidecar and Coach Building Company, which by the late 1920s was constructing the bodywork for the Austin Swallow two-seater car and the Morris Cowley Swallow.

In 1928, having outgrown their works in Blackpool, the Swallow firm moved to Coventry, and in 1929 the first Swallow Coachwork stand appeared at the Olympia Motor Show. Two years later the company's first complete car, the Swallow Sports, its chassis and engine supplied by the Standard Motor Company, made its appearance at the show, and by 1934 SS cars had their own stand in the main hall at Olympia.

Early in 1935 William Walmsley retired, and Lyons became chairman and managing director of SS Cars Ltd., and produced the first Jaguar car, which, during the next four years, made its mark in the motor-racing world. The outbreak of war in 1939 put a stop to this story of success, and the Coventry factory was turned over to the manufacture and repair of parts for bomber and fighter aircraft.

As soon as the war ended Lyons, despite the shortage of steel, set about the task of resuming car production; the company's name was changed to Jaguar Cars Ltd., and great efforts were made to extend the sales of its product overseas, especially in the USA. Lyons was careful to elicit the fullest co-operation from his work people, who soon had their own club room and the company its own magazine, the *Jaguar Journal*. In 1948 a new sports Jaguar was produced, and in 1951 this won the Le Mans 24-hours race, a success which was repeated in 1953, 1955, 1956, and 1957. In 1953 the Jaguar C type cars finished first, second, and fourth. In 1956 Lyons was knighted in recognition of his outstanding contribution to Britain's export business.

The works in Coventry had by this time been moved to a much larger site, and in January 1957 fire gutted that part of the building where new sports racing models were being constructed. Lyons personally helped in fighting the fire, and, within two days cars were being produced again; but it was decided to withdraw from motor racing. During the next ten years the Jaguar Company continued to expand. In 1960 they took over the name and business of the prestigious Daimler Company; then in 1961 they acquired Guy Motors of Wolverhampton and Coventry Climax Engines Ltd., and entered the market for buses, commercial vehicles, fire-engines, and fork-lift trucks.

Lyons, however, despite the profitability of Jaguar cars, had some doubt whether such a comparatively small firm could hold its own, and in 1966 he and Sir George Harriman agreed that Jaguar Cars Ltd. would merge with the British Motor Corporation Ltd. in a new company, British Motor Holdings, of which Lyons became chairman and managing director. Two years later BMH was merged in British Leyland, a step which Lyons had tried to avoid, fearing that Jaguar cars would cease to exist as a separate entity. He remained chairman and chief executive of the Jaguar Group whilst taking a seat on the board of British Leyland. His fears for Jaguar cars were realized in 1972, and in that year he retired from his executive position; but he continued up to the time of his death to take a keen interest in the company as its honorary president, and was very happy when, in 1984, Jaguar Cars returned to private ownership under his successor, Sir John Egan.

Throughout his remarkable career Lyons worked enthusiastically to enhance the reputation and success of the British motor industry; he became president of the Society of British Motor Manufacturers and Traders and of other similar organizations; he also held an honorary degree from Loughborough University (1969). He became RDI in 1954.

In 1924 he married Greta, daughter of Alfred Jenour Brown, a schoolmaster, of Cuddington, near Thame. They had two daughters and one son, who was killed in a car accident in 1956. Lyons himself died at his home, Wappenbury Hall, Warwickshire, 8 February 1985.

[Andrew Whyte, *Jaguar. The Definitive History of a Great British Car*, 1980, 2nd edn. 1985; *The Times*, 9 February 1985.]

H. F. OXBURY

M

McALPINE, (ARCHIBALD) DOUGLAS (1890-1981), neurologist, was born in Garscadden, Glasgow, 19 August 1890, the eldest of three children and only son of the second marriage of (Sir) Robert McAlpine, first baronet, civil contractor and founder of the firm of Sir Robert McAlpine & Sons, and his wife, Florence Evaline Margaret Palmer. There were five sons and three daughters of the first marriage. He was educated in Edinburgh at Kirton College, at Cheltenham, and later at Glasgow University, where he qualified with distinction MB, Ch.B. in 1913. He joined the Royal Army Medical Corps in August 1914, was posted to the 13th Field Ambulance in France, and later became a regimental medical officer with the Queen's Own Scottish Borderers. In 1915 he joined a hospital ship and then transferred to the Royal Navy. He served in HMS *Falmouth* of the 3rd light cruiser squadron at the battle of Jutland and was aboard in August 1916 when she was sunk off Flamborough Head. Subsequently in Scapa Flow he boxed and played rugby for the squadron. He was mentioned in dispatches.

In March 1918 he moved to the RN sick quarters in Plymouth and treated nervous ailments with hypnosis. Soon after demobilization in November 1918 he decided on a career in neurology and worked on post-encephalitic Parkinsonism with J. G. Greenfield at the National Hospital, Queen Square. He became MRCP in 1921 (FRCP, 1932) and MD in 1923. After a short period spent with L'Hermitte at the Salpetrière in Paris, he was appointed in February 1924 as physician for nervous diseases at the Middlesex Hospital and assistant physician and pathologist at Maida Vale.

At the Middlesex he was an out-patient physician without beds, but in 1930 a neurological unit was created through a gift from his family firm (Sir Robert McAlpine) which made it possible to convert a derelict group of small rooms into two pleasant wards holding twenty-four patients; neurosurgical beds were added later. There he taught and practised with such enthusiasm and a firm but gentle discipline that he inspired many of his juniors to practise neurology.

In World War II he was successively adviser in neurology in the Middle East, in India Command, and in South East Asia. During the battle of El Alamein, when over 200 head-injured casualties were admitted to a mobile neurosurgical unit within a week, he assisted, even though a brigadier, in examining and treating them, subsequently telling Major Michael Kremer, neurological specialist (formerly McAlpine's house physician at the Middlesex), 'It was nice to be your house physician for a week.' He was again mentioned in dispatches.

McAlpine's major interest was the study of multiple sclerosis. The massive clinical data he collected was analysed in his book *Multiple Sclerosis*, published in 1955. In the next edition he recruited two notable co-authors and the book has since reappeared under new editorship. In 1951 he initiated the formation of the Multiple Sclerosis Society of Great Britain (finally established in 1953). He recruited the first medical panel and the medical advisory research committee, of which he was the first chairman. His enthusiasm initiated and nurtured research programmes and encouraged many others to work in the field. He was the first recipient of the Charcot award of the International Federation of Multiple Sclerosis Societies and no similar award was made until twelve years after he received it in 1969.

McAlpine was sometimes thought to be shy and at times a trifle abrupt, but those who knew him well appreciated his innate kindness, generosity, and above all his empathy. He was an avid sportsman, an expert fly-fisherman, and also interested in shooting, golf, skiing, photography, and ornithology. In 1917 he married (Elizabeth) Meg, daughter of Isaac Sidebottom, textile merchant. They had a son and a daughter. After his first wife died in 1941, he married in 1945 Diana Christina Dunscombe, daughter of Bertram Plummer, a solicitor, of Leicester. They had one son. Sadly in later years his wife was often ill but he cared for her with total cheerful dedication. His occasional brusqueness, largely resulting from impatience fuelled by boundless enthusiasm, is recalled by his great-nephew, Ian McAlpine, who believes that he is the only member of the family to have consulted Douglas McAlpine as a patient; nevertheless, the consultation left a lasting impression on him.

When twenty of his former colleagues joined him in a birthday lunch in September 1980 to celebrate his ninetieth birthday, he showed the same lively interest and genial bonhomie which had endeared him to friends and colleagues. He died in his sleep at Marnhull, Dorset, 4 February 1981; his wife died a few days later.

[Michael Kremer in *Munk's Roll*, vol. vii, 1984; *British Medical Journal*, 7 March 1981; private information; personal knowledge.]

JOHN WALTON

McCANCE, SIR ANDREW (1889-1983), industrialist and scientist, was born at 12 Adder

Steps, Cadder, near Glasgow, 30 March 1889, the third of four children (the youngest of whom died before the age of two) and younger son of John McCance and his wife, Janet Ferguson McGaw. His father was at the time a cloth buyer but later worked for the Graham Joint Stock Shipping Company of Glasgow on trade with the East. McCance was given the best affordable education first at Morrison's Academy in Crieff and subsequently at Allan Glen's School in Glasgow. He graduated in mining in 1910 at the Royal School of Mines in London.

McCance started unpaid work in the same year on the open hearth furnaces at W. Beardmore & Co. in Glasgow. Principally through his contributions to the prevention of cracking in armour plate ingots and to the softening heat treatment of armour plate he became assistant armour manager until 1919. In 1913 he was injured in a major crane accident from which he fully recovered. In 1916 he gained a London D.Sc. In 1919 McCance started the Clyde Alloy Steel Company in disused premises at Craigneuk Motherwell supported financially mainly by Colvilles Ltd. During the 1920s he worked long and hard to achieve deserved success.

In 1930 (Sir) John Craig [q.v.], chairman of Clyde Alloy and Colvilles, invited McCance to become general manager of Colvilles. The decade saw many changes—political, industrial and technical. Of these there was the merger at different times of Colvilles and James Dunlop & Co., the Beardmore Mossend Works, the Steel Company of Scotland and Lanarkshire Steel; there was the closure of the uneconomic blast furnaces at Glengarnock; and the building of new blast furnaces and coke ovens at Clyde Iron and their integration with the Clydebridge steelworks. If there is any criticism to be made of McCance it is that he preferred to build on the past and not concern himself industrially with the future. He therefore rationalized the existing works at Clydenbridge, Dalzell, Lanarkshire, and Glengarnock, converted Hallise to electric arc melting and started the Ravenscraig complex. In 1956 McCance succeeded Craig as chairman, introduced the wide strip mill to Ravenscraig, built the third blast furnace there, and replaced the open hearth furnaces by oxygen steelmaking converters. In 1965 he retired.

Whatever demands were made on him as an industrialist, McCance found time throughout his life to view the problems of steel and its manufacture using scientific methods and techniques. Of the twenty-three papers which bear his name, at least half are original contributions to the scientific understanding of the 'mysteries' of a solid iron exhibiting two phases, one above and one below 600°C, and of the interaction between slag composition and the metal it was intended to refine. On the former there is

his work on troostite, the hardening of steel by quenching, and the association of the non-crystalline change with specific heat; and on the latter he showed the importance of non-metallic inclusions (manganese sulphide and/or silicate) on the properties of steel and of time for aggregation and the larger particles to rise to the surface measures related to suitable slag basicity and temperature. McCance was elected FRS in 1943 and was knighted in 1947.

McCance was a Scot above all. He created and ran the Scottish steel industry and was opposed to its nationalization in 1955 and again in 1967. Many of his scientific papers were published in the *Journal of the West of Scotland Iron and Steel Institute*, of which institute he was president from 1933 to 1937. In 1950 he was chairman of the Royal Technical College, which was renamed the Royal College of Science and Technology in 1956 and then became the University of Strathclyde which awarded him an honorary D.Sc. in 1965 and named a building after him. He also had an honorary LLD from Glasgow (1948). He was president of the Institute of Engineers and Shipbuilders in Scotland in 1951-2 and a member of the board of the Royal Bank of Scotland. Nationally his merits were recognized. He was Bessemer gold medallist of the Iron and Steel Institute in 1940 and president in 1948-50; president of the British Iron and Steel Federation in 1957 and 1958; chairman of the mechanical research board, Department of Scientific and Industrial Research, from 1952 to 1958 during which time the National Engineering Laboratory was located at East Kilbride; and DL for Lanarkshire (1950).

McCance was always impeccably dressed, an establishment figure if ever there was one. He was very inquisitive and a good listener but, having listened, he made up his own mind. In 1936 he married Joya Harriett Gladys (died 1969), daughter of Thomas B. Burford, licensee; they had two daughters. McCance died 11 June 1983 in Davidson Hospital, Girvan.

[N. J. Petch and L. Barnard in *Biographical Memoirs of Fellows of the Royal Society*, vol. xxx, 1984; conversation with Nathalie McCance (daughter); personal knowledge.]

MONTY FINNISTON

MacDONALD, MALCOLM JOHN (1901-1981), politician, diplomat, and writer, was born 17 August 1901 at Lossiemouth, Scotland, the second of the three sons and six children of James Ramsay MacDonald [q.v.], journalist, politician, and later prime minister, of 3 Lincoln's Fields, London, and his wife, Margaret Ethel, daughter of John Hall Gladstone, FRS, a distinguished scientist and an active social and religious worker.

His mother died in September 1911 at the early age of forty-one, eighteen months after her third and youngest son, David. Thereafter Malcolm, his elder brother Alister, and his three sisters were drawn even closer to their father. Malcolm's career in particular was crucially affected by that of his father right up until the latter's retirement from the premiership in June 1935, and he was always a stalwart champion of his father's reputation.

He was educated at the coeducational Bedales School in Hampshire and at Queen's College, Oxford. He took a second class honours degree in modern history (1923), and then stayed on for a further postgraduate year reading for a diploma in economics and politics in which he was awarded a distinction in 1924.

He toured the United States, Canada, and Australia in 1924-5 as a member of a three-man Oxford debating team, and he visited Hawaii and Kyoto, in 1927 and 1929 respectively, as secretary to the British delegation to conferences of the Institute of Pacific Relations. Although unsure as to whether he wished to pursue a lifetime career in politics, he became a member of the London County Council from 1927 to 1930 and in 1929 at his third attempt (having been defeated in 1923 and 1924) he was elected to the House of Commons as Labour member for the Bassetlaw division of Nottinghamshire.

In August 1931 the formation of the national government split the Labour Party and he followed his father as one of a small group of National Labour MPs. He was immediately appointed parliamentary under-secretary in the Dominions Office and in this capacity moved the third reading of the Statute of Westminster. Between 1935 and 1940 he was a cabinet minister, holding one or both of the offices of secretary of state for dominion affairs and secretary of state for the colonies. He was thus intimately involved in the controversies in these years over Eire, Palestine, the West Indies, and the promotion of colonial development policy, becoming the first active ministerial advocate of Commonwealth as a desirable outcome of empire. Although defeated at the 1935 general election he was re-elected to parliament at a by-election in 1936, for Ross and Cromarty, whose MP he remained, even through his war years in Canada, until 1945. In (Sir) Winston Churchill's government of 1940 he was appointed minister of health, a post which involved him in such matters as food rationing, hospitals, and air raid shelters during the blitz.

In 1941 he was appointed Britain's high commissioner to Canada, and his youngest sister Sheila went with him as his hostess. During his five years there, he played a significant and characteristically unpublicized role in smoothing relations between the prickly Canadian prime minister, W. L. Mackenzie King [q.v.], and Winston Churchill, through a time when Canada's contribution to the war effort and to post-war planning was of the utmost importance.

From Ottawa MacDonald finally abandoned a political career when he was appointed governor-general of the Malayan Union and Singapore in 1946, refusing the offer by C. R. (later Earl) Attlee of an office in his government or some other prestigious post. British Borneo was added to his responsibilities later that year. In 1948 his direct responsibilities in what was later to become Malaysia were superseded by the more general and somewhat ambiguous role of being in charge of overseeing Britain's relationships with the whole of South East Asia, including the former French Indo-China, as UK commissioner-general for South East Asia. This was a post in which he was reconfirmed several times, by Conservative as well as Labour governments. He was adopted as 'son' of the Iban paramount chief Temonggong Koh. He was high commissioner in India from 1955 to 1960, nursing Indo-British relations through the strains which followed the Suez war of 1956.

It was a tribute to his unique standing and experience that he was called out of a short-lived semi-retirement in England to be co-chairman and leader of the British delegation at the international conference on Laos held in Geneva in 1961-2, where his tact, patience, and diplomatic skills were fully stretched—not least in helping here, as at other times, in associating China constructively with the enterprise, through his friendship with Zhou Enlai.

In 1963 he accepted the difficult task of being Britain's last governor and commander-in-chief in Kenya in the period leading to independence, a process he was keen to hasten. He remained as governor-general in 1963-4 and subsequently, in what was a clear testament to the trust and standing he enjoyed with Jomo Kenyatta [q.v.], he returned as high commissioner to Kenya, 1964-5. Then for three more years (1967-9) he was employed by the government of Harold Wilson (later Lord Wilson of Rievaulx) as a special representative of the British government in Africa, undertaking a variety of sensitive missions, especially regarding Britain's difficulties over Rhodesia and Nigeria.

His nominal retirement in 1969 was but the prelude to assuming new offices and responsibilities, in continuation of a prodigiously active and varied life. Right up until his death, and as a testimony to each society's reluctance to let him go once he had become their ever helpful figurehead he was president of the Royal Commonwealth Society from 1971, of the Great

Britain–China Centre from 1972, of the Federation of Commonwealth Chambers of Commerce from 1971, of VSO from 1975, of the Caribbean Youth Development Trust from 1977, and of the Britain–Burma Society from 1980. He was markedly reluctant to accept 'social' honours and several times refused overtures regarding a knighthood or a peerage. The notable exception came in 1969 when he was accorded the sovereign's accolade of the Order of Merit. He had been sworn of the Privy Council in 1935. He received, however, many academic and some civil honours: he became an honorary fellow of his old college Queen's, Oxford; an honorary doctor of laws and also of letters of the University of Durham, and had honorary doctorates conferred on him by various North American universities and by the universities of Hanoi, Hong Kong, Singapore, and Malaya. He was made freeman of the city of Singapore in 1955 and of the burgh of Lossiemouth in 1969.

His continuing devotion to young people and to education is illustrated by the fact that he was a Rhodes trustee from 1948 to 1957, chancellor of the University of Malaya from 1949 to 1961, visitor for the University College of Kenya (1963–4), and a senior research fellow at the University of Sussex (1971–3).

MacDonald was small, wiry, and remarkably friendly, with something of a reputation for being a ladies' man. He wore spectacles from youth to middle age and then was able to dispense with them during World War II. In later life he ate one main meal a day (though his diary sometimes registered two or three 'tea-times' in a single afternoon, usually at the Royal Commonwealth Society) and he required no more than a few hours sleep each night. His painstaking application would often keep him working late into the night, but he would usually be up at dawn to renew his passionate interest in ornithology, which led him to produce books on the birds of Ottawa, Delhi, and Kenya. His other published writings were either reminiscences and pen portraits of people he had known or travelogues. For years he listed his recreations in *Who's Who* as 'ornithology, collecting, skiing', the first two undoubtedly occupying more of his time than the third—his collection of Chinese porcelain is now at Durham University—but he met his future wife on a ski slope in Canada.

He married a young Canadian war widow, Audrey Marjorie Rowley (daughter of a civil servant, Kenyon Fellowes) in Ottawa in December 1946. She had a son and a daughter by her first marriage, and by her marriage to MacDonald a daughter, Fiona, born 20 January 1949. MacDonald died at his home, Raspit Hill, near Sevenoaks, in Kent, 11 January 1981.

A service of thanksgiving for his life and work was held in Westminster Abbey on 3 March 1981 and in his commemorative address Sir Shridath Ramphal, the Commonwealth secretary-general, summarized his richly varied career as a quintessentially Commonwealth man and stressed how he had brought to his extensive representational work the highest qualities of intermediation, describing him as 'the supreme interlocutor'.

An extensive collection of photographs by and relating to MacDonald was given to the Royal Commonwealth Society for its library in London.

[*The Times*, 12 January 1981; the Malcolm MacDonald papers at Durham University, including an unpublished autobiography; private information; personal knowledge.]

PETER LYON

McEVOY, HARRY (1902–1984), industrialist and leading figure in British food manufacture, was born 16 August 1902 in Bradford, the younger of twin sons (there were no other children) of Thomas McEvoy, grocer, and his wife, Polly Taylor, daughter of a Norwegian sea captain. He was educated at Bradford Grammar School, and as a young man worked in his father's grocery business in Yorkshire. In his late twenties he studied at Columbia University, USA, where in 1930 he obtained a degree with distinction in American business methods and administration. In the following year he joined the export division of the American Kellogg Company, manufacturers of cereal foods, and after a year learning the business in Canada and the USA, came to England and worked as assistant to the manager of the new Kellogg Company of Great Britain. In 1934 he was appointed managing director of the company. During the next four years he did much to persuade the British public, normally conservative in their eating habits, to accept packaged cereal foods as a healthy constituent of the British breakfast, not an easy task as these products were at the outset treated with some suspicion. By 1936 some 400,000 packets of Kellogg's cereals were being sold each week, and it was decided that, in preference to importing them from America, a factory should be opened in Britain.

In 1938 McEvoy chose Stretford, Manchester, as the site for what would become the largest food processing factory outside the USA. During World War II he worked with Frederick James Marquis, Earl of Woolton [q.v.] at the Ministry of Food and was responsible for setting up the committee which dealt with cereal breakfast foods as part of a policy in furthering the war effort to ensure that the civilian population was adequately fed in spite of the difficulties caused by the destruction of British shipping by German U-boats. The Stretford works continued in production despite near misses from bombing and fires. Output doubled; vast quantities of breakfast foods went to the armed forces, and civilians had to be rationed. When it became impossible to import maize and rice from America and Canada wheat flakes and all-bran had to be manufactured from home-grown wheat.

When the war ended McEvoy planned and supervised the extension of the factory in Manchester, overcoming serious difficulties in procuring cement and steel. In 1938 there had been a work-force of 250; by 1968 it had risen to 1,700, and the turnover of products being sold in supermarkets and grocery shops throughout Britain had increased eightfold.

McEvoy saw that the breakfast cereals industry needed some centralized organization to deal with problems of mutual interest, and in particular, to provide a forum for consultation and discussion with the government when post-war legislation was under consideration. Together with the directors of other firms dealing in breakfast foods, he was a founder member of the Association of Cereal Food Manufacturers, and was its chairman from its inception in 1955 until his retirement in 1967. In pursuing its interests he played an important part in negotiations between the Association and the Board of Trade which led to the inclusion of special provisions covering cereal foods in the Weights and Measures Act of 1963.

During his thirty-three years as managing director of the Kellogg Company of Great Britain McEvoy was the driving force in the extension of the cereal foods industry to western Europe, Scandinavia, and South Africa; to quote his obituary in The Times, 'more than any man he influenced the present trend towards a natural cereal based diet'. Throughout those years he acquired a reputation as a man of dynamic energy whose sense of humour and consideration for others earned him the affection and respect of his colleagues and staff and all those with whom he conducted business.

In 1926 he married Hilda (died 1979), daughter of Enoch Wood, a Bradford mill owner. They had two daughters. In 1968 the McEvoys retired to Sydney, Australia. In 1983 Harry McEvoy returned to Britain and settled in Douglas, Isle of Man. He died at his home there 3 November 1984.

[The Times, 10 November 1984; private information.]   H. F. OXBURY

MACINTYRE, DONALD GEORGE FREDERICK WYVILLE (1904–1981), naval officer and historian, was born in Dehra Dun, India, 26 January 1904, the younger son (there were no daughters) of Lieutenant-Colonel (later Major-General) Donald Charles Frederick Macintyre, Indian Army, and his wife Maud, daughter of Colonel George Strahan, of the Royal Engineers. Macintyre entered the Royal Navy on 15 September 1917 and was educated at the Royal Naval colleges, Osborne and Dartmouth.

After initial sea service in destroyers he became a Fleet fighter pilot in 1927 and for the next six years served in squadrons based in aircraft carriers. This phase of his service was ended by medical unfitness for flying in 1935 and he returned to small ships, gaining his first command, HMS Kingfisher, soon afterwards. The outbreak of war found him in home waters in command of HMS Venomous.

Macintyre's experience as a destroyer officer made him a natural choice for fleet tasks but his instinct was more for the convoy war developing in the Atlantic and it was here that he spent most of his immensely distinguished war career. He was an early leader in the development of the escort group concept, where a mixed group of destroyers, frigates, and corvettes worked up and worked together under an experienced senior officer and, by their well-practised procedures and knowledge of each other's characteristics, were able to give much better support to a convoy than any ad hoc formation. Macintyre's first success against the U-boats came on 16 March 1941, and it was a highly dramatic one, for in HMS Walker, with Vanoc in company, he sank the submarines of two noted German aces—Schepke and Kretschmer—in a single action. For this he was appointed to the DSO.

After a brief spell ashore Macintyre returned to sea in HMS Hesperus. He measured his success in this ship and the escort group he commanded in two ways: by the number of U-boats sunk by the group but, equally important in his eyes, the fact that he lost only two ships from convoy in nearly two years. Macintyre regarded this trade-off as a vindication of convoy as both an offensive and a protective measure. Notable successes during this period were the destruction of U-357 by ramming in December 1942 and of U-191 and U-186 in April and May 1943 respectively. For these actions Macintyre was awarded two bars to his DSO. The tide had now turned decisively against the U-boats. Macintyre remained at sea until mid-1944, with one further success in May of that year when HMS Bickerton under his command sank U-765, HMS Bligh in company. He was awarded the DSC for this action. He was also, for his service to the USA during the war, made an officer of the Legion of Merit.

Macintyre had been promoted commander at the end of 1940 and spent the rest of the war in that rank. He was promoted captain at the end of 1945 and his service for the next decade was mainly ashore, though he commanded the third training flotilla between 1948 and 1950. He was placed on the retired list in July 1955.

He then turned his energies to writing, beginning with an account of his own war in *U-Boat Killer* (1956) and following up with almost one book a year for the next decade and a half. Most of his books were on the recent history of maritime war, though from time to time he went further back with, for example, a biography of Admiral Lord Rodney [q.v.]. With their lack of footnotes, references, and bibliography, and their graphic and sometimes colourful language, Macintyre's books are at the 'popular' end of history and not in the style of heavy scholarship fashionable twenty years later. But on analysis they are admirably accurate and show tremendous grasp of the essentials of sea power and of the naval art. Macintyre's researches were undoubtedly aided by his association with the Naval Historical Branch, which he joined in August 1964 and from which he retired in December 1972.

Burly and somewhat bear-like in appearance, gruff in manner, Macintyre could be daunting on first acquaintance, but it needed only the briefest of conversations to convince him one knew one's job and he was then the most kind and helpful of men. Considering his analytical powers and skill with words, it is astonishing that use was not made of him in the Admiralty during his service as a captain; but perhaps his outspokenness was not regarded as an asset in those years.

In 1941 he married Monica Josephine Clifford Rowley, daughter of Roger Walter Strickland, gentleman. They had a son and a daughter. Macintyre died in Ashford, Kent, 23 May 1981.

[Naval secretary's records; *The Times*, 11 June 1981; Ripley Registers; Donald Macintyre, *U-Boat Killer*, 1956; private information; personal knowledge.]   RICHARD HILL

MCKENZIE, ROBERT TRELFORD (1917–1981), political scientist and broadcaster, was born 11 September 1917 in Vancouver, the only child of William Meldrum McKenzie, grocer, and his wife, Frances. He was educated at King Edward High School, Vancouver, and at the University of British Columbia (BA, 1937), where in 1937 he became a lecturer, and where he developed what became a lifelong passion for politics. He joined the CCF, the Canadian equivalent of the British Labour Party, but later

assumptions that this marked him down as 'left wing' were exaggerated. He was careful to conceal his own views but they were in fact only mildly to the left of centre, and firmly anti-Marxist.

In 1943 McKenzie joined the Canadian Army, and a year later war service (he was seconded from the artillery to current affairs education) took him for the first time to London, where he was destined to spend the rest of his working life. He never ceased to be a Canadian, both in outlook and accent, and his affection for Britain was tinged with exasperation at what he regarded as its social affectations, but he came to love the country and to be fascinated by its politics.

When he left the Canadian Army (as a captain) in 1946, he was persuaded by Professor Harold Laski [q.v.] to join the London School of Economics and work for a doctorate. The result was of importance to any study of twentieth-century British politics. His doctoral thesis was the basis of a book published in 1955—*British Political Parties: the Distribution of Power within the Conservative and Labour Parties*. McKenzie's meticulous analysis led him to see the Conservatives as more professional, and more ruthless, than popularly supposed, while the Labour Party he found to be crucially skilled in circumventing the extra-Parliamentary bias in its constitution. McKenzie's judgements became widely accepted, and remained influential, although his work was to some extent outdated by such changes as the emergence of a strong third party, and the shift of power within the structure of the Labour Party.

In 1968 McKenzie published, with Allan Silver, a further book—*Angels in Marble: Working Class Conservatism in Urban England*. This did not attract the same interest as the earlier work, but was acknowledged to be a respectable contribution to political analysis.

McKenzie's academic work was of lasting importance, but paradoxically the wider public knew him largely through his appearances on television. The viewers identified him above all with a device he called the 'swingometer'. This was shaped like a large, inverted metronome and, standing alongside it every election night, McKenzie pushed it to left or right, marking the percentage swing in votes to one side or the other. McKenzie became in time rather too obsessed with his invention, the function of which was overtaken by computer analysis, but it served its turn as a device for demonstrating to an uninformed public how the swing in votes translates to a shift in parliamentary seats.

It is perhaps typical of television that McKenzie should be remembered for a few bravura performances with the 'swingometer' rather than for his many, more substantial, appearances, particularly his political commentary and interviewing on radio and television, which was generally regarded as the best informed and most penetrating heard or seen up to that time. He was equally skilled when interviewing a truculent Viscount Hailsham (later Lord Hailsham of St Marylebone) at the height of the Profumo affair in 1963, or gently leading a benign Harold Macmillan (later the Earl of Stockton) through a series of reminiscences.

McKenzie sometimes felt that the academic world did not give him sufficient credit for his television work, because to him they were different sides of the same coin, spreading political understanding to divergent audiences. Since 1949 he had been a lecturer in the LSE sociology department and he was particularly pleased when in 1964 the LSE made him professor of sociology (with special reference to politics). He had earlier been a visiting lecturer on politics at Harvard and Yale (1958-9), and in 1969 he received an honorary degree from Simon Fraser University in his native Vancouver.

To television viewers, McKenzie appeared a chubby figure, rather intimidating behind heavy spectacles, but with a lurking sense of humour. Politicians respected his wide knowledge and recognized that however sharp his questioning, he was never discourteous.

Although rather a private man—he never married—McKenzie had a wide circle of friends, with whom he delighted to argue, and to entertain at his little house on an island in the Thames, where he was able to exercise his love of gardening.

McKenzie died of cancer 12 October 1981 at University College Hospital, London. The next day Harold Macmillan wrote: 'Apart from his great charm he had a quite remarkable knowledge of political history over the last hundred years or more. The depth of his learning he concealed under a light touch.'

[*The Times*, 14 October 1981; private information; personal knowledge.]

IAN TRETHOWAN

McKIE, SIR WILLIAM NEIL (1901-1984), organist, choirmaster, and church musician, was born 22 May 1901 in Melbourne, Australia, the second son (the first died in infancy) in the family of three sons and three daughters of the Revd William McKie, vicar of St Philip's, Collingwood, and his wife, Mary, daughter of James Doyle, journalist. His mother was one of the earliest women graduates of Melbourne University, which gave degrees to women before Oxford and Cambridge began to do so.

Both parents were Australian-born, but all four grandparents had emigrated from Ireland in the 1850s, after one of McKie's grandfathers had been imprisoned, with other members of the Young Ireland Party, on a charge of sedition, of which he was acquitted.

At Melbourne Grammar School McKie was taught by Arthur Nickson, a pupil of Sir Walter Parratt [q.v.]. In 1921, after a year at the Royal College of Music under Henry Ley [q.v.], he won the organ scholarship at Worcester College, Oxford, of which he was later to be an honorary fellow (1954). He gained his B.Mus. in 1924.

During vacations McKie used sometimes to act as assistant to Noel Ponsonby at Ely Cathedral, where, under that dedicated man, he gained his first insights into the disciplines of cathedral music. An even stronger influence, however, was that of Sir Hugh Allen [q.v.], at the time sceptical about cathedral music, and inclined to channel his abler apprentices into school music, or administration, rather than church music. McKie, accordingly, after a short probationary period at Radley College (1923-6), went to Clifton College (1926-30) for five years as director of music. By nature, however, he was a performer rather than an administrator, and in 1931 he decided to return as city organist to Melbourne, where he soon won recognition as an outstanding player and personality.

In 1938, on the retirement of H. C. Stewart, McKie was appointed to succeed him at Magdalen College, Oxford, where he stayed for two years, devoting himself almost exclusively to college music and to the development of his own skills, particularly as choir trainer, where he excelled. His standards were exacting and his methods rigorous, but he won the affection of his choristers and was happy in the work, except for occasional brushes with college dons, notably C. S. Lewis [q.v.], who disapproved of the choral services and resented the cost of their upkeep.

In 1941, after (Sir) Ernest Bullock [q.v.] was bombed out of Westminster, McKie was called to take his place. It was a tragic time for Westminster Abbey. The music establishment had been dispersed; its future was uncertain; and McKie himself was of military age, due to be absent until further notice with the RAFVR. Not until 1946 was he able to begin the great task of rebuilding the Abbey's musical tradition. His main aim, in this single-minded effort, was to maintain unvaryingly high standards in the daily services, and to modernize the technique, style, and repertory of the choir. In all this he was successful, in spite of some opposition and frequent disruption of his plans by the obligation to provide, sometimes at short notice, music for royal and state occasions. Of these the most demanding was the coronation of 1953, whose music was well chosen, brilliantly performed, and highly praised. It was largely in recognition of this achievement that McKie was knighted that year. Some years earlier, in 1944, Oxford had made him an honorary D.Mus., and in 1948 he was appointed MVO. He was also FRSCM, FRCM, FRCO, FTCL, and honorary RAM.

The claims of his work at the Abbey did not prevent McKie from taking an active part in London music. He taught at the Royal Academy of Music (1946–62), and in 1956–8 served as president of the Royal College of Organists. He also composed some unpretentious but distinguished church music, and greatly helped many young musicians who came within his influence. To this Dictionary he contributed the notice of Sir Sydney Nicholson.

McKie was a fine-looking man, masterful in bearing, with strong features and a jutting chin that gave the impression of formidable authority and self-confidence. But the cost of sustaining this appearance was high, for, especially in early life, McKie was diffident, anxious, and liable to severe depression. More than once, in his time at the Abbey, he admitted to being close to breakdown. But his unexpected marriage in 1956, at the age of fifty-six, greatly strengthened him. His wife, Phyllis, daughter of John W. Ross, LLD, of Montreal, and widow of Gerald Walker Birks, a Canadian businessman, was a lady of great charm and character, under whose influence his social life broadened. He became more confident and relaxed, better able to believe that he was admired and indeed loved. Even so, he remained what he had always been, a shy and reserved man, most at home in the company of a few friends, preferably male, as critical as himself, and not necessarily musicians.

After retirement from the Abbey in 1963 the McKies went to live in Ottawa where Lady McKie had many connections, and for a time William continued to take part in musical activities. But his health was declining, and some time before his wife's death in 1983, he was admitted to hospital suffering from Alzheimer's disease, from which he died in Ottawa 1 December 1984. There were no children.

[Private information; personal knowledge.]
THOMAS ARMSTRONG

MACLEAN, DONALD DUART (1913–1983), British diplomat and Soviet spy, was born in London 25 May 1913, the third of four sons and five children of (Sir) Donald Maclean [q.v.], Liberal politician and cabinet minister, and his wife, Gwendolen Margaret, eldest daughter of Andrew Devitt JP, of Oxted,

Surrey. He, like his two elder brothers, was educated at Gresham's School, Holt, where the headmaster, J. R. Eccles, enforced the so-called 'honour system' with the aim of maintaining the highest moral standards. It may well be supposed that this system, allied to a strict upbringing at the hands of Sir Donald, a non-smoker, temperance advocate, and severe Sabbatarian, may have brought out in his son the tendency both to rebel and to deceive authority.

In October 1931 Maclean went up to Trinity Hall, Cambridge, with an exhibition in modern languages. He soon joined those on the extreme left, who aimed to reanimate, and dominate, the University Socialist Society; this group included Guy Burgess, Anthony Blunt [q.v.], and H. A. R. ('Kim') Philby. Maclean, handsome, standing six feet four inches tall, was prominent both physically and intellectually among his contemporaries. He at first made no secret of his communist sympathies; but in mid-1934 he abandoned open political activity and announced his intention to enter the Diplomatic Service. This move coincided with his recruitment by the NKVD (later KGB) of the Soviet Union. In June 1934 he graduated with first class honours in part ii of the modern languages tripos, having gained a second class in part i in 1932. He entered the Diplomatic Service in 1935, serving in the League of Nations and Western department of the Foreign Office.

In 1938 he was appointed third secretary at the Paris embassy. There he met an American student, Melinda Marling, eldest daughter of Francis Marling, a Chicago businessman, whose wife divorced him in 1928 to marry a New England property owner, Hal Dunbar. Donald and Melinda were married in Paris on 10 June 1940 at the time of the evacuation of the city. Back in London he was promoted second secretary and employed in the General department until April 1944, when he was transferred to the Washington embassy, and soon after promoted first secretary. An indication that his credit with the NKVD was as high as with the Foreign Office is provided by the Soviet decision to transfer to Washington Maclean's London case officer ('control'). For some months in 1946 Maclean was acting head of Chancery; but his most important duties, from the NKVD's viewpoint, were connected with the development of the atom bomb. Early in 1947 he became joint secretary of the Anglo-American-Canadian combined policy committee, a post that gave him access to the American Atomic Energy Commission at a time when the US government was making maximum efforts to prevent leakage of information about nuclear weapons.

Two sons were born to the Macleans in the Washington years. When Maclean left in September 1948, on promotion to counsellor and head of Chancery in Cairo, he was the youngest officer in his new grade. In Cairo, however, an all-round deterioration set in, culminating in a drunken spree that caused him to be sent back to London and subjected to psychiatric examination for his homosexuality and alcoholism. The Foreign Office, believing he had recovered, appointed him in November 1950 to be head of the American department. This was a sensitive post, because of tensions resulting from the Korean war. Meanwhile the investigation of earlier leakages in Washington began to point to Maclean as prime suspect. Through Philby and Burgess he became aware of this and on 25 May 1951 defected with Burgess to the USSR. Soon afterwards his wife gave birth to a daughter in London. She later moved with her children to Switzerland and in September 1953 left secretly to join her husband. In 1967 she had an affair with Philby in Moscow.

In Moscow Maclean taught graduate courses in international relations and published *British Foreign Policy since Suez* (London, 1970). An expanded Russian edition led to his award of a doctorate of the Institute of World Economics and International Relations. By 1979 his former wife and children had left and gone to the West. He died 6 March 1983 in Moscow and was cremated there. His ashes were taken by his elder son and buried in the family plot at Penn, Buckinghamshire.

There has never been any official assessment on the British side of the damage done by Maclean as a spy; opinions differ concerning the benefit derived by the USSR from his undoubted insight into US nuclear capacity in the crucial years before the first Soviet atomic test in 1949. A report prepared in 1955 by US Military Intelligence, however, is unambiguous in attributing very grave damage to the combined activities of Maclean and Burgess. In the long term the most lasting damage was probably that suffered by Anglo-American relations, much of it arising from US criticism of the laxity of British security.

[Robert Cecil, *A Divided Life: A Biography of Donald Maclean*, 1988; official and private sources; personal knowledge.]

ROBERT CECIL

McNEE, SIR JOHN WILLIAM (1887–1984), professor of medicine, was born at Murieston, Mount Vernon, Lanarkshire, 17 December 1887, the son of John McNee, insurance agency inspector, and his wife, Agnes Caven. When the family moved to Newcastle upon Tyne he was educated at the Royal Grammar School there and later went to Glasgow University graduating MB, Ch.B. with honours in 1909.

After appointments held in pathology with (Sir) Robert Muir [q.v.] and in medicine he worked on a two-year research scholarship with the distinguished pathologist Professor L. Aschoff at Freiburg in Germany. He made important observations on haemolytic jaundice in geese which formed the basis of modern views on the functions of the liver and reticulo-endothelial cells in the formation of bile, work for which he obtained an MD with gold medal in 1914.

During World War I he served in the Royal Army Medical Corps, attaining the rank of major. He was assistant adviser in pathology to the First Army in France and wrote several articles on gas poisoning, gas gangrene, and trench fever. He was appointed to the DSO in 1918 for his war services and mentioned in dispatches. On the strength of his war work he became a member of the new teaching medical unit at University College Hospital, London, and later the deputy director of that unit and physician to the hospital. While there he did major research, which resulted in the third edition of the textbook (with Sir Humphry Rolleston, q.v.) *Diseases of the Liver, Gall-Bladder, and Bile-Ducts* (1929). His work on cholesterin metabolism obtained a D.Sc. for him in 1920. In 1924 he obtained a Rockefeller medical fellowship to Johns Hopkins University at Baltimore where he investigated bile acids. He will be mainly remembered for his work in the USA by the publication of 'The Clinical Syndrome of Thrombosis and the Coronary Artery' published in the *Quarterly Journal of Medicine of Britain* (1925), a disorder which had not previously been described in the United Kingdom. While in London he gave several prestigious lectures, particularly the Lettsomian lecture in 1931 and the Croonian lectures in 1932. He became FRCP (Lond. 1925, Edin. 1943) and FRSE (1940).

In 1936 he took up the regius professorship of the practice of medicine at Glasgow University. A new Gardiner Institute of Medicine was purpose-built adjacent to the wards in the Western Infirmary, a combined facility for clinical work, ward teaching, and research. He swung himself into the work of developing the department but there was no question that his ambitions were delayed by the onset of the war in 1939. Since 1935 he had been a consultant physician to the Royal Navy and throughout World War II he served with the rank of surgeon rear-admiral to the Royal Navy in Scotland and the Western Approaches. He organized the medical personnel and equipment of the rescue ships which accompanied the convoys during the war and played an important part in instituting a convoy medical code, designed particularly for merchant ships which did not carry a medical officer. During the war he still lectured to students in the uniform of the rear-admiral, a figure of elegance and panache. After the war he encouraged a group of young men in their researches and also attracted a Medical Research Council unit on clinical chemotherapy to the Western Infirmary.

He was involved in several major publications, not only in his work on the liver and spleen but also the textbook of medical treatment (1st edn. 1939, 6th edn. 1955) where he joined forces with (Sir) Derrick Dunlop and (Sir) L. Stanley P. Davidson [qq.v.]. In 1952 he delivered the Harveian lecture on infective hepatitis. Many honours came his way. He was physician to the King in Scotland (1937–52) and the Queen (1952–4), and was knighted in 1951. He was president of the Royal Medico-Chirurgical Society of Glasgow and of the Gastro-Enterological Society of Great Britain in 1950–1 and of the Association of Physicians of Great Britain and Ireland in 1951–2 when it met in Glasgow. He was also president of the British Medical Association in Glasgow in 1954–5. He had honorary degrees from the National University of Ireland, Glasgow, and Toronto.

In 1923 he married Geraldine Zarita Lee, daughter of Cecil Henry Arthur Le Bas, stock-broker's clerk, of Charterhouse, London, who was herself at that time a university research worker at University College Hospital Medical College. They had no children. After McNee's retirement in 1953 they moved to Winchester, Hampshire, where they had a lovely garden containing many rare plants. They were devoted to the countryside and country sports; McNee's favourite sports were fishing and shooting. In his long and happy retirement he was able to express his love of the woods, rivers, and hills by active participation in the Council for the Protection of Rural England, of which he became chairman of the Winchester district branch. He retained his memory and intellectual acuity to the end despite a degree of Parkinsonism.

McNee was the prototype of the modern professor of medicine. He had a remarkably strong base in experimental pathology and worked as a young man with two great pathologists of his generation. He combined this also with the charm of the traditional physician. After his wife's death in 1975 he had an increasing sense of loneliness, but he kept a sharp eye on his old department in Glasgow and wished to be kept up-to-date on research and progress. He died 26 January 1984 at his home in Winchester.

[Personal knowledge.]   A. GOLDBERG

MacTAGGART, SIR WILLIAM (1903–1981), painter, was born 15 May 1903 at Loan-head, Midlothian, the elder son and third of four children of Hugh Holmes MacTaggart, mechanical engineer, and his wife, Bertha, daughter of Robert Little, businessman, of Edinburgh. His father was the eldest son of the Scottish landscape painter William McTaggart [q.v.]. Although he always revered his grandfather's memory MacTaggart established his artistic personality from the beginning and even his earliest pictures owe nothing to his illustrious namesake. As a child MacTaggart suffered from ill health and was educated privately. From an early age he was determined to become a painter. In this he received every encouragement from his father whose collection of pictures not only contained an excellent representation of the elder McTaggart but also pictures by Boudin and Le Sidaner. Between 1918 and 1921 he attended the Edinburgh College of Art as a part-time student. It was at this time he met the painters (Sir) W. G. Gillies, Anne Redpath [qq.v.], and John Maxwell who were to remain lifelong friends. The person who had the greatest influence on MacTaggart at this period, however, was the artist William Crozier (1893–1930). He had studied in Paris with André Lhote, had travelled extensively, and brought an intellectual approach to painting which was important to MacTaggart.

Between 1922 and 1929 MacTaggart went regularly to the South of France for the sake of his health, sometimes in the company of Crozier: to Cannes, Le Cannet, Cassis, Bormes, and Grimaud. He took advantage of these visits to paint, and a six-week period at Cannes in 1923 was particularly important. The pictures from this period are bright and strong in colour and show an affinity with those of the Scottish colourists, particularly S. J. Peploe. It was appropriate therefore that in 1927 MacTaggart joined the Society of Eight, of which Peploe and F. J. Cadell were also members. MacTaggart held an exhibition of his work at St Andrews church hall, Cannes, in 1924. His first one-man show in Edinburgh took place at Aitken Dott & Son in 1929.

The inclusion of twelve pictures by Edvard Munch in the annual exhibition of the Society of Scottish Artists in 1931–2 was the occasion for MacTaggart meeting the person who was largely responsible for bringing them to Edinburgh: Fanny Margarethe Basilier, daughter of General Ivar Aavatsmark, of Oslo. She was perhaps the most important influence in his life. She gave him confidence in his work and broadened his outlook. Together they visited Matisse in the summer of 1936 and after their marriage in July 1937 they paid regular visits to Norway. MacTaggart was president of the SSA between 1933 and 1936, during which period paintings by Klee and Braque were shown at their annual exhibitions in Edinburgh. In 1933 he began his teaching career at the College of Art, Edinburgh, an association that lasted until 1956. In 1937 he was elected as associate of the Royal Scottish Academy. (He became a full member in 1948 and was president between 1959 and 1969.)

Between 1939 and 1945 MacTaggart again turned his attention to the landscape of East Lothian which remained an important source of inspiration for the remainder of his life. Visits to Scandinavia resumed after the war and between 1947 and 1952/3 the MacTaggarts stayed regularly at Orry-la-Ville, just north of Paris. The Rouault exhibition at the Musée d'Art Moderne in 1952 had a profound effect on MacTaggart's technical approach to painting. The palette becomes more sombre and the paint surface more richly worked. In the landscapes and still lifes of this period, especially those of flowers seen through an open window against a night sky, colours take on an inner glow. The most important example of the 'window' theme which he often repeated throughout the 1960s is the 'Starry Night, the New Town', 1955 (Coll. Laila Aavartsmark, Oslo). MacTaggart gave up painting directly from nature in the mid-1950s and increasingly worked from sketches boldly executed on the spot in black chalk.

The MacTaggarts were a focal point of social life in Edinburgh and entertained constantly at 4 Drummond Place, which was their home from 1938. In the later 1960s and 1970s the routine of work, parties, and official functions was broken by the annual visit to the spa town of Skodsborg in Denmark. MacTaggart was a man of great personal charm with a gift for friendship. Position and honours were important to him, if only to prove that he had been able to conquer adversity and ill health. He had an honorary LLD from Edinburgh University (1961), was knighted in 1962, and became a chevalier of the Legion of Honour (1968). He was elected ARA in 1968 and RA in 1973.

MacTaggart's range as an artist was limited and he was no innovator. But he spoke with a distinctive voice and the East Lothian landscapes of the 1960s in which harvest fields and ploughed land glow under an incandescent sun belong to the visual imagery of the Scottish scene and make a small but distinctive contribution to a larger Nordic tradition in painting. He was very popular in his lifetime and his pictures always sold well. MacTaggart stopped painting about 1976. He died in Edinburgh 9 January 1981, and his wife died nine days later. They had no children. An unfinished portrait by W. G. Gillies (c.1935) is in the Scottish National Portrait Gallery (2496). A bronze bust by Benno Schotz, RSA, belongs to the Royal Scottish Academy who commissioned it in 1970.

Both are good likenesses.

[The MacTaggart Papers, National Library of Scotland, MS Acc. 8636, 8416, 8755; catalogue (by Douglas Hall) of the retrospective exhibition at the Scottish National Gallery of Modern Art, Edinburgh, 1968; H. Harvey Wood, *W. MacTaggart* (Modern Scottish Painters 3), Edinburgh, 1974; catalogue of the Studio Sale, Christie's & Edmiston's at 4 Drummond Place, Edinburgh, 2 July 1981, pp. 45-74; personal knowledge.]

HUGH MACANDREW

MANNIN, ETHEL EDITH (1900-1984), writer, was born 11 October 1900 in Clapham, the eldest of three children (the middle one a boy) of Robert Mannin, a postal sorter, and his wife, Edith Gray, a farmer's daughter. Robert Mannin was a Londoner, with Irish ancestry, who, from his youth, spent much of his earnings on books. Ethel was educated at the local council school, and left at fifteen to become a stenographer in the London advertising agency of Charles F. Higham. Two years later she was appointed associate editor of the *Pelican*, a theatrical and sporting periodical. In 1919 she married John Alexander Porteous (died 1956), a Scotsman, thirteen years her senior, who worked as a copy-writer at Higham's, and later became general manager. He was the son of William Porteous, who had a grocery business. Their only child, a daughter, was born within a year.

By then, Ethel Mannin was already selling articles to women's magazines, and had embarked on her career as a compulsive writer of novels, travel books, children's stories, and volumes of autobiography. Her first novel, *Martha* (1923), was runner-up in a competition for first novels. Her first success came with *Sounding Brass* (1925) in which she satirized the world of advertising, using her intimate knowledge of its practices. This she thought to be her best novel, but, such was her prolific output, that she could not hope to maintain the same standard in all her later novels.

During the next fifty years she published nearly one hundred books, but, as she herself admitted in *Brief Voices* (1959), she had 'only a limited talent' and was well aware of her limitations.

In her first autobiographical volume *Confessions and Impressions* (1930), she set out to shock an older generation since, to use her own words, she was 'angry with the existing social system, angry with the humbug of conventional morality, angry with the anti-life attitude of orthodox religion and the futility of orthodox education'. In 1930 she also published *Children of the Earth*, a novel with a Channel Islands' background, and from then onwards she deliberately set out to produce one novel and one non-fiction book every year.

In 1929 she had been able to buy Oak Cottage, Wimbledon, a house to which she had been attracted since as a young girl she had walked across Wimbledon Common. She lived there for nearly forty years and there she entertained her many friends, including (Sir) Allen Lane [q.v.], Christina Foyle, and (Dame) Daphne du Maurier. As her second husband, Reginald Reynolds, wrote in his autobiography *My Life and Crimes* (1956) she also gave hospitality to numbers of visitors from overseas for whose causes she worked tirelessly, 'from Spanish anarchists to Arabs'.

Her interests were in the main the theatre, cinema, ballet, and the best-selling novels of the period such as *The Green Hat* (1924) by Michael Arlen [q.v.]. She admitted quite honestly that she was not attracted by the more esoteric works such as those of Virginia Woolf [q.v.], published in that decade.

She held certain opinions very strongly. She was a confirmed pacifist from the time of the Spanish civil war; she was fiercely anti-Zionist; and she regarded monarchy as an anachronism. In 1933 she joined the Independent Labour Party but in later life became disenchanted with the policies of the left, and particularly with those of the USSR.

Her marriage to John Porteous broke down (they were divorced in 1938), though she continued to maintain amicable relations with him, and when he died she hastened to Devon to be with her daughter at his funeral. In *Brief Voices* she wrote: 'That we were able to be friends for the last fifteen years of his life was absolution for me for the wrong I had done him in my youth.' In 1938 she married Reginald Reynolds (died 1958), with whom she had been friendly since meeting him at an ILP dance in 1935. The son of Bryant Reynolds, market gardener, of Sanderstead, he was a Quaker, a pacifist, and a friend of Mahatma Gandhi [q.v.]. He and his wife were both great travellers, but they never travelled together.

However, she and her daughter frequently travelled abroad together and spent many happy weeks walking and climbing in the Lake District and boating in Connemara. Whatever she did and wherever she went Ethel Mannin described in her travel books and novels. In 1934 she published *Men Are Unwise*, a novel about mountaineering; her *Connemara Journal* (1948) is the story of her cottage in that corner of Ireland near Mannin Bay; and her novel *Late Have I Loved Thee* (1948) is also set in Ireland. Her early visits to the Continent are described in such books as *Bavarian Story* (1950) and

German Journey (1948). After her visit to India she published *Jungle Journey* (1950) and a novel *At Sundown, the Tiger . . .* (1951); Morocco is described in *Moroccan Mosaic* (1953) and Burma in *Land of the Crested Lion* (1955) and a novel *The Living Lotus* (1956).

Ethel Mannin's father died in 1949. When she was a young girl he had encouraged her to write and they had always been very close. In *This Was a Man* (1952) she paid tribute to him in a little memoir which she herself believed to be 'perhaps the best thing I have written'.

She had a distinctive hairstyle; her hair, parted in the middle, was brushed smoothly aside from her broad forehead. In *Sunset over Dartmoor, a Final Chapter of Autobiography* (1977) she describes how she decided to leave Oak Cottage and retire to Overhill, a bungalow which her daughter Jean Faulks had found for her at Shaldon, overlooking the Teign estuary. She found old age hard to bear. In looking back on her career as an author she said, 'I did not write to any pattern, merely accepting and using material as it came to hand', and of her early work she admitted: 'I had too much too soon; too much facile success, too much money, too much unassimilated experience.' She realized that her books might have been of higher quality had she been prepared to restrict herself to a lower output. In July 1984 she was hurt in a fall at her home and was taken to Teignmouth hospital where she died 5 December 1984.

[*The Times*, 8 December 1984; Ethel Mannin, *Confessions and Impressions*, 1930, *Privileged Spectator*, 1939, *Brief Voices*, 1959, *Young in the Twenties*, 1971, *Stories from my Life*, 1973, and *Sunset over Dartmoor*, 1977 (autobiographies); information from Jean Faulks (daughter).]

H. F. OXBURY

MARRIAN, GUY FREDERIC (1904-1981), biochemist, was born 3 March 1904 in London, the only son and youngest of the three children of Mary Eddington Currie and her husband, Frederic York Marrian, civil engineer, of London, who also had two sons and a daughter by his first wife, who died. He was educated at Tollington School, London, but his subsequent entry into University College, London, was delayed by one year in order to meet county scholarship conditions. This administrative problem proved beneficial as he took a technician's job at the National Institute for Medical Research in London, where he was introduced to the topic that formed the core of his future work—endocrinology. This was the era when, what subsequently became known as insulin, was being isolated and characterized. Marrian appreciated the importance of that work and utilized the principles of large-scale isolation and biological testing learned during that period to great effect when he eventually set up his own research team.

He graduated with a B.Sc. in chemistry in 1925 and remained at the college for a further eight years during which he initiated the work for which he became famous, namely the identification of two chemicals, pregnanediol and oestriol, involved in female sexual function. His ability was acknowledged by the award of a D.Sc. by London University (1930) at the young age of twenty-six without bothering with the more conventional Ph.D. He became FRIC in 1931. In 1933 Marrian moved to Toronto University, Canada, where he further enhanced his scientific reputation with pioneering work on oestrogen assay and metabolism.

Marrian considered this to be his most productive period and his research activity was recognized by his election as a fellow of the Royal Society of Canada (1937). He returned to the United Kingdom in 1939 to occupy the chair of chemistry in relation to medicine at Edinburgh University and the ensuing war years were spent studying the poison gas, arsine. With the experiences of World War I in mind, the authorities were concerned that the Germans would again use poison gas and arsine was thought to be the most likely candidate. He was elected a fellow of two further Royal Societies, Edinburgh (1940) and London (1944).

The post-war years saw Marrian re-establish himself in the forefront of steroid-hormone research, playing an important role in persuading the Medical Research Council to establish the clinical endocrinology research unit, the forerunner of the present reproductive biology unit in Edinburgh. However, he moved yet again, this time to become director of research at the Imperial Cancer Research Fund (1959) where he remained until his retirement in 1968. Although taking a major role in rapidly expanding the activities of that charity, his period there was less than happy, in part due to personality clashes with other personnel.

There is no doubt that Marrian made a lasting contribution to steroid endocrinology, both through his own work and the many top-class students he trained. What remains debatable is whether his full potential was truncated by the move from academia to that of running a large institute. Perhaps his best environment was the laboratory and not the office. The continuing high international status of the Medical Research Council unit in Edinburgh, and Imperial Cancer Research Fund in London, remain fitting memorials to Marrian's scientific and medical foresight.

His personal behaviour would, in every respect, be identified as 'gentlemanly' with oc-casional lapses when he thought that scientific principles were being undermined. His characteristic manner of speaking could, and was, used to devastating effect on those rare occasions. At the scientific level, he generated strong loyalties in his co-workers and students but this was rarely translated into personal associations. Not that life was all science, for being no mean athlete in his youth he retained his sporting interests and took particular pride in his daughter's international achievements in this sphere.

Marrian was a fellow of University College, London (1946), was appointed CBE (1969), became honorary DM of Edinburgh (1975), and was the Meldola medallist of the Royal Institute of Chemistry (1931), a recipient of the Francis Amory prize of the American Academy of Arts and Sciences (1948), and Sir Henry Dale medallist of the Society of Endocrinology (1966).

In 1928 he married Phyllis May, a fellow chemist, daughter of Albert Robert Lewis, pharmacist. She provided him with valuable practical and moral support. They had two daughters. Marrian died in Ickham, Kent, 24 July 1981.

[J. K. Grant in *Biographical Memoirs of Fellows of the Royal Society*, vol. xxviii, 1982; personal knowledge.]　　　　R. J. B. KING

MARSH, DAME (EDITH) NGAIO (1899–1982), detective novelist and theatre director, was born in Christchurch, New Zealand, 23 April 1899, the only child of Henry Esmond Marsh, an English immigrant to New Zealand, holder of a bank clerkship in the Bank of New Zealand, and his wife, Rose Elizabeth Seager, a New Zealander of longer standing. Given the Maori name of Ngaio, pronounced 'Nyo' (the ngaio is a sturdy New Zealand coastal tree which Maori legend says also grows on the moon), she was educated at St Margaret's College and Canterbury University College School of Art, Christchurch.

Her original aim to become a painter was gradually diverted to the theatre through amateur acting and eventually an invitation to join (1920–3) the professional Allan Wilkie Shakespeare Company, then touring the antipodes. From 1923 to 1927 she was a theatrical producer.

It was during a first stay in England, from 1928 to 1932, when she worked as an interior decorator and helped run a gift shop, that almost by chance her life took on the aspect through which she was to become known wherever English is read. In her reticent and evocative autobiography, *Black Beech and Honeydew* (1966, revised 1981), she told how on a rainy London Sunday, having devoured a detective story, the notion entered her head of writing one herself and how she 'went to the newspaper shop . . . and bought several penny exercise books and some sharpened pencils'. The novel that resulted, *A Man Lay Dead* (1934), was to be the first of thirty-one.

It is to the creation of the detective of her first and all subsequent books, Roderick Alleyn of Scotland Yard, that her success can chiefly be attributed. Named after the founder of her father's old school, Dulwich College, Alleyn is above all a gentleman. How this Old Etonian, a cross 'between a grandee and a monk', in his creator's words, ever became a policeman is not recounted. But a detective in the full literary tradition he was. Seldom involved in physical encounter, he is seen, largely through the words he speaks, as invariably polite, always authoritative, free from affectation, and quietly humorous. In short, as 'that nice chap, Alleyn'. But his interviews, often prolonged, never failed to discover the culprit he sought.

It is, paradoxically, the Englishness of this New Zealand author's work that is the second factor accounting for its success. As a girl Ngaio Marsh was frequently told by her mother, anxious for her gentility, that 'rude is never funny' and in her books, despite the bizarreness of many of her murders, she does nothing to disturb the reader. They are notable, indeed, for the agreeable characters who people them and the agreeable settings they take place in. Pleasantness in the people did not, however, prevent many of them displaying a delightful eccentricity, most notably in the almost direct portrayal in *Surfeit of Lampreys* (1941) of an aristocratic and somewhat feckless family met first in New Zealand and subsequently joined in England.

A further factor accounting for the books' success is the author's inherent modesty which kept her, a craftsman, always within the bounds of her own knowledge. Finally, there is the writing itself. Nothing, she once said, prevents a detective-story author from using 'as good a style as he or she can command'. She wrote with a quiet fidelity that is the salt which has preserved her books when others in the genre are forgotten.

Ngaio Marsh divided her time between England and detective fiction and New Zealand and the theatre. As well as being a driver for a Red Cross transport unit in New Zealand in World War II she was also from 1944 until 1952 a producer in the D. D. O'Connor Theatre Management. In the early 1940s, Shakespeare had not been played in New Zealand since the Allan Wilkie tour of the 1920s. Ngaio Marsh cajoled her untried students into tackling *Hamlet* and, despite fearful prognostications, achieved a sell-out success. Thenceforth a considerable part of her life was devoted in New Zealand

to Shakespeare and other dramatists of repute. Eventually she became an honorary lecturer in drama at Canterbury university college.

In 1950 she visited London to found a British Commonwealth Theatre Company. It toured New Zealand and Australia with success but did not prove viable. While in London she directed Pirandello's *Six Characters in Search of an Author* at the Embassy Theatre. An actor member of her company has described her as 'a perfect coach', a director excelling in inspiring the young. In 1948 she was appointed OBE for services to New Zealand drama and literature and she was advanced to DBE in 1966. In 1956 she became an honorary Dr. Litt. of Canterbury University, New Zealand, and in 1962 the Ngaio Marsh Theatre was founded at that university.

Ngaio Marsh was unmarried. She died, in the wooden house her father had built overlooking her native Christchurch, 18 February 1982.

[Ngaio Marsh, *Black Beech and Honeydew*, 1966, revised edn. 1981; *The Times*, 19 February 1982; private information; personal knowledge.]   H. R. F. KEATING

MARSHALL, Sir STIRRAT ANDREW WILLIAM JOHNSON- (1912–1981), architect. [See JOHNSON-MARSHALL.]

MARSHALL, THOMAS HUMPHREY (1893–1981), sociologist, was born in London 19 December 1893, the younger son and fourth of six children of William Cecil Marshall, architect, of London, and his wife, Margaret, daughter of Archdeacon J. F. Lloyd of Waitemata, New Zealand. He was educated at Rugby School and Trinity College, Cambridge, where he obtained a first class in part i of the historical tripos in 1914. While visiting Germany that year he was interned on the outbreak of hostilities. He spent the next four years as a civilian prisoner of war. After his release he returned to Cambridge, where in 1919 he was elected to a fellowship in history at Trinity.

At Cambridge he seemed to be settling into an orthodox academic career as an economic historian. He revised and extended George Townsend Warner's *Landmarks in English Industrial History* (1924) and wrote a short life of James Watt (1925). Marshall, however, was also a musician of remarkable talent and he seriously considered becoming a professional violinist. In the event his change of career came in a different form when in 1925 he moved to an assistant lectureship in the social science department at the London School of Economics.

The LSE in the 1920s was already an established centre of teaching and research in theoretical sociology and social policy, and the growing range of Marshall's scholarly interests is reflected in his subsequent progress at the school. Initially engaged to teach social work students, he transferred to the sociology department under Morris Ginsberg [q.v.] in 1929. Marshall's growing reputation as a sociologist received formal recognition in the shape of a readership in 1930. Thereafter he played an essential part in launching the *British Journal of Sociology* and in editing major publications such as *Class Conflict and Social Stratification* (1938) and *The Population Problem* (1938), thereby laying the institutional foundations for much of the pioneering work on social class and population studies which was carried out at the LSE after 1945.

During World War II Marshall was in charge of a section in the research department of the Foreign Office until 1944, when he rejoined the LSE as professor of social institutions and head of the social science department. On his second return to the school, however, after a period of secondment as educational adviser to the British High Commission in Germany from 1949 to 1950, he was made head of the sociology department, succeeding to the Martin White chair of sociology in 1954. Only two years later he was appointed director of the social sciences department at Unesco, where he remained until his formal retirement in 1960.

The period of intense and varied activity after the war coincided with Marshall's most productive years as a scholar. His first collection of essays, *Citizenship and Social Class* (1950), later reprinted in an extended edition as *Sociology at the Crossroads* (1963), became a seminal text in the history of British sociology. These essays contain a brilliant analysis of the changing relationship between the institutions of citizenship and social class, set in historical and comparative perspective. They also put forward Marshall's distinctive view of sociology as a synthesizing discipline encompassing not only theoretical issues but a wide range of practical applications in the field of social policy.

This authorship alone vindicates Marshall's eminence as a sociologist; his equally definitive studies in social policy were not written until after his retirement. In particular *Social Policy in the Twentieth Century* (1967) and the last collection of essays, *The Right to Welfare* (1981), demonstrate Marshall's unique ability to relate the sociological aspects of social institutions to issues of social policy. Marshall believed that the value conflicts generated by the interaction of competitive economic markets, representative democracy, and statutory social services are an indication of the resilience rather than the weakness of democratic welfare-capitalism in the context of social change. He also challenged the convention that the abolition of poverty is contingent on strictly egalitarian policies, arguing that certain inequalities which facilitate economic growth are a precondition of the elimination of poverty, provided that the right to a basic level of social services is guaranteed by the state.

Marshall's writings are characterized by their intellectual integrity, impeccable logic, and stylistic elegance. His grasp of essentials and his talent for precise exposition made him an essayist of the first rank.

In 1925 he married Marjorie, daughter of Arthur Tomson, artist. She died in 1931. There were no children of the marriage. In 1934 Marshall married Nadine, daughter of Mark Hambourg, pianist; their son Mark was born in 1937. Throughout their marriage they shared with relatives and close friends a deep and informed interest in music. At home, in Cambridge, Marshall pursued, with considerable skill, his enthusiasm for gardening, and he rediscovered the countryside of his childhood holidays during visits to their house in the Lake District. He was by disposition a retiring man but beneath his outer reserve there was a warmth and generosity of spirit which endeared him to his friends and colleagues. He had a finely handsome appearance and even in his last years he conveyed a sense of intellectual and physical vitality. He was by all accounts an inspiring teacher.

Appointed CMG in 1947, he received other honours and awards in his retirement, including the presidency of the International Sociological Association from 1959 to 1962, an honorary fellowship at the LSE and honorary degrees from Southampton (D.Sc. 1969), Leicester (D.Litt. 1970), York (D.Univ. 1971), and Cambridge (Litt.D. 1978). In effect Marshall never retired, and his remarkable intellectual vigour continued to the very end. His last book *The Right to Welfare* was published a few months before his death, which occurred at his home in Cambridge 29 November 1981.

[T. H. Marshall, 'A British Sociological Career', *International Social Science Journal*, vol. xxv, no. 1/2, 1973; *The Times*, 3 and 7 December 1981; Donald G. MacRae in *L.S.E.*, the magazine of the London School of Economics and Political Science, no. 64, November 1982; personal knowledge.]

ROBERT PINKER

MARSHALL-CORNWALL, SIR JAMES HANDYSIDE (1887–1985), soldier, linguist, and author, was born in India 27 May 1887, the only son and elder child of James Cornwall, postmaster-general of the United Provinces, India, and his wife, Agnes Hunter. He was educated at Rugby where his contemporaries included Rupert Brooke, Rufus Isaacs (later Marquess of Reading) [qq.v.], and Philip Guedalla. After Rugby, where he was not happy, he went in 1905 to the Royal Military Academy at Woolwich, which at the time was known as 'The Shop' and had an atmosphere more suited to his temperament. In 1907 he was commissioned into the Royal Field Artillery and posted to Edinburgh, whence, during his first spell of annual leave, he travelled to Germany to study German. He later passed the Civil Service Commission examination as a first class German interpreter. This was the first of the eleven interpreterships he was to gain. Before World War I he passed as first class interpreter in French, Norwegian, Swedish, Hollander Dutch, and Italian.

On the outbreak of war he was ordered to join the Intelligence Corps, 'a formation of which I had never heard', and was sent to Le Havre. After the débâcle at Le Cateau he was, rather surprisingly, given command of a squadron of the 15th Hussars. In February 1915 he was appointed G-503 (Intelligence) at II Corps headquarters in the Second Army, with the rank of captain. In January 1916 he was promoted to the rank of temporary major and sent to the general headquarters of the British Expeditionary Force as G-502 (Intelligence). This was his first encounter with Sir Douglas (later Earl) Haig [q.v.], who had recently taken over command of the BEF from Sir John French (later Earl of Ypres, q.v.). Cornwall soon realized that the information given to Haig by his chief intelligence staff officer, Colonel John Charteris, was based more on what he thought his chief would be pleased to hear than on a realistic assessment of the situation. He spent two frustrating years at GHQ from which Charteris was eventually transferred. In January 1918 Cornwall was posted to the War Office as head of the MI3 section of the Military Intelligence Directorate, where he remained until the armistice. He won the MC in 1916 and was appointed to the DSO in 1917.

In January 1919 he was sent to Paris as ex-officio member of the general staff delegation at the peace conference where he struggled, along with (Sir) Reginald W. Allen Leeper and (Sir) Harold Nicolson [q.v.], with the knotty problems of the new boundaries of Europe. Then followed a spell as student at the Staff College. From 1920 to 1925 he was employed in a number of jobs in the Middle East, which gave him the opportunity to polish up his Turkish and modern Greek. While encamped in the Ismid peninsula he met Marjorie Coralie Scott Owen (died 1976) who was driving an ambulance for a Red Cross mission to White Russian refugees. They were married in Wales in April 1921. She was the daughter of William Scott Owen.

In 1927 the Baldwin government sent a force to China to protect British life and property in the Shanghai international settlement. Cornwall's position as brigade major in the Royal Artillery (Shanghai defence force) enabled him to learn Chinese and to travel extensively in the Far East with his wife. In the same year he inherited a small estate in Scotland from his maternal uncle, William Marshall, on condition that he should assume the surname of Marshall. As Marshall was one of his forenames this was achieved by the simple insertion of a hyphen. From 1928 to 1932 he held the post of military attaché in Berlin where Sir Horace Rumbold (whose notice he later wrote for this Dictionary) was ambassador. He was therefore well placed to observe the rising power of the Nazis and the nascent rearmament of Germany.

In 1934, after two years as commander of the 51st Highland division, Royal Artillery, based at Perth, Scotland, he was promoted major-general, which, under the peculiar system of those days, left him without a job. The next four years were mainly spent travelling in Europe, India, and the United States. There followed two years in Cairo as head of the British Military Mission to Egypt, where, not surprisingly, he qualified as an interpreter in colloquial Arabic. In 1938 he was promoted lieutenant-general and was put in charge of the air defence of Great Britain. The year 1940 found him in France helping to evacuate British troops from Cherbourg. He himself boarded the last ship to leave that port when Rommel was only five kilometres from the harbour. In 1941, as the only senior serving officer who spoke Turkish, he was sent by (Sir) Winston Churchill to Turkey in an attempt to persuade the Turks to join the war. He was not unhappy at the failure of this mission as he was certain that at that time the Turks would have been more of a liability than an asset to the Allied cause.

He took over Western Command in November 1941, but was dismissed in the autumn of 1942 for going outside the proper channels to secure the safety of the Liverpool docks. He spent the rest of the war with SOE and then MI6.

Retired from the army in 1945 at the age of fifty-eight with the rank of full general, he was nevertheless still very much in the prime of life and his later years were as active as his military career had been. Between 1948 and 1951 he was editor-in-chief of captured German archives at the Foreign Office. Arms dealing, writing military history, and presidency of the Royal Geographical Society (1954-8) were among the various activities which crowded the many years left to him.

Short of stature and slight of build, Marshall-Cornwall's outstanding intellect and phenomenal memory more than compensated for his modesty and lack of self-advertisement. He said that towards the end of his life his memory began to fail him and was annoyed to find on revisiting Turkey in 1982 that 'all my Turkish was completely outdated'. But anyone reading his autobiography *Wars and Rumours of Wars* (1984), written in his ninety-seventh year, will deduce that he retained near perfect recall until the end.

The Marshall-Cornwalls had three children. Their only son was killed in France in 1944. He is buried on the spot where he fell in a Normandy orchard subsequently purchased by Marshall-Cornwall to protect his son's grave which he visited each year on the anniversary of the death. Their elder daughter died aged fourteen in 1938 after an operation for appendicitis in Switzerland.

Marshall-Cornwall was appointed CBE in 1919, CB in 1936, and KCB in 1940. He died 25 December 1985 at Birdsall Manor, Malton, north Yorkshire, the home of his younger daughter.

[James Marshall-Cornwall, *Wars and Rumours of Wars*, 1984; private information; personal knowledge.]

LEO COOPER and T. R. HARTMAN

MASON, JAMES NEVILLE (1909–1984), actor, was born 15 May 1909 at Huddersfield, the youngest of three sons (there were no daughters) of John Mason, a textile merchant of that town, and his wife, Mabel Hattersley, only daughter of J. Shaw Gaunt, also of the west riding of Yorkshire. He was educated at Marlborough College and Peterhouse, Cambridge, where he took a first in architecture in 1931. At Cambridge he discovered a taste for acting and the theatre. His performance as Flamineo in a Marlowe Society production of *The White Devil* by John Webster [q.v.] was well reviewed by the theatre critic of the London *Daily Telegraph*. Thus encouraged, he began to reconsider his decision to become an architect (he was skilful at drawing for the rest of his life). He was stage-struck, of course, but also shrewd about himself. He knew that he had a good voice, a true ear, and other graces of body and mind. He also knew that in acting, as in the other performing arts, a broad gulf divides the talented amateur from the employable professional. He believed, though, that he could bridge it. His interest in films was that of a young intellectual. The thought of working in them (except, possibly, as an avant-garde director) had not yet occurred to him.

He had no formal training as an actor, but served an older, informal kind of apprenticeship: that of answering advertisements in the *Stage*, of presenting himself for auditions, of living cheaply, of taking ill-paid jobs in provincial touring and repertory companies, of making friends in the theatre, of doing the best he could with unsuitable parts and of making the suitable ones seem better. He made his professional début at the Theatre Royal, Aldershot, in 1931 playing the Grand Duke Maritzi in a play called *The Rascal*. Two years later he made his first London appearance in *Gallows Glorious* at the Arts Theatre. Between 1934 and 1937 he continued his stage education with the Old Vic Company and at the Gate Theatre, Dublin. He played his first film part in 1935. This was in *Late Extra*, a low-budget 'quickie' of the kind then being made by the dozen in England to enable film exhibitors to comply with the Quota Act.

He played in more quota films and in doing so began to identify and acquire the special skills needed to act effectively for the camera. He also made new friends, among them Roy Kellino, a cameraman turned director, and his wife Pamela. She was the daughter of Isidore Ostrer who, with his brothers, then controlled half the British film industry, including the Gaumont-British cinemas and the Shepherd's Bush and Islington studios. He was not, however, an indulgent father. When, in 1937, Mason and the Kellinos decided to make a film of their own, using their own pooled savings and a script written by the three of them, they were unable to get it properly distributed. *I Met a Murderer* (1939) was an intelligent little crime thriller and well received by the better critics, but the Ostrers were reluctant to exhibit in their cinemas a British film that had not been made in their studios. Mason returned to the stage to repair his fortunes and was in rehearsal for the BBC at Alexandra Palace when television production was halted there in August 1939.

He had an eventful but confusing war. He was estranged from his family who disapproved of his living with Mrs Kellino, not yet divorced. His attempt to register as a conscientious objector was frustrated by a tribunal which directed him to non-combatant military service. His appeals against this ruling, however, became in the end irrelevant. After he and Pamela were married in 1941, he found that work in the film industry had been declared of national importance. As long as he worked in films his call-up would be deferred and he would remain a civilian. He had worked his way through some very bad films when, at the Islington studios of Gainsborough Pictures, he played the wicked Lord Rohan of *The Man in Grey* (1943). It was the first of a series of costume melodramas which had a phenomenal popular success. *Fanny by Gaslight* (1944), *They Were Sisters* (1944), and *The Wicked Lady* (1945, with Margaret Lockwood) followed. In 1944 he was polled by the New York *Motion Picture Herald* as Britain's top box-office star. The following year he appeared, with Ann Todd, in *The Seventh Veil*, the film that introduced him to American audiences. *Odd Man Out* (1946, directed by (Sir) Carol Reed, q.v.), an exceptionally good film made shortly after the war ended, established Mason as a fine actor as well as a star. In less than four years he had become what Hollywood then called 'a hot property'.

In post-war England the problems of managing a success of that sort were unfamiliar; and, perhaps inevitably, the solutions were decided upon by the property himself, assisted by his wife. Determined at that time to produce films as well as act in them, he proceeded to dissipate much of his potential influence as a star by writing newspaper articles and open letters denigrating the British film industry in general (lacking in glamour, third-rate) and J. Arthur (later Lord) Rank [q.v.], by then its major proprietor, in particular. He made other mistakes. In deciding which of his Hollywood suitors to accept, he used his own judgement rather than his agent's. As a result he spent most of his first year in America preoccupied with an expensive lawsuit. He won the suit, but not all the costs of it. He needed work and was glad to play in two minor Hollywood films directed by Max Ophüls. It was his successful portrayal of a German field-marshal in *Rommel, Desert Fox* (1951) that re-established him as a box-office attraction, and his Brutus in the 1953 film version of *Julius Caesar* reminded the public of his qualities as an actor.

In his fifty years as a screen actor he appeared in over a hundred films. Those of his middle years were perhaps the best. His fine performance in George Cukor's version of *A Star is Born* (1954) brought him an Oscar nomination; his portrait of a middle-aged man infatuated with a teenage girl in the film of Vladimir Nabokov's *Lolita* (1962) was a triumph. He was remarkably versatile and could always make even a small part memorable. *Georgy Girl* (1966) and James Ivory's *Autobiography of a Princess* (1975) are examples.

His marriage to Pamela was dissolved in 1965, when he left Hollywood. They had a daughter, Portland, and a son, Morgan. In 1971 he married the Australian actress Clarissa Grace Kaye, daughter of Austin Knipe, racehorse training manager, of Sydney, Australia. There were no children of this marriage. The Masons settled in Corseaux s/Vevey in Switzerland where they were near neighbours of Sir Charles Chaplin [q.v.] and his wife Oona. Corseaux was their

base from which they travelled and went to work in other places. They took up birdwatching. Their favourite subjects were a family of crows which occupied the large pine on the lake side of their house. Mason died in hospital in Lausanne 27 July 1984.

[*The Times*, 28 July 1984; *Who's Who in the Theatre*, 14th edn., 1976; James Mason, *Before I Forget* (autobiography), 1981; Sheridan Morley, *Odd Man Out: James Mason*, 1989; Diana de Rosso (sister-in-law), *James Mason*, 1989; personal knowledge.]   ERIC AMBLER

MASSEY, SIR HARRIE STEWART WILSON (1908–1983), physicist, was born 16 May 1908 in St Kilda, a suburb of Melbourne, Australia, the only child of Harrie Stewart Massey, a self-taught engineer who had worked in the gold fields, and his wife, Eleanor Wilson. His father was active in setting up the local school, at Hoddles Creek, near Melbourne. Its sole master was delighted to find that the young Massey was something of an infant prodigy, progressing up the school at twice the normal rate. From there, Massey won a scholarship to the University High School, Melbourne, where after a slow start he climbed to the top of the school and was also vice-captain for cricket. In 1925 he entered the University of Melbourne, studying physics and chemistry and playing cricket. He obtained a degree in physics and chemistry, an M.Sc. in physics, and simultaneously a BA in pure and applied mathematics. In 1926 the work of Heisenberg and Schrödinger on quantum or wave mechanics was published, revolutionizing our understanding of the atom. For his M.Sc. thesis Massey's supervisor suggested 'wave mechanics', which he had to master from the original papers, many in German, a language he did not know. This 400-page thesis is preserved in Melbourne, and it was this which determined his future career. Though he carried out experiments, he was a theorist at heart, a scientist who used wave mechanics to explain natural phenomena.

The Cavendish Laboratory in Cambridge, under Sir Ernest Rutherford (later Lord Rutherford of Nelson, q.v.) had to be the next place for Massey. He was awarded the Aitchison scholarship to study there. In Cambridge, from 1929 to 1933, he was present during perhaps the most exciting time through which the Cavendish has passed, with the discovery of the neutron, the positron, and much else. The atomic nucleus, put on the map by Rutherford, had been proposed to explain the way very fast particles were deflected while passing through matter. Theories leading to an understanding of this are called scattering or collision theories, and Massey in Cambridge resolved to work in this field. Knowing this, the author of this article, then a young university lecturer, asked him to join in writing a book about it. The result, *Theory of Atomic Collisions*, appeared in 1933 and two subsequent and much enlarged editions were entirely Massey's work. Massey, attached to Trinity College, obtained a Cambridge Ph.D. in 1932. His whole career in research was based on collision theory; it led him to the upper atmosphere where molecules collide, then to balloons and rockets to see what happens up there, and eventually to presidency of the council of the European Space Research Organization (from 1964). In 1959 he was appointed first chairman of the British national committee on space research and he also maintained contacts with Australian scientists, helping with work on the Woomera rocket range.

Massey's great scientific achievements came through his capacity to understand things quickly, and through immensely hard work and the ability to inspire his colleagues. He was the kind of man who makes things happen. After Cambridge his first tenured job was in Belfast, at Queen's University as a lecturer in mathematical physics (1933-8) and then in 1938 he became Goldsmid professor of mathematics at University College, London. After that came the war with work at first in the Admiralty research laboratory in 1940 and then in the Admiralty mine design department, working on magnetic mines (he became deputy chief scientist in 1941 and chief scientist in 1943). To Massey should be given some of the credit for the setting up of the MX organization in which a team of scientists co-operated with a team of naval officers to counter enemy mine-sweeping action. He subsequently worked in America (1943-5) on the fission bomb in the Manhattan Project at Berkeley, California. In 1950 Massey became Quain professor of physics at University College, London, of which he was vice-provost (1969-73). Under his leadership his department was active in almost all branches of physics.

Of more than 200 scientific papers and many books, Massey was sole or part author. Later he became a leading figure in science policy making. In 1965, a Council for Scientific Policy was set up, with Massey as its first chairman (until 1969). He was also physical secretary of the Royal Society from 1969 to 1978. And yet his production of scientific papers did not dry up. He was elected FRS in 1940 and received the Society's Hughes and Royal medals (1955 and 1958). He was knighted in 1960, and was an honorary doctor of twelve universities.

As a cricketer, he was judged good enough to have been a professional. His sense of proportion often did not allow him time to study

the minutes of meetings he chaired. But he always had time for people. He loved his native Australia, and went back twenty times. In 1925 he married Jessica Elizabeth, daughter of Alexander Barton-Bruce, manager of timber mills, near Perth, Australia. They had one daughter. Massey died after a long illness in his home 'Kalamunda', Pelham's Walk, Esher, Surrey, 27 November 1983.

[Sir David Bates in *Biographical Memoirs of Fellows of the Royal Society*, vol. xxx, 1984; correspondence with Sir David Bates; personal knowledge.]   NEVILL MOTT

MASTERS, JOHN (1914-1983), soldier and author, was born 26 October 1914, at Calcutta, the elder son (there were no daughters) of Captain John Masters of the 16th Rajput Regiment and his wife, Ada Coulthard. Captain Masters had been born at Midnapore, Bengal, and the Masters family had served in the Indian army since 1805. John was educated at Wellington and Sandhurst, and in 1934 joined the Duke of Cornwall's Light Infantry but, within a year, was transferred to the 4th Prince of Wales's Own Gurkha Rifles. In 1936-8 he fought on the North-West Frontier; he served in Baluchistan in 1939, and in 1941 in Iraq, Syria, and Persia (Iran).

After a year at the Staff College, Quetta (1942-3) he commanded a brigade of the Chindits raised by Major-General Orde Wingate [q.v.], fighting behind the Japanese lines in Burma (1944-5). As the Fourteenth Army advanced into Burma from Manipur under General W. J. (later Viscount) Slim [q.v.], Masters became chief of staff to the 19th Indian Infantry division, which captured Mandalay and Maymyo. In 1944 he was appointed to the DSO and in 1945 appointed OBE, and, when the Japanese war ended, was posted to the directorate of military operations at GHQ, Army in India. Since 1942 he had served continuously with troops in the field and he wished to continue his career with the Gurkhas in India. But when India became independent Masters was disappointed to learn that the 4th Gurkhas would be merged with other Gurkha regiments for service in South East Asia and the Far East. He held an Indian passport and considered remaining in India as an Indian citizen, but his wife, Barbara Allcard, whom he had married in 1945, persuaded him that neither he nor she could make themselves Indians. Barbara was the daughter of Victor Allcard, stockbroker. Masters applied for and obtained in 1947 the post of instructor in mountain and jungle warfare at the Staff College, Camberley.

Masters did not feel at home in England, and at the end of 1948 he resigned his commission and retired from the army as a lieutenant-colonel. He later wrote in his autobiography *Pilgrim Son* (1971): 'After the geographical space and freedom of India I felt (in England) I was in a prison.' He was now thirty-four years old with a wife and a son and a daughter; a second daughter had been born with brain damage and only lived a few days. Barbara also had a son and a daughter by her former husband, from whom she obtained a divorce. They were not brought up by Masters. Masters had fourteen years' military service and was granted a loss-of-career gratuity of £3,000. He had no professional qualifications other than those of a fighting soldier. He decided to emigrate to the United States which he had visited in 1938.

Masters soon discovered, however, that because he had been born in Calcutta and had an Indian passport, the US immigration authorities regarded him as an Indian, and for Indians there was a waiting list delaying entry for over four years. After some time and much anxiety he obtained a visitor's visa, valid for six months.

On arrival in America, he tried his luck as a salesman, and as an agent for conducted tours to the Himalayas, but in neither was he successful. However he was fortunate in getting an article accepted by the *Atlantic Monthly*. He decided that he would be an author writing novels about the British in India. His first, with a background of the Indian mutiny of 1857, he entitled *Night-runners of Bengal*. This was rejected by many American publishers, but he managed to avoid insolvency by writing a booklet for the American Army Film Unit. Meanwhile, his visa expired and he became liable to deportation, but, with the help of an American friend, he delayed this catastrophe until eventually his wife, who had an English passport, legally entered the country, and he was permitted to stay on the grounds that he was her sole financial support. Then, at last, Viking Press accepted *Nightrunners of Bengal* subject to amendment by a professional editor, Helen Taylor. The novel was published in 1951 and was an immediate success.

Masters called his next novel with an Indian background *The Deceivers* (1952). A British officer, disguised as an Indian, joined a band of Thugs, and was able to prove to the authorities the widespread nature of this conspiracy of murder and extortion. The third of the Indian trilogy, *The Lotus and the Wind* (1953), was based on an idea culled from *The Riddle of the Sands* (1903) by R. Erskine Childers [q.v.]. Masters transferred the setting of espionage from the German coast to the Himalayas with Russia as a threat to Afghanistan. Careful research gave authenticity to these novels and in their composition Masters demonstrated his own military experience. All were good stories

with no pretensions to any literary merit.

In 1954 Masters became an American citizen and published *Bhowani Junction*, a novel showing keen insight and sympathetic understanding of the dilemma faced by Eurasians in India when that country became independent. This was made into a film (1955) with Ava Gardner and Stewart Granger in the principal roles.

At the outset of his writing career Masters had decided not to write about the British in Britain, but, later in life, he changed his mind and published his 'Loss of Eden' trilogy—*Now God be Thanked* (1979), *Heart of War* (1980), and *By the Green of the Spring* (1981). These books about the 1914–18 war display his knowledge of the military scene but are less convincing in depicting civilian characters. His last novel *Man of War* (1983) is obviously based largely on his own experience of life in the army. He also wrote two autobiographical volumes about his military career, *Bugles and a Tiger* (1956) and *The Road Past Mandalay* (1961). In several of his novels such as *Far, Far the Mountain Peak* (1957) he revealed his love of mountains, in others his fascination with railways. His publisher in England was Michael Joseph Ltd. Masters died 7 May 1983 in hospital at Albuquerque, New Mexico, where he had received heart surgery.

[John Masters, *Bugles and a Tiger*, 1956, *The Road Past Mandalay*, 1961, and *Pilgrim Son*, 1971 (autobiographies); *The Times*, 9 May 1983.]   H. F. OXBURY

MATTHEWS, JESSIE MARGARET (1907–1981), actress, was born in Soho, London, 11 March 1907, the seventh of eleven surviving children of George Ernest Matthews, owner of a greengrocery stall in Berwick Street market, and his wife, Jane, daughter of Charles Henry Townshend, a timber porter. She went to Pulteney Street School for Girls, Soho, and showed such promise as a dancer that her oldest sister, Rosie, arranged for her to be trained in classical ballet by Mme Elise Clerc. When Mme Clerc died suddenly, Rosie determinedly arranged for Jessie to train as a chorus girl with Miss Terry Freedman of Terry's Juveniles.

Jessie Matthews made her first London appearance in 1919 in *Bluebell in Fairyland* produced by (Sir) E. Seymour G. Hicks [q.v.]. Four years later she played in Irving Berlin's *Music Box Revue* presented by (Sir) Charles Cochran [q.v.]. In his book *I Had Almost Forgotten . . .* (1932) Cochran described her as 'an interesting looking child with big eyes, a funny little nose, clothes which seemed a bit too large for her, and a huge umbrella'.

At the age of sixteen she made her New York

début in the chorus of *André Charlot's Revue of 1924*. Gertrude Lawrence [q.v.] was the leading lady in that show, and when she fell seriously ill with pneumonia in Toronto, Jessie Matthews took over her part.

She reached full star status in *The Charlot Show of 1926* when she danced in ballet numbers with Anton Dolin [q.v.] and in musical comedy items with Henry Lytton Jun. (Lord Alva Lytton, died 1965), son of (Sir) Henry Alfred Lytton [q.v.], actor. She married Henry Lytton in 1926 but from the outset the marriage was a failure, and in 1929 it was dissolved. Meanwhile, Jessie Matthews had obtained a £25,000 contract from Cochran, and in 1927 she starred with John Robert Hale Monro ('Sonnie Hale') in *One Dam Thing After Another*, finding in him the perfect dancing partner. In the next year they appeared together in *This Year of Grace* by (Sir) Noël Coward [q.v.], in which they sang Coward's romantic duet, 'A Room with a View'. The critics acclaimed her performance, which was followed by similar triumphs in Cole Porter's *Wake Up and Dream* (1929) and *Ever Green* (1930). She had now reached the peak of her theatrical career.

Sonnie Hale (died 1959), son of Robert Hale, actor, was married to Evelyn Laye, another highly successful actress. In 1930 they divorced, and Jessie Matthews received much unwelcome publicity as the woman responsible for the break-up of the marriage. Her own divorce had been finalized and in 1931 she and Hale married. In that year she made her first sound film, *Out of the Blue*, which was a failure, but her second, *There Goes the Bride* (1933), was a triumph, and led to her becoming Britain's first international film star. During the 1930s she starred in fourteen films, including *The Good Companions* (1933) opposite (Sir) John Gielgud, *Friday the Thirteenth* (1933) opposite (Sir) Ralph Richardson [q.v.], and *Evergreen* (1934), all directed by Victor Saville.

During the filming of *Evergreen* Jessie Matthews had her first nervous breakdown; many, more serious, were to follow. In 1934 her first baby, a son, only lived four hours; the doctors advised the desolate mother to adopt a child, and early in 1935 she and her husband adopted a baby girl, Catherine. In 1936 there was another serious nervous breakdown. In spite of Jessie Matthews's spectacular successes she was always beset by feelings of insecurity; at the beginning of her autobiography, *Over My Shoulder* (1974), she wrote: 'All my life I had been frightened.' She was now directed by her husband in *Head Over Heels* (1937) and feared it would be a failure; but it made money. *Gangway* (1937) and *Sailing Along* (1938), however, were disappointments, and relations with Sonnie Hale were becoming more and more strained.

Her only Hollywood film was *Forever and a Day* (1943).

The Hales returned to the stage in 1939 in their own musical production *I Can Take It*. Its provincial tour was a great success. It was due to open at the London Coliseum on 12 September 1939; war broke out on 3 September, and cancellation of the show meant financial disaster. In 1941 Jessie Matthews had an offer to appear on Broadway in *The Lady Comes Across*, and her husband urged her to accept. She reluctantly left him and Catherine, and set off alone for New York, but, before the show could open she was ill again and the play flopped. At the age of thirty-four her doctors predicted that her theatrical career was over. During her absence in America her husband was having an affair with Catherine's nurse, Mary Kelsey, and in 1942 he and his wife parted company; two years later they divorced.

Jessie Matthews resumed her stage career in the West End in Jerome Kern's *Wild Rose* (1942). While appearing in concerts with ENSA, she met Lieutenant (Richard) Brian Lewis, of the Queen's Royal Regiment, who was twelve years her junior; in 1945 they married. Brian Lewis was the son of Norman Percy Lewis, a schoolmaster, from West Hartlepool. Four months later Jessie Matthews had a stillborn son and her doctors warned her that another pregnancy would threaten her life. In 1948, after six years' absence, she reappeared on the London stage in *Maid to Measure*, followed in 1949 by the revue *Sauce Tartare*. She also appeared in *Pygmalion* (1950) and *Private Lives* (1954). She and Brian Lewis divorced in 1958.

She returned to films in *Tom Thumb* (1958), and demonstrated that she could still command an audience when she sang one of her well-known songs, 'Dancing on the Ceiling', in the 1960 *Night of One Hundred Stars*. By this time she had lost her sylphlike figure but not her charm. In 1963 the BBC invited her to take over the matronly role of Mrs Mary Dale in the radio serial *The Dales*. She played this part for the next six years. She also appeared frequently in television drama and returned to the stage in such plays as *The Killing of Sister George* (1971) and *Lady Windermere's Fan* (1978). In 1979 her one-woman show *Miss Jessie Matthews in Concert*, produced in Los Angeles, won the US Drama Critics award. Her last appearance was at the National Theatre, London, in *Night of One Hundred Stars* on 14 December 1980. She was appointed OBE in 1970. She died at Eastcote, Middlesex, 19 August 1981.

[*The Times*, 21 August 1981; Jessie Matthews and Muriel Burgess, *Over My Shoulder*, 1974; Michael Thornton, *Jessie Matthews*,

1974; David Shipman, *The Great Movie Stars: The Golden Years*, 1979; Jeffrey Richards, *The Age of the Dream Palace: Cinema and Society in Britain 1930-1939*, 1984.]

H. F. Oxbury

MAUD, JOHN PRIMATT REDCLIFFE, BARON REDCLIFFE-MAUD (1906-1982), public servant, was born 3 February 1906 at Bristol, the younger son and last of the six children of the Revd John Primatt Maud, then vicar of St Mary Redcliffe, Bristol, and later bishop of Kensington, and his wife, Elizabeth Diana, eldest daughter of Canon Charles Wellington Furse, archdeacon of Westminster, and rector of St John's, Smith Square.

He was educated at Summer Fields, Oxford; as a King's scholar at Eton (of which he was a fellow, 1964-76); and as a scholar at New College, Oxford, where he won a second in classical honour moderations (1926) and a first in *literae humaniores* (1928). In 1929, after a year studying economics at Harvard, he was appointed a junior research fellow in politics at University College, Oxford—the first full-time politics don to be appointed in Oxford—and made his speciality the subject of local government. Johannesburg invited him to write a history of its local government, and a visit in 1932 was to establish a lifelong link with South Africa. He became a tutorial fellow at University College in 1932, and dean in 1933.

In 1939 he was appointed master of Birkbeck College, London. The war broke out just as he took up his appointment, but he managed until 1943 to combine the mastership with work as a temporary civil servant, for a few months in the office of the regional commissioner for the Southern Region of England, and then in the Ministry of Food. When Lord Woolton (whose notice Maud later wrote for this Dictionary) became minister of food in April 1940, he chose Maud to be his principal private secretary. Maud's rise thereafter was rapid: to deputy secretary in 1941 and second secretary in 1944, when he moved with Woolton to the newly formed office of the Ministry of Reconstruction, at first as its deputy head and soon, when Norman Brook (later Lord Normanbrook, q.v.) became secretary of the cabinet, its head.

Maud was so evidently fitted for the public service that he was invited to continue in it after the war ended. Still only thirty-eight, he was appointed permanent secretary to the Ministry of Education in 1945. This appointment engaged his concern for education as well as his knowledge of local government, and called for all his qualities of administrative skill, persuasiveness, and enthusiasm. In this period he was one of the founding fathers of Unesco.

In 1952 Maud became permanent secretary

at the Ministry of Fuel and Power. The main thrust of policy during his time was to reduce dependence on coal, by increasing the substitution of oil for coal and developing alternative energy sources: for Maud the prospects of nuclear power were particularly exciting. In 1956 his attention was diverted for a time to the response to the immediate energy problems created by the closure of the Suez canal and the need to re-establish the administrative machinery for petrol rationing.

At the beginning of 1959 Harold Macmillan (later the Earl of Stockton) asked Maud to go to South Africa as British high commissioner. After South Africa became a republic and left the Commonwealth in 1961, he became the first British ambassador to the new republic. It was the time of Macmillan's 'wind of change' speech. Wholly out of sympathy with the South African government's apartheid policies, Maud remained scrupulously correct in his behaviour and pronouncements, without compromising his integrity or betraying his convictions, and came to be widely respected in that country. He combined with his diplomatic appointment that of high commissioner for Swaziland, Basutoland, and Bechuanaland, and worked hard to promote the constitutional, economic, and educational advancement of those territories.

In 1963 the wheel came full circle: Maud returned to University College, to serve as its master until his retirement in 1976. During this time he undertook the chairmanship of a succession of public inquiries on the organization, structure, and management of local government, above all the royal commission on English local government (1966-9); its recommendations for changing the structure of local government proved to be too radical for the 1970-4 government, whose own 'reorganization' Maud regarded as a child of no great beauty. But the college came first, and he devoted himself wholeheartedly to its interests and to the welfare of its members. It was not by mere coincidence that under his mastership University College became pre-eminent among Oxford colleges in reputation and success.

Maud's publications included *Local Government in Modern England* (1932), *City Government, the Johannesburg Experiment* (1938), *English Local Government Reformed* (with Bruce Wood, 1974), *Support for the Arts in England and Wales* (1976), *Training Musicians* (1978), and *Experiences of an Optimist* (1981).

Maud was (as he said the ideal permanent secretary should be) discreet, wise, entertaining, and incorruptible; and he had in high degree the traditional attributes of the distinguished public servant: intelligence, high-mindedness, dedication, and a capacity for sustained hard work. He believed in the power of enlightenment to hold back the tide of barbarism and in the capacity and duty of individual men and women to bring about progress. In these virtues and beliefs he was true to his Christian faith and to the value which he derived from his education. There were some who found all this, combined with his enjoyment of the rewards of his achievements, with his effortless charm and ease of manner, and with his seemingly inexhaustible zest, old-fashioned and even a little too good to be true. There were many others, more responsive to his qualities and in tune with his values, who looked to him for friendship, help, and inspiration, and were never disappointed.

He was a fine-looking man: tall, elegant, and distinguished. In his youth he loved and practised the art of acting, and his skill as a performer showed in his later facility as a public speaker and as a broadcaster: he had a gift for combining wisdom and wit in his matter with grace and felicity in his manner.

Music also meant much to him—as performer in his younger days as well as listener—though in this part of his life he played a supporting role to his wife Jean (the younger daughter of John Brown Hamilton, headmaster of St Mary's School, Melrose, Roxburghshire) who was, and continued to be, a professional pianist. They married in 1932; and from then on his success and his serenity were sustained by fifty years of close and happy family life. There were four children of the marriage: three daughters (one of whom died in 1941, the day before her fifth birthday) and a son, Humphrey, who had a distinguished career in the Diplomatic Service.

Maud was appointed CBE (1942), KCB (1946), and GCB (1955). He was given honorary degrees by the universities of Witwatersrand (1960), Natal (1963), Leeds (1967), Nottingham (1968), and Birmingham (1968). He was FRCM (1964) and an honorary fellow of New College (1964) and University College, Oxford (1976). He was created a life peer in 1967, taking the title of Redcliffe-Maud. He took an active part in the work of the House of Lords, particularly relishing the committee work, until very shortly before his death in Oxford 20 November 1982.

[Lord Redcliffe-Maud, *Experiences of an Optimist*, 1981 (memoirs); *The Times*, 22 November 1982; private information; personal knowledge.]   ROBERT ARMSTRONG

MAYER, SIR ROBERT (1879–1985), businessman, patron of music, and philanthropist, was born 5 June 1879 at Mannheim, Germany, the third of the four sons (there were no daughters) of Emil Mayer, hop merchant and later brewer, of Mannheim, and his wife Lucie Lehmaier, of Frankfurt. He was educated at Mannheim Gymnasium and Conservatoire. He displayed musical gifts from his earliest years, and was encouraged by an encounter with Johannes Brahms. Increasing distaste for Prussian militarism led Mayer's father to send him in 1896 to settle in Britain, where his first job was with a firm of stockbrokers. On leaving that he went into the non-ferrous metal business, in which he remained until 1929. He became a naturalized British citizen in 1902, and from 1917 to 1919 served in the British army.

In 1919 he married Dorothy Moulton, the daughter of George Piper, OBE, civil servant at the War Office, of London. They had a daughter and two sons, the elder of whom died in 1983. Dorothy was a soprano singer of considerable distinction who was notable for introducing to the public the work of young composers (particularly English ones) while they were still unknown. She encouraged Mayer to support music and in particular to promote the musical development of children. Following the example of Walter Damrosch's special concerts for children in America, they instituted the Robert Mayer Concerts for Children, the first of which was given in the Central Hall, London, on 29 March 1923. Mayer chose his conductors well: the first season was conducted by (Sir) Adrian Boult [q.v.]; and most of the seasons thereafter until 1939, when the concerts had to be suspended, were directed by (Sir) H. Malcolm W. Sargent [q.v.]. The combination of Sargent's musicianship and skill with the young audience, Dorothy's enthusiasm, and Robert's generosity and determination ensured that the Robert Mayer concerts became and remained an important institution in the musical life of London. They spread to a large number of provincial centres in the 1930s and made a significant contribution to the renaissance of music in England, as well as later in Ireland where Mayer supported his wife's foundation for the promotion of music.

In 1929 Mayer retired from a formal business career, his means being by now sufficient to fund his work for music. He was co-founder with Sir Thomas Beecham [q.v.] of the London Philharmonic Orchestra in 1932. In 1939 he was knighted for his services to music.

The concerts for schoolchildren started again after the war, which Mayer spent in the United States, and in 1954 Mayer established Youth and Music, an organization modelled on the continental Jeunesses Musicales and catering for young people from fifteen to twenty-five. Its main activity was to take blocks of seats at concerts and opera performances and make them available at affordable prices to groups formed in places of education and work. Thus many who had been introduced to orchestral music by the children's concerts were enabled to develop their appreciation of music in the years after they had left school.

In addition to these activities Mayer supported talented musicians in various ways: he would, for example, assist groups of players or singers to undertake concert tours abroad, or help promising students to continue a course of training when other support was not available.

In later years Mayer's philanthropy was not confined to musical causes. There were three threads that ran through it—music, young people, and the improvements of relations with citizens of other countries—and they were related in his mind: he saw music as a civilizing force in society and in international relations. He became interested in the problems of juvenile delinquency, and in 1945 published a book *Young People in Trouble*. He supported the Elizabeth Fry Fund, the International Student Service, the Children's Theatre, the Transatlantic Foundation, the Anglo-Israel Foundation, and many other such causes. In his nineties he was a strong supporter of the movement for British membership of the European community.

His wife died in 1974, when he was ninetyfive. It seemed at first as if all the light had gone out of his life. But his irrepressible energy and vitality triumphed over age and bereavement, and within a few months his small, brisk, neat—almost dapper—figure was to be seen in London concert halls and opera houses as often as ever, and his imagination was once again at work on plans for expanding the scope of Youth and Music. His hundredth birthday was celebrated by the publication of an autobiography confidently entitled *My First Hundred Years* (1979) and by a gala concert at the Royal Festival Hall in the presence of the Queen, who afterwards bestowed upon him the insignia KCVO.

Mayer was appointed CH in 1973. He was an honorary fellow or member of the Royal Academy of Music, the Royal College of Music, the Guildhall School of Music, and Trinity College London; and was given honorary doctorates at Leeds (1967), the City University, London (1968), and Cleveland, Ohio (1970). He was awarded the Albert medal of the Royal Society of Arts in 1979. The international dimension of his activities was recognized by the award of the grand cross in the Order of Merit in the Federal Republic of Germany (1967) and membership of the Order de la Couronne in Belgium (1969).

In 1980 he married Mrs Jacqueline Noble (née Norman), who cared for him with devotion through his last years of increasing frailty and withdrawal from public activity, until he died in London 9 January 1985 at the age of 105.

[*The Times*, 15 January 1985; private information; personal knowledge.]

ROBERT ARMSTRONG

MIERS, SIR ANTHONY CECIL CAPEL (1906–1985), submariner, was born at Birchwood, Inverness, 11 November 1906, the younger son and second of three children of Captain Douglas Nathaniel Carleton Capel Miers, Queen's Own Cameron Highlanders, who was killed in France in September 1914, and his wife, Margaret Annie Christie. He was educated at Stubbington House, Edinburgh Academy, and Wellington College whence he entered the navy as a special entry cadet in 1924.

He joined the Submarine Service in April 1929 where he made his mark as totally loyal, outstandingly keen, fearless, hot-tempered and incautiously outspoken. A prescient training officer wrote that he would either be awarded the VC or a court martial: in the event he got both, the latter in 1933 for self-confessedly striking a rating. Miers, who came to be known as 'Gamp' on the lower deck and 'Crap' by officers for reasons that have never been convincingly explained, was fiercely competitive at sea, ashore, and on the playing field. He was a good athlete, a tennis and squash player, and a fine rugby footballer. His vigorous singleminded aim at all times was to win. He was not a good loser; but his sheer determination to beat the opposition was to overcome all obstacles in war. The men who served him had complete confidence in him and forgave his impetuous outbursts; but, undeniably, those who sought a calmer life avoided his company. Nobody could be indifferent to his presence. He was wholeheartedly supportive to subordinates, provided they were efficient and stood up to him, albeit at some risk of fisticuffs.

His first submarine command was *L54* (1936–7). Then, after a spell of general service in the battleship *Iron Duke*, he qualified on the staff course (1938) before joining, as a lieutenantcommander, the seagoing staff of the C-in-C Home Fleet (1939–40), where he was mentioned in dispatches. He returned to submarines in 1940 to command the new HMS *Torbay* on 12 November. It was in this submarine that Miers gained lasting fame for varied patrols in the Mediterranean over the next two intensely active years. He was promoted to commander in December 1941, again mentioned in dispatches, decorated twice with the DSO (1941 and 1942), and finally invested with the VC (1942).

The VC was richly deserved. On 20 February 1942 *Torbay* sailed from Alexandria for her tenth patrol. Bad luck dogged Miers for several days and his frustration reached full fury when, out of position due to an earlier fruitless hunt, he was unable to close on an important troop ship convoy. Angry with himself, he coldly decided to attack the convoy at its destination—within the well-protected enemy harbour of Corfu Roads. Accordingly, he took *Torbay* right inside the harbour and remained in this highly hazardous situation for some seventeen hours. In fact, by the time he got there, the convoy had departed; but two supply ships lay at anchor and these went down to his torpedoes.

In 1941, off Crete, Miers ordered the machine-gunning of German soldiers who had taken to a rubber float while their caique was sunk by *Torbay* crewmen with a demolition charge. Accounts of this controversial incident vary; but, whether the shooting can be justified in retrospect or not, Miers was always determined to neutralize as many of the enemy as possible; and, in this case, he reported that he was preventing them from regaining their ship. It has otherwise been said that he was intent on the soldiers taking no further part in the war—either in some kind of action against his own submarine or on the nearby shore against the remaining Allied troops there.

In December 1942 Miers was sent as submarine liaison officer to the American Pacific Fleet where he was made an officer of the Legion of Merit. In July 1944 he became commander S/M of the eighth submarine flotilla based on Perth, Western Australia; and here, on 20 January 1945, Miers married Patricia Mary, daughter of David McIntyre Millar, of the Chartered Bank of India, Australia, and China, who was serving in the WRANS. They had a daughter and a son.

Miers was promoted to captain in December 1946 and, gaining a pilot's A licence, went on to command the naval air station HMS *Blackcap* (1948–50), HMS *Forth* and the first submarine squadron (1950–2), the RN College at Greenwich (1952–4), and the aircraft carrier HMS *Theseus* (1954–5). In 1955 he was made a burgess and freeman of Inverness and in 1966 a freeman of the City of London. He was promoted to rear-admiral on 7 January 1956 and became flag officer, Middle East, until he retired on 4 August 1959. He was appointed CB in 1958 and KBE in 1959.

Thereafter Miers maintained a large number of business, sporting, and charitable interests. Most notably, he joined National Car Parks as director for development co-ordination in 1971; he became president of the RN Lawn Tennis Association and of the RN Squash Racket Association; and for many years he was national president of the Submarine Old Comrades' Association. He died at his home in Roehampton 30 June 1985.

[William Jameson, *Submariners VC*, 1962; Paul Chapman, *Submarine Torbay*, 1989; records in RN Submarine Museum, Gosport; private information; personal knowledge.]

RICHARD COMPTON-HALL

MILFORD, DAVID SUMNER (1905–1984), world open rackets champion and international hockey player, was born in Oxford 7 June 1905, the second of three children and younger son of (Sir) Humphrey Sumner Milford [q.v.], publisher, and his first wife, Marion Louisa, daughter of Horace Smith, metropolitan police magistrate. He was educated first at West Downs School, Winchester, where it was said he never missed the chance of hitting a ball against a wall with whatever weapon came to hand; then at Rugby, where he not only began his remarkable career as a games player but also won the public schools rackets championship with C. H. Goodbody. He went on to New College, Oxford, where he obtained a third class in classical honour moderations (1926) and a fourth in *literae humaniores* (1928). In 1928 he joined the staff at Marlborough, where he taught Latin and geography for thirty-five years.

In 1930 he first became British amateur rackets champion; he went on to win this championship seven times. In all he won the British amateur doubles eleven times, ten of them partnered by John Thompson, a colleague of his at Marlborough. He also won the Noel Bruce old boys championship on twelve occasions: four times with Cyril Simpson before World War II and eight times afterwards with Peter Kershaw. But his real triumph was in 1937 when he won the world championship in New York and at the Queen's Club, London, defeating the American challenger Norbert Setzler by 7 games to 3. He held the title until 1947 when he retired.

At rackets he was outstanding. He had a good eye and fleetness of foot, and extraordinary anticipation. He could fashion unorthodox strokes, in particular a deadly angled dropshot, which is most difficult to achieve when the ball moves as fast as it does at rackets. He often paid tribute to Harry Gray, his coach at Rugby, and Walter Laurence, the Marlborough professional.

When Milford arrived at New College, he played hockey and was at once recognized as a player of promise. He first got his blue in 1927, playing in three successive university matches against Cambridge. In 1930 he was chosen to play for England at inside left; later occasionally he played at centre forward, and became an automatic choice for the side. The qualities he had at rackets he showed too on the hockey field. He was extremely energetic and hard working: he seemed to be all over the field and available to take a pass from which he usually scored a goal. His splendid stickwork and ball control were confusing to his opponents and he was almost at his best when the ground was wet or bumpy. When at Marlborough he wrote two books on hockey which in their day were standard works on the game.

He was a good cricketer—a slow left-arm bowler—and a very useful middle-order batsman. Once when playing for Wiltshire against Dorset he made 150 on a dangerous wicket, when the rest of the side was struggling to reach double figures. He also played lawn tennis for Wiltshire for many years.

Milford was five feet eleven inches in height, very slim and wiry, with good hands and wrists. His lack of physique often caused comment. Once while visiting an osteopath about a strained shoulder, when he was at the height of his powers, the man looked him over and suggested that he should think of taking up some game to improve his strength. On another occasion a club secretary watching him play golf remarked: 'I don't believe that boy has a joint in his body.'

He was a person of many interests. He was an excellent bridge player because he was quick-minded and also enjoyed a gamble. He sang madrigals with local enthusiasts and later in life took up bird-watching. Like many good games players he was modest. He was also rather shy: he disliked pushing himself forward, or speaking in public, which was perhaps unusual for a schoolmaster. Nevertheless he was friendly and good-tempered.

In 1930 he married Elizabeth Mary, a granddaughter of the composer Sir John Stainer [q.v.] and daughter of John Frederick Randall Stainer, chief inspector at the Passport Office. They had a daughter (Marion, a professional singer) and three sons. Milford died 24 June 1984 after playing three sets of tennis.

[Information from John Thompson, Peter Kershaw, J. B. H. Bisseker, and Milford's family; personal knowledge.]

D. M. Goodbody

MITCHELL, Sir GODFREY WAY (1891–1982), leading figure in the construction industry, was born at New Cross 31 October 1891, the younger son (there were no daughters) of Christopher Mitchell, stone mason and quarry owner, and his wife, Margaret Way, of Weymouth. He was educated at Haberdashers' Aske's School, Hatcham. Before World War I Mitchell worked in Rowe & Mitchell, his father's quarry business in Alderney, from which stone was exported to Britain. At the outbreak of war he joined the Royal Engineers; he later commanded a quarry company in the Pas de Calais, and was demobilized with the rank of

captain. When he returned to the business, the terms of trade had changed adversely, causing him to seek other opportunities. With his war gratuity and a £3,000 loan from his father, in 1919 he purchased a small, insolvent, general works contracting business from George Wimpey, of Hammersmith.

He built up the civil engineering side of George Wimpey & Co. Ltd. and in 1927 started private house building, quickly becoming one of the largest London-based house builders. With the advent of rearmament, the company's contracting organization further considerably expanded. During World War II the company carried out successfully some of the largest of building and civil engineering projects. It built ninety-eight airfields throughout Britain, besides barrage balloon stations, docks, and army camps. It was this for which Mitchell was knighted in 1948. He had also served on several ministry committees during the war. For a time he was a controller of building materials at the Ministry of Works.

At the end of the war Mitchell was concerned about how he was to re-employ several thousand ex-service employees on demobilization, and this was a major consideration in both the setting up of a number of regional offices and the creation of a highly efficient overseas contracting organization which also embraced electrical and mechanical engineering and plant erection. At its peak, overseas work accounted for over 33 per cent of total turnover.

Mammoth projects were carried out in the early 1950s in various parts of the world (particularly the Middle East), probably the most renowned being the BP Aden oil refinery, in partnership with Bechtel of California. Concurrently the regionalization of the home-based construction activity proved to be most successful, and the company developed the 'no fines' system of house building for local authorities. It built more than three hundred thousand dwellings throughout Britain.

Moreover, its road surfacing activity became Wimpey Asphalt and grew apace. The problem was not how to re-employ ex-service personnel, but how to obtain a sufficiently able and mobile labour force. Sophisticated but highly practical training at all levels up to graduate engineering was undertaken. Many thousands of people in construction were justly proud of being 'Wimpey trained'. As demand for local authority housing dropped, the company increased its private housing activities and, by the time of Mitchell's death, it was the major private house builder in Britain. Mitchell was chairman of his company from 1930 to 1973, executive director from 1973 to 1979, and life president from 1979. In 1948 he was chairman of the Federation of Civil Engineering Contractors, and also master of the Worshipful Company of Paviors. He did much to revive this livery company after World War II. He was an honorary fellow of the Institution of Civil Engineers (1968) and the Institute of Building (1971).

In any discussion Mitchell, with his quick incisive mind, identified the problem, the causes, and the potential solution. He was very well read, particularly in philosophy, and enjoyed discussions in depth on social, economic, philosophical, and religious matters. Although an admitted agnostic, he endeavoured quietly to live up to the highest of Christian principles. If he thought any friend, acquaintance, or employee was not 'playing the game', he told them so. One of his regular sayings was: 'You must always put more into the pot than you take out, for the sake of the other fellows.' He was an ascetic; a non-gregarious, but humorous family man, who was much at home with the young, whom he encouraged. He was modest and unostentatious and was uninterested in personal power and wealth. In particular he could not abide humbug and cant. He lived simply, for the love of his family, his business, his country, and, in his latter years, his charitable trust (the Tudor Trust) which at the time of his death controlled the company which, together with a band of loyal colleagues, he had built.

In 1929 he married Doreen Lilian (died 1953), daughter of Ernest Mitchell, a Melbourne (Australia) civil servant. They had two daughters. Mitchell died at his home in Beaconsfield 9 December 1982.

[*The Times*, 10 December 1982; private information; personal knowledge.]

MAURICE LAING

MITCHELL, GRAHAM RUSSELL (1905–1984), deputy director-general of the Security Service, was born 4 November 1905 in Broom House, Warwick Road, Kenilworth, the only son and elder child of Alfred Sherrington Mitchell, a captain in the Royal Warwickshire Regiment, and his wife, Sibyl Gemma Heathcote. He was educated at Winchester, of which he was an exhibitioner, and at Magdalen College, Oxford, where he read politics, philosophy, and economics. In spite of suffering from poliomyelitis whilst still at school he excelled at golf and sailed for his university. He was also a very good lawn tennis player, and won the Queen's Club men's doubles championship in 1930. He played chess for Oxford, and was later to represent Great Britain at correspondence chess, a game at which he was once ranked fifth in the world. He obtained a second class honours degree in 1927.

His first job was as a journalist on the *Illustrated London News* but his only credited article, which appeared in the 12 October 1935 edition, was entitled 'What Was Known About Abyssinia in the Seventeenth Century—A Detailed Account in a Geography of 1670'. Thereafter he joined the research department of Conservative Central Office which was then headed by Sir G. Joseph Ball [q.v.].

Mitchell's bout of polio had left him with a pronounced limp and when war broke out in September 1939 he was considered unfit for military service. Instead he joined the Security Service, MI5, in November. Exactly who sponsored his recruitment is unknown although Sir Joseph Ball, who was later to be appointed deputy to Lord Swinton [q.v.] on the top secret Home Defence Security Executive, was sufficiently influential and well connected in security circles to have assisted his entry into the organization.

Mitchell's first post in MI5 was in the F3 sub-section of F division, the department responsible for monitoring subversion headed by (Sir) Roger Hollis [q.v.]. F3's role was to maintain surveillance on right-wing nationalist movements, the British Union, German and Austrian political organizations, and individuals suspected of pro-Nazi sympathies. One of Mitchell's first tasks was to assist his immediate superior, Francis Aiken-Sneath, to investigate the activities of Sir Oswald Mosley [q.v.] and collate the evidence used to support his subsequent detention.

At the end of the war Mitchell was offered a permanent position in the Security Service and was promoted to the post of director of F division, where he remained until 1952 when he was switched to the counter-espionage branch. While in charge of D branch Mitchell led the team of case officers pursuing the clues of Soviet penetration left by Guy Burgess and Donald Maclean [q.v.], the two diplomats who defected to Moscow in May 1951. At the same time he was 'one of the chief architects of positive vetting', the screening procedure introduced in Whitehall to prevent 'moles' from penetrating the higher echelons of the Civil Service.

In 1956 Roger Hollis succeeded Sir Dick White as director-general of MI5 and selected Mitchell as his deputy. He remained in this post until September 1963 when he unexpectedly took early retirement. It was later revealed that at the time of his departure Mitchell was himself under investigation as a suspected Soviet spy. The evidence accumulated against Mitchell was all very circumstantial, and centred on the poor performance of MI5's counter-espionage branch during the 1950s. During this period MI5 experienced a number of set-backs, failed to attract a single Soviet defector, and only caught one spy on its own initiative.

During the last five months of his career

Mitchell was the subject of a highly secret and inconclusive 'molehunt' which was eventually terminated when he was brought back from retirement to face interrogation. This gave him the opportunity to answer his accusers, but did little to end the debilitating atmosphere of suspicion that at one point threatened to paralyse the entire organization, and resulted in Roger Hollis himself being accused of having spied for the Russians. Both Mitchell and Hollis strenuously denied having been traitors, leaving the whole question of the identity of the KGB's master spy, if indeed there was one, unresolved.

Tall, stooped, and habitually wearing tinted glasses, Mitchell cut a lonely figure with solitary interests, like chess puzzles and *The Times* crossword. In 1934 he married Eleonora Patricia, daughter of James Marshall Robertson, gentleman. They had a son and a daughter. Mitchell died at his home in Sherington, Buckinghamshire, 19 November 1984.

[*The Times*, 3 January 1985; *Illustrated London News*, 12 October 1935; Nigel West (Rupert Allason), *Molehunt: The Full Story of the Soviet Spy in MI5*, 1987.] Nigel West

MITCHELL, JAMES ALEXANDER HUGH (1939–1985), publisher, was born 20 July 1939 at Great Hallingbury in Essex, the youngest of three sons (there were no daughters) of William Moncur Mitchell, a solicitor, and his wife, Christine Mary Browne. The family subsequently moved to Cirencester, Gloucestershire, from where Mitchell and his brothers Julian (subsequently a novelist and playwright) and Anthony went to Winchester College. At Winchester James's principal achievement, as he recalled, was to found the Winchester College Astronomical Society and to persuade no less than 400 of the 480 boys to join. He was a highly persuasive person.

Mitchell went to Trinity College, Cambridge, and obtained a third class in both part i of the economics tripos (1958) and part iA of the theological tripos (1960). At that time, and subsequently, he considered taking holy orders, but was dissuaded by an entrepreneurial instinct that led him, via a short spell as a sales assistant at Hatchards bookshop in Piccadilly, into publishing. In 1961 he became an editor at Constable & Co., and in 1967 he was appointed editorial director of Thomas Nelson & Sons in Park Street, London. At Nelsons he met his future business partner, the production director, John Beazley, and several authors and editors who were to become important to him when he started his own publishing firm, notably Nicolas Bentley [q.v.], John Hedgecoe, and the present writer.

In 1969 Mitchell and Beazley opened a tiny office in Goodwin's Court off St Martin's Lane with the notion that they would publish only bestsellers. One a year, they reckoned, would do. The press was full of excitement about the imminent American moon flight programme. Mitchell contacted the astronomer Patrick Moore (whom he had met at Winchester) and asked him to write the *Moon Flight Atlas* (1969). Moore's text combined with the United States Space Agency pictures to make a historic and authoritative account which was ready for the printer by the time the astronauts returned to earth. It sold 800,000 copies. The next Mitchell Beazley bestseller was Patrick Moore's *The Atlas of the Universe* (1970), and the next Hugh Johnson's *The World Atlas of Wine* (1971). The firm's bias towards atlases was partly due to the financial backing of George Philip & Son.

The design principle of these books was largely the work of Peter Kindersley (who had also worked at Thomas Nelson & Sons and who left Mitchell Beazley in 1974 to found his own firm). Their inspiration, and the sales technique that made them possible, were Mitchell's own. He offered each book as a completely designed package intended for international consumption. A foreign publisher only had to translate the text (and an American not even that). All the colour printing was done together, at considerable saving. Such was Mitchell's persuasiveness and Beazley's organization that initial international orders for apparently specialist reference books made them bestsellers before publication. A very high standard of illustration and graphics was supported by densely factual text. *The World Atlas of Wine* became the most discussed and most imitated reference book of the 1970s among rival publishers. By the end of the decade it had sold one million copies. In 1975 Mitchell Beazley Ltd. was given a Queen's award to industry for success in exporting.

Successful reference book publishing (particularly in the fields of astronomy, gastronomy, and photography) led Mitchell to conceive an illustrated encyclopaedia. *The Joy of Knowledge* was a vastly ambitious project, for which he assembled a formidable team of experts and commissioned an 'art-bank' of illustrations without parallel. It was published, in eight thematic volumes, with a two-volume 'Fact Index', in 1977–8. Mitchell was a principal editor, and sold the rights for translation into twenty-three languages, among them Chinese and Arabic.

For this huge undertaking he had to expand Mitchell Beazley to employ nearly 200 people. It was an unsustainable effort, for all the success of the encyclopaedia and despite Mitchell's further successful reference book innovations (of which the 'pocket book' has proved the most durable) and such bestsellers as Alex Comfort's *The Joy of Sex* (1974). John Beazley died of cancer in 1977. In 1980 Mitchell sold the firm to American Express Publishing, retaining the chairmanship. The experience was not a happy one and in 1983 he bought it back again.

Throughout his career Mitchell remained an essentially boyish character, given to fits of boisterous, even irresponsible, enthusiasm. He became bald in his thirties, and his plump and shiny face exuded a charm that was sometimes wily, but always high-spirited. His best friends, among them the Revd Harry Williams, whose devotional books he published, knew him to be far more introspective than he appeared. His favourite recreation was fly fishing on the Avon near his greatly enlarged 'cottage' at Wilsford.

In 1962 Mitchell married Janice Page, daughter of Jack Gunn Davison, a Lloyd's broker. They had three sons, Charles, Alexander, and Edward. Mitchell died 12 March 1985 at 110 Regent's Park Road, London NW1.

[Personal knowledge.]   HUGH JOHNSON

MITCHELL, LESLIE SCOTT FALCONER (1905–1985), actor, pioneer television announcer, and commentator, was born 4 October 1905 in Edinburgh, the only child of Charles Eric Mitchell, caterer, of Edinburgh, and his wife, Leslie Florence Whittington, née Lowe. He had a clouded childhood. His parents separated. His beautiful but wayward mother (whose three marriages were all to end unhappily and who twice attempted suicide) went on a holiday to the United States early in World War I and found herself unable to return. He was told that his absent father had died in battle. He was left to be brought up by the novelist W. J. Locke [q.v.] and his wife. Locke became a trusted mentor, and arranged for Mitchell to be educated at King's School, Canterbury. He was destined for the navy, but ill health, which dogged him for most of his life, thwarted that plan. He completed his education at Chillon College on Lake Geneva.

In 1923 the first of two unsympathetic stepfathers found him a job as a trainee stockbroker. However, after a short period he tried the stage. He had not had any professional dramatic training, but striking good looks and a strong, clear voice, aided by some useful introductions from a number of theatrical friends, secured him some small parts in which he rapidly demonstrated natural ability as an actor. Between 1923 and 1925 he toured Britain with the Arts League and for the next three years acted in various West End productions. In 1928, after a successful tour in *Flying Squad* by R. H. Edgar Wallace [q.v.] he suffered a major motor-cycle

accident. His multiple injuries were treated by many operations. Plastic surgery to his face, a rebuilt jawline, a damaged eye, and the fitting of a brace to his leg kept him away from work for more than a year.

Eventually he returned successfully to the West End stage. He narrowly missed being the first to play Captain Stanhope, the leading part in *Journey's End* by R. C. Sherriff [q.v.], but he later acted the part in London, in a BBC broadcast on armistice day, and on tour in South Africa. In Johannesburg he happened to meet the father he believed had been killed seventeen years earlier.

In 1934, after more than a decade of acting, Mitchell joined the BBC as a general announcer and soon transferred to compèring dance-band music. He was chosen from 600 applicants as the male announcer to launch the world's first public service of high definition television at Alexandra Palace on 2 November 1936. *The Times* referred to Mitchell's 'very successful transmissions'. He was charming and handsome, despite the injuries of the motor-cycle accident, and always immaculately groomed, usually with a red carnation in his buttonhole. He soon became a favourite with the pre-war television audience.

Mitchell conducted some twenty interviews a week for *Picture Page*, a television magazine. Before the outbreak of World War II, which was to suspend television for the duration, he resigned from the BBC to become the British Movietone news commentator and a free-lance broadcaster. He enlisted in the Home Guard and in addition to being the encouraging voice of the cinema news-reels he worked on the Allied Expeditionary Forces Radio. His innumerable wartime broadcasts ranged from the *March of the Movies* to the *Brains Trust*. After the war he visited the United States to study publicity methods, and on his return he was appointed publicity director to the film-maker Sir Alexander Korda [q.v.].

In 1948 he returned to his free-lance activities as a writer, commentator, and producer. He was a television commentator for the wedding of Princess Elizabeth and the Duke of Edinburgh, and for the silver wedding of King George VI and Queen Elizabeth. In 1951, when the political parties first tentatively used television for publicity, he interviewed R. Anthony Eden (later the Earl of Avon, q.v.) in the Conservatives' one general election programme. He was prominent in the launch of commercial television in London in 1955, and in Birmingham a year later. In 1981 he produced an entertaining autobiography *Leslie Mitchell Reporting . . .* . In 1983 he was made the first honorary member of the Royal Television Society and in 1984 a freeman of the City of London in recognition of his wartime contribution to the morale of Londoners.

In 1938 Mitchell married Phyllis Joan Constance, a young widow whose husband Anthony Wood had died after less than a year of marriage. She was the daughter of the London impresario Firth Shephard. In 1965 Phyllis died and in 1966 Mitchell married Inge Vibeke Asboe, daughter of Niels Andreas Jørgensen, a merchant, of Aarhus, Denmark. There were no children of either marriage. Deteriorating health cast a shadow over Mitchell's last years and he died in Paddington Community Hospital, London, 23 November 1985.

[*The Times*, 3 November 1936 and 25 November 1985; Leslie Mitchell, *Leslie Mitchell Reporting . . .* , 1981; private information; personal knowledge.]   LEONARD MIALL

MONCREIFFE OF THAT ILK, SIR (RUPERT) IAIN (KAY), eleventh baronet (1919–1985), herald and genealogist, was born 9 April 1919, in the parish of Hampton Court, Middlesex, the only child of Lieutenant-Commander (Thomas) Gerald (Auckland) Moncreiffe, RN, and his wife, Hinda, daughter of Frank Meredyth, styled Comte François de Miremont. His father, who had settled in Kenya as a coffee planter before World War I, died in 1922; his mother left her son's upbringing to nurses and later to uncles, aunts, and elderly relatives in London and Scotland. He was educated at Stowe School, briefly at Heidelberg University, and at Christ Church, Oxford. He served in the Scots Guards throughout World War II, saw much active service, attained the rank of captain, and was injured in Italy. After the war he became an attaché in the British embassy in Moscow, but soon returned to study Scots law at Edinburgh University (LLB, 1950), being admitted to the Faculty of Advocates in 1950. His legal practice was small but his distinction in peerage and related matters enabled him to take silk in 1980.

From an early age he had displayed a strongly genealogical and heraldic cast of mind, which derived in part from the great antiquity of his own family and partly from wide historical interests that took his researches far beyond Scots ancestry and gave him an enviably synchronous view of historical development. These instincts were emphasized by his marriage, in 1946, to Diana Denyse ('Dinan' or 'Puffin') Hay (died 1978), in her own right Countess of Erroll and hereditary lord high constable of Scotland, and later by his inheritance in 1957 from a cousin of the baronetcy of Moncreiffe of that Ilk ('of that Ilk' means 'of that same place').

He joined the Court of Lord Lyon King of Arms as Falkland Pursuivant in 1952, and

became Kintyre Pursuivant (in ordinary) from 1953, Unicorn Pursuivant from 1955, and Albany Herald from 1961. Although the chief office of the Lyon Court eluded him, to his disappointment, he was widely recognized as an expert authority. He was appointed CVO in 1980.

Such resounding titles appealed to the popular press, which increasingly came to regard him as an (eminently quotable) super-snob, and he teasingly played up to this designation in an unguarded way that did little justice to his wide view (based on his deep understanding of Scottish clan history) of the importance of individual ancestry and of the universal bonds of tradition and continuity, themes which were well displayed in his writings.

In collaboration with the heraldic artist Don Pottinger he published *Simple Heraldry* (1953), a considerable success that was followed by *Simple Custom* (1954) and *Blood Royal* (1956). Each was 'cheerfully illustrated', but their lightness of manner disguised a great deal of sound learning that was also shown in later writings, including the excellent *The Highland Clans* (1967) and many discursive book reviews. To this Dictionary he contributed the notice of Sir Thomas Innes of Learney. Most of his best work, however, remains unpublished, including his Edinburgh University Ph.D. thesis (1958) on 'Origins and background of the law of succession of arms and dignities in Scotland'. He gave many expert opinions in peerage cases, and was frequently consulted, formally and informally.

Moncreiffe was a well-known figure in London clubland, and in Edinburgh founded his own club, Puffin's (named after his first wife), as a resort for Scottish country gentlemen. Quite short, with hair latterly almost white and a bristling moustache curved like a tilde, there was nothing overbearing in appearance or manner about the quietly spoken baronet and chieftain. His neat little handwriting was described by a friend as 'like the footprints of a wren'. In conversation, as well as in his published work, he relied on a marvellously retentive memory that was unimpaired even by a considerable intake of alcohol. Conviviality that could prostrate others left him in full command of detail and only increased his extraordinary range of allusiveness.

His first marriage was dissolved in 1964, and in 1966 he married Hermione Patricia, daughter of Lieutenant-Colonel Walter Douglas Faulkner, Irish Guards, and his wife, Patricia Katherine, later Countess of Dundee. He had long lived at Easter Moncreiffe, near Perth, on lands held by his family from time immemorial, but it was at his London flat, 117 Ashley Gardens, SW1, that he died 17 February 1985. By his first marriage he had two sons and a daughter. His elder son, Merlin Sereld Victor Gilbert (born 1948), twenty-fourth Earl of Erroll, succeeded to the baronetcy; to the younger, Peregrine, devolved the style of 'Moncreiffe of that Ilk', which his father had done so much to enhance in the memory of a wide public.

[John Jolliffe (ed.), *Sir Iain Moncreiffe of that Ilk, an Informal Portrait*, 1986; Hugh Montgomery-Massingberd (ed.), *Lord of the Dance: a Moncreiffe Miscellany*, 1986; personal knowledge.]   ALAN BELL

MONRO, MATT (1930-1985), singer of popular music, was born Terence Parsons 1 December 1930 in Shoreditch, London, the youngest in the family of four sons and one daughter of Frederick Parsons, druggist packer, and his wife, Alice Mary Ann Reed. He began singing when he joined the army in 1947, at the age of seventeen. He became a tank instructor and divided his time between tanks and talent contests while serving in Hong Kong. It was there that he decided to become a professional singer. After demobilization in 1953 he became a long-distance lorry driver, electrician, coalman, bricklayer, stonemason, railway fireman, layer of kerbstones, milkman, baker, offal boy in a tobacco factory, plasterer's mate, builder's mate, and general factotum in a custard factory. Using the name Al Jordan, he took a semi-professional job with Harry Leader and his orchestra. It meant months of travelling from town to town. Eventually he decided to abandon the work and became a London bus driver instead. However, the desire to become a fully professional singer persisted and he recorded a demonstration disc of 'Polka Dots and Moonbeams' with a small rhythm section. One of its members was so impressed it was forwarded to the pianist Winifred Atwell who arranged a number of important meetings. At this point he decided on a change of name. Terence Parsons became Matt Monro—'Matt' from Matt White, the first journalist to write about him, and 'Monro' from the first name of Winifred Atwell's father.

For the newly named Matt Monro there followed a series on Radio Luxembourg in 1956, a regular singing spot with Cyril Stapleton's show band, and a recording contract with Decca. However, his career took off almost by accident. The record producer George Martin, of EMI/Parlophone, was at this time making a name for himself in the comedy record field, and was looking for someone to sound like Frank Sinatra for an LP for Peter Sellers [q.v.], *Songs for Swingin' Sellers*. He chose Matt Monro with his rich, clean-cut baritone voice. Everyone connected with the recording was impressed,

especially Martin who asked if he would like to record under his new professional name (on the LP he had used the name Fred Flange).

Monro's first single, 'These Things Happen', was followed by 'Love Walked In'. His third recording, 'Portrait of my Love', which he thought one of the most uncommercial songs he had ever heard, entered the list of British best-selling records in December 1960 and reached no. 3. This was followed by 'My Kind of Girl' which climbed to no. 5 in 1961, 'Why Not Now?/Can This Be Love?', 'Gonna Build a Mountain', 'Softly as I Leave You', 'When Love Comes Along', 'My Love and Devotion', 'From Russia with Love', 'Walk Away', 'For Mama', 'Without You', and 'Yesterday'. He was to achieve his final singles record success in Britain in 1973 with 'And You Smiled'. Oddly, 'Born Free' and 'We're Gonna Change the World', two of his most requested records on radio programmes, never achieved success in the British best-seller lists, although the former won an Academy Award for the best song in a motion picture in 1965. Amongst his best known albums were *Walk Away, I Have Dreamed, My Kind of Girl, The Late Late Show*, and *Softly*. Monro came second in the Eurovision Song Contest in 1964 and was voted best male singer in England in 1965.

Monro spent a considerable amount of time in the United States of America, where he turned increasingly to cabaret. His first visit was in 1960 on a special exchange agreement. He sang at the Pentagon in Washington, while Ella Fitzgerald appeared in Great Britain. In 1966 Monro signed with Capitol Records and he resided in the United States during 1967. He was a constant traveller, appearing in cabaret and concerts in Australia, New Zealand, Japan, Hong Kong, the Philippines, Malaysia, Canada, South Africa, Scandinavia, and most European countries. He travelled approximately 150,000 miles a year.

Monro's hobbies were golf, the cinema, and watching television programmes, especially westerns. He once said 'the worst fate that can befall me is to be stranded in a town without a television set'. His favourite film was *The Magnificent Seven*. He was a man of great natural charm and was highly respected by his colleagues, including Frank Sinatra, Tony Bennett, and Bing Crosby, who all regarded him as 'a singer's singer'. According to his old friend, George Martin, he 'had the rare gift of getting to the heart of a lyric and delivering it in such a way that it became a personal message to his audience'. Monro is rightly considered one of the best singers of popular music Britain has ever produced.

In 1955 Monro married Iris, daughter of Frederick Jordan, factory wallpaper dyer. A son,

Mitchell, was born the same year. The marriage was dissolved in 1959 and in the same year Monro married Mickie, daughter of Adolph ('Dolly') Schuller, dentist. They had a daughter Michele in 1959 and a son Matthew in 1964. Monro died in the Cromwell Hospital, Kensington, London, 7 February 1985.

[Private information; personal knowledge.]
DAVID JACOBS

MONTAGU, EWEN EDWARD SAMUEL (1901–1985), judge and deception planner, was born 29 March 1901 in Kensington, the second of the three sons (there was also a younger daughter) of Louis Samuel Montagu, banker, who succeeded his father Samuel Montagu [q.v.], founder of the family firm, as second Baron Swaythling in 1911. His father's younger brother was E. S. Montagu [q.v.]; his own younger brother Ivor Montagu [q.v.] became a leading communist intellectual. His mother was Gladys Helen Rachel, daughter of Colonel Albert Edward Williamson Goldsmid. The children were brought up at Stoneham House, now swallowed by the suburbs of Southampton, in lavish circumstances. He was at Westminster School during the war of 1914–18, and spent a year at Harvard before going up to Trinity College, Cambridge, in 1920. There he gained second classes (division II) in both part i of the economics tripos (1922) and part ii of the law tripos (1923).

Just before he went down in 1923 he married Iris Rachel, daughter of Solomon Joseph Solomon [q.v.], the portrait painter; her first cousin had married (Sir) Victor Gollancz [q.v.]. They had a son and a daughter; all three survived him. In 1924 he was called to the bar from the Middle Temple, and after fifteen years' hard work on the Western circuit took silk in 1939.

He spent many holidays yachting, and his interest in small boats led him to a commission in the Royal Naval Volunteer Reserve that September. As a naval intelligence officer at Hull he attracted the attention of Rear-Admiral J. H. Godfrey [q.v.], who summoned him in 1941 to join the naval intelligence division at the Admiralty. There Montagu ran a highly secret sub-branch, NID 17(M), which handled counter-espionage. He accompanied Godfrey to the W board, the informal committee that ran the most secret intelligence war, and sat as the naval member of the XX committee which supervised the playing-back of captured Abwehr agents. Extreme caution and extreme daring had to be combined for this work to be effective; Montagu displayed both.

The best known of several coups he organized was Operation Mincemeat. This involved the floating ashore on 30 April 1943 in south-west Spain of what appeared to be the body of a Royal Marine officer, carrying documents which indicated an imminent Allied attack on Sardinia rather than Sicily. Through the deciphering service, Montagu was able to trace enemy reactions; this ploy reached at least as high as Admiral Canaris, head of the Abwehr, and the chiefs of staff telegraphed to the absent prime minister 'Mincemeat swallowed whole.' No one can quantify the results of such work, but it undoubtedly saved thousands of lives.

At the end of the war Montagu returned to law, becoming judge advocate of the fleet in 1945, a post he held till 1973. Even before the war ended, he had been appointed recorder of Devizes (1944–51); he was then recorder of Southampton for nine years more, becoming deputy lieutenant in that county in 1953. He served also for fourteen years as assistant chairman, deputy chairman, and chairman of quarter-sessions for Middlesex, and became a bencher of the Middle Temple in 1948 and treasurer in 1968.

This work was interrupted in 1952. A novel by A. Duff Cooper (later Viscount Norwich, q.v.) was based on Operation Mincemeat, of which Cooper had had official knowledge. Word leaked out that the story might be true; important secrets were held to be at risk. The government was worried that too much would come out, both about the existence and about the methods of the deception service. The Admiralty appealed to Montagu. In forty-eight hours, from a Friday evening to a Sunday evening, he dictated an account that was at once accurate, exciting, and circumspect: it revealed nothing the government wanted kept secret. As The Man Who Never Was (1953) it sold over two million copies and inspired a film of the same name (1955). At more leisure, after the principal secret he protected had been released, Montagu wrote Beyond Top Secret U (1977), a war autobiography in which he explained what he had left out of his bestseller—how much of his work had depended on the deciphering done at Bletchley Park.

He was a devout Jew, long active in charitable works, and president of the United Synagogue from 1954 to 1962. He was appointed OBE in 1944, for his work on Operation Mincemeat, and advanced to CBE in 1950; he also received the Order of the Crown of Yugoslavia in 1943. He was a tall, spare, handsome man, who kept his good looks and gentle manners into old age. He died, at his flat in Exhibition Road, Westminster, 19 July 1985.

[The Times, 20 and 30 July 1985; Ewen Montagu, The Man Who Never Was, 1953, and Beyond Top Secret U, 1977; private information.]
M. R. D. FOOT

MONTAGU, IVOR GOLDSMID SAMUEL (1904-1984), film producer and writer, was born 23 April 1904 in Kensington, the third of four children, and the youngest of three sons of Louis Samuel Montagu, second Baron Swaythling, banker, and his wife, Gladys Helen Rachel, daughter of Colonel Albert Edward Williamson Goldsmid. He was educated at Westminster School, the Royal College of Science, London, and King's College, Cambridge, where he gained a pass degree in 1924. From an early age he cultivated the widely different interests that he pursued for over sixty years—zoology, sport, most of all table tennis, film, and socialism. By the time he was twenty-two he had already published his first book (*Table Tennis Today*, 1924), founded the English Table Tennis Association, made two zoological expeditions to the Soviet Union, and created the Film Society. The society was the crucial first recognition of film as an art form in Britain, numbering among its sponsors Augustus John, J. M. (later Lord) Keynes, G. B. Shaw, and H. G. Wells [qq.v.]. Montagu remained its chairman until 1939.

He wrote some of the earliest London newspaper film reviews for the *Observer* and in 1926 was invited to re-edit *The Lodger*, directed by (Sir) Alfred Hitchcock [q.v.]. Two years later, with his partner Adrian Brunel, he directed his first films, *Bluebottles* (1928), *Daydreams* (1929), and *The Tonic* (1930), from stories by H. G. Wells, starring for the first time together Charles Laughton and Elsa Lanchester. Although he continued to live at this pace for the next thirty years, Montagu never quite fulfilled the promise of this tumultuous beginning to his career.

His first visits to the USSR confirmed his early enthusiasm for socialism, and he moved from the Fabian Society and the British Socialist Party into the British Communist Party, speaking and writing vividly on its behalf. This move and his marriage in January 1927 to a secretary, Eileen Hellstern (affectionately known as 'Hell'), the daughter of a south London boot maker (Francis Anton Hellstern), completed the abandonment of the patrician life, liberal politics, and Jewish orthodoxy of his family background. Until they died within a month of each other in 1984, Ivor and Hell were inseparable partners, together with Hell's daughter, Rowna. There were no children of the marriage.

On a visit to Switzerland in 1929 Montagu was captivated by the Russian film director Sergei Eisenstein. He travelled to Hollywood and prepared the way for a contract for Eisenstein to work on a film at the Paramount Studios. Eisenstein arrived with his assistant Alexandrov and the cameraman Tisse, and moved into a house with the Montagus, where Ivor Montagu acted as manager, script writer, mentor, and disciple, as recorded in his book *With Eisenstein in Hollywood* (1968). After six confused months Paramount terminated the contract; Montagu disagreed with Eisenstein's plans to film in Mexico. They parted company, friends but no longer partners, with their great projects 'Sutter's Gold' and 'An American Tragedy' unmade.

On his return to England Montagu began the move from film to a mixture of film and political activity that led later to his main concentration on politics and writing. He worked as an associate producer for (Sir) Michael Balcon [q.v.] on five of Hitchcock's British feature films of the 1930s, including *The Man Who Knew Too Much* (1934) and *The Thirty-Nine Steps* (1935). He also produced left-wing documentary films *Free Thälmann* (1935), *In Defence of Madrid* (1938), *Spanish ABC* (1938), and *Behind the Spanish Lines* (1938), after a visit to the Spanish Republican front. *Peace and Plenty* in 1939 was a brilliantly satirical attack on the Chamberlain government made for the Communist Party.

During World War II Montagu was on the editorial staff of the *Daily Worker*, an adviser to the Soviet Film Agency, and producer of a film for the Central Office of Information, *Man, One Family* (1946) on the theme of racial harmony. A short post-war reunion with Balcon at the Ealing Studios, where he co-wrote *Scott of the Antarctic* (1948), was almost the end of his association with commercial cinema. After the Wroclaw peace conference in 1949 he was caught up in the Peace Movement, as a writer and speaker, travelling extensively in Communist Europe, China, and Mongolia (for which he had a special affection). Montagu remained steadfast to the Communist Party to the end. He was awarded the Lenin peace prize in 1959, many other socialist decorations, and was a tireless, colourful dignitary at conferences and film festivals in eastern Europe.

In the late 1960s he gave up active office in many of the organizations he had supported, even established, such as the International Table Tennis Association, the Association of Ciné and Television Technicians, and the World Council of Peace, but he stayed in the wings as a respected elder statesman. From his home in Watford and his retreat in Orkney, he translated plays, novels, and works by Soviet film makers, wrote political pamphlets, and produced his own major books, *Film World* (1964), *With Eisenstein in Hollywood* (1968), and *The Youngest Son* (1970). Unfortunately these autobiographical sketches about his family and intellectual adventures up to the late 1920s were never complemented by a second volume of autobiography that he worked on up to his death.

Montagu listed his recreations in *Who's Who* as 'washing up, pottering about, sleeping through television', modest occupations for a man who had started life with so many social and intellectual advantages. He did 'potter about' in a wide range of activities but always with a passionate energy and often with striking effect. A tall, strongly built man, with the features of a Jewish intellectual, Montagu dressed carelessly but distinctively, often in a shaggy pullover, buried beneath the long black leather overcoat he had picked up in Mongolia. He played tennis with more energy and craft than skill at Cambridge, the Queen's Club, and in Hollywood with Eisenstein and (Sir) Charles Chaplin [q.v.]. His causes included cricket (he was a member of the MCC and Hampshire Cricket Club) as well as communism; Southampton Football Club as well as the Zoological Society of London. With his friends in all social classes he communicated vividly in many languages and endlessly over the telephone from Watford, Orkney, Moscow, or Peking. He died in Watford 5 November 1984, shortly after his wife.

[Ivor Montagu, *The Youngest Son*, 1970; articles in *Sight and Sound;* private information.]   D. J. WENDEN

MOOREHEAD, ALAN McCRAE (1910–1983), author and journalist, was born 22 July 1910 in Melbourne, Australia, the youngest of three children and second son of Richard McCrae Moorehead, journalist, and his wife, Louise Edgerton. Educated at Scotch College, Melbourne, and Melbourne University, where he gained a BA in law, he displayed an early interest in newspapers as a student contributor to the *Melbourne Herald*. He later claimed that his awareness of journalism as a profitable career was awakened by the discovery that the *Herald* would pay lineage rates even for the interminable lists of finals results that he supplied to them. After six years as a reporter for the *Herald*, ambition and his intense curiosity about the wider world persuaded him in 1936 to sail for England, where he gained a job with the *Daily Express*. In the next ten years, although Moorehead's relations with the Beaverbrook press proved as stormy as those of many other writers, he became one of its star reporters. He made his reputation as a war correspondent in the Western Desert and later in north-west Europe. He was mentioned in dispatches 1939 and 1945. His series of books on the desert campaign were collected in a single volume, *African Trilogy*, in 1944. In 1946 he produced an early biography of Montgomery.

As Moorehead confessed with engaging honesty in his fragment of autobiography, *A Late Education* (1970), he arrived in England before the war as a gauche and lightly educated provincial. He developed his cultural interests avidly, not least through his friendship with his talented fellow correspondent Alex Clifford, with whom he enjoyed one of the closest relationships of his life. His charm and intelligence gained him many friends, not least among women. Jealousy of his success and dislike of his unembarrassed ambition also made him a controversial figure among Fleet Street colleagues. But even his enemies could not dispute his brilliant gifts as a descriptive writer and military analyst, which also won him respect among senior soldiers. Most journalism is immediately forgettable, but Moorehead's wartime dispatches still seem models of vividness and insight half a century later.

In 1946, weary of daily journalism, Moorehead defied the imprecations of Lord Beaverbrook [q.v.] and resigned from the *Daily Express*. His enthusiasm for Europe was undiminished, but he had little patience with English weather. He departed for Italy with his young children and wife Lucy, a former womens' editor of the *Daily Express*. In the next four years, at the house they had rented outside Florence, he wrote two forgettable novels which taught him his own limitations as a writer of fiction. He was on the verge of returning to England, nursing failure, when he received a cable from the *New Yorker* announcing its acceptance of a long extract from a non-fiction book he had written about the occupants of his Florence house in the fifteenth century, *The Villa Diana* (1951). The generous American cheque that followed was insufficient to save him from returning to England and accepting an improbable job for some months as a Ministry of Defence press officer. But it restored his confidence in his own abilities. In 1952 he achieved his first substantial commercial success with *The Traitors: The Double Life of Fuchs, Pontecorvo, and Nunn May*. The pattern of his career was then set, founded upon intensely disciplined work on a succession of historical books. In 1957 he built a house at Porte Ercole where the family spent several months every year thereafter. Yet he remained a restless, almost compulsive traveller, with little interest in developing permanent roots.

Moorehead's reputation as a major historical writer was made by *Gallipoli* in 1956. *The Russian Revolution* and *No Room in the Ark* followed in 1958 and 1959. In 1960 he published what is generally regarded as his finest work, *The White Nile*, a study of the nineteenth-century search for the sources of the greatest river in Africa, based upon archival research and extensive personal travel in the heart of the continent. Written with Moorehead's customary precision and literary grace, the book sold 50,000 hardback copies in its first British edition. *The Blue Nile* followed in 1962. In 1963 he produced *Cooper's Creek*, a study of R. O'H. Burke's and W. J. Wills's journey across Australia in 1860, which reflected his reawakening interest in his own country. In 1966 came *The Fatal Impact*, a study of the coming of westerners to the south Pacific. But in that year, while still in his mid-fifties, Moorehead suffered a stroke. This deprived him of the power to write or speak coherently. He retained sufficient comprehension to read widely by the simultaneous use of text and aural tape. He mitigated the terrible frustration of his condition by painting, and still found some pleasure in travel. But his tragedy was compounded by the death in a car accident in 1979 of his wife, whose role as typist, editor, critic, and business adviser had been essential to his career.

Moorehead was an outstanding example of the writer as professional. Some colleagues in his early years rebuked his indifference to ideology. He was single-mindedly dedicated to the job in hand, whatever this might be. Though he lost his Australian ‚accent at an early age, there remained about him an underlying raw, outsider's quality. He used his gifts—above all as a writer of descriptive prose—to earn a living rather than to pursue any personal passion. In his youth, he seemed somewhat careless of personal relationships, although he valued those with other successful men and women: Ernest Hemingway, the Australian painters Sidney Nolan and (Sir) G. Russell Drysdale, the television tycoon Sidney (later Lord) Bernstein. He mellowed towards his fellow men as he grew older, above all after his stroke. He was one of the most admired non-fiction writers of his generation, who also achieved great popularity in America. He was appointed OBE in 1946 and CBE in 1968. He became an officer of the Order of Australia in 1978. He was awarded the *Sunday Times* gold medal and Duff Cooper memorial award for *Gallipoli*, and received the Royal Society of Literature award in 1964.

He married 29 October 1939, in Rome, Lucy Martha, daughter of Vincent Milner, a medical doctor, of Torquay, and Isabella Mary Milner. They had two sons and a daughter. Moorehead died in London 29 September 1983.

[Alan Moorehead, *A Late Education*, 1970; private information.]   MAX HASTINGS

MORECAMBE, ERIC (1926–1984), actor and comedian, was born (John) Eric Bartholomew in Morecambe, Lancashire, 14 May 1926, the only child of George Bartholomew, a manual worker for the Morecambe and Heysham Corporation, and his wife, Sarah ('Sadie') Elizabeth Robinson. Educated at Euston Road Elementary School, he was prompted by his mother to leave school at thirteen and embark on the professional stage as a child act.

In 1940, as a result of winning one of the many juvenile talent contests then in vogue, Eric Bartholomew earned a place in a touring 'discovery' show, *Youth Takes a Bow*, presented by Bryan Michie under the emergent impresario Jack Hylton [q.v.]. Coincidentally, Ernie Wise (Ernest Wiseman) was a fellow juvenile already engaged on the show. In 1941 Sadie Bartholomew inspired the pair to form a double act (at the Empire Theatre, Liverpool), but it was to be some years before they adjusted their roles with Eric predominantly as the comedian. The war intervened and Eric was drafted as a 'Bevin Boy' in the mines; he was discharged unfit eleven months later. Upon Ernie's release from the Merchant Navy, the couple resumed their partnership in 1947 in *Lord George Sanger's Variety Circus*. Over the next five years, now billed as Morecambe (the name taken from Bartholomew's birthplace) and Wise, they gradually broke into radio and television.

The critics' response to their first television series, *Running Wild* (BBC, 1954), indicated that they had not yet discovered a winning combination. Soul-searching and experimentation filled their subsequent summer seasons and pantomimes. In 1958 they undertook a six-month variety tour of Australia. Upon their return, their act was purged of its brashness and relied more on subtlety.

They became regular guests on TV variety shows, being at their most effective on Val Parnell's *Sunday Night at the London Palladium*. They felt at home on the ATV commercial network. By this time Morecambe's burgeoning genius as a comedian was making a mark. The first period of the long-running *The Morecambe and Wise Show* ran uninterruptedly on ATV between 1961 and 1968 and undoubtedly established them as national favourites. Their writers, Sid Green and Dick Hills, stimulated by Morecambe's limitless range, were inexhaustibly inventive. The first of many distinguished awards came in 1963 when Morecambe and Wise were chosen by the Guild of TV Producers and Directors as Top Television Light Entertainment Personalities of the Year.

In 1968 Eric Morecambe suffered a major heart attack, but he recovered sufficiently to continue the outstanding run of *The Morecambe and Wise Show*, which had now moved to the BBC and was aided by its more opulent budgets and the new scriptwriter, Eddie Braben.

Meanwhile the pair had appeared many times on the *Ed Sullivan Show* in New York, and in several royal command performances. At Windsor, they were favourites at royal family Christmas parties. Their three films for the Rank Organization—*The Intelligence Men* (1964), *That Riviera Touch* (1965), and *The Magnificent Two* (1966)—made noble attempts to translate their comedy on to the large screen. In 1979 Morecambe had to undergo open-heart surgery. During his convalescing periods he turned author. *Mr Lonely*, a novel, was published in 1981. Then followed two children's books—*The Reluctant Vampire* (1982) and *The Vampire's Revenge* (1983), which were subsequently translated into several languages. *Eric Morecambe on Fishing* (his main hobby) was published posthumously.

Morecambe was a director of Luton Town Football Club from 1969, retiring in 1975 to become a vice-president. He served as president of the Lord's Taverners between 1976 and 1979. He actively supported the Variety Club's charities, the Stars Organization for Spastics, the Sport Aid Foundation, and (among others) The British Heart Foundation. In 1976 he was appointed OBE and in the same year became a freeman of the City of London. Lancaster University conferred upon him an honorary D.Litt. in 1977.

Morecambe's eyebrows arched over the upper rims of his spectacles like twin circumflexes of bewilderment and surprise; his chirpy head movements, like a sparrow's, showed him as always qui vive, while his baggy underlids hinted at the world-weariness and sadness of the clown. Apart from various running jokes, among them 'What d'you think of the show so far?' (audience: 'Rubbish!'), Morecambe did not cultivate catchphrases for the sake of them; although some of his stock comments—'There's no answer to that!', 'My buddy Ern with the short fat hairy legs', 'What do I think of it? *Not a lot* . . . '—and his habit in moments of endearment towards Wise of clapping both hands on his partner's cheeks, will evoke memories of his style.

Morecambe and Wise, like Laurel and Hardy and Hope and Crosby, gave the double act a new dimension, moulding the archetypal 'straight man and comic' into a more complex but totally believable human relationship, and shared their art of being able to involve mass audiences in an enormous sense of fun. Morecambe possessed a needle-sharp awareness of the comic potentialities of any situation, and a comedic flexibility which enabled him, more so even than past masters like Robb Wilton and Jimmy James, to rebound off any number of characters on stage and so widen the comedy impact. No comedian can achieve greatness however unless he embodies those endearing human weaknesses which are the essence of laughter. This Morecambe did in his classic roles with Wise, as the one whose unquenchable exuberance often caused embarrassment, whose performance never quite matched his self-proclaimed ability, and whose sharp wit was used to cover up his ignorance or as a Parthian shot to rescue his self-esteem. Later in life, he became more the nation's humorist, with prestigious people queuing up to be the buttress of his quips. Whatever the role, Eric Morecambe shared the talent of, for example, Will Fyfe and Gracie Fields, to inspire as much love and affection as laughter. By the time of his death he was a national institution.

In 1952 he married a dancer, Joan Dorothy, daughter of Harold Bartlett, a captain in the Royal Army Medical Corps. They had a son and a daughter of their own and they later adopted another son. Morecambe died from a heart attack 28 May 1984 in Cheltenham General Hospital, to which he had been taken after performing at the Roses Theatre, Tewkesbury.

[E. Morecambe and E. Wise, *Eric and Ernie*, 1973; Gary Morecambe (son), *Funny Man*, 1982; private information; personal knowledge.]   DICK HILLS

MORRIS, JOHN HUMPHREY CARLILE (1910-1984), academic lawyer, was born in Wimbledon 18 February 1910, the elder son and eldest of three children of Humphrey William Morris, a solicitor, and his wife, Jessie Muriel Vercoe. Educated at Charterhouse, he was awarded a Holford history scholarship to Christ Church, Oxford. He obtained first classes in jurisprudence (1931) and in the BCL (1932), and was elected Eldon law scholar (1933). He was called to the bar in 1934 (Gray's Inn), spending the next two years in practice. In 1936 he was elected a fellow of Magdalen College, Oxford. In 1940 he joined the Royal Naval Volunteer Reserve, serving in the Faroe Islands, and in Ayrshire to assist with the preparations for the Normandy invasion. Released from naval service in 1945 as lieutenant-commander, he returned to Oxford.

His career as an academic lawyer was centred in Oxford: fellow of Magdalen College (1936-77); All Souls lecturer in private international law (1939-51); reader in the conflict of laws (1951-77). The one period of absence was 1950-1, spent as a visiting professor at Harvard. He declined all similar invitations until his retirement. In Magdalen, with (Sir) (A.) Rupert (N.) Cross [q.v.], he transformed the study of law. He was an exacting tutor; but he inspired great affection and loyalty in his pupils, several of whom reached high office. He was proud of their success and reciprocated their affection. His impact as a teacher was not restricted to Magdalen. In his university teaching of the conflict of laws, he had a major influence on generations of legal scholars, expounding with extraordinary clarity some of the most complex aspects of that subject.

Morris was a big man, physically and intellectually, and few who attended his seminars can forget the sense of expectation (and trepidation) as he swept into the room, gown billowing, attaché case in hand. It would have surprised those graduate students to know that Morris was often as concerned after the seminar as they had been before it, anxious as to whether all the issues had been properly explored. Morris was a scholar of great intellectual power and breadth of knowledge which impressed all who knew him; few realized that he could be confronted with real doubts on legal matters. When, however, principle demanded a resolute approach, he would be vigorous in its defence, crossing swords with many colleagues at one time or another, though usually quick to make up any differences.

He was author of or contributor to twenty-seven different volumes of legal works, extending over four major fields, and the whole gamut of types of publication. His monograph, *The Rule Against Perpetuities* (1956, 2nd edn. 1962, with W. Barton Leach) is a model of clarity and precision; he was the editor of three editions (1939–54) of *Theobald on the Law of Wills*, and general editor of the 22nd edition of *Chitty on Contracts* (1961). It is, however, as a profound influence on the conflict of laws that his scholarship will best stand the test of time. The list of publications is impressive: *Cases on Private International Law* (1939–68), *Cases and Materials on Private International Law* (with P. M. North, 1984); and three editions of his students' textbook, *The Conflict of Laws* (1971–84). However, his outstanding work was as general editor of five editions (1949–80) of A. V. Dicey and Morris, *The Conflict of Laws*. This became probably the most influential English practitioners' work in any area of law, which

was due in the main to the incisiveness, subtlety, and elegance of Morris's legal writing.

His influence on the development of the law was also to be found in his relations with judges and practitioners, by whom his advice was sought. It was sought more formally by the Law Reform Committee and the Law Commission on whose work on perpetuities and the conflict of laws he had a significant impact.

Morris's scholarly distinction achieved wide public recognition. He was awarded a DCL at Oxford in 1949; was elected an associate member of the Institute of International Law (1954), associate member of the American Academy of Arts and Sciences (1960), honorary fellow of Magdalen College (1977), fellow of the British Academy (1966), and honorary bencher of Gray's Inn (1980); and was appointed QC in 1981. He spent 1978–9 in Cambridge as Arthur Goodhart visiting professor of legal science and fellow of Gonville and Caius College. In 1964 he had declined the Vinerian chair of English law at Oxford, a chair held at All Souls, a college for which he had no affection.

An ardent sailor for most of his life he cruised extensively in north European waters and the accounts of his cruises, elegantly written as ever, are to be found in the *Journal* of the Royal Cruising Club. In *Thank You, Wodehouse*, a set of essays published in 1981, Morris applied both his analytical and stylistic skills to produce a self-parody which informs as it entertains. Who else would have tackled the problem of determining 'The Domicile of Agnes Flack'? Of his further work on Dorothy L. Sayers [q.v.] only some of the essays were published.

He married in 1939 (Mercy) Jane, daughter of Stanley Asher Kinch, civil servant. They had no children. In retirement Morris lived in Suffolk. He died 29 September 1984 in Ipswich.

[Guenter Treitel in *Magdalen College Record*, 1985, pp. 34–41; A. V. Dicey and J. H. C. Morris, *The Conflict of Laws*, 11th edn., 1987, pp. xxi–xxiii; Peter North in *Proceedings* of the British Academy, vol. lxxiv, 1988; personal knowledge.]   PETER NORTH

N

NAIPAUL, SHIVADHAR SRINIVASA (SHIVA) (1945-1985), writer, was born 25 February 1945 at a kinsman's house in Woodbrook, a quarter of Port-of-Spain, but was soon taken to Nepaul Street in the neighbouring quarter of St James, where he grew up. He was the sixth of the seven children, two of them sons, of Seepersad Naipaul, journalist, and his wife Droapatie Capildeo; his elder brother by more than twelve years the novelist V. S. Naipaul. The family belonged to the Indian community which had originally migrated to Trinidad as indentured labourers in the nineteenth century.

Family life was close and essentially feminine after the death of his father when Shiva was seven and his brother's departure to England. Naipaul himself wrote that he enjoyed the company of women and was 'responsive to the tidal motions of their moods—their curious gaieties and darknesses; and without consciously intending it, I see that they have had a major role in my fiction'. He was educated at Queen's Royal and St Mary's Colleges in Port-of-Spain whence he won an Island scholarship to Oxford. It was a release. Naipaul later wrote that he never revisited Port-of-Spain without a sense of panic 'that, having arrived there, I may never be able to get out again'.

He sailed to England in 1964 and went up to University College, Oxford, where he read psychology, philosophy, and physiology before changing capriciously to Chinese; he took a third class in 1968. Naipaul was a striking figure at Oxford—he wore long black boots and long black hair in those days—with an affectionate and bibulous circle of friends. As an undergraduate he met Virginia Margaret, daughter of Douglas Stuart, a BBC journalist. 'Jenny' was to be his helpmeet for eighteen years. They were married in Oxford 17 June 1967 and their son Tarun was born in 1973.

Before he left university, Naipaul had begun a novel. It was continued in a bedsitter off Ladbroke Grove in London, first of several modest accommodations before he found a flat in Warrington Crescent in 1975. The result was *Fireflies*, published in 1970 and recognized as a work of very high talent. It was set in Indian Port-of-Spain and told the story of Ram Lutchman, his long-suffering wife, and their two sons; told it in limpid prose, with a crystal ear for dialogue, with electric comedy, but also with deep feeling for the struggle of human beings to come to terms with life and one another. The book won three awards, the John Llewellyn Rhys memorial prize, the Jock Campbell *New Statesman* award, and the Winifred Holtby me-

morial prize of the Royal Society of Literature. It was followed in 1973 by *The Chip-Chip Gatherers*, also set in Trinidad, tragi-comic again though palpably more sombre in tone. It was enthusiastically received and won the Whitbread award.

The enthusiasts were perplexed that no novel followed for ten years. During those years Naipaul travelled and wrote a good deal of outstanding journalism, much of it collected in *Beyond the Dragon's Mouth* (1984), an anthology of articles and short stories. He also published two works of non-fiction. *North of South* (1978) was a bleak account of a journey through Kenya, Tanzania, and Zambia. It was too bleak for those *bien pensants* who wanted to see only the best and the most hopeful in independent Africa. Naipaul was in fact a painfully honest observer, reacting with savage indignation to this 'hopeless, doomed continent . . . swaddled in lies—the lies of an aborted European civilisation; the lies of liberation. Nothing but lies.'

He had a specific advantage over the *bien pensants* as a non-European quite free from 'white liberal' hang-ups of colonial and racial guilt. This advantage was put to further use. In 1979 80 the Naipauls spent more than a year intermittently in San Francisco and Connecticut as he worked on a bleaker subject still. *Black and White* (1980) recounted the background to a ghastly story, the mass suicide at a jungle camp in Guyana by the deluded followers of a Californian heresiarch called Jim Jones. If *North of South* had been called 'anti-African', the unflinchingly sharp-eyed *Black and White* was called 'anti-American'; again unjustly: it was another cry of rage at folly and cruelty. Naipaul's third novel appeared at last in 1983. He had evidently had some block or inhibition, but at the same time *A Hot Country* is a tribute to his fastidiousness and perfectionism, which were always as much a blessing to his readers as they were a trial to his publishers and editors: he was a writer who could spend a morning working on one sentence, to delete it in the afternoon. Direct and intense, his style is now pared down to a minimum. Maybe for that reason the book was underrated at the time.

In 1984 Naipaul visited Australia to write a book about the country. His work in progress on that book was included in his posthumous collection *An Unfinished Journey* (1986). He had just moved to an airy flat in Belsize Park Gardens where in a small work-room he wrote in longhand standing at an old-fashioned lectern.

Naipaul was above medium height but stooping and thickly built. He was bear-like in looks and also in manner. Prickly, acutely sensitive, difficult at times, he provoked hostility without trying; or perhaps without trying much. Some found him arrogant or truculent, and by his own hilarious account his visit to Australia ended in a succession of socially disastrous evenings. Others relished his company. A group of friends, a number of them writers associated with the *Spectator* where his wife was the editor's secretary, regarded him with amused affection and awed respect. Sardonic and world-weary, over a long wine-fuelled lunch or a long whisky-fuelled evening he was the most engaging and rewarding of companions.

He celebrated his fortieth birthday at home in February 1985. He had always been afraid of death. In that sense alone it came to him mercifully six months later when he was struck by a thrombosis at his flat on 13 August 1985 and died instantly. He was cremated by Hindu rites at Golders Green crematorium.

[Shiva Naipaul, *Beyond the Dragon's Mouth*, 1984, and *An Unfinished Journey*, 1986; private information; personal knowledge.]

GEOFFREY WHEATCROFT

NEEL, (LOUIS) BOYD (1905–1981), conductor, teacher, and administrator, was born 19 July 1905 at 30 Ulundi Road, Blackheath, the only child of Louis Anthoine Neel, a paint manufacturer, and his wife Ruby Le Coureur—her family came from Jersey. He was educated at the Royal Naval College of Osborne and Dartmouth, which he left to study medicine at Gonville and Caius College, Cambridge (BA, 1926). He then went to St George's Hospital, London, and became MRCS Eng. and LRCP Lond. (1930). Meanwhile he grasped every chance to conduct amateurs, took lessons at the Guildhall School of Music, and listened to Sir Thomas Beecham [q.v.], Wilhelm Furtwängler, and Bruno Walter. While still in medical practice, he founded the Boyd Neel Orchestra, which made its début at the Aeolian Hall, London, on 22 June 1933; Neel delivered a baby later that night.

For reasons of economy and repertory the group was small and at first consisted mostly of students. At the time little romantic, let alone baroque, string music was played, and then only by symphony orchestras; Neel saw a gap waiting to be filled. He brought forward works by Dvořák, Sir Edward Elgar [q.v.], Grieg, Gustav Holst [q.v.], and Tchaikovsky. The *Tallis Fantasia* of Ralph Vaughan Williams [q.v.], Bloch's *Concerto Grosso*, and Stravinsky's *Apollon Musagète* were in the orchestra's twentieth-century repertory; and *Variations on a Theme by Frank Bridge*, composed by E. B. (later Lord) Britten [q.v.] for them to play at the 1937 Salzburg

Festival, established the composer's international reputation and their own. Based on eighteen strings who worked together regularly, the BNO developed a true and distinctive chamber style, finely suited to the concertos of Mozart (often with Kathleen Long or Frederick Grinke) and to revivals of composers such as Torelli, Vivaldi, and Geminiani. A debonair and restrained figure on the podium, Neel had an instinctive gift for just tempos and lucid textures in Bach and Handel. Among the orchestra's many fine recordings, those of Handel's Concerti Grossi Op. 6, pioneering when they were made, held their own when they were reissued in the more critically informed 1970s.

During World War II Neel returned to medicine but also performed at the National Gallery concerts in London. He then branched out into opera, conducting briefly for Sadler's Wells and D'Oyly Carte, and also took the Sir Robert Mayer [q.v.] Children's Concerts. Even during the war, the Boyd Neel Orchestra managed to celebrate its tenth anniversary, for which Britten composed his Prelude and Fugue. Then came their widespread tours, in Britain and elsewhere in Europe, in Australia and New Zealand in 1947, in Scandinavia in 1950, in Canada and the USA in 1952. In 1950 Neel published *The Story of an Orchestra*; he also contributed the chapter on string music to *Britten: A Commentary* (ed. Donald Mitchell and Hans Keller, 1952).

The success of the 1952 Canadian tour led to his move to Toronto in 1953, as dean of the Royal Conservatory of Music and head of the university faculty of music. Without academic musical training, Neel might have seemed a figurehead; but, finding the premises inadequate, he immediately turned his energy to the planning of the Edward Johnson Building, named after the great Canadian tenor. Opened in 1962, this provided the reorganized faculty with an opera theatre, a concert hall, rehearsal, lecture, and practice rooms, a library, and many other facilities, all air-conditioned and soundproofed. Neel's good humour and skill in communication stood him well in his relationship with both the university and the community (where he was highly regarded), and found another outlet in his work as a popular lecturer and broadcaster.

In 1955 he founded the Hart House Orchestra, a chamber group similar to his London one (which, directed by R. Thurston Dart, q.v., who had played continuo for Neel, became the Philomusica). Neel had quickly realized that Canada trained more performers than it could then employ, and that a professional orchestra based on the university would stimulate the community; and he himself needed active music-making. The Hart House Orchestra toured widely over North America, and visited the Brussels World Fair in 1958 and Aldeburgh, at Britten's invitation, in 1966. Neel was also in demand as a guest conductor, particularly after he retired in 1971 from his academic post, where his work had substantially raised the prestige of music in the university. A relaxed, buoyant figure, and a convivial homosexual, he became one of the best-known, most influential musicians in his adopted country. He was appointed CBE (1953) and a member of the Order of Canada (1973). He was honorary RAM. He died in Toronto 30 September 1981. He was unmarried.

[L. Boyd Neel, *My Orchestras*; private information.]   DIANA McVEAGH

NEILL, STEPHEN CHARLES (1900-1984), church historian and ecumenical worker, was born 31 December 1900 at Edinburgh, the third child in the family of four sons and two daughters of northern Irish evangelical parents, Charles Neill and his wife, Margaret Penelope, daughter of James Monro, CB. His father was ordained as a minister in the (Anglican) Church of Ireland, and was a low churchman and fundamentalist. Both parents had a strong interest in the conversion of India and Neill's sister spent all her life as a missionary in India. Neill was sent to be educated at Dean Close School at Cheltenham. He went on to Trinity College, Cambridge, and was placed in the first class of part i of the classical tripos (1920) and part ii of both the classical tripos (1922) and the theological tripos (1923). Cambridge classical dons later regarded him as one of the ablest classical students they met between the two world wars. He then wrote a thesis on the Cappadocian Fathers of the fourth century AD and their relationship with the neo-Platonic philosophy. On this thesis he won a fellowship at Trinity College.

He held the fellowship only a short time, for in 1924 he decided to go as a missionary to south India. There his astonishing ability at languages soon gave him a mastery of Tamil in which he could preach and write fluently. He also learned Sanskrit and made himself an authority on the early Hindu religious texts. He wrote (in English) some modest but charming little books about his parish life in India. In 1930 he became warden of a theological college at Tirumaraiyur, and in 1939 was elected bishop of Tinnevelly. He was full of energy, often on a bicycle. He preached brilliantly in English or Tamil and had high standards for his clergy. Here he began his ecumenical interest. The various Christian denominations in south India were wondering how it would be possible to have a united south India church. Neill understood the theological as well as the practical problems and took a prominent part in the delicate negotiations which led up to the formation of the Church of South India in 1947.

But before the achievement of unity a disaster of health struck him. He began to suffer the incidence of mental aberration and a continuous weariness through insomnia and pain through headaches. The other Indian bishops believed that it was impossible for him to go on as a bishop and in 1944 asked him to resign his see.

He returned to Cambridge and became chaplain to his college and a lecturer in the faculty of divinity. He also became well known as a preacher and lecturer and conductor of missions throughout England. He could talk at length without notes, his voice was attractive, his experience rare, and his capacity to illustrate his points wide-ranging. He fascinated the young, and not only the young, with his tale of Christianity spreading across the globe at a time when in Europe it seemed to be in recession.

In 1947 G. F. Fisher (later Lord Fisher of Lambeth, q.v.), the archbishop of Canterbury, made him his assistant bishop for ecumenical work. He moved to Geneva to be near the World Council of Churches. But the problems of health recurred. Part of 1950 he spent in hospital, and in February 1951 the archbishop asked him to resign.

He now became a writer and a traveller for the World Council of Churches. He lived in Geneva; helped to edit the *Ecumenical Review*; edited (with Ruth Rouse) *The History of the Ecumenical Movement* (1954), an important and original achievement; and started a series called World Christian Books which published between 1952 and 1970 seventy books under his editing or direction. The intention was to help the education of the developing ministries in the churches of the third world. He wrote several of the books himself. This post at Geneva, the publications, and the travel on behalf of the World Council or for the International Missionary Council, made him a well-known figure in the world church. The best known of his books to the English public was a Pelican on *Anglicanism* (1958).

From 1962 to 1967 he was a lecturer in Christian missions and ecumenical theology, with the title of professor, at the University of Hamburg. His interest was the question how far the expansion of Christianity into Asia and Africa depended on the European imperial mission, and how it would free itself from that inheritance once the imperial mission ended. This phase was marked by *A History of Christian Missions* (1964), *Colonialism and Christian Missions* (1966), and *The Church and Christian Union* (1968). For these and other works he was elected a fellow of the British Academy in 1969.

From 1969 to 1973 he was the first professor of philosophy and religious studies at the University of Nairobi. Then he retired to Wycliffe Hall at Oxford to write the history of Christianity in India. The first volume was published in 1984, the second a year after his death which occurred 20 July 1984 at Wycliffe Hall, Oxford. He never married. He held eight honorary degrees, and took the Cambridge DD when he was seventy-nine.

[Kenneth Cragg and Owen Chadwick in *Proceedings* of the British Academy, vol. lxxi, 1985; *The Times*, 24 July and 3 August 1984; unpublished autobiography; private information; personal knowledge.]

OWEN CHADWICK

NEMON, OSCAR (1906–1985), sculptor, was born in Osijek in East Croatia 13 March 1906, the elder son and second of three children of Mavro Nemon, pharmaceutical manufacturer and a member of the Jewish community in Osijek, and his wife, Eugenia Adler. As a youth he practised modelling in clay at a local brickworks and he participated in local exhibitions in 1923 and 1924. After his unsuccessful application for admission to the Akademie der Bildenden Kunst in Vienna (where one of his uncles owned a bronze foundry) he spent many months there, long enough to meet Sigmund Freud of whose dog Topsy he made a portrait, on the completion of which he was allowed to make his great brooding over-life-size seated figure of Freud himself (bronze, 1930–1, Hampstead Public Library). He also modelled a portrait of the Princess Bonaparte in Vienna.

He obtained a bursary from his native city in 1925 to study at the Académie des Beaux-Arts in Brussels, where he won the gold medal for sculpture. A one-man exhibition of his portrait heads was mounted there in the Palais des Beaux-Arts. His sculptures were also shown by the Galerie Monteau there in December 1934 and January 1939. He shared a house with the Surrealist painter René Magritte for some years during this period. Whilst in Brussels he did not lose touch with Osijek for he made the monument 'June Victims' in 1928 for the city.

To avoid the anti-Semitic threats of the Nazis, he decided, on the advice of his lifelong friend and confidante Madame Simone Speak, to take refuge in England just before war broke out in 1939 (with the exception of his sister, his entire family perished during the Holocaust). He was naturalized in 1948. He was fortunate to be given English lessons by Sir H. Max Beerbohm [q.v.] (bust, c.1941 in Merton College, Oxford). Having had to abandon most of a dozen years of studio work, including a

twenty-foot clay model, 'Le Pont', of which only photographs have been preserved, he had to start again in a new country, with but a few introductions, some sympathy, and little money. He settled in Oxford, set to work, and in 1942 a small exhibition of portraits was arranged at Regent's Park College. Through Albert Rutherston [q.v.], Ruskin master of drawing in the university, he came to the notice of (Sir) John Rothenstein, director of the Tate Gallery (bust, Tate Gallery), and of Sir Karl Parker, keeper of the Department of Fine Art in the Ashmolean Museum (reduced half-length terracotta, Ashmolean Museum). He was befriended by G. Gilbert A. Murray [q.v.], not far from whose house he built Pleasant Land, his studio-residence on the north slope of Boars Hill overlooking the city.

For the convenience of his royal sitters he was allotted a lofty studio in St James's Palace where he modelled the busts of Queen Elizabeth II (in the hall, Christ Church, Oxford), Queen Elizabeth the Queen Mother (Grocers' Hall, London), Earl Mountbatten of Burma, and Prince Philip. There he also created what must be looked on as his most successful series, the monumental effigies of Sir Winston Churchill of which more than a dozen larger than life busts and compositions are in public places. Other wartime heroes who sat for him were Dwight D. Eisenhower, Earl Alexander of Tunis (Imperial War Museum), Viscount Montgomery of Alamein (Whitehall, 1955, and Brussels), Lord Freyberg, Lord Portal of Hungerford, and Lord Beaverbrook (Brunswick Museum, Nova Scotia). Notable personalities of the post-war period whom he portrayed include Harold Macmillan (c.1959, Oxford Union Society) and Margaret Thatcher (1978, Conservative Central Office). Besides many other portraits in clay (terracotta), plaster, bronze (his chosen material for finished works), and stone, he designed many monuments and a series of brilliantly handled small reliefs in plasticine, a medium which allowed him to show bold outlines in fine modulated wiry relief to accentuate the low relief of the figures; these were cast in plaster to give them permanence. His last major work was a monumental memorial to the Royal Canadian Air Force, which was unveiled by the Queen in Toronto in 1984.

Throughout his career, Nemon modelled in clay directly from life, rapidly making many small studies in the tradition of Bernini, though unlike the great baroque master, he never made preliminary drawings. He animated his sitters with a sequence of droll stories and observations told with turns of phrase that betrayed his earlier fluency in Serbo-Croat, German, and French. He was thus able to evoke in them a variety of moods from which he essentialized both likeness and typical poses. Though deft in modelling he never hurried. If he talked on with an accentuated drawl, amusingly, and with originality of thought and phrase, he pondered long and with penetration towards the final form of the portrait on a larger scale. He seemed to feel that the form, capable of modification by his sitter's changes in mood and age and physique, was ever elusive. He was a perfectionist, always aware of the eternal dilemma of the portraitist: how to catch in a single finite form the unceasing mobility of feature and mood. His preferred foundries were those of Morris Singer and occasionally, for smaller pieces, Burleigh Field Art Foundry of High Wycombe.

An exhibition of his work was held in the Ashmolean Museum, Oxford, in 1982. Shortly afterwards he was honoured by the 10th Biennale Slavonaca, which in the event became a memorial exhibition in the Galerija Likovnih Umjetnosti in the city of his birth.

He married in 1939 Patricia, daughter of Lieutenant-Colonel Patrick Villiers-Stuart. They had a son and two daughters. He died 13 April 1985 in the John Radcliffe Hospital, Oxford, having been working on models for a portrait of Princess Diana up to the last. His studio effects and models and several cases of assorted photographs, letters, and other papers, including a manuscript autobiography, have been preserved.

[Catalogues of exhibitions; 'Sculptures of our Time', Ashmolean Museum, Oxford, April/May 1982; Oskar Nemon, 1906-1985, Galerija Likovnih Umjetnosti, Osijek, November/December 1985; *Oxford Times*, 19 April 1985; *The Times*, 16 April 1985; private information; personal knowledge.]

GERALD TAYLOR

NEWMAN, MAXWELL HERMAN ALEXANDER (1897-1984), mathematician, was born in Chelsea 7 February 1897, the only child of Herman Alexander Neumann, a German working in England as secretary to a small company, who lived in Dulwich, and his wife, Sarah Ann Pike, who was of farming stock and an elementary schoolteacher. In 1908 he went to the City of London School, and in 1915 he gained an entrance scholarship to St John's College, Cambridge. He got a first class in part i of the mathematical tripos in 1916, the year in which he changed his name to Newman by deed poll. Much of the next three years he spent in the army, returning to Cambridge in 1919. In 1921 he was a wrangler in the mathematical tripos (part ii) with a distinction in schedule B. He was elected to a fellowship at St John's in 1923 and appointed a university lecturer in 1927.

Newman was among the pioneers of combinatory (later called geometric) topology and wrote a number of important papers on it in the late 1920s. Earlier definitions of combinatory equivalence, based on subdivision, had hit snags. Newman had the bold idea of using only three elementary 'moves' for defining equivalence, none resembling subdivision. He succeeded in developing all the desirable definitions and theorems for combinatorial manifolds, and also showed that his definition of equivalence encompassed earlier definitions and resolved their difficulties. He spent the year 1928-9 on a Rockefeller research fellowship at Princeton University in fruitful collaboration with J. W. Alexander.

In the 1930s, apart from continued work on combinatory topology, he wrote a seminal paper on periodic transformations in Abelian topological groups; he collaborated with J. H. C. Whitehead [q.v.] in producing an intriguing topological counter-example; and he wrote his admirable book, *Elements of the Topology of Plane Sets of Points* (1939). In the early 1940s he wrote a number of papers on logic and Boolean algebras.

In 1942, during World War II, he joined the Government Code and Cipher School at Bletchley Park, starting in the research section. There he became familiar with the recent diagnosis of an important German cipher system. Newman found a way to exploit this breakthrough. His attack involved designing and building some electronic machines, and forming a substantial section. He soon realized that he could improve on the design of the critical machine and started on the logical plan of another, called Colossus. He used his mastery of formal logic to give Colossus maximum flexibility, which, gratifyingly, later paid off in an unforeseen way. Although he can have had little previous experience of discussing designs with engineers or of managing staff, he was successful in both. He had a natural authority combined with an engaging respect for other people that made him easy to work with and for; in his section originality flourished.

From 1945 to 1964 he was Fielden professor of mathematics at Manchester University. He went there convinced that general purpose computers were on the horizon, and he was active in persuading the authorities to build one at Manchester. His main personal involvement in the project was the notion of the B-register. He ran the mathematics department effortlessly, attracting a formidable succession of fine mathematicians and getting the best out of them.

When he retired in 1964 he embarked on a remarkable second burst of mathematical research in geometric topology, culminating in an engulfing theorem for topological manifolds, published in 1966. This extended to topological manifolds what was already known of combinatorial manifolds, namely that the Poincaré Conjecture is true for manifolds of dimension >4.

Newman was a very gifted pianist and a strong chess player. He also enjoyed reading, claiming once to have read everything, 'including Thoreau's *Walden*'. At first contact perhaps austere, he was in fact a splendid companion, with a great sense of humour and a quite delightful turn of phrase.

He was elected FRS in 1939. In 1959 he was awarded the Sylvester medal and in 1962 the de Morgan medal. He was given an honorary D.Sc. by the University of Hull in 1968, and in 1973 St John's made him an honorary fellow.

In 1934 he married Lyn Lloyd, daughter of John Archibald Irvine, a Presbyterian minister; she was a writer. They had two sons. After her death in 1973 he married in the same year Dr Margaret Penrose (died 1989), daughter of John Beresford Leathes, professor of physiology, and widow of Professor Lionel Sharples Penrose [q.v.], physician. Newman died in Cambridge 22 February 1984.

[J. F. Adams in *Biographical Memoirs of Fellows of the Royal Society*, vol. xxxi, 1985; private information; personal knowledge.]

SHAUN WYLIE

NICHOLS, (JOHN) BEVERLEY (1898–1983), author, composer, and playwright, was born 9 September 1898 at Long Ashton, Bristol, the third and youngest son (there were no daughters) of John Nichols, a wealthy solicitor, and his wife, Pauline Zoe Lilian Shalders. He was educated at Marlborough following preparatory school in Torquay. After a term at Balliol in 1917 he obtained his commission in the army, as a second lieutenant in the Labour Corps. While attached to the War Office his unconventional and indiscreet sex life caused consternation and he was hastily transferred to Cambridge to instruct officer cadets in military strategy. Here he was befriended by (Sir) A. E. Shipley [q.v.], vice-chancellor of the university, and was seconded to him as aide-de-camp when he headed the Universities Mission which toured America for the last three months of 1918. Nichols, strikingly handsome in uniforms of his own design, was hailed as a war hero and acted the part with aplomb and charm. Such was his success that he extended his stay in New York by two weeks to cope with social engagements. He also earned his first forty dollars for a magazine article. During the long train journeys across America he completed his novel *Prelude*, based on his school-days. It was published in 1920 to excellent reviews.

He returned to Balliol in 1919 determined to become a celebrity. As editor he revitalized *Isis*, the student newspaper, launched *Oxford Outlook*, a leftish magazine, and cajoled the Liberal Party to finance a new club. After an initial and humiliating defeat engineered by his Tory detractors he was elected president of the Union for Michaelmas term 1920 and used the position to promote himself unceasingly. Much to his delight, when he left Oxford in 1921 with a degree in modern history (short course), having nearly been sent down for neglecting his studies, his controversial opinions were already the subject of indignant letters to the national press.

After flirting with politics and abandoning a career in law he drifted into journalism. His shrewd and refreshingly impudent interviews with the famous, from the US president to (Sir) Charles Chaplin [q.v.] won readers on both sides of the Atlantic. He became well known for his diary on page two of the *Sunday Chronicle*, which he wrote for fourteen years from September 1932. It was filled with sweet and sour comment and name-dropping anecdotes. His second novel, *Patchwork* (1921), set in Oxford, audaciously stated that the characters pleasant and otherwise were drawn from life. His next, *Self* (1922), based on Thackeray's *Vanity Fair*, was a bestseller. In 1924 he ghosted the memoirs of Dame Nellie Melba [q.v.], mistakenly hoping they would be as uninhibited as her conversation. After her death he wrote *Evensong* (1932), highlighting her faults; the dramatized version gave (Dame) Edith Evans [q.v.] her first commercial stage success in London. His precocious autobiography *Twenty-Five* appeared in 1926 but its elegance and wit were marred by sentimentality. This tendency bedevilled his satire of the bright young people, *Crazy Pavements*, autobiography posing as fiction, which startled readers in 1927 with scenes of drug taking and sexual ambivalence. Lawsuits were threatened but he blandly denied any intended resemblance to anyone and no action transpired.

His work for the theatre comprised music and words for six shows including *C. B. Cochran's 1930 Revue* and five plays covering social issues such as blood sports, abortion, and alcoholism. He was an eager champion of causes; his slogan 'Peace at any price' became a rallying cry and *Cry Havoc!* (1933), was required reading for disarmers and pacifists. He was not a Fascist but he believed Sir Oswald Mosley [q.v.] was the only politician strong enough to prevent war with Germany and said so. He regretted this. By contrast he never regretted his support for Indian partition and the Muslim cause or his vitriolic criticism of M. K. Gandhi [q.v.] in *Verdict on India* (1944). He wrote two other headline-catching bestsellers: *A Case of Human*

Bondage (1966) condemned W. Somerset Maugham [q.v.] for the vilification of his wife Syrie in Maugham's *Looking Back* (1962), and he attacked his own father for cruelty to the family in *Father Figure* (1972). It contained three melodramatic accounts of attempted patricide, probably tongue-in-cheek for he was an accomplished spoofer and a consummate actor on the page and off it. Even at his most impassioned his capacity for mockery was not far away.

He saw himself as a frustrated composer forced by circumstance to become a hack, albeit a brilliant and versatile one. He wrote over fifty books including reminiscences, works on mysticism, cookery, religion, travel, and cats, five detective novels, four novels for children, and finally in 1982 *Twilight*, a volume of poetry. He is best remembered for *Down the Garden Path* (1932), a witty, fictionalized account of his experience as an amateur gardener at his weekend cottage. The thought of this glamorous cosmopolitan getting his hands dirty captivated the public. He made a fortune and nine similar books followed. The last was *Garden Open Tomorrow* (1968), set in the garden he designed at Sudbrook Cottage, Ham Common, his home until his death 15 September 1983 in Kingston Hospital following a fall in the music room where he was working. He was unmarried.

[Beverley Nichols, *Twenty-Five*, 1926, *All I Could Never Be*, 1949, *Father Figure*, 1972, and *The Unforgiving Minute*, 1978 (autobiographies); private information; personal knowledge.]   BRYAN CONNON

NICHOLSON, BENJAMIN LAUDER (BEN) (1894-1982), painter, was born 10 April 1894 at Denham, Buckinghamshire, the eldest in the family of three sons and one daughter of the painter (Sir) William Newzam Prior Nicholson [q.v.] and his wife, Mabel Pryde, also a painter and sister of the painter James Pryde [q.v.]. He was educated at Heddon Court, Hampstead, and Gresham School, Holt. He attended the Slade School of Art for about a year, 1910-11, though he always maintained that he stayed only one term and said he learned more from his billiards games with Paul Nash [q.v.] at a nearby hotel. There followed periods abroad between 1911 and 1918 at Tours, Milan, and, for his health, Madeira and Pasadena, California. The chronic asthma from which he suffered all his life spared him military service in World War I.

In reaction against the sophisticated urbanity of his father's world, Nicholson developed a distinctive lyrical style during the 1920s. Working alongside his first wife, (Rosa) Winifred Roberts (also known as Dacre), he painted largely still lifes and landscapes, with the occasional portrait or abstract picture (the first being in 1924). Winifred was the daughter of Charles Roberts, and granddaughter of George James Howard, ninth Earl of Carlisle [q.v.]. They were married in 1920 and had two sons and a daughter. They divided their time between Cumberland, London, and Lugano in Switzerland. The marriage was dissolved in 1938.

Nicholson's encounter with Cubism during his visits to Paris in these years was eventually to be decisive in the development of his work. The discovery in St Ives, Cornwall, in 1928, with his friend the painter Christopher Wood, of the retired fisherman-turned-painter Alfred Wallis, encouraged the 'primitive' streak in Nicholson's painting. He was not a quick developer, however, and needed time to absorb these influences.

Nicholson had his first one-man show at the Adelphi Gallery in London in 1922 and he was elected a member of the Seven and Five Society in 1924, becoming chairman in 1926. As he became more avant-garde himself, he took the group into more adventurous territory. In 1933 he was a founder member of Unit One, a group which included Paul Nash and the sculptors Henry Moore and (Dame) Barbara Hepworth [q.v.].

Nicholson had begun to share a studio in Hampstead with Hepworth in 1932 and a community of style was evident in their joint exhibition at Tooth's of London in November 1932. Hepworth (died 1975), the daughter of Herbert Raikes Hepworth, CBE, civil engineer, became Nicholson's second wife in 1938. Their triplets, a son and two daughters, were born in 1934. The marriage was dissolved in 1951.

In 1933 Hepworth and Nicholson joined the Paris-based group Abstraction-Création. On their trips together to France in the early thirties, it was the contacts made with Braque and Miró which were particularly important for Nicholson's increasingly abstract work of the time. In late 1933 he made his first completely abstract relief on board and he did his first all-white reliefs the following year. Hand-crafted and highly personal works, the imprint of the Dutch painter Mondrian, whom Nicholson met in 1933, is to be found in their geometric vigour.

The white reliefs are today generally agreed to be Nicholson's major contribution to English and European Modernism. They gave Nicholson a leading position in the avant-garde in art of the time, at the centre of a group which was based in Hampstead and included Hepworth, the poet and critic (Sir) Herbert Read [q.v.], and the Russian constructivist sculptor, Naum Gabo. Mondrian joined them in Hampstead in 1938. With Gabo and the architect (Sir) (J.) Leslie Martin, Nicholson edited *Circle, an International Survey of Constructive Art* in 1937.

Just before the outbreak of war, in August 1939, Nicholson and Hepworth, with their family, left London for St Ives, Cornwall, at the invitation of the writer and critic Adrian Stokes [q.v.], to be joined later by Gabo. The landscape of the Penwith peninsula, often seen through a window, soon found its way into Nicholson's work. These more naturalistic paintings and drawings of the 1940s run parallel with completely abstract works in which simply coloured geometric forms are harmoniously juxtaposed. At the same time Nicholson was pursuing what might be described as post-Cubist still-life painting, in which the shapes and colours of jugs and bottles on a table top were abstracted into grand pictorial compositions.

It was these in particular that gave Nicholson an international reputation in the late 1940s. He won numerous awards, including first prize at the Carnegie International, Pittsburgh, in 1952; the Ulisse prize at the Venice Biennale in 1954; the Governor of Tokyo prize at the 3rd International Exhibition, Japan, in 1955; the grand prix at the 4th Lugano International in 1956; the first Guggenheim International award in 1957; and the first International prize for painting at the São Paulo Bienal, also in 1957.

In 1957 Nicholson married a German journalist and photographer, Dr Felicitas Maria Vogler, daughter of Kurt Vogler, a bank director. In 1958 he left St Ives and moved to the Ticino in Switzerland. He concentrated on carved reliefs which became increasingly severe in composition and in colour, but he continued to make many drawings of landscapes and his favourite still-life objects which were lighter in mood and delicate and sensitive in feeling. He returned alone to England in 1972, and lived first in Cambridgeshire and then in Hampstead.

In his final years Nicholson was recognized as one of the major painters of his age, and a leading figure in English art. He was admitted to the Order of Merit in 1968 and received the Rembrandt prize in 1974. He had the rare distinction of two major retrospective exhibitions at the Tate Gallery, in 1955 and 1969. Examples of his work can be seen in museums throughout the world.

Nicholson devoted his life to his art, which he took with intense seriousness. He was a practical man, with a suspicion of intellectual arguments and a dislike of abstract theory. It was entirely in character that he should list his recreations in *Who's Who* as 'painting, tennis, golf, table tennis etc.'. In his youth a champion diabolo player, he often made an analogy be-

tween the perfect poise needed for both drawing a tree and playing a ball game well. He was something of a practical joker and a maker of puns, both verbally and in his witty and copious correspondence. Of small height, Nicholson was of scrupulous though casual appearance. He was a determined avoider of formality and convention, in his life as in his art. He died at his Hampstead home 6 February 1982.

[Sir John Summerson, *Ben Nicholson*, 1948; Sir Herbert Read (ed.), *Ben Nicholson*, 1955; J. P. Hodin, *Ben Nicholson, the Meaning of his Art*, 1957; Maurice de Sausmarez (ed.), *Ben Nicholson*, 1969; *Ben Nicholson, Drawings, Paintings, and Reliefs* (introd. John Russell), 1969; personal knowledge.]

ALAN BOWNESS and SOPHIE BOWNESS

NIVEN, (JAMES) DAVID (GRAHAM) (1910-1983), actor and author, was born 1 March 1910 in Belgrave Mansions, London, though in his best-selling autobiographies he later followed the example of his own Hollywood studio publicists by listing the more romantic and picturesque birthplace of Kirriemuir in Scotland. He was the youngest of four children, two sons and two daughters, born to William Edward Graham Niven, a landowner, and his wife Henrietta Julia, daughter of Captain William Degacher, of the South West Borderers. At the outbreak of World War I, Niven's father enlisted in the Berkshire Yeomanry and was killed in action at Gallipoli on 21 August 1915, leaving a widow to bring up their children in somewhat reduced circumstances until she remarried, in 1917, (Sir) Thomas Platt (from 1922 Comyn-Platt), a businessman who contested, for the Conservative Party, Southport (1923) and Portsmouth Central (1929).

Niven neither knew his father well nor cared for his stepfather at all, his childhood being largely spent at a succession of preparatory boarding schools (from one of which, Heatherdown in Ascot, he was summarily expelled for stealing) and then at Stowe where he at last found in the pioneering headmaster J. F. Roxburgh [q.v.] the father figure he so lacked at home. It was at Roxburgh's urging that Niven was taken into the Royal Military College at Sandhurst in 1928, and his final school report on Niven was unusually prescient: 'Not clever, but useful to have around. He will be popular wherever he goes unless he gets into bad company, which ought to be avoided because he does get on with everybody.'

It was while at Sandhurst that, in a college production of *The Speckled Band*, Niven made his first notable stage appearance, though there was as yet little indication of any desire to enter the acting profession. Instead he was dispatched from Sandhurst in 1929 into the Highland Light Infantry as a junior officer and stationed on Malta, which conspicuously lacked the social and night life to which he had now become accustomed as a young man about London. After several military pranks born of tedium had misfired, and his army future looked extremely bleak, he sent a telegram to his commanding officer in the summer of 1933 reading simply 'Request Permission Resign Commission', a request which was met with evident relief and almost indecent haste.

With no immediate job prospects in England, his mother recently deceased, and only a vague idea that he might perhaps quite like to be an actor, Niven set sail for Canada: he was just twenty-three and it seemed as good a place as any to start out on a new life. Within a matter of weeks he had travelled south to New York and found work as a whisky salesman before joining a dubious pony-racing syndicate in Atlantic City. From there he travelled on to Los Angeles and began to seek employment as an extra in minor westerns. His Hollywood fortunes distinctly improved when he formed a romantic attachment to Merle Oberon however, and by 1939 as a contract artist at the Goldwyn Studios he had made starring appearances in *The Charge of the Light Brigade* (1936), *The Prisoner of Zenda* (1937), *The Dawn Patrol* (1938), *Raffles* (1939), and *Wuthering Heights* (1939), among a dozen other and lesser films. His Hollywood image was that of the 'grin and tonic' man, a veneer actor who traded in a kind of jovial good fortune, that of the happy-go-lucky adventurer who once shared a beach house with Errol Flynn known locally as Cirrhosis-by-the-Sea on account of its constant stock of alcohol.

In truth, Niven was a considerably more serious, astute, and talented man, one whose behaviour at the declaration of World War II showed characteristic courage: abandoning a lucrative studio contract and a career which was at last successful, he was the first of the few English actors to return from California to enlist. He rejoined the army as a subaltern in the Rifle Brigade, was released to make three of his best films (*The First of the Few*, 1942, *The Way Ahead*, 1944, and *A Matter of Life and Death*, 1945), and returned to Hollywood in 1946 accompanied by his beloved first wife Primula Susan, whom he had married in 1940, and their two young sons, David and James. Primula was the daughter of Flight-Lieutenant William Hereward Charles Rollo, solicitor, grandson of the tenth Baron Rollo. Within a few weeks of the Nivens' arrival in California however, 'Primmie' was killed in a fall down a flight of cellar stairs, and although Niven was to marry again in 1948 (the Swedish model Hjördis

Paulina Tersmeden, who survives him and with whom he adopted two daughters, Kristina and Fiona) a certain sadness was now discernible behind the clenched grin of the gentleman player.

Niven's post-war career as an actor was remarkably undistinguished, coinciding as it did with the collapse of the Hollywood Raj of expatriate British officers and gentlemen on screen. By 1951, however, with the publication of a first novel (*Round the Rugged Rocks*) Niven had discovered a second career as a writer, and in the 1970s he was to publish two anecdotal volumes of memoirs (*The Moon's a Balloon*, 1971, and *Bring on the Empty Horses*, 1975) which were the most successful ever written by an actor and ran into many millions of paperback reprints. Shortly before his death he also published a second novel (*Go Slowly, Come Back Quickly*, 1981) and had become a regular guest on British and American television chat shows where, as himself, he gave some of his best performances.

In 1958 Niven deservedly won an Oscar for *Separate Tables*, in which he played an army officer who invented a private life when his own proved unsatisfactory, a habit often endorsed by Niven himself in his autobiographies. His later films of note included *The Guns of Navarone* (1961), *Paper Tiger* (1974), and *Murder by Death* (1976), but during an author tour for his last novel in 1981 he was stricken by motor neurone disease which condemned him to a lingering and painful death, one he approached with all the courage and good humour that were the hallmarks of his life. Niven died 27 July 1983 at his home in the Swiss village of Château d'Oex where he spent many of his later years skiing. He was buried there. There was a memorial service at St Martin-in-the-Fields in London attended by more than five thousand people.

[David Niven, *The Moon's a Balloon*, 1971, and *Bring on the Empty Horses*, 1975 (autobiographies); Sheridan Morley, *The Other Side of the Moon*, 1985 (biography); private information; personal knowledge.]

SHERIDAN MORLEY

NOEL-BAKER, PHILIP JOHN, BARON NOEL-BAKER (1889–1982), Labour politician and winner of the Nobel peace prize, was born at Brondesbury Park, London, 1 November 1889, the sixth child in the family of three sons and four daughters of Joseph Allen Baker, Liberal MP and engineer, who came to England from Canada in 1876, and his wife, Elizabeth Balmer Moscrip, of Roxburghshire, New Brunswick, Canada. Joseph Baker had much to do with the development of London's tramway system. After the Quaker school, Bootham in Yorkshire, Philip went to Haverford College in Pennsylvania, USA. He entered King's College, Cambridge, in 1908 where he distinguished himself academically with a second in part i of the historical tripos (1910) and a first in part ii of the economics tripos (1912). He was university Whewell scholar in international relations in 1911 and was president both of the Cambridge Union (1912) and the University Athletic Club (1910-12), at that time a unique combination. He became vice-principal of Ruskin College, Oxford, in 1914.

During World War I, as a Quaker and conscientious objector, he served with the Friends' Ambulance Unit and received the Mons star and silver medal for valour (Italy, 1917) and the croce di guerra (1918). His first experience of public life came after the war, when he was a member of the British delegation to the Paris peace conference and then served with the secretariat of the League of Nations until 1922. He was an admirer of the League's creator, Viscount Cecil of Chelwood, and he later contributed the notice of him to this Dictionary. He returned briefly to the academic world as Sir Ernest Cassel professor of international relations at the University of London from 1924 to 1929. He then became a Labour MP for Coventry in 1929 and was made parliamentary private secretary to Arthur Henderson [q.v.], the secretary of state for foreign affairs. He lost his seat in 1931 but Henderson chose him as his principle assistant when he presided over the disarmament conference of 1932-3. His academic and personal interest in disarmament was reflected in the efforts to end the private trade in arms which he described in a book *The Private Manufacture of Armaments* (1936).

In 1936 he was elected MP for Derby, a constituency (which became Derby South) which he held for thirty-four years. His ministerial career started with the wartime coalition government under (Sir) Winston Churchill in 1942 when he was appointed joint parliamentary secretary to the Ministry of War Transport. In 1945 he became minister of state to the Foreign Office under Clement (later Earl) Attlee and in the same year he was admitted to the Privy Council. He was then secretary of state for air in 1946-7 and for commonwealth relations from 1947 to 1950. His last government post was as minister for fuel and power in 1950-1. He was a very popular chairman of the Labour Party in 1946-7 and was chairman of the foreign affairs group of the Parliamentary Labour Party from 1964 until 1970.

After 1951 he was a member of the shadow cabinet for the next eight years in opposition. Slight and white-haired, sincere and eloquent, he dominated disarmament debates in the House of Commons with his total mastery of all the facts and figures. His work for disarmament was summarized in his major book, *The Arms Race — a Programme for World Disarmament* (1958), which with his other work was recognized by the award of the Nobel peace prize in 1959, the proceeds of which were used for the cause of disarmament. He received many subsequent honours, decorations, and honorary degrees and was made a life peer in 1977. He was a man of great charm and impressive intellect, who was widely liked, even by those who might have disagreed profoundly with some of his political views. It is likely that his uncompromising stand on disarmament prevented him from achieving the very high office to which his intelligence and political skills might otherwise have propelled him. The Labour governments of the time were reluctant to endorse the type of pacifism and multilateral disarmament which he saw as the only hope for the avoidance of future conflicts.

His life as sportsman and sports administrator spanned a century of transition in sport. Throughout his life he was closely associated with the Olympic Games, in which he ran twice (in 1912 and 1920, in the 1,500 metres race). In both races he self-sacrificingly arranged the way he ran in the final to make the victory of his team-mate more certain. Yet towards the end of his long life he saw many of his ideals in sport flouted; success in sport too often became the justification of political ideology. Perhaps in contrast to his international committees on disarmament, the results of which were disappointing, his International Council for Sport and Physical Recreation (of which he was the founder and president for sixteen years) seemed to bear more fruit. He had an immense belief in the importance of sport both for the individual and as a means of international understanding. At the age of ninety-one he made one of the last major speeches of his life at the Olympic Games in Baden-Baden, and received a standing ovation.

On 12 June 1915 at Worth, he married Irene (died 1956), the daughter of Francis Edward Noel, a British landowner, of Achmetaga in Greece. He first met his wife when they worked together in the Friends' Ambulance Unit in France. He added Noel to his name on marriage. There was one son of the marriage, Francis, who was for some years Labour MP for Swindon. Noel-Baker died at his Westminster home 8 October 1982.

[Personal knowledge.]   ROGER BANNISTER

NORRINGTON, Sir ARTHUR LIONEL PUGH (1899-1982), publisher and vice-chancellor, was born 27 October 1899 at Kenley in Surrey, the only son and elder child of Arthur James Norrington, a businessman in the City of London, and his wife, Gertrude Sarah Elizabeth, daughter of William Pugh, a merchant in China, from Montgomeryshire, Wales. In 1913 he went to Winchester as a scholar, where his general scepticism of received lore earned him the nickname of 'Thomas' which remained with him for life. In 1918 he joined the Royal Field Artillery; the end of the war denied him service overseas, but he lost a little finger in an accident. In 1919 he went to Trinity College, Oxford, as a scholar and achieved a first class in classical honour moderations (1920) followed by a second in *literae humaniores* in 1923. On going down he joined the Oxford University Press in London and was forthwith sent out to India; in later years he would recall his life there in almost Kiplingesque phrases. In 1925 he returned to Oxford as junior assistant secretary to the delegates of the Press. Henceforward Oxford was his home.

In 1942 he became assistant secretary when Kenneth Sisam succeeded R. W. Chapman [q.v.] as secretary and in 1948 he followed the former as secretary to the delegates or, in practice, the senior administrative officer of the OUP. His old college at once elected him a professorial fellow. Norrington was deeply concerned with the expansion in the range of educational books in the Press. In particular he was closely associated with the great wartime series of Oxford Pamphlets on World Affairs and on Home Affairs which ultimately sold almost six million copies. He also made the Press better known in the outside world. His delight in music contributed much to the Press's music publishing.

To his classical learning he added a widespread knowledge of English literature and a quick capacity for mathematical evaluation. His manner was kindly and he established friendly contacts everywhere. His searching blue eyes and quizzical glance would confront a problem and then a ready smile would enliven his taut almost arid skin and an apt solution or verdict would appear. Gardening he loved, and he was a keen ornithologist. His devotion to music was a feature of family and public life alike. He served as a JP and rose to be chairman of the Oxford bench, but his innate sense of justice was also manifested in college and university affairs.

In 1952 the fellows of Trinity invited him to become their next president and he took up office in 1954, in time to preside over the quatercentenary celebrations in 1955. These were followed by an appeal and also by an extensive and successful programme of refacing the crumbling walls of the college. Norrington was an admirable head of house and also played a full part in the affairs of the university, both on council and on the general board. Further afield he was on the revising committee for the New English Bible and from 1960 was the first chairman of the government committee for the publication of cheap books abroad. His name became attached to the Norrington table which assessed the performance of the various Oxford colleges in final examinations.

In 1960 he succeeded T. S. R. Boase [q.v.] as vice-chancellor. He devoted much attention to the status of dons who were not fellows of colleges and to the need for planning the future development of the science area. His college was delighted to welcome him back in 1962; and he resumed his close contacts with its members at all levels. In 1968 he received a knighthood. By a happy conjunction of interests, Blackwell's were enabled to create a large room of their bookshop under Trinity land which was given his name. His more private interests were reflected in the publication (with Professors H. F. Lowry and F. L. Mulhauser) of the poems of A. H. Clough [q.v.] in 1951. *Blackwell's 1879–1979* was published posthumously in 1983. To this Dictionary he contributed the notices of J. T. Christie, Sir Humphrey Milford, and A. T. P. Williams. In 1970 he retired from Trinity but became warden of Winchester until 1974, a position which he filled with distinction and zest. He was an honorary fellow of Trinity, St Cross, and Wolfson colleges, Oxford, and an officer of the Legion of Honour (1962).

In 1928 he married Edith Joyce, daughter of William Moberly Carver, electrical engineer. It was a happy and Christian marriage; her early death in 1964 was a shattering blow to Norrington. There were two sons, of whom one, Roger, became a conductor, and two daughters. In 1969 he married secondly Ruth Margaret, widow of (Peter) Rupert Waterlow and daughter of Edmund Cude, architect, surveyor, and estate agent. There were no children. There is a portrait of him in Trinity by John Ward (1967). He died in the John Radcliffe Hospital, Oxford, 21 May 1982.

[*The Times*, 24 May 1982; private information; personal knowledge.]

MICHAEL MACLAGAN

NORTHCHURCH, BARONESS (1894–1985), politician. [See DAVIDSON, (FRANCES) JOAN.]

O

OAKLEY, KENNETH PAGE (1911-1981), anthropologist, was born 7 April 1911 at Amersham, Buckinghamshire, the only child of Tom Page Oakley, schoolmaster and later headmaster of Dr Challoner's Grammar School, Amersham, and his wife, Dorothy Louise Thomas. He was educated at Challoner's Grammar School, University College School in Hampstead, and University College, London, where in 1933 he took a B.Sc. in geology (with anthropology as a subsidiary subject), gaining first class honours and the Rosa Morison memorial medal. He then began Ph.D. research at London University, successfully completing his thesis on Silurian pearl-bearing Bryozoa (Polyzoa) in 1938. That his research took five years to complete was due to his appointment in 1934 to a post with the Geological Survey and a year later to an assistant keepership in geology (palaeontology) at the British Museum (Natural History). Here he spent the rest of his working life, with the exception of secondment for war service back to the Geological Survey to work on water supply and mineral resources in Britain.

In 1955 the Museum created a sub-department of anthropology within its department of palaeontology, and Oakley became its head, holding the title deputy keeper (anthropology) from 1959 to 1969. In 1969 the relentless progress of multiple sclerosis forced his premature retirement. Though he suffered much pain for the rest of his life, and was eventually confined to a wheelchair, his mental powers and his appetite for anthropological research remained strong and he continued to study and to publish his work right to the end.

While it is correct to describe Oakley as an anthropologist, this somewhat masks the breadth of his further expertise in geology, archaeology, and other disciplines within Quaternary studies. In the 1930s he did work of lasting importance on the Palaeolithic and Pleistocene successions of the Thames Valley, and was secretary and an influential member of the research committee set up in 1937 by the Royal Anthropological Institute to superintend work at Barnfield Pit, Swanscombe, following discovery of the famous Swanscombe skull fragments. Early hominid fossils fascinated Oakley, and his principal research throughout his career concerned them in one way or another. Soon after the war he began work with various colleagues on methods of dating bone, notably by analysis of its fluorine content. An early result was his demonstration (with M. F. Ashley Montagu) that a supposedly Middle Pleistocene human skeleton from Galley Hill, Swanscombe, was in fact much younger than the gravels in which it was found. The fluorine method was also a starting point for his magnificent exposure, with C. R. Hoskins, J. S. Weiner, and (Sir) W. E. Le Gros Clark [q.v.], of the Piltdown skull hoax, a revelation which caught the public's imagination in the mid-1950s and has held it ever since.

Oakley later became involved with radiocarbon dating of hominid remains, and for many years was widely consulted on problems concerning the age and status of hominid fossils all over the world. He travelled extensively to study material and attend conferences. He also became deeply interested in the mental capacities and cultural attainments of our early ancestors, for example when they first made tools or used fire, and in evidence for their collecting decorative curiosities such as fossil shells. His popular handbook *Man the Toolmaker*, first published in 1949, went to six editions and a Japanese translation. His *Frameworks for Dating Fossil Man* (1964 and two later editions) combined his palaeoanthropological and archaeological interests into something approaching a textbook of early prehistory, which delighted many students. With B. G. Campbell and T. I. Molleson he produced the three-part *Catalogue of Fossil Hominids* (1967, 1971, 1975), a scholarly and invaluable reference work. Oakley's many dozen journal articles, spread over forty-five years, are distinguished by their clarity and readability, regardless of subject. His lectures were equally clear and always well received, and he was a stimulating and incisive contributor at conference discussion sessions.

Many academic honours marked Oakley's career, including the Wollaston Fund award (1941), the Henry Stopes memorial medal of the Geologists' Association (1952), fellowship of the Society of Antiquaries (1953), a London University D.Sc. (1955), election as a fellow of the British Academy (1957), a fellowship of University College, London (1958), presidency of the Anthropological Section of the British Association for the Advancement of Science (1961), and the Prestwich medal of the Geological Society of London (1963).

In 1941 Oakley married Edith Margaret (died 1987), daughter of Edgar Charles Martin, OBE, adjutant to the assistant comptroller at the Patent Office. They had two sons. Edith was a psychiatric social worker. Particularly during the last ten years of his life, Oakley fought a gallant and uncomplaining battle against his illness, sometimes seeming to take a scientist's dispassionate interest in its progress. He spent his final years living at Oxford, where friends from all over the world flocked to visit him and bring him news, a striking tribute to the esteem and affection in which he was held. He retained his interest in art (having been, like his father, an accomplished amateur painter) and in listening to music, and he never lost his charm of manner or his quiet sense of humour. He died 2 November 1981 at Oxford, having worked up to the final weeks on publications which appeared posthumously.

[Information from Theya I. Molleson, British Museum (Natural History); *The Times*, 5 November 1981; F. W. Shotton in *Proceedings* of the British Academy, vol. lxviii, 1982; personal knowledge.]   DEREK A. ROE

O'CONNOR, SIR RICHARD NUGENT (1889–1981), general, was born in Srinagar, Kashmir, India, 21 August 1889, the only child of Maurice Nugent O'Connor, a major in the Royal Irish Fusiliers, and his wife, Lilian, daughter of Sir John Morris of Killundine, Argyll. After education at Wellington College and the Royal Military College, Sandhurst, O'Connor was commissioned into the Scottish Rifles (Cameronians) in 1909. In 1914, as signals officer of 22nd brigade in the 7th division, he fought in the first battle of Ypres, in the battles of Neuve Chapelle and Loos in 1915 (he was awarded the MC), and in the battle of the Somme in 1916. In 1917 he was given command, as a lieutenant-colonel, of the 2nd battalion of the Honourable Artillery Company, and with them he took part in the third battle of Ypres (Passchendaele). Subsequently he went with them to Italy. He was appointed to the DSO (1917), gaining a bar (1918). He was mentioned in dispatches nine times.

O'Connor attended the Staff College at Camberley during 1920, and later was an instructor there, and held staff appointments. In 1935 he attended the Imperial Defence College and then was posted to command the Peshawar brigade on the North-West Frontier of India. In 1938 he was promoted to major-general, and appointed to command the 7th division, responsible for operations against Arab terrorists in southern Palestine, and to the post of military governor of Jerusalem. He was there when war broke out in September 1939. In June 1940 O'Connor, promoted temporary lieutenant-general, was appointed to command the Western Desert Force in Egypt.

Sir A. P. (later Earl) Wavell [q.v.], commander-in-chief Middle East, instructed O'Connor secretly to plan the destruction of the Italian Tenth Army in Egypt. O'Connor's attack, launched on 9 December 1940, was brilliantly successful. Early in 1941 he received reports of an Italian withdrawal from Benghazi towards Tripolitania, and obtained authority to try and cut it off by a direct move across the desert to the Gulf of Sirte. His forces succeeded in forestalling the Italian withdrawal at Beda Fomm on 6 February, 25,000 Italians surrendering with 100 guns and an equal number of tanks. In two months O'Connor's force had advanced 350 miles, capturing 130,000 prisoners, nearly 400 tanks, and 845 guns at a cost to itself of 500 killed, 1,373 wounded, and 55 missing, a victory on which his reputation was to rest. O'Connor was then sent to Cairo to command British troops in Egypt, being replaced by Sir Henry Maitland (later Lord) Wilson [q.v.] and later by (Sir) Philip Neame, who faced the Germans under Rommel. When the latter's attack at the end of March 1941 threw Neame's force into confusion, Wavell sent O'Connor up from Cairo, intending that he should replace Neame, but O'Connor persuaded Wavell to leave Neame in command with himself as an adviser. Benghazi had to be abandoned and the decision was taken to withdraw from Cyrenaica. O'Connor and Neame were both captured and sent to Italy. When Italy surrendered in September 1943, an Italian general helped them get away before the Germans arrived. With the help of Italian partisans they finally reached the Eighth Army's lines, and O'Connor returned to Britain.

In January 1944 O'Connor assumed command of VIII Corps, which, after the landing in Normandy in June, was intended to effect the break-out from the beach-head. It was met by fierce opposition from the German forces in the area of Caen. Two major attacks failed to achieve a break-out, causing recriminations at many levels. VIII Corps was then switched to the western end of Second Army's sector and, from 28 July to 4 August, drove a deep wedge south to Vire on the boundary with US First Army.

O'Connor was given only a subsidiary role in the fighting which followed the liberation of Belgium, and was disappointed when, in December, Sir B. L. Montgomery (later Viscount Montgomery of Alamein, q.v.) informed him that he was to go to Calcutta to take over India's Eastern Command, a post which involved no responsibility for operations. Soon after he arrived there, he was transferred to command North West Army in India and promoted general in April 1945.

In the following year, when Montgomery had become chief of the imperial general staff, O'Connor was recalled to become his colleague on the Army Council as adjutant-general. He spent much time visiting overseas commands to discuss an orderly demobilization, and proffered his resignation in summer 1947 rather than agree to an Army Council decision to reduce the numbers returning from the Far East, owing to a shortage of shipping. He was held to his offer on the grounds that he was not up to the job, an accusation he greatly resented. He retired on 30 January 1948.

In 1935 O'Connor married Jean, daughter of Sir Walter Charteris Ross of Cromarty. They acquired a house at Rosemarkie, from which O'Connor played a full part in Scottish public life. He was colonel of the Cameronians from 1951 to 1954, and in the following year became lord lieutenant of Ross and Cromarty, a post he held until 1964, in which year he was lord high commissioner of the general assembly of the Church of Scotland. He was appointed CB (1940), KCB (1941), GCB (1947), and KT (1971). He became an honorary DCL (St Andrews, 1947), was a member of the Legion of Honour, and held the croix de guerre with palm. His wife died in 1959. In 1963 he married Dorothy, widow of Brigadier Hugh Russell and daughter of Walter Summers, a steel merchant, of Dublin. In 1978 they moved from Rosemarkie to London, where O'Connor died 17 June 1981. He had no children, the son of his first wife by a previous marriage taking the surname of O'Connor by deed poll in 1944.

[I. S. O. Playfair, *The Mediterranean and the Middle East* (official history), vols. i (1954) and ii (1956); Barrie Pitt, *The Crucible of War*, 1980; Max Hastings, *Overlord*, 1984; O'Connor's personal papers at King's College, London; private information.]

MICHAEL CARVER

OLDFIELD, SIR MAURICE (1915–1981), head of SIS (Secret Intelligence Service), traditionally known as 'C', was born 16 November 1915 in the village of Over Haddon, near Bakewell, Derbyshire, the eldest of eleven children of Joseph Oldfield, tenant farmer, and his wife, Ada Annie Dicken. He was educated at Lady Manners School in Bakewell, where he learned to play the organ and began his lifelong devotion to the Anglican Church. In 1934 he won a scholarship to Manchester University and specialized in medieval history. After the award of the Thomas Brown memorial prize, in 1938 he graduated with first class honours in history and was elected to a fellowship. The war upset his plans for an academic career.

After joining the Intelligence Corps, his service was spent mostly at the Cairo headquarters of SIME (Security Intelligence Middle East) where his talent was spotted by Brigadier Douglas Roberts. Oldfield finished the war as a lieutenant-colonel with an MBE (1946). When Roberts joined SIS at the end of 1946 as head of counter-intelligence, Oldfield became his deputy from 1947 to 1949. There followed two

postings to Singapore from 1950 to 1952 and from 1956 to 1958, first as deputy and later as head of SIS's regional headquarters covering South East Asia and the Far East. It was here that he established himself as a flyer. In 1956 he was appointed CBE. Throughout his life he never lost interest in the family farm and kept up his organ playing and regular attendance in church, both at home and abroad. Although there was never any hint of indiscretion in his private life, a number of his colleagues put the standard interpretation on why he had reached middle age without getting married.

Following a short spell in London from 1958 to 1959, he was selected for the key post of SIS representative in Washington, where he remained for the next four years, with the main task of cultivating good relations with the CIA. In 1964 he was appointed CMG. His close ties with James Angleton, the head of the CIA's counter-intelligence branch, were reinforced by their shared interest in medieval history. But Angleton also persuaded Oldfield to swallow the outpourings of the KGB defector, Anatoli Golitsyn, who was claiming, *inter permulta alia*, that the Sino-Soviet conflict and President Tito of Yugoslavia's breach with Moscow were clear cases of Soviet disinformation. Soon after leaving Washington, Oldfield withdrew his belief in most of Golitsyn's fairy stories. If, however, he confessed his errors when on his knees, there was, understandably, no overt explanation of how someone of his calibre had been led up the garden path.

On his return to London he became director of counter-intelligence and in 1965 'C's' deputy. He therefore had reason to feel aggrieved when he was passed over in 1968 in favour of Sir John Rennie [q.v.] from the Foreign and Commonwealth Office, whom he later succeeded as 'C' in 1973. This made Oldfield the first member of the post-war intake to reach the top post. Under his leadership, SIS benefited from the good relations he cultivated with both Conservative and Labour ministers at home and from its improved standing with friendly foreign intelligence services with which he kept in personal touch. Oldfield was appointed KCMG in 1975 and GCMG on his retirement in 1978: the only 'C' so far to have received this award. He was also the first to cultivate chosen journalists at meetings in the Athenaeum. This led to the smile on his pudgy face behind horn-rimmed glasses appearing in the press.

All Souls made him a visiting fellow in 1978, where he began a study of Captain Sir Mansfield Cumming, the first 'C', but soon lost interest in it through lack of material. He therefore welcomed Margaret Thatcher's proposal in October 1979 to appoint him co-ordinator of security intelligence in Northern Ireland. In Belfast he did his best to improve relations between the chief constable and the new GOC, but the strains of office soon told on him. It was not only incipient cancer, but also alleged evidence on his unprofessional contacts that caused his return to London in June 1980. Subsequent interrogation resulted in the withdrawal of his positive vetting certificate, after he confessed he had lied to cover up his homosexuality. There is, however, no evidence that his private life had prejudiced the security of his work at any stage in his career. He died in London 11 March 1981. He never married.

[Richard Deacon, '*C*': *A Biography of Sir Maurice Oldfield*, 1984; private information; personal knowledge.]   NIGEL CLIVE

OPIE, PETER MASON (1918-1982), historian of the lore of childhood, the only child of Major Philip Adams Opie, of the Royal Army Medical Corps, and his wife, Margaret Collett-Mason, was born 25 November 1918 in Cairo, Egypt, where his father was serving as an army doctor and where his mother had been a volunteer nurse. His father was later posted to India and was to die there from a fall while horse-racing. Peter was educated at Eton, and his schooldays and life in India are recalled in *I Want to be a Success* (1939), written when he was eighteen, which opens: '[This story] is about a boy who dreamed of success . . . [but] he could not conceive how the Neon lights were to be induced to take notice of him.' He had also stated: 'I wanted to *do something real now*, not just drift on for a further three years at a University' and in the event the 'something real' was the army. At the outbreak of World War II he joined the Royal Fusiliers, was commissioned in the Royal Sussex Regiment in 1940, and after a fall during an assault course was invalided out in 1941. He took up work with the BBC and then, after his marriage in 1943 to Iona Margaret Balfour Archibald, born 13 October 1923 (daughter of Sir Robert George Archibald, a specialist in tropical diseases), joined a firm of reference book publishers (Todd Publishing). Iona's mother was Olive Chapman, the only child of Arthur Cant of Claremont House, Colchester. Iona was educated at Sandecotes School, Parkstone, and had joined the meteorological section of the WAAF.

Opie was still ambitious to be a professional writer and a volume of impressions of wartime life, *Having Held the Nettle* (1945) and a second instalment of autobiography, *The Case of Being a Young Man* (1946) were written during those years. The latter won the Chosen Book competition, and with the proceeds, £337 10s., he decided to abandon his job and settle down

in partnership with Iona to study the lore of childhood, a subject which had been suggested by the birth of their first child in 1944. The first fruit of this was *I Saw Esau* (1947), a small compilation of rhymes of British schoolchildren, but their sights were set on something far more ambitious, a definitive study of nursery rhymes, their histories and variants. The Bodleian Library's keeper of western manuscripts, seeing them at work in 1946, suggested that the Oxford University Press might be interested in the projected work, and thus initiated the Opies' lifelong connection with Oxford. *The Oxford Dictionary of Nursery Rhymes* finally appeared in 1951 and had instant popular success as well as the approval of scholars. 'Unprecedented in the care which it devoted to printed sources', as *The Times* obituary was to comment, 'it nonetheless carried its factual burden with an appropriate cheerfulness and a constant awareness that it was dealing with a living subject.' *The Oxford Nursery Rhyme Book*, a collection for the use of children, followed in 1955 and *The Puffin Book of Nursery Rhymes* in 1963.

From the outset they had realized that children's own lore and their games was a separate subject from nursery rhymes, and three books finally came out of this: *The Lore and Language of Schoolchildren* (1959), *Children's Games in Street and Playground* (1969), and *The Singing Game* (1985), completed by Iona after Peter Opie's death. In all these Iona was the field worker, standing in playgrounds taking notes which her husband subsequently wrote up, and later taking over the research work at the libraries. Iona also wrote the notice of Rose Fyleman for this Dictionary. Even in the early lean years they, and particularly Peter, had been avid collectors of children's books and all objects connected with childhood, from toys and games to baby clothes and feeding bottles, and were to amass one of the most important collections of children's books in private hands, which in 1988 was secured for the Bodleian Library by a public appeal. They had two sons and one daughter. Both the sons inherited the collecting instinct, James becoming a leading expert on toy soldiers and Robert founding his own museum of packaging in Gloucester. From their historical children's books were derived *The Oxford Book of Children's Verse* (1973); *The Classic Fairy Tales* (1974), which contains the earliest published English texts of the twenty-four selected stories together with notes on their history; *A Nursery Companion* (1980), a compilation of some thirty facsimiles of early nineteenth-century picture books, and *The Oxford Book of Narrative Verse* (1983). Oxford University made both Opies honorary MAs in 1962.

Opie died at home in West Liss 5 February

1982, with many books still projected. Among the works he had in mind was a dictionary of folklore (this became *A Dictionary of Superstitions*, edited by Iona Opie and Moira Tatem, 1989), for his interests extended well beyond the study of childhood. He had been president of the anthropology section of the British Association in 1962–3 and president of the Folklore Society (1963–4) and always hoped when he and Iona had completed their work on childhood to be able to return to folklore studies.

Slight and boyish in appearance, he was a compulsive and dedicated worker who saw friends and recreation as an interruption to his life's purpose. Although in *Who's Who* he gave 'blackberry picking' (with book collecting) as his hobby, latterly he rarely could be persuaded to leave the house in West Liss, Hampshire, which they bought in 1959 (and had chosen for the space it provided to accommodate their books). Indeed it is recorded that he only once made the perambulation of the boundaries of the house's grounds.

[*The Times*, 8 and 16 February 1982; *New Yorker*, 4 April 1983; *The Oxford Companion to Children's Literature*; private information.]

GILLIAN AVERY

ORMSBY GORE, (WILLIAM) DAVID, fifth BARON HARLECH (1918–1985), politician and ambassador, was born 20 May 1918 in London, the second son and third child in the family of three sons and three daughters of William George Arthur Ormsby Gore, fourth Baron Harlech [q.v.], politician and banker, and his wife, Lady Beatrice Edith Mildred, daughter of James Edward Hubert Gascoyne-Cecil, the fourth Marquess of Salisbury [q.v.]. His elder brother died at the age of nineteen. He was educated at Eton and New College, Oxford (of which he became an honorary fellow in 1964). He obtained a third class in modern history in 1939 and joined the Berkshire Yeomanry in the same year, becoming a major (general staff) by the end of the war.

In 1950 he became Conservative MP for the Oswestry division of Shropshire and held the seat until 1961. After a few months as parliamentary under-secretary, he was appointed minister of state for foreign affairs at the beginning of the administration led by Harold Macmillan (later the Earl of Stockton) in January 1957. In this office much of his attention was devoted to disarmament negotiations, partly in Geneva and partly in New York during successive sessions of the United Nations General Assembly.

In November 1960 John F. Kennedy was elected president of the United States. Ormsby Gore, two years his junior, had been a close friend since Kennedy's pre-war years in London during his father's embassy. Macmillan, anxious to achieve the closest relations with the new president, who was a brother-in-law of his late nephew by marriage, the Marquess of Hartington, decided to send as British ambassador to Washington another nephew by marriage. Ormsby Gore resigned from the House of Commons and arrived in Washington in May 1961.

Until that point the new ambassador's career could be fairly described as remarkably nepotic. The success which he then made of his mission, however, sprang from qualities which could not be bestowed by family connection. Ormsby Gore was almost perfectly attuned to the new US administration. His friendship with the president strengthened rather than wilted under the strains of office and official intercourse. It was buttressed by the fact that he was also on close terms with Jacqueline Kennedy, as were the Kennedys with Lady Ormsby Gore (Sylvia, or Sissy, daughter of Hugh Lloyd Thomas, diplomat and courtier, whom he had married young in 1940, and by whom he had two sons and three daughters), whose shy but elegant charm made her an addition to the embassy and easily at home in the White House. President Kennedy much liked to have small dinner parties organized at short notice. The Ormsby Gores were probably more frequently invited on this basis than was anybody else, including even the president's brother and attorney-general. It was a wholly exceptional position for any ambassador to be in. It made him almost as much an unofficial adviser to the president as an envoy of the British government—although there was never any suggestion that British interests were not firmly represented in Washington during these years. His position was particularly influential during the Cuban missile crisis in October 1962.

Gore's special relationship may have caused some jealousy amongst other ambassadors but it in no way weakened his position in official Washington outside the White House. Other members of the administration—Robert McNamara, Robert Kennedy, McGeorge Bundy, Arthur Schlesinger—became his close and continuing friends, and after Kennedy's assassination he was even able to be a more than averagely effective ambassador for the first seventeen months of Lyndon Johnson's presidency. But the *raison d'être* of his embassy had gone. In February 1964 his father died and he became Lord Harlech. Later that year the Conservative government which had sent him as a political appointment to Washington was replaced by a Labour one. The new government was in no hurry to remove him. Nor should it have been. Apart from his effectiveness on the spot, his Kennedy years had shifted Harlech to the centre or even the left-centre of politics. After his return to England (in the spring of 1965) he was briefly (1966-7) deputy Conservative leader in the House of Lords, but he had lost any taste which he ever possessed for political partisanship and resigned after a year. This apart, all his subsequent semi-political activities were firmly centrist: the presidency of Shelter, the chairmanship of the European Movement (1969-75) and of the National Committee for Electoral Reform (from 1976). He was also twice concerned in a semi-official capacity with trying to find a multi-racial solution to the Rhodesian problem. In addition he was president of the British Board of Film Censors from 1965, the initiator and chairman of Harlech Television from 1967, and a director of a few other companies, although never centrally occupied with business. He was chairman of the Pilgrim Trust (1974-9) and of the British branch of the Pilgrims (a quite separate organization) in 1965-77. He was also the leading British figure in all Kennedy commemorative activities.

Sadly his own life then came to be almost as marked by tragedy and violent death as was that of the Kennedy family itself. In 1967 Sissy Harlech was killed in a car crash almost at the gates of their north Wales house. In 1974 his eldest son committed suicide. He surmounted these vicissitudes with fortitude and buoyancy, greatly assisted by his second marriage in 1969 to Pamela, the daughter of Ralph Frederick Colin, a New York lawyer and financier. They had one daughter. Pamela Harlech, a Vogue editor and talented compiler of books, brought vitality and verve to the marriage and had the gift of keeping her husband young. David Harlech at sixty-six looked no different from the way he had looked at fifty-six. He had no mountains left to climb, but he lived a life of style and grace, in which a large part was played by pleasure, tempered by high public spirit, good judgement, and unselfish instincts on all the main issues of the day. There seemed little reason why he should not continue for many years as an easy-going public figure of good sense and high repute. But tragedy struck again. On the evening of 25 January 1985, driving from London to Harlech, he was involved in a car crash (the third major one of his life) near to his constituency of the 1950s and the Shropshire homes of his earlier life. He died in hospital in Shrewsbury early the next morning. He was succeeded in the barony by his second son, Francis David (born 1954).

Ormsby Gore was admitted to the Privy Council in 1957 and appointed KCMG in 1961. He had honorary degrees from several American universities and from Manchester University (LLD, 1966).

[Personal knowledge.]   ROY JENKINS

P

PARKES, JAMES WILLIAM (1896–1981), historian and theologian, was born at Les Fauconnaires, Guernsey, 22 December 1896, the younger son and second of three children of Henry Parkes, an English civil engineer who had settled in Guernsey to grow fruit, and his wife, Annie Katharine Bell. He was educated at Elizabeth College, Guernsey, and, after World War I, at Hertford College, Oxford, where he was an open classical scholar. He took a second in classical honour moderations in 1921 and an *aegrotat* in theology in 1923. He returned to Oxford for a term of postgraduate work in 1931 (and for a term in both 1932 and 1933) as a closed scholar of Exeter College, and was awarded the D.Phil. in 1934 for a thesis which was immediately published and became one of his most influential and durable works, *The Conflict of the Church and the Synagogue* (1934).

Parkes was permanently marked by his experiences during World War I, in which he was gassed in the Ypres salient, and encountered (as he later put it) the 'alternating black and white' of hierarchical incompetence and human solidarity, as well as ample outlets for his own brand of good-natured and often unconventional practicality. At Oxford he became involved in the League of Nations Union, and took the decision to enter the church (he was ordained a priest in 1926). In March 1923 his life took a decisive turn with an invitation to join the staff of the Student Christian Movement, and from then on he devoted himself body and mind to the cause of fighting against prejudice and conflict. In addition to his work for the SCM, he was actively involved in the British (later Royal) Institute of International Affairs, the National Union of Students, the League of Nations Union, and the Committee for European Student Relief of the World's Student Christian Federation. In 1926 he became warden of Student Movement House, a club for (mainly foreign) students in London.

In March 1928 he joined the staff of the International Student Service in Geneva, and with the rise of Nazism he found his work increasingly dominated by the problem of antisemitism (Parkes insisted that the word should be spelt in this way, the form 'anti-Semitism' being itself antisemitic). He remained in Geneva until 1935; just before he left he narrowly escaped an assassination attempt by Swiss Nazis. From then until 1964 he lived in the village of Barley, a few miles south of Cambridge, and it was here, with the financial backing of Israel M. (later Lord) Sieff [q.v.] and others, that his library evolved into a research centre on antisemitism and the history of Christian–Jewish relations which attracted scholars from all over the world. It was also in Barley that the first steps were taken towards the foundation of the Council of Christians and Jews, with which Parkes remained closely associated. In 1964, when Parkes and his wife 'retired' from Barley to a small cottage in Dorset, a home was found for the Parkes Library in the University of Southampton, which agreed to keep the library up to date and also established a Parkes Library fellowship in the field of Christian–Jewish relations.

Parkes was a prolific writer. His solidly based investigations of antisemitism and of the history of Christian–Jewish relations opened a new chapter in the study of this difficult and sensitive subject. Parkes insisted that antisemitism, although a modern political phenomenon, cannot be detached from centuries of hostile Christian preaching which wilfully misrepresented the true nature of Judaism and directly contributed to the group hatred which is at the root of the isolation and persecution of the Jews. This view, which later seemed banal, was novel in the 1930s and was deeply resented in certain quarters. Side by side with a historical approach, Parkes developed a theological perspective which found a positive place for the Jews within the divine economy and reinterpreted the Christian Trinity as a functional description of God's activity, rather than a definition of his nature. In 1940 he published *Good God*, a kind of popular biography of God, under the pseudonym John Hadham; it was followed by other theological writings and broadcasts under the same name. To this Dictionary he contributed the notice of Selig Brodetsky.

Parkes was a soft voiced and deceptively gentle man, with a relentless inner energy and an impish sense of humour. His voice he lost completely at the end of his life, and also the use of his hands, crippled by Dupuytren's contracture. He had previously derived great pleasure from tapestry making and from gardening. He was a keen student of architecture, and assembled a remarkable collection of architectural photographs. He also collected English brass candlesticks, on which he became a recognized authority.

Parkes was president of the Jewish Historical Society of England (a rare honour for a Gentile) in 1949–51. He was made an honorary fellow of the Hebrew University of Jerusalem in 1970, and received honorary doctorates from the Jewish Institute of Religion, New York, and the University of Southampton.

In 1942 he married Dorothy, daughter of

Frank Iden Wickings, an agricultural merchant, of Hildenborough, and his wife Emily. She had been helping him secretarially and continued to support him in his work to the end of his life. He died in Bournemouth 6 August 1981.

[James Parkes (John Hadham), *Voyage of Discoveries* (autobiography), 1969; *A Bibliography of the Printed Works of James Parkes with selected quotations*, compiled by S. Sugarman and D. Bailey, ed. by D. A. Pennie, 1977; personal knowledge.]

NICHOLAS DE LANGE

PARSONS, TERENCE (1930–1985), singer of popular music. [See MONRO, MATT.]

PAUL, LESLIE ALLEN (1905–1985), author, was born 30 April 1905 in Dublin, the second son and second child in the family of three sons and two daughters of Frederick Paul, an advertising manager for a chain of provincial newspapers, who lived for most of his working life at Honor Oak in south-east London, and his wife, Lottie Burton, a state qualified nurse. He was educated at a London central school, during and immediately after World War I. Following his father's footsteps in the world of journalism, he entered Fleet Street, and, after editing a newspaper which failed, earned his livelihood as a free-lance journalist.

As a boy he had been a keen Scout, and this enthusiasm led him to an interest in youth movements. In 1925, after some time with the Kibbo Kift (a pacifist movement with an emphasis on family and tribal ritual, begun by John Hargrave, q.v.), he founded his own movement, the Woodcraft Folk. This included among its aims 'the communal ownership of the means of production'. Six years later he led a delegation to the USSR, and published his findings in a book, *Co-operation in the U.S.S.R.* (1934). This was not the first of his large and immensely varied list of publications, which poured from his pen almost to the end of his life. His first book was a volume of poems, *Pipes of Pan*, published in 1927.

During the 1930s he was occupied in educational and social work, mainly in London, but for a time on the Continent with refugees: he was a tutor for the London county council and the Workers' Educational Association. In 1936 he published a novel, *Men in May*, about the general strike. The rise of Fascism was abhorrent to him, and he outgrew his earlier pacifism. When war broke out he was called up and served in the Army Educational Corps, mainly in the Middle East, and also as a staff tutor at Mount Carmel College. It was during his war service that Paul returned to his boyhood Christian faith, as recorded in his *The Annihilation of Man* (1944), for which he received the Atlantic award in literature (1946).

The post-war period was filled with writing and teaching. Two books were autobiographical—*The Living Hedge* (1946) and *Angry Young Man* (1951), the latter being perhaps his best-known work. Two years later, *The English Philosophers* and a life of Sir Thomas More appeared. A short spell of teaching at the Ashridge College of Citizenship (1947–8) was followed by a longer, more fruitful period (1953–7) as director of studies at Brasted Place, near Sevenoaks, a college for pre-theological training of non-graduate Church of England ordinands. More public recognition was shown by his appointments as a member of the departmental committee on the Youth Service (1958–60) and as a research fellow of King George's Jubilee Trust (1960–1).

In July 1960 the Church Assembly resolved that 'a Commission be appointed to consider, in the light of changing circumstances, the system of the payment and deployment of the clergy, and to make recommendations'. The purpose was to examine the interrelated problems of clergy stipends, pensions, and appointments; and the task was given to the central advisory council for the ministry, which decided that a single person should be appointed so that the totality of the problem could be grasped by one mind. The person chosen was Leslie Paul. He began work in February 1962 and submitted his massive report in November 1963; it was published in 1964 as *The Deployment and Payment of the Clergy*. It was a remarkable achievement: the 135,000-word report, with sixty-two recommendations, was written virtually singlehanded by Paul, aided by one omni-competent secretary, Jean Henderson. The report was considered at three successive sessions of the Church Assembly, which appointed a commission under W. Fenton Morley (then vicar of Leeds) to consider the implementation of its main recommendations about the parson's freehold, patronage, and a fairer system of remuneration. Radical changes were made. The Church of England as an institution is not easy to change, but the Paul report was a landmark in its rationalization.

This was the peak of Paul's achievement, and he received widespread recognition, including a FRSL and an American D.Litt. From 1965 to 1970 he was lecturer in ethics and social studies at Queen's College, Birmingham, a post followed by five years (1970–5) as a member of the General Synod. During that decade and the ensuing five years, he continued to write extensively—poems, a novel, and works of theology and philosophy. His last teaching post was an *ad hoc* appointment as writer-in-residence at the teachers' training college of St Paul and St Mary, Cheltenham. During the closing years of his life he lived in the village of Madley outside Hereford, where he could indulge his recreation of bird-watching. He died there 8 July 1985, just two months after his eightieth birthday. He was unmarried.

[Leslie Paul, *Angry Young Man* (autobiography), 1951; personal knowledge.]

W. H. SAUMAREZ SMITH

PENROSE, SIR ROLAND ALGERNON (1900–1984), artist, writer, and exhibition organizer, was born in London 14 October 1900, into a Quaker family, the third of the four sons (there were no daughters) of (James) Doyle Penrose, and his wife, Elizabeth Josephine, daughter of Alexander Peckover, later first Baron Peckover and lord lieutenant of Cambridgeshire. His Anglo-Irish father was a painter, a Royal Hibernian academician who disapproved of the modern art which would claim Penrose's passionate loyalty. But he stimulated Penrose's love of the arts, an interest reinforced by his maternal grandfather. The young Penrose gained a great deal from visits to his grandfather's Wisbech house where an extensive library and collection was available. After studying at Leighton Park School, Reading, he decided to become a painter. At Cambridge, where he attended Queens' College and graduated in 1922, he studied architecture. But he painted in his spare time, designed décor for and performed in plays by the Marlowe Society, and met in the rooms of J. Maynard (later Lord) Keynes [q.v.] Bloomsbury figureheads like A. Clive H. Bell, Duncan Grant, and an especially encouraging Roger Fry [qq.v.]. Soon after going down from Cambridge, Penrose set off for Paris to become a painter.

It was a profoundly liberating experience. Studying at André Lhôte's academy, he quickly acquainted himself with the manifold pleasures of Parisian life in the early 1920s. He got to know Georges Braque, Man Ray, Max Ernst, and André Derain, and in 1925 he married the beautiful young Gascon poet Valentine Andrée Boué, daughter of Maxime Boué, colonel, of Condom, France. Close friendship with Ernst ensured that he became familiar with other Surrealists, André Breton and Joan Miró prominent among them. In 1928 he held his first one-man show at the Galerie Van Leer in the rue de Seine. But he also worked with other artists, organizing the publication of Ernst's *Une Semaine de Bonté* in 1934 and securing a small role in Luis Bunuel's *L'Age d'Or* (1930). Penrose met the young English poet David Gascoyne, who shared his enthusiasm for Surrealism. The two men agreed that they would convert Britain

to the Surrealist faith, and in 1935 Penrose returned to England leaving his estranged wife to travel alone in India.

Wasting little time, he organized the International Surrealist Exhibition at the New Burlington Galleries in the summer of 1936. Aided by a committee which included Henry Moore, Paul Nash, and (Sir) Herbert Read [qq.v.], he succeeded in assembling a major survey provocative enough to establish Surrealism as a lively cultural force in England. The onset of the Spanish civil war a fortnight after the exhibition closed did not prevent him from working with Christian Zervos in Spain on a book called *Catalan Art* (1937). Nor was he slow in raising money for Spain and putting pressure on the British government to back the republican cause. As for Surrealism, he was tireless in its promotion. In 1937 he established the London Gallery in Cork Street under the directorship of the Belgian Surrealist E. L. T. Mesens, and the following year launched the *London Bulletin* where he acted as assistant editor to Mesens.

While continuing to produce his own Surrealist art, Penrose ensured that Picasso's 'Guernica' toured several British cities after the Paris International Exhibition had finished. He began to assemble a distinguished collection as well, including Picasso's great 'Weeping Woman' which was later acquired by the Tate Gallery. In 1937 he met the American photographer Lee Miller, who had been photographed so memorably by Man Ray and starred in Jean Cocteau's film *Le Sang d'un Poète*. They toured the Balkans together by car in the summer of 1938, and Penrose dedicated his experimental book *The Road is Wider than Long* (1939) to her. They were finally married in 1947, when their son Tony was born. They had no other children. Lee Miller (died 1977) was the daughter of Theodore Miller, of Poughkeepsie, New York.

Although Penrose staged a one-man show at the Mayor Gallery in 1939, the war interrupted his art. He served as an air-raid warden in Hampstead and then as a camouflage instructor, but as soon as hostilities ceased he resumed all his old activities. In 1947 they bore ambitious fruit in the founding of the Institute of Contemporary Arts, of which Penrose became the first chairman. Dedicated to promoting the adventurous spirit of modern art on an international scale, but also extending a welcome to writers and musicians, it quickly established itself as an important meeting-place for anyone with innovative ideas.

The post-war period also saw the emergence of Penrose the biographer and art historian. His biography of Picasso (*Picasso: His Life and Work*), published in 1958, is an outstanding work, and he followed it with studies of Mirò (1970), Man Ray (1975), and Tàpies (1978). At the same time Penrose organized an impressive sequence of retrospectives at the Tate Gallery. They commenced with a vast survey of Picasso's work in 1960, and continued with Ernst (1962), Mirò (1964), and Picasso's sculpture (1967). His close friendship with each of these artists enabled him to select the shows with great authority, and his services to art in Britain were recognized by appointment as a CBE in 1961 and a knighthood in 1966. To the end of his long life he delighted in making inventive collages, and during his last years he produced some of his most exuberant images. He died at Chiddingly 23 April 1984. 'What matters to him is life, openness, receptivity, involvement, appetite, energy', wrote Norbert Lynton, who saluted Penrose 'as one whose very life has been an unconscious work of art performed to the benefit of all'.

[Roland Penrose, *Scrapbook 1900-1981*, 1981; *Roland Penrose*, Arts Council catalogue, 1980; *Roland Penrose: Recent Collages* (*A Commemorative Exhibition*), Gardner Art Centre catalogue, 1984; *Surrealism in Britain in the Thirties*, Leeds City Art Galleries catalogue, 1986; *Dada and Surrealism Reviewed*, Hayward Gallery catalogue, 1978; personal knowledge.]   RICHARD CORK

PERHAM, DAME MARGERY FREDA (1895-1982), writer and lecturer on African affairs, was born 6 September 1895 at Bury, Lancashire, the youngest in the family of five sons and two daughters of Frederick Perham, a wine merchant, and his wife, Marion, daughter of Mrs John Hodder Needell, a novelist. The family soon moved to Harrogate, but Margery was sent to the Anglican school, St Anne's, Abbots Bromley, where she won an open scholarship to St Hugh's College, Oxford, in 1914. After taking a first class in modern history in 1917 she was appointed assistant lecturer at Sheffield University where she also acted and wrote a play on early Saxon England.

A nervous breakdown in 1922 gave her the opportunity to visit her only sister in Somaliland where her brother-in-law was a district commissioner. The romance of riding and camping with the mountains of Ethiopia on the horizon captured her imagination and was the beginning of a sixty-year love affair with Africa. Its first fruits was a novel, *Major Dane's Garden* (1925)—'the outlet', she said somewhat defensively later, 'for my deeply felt impressions' at a time when there was no routine D.Phil. course.

Returning to St Hugh's in 1924 as official fellow and tutor in modern history and in the school of philosophy, politics, and economics, she lectured on the League of Nations, the mandates, and British 'native' policy. But her heart was not in undergraduate teaching and in 1929 she ecstatically accepted a year's travel grant by the Rhodes trustees, through their secretary Philip Kerr (later the Marquess of Lothian, q.v.), to study 'native problems'. A year later two cables arrived in Tanganyika: one from Lothian offering an extension of travel, the other from her college asking her to return or resign her official fellowship. Long afterwards she would tell how she dispatched two one-word replies: 'Accept' and 'Resign'. Then she wrote 'I now had to sit down and consider . . . how best to use this glorious gift of time.' But already there was only one answer: Africa. In 1932 the Rockefeller Foundation made generous provision to make that single-minded devotion possible. So it was that for over five years, now a research fellow of St Hugh's (1930-9), she was primarily but not exclusively on safari in East, West, and South Africa, moving easily from government house to district station and back. She was now writing regularly letters to and leaders in *The Times* and actively lobbying MPs, the Colonial Office, and many governors. Through Lionel Curtis [q.v.] she became involved in defending the High Commission territories, and through Lord Lugard [q.v.] and his followers, became a champion of 'indirect rule'. Her *Native Administration in Nigeria* (1937), a skilful and perceptive interpretation of its practice and problems in that period, remained a classic analysis of empire-on-the-spot.

In 1939 she became the first official—and first woman—fellow of the new Nuffield College in Oxford. At the same time she was elected reader in colonial administration, the creation of which post was a tribute to her reputation in that field. She held the post until 1948. She had made colonial studies a respected academic subject and African affairs very much her own. In high places she had become an influential publicist and a name to conjure with, even to evade. Arthur Creech Jones [q.v.], secretary of state for the colonies (1946-50) treated her as an oracle. When not in Africa, she travelled tirelessly to London, giving interviews, listening, broadcasting, attending committees on colonial higher education or on missions or on aborigines' protection, or to Cambridge to repeat lectures delivered at Oxford to the Colonial Service probationers. Her teaching indeed was almost entirely devoted to the first and second Devonshire courses for colonial servants. She was a professional teaching professionals, analysing 'indirect rule' as the cushion to European impact. Before the war she did not favour the Africanization of the civil service in African colonies, nor did she think Africa was ready for democracy. But later she maybe adjusted her timescale a little and busied herself with developing universities for the new African leaders and experts. She often worked best in a team, as when earlier at Blickling Hall with Lothian and others she helped plan the project which emerged as *An African Survey* (1938) by Lord Hailey [q.v.] or much later in the initiation of a scheme to rescue the records of decolonization then in jeopardy in former colonies—the Oxford University Colonial Records Project.

In Oxford she became during World War II the nucleus and memoranda-writer for a group (including Sir Reginald Coupland and (Sir) Douglas Veale, qq.v.) which secured government funds and asserted the university's role in colonial studies. Her books, reports, and papers provided the base for an Institute, the Oxford Institute of Colonial Studies, of which in 1945 she was appointed director (until 1948). Rarely did she formally supervise graduate students other than Africans, for whom in positive discrimination she waived her own exacting standards. She took meticulous care in editing her series on colonial legislatures. In Nuffield College she was always a little remote from students other than those who attended her occasional readings of poetry or the chapel regularly. She was largely instrumental in securing John Piper to design and furnish the 'upper room' chapel. For many years she lived in the college with her sister and her dog Guerda. She was a sympathetic colleague, but as a scholar very much involved in her own work and her massive correspondence. Her major work now was the two-volume life of Lugard (1956 and 1960). She gave the Reith lectures, *The Colonial Reckoning*, in 1961 and two delightful series of travel reminiscences on the BBC, 'Time of My Life' and 'Travelling on Trust', after she had retired in 1963.

Margery Perham was a woman who won respect among scholars and administrators: she was an anti-feminist who found it easy to succeed in what were essentially men's worlds. She was devoted to quality and the grand gesture and deplored the shoddy or makeshift or complacent. She was a ruthless croquet player, resolved to play by the rules but rules interpreted in her own way. She was robust with striking features and with more than a hint of masculine strength. She was six feet tall and looked down on many but her eyes were kindly. Her charity embraced many causes, including animals. Though single-minded and often formidable she was also genuinely humble and warm-hearted. Her Christian faith, which was strong in childhood, was renewed by *The Man Born to be King* by Dorothy L. Sayers [q.v.], a radio drama

broadcast from December 1941 to October 1942.

She was the first president of the African Studies Association and was appointed CBE in 1948 and DCMG in 1965. She became FBA in 1961, an honorary fellow of Nuffield College (1963), St. Hugh's (1962), Makerere (1963), and the School of Oriental and African Studies (1964). She held honorary degrees at the universities of St Andrews (1952), Southampton (1962), London (1964), Birmingham (1969), Cambridge (1966), and Oxford (1963). There is a portrait drawing by Sir William Rothenstein in Nuffield College. She was unmarried. She died at a nursing home near Dorchester on Thames 19 February 1982.

[Perham papers in Rhodes House library, Oxford; K. Robinson and A. F. Madden (eds.), *Essays in Imperial Government presented to Margery Perham*, 1963; A. H. M. Kirk-Greene, 'Margery Perham and Colonial Administration' in *Oxford and the Idea of Commonwealth* (A. F. Madden and D. K. Fieldhouse, eds.), 1982; personal knowledge.]   FREDERICK MADDEN

PERKINS, JOHN BRYAN WARD- (1912–1981), archaeologist. [See WARD-PERKINS.]

PERY, ANGELA OLIVIA, COUNTESS OF LIMERICK (1897–1981), leader of the British and International Red Cross movements, was born in Folkestone 27 August 1897, the younger daughter (there were no sons) of Lieutenant-Colonel Sir Henry Trotter, KCMG, soldier, explorer, and diplomat, and his wife, Olivia Georgiana, daughter of Admiral Sir George Wellesley [q.v.] and a great-niece of the first Duke of Wellington [q.v.].

She was educated at North Foreland Lodge at Broadstairs, which she left in 1915 at the age of seventeen to train as a Red Cross VAD. Being too young to serve overseas she got to France by falsifying her age, and there nursed the wounded in both French and British hospitals. After the war she took a diploma in social science and administration at the London School of Economics and travelled extensively in Europe, and also areas of the Middle East where European women were almost unknown.

In 1926 she married Edmund Colquhoun ('Mark') Pery, who succeeded in 1929 as fifth Earl of Limerick. They had two sons and a daughter. After marriage she worked in the London branch of the British Red Cross Society and was also a Poor Law guardian (1928–30). She served on the Kensington borough Council (1929–35) and was chairman of its maternity and child welfare, and public health committees. She represented South Kensington on the London county council (1936-46). She was also a pioneer campaigner for family planning, to the extent of being stoned at a meeting in Glasgow.

By 1939 she was president (having been director) of the London branch of the BRCS, and thus in charge of its services during the blitz. As a deputy chairman of the joint war organization of Red Cross and St John (1941-7) she visited fourteen countries, including several battle fronts. In addition she served on several government committees and was Privy Council representative on the General Nursing Council (1933-50). Immediately after the war she was closely involved with Red Cross rehabilitation work overseas.

In the years 1946-63, as vice-chairman of the executive committee of the BRCS in charge of its international operations, she visited most of its branches in Africa, the Far East, and the Caribbean, encouraging the expansion of services and their preparation for transition into independent national societies. She also visited twenty-six other national Red Cross societies, including those in the USSR and China. She was a most active and articulate leader of the British delegation at all International Red Cross conferences and meetings; uncompromising in support of the fundamental principles and integrity of the movement, she was heard with ever-growing respect.

Her increasing contribution to the interternational work of the Red Cross was recognized by her election as a vice-chairman of the League of Red Cross Societies (1957-73), then in 1965 to the standing commission, the supreme co-ordinating committee of the International Red Cross, which promptly made her its chairman. Exceptionally, she was invited to preside at the quadrennial international conference in Istanbul in 1969, in which year she was re-elected chairman of the standing commission for a second four-year term following joint persuasion from the Americans and Russians; in 1973, at the age of seventy-six, she refused further nomination.

Age notwithstanding, Lady Limerick was persuaded to succeed the Duke of Edinburgh in 1974 as the first non-royal chairman of council of the BRCS (she was a vice-chairman from 1963) to preside over the transition to a more democratic constitution. When she retired in 1976 the Queen approved her appointment as the first non-royal vice-president of the BRCS.

Angela Limerick was a much loved figure with great breadth of vision, humour, a remarkable memory and grasp of detail, and an endearing ability to establish close and lasting relationships after brief acquaintance; above all, she had the gift of bringing out the best in others by her encouragement and inspiration.

Her role in the many organizations with which she was involved was invariably active. She was president or vice-president of the Multiple Sclerosis Society, St Giles' Hospital for Leprosy Patients, the International Social Service of Great Britain, the Family Planning Association, the Family Welfare Association, the Star and Garter Home, and Trinity Hospice. She received honorary LLDs from Manchester (1945) and Leeds (1951) universities, and in 1977 was appointed a deputy lieutenant for West Sussex.

Lady Limerick was appointed CBE in 1942, DBE in 1946, GBE in 1954, and CH in 1974. The last two awards gave her particular pleasure because, uniquely in each case, they matched awards to her husband, who died in 1967 after equally distinguished public service. Their harmonious partnership was exemplified in the beautiful garden they created at their Sussex home, which offered a happy haven for family and friends alike. Lady Limerick was fully active until a few days before her death at home in Sussex 25 April 1981.

[Personal knowledge.]   ANNE M. BRYANS

PETERS, SIR RUDOLPH ALBERT (1889-1982), biochemist, was born 13 April 1889 in Kensington, the only son and elder child of Albert Edward Duncan Ralph Peters, a general practitioner in Midhurst, and his wife, Agnes Malvina Watts. He was educated at Wellington College, which he entered as a classical scholar, though intending to study medicine. His early interest in science was maintained by private reading and by helping a mathematics master with a telescopic survey of the moon. From another master he received encouragement in music. He became a competent violinist and throughout his life enjoyed playing chamber music with friends.

In 1908 he entered Gonville and Caius College, Cambridge, where, under the influence of (Sir) Joseph Barcroft, A. V. Hill [qq.v.], and others, he obtained first class honours in part i of the natural sciences tripos (1910). He went on to spend two years on part ii, in physiology including 'physiological chemistry' taught by (Sir) F. Gowland Hopkins [q.v.], but he was unable to take the examination because he had typhoid fever. Peters assisted Barcroft in research on haemoglobin and, in 1912, joined A. V. Hill in work on the chemistry of muscular contraction.

On the outbreak of war in 1914, though awarded a fellowship at Caius, Peters decided to complete his medical studies at St Bartholomew's Hospital, qualifying MB, B.Chir. in 1915. In November 1915 he joined the Royal Army Medical Corps and left for France to serve for six months with a field ambulance, then transferring to the 60th Rifles as medical officer. He was awarded the MC and bar (1917) and was mentioned in dispatches. Early in 1917 he was recalled to the physiological laboratory at Porton, to research under Barcroft on problems of chemical warfare. He remained there for the rest of the war. He gained his MD in 1919.

In 1918 Peters took up his fellowship at Caius, together with a lectureship in biochemistry. He maintained an interest in work begun at Porton but also started what was to become a major line of research, on the vitamin B complex. In 1923 Peters was appointed to the recently established Whitley chair of biochemistry in Oxford and occupied it until 1954. He was a fellow of Trinity College, Oxford, from 1925 to 1954. Then, immediately following his retirement, he accepted appointment, under Ivan de Burgh Daly, to the Agricultural Research Council's Animal Physiology Institute where he built up a strong research group while continuing his own work. Following a second retirement in 1959, Peters settled in Cambridge where he was given facilities for laboratory work. He continued this until 1976, publishing his last paper in 1981.

Peters's researches fall into three periods. Throughout these he was interested in general and cellular metabolism. In the first, 1923–39, he began a search for the anti-neuritic component of vitamin B which led him to the first isolation of thiamine. He made important contributions to understanding its metabolic function as co-carboxylase. In 1939–45 he directed his own and colleagues' attention to the biochemistry of chemical warfare agents. This led to the discovery (by L. A. Stocken and R. H. S. Thompson) of 'British anti-Lewisite'. Peters was active also in advising on research into human nutritional requirements. Finally, following 1945, he extended work begun during the war on the toxicity of fluoro compounds showing how, by what he termed 'lethal synthesis', they become converted into metabolically inert analogues of normal metabolites.

Peters was a kind man, though capable of beneficial severity. He inspired trust and affection. His manner was friendly. His speech, apt to come in rapid bursts, expressed enthusiasm, alertness, and a sense of fun. Though his first love was biochemistry, he was more than merely an able biochemist. He was careful of the interests of all his associates, from first-year students to senior colleagues. He was as much concerned with his department's teaching as with its research. He saw his subject not in isolation but in relationship with other sciences, physical, biological, and social, and its study as part of the whole matrix of academic fields of learning.

Peters was elected a fellow of the Royal Society in 1935, received its Royal medal in 1949 and delivered its Croonian lecture in 1952. In that year he was knighted and became FRCP. He was elected honorary fellow of Caius College, of Trinity College, Oxford, and of a number of professional organizations. He was awarded eight honorary degrees and gave many invited lectures. As well as his university duties in research, teaching, and administration, he was adviser to the Medical Research Council, the Ministry of Supply, and St Bartholomew's Hospital. From 1958 to 1961 he served as president of the International Council of Scientific Unions, travelling world-wide on its business.

In 1917 Peters married Frances Williamina, daughter of Francis William Vérel, of Glasgow. They had two sons. Peters died in Cambridge 29 January 1982.

[R. H. S. Thompson and A. G. Ogston in *Biographical Memoirs of Fellows of the Royal Society*, vol. xxix, 1983; personal knowledge.]
A. G. OGSTON

PEVSNER, SIR NIKOLAUS BERNHARD LEON (1902–1983), architectural historian, was born in Leipzig 30 January 1902, the younger son (there were no daughters) of Hugo Pewsner, a successful fur trader, settled in Leipzig, and his wife, Annie Perlmann, both of Jewish extraction. The elder brother died in 1919, the father in 1940, his widow committing suicide two years later to avoid internment in a concentration camp. Pevsner was educated at St Thomas's School, Leipzig, and attended the universities of Leipzig, Munich, Berlin, and Frankfurt. His doctoral dissertation focused on the baroque merchant houses of Leipzig and led to the writing of his first book, *Leipziger Barock* (1928). Meanwhile in 1924, he joined the staff of the Dresden gallery as a 'Voluntär' or unsalaried assistant, remaining there till 1927 and contributing the volume on Mannerist and baroque Italian painting to the *Handbuch der Kunstwissenschaft* series (1928). Also in the Dresden years he was attracted to the social history of art, on which nothing yet had been written. An eventual outcome was his *Academies of Art, Past and Present*, published in England in 1940.

In 1928 Pevsner was attached to the University of Göttingen as a *Privatdozent*. In 1930 he visited England for the first time and thereafter lectured on English architecture in the art history department and occasionally in the philosophy faculty. His experience of the arts in England induced a profound curiosity about the contrast between the English arts and their equivalents on the Continent. 'The contrast was complete and it was, against all expectations, agreeable too!'

In 1933, notwithstanding that he had become a Lutheran when he was nineteen, Pevsner was forced to stop teaching under the Nazi race laws. He decided to leave Germany and move to England, making the acquaintance of Philip Sargant Florence of Birmingham University. Florence suggested that he should undertake an investigation into English industrial design. From 1934 Pevsner worked mainly in Birmingham, his findings eventually crystallizing in *An Enquiry into Industrial Art in England* (1937). In the course of his work he met (Sir) S. Gordon Russell [q.v.] and became his firm's adviser on modern furniture design.

Pevsner had early recognized the validity of the Modern movement in architecture. He was conscious, however, of the need for clarification of the movement's sources and in 1936 appeared his *Pioneers of the Modern Movement from William Morris to Walter Gropius*. The book was widely influential and was translated into many languages.

With the outbreak of hostilities in 1939, Pevsner's status became technically that of 'enemy alien' and in 1940 he was interned for a short period in the Isle of Man under the wartime regulation 18B. His friends in academic circles quickly secured his release and he briefly took employment as a labourer clearing rubble from bombed London streets. From this he was rescued by an offer of employment by the Architectural Press and from 1942 to 1945 the editorship of the *Architectural Review* was almost entirely in his hands. He was naturalized in 1946.

Also in 1942 he was appointed to a part-time lectureship at Birkbeck College, University of London. He lectured there till his retirement in 1969, having in the meantime been appointed (1959) professor of the history of art. His approach to architectural history is reflected in the Pelican Book written in 1941, and first published in 1942, *An Outline of European Architecture*. Consisting only of 160 pages and selling at a very modest price, it placed the history of architecture in a new perspective. 'The history of architecture', Pevsner wrote in the introduction, 'is the history of man shaping space.' By the adroit selection and analysis of only a few buildings in each period, he penetrated deeply into the social and philosophic roots of architectural form.

For Penguin Books Ltd. Pevsner edited, from 1941, the series of 'King Penguins', miniature picture books with short texts by distinguished scholars; of these his own study of *The Leaves of Southwell* (the carved foliage in the chapterhouse), 1945, was one of the most original.

Following the success of the Pelican *Outline*, the publisher, Sir Allen Lane [q.v.] asked Pevsner what he would like to tackle next.

Pevsner made two suggestions. One was for a comprehensive series of histories covering the whole world of art and architecture, of which he would be the editor. The other was for a series of county-by-county guides to the architecture of Britain. This he would undertake to write entirely himself. Lane reacted promptly in favour of both proposals. In the event, the guides, *The Buildings of England*, were launched first, in 1951, and completed, in forty-six volumes, in 1974. The 'Pelican History of Art', launched in 1953, is still in progress.

If *The Buildings of England* was Pevsner's most conspicuous and widely acclaimed achievement, he was constantly engaged in writing and lecturing on a wide range of subjects. He was Slade professor of fine art at Cambridge in 1949-55 and at Oxford in 1968-9. In 1955 he delivered the annual Reith lectures on the BBC, taking as his subject 'The Englishness of English Art'. This had always been for him an absorbing theme. The talks were made into a book, published in 1955.

In 1964 he produced *Sources of Modern Art* (reissued as *Sources of Modern Architecture and Design* in 1968). *Some Architectural Writers of the Nineteenth Century* followed in 1972. His last published work was *A History of Building Types* (1976), a valuable collection of notes made over the whole period of his career.

Pevsner's continuous production of books, articles, and lectures left him little leisure for social activities which indeed, he constantly declined. His only recreations were walking and swimming. He participated, however, in the work of commissions and committees which he considered important, serving on the Royal Fine Art Commission, the Historic Buildings Council, the Advisory Board on Redundant Churches, and the National Council for Diplomas in Art and Design. He was a founding member of the Victorian Society and chairman from its inception in 1958 till 1976, after which he was president for life.

He received many honours and awards. He was appointed CBE in 1953, elected FBA in 1965, and knighted 'for services to art and architecture' in 1969. In 1967 he received the RIBA Royal gold medal for architecture. Honorary doctorates were awarded to him by the universities of Oxford, Cambridge, East Anglia, Keele, the Open University, the Heriot-Watt University of Edinburgh, Pennsylvania, and Zagreb. In 1969 he was decorated with the grand cross of the Order of Merit of the Federal Republic of Germany.

In appearance Pevsner was tall, slim, and fresh complexioned but with the somewhat 'owlish' mask of the scholar. In conversation he was lively and responsive, with a keen sense of humour; his letters were always concise and sometimes witty. He lived simply, worked incessantly, and was wholly dedicated to scholarship and the arts, especially to the architecture of his adopted country, to the study of which he may be said to have contributed more than any man of his time.

Pevsner married in 1923 Karola ('Lola') Kurlbaum, who has been described as 'the most important influence on his life'. She was the daughter of Alfred Adolf Kurlbaum, by his first wife, who was of Jewish descent, though Kurlbaum himself was not. He was a lawyer who practised at the Supreme Court in Leipzig. By her Pevsner had two sons and a daughter, all of whom survived him. She died in 1963. Pevsner's last years, during which he was incapacitated by Parkinson's disease, were spent at 2 Wildwood Terrace, Hampstead, which he had occupied since 1941. He died there 18 August 1983.

[*The Times*, 19 August 1983; Peter Murray in *Proceedings of the British Academy*, vol. lxx, 1984; information from Stephen Games; personal knowledge.]   JOHN SUMMERSON

PHILLIPS, JOHN BERTRAM (1906–1982), biblical translator and Christian apologist, was born in Barnes, London, 16 September 1906, the second of three children and elder son of Philip William Phillips, OBE, a civil servant in the Ministries of Health and Labour, and his wife, Emily Maud Powell, formerly a post office assistant. His mother died of cancer in 1921, after a long and painful illness. He was educated at Emanuel School, Wandsworth, and from there progressed to Emmanuel College, Cambridge. He obtained a third class in part i of the classical tripos (1926) and a second in part i of the English tripos (1927). He then went to Sherborne Preparatory School, Dorset, to teach for a year. At Cambridge he had passed from atheism to Christianity and a conviction that he must become a clergyman in the Church of England. In 1928 he was accepted as a student at Ridley Hall, Cambridge, and after a year of study was ordained deacon in St Katharine's church in London's dockland in 1930 (he became a priest in the same year).

Phillips had already agreed to go to Penge, to assist Sidney Ford as his curate at St John's. In his third year he was taken ill with a nervous complaint and resigned his curacy. He was a patient of Leonard Browne, a Christian psychiatrist, who greatly helped him. With recuperation and then a period of journalism, he was able to return to the ministry.

In 1936 Canon Frank Hay Gillingham invited him to be his curate at St Margaret's, Lee, in south-east London. Then he became priest-in-charge of the Church of the Good Shepherd, Lee, and later vicar. There in 1940 he began to translate the Letters of St Paul. They were intended for his youth club, the King's Own. He was encouraged by C. S. Lewis [q.v.] to publish, and eventually a translation of all the letters in the New Testament appeared as *Letters to Young Churches* (1947).

Before then, in the latter part of 1944, Phillips was offered the living of St John's, Redhill. He and his wife went there in January 1945 and remained for ten years until he found the daily grind of parish work too much if he were to continue his now very successful writing and translation. Because of the astonishing success of *Letters to Young Churches* he received invitations to lecture and broadcast in many parts of Britain and America. In January 1955 he moved to Swanage to concentrate upon his writing and now voluminous pastoral correspondence.

His translation work included eventually the whole of the New Testament and four prophets from the Old. *Letters to Young Churches* was followed by *The Gospels in Modern English* (1952), *The Young Church in Action*—a translation of the Acts of the Apostles (1955), *The Book of Revelation* (1957), and *Four Prophets* (1963).

Before leaving Redhill, he had already published his first four books of Christian apologetics— *Your God is too Small* (1952), *Plain Christianity* (1954), *Appointment with God* (1954), and *When God was Man* (1954). At Swanage he continued this writing. *New Testament Christianity* appeared in 1956 and a book for the Church Missionary Society, *The Church under the Cross* also in 1956. He did this extensive writing and translating while also being very active in broadcasting, and undertaking lecture tours and missions. The BBC schools broadcasting department persuaded him to write a series of plays which were broadcast and published as *A Man Called Jesus* (1959). A year later he published *God our Contemporary* (1960).

Early in 1960 his former nervous disorder returned and he fell victim to severe clinical depression. In his own words, 'Without any particular warning the springs of creativity were suddenly dried up; the ability to communicate disappeared overnight and it looked as if my career as a writer and translator was over' (*The Price of Success*, 1984). Phillips cancelled all engagements and refused many attractive invitations. He struggled with great courage against a persistent and ever returning depression. He greatly underestimated his achievements during his twenty years of depression. From 1964 to 1969 he was canon of Salisbury Cathedral. He also continued his voluminous correspondence, dealing with pastoral problems throughout the world. That correspondence formed the basis of a book published after his death, *The Wounded Healer* (1984). The controversies that followed the publication in 1963 of *Honest to God* by John A. T. Robinson [q.v.] moved him to write *Ring of Truth* (1967) and to issue a second edition with reference to his correspondence in the same year. During those years he also revised and brought up to date his *New Testament in Modern English* (revised edn., 1972). He also participated in a project to publish a series of commentaries based upon his revised translation and wrote the commentary on Mark's Gospel, with the title *Peter's Portrait of Jesus* (1976).

In 1939 he married Vera May, a teacher of dancing, the daughter of William Ernest Jones, bank manager. They had one daughter. Phillips became DD (Lambeth) in 1966 and was awarded an honorary D.Litt. by Exeter University (1970). After a long illness he died at home in Swanage, 21 July 1982.

[J. B. Phillips, *The Price of Success*, 1984, and *The Wounded Healer*, 1984 (autobiographies); private information; personal knowledge.]   EDWIN ROBERTSON

PIKE, SIR THOMAS GEOFFREY (1906-1983), marshal of the Royal Air Force, was born at Lewisham 29 June 1906, the youngest of three sons (there were no daughters) of Captain Sydney Royston Pike, of the Royal Artillery, and his wife, Sarah Elizabeth Huddleston. He was educated at Bedford School, where he was a member of the Officers' Training Corps, and entered the Royal Air Force through the RAF College, Cranwell, on 1 January 1924.

After qualifying as a pilot in December 1925 he served with No. 56 Fighter Squadron, equipped with the Gloster Grebe and then the Siskin, based at Biggin Hill. During the next three years, having trained as a flying instructor, he taught at the Central Flying School and undertook a two-year engineering course, after which he served in the Middle East, partly on engineering duties. In 1937, now a squadron leader, he attended No. 15 course at the RAF Staff College, Andover, and he spent 1939 as chief flying instructor at No. 10 Flying Training School, Ternhill.

Throughout 1939 and 1940 he worked in the Air Ministry in the directorate of organization before taking command of No. 219 Squadron as a wing commander in 1941. Here, flying Beaufighters from Tangmere, he took part in the night air defence of the United Kingdom and shot down several enemy aircraft, winning the DFC (1941) and bar (1941). From September 1941 until June 1943, now a group captain (1941), he served at Headquarters No. 11

Group and as officer commanding North Weald before moving to the Middle East on further promotion to become senior air staff officer at Headquarters Desert Air Force during the Italian campaign; in 1945 he was made officer of the US Legion of Merit, and this initial experience of working with the army and the Americans was later to stand him in good stead. He was promoted to air commodore in 1944.

The war over, after a short period commanding the Officers' Advanced Training School, he returned to the Air Ministry as director of operational requirements where he played a key role in deciding the types of aircraft needed by the post-war air force. In 1949 he attended the Imperial Defence College, in 1950 became air officer commanding No. 11 Group in Fighter Command (having been promoted to air vice-marshal in 1950), and in 1951 went to a NATO appointment at Fontainebleau as deputy chief of staff (operations) under General Norstad. Two years later he returned to the Air Ministry and on the sudden death of W. A. D. Brook was promoted to air marshal to join the Air Council as deputy chief of the air staff, responsible for all aspects of the RAF's equipment programme at a time when critical decisions had to be taken concerning the V-bombers, the RAF's future fighter aircraft and its long-range transport force. From 1956 to 1959, as air officer commanding-in-chief Fighter Command, he had to cope with the major contraction of the UK fighter force that followed the defence review of Duncan Sandys (later Lord Duncan-Sandys).

He became chief of the air staff on 1 January 1960 and over the next three and a half years led the RAF through a number of difficult debates. The abandonment of Blue Streak he accepted as unavoidable in view of its vulnerability, and the cancellation of the Skybolt system and the decision to turn to Polaris for the British deterrent he regarded as disappointing for the RAF but inevitable. He fought hard to protect his Service at a time when there were strong pressures for greater centralization, and in endless discussions about Britain's overseas role he firmly advocated the retention of fixed bases east of Suez. In all his work as CAS his approach was quiet, honest, and straightforward; he had little time for the machinations and deviousness of the politicians and some of his military colleagues, and his own views—most carefully considered—were always firmly stated. Despite the problems of the day, which were accentuated by the ending of National Service, the RAF remained under his leadership a highly efficient force, with its central role the provision of the UK nuclear deterrent. In 1962 he was promoted to marshal of the RAF.

In January 1964, uniquely for a former CAS,

he moved with some misgivings to Paris as deputy supreme allied commander, Europe, under General Lemnitzer. When he left there in March 1967 his doubts had been confirmed: his time had been interesting and enjoyable but he felt he had been given too little real responsibility. He retired to Harlow where he devoted much time to the local community, and served as a respected president of the Royal Air Forces Association from 1969 to 1978. He was appointed CBE (1944), CB (1946), KCB (1955), and GCB (1961).

He married Kathleen Althea, daughter of Major Herbert Elwell, in 1930. They had a son and two daughters. Pike died 3 June 1983 at Princess Mary's RAF Hospital, Halton.

[Official records; personal knowledge; advice of relatives and friends.] HENRY A. PROBERT

PILKINGTON, (WILLIAM) HENRY, BARON PILKINGTON (1905-1983), industrialist, was born at Eccleston Grange, St Helens, Lancashire, 19 April 1905, the eldest in the family of three sons and two daughters of Richard Austin Pilkington, a director of the family glass business, and his wife, Hope, daughter of Sir Herbert Hardy (later first Baron) Cozens-Hardy, master of the Rolls from 1907 to 1920. His childhood years from 1909 until shortly before World War I were spent in Colorado where his father, ill with tuberculosis, was sent to recover. He was educated at Rugby School and Magdalene College, Cambridge, where, apart from rowing for the college, he did not distinguish himself, taking a third class in both parts of his tripos: history (1925) and economics (1927).

His memory, outstanding head for figures, and appetite for long hours of work, all of which he was to display after entering the family company in 1927, makes this uninspiring educational performance all the more remarkable. The Pilkington glass business was then going through a particularly difficult phase—his father and uncle both resigned in 1931. In 1934 'Harry' Pilkington became a full member of the board, but in the following year he found himself, quite unexpectedly, senior commercial director. He travelled extensively and took advantage of the spread of air services to meet glass manufacturers and merchants outside Europe. He was chairman of the company from 1949 to 1973, leading Pilkington's through a period of growth and prosperity until it was transformed in 1970 into a public company.

The building boom and increased motor vehicle production demanded greater glass-making capacity not only from Pilkington factories in Britain but also from new ones which were opened in South Africa, Canada, and India as well as from processing plants elsewhere. Pilkington's revolutionary float glass process was invented in the 1950s. Its ultimate success, after years of costly development, owed much to the chairman's unflagging support of (Sir) L. A. B. Pilkington, the inventor, and the company's engineers against its sceptical scientists. The licensing of the process to manufacturers abroad after 1963 not only generated large annual incomes but also greatly strengthened Sir Harry's hand in his negotiations with these powerful business interests and eventually led to his company becoming, by growth and purchase, the world's largest glass manufacturing concern. Pilkington's greatest achievement during the 1960s, however, was the successful presentation to the monopolies commission of the company's case, for which he was personally responsible.

From 1944 to 1952 he chaired the national council of building material producers and in May of the latter year he attracted attention in Whitehall by a report on the methods and costs of school building. A few months later the Pilkington board allowed him to accept the presidency of the Federation of British Industries (1953-5) on the grounds that it was not only an honour but also a national duty, but warned him against the further distracting invitations which would follow. Knighted in the New Year honours in 1953, he served on the important Crichel Downs inquiry (1954) and chaired the royal commission on doctors' and dentists' remuneration (1957-60) and, immediately afterwards, the committee on broadcasting (1960-2), for which the Pilkington board gave permission only 'after considerable discussion'. He also served as a director to the Bank of England (1955-72) and on the council of European industrial federations (1954-7). He had an abiding interest in education which occupied more of his time as he grew older. He served as chairman of the national advisory council for education for industry and commerce (1956-66), and was a member of the Manchester Business School's council (1964-72), president of the association of technical institutions (1966-8), and first chancellor of Loughborough University of Technology (1966-80). He received honorary degrees from Loughborough (1966), Manchester (1959), Liverpool (1963), and Kent (1968). He was made a life peer in 1968.

A lifelong Congregationalist, he believed in the gospel of work and strict self-discipline. His home remained on the outskirts of St Helens itself, in a house which the family had occupied since 1826. He was a keen rose grower and he would sometimes surprise his employees by turning up to prune their roses if his gardener was unable to do so. His devotion to the bicycle became legendary, particularly in the City.

Much time in his last years was spent writing in his own hand to company pensioners who reached the age of eighty and sometimes delivering these letters to their homes.

In 1930 he married Rosamond Margaret ('Penny'), daughter of Colonel Henry Davis Rowan, of the Royal Army Medical Corps. They had a son and two daughters. His wife died of a heart attack in 1953 and his younger daughter was drowned in Iraq in 1960. In 1961 he married Mavis Joy Doreen, daughter of Gilbert Caffrey, cotton manufacturer, and former wife of John Hesketh Wilding, radiologist. Pilkington died in St Helens 22 December 1983.

[T. C. Barker, *The Glassmakers*, 1977; *Dictionary of Business Biography* (ed. D. J. Jeremy), vol. iv, 1985; Pilkington archives; personal knowledge.]   T. C. Barker

PITMAN, Sir (ISAAC) JAMES (1901-1985), educationist and publisher, was born in London 14 August 1901, the eldest in the family of four sons and two daughters of Ernest Pitman, businessman, and his wife, Frances Isabel Butler, and grandson of Sir Isaac Pitman [q.v.], the inventor of Pitman shorthand. His early childhood was spent in the neighbourhood of Bath, and in country pursuits, while his enquiring mind was stimulated by his father's scientific experiments and innovations. He showed early promise as an athlete, setting up at Summer Fields, his preparatory school in Oxford, records which have held for three-quarters of a century. At Eton and Oxford University this promise was more than fulfilled. At Eton he gained the title of victor ludorum in 1918 and 1920, but, by a special rule, was not permitted to compete in 1919. He won the middleweight public school boxing in 1919. He played rugby football for Oxford University against Cambridge in 1921, and he was capped for England against Scotland in the Calcutta Cup match of 1922. He also ski'd and ran for Oxford that same year. He went to South Africa with the combined Oxford and Cambridge athletics team, but had the misfortune to break a leg which put him out of contention. Much later in life he competed in the Anglo-Swiss interparliamentary skiing, which he personally won several times.

Always of independent thought, he had a very fertile brain which turned towards high finance. At Christ Church, Oxford, he obtained second class honours in modern history (1923). He then entered the family business, whose three sections—office training, publishing, and printing—had been developed by his Uncle Alfred and his father. He worked in close association with his father-in-law, the first Lord Luke, who helped him to gain control of the

Pitman business for his branch of the family, and he became chairman in 1934.

Having a paramount interest in youth and education, he became the enthusiastic bursar of the Duke of York's and King's Camp (1933-9), which was set up at the instigation of the Duke of York (later King George VI) as an annual camp at Southwold where boys from public schools mixed with boys from less fortunate backgrounds.

With the coming of World War II, he served as a squadron leader in the RAFVR from 1940 to 1943. In 1943 he was appointed to the Treasury as director of organization and methods. Meanwhile in 1941 he had become the youngest director of the Bank of England, an appointment he held until he entered Parliament. He was asked to stand as Conservative candidate for Bath in the 1945 general election, and was elected. Although he never attained ministerial rank, he nevertheless played a prominent part in politics for nearly twenty years, until 1964, being a popular and most active member in his constituency. Besides his full-time job as chairman and joint managing director of Pitman, he served for many years on the boards of several public companies, notably Boots, Bovril, and Equity & Law.

His later years were much occupied by his efforts to establish the Initial Teaching Alphabet (ita) as a method of teaching children, especially those who had difficulties in learning to read. He had been considerably influenced by the spelling reform ideas of his grandfather, and became increasingly involved in spelling reform after he entered Parliament. He was invited to become a trustee under the terms of the will of George Bernard Shaw [q.v.] for the development of a new alphabet. This influenced him further and he was then to turn to his ita. In this he augmented the twenty-six letter alphabet with sufficient new characters to achieve a system which was phonetic but by which the superficial appearance of most words altered little. It was found that fluency in reading in the ita could, by a simple transition, lead to fluency in reading conventionally. It was in practice largely successful, but he worked tirelessly for the remainder of his life endeavouring to persuade education authorities and teachers to adopt this new means to successful reading. The reluctance of many was to leave him puzzled and disappointed. After his retirement from Pitman in 1966, he was able to devote all his time and his considerable energy to its promotion.

A great joy of his later life was his appointment as charter pro-chancellor of the University of Bath. He took his duties conscientiously, and paid frequent visits on ceremonial and other occasions over a number of years. This, like everything else he did, was associated with his desire for the improvement and expansion of education worldwide.

Pitman was appointed KBE in 1961 and had honorary degrees from the universities of Bath and Strathclyde. In 1927 he married Margaret Beaufort ('Beau') (died 1983), second daughter of Sir George Lawson Johnston (later first Baron Luke, q.v.), chairman of the firm of Bovril Ltd.; they had three sons and a daughter. Pitman died 1 September 1985 at 58 Chelsea Park Gardens, London.

[Personal knowledge.]

HUGH LAWSON JOHNSTON

PLOMLEY, (FRANCIS) ROY (1914–1985), deviser and presenter for forty-three years of *Desert Island Discs*, was born 20 January 1914 at Kingston upon Thames, Surrey, the only child (two previous sons had been stillborn) of Francis John Plomley, pharmaceutical chemist, and his wife, Ellinor Maud, daughter of Wright Heyhoe Wigg, agricultural engineer, of Norfolk. Reading, cinema, and the theatre were early passions, and his formal education at King's College School, Wimbledon, proved less interesting than taking part in a locally produced operetta, an experience which left him 'stage-struck'.

An eager ambition to work in the theatre, or 'in one of the ancillary forms of theatre, such as films or broadcasting' was not quickly fulfilled; Plomley was employed successively by an estate agent, an advertising agency, a publishing company, and a man who sold horoscopes by post from an office in Jersey. From Jersey, Plomley paid his first visit to France, thus inaugurating a lifetime of Francophilia. When phoney astrology palled, he returned home to take spasmodic employment as a film extra, and short engagements in musical comedy and pantomime. He even busked briefly in the streets of Guildford.

Early in 1936 Plomley sent a programme idea to the International Broadcasting Company which ran English programmes from Radio Normandy. The idea was turned down, but Plomley was engaged in April 1936 as an announcer and producer, working from the company's makeshift station at Fécamp, at Poste Parisien in Paris, and at IBC's London headquarters in Portland Place. Sent back to wartime Paris in January 1940, he took with him his fiancée, the actress Diana Beatrice Wong, and they were married by the pastor of the Church of Scotland in the French capital. She was the daughter of Siong Yew Wong, a doctor. When Hitler's army advanced to occupy Paris, the couple had a narrow escape across the channel. They were to have one daughter.

One cold November night in 1941, sitting by the embers of a dying fire, Plomley had the idea which was to make him a household name. Even the title *Desert Island Discs* came to him at that moment and, 'with a rare spurt of animation', he wrote at once to the BBC. In January 1942 he interviewed his first 'castaway' in a projected series of eight programmes, which was extended to fifteen. Not long before his death he recorded his 1,791st edition of *Desert Island Discs*. It was by far the longest run of any radio programme with the same presenter.

Roy Plomley asked his celebrity guests which eight gramophone records they would take with them to a desert island, 'assuming, of course, that they also had a gramophone'. As the series progressed, they were also allowed to take with them a favourite book 'apart from Shakespeare and the Bible', and a luxury object. The formula proved highly effective as a means of revealing character, thanks to Plomley's skill as an interviewer. With a blend of meticulous research and disarming courtesy, he got people to speak frankly by putting them completely at ease.

Plomley presented many other radio and television programmes, including a TV variant of *Desert Island Discs*, *Favourite Things*, and his broadcasting achievements were recognized by numerous awards. But it was perhaps success as a writer, especially for the theatre, that he most coveted. When he himself was the 'castaway' in 1958, he told Eamonn Andrews that on his island at about half past seven, he would like to think of 'the lights going up in Shaftesbury Avenue'. Plomley wrote sixteen plays, though only one (*Cold Turkey*) achieved a West End run, and he had a remarkable collection of theatrical memorabilia. A compulsive scribbler, he was also fascinated by history and worked on a number of historical novels, including an account of his ancestors' sufferings in the Monmouth rebellion. Among his published books is a delightful volume of reminiscences which reflects his whimsical sense of humour: *Days Seemed Longer* (1980).

Plomley compensated for a lonely childhood by becoming a very clubbable person—'an inglenook man' was one apposite description. He was a beloved member of the Savage Club and of the Garrick, where he lunched his male castaways before recording *Desert Island Discs*.

He was appointed OBE in 1975. He died 28 May 1985 at his home in Putney.

[Roy Plomley, *Days Seemed Longer*, 1980; BBC scripts; information from Mrs Diana Plomley; personal knowledge.]

RICHARD BAKER

POLUNIN, OLEG (1914–1985), botanist and traveller, was born 28 November 1914 at Reading, the second son of four children (he had two brothers and one sister) of Vladimir Polunin, who was Russian, a lecturer at the Slade School of Art, and his British wife, Violet Hart, daughter of a Kent farming family. Both parents were artists who worked with Diaghilev and the Ballets Russes. His home was filled with their paintings, supplemented by a rich variety of objects collected in his travels: witness to an inherited love of visual detail in man-made artefacts equalling a sharp eye for the natural world.

Polunin was educated at St Paul's School and read botany as an exhibitioner at Magdalen College, Oxford, obtaining a second class degree in 1937. He served in the army throughout the war of 1939-45, in the Education Corps and then Intelligence Corps, being demobilized with the rank of captain following service in India and Java. After the war he returned to Charterhouse School where he had begun teaching botany and biology in 1938. He retired early in 1972 to concentrate on writing, having inspired generations of field botanists, many of whom had joined him and his wife on excursions to many parts of Europe. However, a love of travel, school holidays, and a willingness by the school to release him for six-month sabbaticals, enabled Polunin to journey to more distant places and it was these botanical travels and the resultant books that became the main thrust of his career.

In 1949 (Sir) George Taylor, deputy keeper of botany at the British Museum of Natural History, helped get Polunin attached—as botanist—to the first British mountaineering expedition into Nepal, led by H. W. Tilman [q.v.], and in 1952, on a wholly botanical expedition, organized jointly by the Museum and the Royal Horticultural Society, to the little known region of western Nepal. Later botanical exploration and collecting took him to Kashmir, Iraq, Turkey in 1954-6 working on the flora of Turkey, and in 1960 on the Anglo-American expedition to the Karakoram, led by Wilfred Joyce.

Communication was as important as travel and in 1960 he took part in what was to become the first of many tours as a lecturer and guide for the Society for Hellenic Travel. The circumstances leading to this pivotal redirection of Polunin's career had its origins in the fact that a brother of one of the two women owners of the travel firm of Fairways & Swinford was an archaeologist, and from this fraternal connection sprang the idea for (and later astonishing post-war growth of) guided tours to archaeological sites. As a further development of these specialist tours, Anthony Huxley and Oleg Polunin were invited to lead botanical tours to Greece. From this experience, the need for a plant book of the region was recognized

and later realized in their joint authorship of *Flowers of the Mediterranean* (1965). Many years of travel followed between 1965 and 1984. Polunin travelled to the remoter parts of Europe with his wife in their Volkswagen camper, photographing and collecting plants for his books, adventures which culminated in their drive to India in 1977.

As a major consequence of these private expeditions Polunin was inspired to produce a trilogy of related field guides to the flora of Europe: *Flowers of Europe: A Field Guide* (1969), *Flowers of South-West Europe* with B. E. Smythies (1973), and *Flowers of Greece and the Balkans* (1980), supplemented by *The Concise Flowers of Europe* in 1972 with B. Everard, *Trees and Bushes of Britain and Europe* in 1977, and, with Martin Walters, *A Guide to the Vegetation of Britain and Europe* in 1985. A felicitous style of writing and high standard of plant photography, much of which illustrated Polunin's own skills with the camera, led to most of his books being translated into several languages. In using photographs rather than drawings, Polunin pioneered the use of small cameras and good quality colour film for flower illustration. He also, in his concise floras, pioneered the use of symbols for flower form, habitat, and geographical distribution.

In 1984 *Flowers of the Himalaya*, which Polunin wrote with Adam Stainton, was published, thereby bringing to fruition a long-standing ambition since Polunin's first journey to Nepal, thirty-five years earlier. A concise version was published after his death in 1987; also posthumously there appeared *Collins Photoguide to Wild Flowers of Britain and Northern Europe* in 1988, edited by John Akeroyd—one of Polunin's pupils—who wrote of him that he was a kindly generous man, giving freely of his botanical knowledge; as much respected by the professional botanist as by his large non-professional readership.

Saxifraga poluninii and *Primula poluninii* were named after him as also was an aphid collected at 5,500 metres. He discovered the type specimen of *Rhododendron cowanianum* and a new species of lousewort. In 1962 he was awarded the Veitch memorial medal by the Royal Horticultural Society for his success in bringing back good garden plants, and in 1983 the Linnean Society conferred on him the H. H. Bloomer award for his contribution to biological knowledge. He was also joint recipient with Mrs B. Everard of the Grenfell medal for the photographs and line drawings in *Flowers of Europe*. Polunin was active in local conservation, becoming a founder member of the Surrey County Naturalists Trust, and later its secretary and chairman.

In 1943 he married Lorna Mary, daughter

of John Venning, sanitary engineer, and Ethel Knight. They had a son and a daughter. His wife was an active participant in many of his explorations, guided tours, and publications. Polunin died of motor-neurone disease at his daughter's house near Bradford-on-Avon 2 July 1985.

[*The Times*, 5 July 1985; family information.]

THOMAS HUXLEY

PORTER, RODNEY ROBERT (1917-1985), biochemist, immunologist, and Nobel prize-winner, was born at Newton le Willows in Lancashire 8 October 1917, the younger son (there were no daughters) of Joseph Lawrence Porter, chief clerk at the Railway and Carriage Works at nearby Earlestown, and his wife, Isobel Mary Reese. From Ashton-in-Makerfield Grammar School he entered Liverpool University where he qualified with a first class honours degree in biochemistry in 1939. He was awarded the Johnston colonial fellowship to enable him to undertake postgraduate work at Liverpool. Soon after starting research in the biochemistry department he volunteered for war service and in the late summer of 1940 joined the Royal Artillery. Subsequently he was commissioned in the Royal Engineers and saw service in the Mediterranean theatre where later as a major in the Royal Army Service Corps he ended his military service working as a War Department analyst.

A very formative period in his career began after demobilization in 1946 when he joined Frederick Sanger as a postgraduate student in the department of biochemistry at Cambridge. Sanger was then developing methods that were later to enable him to determine the structure of insulin, work which led to the award of his first Nobel prize. Inspired by the writings and the work of the immunologist, Karl Landsteiner, Porter was struck by the remarkable properties of the immunoglobulins, proteins that were not widely studied by biochemists at the time. He was so astounded by the wide range of antibody specificity exhibited by what appeared to be a homogeneous protein molecule that he devoted the whole of his scientific life to the study of the chemistry of the immune process.

In 1949 he moved to the National Institute for Medical Research at Mill Hill where he continued his work on the immunoglobulins and built up an international reputation as an authority on protein structure and function. By repeating in 1958 an experiment that he first carried out in 1948, but using the improved techniques then available, he made the important discovery which was later to lead to the elucidation of the immunoglobulin structure. He was able to cleave the molecule into two fragments, each possessing different unique properties of the original protein and thus greatly simplifying the study of the immunoglobulins. With his scientific reputation rapidly rising he was appointed in 1960 to the first chair of immunology to be established in Britain, at St Mary's Hospital Medical School. His discovery stimulated intense activity throughout the world and Porter, always an internationalist in his work, was one of the founder members of the Antibody Workshop. This was a loose association of international scientists who were leaders in the field and who did much to promote the advancement of the subject. In 1962 at a meeting in New York Porter was the first to propose the four-chain structure of immunoglobulin, a model that stands today with little modification. The importance of these advances for medical science was clear and appropriately in 1972 he shared the Nobel prize for physiology and medicine with G. M. Edelman.

At Oxford, where he occupied the Whitley chair of biochemistry from 1967 (with a fellowship at Trinity College), he extended his interests to another aspect of the immune response, the complement system. Aided by colleagues in the immunochemistry unit set up in Oxford under his direction by the Medical Research Council, he established himself within a few years as an outstanding international authority on complement. This was accomplished by supplementing his skill as a protein chemist with the powerful new techniques of immunogenetics and molecular biology. In view of the fact that Porter was rapidly approaching the retiring age for university staff this was an impressive achievement. Even at the age of sixty-seven his scientific vigour was such that the Medical Research Council, quite against their normal practice, asked him to continue to direct the immunochemistry unit for a further four years. Always mindful of the medical significance of his discoveries, towards the end of his life he was beginning to speculate on the relationship between the expression of the various forms of the complement genes and autoimmunity.

The quality of his research was such that he has been described by some as the father of modern immunochemistry. His achievement was to have the foresight, confidence, and persistence to attempt to explain in biochemical terms, at a time when knowledge of protein structure was extremely rudimentary, the remarkable properties of the immunoglobulins. Essentially an individualist in research he had the capacity to engender respect and develop continuing friendships with all those who worked with him. Collaborators came from all over the world to work in his laboratory and many went on to become leaders in their fields. The impact of Porter's work was such that honours were showered on him, but about these he characteristically rarely spoke. He was elected FRS in 1964 and received the Royal (1973) and Copley (1983) medals. His work was particularly highly regarded in the United States where in addition to election to foreign membership of the National Academy of Sciences in 1972 he was honoured by a number of other societies. Honorary doctorates came from universities in Britain and continental Europe and he was elected honorary FRCP (1974) and FRSE (1976). A few months before his death he was made a Companion of Honour.

Porter was a tall almost gaunt figure with a down-to-earth manner. His remarks were often blunt, usually brief and to the point but on acquaintance the attractive personal qualities that made him so many friends soon became obvious. He was well known for his dry, laconic humour which was liable to emerge on almost any occasion irrespective of its nature. Research was the main passion in his life but much of his relaxation was spent in the open air. He was a keen and expert rock climber in his youth and continued to walk in the mountains right up to his death. He had a great love of gardens and gardening which became his major interest outside his work when he moved to Oxford where he lived near Witney in a farmhouse with a large garden. This was probably the happiest time of his life for here he enjoyed the intellectual life of Oxford and the prestige of his scientific eminence combined with the simple pleasures of country pursuits.

In 1948 he married Julia Frances, daughter of George Francis New, an industrial physicist. They had two sons and three daughters. Porter was killed in a road accident near Winchester 6 September 1985, a few weeks before he was due to retire from the Whitley chair.

[S. V. Perry in *Biographical Memoirs of Fellows of the Royal Society*, vol. xxxiii, 1987; private information; personal knowledge.]

S. V. Perry

POSTAN, Sir MICHAEL MOÏSSEY (1899–1981), economic historian, was born 24 September 1899 at Tighina in Bessarabia, the eldest of three children (two sons and a daughter) of Efim Postan, a property owner, and his second wife, Elena. His early education was at the local high school and in Odessa; but war, his own army service, and revolution combined to frustrate attempts to continue his studies successively at the universities of St Petersburg, Odessa, and Kiev, and by late 1919 he was

sufficiently out of sympathy with the trend of events in Russia to decide to leave that country. Following a year spent mainly in the Balkans he arrived in England at the end of 1920. He became a British citizen in 1926.

If these early years gave Postan his cosmopolitan quality, his enrolment for a first degree course at the London School of Economics in October 1921 was decisive in his development as an economic historian. He graduated in 1924 and, after postgraduate study, became research assistant to Eileen Power [q.v.], an appointment which confirmed the medieval bias of his interests. He went on to establish himself as a teacher in academic posts at University College, London (1927–31) and the London School of Economics (1931–5), years during which he also built up the circle of friends engaged in many branches of the social sciences (it included Hugh Gaitskell, T. H. Marshall [qq.v.], Evan Durbin, and (Sir) Raymond Firth) from which he derived so much intellectual stimulus. Then in 1935 he moved to Cambridge where in 1938, when still only thirty-nine, he was elected to the professorship of economic history. The rest of his life was spent in Cambridge except for the years of World War II when uses were found for his specialist skills. From 1939 to 1942 he was head of the Russian section of the Ministry of Economic Warfare, and thereafter took charge of the history of war production for the official civil history of the war being prepared under the direction of (Sir) W. Keith Hancock.

Work on this history continued for some time after Postan's return to Cambridge in 1945, and it was during the Cambridge years before and after the war that he emerged as beyond doubt one of the most influential economic historians of his generation. His influence derived in part from the fact that his conception of what economic history ought to be, as he worked it out in many of the essays republished as *Fact and Relevance* (1971), was more coherent than that of many scholars. This task of definition appealed to his taste for debate and speculative thought and engaged an interest in the social sciences which was exceptionally wide. Even more, however, his influence depended upon his qualities as a teacher, whether as a lecturer or in seminars or in his ability to feed the enthusiasm of individuals; and its scope was enlarged by the unusually positive way in which he interpreted his role as editor of the *Economic History Review* (1934–60) and *The Cambridge Economic History of Europe*. Most important of all were his own writings. He followed the advice R. H. Tawney [q.v.] gave him as a student and engaged in some modern history, for his books included *British War Production* (1952) and *An Economic History of Western Europe*,

1945-64 (1967); but his principal concerns were always the trade and the rural society of the middle ages. His numerous essays on these topics were collected in two substantial volumes in 1973 and many of his conclusions were summed up in chapters on agrarian society in medieval England and on the trade of northern Europe contributed to *The Cambridge Economic History of Europe*, vol. i (2nd edn., 1966) and vol. ii (1952). In sum these writings constituted a wide-ranging reassessment of the economy of the middle ages.

Postan was single-minded in his dedication to his work, but he had time for much else, including opera, climbing and walking, and collecting porcelain. He was also a welcoming host and his circle of friends, which was characteristically international, grew wider with the years. Perhaps most of all he delighted in discussion and debate, in which, as in his lectures, his effectiveness was often enhanced by his accent, an occasional individuality of phrasing, and a pervading sense of humour.

Postan's standing as a scholar was widely recognized. He was a fellow of Peterhouse from 1935 to 1965 and an honorary fellow from 1965, an honorary fellow of the London School of Economics from 1974, president of the Economic History Society (1963-6) and of the International Economic History Association, and the recipient of honorary degrees from the universities of Birmingham, Edinburgh, York, and the Sorbonne. He was elected a fellow of the British Academy in 1959 and knighted in 1980.

In 1937 Postan married Eileen Edna le Poer Power, historian, daughter of Philip Ernest le Poer Power, a London stockbroker. They had no children and she died in 1940. He married, secondly, in 1944 Lady Cynthia Rosalie, second daughter of Walter Egerton George Lucian Keppel, ninth Earl of Albemarle; they had two sons. Postan died in Cambridge 12 December 1981.

[Edward Miller in *Proceedings* of the British Academy, vol. lxix, 1983; private information; personal knowledge.] EDWARD MILLER

PRIESTLEY, JOHN BOYNTON (1894-1984), novelist, playwright, and essayist, was born 13 September 1894 at 34 Mannheim Road, Toller Lane, Bradford, the only child of Jonathan Priestley, schoolmaster and Baptist layman, and his wife, Emma Holt, who died when John was an infant. He was brought up by a stepmother, Amy Fletcher, 'who defied tradition by being always kind, gentle, loving'. He had one stepsister. He attended Belle Vue Grammar School, Bradford, but left at sixteen by his own choice, and worked as a junior clerk

at the wool firm of Helm & Co. When G. B. Shaw [q.v.] later praised Stalin's Russia 'because you meet no ladies and gentlemen there', Priestley retorted: 'I spent the first twenty years of my life without meeting these ladies and gentlemen.' But he made the most of Bradford's two theatres, two music-halls, flourishing arts club, and vigorous musical life, as well as the Bradford Manner, 'a mixture of grumbling, irony and dry wit'. He sported Bohemian dress, including a jacket 'in a light chrome green', enjoyed an attic study, where he produced poetry and articles, had them typed by a professional, 'a saucy, dark lass' who was 'paid in kisses, for I had no money', and got a few printed in the local Labour weekly, the *Bradford Pioneer*, and even in *London Opinion*.

Priestley later portrayed his Bradford life as idyllic, asserting 'I belong at heart to the pre-1914 North Country', but at the time he was bored by it and when war came promptly enlisted in the Duke of Wellington's Regiment. He had two long spells in the Flanders front line, was wounded twice, commissioned in the Devon Regiment in 1918, and demobilized the following March. He always refused to be unduly impressed by any event, however momentous, and put his war experiences quietly behind him, never collecting his medals and declining to write about the war, except briefly in *Margin Released* (1962). But it left its mark. Half a century later, when a young guest told him she never ate bread at meals, he snorted: 'I can see *you* never served in the trenches.'

With an ex-serviceman's grant he went to Trinity Hall, Cambridge, where he refused to be enchanted by the atmosphere, let alone to acquire what he termed 'a private income accent', but did a vast amount of reading, laying the foundation for his later *tour de force, Literature and Western Man* (1960). He obtained a second class in English (1920) and a second in division I of part ii of the history tripos (1921). He also produced there his first volume of essays, *Brief Diversions* (1922), which brought him the patronage of (Sir) J. C. Squire [q.v.] at the *London Mercury*, reviewing from Robert Lynd [q.v.] at the *Daily News*, and a job as reader for the Bodley Head. For the rest of the decade he led the life of a London literary journalist, producing reviews, articles, and books, including two novels, workmanlike biographies of George Meredith and Thomas Love Peacock [qq.v.] for Macmillan's English Men of Letters series, and a little volume, *The English Comic Characters* (1925), which became a particular favourite of actors. Such work gave him a living but no leisure and in 1928 (Sir) Hugh Walpole [q.v.], always eager to assist new talent, collaborated with him in a novel, *Farthing Hall* (1929), so that the advance Walpole's fame commanded would give Priestley the time to write the major picaresque story he was plotting.

The Good Companions was begun in January 1928 and its 250,000 words finished in March 1929. Heinemann's, who had daringly printed 10,000 copies, brought it out in July and by the end of August it had sold 7,500. Thereafter it gathered pace and, to the accompaniment of the Wall Street collapse, became one of the biggest sellers of the century. By Christmas the publishers were delivering 5,000 copies a day by van to the London bookshops. Priestley, typically, did not allow his head to be turned and privately pooh-poohed the merits of his warmhearted tale of a travelling theatrical troupe. He thought his next novel, *Angel Pavement* (1930), set in London, much better. But *The Good Companions*, besides being twice filmed (1932 and 1956), was put on the stage in 1931, where it brought out the talents of the young (Sir) John Gielgud and opened up a new career for Priestley as a dramatist.

While not a natural novelist, always having difficulty with the narrative flow, Priestley was stimulated by any kind of technical challenge and the stage offered plenty. He dismissed his first West End play, *Dangerous Corner* (1932) as 'merely an ingenious box of tricks'. But James Agate [q.v.], then the leading critic, called it 'a piece of sustained ingenuity of the highest technical accomplishment' and it began a decade of theatrical success. In 1937 three Priestley plays opened within a few weeks and for several years his earnings from the theatre alone exceeded £30,000. His were not, like Shaw's, literary plays, at their best when read, but solid pieces of theatrical machinery, dependent on stagecraft and timing, offering rich opportunity for actors. Priestley never turned success into formula: all his plays are different, many of them experimental. *Eden End* (1934) evokes pre-1914 nostalgia, *Time and the Conways* (1937) deals with the theories of J. W. Dunne, *I have Been Here Before* (1937) explores the philosophy of P. D. Ouspensky, *Music at Night* (1937) examines the psychological impact of sounds, *Johnson Over Jordan* (1939) probes life after death, and *When We Are Married* (1938) is mordant Yorkshire comedy.

Priestley's desire never to repeat himself was strength and weakness. 'I am too restless', he told Agate in 1935, 'too impatient, too prolific in ideas. I am one of the hit-or-miss school of artists.' He wrote quickly—his novel *The Doomsday Men* (1938) took only nineteen days—but whatever he did had to be new and this disappointed admirers anxious to typecast him as the provider of provincial warmth. He took a close interest in new media, producing screenplays, studying pre-war TV, and writing

for the BBC, including a novel, *Let the People Sing* (1939), the first instalment of which was broadcast the day war was declared. Priestley had the instincts of an actor, and indeed would act whenever opportunity offered, though his face, which he described as 'a glowering pudding', limited his range. His voice was another matter: it combined unmistakable northern values with mesmeric clarity, 'rumbling but resonant, a voice from which it is difficult to escape', he wrote. In spring 1940, with Hitler triumphant, the BBC had the inspired idea of getting Priestley to broadcast brief 'Postscripts' after the main news bulletin on Sundays at 9 p.m., starting on 5 June and running till 20 October. Throughout that historic summer, his talks, combining light-hearted pleasure in things English with sombre confidence in final victory, and delivered with exceptional skill, formed the perfect counterpoint to the sonorous defiance of (Sir) Winston Churchill's broadcasts. They made him an international figure. Indeed they excited, he believed, Churchill's jealousy and when the BBC, in its mysterious way, dropped him, he thought the prime minister responsible, though it was more probably Conservative Central Office.

Priestley never belonged to a party, but he described his father as 'the man socialists have in mind when they write about socialism' and his own ideas were usually radical. His novels, like Emile Zola's, were often journalistic in choice of subject, taking a topical theme, and such wartime stories as *Black-out in Gretley* (1942) and *Three Men in New Suits* (1945) seemed to place him on the Left. Along with the *Daily Mirror* and the Left Book Club he was credited with the size of Labour's 1945 victory and in 1950 he even made an official Labour election broadcast. He contributed regular essays on current trends to the *New Statesman*, later collected as *Thoughts in the Wilderness* (1957), *The Moments* (1966), and *Outcries and Asides* (1974). But Priestley was incapable of acting in concert with a political group, or indeed any organization which valued 'sound men' (a favourite term of disapproval). He resigned in disgust from the British committee to Unesco and from the boards of both the National Theatre and the *New Statesman*. He contributed to the latter a remarkable article in 1957 which led directly to the Campaign for Nuclear Disarmament. But at a private meeting to plan it, an objection by Denis Healey MP, 'we must be realistic', evoked a characteristic Priestley explosion: 'All my life I have heard politicians tell us to be realistic and the result of all this realism has been two world wars and the prospect of a third.' He was briefly associated with the Aldermaston marches but left the movement when it became an arena for left-wing faction. 'Commitment' was a posture he despised.

Priestley liked to think of himself as a lazy man but there were very few days in his long life when he did not write something, usually in the morning. His output was prodigious in size and variety. In the 1940s he wrote two of his most striking plays, *An Inspector Calls* (1947) and *The Linden Tree* (1948); his post-war novels included *Lost Empires* (1965) and his own favourite, *The Image Men* (2 vols., 1968 and 1969), a sustained attack on the phenomenon he called Admass. He travelled constantly, and both painted (in gouache) and wrote about what he saw. *English Journey* (1934), recording light and shade during the Slump, was constantly revived and imitated, and became a classic; there is fine descriptive writing in his autobiographical works, *Midnight on the Desert* (1937) and *Rain upon Godshill* (1939), while *Trumpets over the Sea* (1968) records an American tour with the London Symphony Orchestra. In 1969-72 he produced a historical trilogy dealing with the period 1815-1910: *The Prince of Pleasure*, *The Edwardians*, and *Victoria's Heyday*. Above all, there were scores of essays, long and short, relaxed and serious. He always wrote clear, unaffected, pure English, but it is his essays which best display his literary skills.

Priestley never claimed genius, another word he despised, merely 'a hell of a lot of talent'. He fought a lifelong battle with the critical establishment: 'I was outside the fashionable literary movement even before I began.' He believed himself to be undervalued after 1945, having been overvalued before it, and often pointed out that his plays were more highly regarded abroad than in Britain. In fact from the 1970s onward they were revived with increasing frequency and success. His work brought substantial material rewards. While making a decisive shot at croquet (a game he relished), he once startled a guest by listing to him the formidable aggregate sums he had paid in income tax and surtax. In 1933 he bought Billingham Manor and estate, where a roof-top study gave him a panoramic view over the Isle of Wight; after the war he moved to an ample Regency house near Stratford-upon-Avon. There, in its splendid library, its bookshelves hiding a bar where he mixed formidable martinis, he would receive a constant stream of guests and interviewers, or switch on monumental gusts of stereophonic music and, when he thought no one was watching, conduct it. He turned down a knighthood and two offers of a peerage but accepted the OM in 1977 and a clutch of honorary degrees, 'as a chance to dress up'. He was never the 'Jolly Jack' of his popular image; rather, a shrewd, thoughtful, subtle, and scep-

tical seer, a great craftsman who put a good deal into life, and a discriminating hedonist who got a lot out of it. In old age he became a little deaf and forgetful but stayed fit and industrious almost to the end, pleased to have got excellent value from his annuities.

In 1919 Priestley married Emily ('Pat'), daughter of Eli Tempest, insurance agent. She died in 1925 after a long distressing illness, leaving him with two daughters. In 1926 he married Mary ('Jane'), the former wife of Dominic Bevan Wyndham Lewis, author, and daughter of David Holland, marine surveyor, of Cardiff. She already had a daughter, who was brought up in the Priestley household, and she and Priestley had a son and two daughters. The marriage ended in 1952 and, after a contested divorce which left him with an abiding dislike of lawyers, especially judges, in 1952 he married the archaeologist Jacquetta Hawkes, daughter of Sir Frederick Gowland Hopkins [q.v.] biochemist, and former wife of Professor (Charles Francis) Christopher Hawkes, archaeologist, by whom she had one son. Priestley died at his home, Kissing Tree House, Alveston, 14 August 1984.

[John Braine, *J. B. Priestley*, 1978; Susan Cooper, *J. B. Priestley*, 1970; private information; personal knowledge.]

PAUL JOHNSON

PRINGLE, JOHN WILLIAM SUTTON (1912-1982), biologist, was born in Manchester 22 July 1912, the eldest of four sons (there were no daughters) of John Pringle, a medical doctor, of Rochester and then of Manchester, and his wife, Dorothy Emily Beney, of Huguenot extraction. One of his ancestors was Sir John Pringle (1707-82), a founder of modern military medicine who became president of the Royal Society. Thus an interest in biology ran in the family for several generations.

Pringle won a scholarship to Winchester in 1926 where in the sixth form he took science subjects and also Greek. He was awarded a major scholarship at King's College, Cambridge, in 1931, and then went on to gain a first in both parts of the natural sciences tripos (1933 and 1934). He held a research studentship at King's in 1934-7 under R. J. Pumphrey, whose notice he later wrote for this Dictionary. His gift for rigorous analysis and immense skill in experimental design were soon in evidence and he produced important papers on the function of campaniform sensilla in 1938-9. He was made a university demonstrator in 1937 and was elected to a fellowship at King's College in 1938.

When World War II broke out in 1939 he joined up with the RAF and was assigned to research on airborne radar with the Telecommunications Research Establishment. He headed a research team which made many valuable contributions to the war effort, particularly during the invasion of Europe. For this service he was appointed MBE and awarded the American medal of freedom (both 1945).

He returned to the Cambridge department of zoology in 1945 remaining there until 1961, acting as lecturer, administrative officer, and finally as reader in experimental cytology. In 1959 he was made a member of the general board of the university. He was appointed to a fellowship at Peterhouse (1945) where successively he became tutor, senior tutor (1948-57), senior bursar (1957-9), and librarian (1959-61). In 1954 he was made FRS, and in 1955 was awarded his Sc.D. in recognition of his outstanding contributions to arthropod biology.

In 1961 he accepted the Linacre chair of zoology at Oxford together with a fellowship at Merton College. The challenge of designing the promised new department and of 'opening up zoology at Oxford' attracted him greatly. However the initial plan for a tower block on a very limited site in the University Parks was rejected by Congregation, and the situation was redeemed only when another and much larger site was made available on Merton ground where ultimately a more satisfactory building was erected, to be opened in 1971. He took immense pains over planning the building, going into every detail with characteristic thoroughness and imagination, and the result, if somewhat unlovely on the outside was exceptionally well planned within. He gave the Royal Society Croonian lecture in 1978, was made president of the Society for Experimental Biology in 1977, and gave the Bidder lecture to that society in 1980. He retired in 1979.

His contributions to research were in the fields of insect physiology and muscle biophysics. He made a full analysis of the haltere mechanisms of flies; he investigated the skeletal and muscular mechanisms of insect flight; and at Oxford he directed an Agricultural Research Council unit which through the use of insect material made many fundamental contributions to the understanding of muscular contraction. Penetrating analysis combined with skill in experimental design were the hallmarks of his research, and he made imaginative use of engineering principles, incorporating them into sophisticated models of mechanisms. His book *Insect Flight* (1957) was of lasting influence, and his numerous papers published in learned journals set a high standard of scientific endeavour. In addition he was able to sustain a productive interest in many other biological fields, making original contributions to such

topics as the origins of life, the two biological cultures, biological responsibility and education, world population, the scientific study of mankind, and conservation.

Pringle took an active interest in educational matters, and his vision of the central unifying role of biology bore fruit when Oxford was persuaded by his vigorous arguments to launch a new honour school in human sciences whose first undergraduates arrived in 1970. He was a major figure in planning the course, which later became a well-regarded feature of the Oxford landscape. He was also much concerned with the development of science in Third World countries and with forming links with tropical universities. He was visiting professor at Nairobi University in 1973 and he gave strong support to a new International Centre for Insect Physiology and Ecology, later acting as chairman of its governing board.

In the laboratory his transparent honesty, sincerity, and clear-minded approach to problems were paramount, while at home he could be a warm, friendly, and genial host. At times, however, these qualities were concealed below a cold and formidable exterior when iron self-discipline and impatience with stupidity were most evident. He was once compared to a *bombe surprise* turned inside out, the thin cold layer of austerity being spread over a warm and generous humanity. He was a deeply religious person who found no incompatibility between his science and his faith. Throughout his life he pursued many other interests, which included gardening, bee-keeping, painting, embroidery, wine-making, woodwork, canals, and gliding (for which he won a gold medal in 1960).

In 1946 he married Beatrice Laura, a widow with one daughter of Captain Martin Wilson and daughter of Humphrey Gilbert-Carter, a well-known Cambridge botanist. They had a son and two daughters. Pringle died in Oxford 2 November 1982.

[V. B. Wigglesworth in *Biographical Memoirs of Fellows of the Royal Society*, vol. xxix, 1983; personal knowledge.]   P. L. MILLER

PRINGLE, MIA LILLY KELLMER (1920-1983), psychologist and first director of the National Children's Bureau, was born in Vienna 20 June 1920, the elder child and only daughter of Samuel Kellmer, a prosperous wholesale timber merchant, and his wife, Sophie Sobel. Her younger brother, Chanan Kella, emigrated to Israel. She attended the State Humanistic Gymnasium at Vienna, VI Rahlgasse, where she matriculated with distinction. She went to Britain with her mother as a refugee in 1938. They arrived virtually penniless, speaking no English, and Mia worked as a primary school teacher to support herself and her mother, whilst studying at Birkbeck College, London. She was awarded a first class honours BA in psychology in 1944. She then qualified as an educational and clinical psychologist at the London Child Guidance Training Centre in 1945. In 1945–50 she worked as an educational psychologist in Hertfordshire, whilst studying at the University of London for a Ph.D., which she was awarded in 1950.

From 1950 to 1963 she taught in the University of Birmingham, first as a lecturer, then senior lecturer and deputy head of the Remedial Education Centre (later the Department of Child Study). During this time she did much to develop the department as a research and postgraduate training centre, and published a large number of research articles concerned with handicapped children, remedial education, and children in care.

In 1963 she was invited to become the first director of the National Children's Bureau, then known as the National Bureau for Co-operation in Child Care. The eighteen years she spent there until her retirement in 1981 were her most productive. Initially the Bureau consisted of herself, two researchers, and a secretary. When she retired she left a staff of sixty-five in a large purpose-built building. She started the Bureau with the four aims of bringing together the different professions concerned with children; publicizing research knowledge about children; improving services for children and pioneering new ones; and carrying out policy-related research about children.

The most important of the Bureau's research projects was the National Child Development Study, a longitudinal study of 17,000 children born in 1958. As well as the major findings on this cohort, published in such books as *Birth to Seven* (1972, by R. Davie and others) and *Britain's Sixteen Year Olds* (1976, by K. Fogelman), there were important studies of special groups within the cohort, such as children in care, one-parent families, and gifted children. Many researchers were involved in this research programme, but its inception, continuation, and success owed much to Mia Pringle's drive. Whilst at the Bureau, she wrote or edited twenty books, as well as many articles. One of her most influential and best-selling books, *The Needs of Children* (1974), was translated into German, Swedish, and French, but some researchers preferred her annotated research summaries on policy issues, including *Adoption: Facts and Fallacies* (1967) and *Foster Home Care: Facts and Fallacies* (1967).

Although a considerable scholar, with an almost obsessive concern for detail, Mia Pringle was only interested in research that had a direct bearing on a practical problem. She was not only, or perhaps essentially, an academic. She was also a great administrator and publicist, tirelessly working to bring different professions together to influence policy on children. She regarded herself above all as a campaigner for children's rights, and to this end wrote widely in the popular as well as the academic press, appeared frequently on television and radio, and served on many government committees. At times she aroused hostility, as when she argued against nursery provision, and the employment of women with under-fives. However, she was listened to with respect by cabinet ministers and senior civil servants.

She was appointed CBE in 1975. In 1970 she was given the Henrietta Szold award for services to children. She also received honorary doctorates from the universities of Bradford (1972), Aston (1979), and Hull (1982). She was made an honorary fellow of Manchester Polytechnic (1972), of the College of Preceptors (1976), and of Birkbeck College, London (1980). After her retirement she acted as a consultant to Unicef.

Mia Pringle was a very hard-working person, who needed little sleep. In personality she was somewhat authoritarian, finding it difficult to share power. At the same time she had great charm, and part of her effectiveness came from a combination of charm, diplomacy, and quick-wittedness. About personal matters she was unusually reserved, and some who knew her well spoke of an underlying melancholy.

In 1946 she married William Joseph Somerville Pringle, an analytical chemist, son of William Mather Rutherford Pringle, barrister and MP. In 1962 William Pringle died, and in 1969 she married William Leonard Hooper, assistant director-general of the Greater London Council, son of Alfred Albert Edward Hooper, insurance inspector. Leonard Hooper died in 1980. Mia Pringle had no children. She died by her own hand in her London flat 21 February 1983.

[*The Times*, 25 February 1983; *Newsletter* of the Association of Child Psychologists and Psychiatrists, summer 1983; address by W. D. Wall at a public memorial meeting at Queen Elizabeth Hall, South Bank; private information; personal knowledge.]

BARBARA TIZARD

PRITTIE, TERENCE CORNELIUS FARMER (1913–1985), journalist and author, was born in London 15 December 1913, the younger son (there were no daughters) of Henry Cornelius O'Callaghan Prittie, fifth Baron Dunalley in the peerage of Ireland, and his wife, Beatrix Evelyn, daughter of James Noble Graham, of Carfin, Lanarkshire. His parents led a peripatetic life, partly in France, until they moved back to the rebuilt family seat at Kilboy in county Tipperary. Prittie had indifferent health as a child and remained short of stature, but he became an excellent player of ball games in adult life, and was a first-class shot. He was sent to Cheam School, where he was not happy, and then on to Stowe. Here he blossomed, partly under the influence of an outstanding history master, and made many friends. Leaving Stowe at Christmas 1932, he spent some months in Germany and began to acquire his mastery of German. In 1933 he entered Christ Church, Oxford, and widely extended his circle of friends; he read modern history, gaining a good second class in 1936, and in his last year was awarded a Boulter exhibition. He narrowly missed blues in both tennis and lawn tennis; in 1935 he was the open champion at lawn tennis of his native province of Munster.

In 1937 he joined Childs Bank in Fleet Street, but not finding banking congenial moved to a firm of stockbrokers. In 1938 he had joined the reserve of the Rifle Brigade; after episodic training in England and Northern Ireland, he embarked in May 1940 for Calais, where on 26 May after the stubborn defence he was taken prisoner. He was mentioned in dispatches (1940). An attempt to escape in France miscarried, and thereafter he succeeded in breaking out of prison on six occasions (described in *South to Freedom* 1946). One escape brought him within sight of the Swiss frontier, and on his final evasion he joined a unit of the advancing Americans in April 1945. While within walls he was agile at goading his captors, and also invented a form of cricket in the moat at Spangenberg. On his return to England in 1945 he was appointed MBE (military).

His knowledge of cricket and his memory of obscure statistics were already outstanding: he joined the staff of the *Manchester Guardian* in February 1946 as sports correspondent in succession to (Sir) J. F. Neville Cardus [q.v.]; he was to publish four books on cricket past and present. But in October 1946 he was sent to Germany as the paper's chief correspondent and made his home with his new wife in Düsseldorf after three difficult years in occupied Berlin. His alert intelligence and potent memory, coupled with a slightly detached viewpoint furnished by his Cromwellian-Irish ancestry, enabled him to become a superb commentator on the German scene. In addition to his lively and well informed reports, he produced biographies of two of the German chancellors whom he met. The true affection which he developed for the country comes out in *My Germans* (1983). An interest in local wine inspired *Moselle* (with Otto Loeb, 1972). When he left Germany in 1963, his industry and his experience rendered him an outstanding and influential figure in press and diplomatic circles.

In London he became the diplomatic correspondent of the *Guardian* and held this post till 1970. In December 1971 the German authorities awarded him the officer's cross (civil) of their Order of Merit. His wife had been so ill that he contemplated retirement to Malta when a new career opened up. He had been interested in the cause of Zionism and Jewry since his early days in Germany, and now became an official advocate of the Israeli cause; he had already published *Israel: Miracle in the Desert* (1967) and a biography of Eshkol (1969), and contributed to the Israeli paper *Ha'aretz*. From a series of small offices he issued a monthly broadsheet 'Britain and Israel', commenting on contemporary problems and expounding the Jewish case. His research was meticulous and his language clear and convincing; his dedication was entire. Frequent visits to Israel and journeys to the United States and elsewhere enlarged his audience. Meanwhile he was frequently at Lord's to watch cricket and play tennis.

Prittie's wonderfully retentive memory enriched his public and his private life. A sleepless night could be alleviated by naming an XI of left-handed clergy in first class cricket. His public manner embodied a robust natural courtesy, but among intimates he loved a private, and occasionally a practical, joke; pomposity he abhorred. Many friends from each facet of his varied career enjoyed his natural gaiety. Relations between both Britain and Germany, and Britain and Israel, were enriched and enhanced by his career and writings. He married on 29 August 1946 Laura (died 1988), only daughter of Gustave Dreyfus-Dundas, an oil engineer, of Colombia, South America. They had two sons. Prittie died in London 28 May 1985.

[Terence Prittie, *Through Irish Eyes*, 1977; *The Times*, 29 May 1985; personal knowledge.]   MICHAEL MACLAGAN

R

RACE, ROBERT RUSSELL (1907-1984), human geneticist and discoverer of numerous blood groups, was born 28 November 1907 at Hull, the eldest of three children, all boys, of Joseph Dawson Race, banker, of York, and his wife May, daughter of Robert Tweddle, a sanitary engineer of West Hartlepool. He was educated at St Paul's School and then at St Bartholomew's Hospital, where he qualified MRCS (Eng.) and LRCP (Lond.) in 1933.

Up to this point his career was entirely undistinguished and he moved to a junior post in pathology. The turning point came in 1937 when he secured the post of assistant serologist at the Galton laboratory after being interviewed by the well-known statistician, (Sir) R. A. Fisher [q.v.]. Race's job was to assist in the laboratory's serum unit which had been set up by Fisher in 1935 with a grant from the Rockefeller Foundation. At the outbreak of World War II the unit moved to Cambridge. At first work was confined to the ABO and MN blood group systems but in 1942 the unit began to study the Rh antigen and antibody which had recently been described in the USA. It soon became clear that there were several different Rh antigens and the picture became bewildering, but Fisher suggested a scheme which brought immediate understanding of the interrelationships. So far, Race's role had been to produce entirely reliable serological results and Fisher's to supply the ideas but very soon Race showed that he had his own powerful contribution to make.

In 1946 Race moved with his small team to the Lister Institute in London and the enterprise became the Medical Research Council's Blood Group Unit. Race now entered on a period of great productivity. As a result of his growing fame and due largely to his open, friendly, and cheerful manner, both face to face and in correspondence, he started to receive blood samples from all over the world, containing antibodies to be identified. When there was no obvious solution, he showed a great flair for thinking of all sorts of possibilities and the products of his imagination were then subjected to rigorous statistical analysis. As each new antigen was identified he was able to show whether or not it was related to existing antigens and in this way he played a leading role in establishing knowledge of the various blood group systems. His unit was soon recognized internationally as the leader in the field. One of its most important discoveries was of an antigen carried on the X chromosome. The potential of this discovery was immediately recognized and in the next few years understanding of the numerous clinical syndromes stemming from abnormal arrangements of the X and Y chromosomes was greatly increased. Race became FRCP (Lond.) in 1959 and FRCPath. in 1972.

During the period from 1942 to the time of his retirement in 1973, Race's work was immensely influential not only in determining the rapid and orderly development of knowledge of human blood group systems but also in initiating the mapping of the human genome.

In 1938 Race married Margaret Monica, daughter of John Richard Charles Rotton, a solicitor. They had three daughters (the first adopted). After the death of his first wife in 1955, Race married in 1956 Ruth Ann, daughter of the Revd Hubert ('Tom') Sanger, headmaster of Armidale School, New South Wales. There were no children of the marriage. Ruth Sanger came to work at the Blood Group Unit in 1946 and stayed there until her retirement. From 1948 onwards she was a co-author of virtually every important paper which Race wrote, and their separate contributions became hard to disentangle. Together they produced six editions (1950-75) of a notable book *Blood Groups in Man*. This work was not only the best source of information in its field for twenty-five years but was written with great clarity, reflecting Race's love of literature, and humour, reflecting his exceptionally amusing personality.

Race was elected FRS in 1952 and appointed CBE in 1970. He received honorary degrees from the universities of Paris (1965) and Turku (1970), and many awards, for example, the Oehlecker medal of the Deutsche Gesellschaft für Bluttransfusion in 1970. In 1973 he was made a member of the Deutsche Akademie der Naturförscher Leopoldina. Jointly with Ruth Sanger he was given the Karl Landsteiner award of the American Association of Blood Banks in 1957 and Gairdner Foundation award in 1972. He was a member of many foreign societies. He died 15 April 1984 in Putney Hospital, London.

[Sir Cyril Clarke in *Biographical Memoirs of Fellows of the Royal Society*, vol. xxxi, 1985; private information; personal knowledge.]

P. L. MOLLISON

RAMBERG, CYVIA MYRIAM (1888-1982), ballet director and teacher. [See RAMBERT, DAME MARIE.]

RAMBERT, DAME MARIE (1888-1982), ballet director and teacher, was born Cyvia Myriam Ramberg 20 February 1888 in Warsaw, Poland, the youngest of the three children, all daughters, of Yakov Ramberg, a bookseller of Jewish descent whose father's surname was Rambam, and his wife, Yevguenia Alapina. For a time she called herself Myriam Ramberg, but when she came to London she assumed the name by which she was thereafter to be known, Marie Rambert. She was educated at the Gymnasium in Warsaw, and in 1905 was sent to Paris with the intention of studying medicine. Instead she began to associate with the artistic world, attracting the attention of Raymond Duncan, brother of Isadora, with whose free style of dancing she first identified. Between 1909 and 1912 she worked in Geneva under Emile Jaques-Dalcroze, whose influence was central to her artistic development.

Towards the end of 1912 she was engaged by Serge Diaghilev to give classes in Dalcroze eurhythmics to the dancers of his Ballets Russes and more especially to assist Nijinsky in the difficult task of choreographing *Le Sacre du Printemps*. Shortly after that ballet's riotous première she accompanied the company on its visit to South America. From her association with the Diaghilev Ballet she received a lasting legacy: a profound interest in classical dance. Her engagement was not renewed when the company returned to Europe, and she went back to Paris, moving to London on the outbreak of war in 1914. There she met the playwright Ashley Dukes (died 1959) [q.v.], whom she married on 7 March 1918, and by whom she had two daughters. Dukes was the son of the Revd Edwin Joshua Dukes, Independent minister. Rambert became a British subject by this marriage. After the war she became an assiduous pupil of the celebrated ballet teacher, Enrico Cecchetti, and was soon teaching ballet on her own account, gathering around her, as time went by, students of exceptional promise, among them (Sir) Frederick Ashton, Harold Turner [q.v.], and Pearl Argyle.

Such talented young dancers needed stage experience to fulfil themselves, and in the later 1920s Rambert began to supply this, first with the ballet, *A Tragedy of Fashion* (1926), which she persuaded Ashton to choreograph: it was his first ballet, and can now be seen as a historic landmark, from which a national ballet tradition was to spring. In the years that followed her students continued to make occasional appearances under the name of the 'Marie Rambert Dancers'. When Diaghilev died in 1929, there was a sudden dearth of ballet in London. In 1931, to fill the void, Rambert formed the Ballet Club with the object of forming a permanent ballet company with a theatre of its own. She even had a theatre—the miniscule Mercury Theatre near Notting Hill Gate, purchased by her and her husband out of their savings. It was to be the home of their ballet until 1939, and

for many years housed their remarkable collection of historic ballet lithographs (now in the Theatre Museum).

Rambert possessed a unique gift for discovering and nurturing young choreographers, and Ashton, Antony Tudor, and Andrée Howard were all greatly in her debt for the cultural enrichment that she brought them. Under her inspired direction, Ballet Rambert (renamed thus in 1934) became part of the fabric of the growing English ballet tradition, although it was to fall to (Dame) Ninette de Valois' Vic-Wells Ballet to be chosen as the national company. Owing to the tiny dimensions of the Mercury Theatre, Ballet Rambert operated as 'chamber ballet', but the absence of spectacle was compensated for by exquisite taste and attention to detail. During World War II the company became larger and outgrew the Mercury. But it never lost its strong interpretative quality, and in 1946 Rambert herself staged a production of *Giselle* that was remarkable for its dramatic content.

Towards the end of her active life Rambert guided the steps of another budding choreographer, Norman Morrice, who succeeded her as director in 1966 and launched Ballet Rambert on a new course with greater emphasis on modern dance. Rambert's support of this bold move was an indication of her extraordinarily active and receptive mind. She enjoyed a long retirement and lived on to 12 June 1982, when she died at her London home in Camden Hill Gardens.

Considered as one of the architects of British ballet, she was appointed CBE in 1953 and DBE in 1962. In 1957 she became a chevalier of the Legion of Honour; she received the Royal Academy of Dancing's Queen Elizabeth II Coronation award in 1956. The University of Sussex awarded her an honorary D.Litt. in 1964.

She had inexhaustible energy (she could turn cartwheels until she was seventy), an infectious sense of fun, and a very retentive memory (displayed in reciting poetry by the page). Above all she gave inspiration to others, without which British ballet would today be much poorer.

[Mary Clarke, *Dancers of Mercury*, 1962; Marie Rambert, *Quicksilver*, 1972 (autobiography); Richard Buckle, *Nijinsky*, 1971, and *Diaghilev*, 1979; Clement Crisp and others (ed.), *Ballet Rambert: 50 Years and On*, 1981; Mary Clarke, obituary in *Dancing Times*, July 1982.]   IVOR GUEST

RANDALL, SIR JOHN TURTON (1905–1984), physicist and biophysicist, was born 23 March 1905 at Newton-le-Willows, Lancashire, the only son and the first of the three children of Sidney Randall, nurseryman and seedsman, of Newton-le-Willows, and his wife, Hannah Cawley, daughter of John Turton, colliery manager in the area. He was educated at the grammar school at Ashton-in-Makerfield and at the University of Manchester where he was awarded in 1925 a first class honours degree in physics and a graduate prize and in 1926 an M.Sc.

From 1926 to 1937 Randall was employed on research by the General Electric Company at its Wembley laboratories, where he took a leading part in developing luminescent powders for use in discharge lamps. He also took an active interest in the mechanisms of such luminescence. By 1937 he was recognized as the leading British worker in the field and was awarded a Royal Society fellowship to study electron processes in luminescent solids in the physics department at Birmingham University. He made important advances in the electron trap theory of phosphorescence. But, when war began in 1939, he transferred in the department to the large group working on centimetre radar. By 1940 he had, with H. A. H. Boot [q.v.], invented the cavity magnetron and obtained an output of centimetre-wave power much greater than any before, thus overcoming the greatest obstacle in the development of radar. The magnetron invention was probably the only wartime scientific advance which was decisive in winning the war.

In 1944 Randall was appointed professor of natural philosophy at St Andrews University where he began planning research in biophysics. In 1946 he moved to the Wheatstone chair of physics at King's College, London, where the Medical Research Council set up the Biophysics Research Unit with Randall as honorary director and a wide-ranging programme of research was begun by physicists, biochemists, and biologists. Use of new types of light microscopes led to the important proposal in 1954 of the sliding filament mechanism for muscle contraction; also X-ray diffraction studies aided the development in 1953 at Cambridge of the double helix model of DNA by Francis Crick and J. D. Watson (who, for that work, were in 1962 awarded a Nobel prize jointly with M. H. F. Wilkins of Randall's laboratory).

In 1951 Randall set up a large multidisciplinary group working under his personal direction to study the structure and growth of the connective tissue protein collagen. The group gave important help to the elucidating of the three-chain structure of the collagen molecule. Randall himself specialized in using the electron microscope, first in studying the fine structure of spermatozoa and then in concentrating on collagen. In 1958 he began to study the structure of protozoa and he set up a new group to use the cilia of protozoa as a model system for the analysis of morphogenesis by correlating the structural and biochemical differences in mutants. In 1970 he retired to Edinburgh University where he formed a group which applied a range of new biophysical methods to study various biological problems. He continued that work with characteristic vigour until his death.

In science Randall was not only original but even maverick. He made extremely important contributions to biological science when, with imagination and foresight, he set up, at the right time, a large multidisciplinary biophysical laboratory where his staff were able to achieve much success. His contributions as an individual worker in biophysics were possibly not so outstanding as those in physics. In science and elsewhere he showed much wisdom and good judgement. He had unusual capacity to see the essentials of an overall situation and had outstanding entrepreneurial skill in obtaining funds and buildings for research. He was also very successful in integrating the teaching of biosciences at King's College.

Randall was ambitious, courageous, and liked power; but his ambition worked very largely for the common good. His friendly, informal, and democratic characteristics could contrast strongly with his Napoleonic self-assertion. He showed great dedication and enthusiasm in his scientific work and also in the extensive gardening he much enjoyed.

In 1938 Randall was awarded a D.Sc. by the University of Manchester. In 1943 he was awarded (with H. A. H. Boot) the Thomas Gray memorial prize of the Royal Society of Arts for the discovery of the cavity magnetron. In 1945 he became Duddell medallist of the Physical Society of London and shared a payment from the royal commission for awards to inventors for the magnetron invention, and in 1946 he was made a fellow of the Royal Society and became its Hughes medallist. Further awards (with Boot) for the magnetron work were in 1958 the John Price Wetherill medal of the Franklin Institute of the state of Pennsylvania and in 1959 the John Scott award, city of Pennsylvania. In 1962 he was knighted and in 1972 became a fellow of the Royal Society of Edinburgh.

He married Doris, daughter of Josiah John Duckworth, a colliery surveyor, in 1928. They had one son. Randall died 16 June 1984 at Edinburgh.

[M. H. F. Wilkins in *Biographical Memoirs of Fellows of the Royal Society*, vol. xxxiii, 1987; private information; personal knowledge.]   M. H. F. WILKINS

RAYNOR, GEOFFREY VINCENT (1913–1983), metallurgist, was born in Nottingham 2

October 1913, the youngest of the three sons, the second of whom died in infancy (there were no daughters), of Alfred Ernest Raynor, a lace dressers' manager, and his wife, Florence Lottie Campion. His family encouraged his early academic studies and in 1925 he won a scholarship to Nottingham High School. Under the enthusiastic guidance of the headmaster, C. L. Reynolds, and academic staff, he developed into an outstanding scholar with special abilities in chemistry and science in general. He developed parallel interests in sport, excelling in rowing, shooting, and swimming. Music always occupied a central place in his family life, centring upon choral singing and piano; in later life, this interest extended into ballet, orchestral performances, and opera. The death of his father in 1927 led to serious difficulties for the family and there were pressures upon him to take a 'safe job'. Advice from his mentors prevailed and in 1931 he won a Nottingham county major scholarship which enabled him to enter Keble College, Oxford, in October 1932. His undergraduate years coincided with a remarkable flowering of inorganic and organic chemistry at Oxford and he benefited greatly from the guidance of his tutors, F. M. Brewer and G. D. Parkes. He read for the final honour school of natural science, taking part i in 1935, and, most significantly, electing to work with the group led by Professor William Hume-Rothery (whose notice he later wrote for this Dictionary) in his part ii research period. In 1936 he gained a first-class honours degree.

The period 1936 to 1939, during which he was a research student, saw the development of the close and fruitful working relationship with Hume-Rothery that was to have a seminal influence upon his future career in science. Exciting ideas on solid-state zone theories had recently appeared and were leading to rapid growth in the science of physical metallurgy. His special interest lay in establishing the influences of atomic size and electronic factors upon the constitution of copper- and magnesium-based alloys and upon the formation of intermediate phases. To his regret, the demands of research interfered with opportunities for serious rowing which, in 1934, had led to him rowing in the university trial eights. Having made his individual mark in the well-known 'Hume-Rothery and Raynor' research group, he gained his D.Phil. in 1939.

During the war of 1939–45 he directed alloy chemistry research at Oxford on behalf of the Ministry of Supply and Ministry of Aircraft Production, supplementing his income by acting as a university demonstrator in inorganic chemistry. This classified research prevented him from joining the army.

In 1945 pressing economic circumstances made him decide to leave Oxford and to accept an ICI research fellowship in the department of metallurgy of the University of Birmingham. The department focused its research effort upon two areas: mechanical properties under the direction of (Sir) A. H. Cottrell and alloy research under Raynor. Promotion was rapid. Raynor became reader in theoretical metallurgy in 1947, professor of metal physics (1949–54), professor of physical metallurgy (1954–5), Feeney professor of physical metallurgy (1955–69), and head of department in 1955. An Oxford D.Sc. was awarded in 1948. His exceptional success in research inevitably drew him into the administrative complexities of university life. From 1966 to 1969 he was dean, faculty of science and engineering. During this period the faculty was expanding and changing greatly; in the metallurgical field it was notable for the installation of one of the first million-volt transmission electron microscopes in Britain. In 1969 a school of metallurgy was formed, combining the previously separate disciplines of physical and industrial metallurgy. In the same year Raynor became deputy principal of the university. Unfortunately his four years of office coincided with the aftermath of the 1968 student revolt; his role as chairman of the academic appointments committee was particularly wearing and difficult. In retrospect it is widely recognized that, despite provocation and some unwarranted vilification, he always acted in a courageous and steadfast manner in accordance with the best long-term interests of the university.

The critical evaluation and annotation of published phase diagrams and the painstaking construction of a systematic corpus of information upon the alloying characteristics of a wide range of metals are fundamental to modern metallurgical development. Their importance to the practising metallurgist is comparable to that of the periodic table to the chemist. After relinquishing the deputy principalship, partly for reasons of ill health, Raynor was able to devote his professorial activities to these objectives, continuing to do so until his death. This work was stimulated by a concurrent world-wide resurgence of interest in phase diagrams and alloy systems. It was therefore a natural and personally satisfying step for him to become chairman of the important alloy phase diagram data committee of the Metals Society (1976) and to join the international council of the Data Program for Alloy Phase Diagrams which operated under the aegis of the American Society for Metals and the US Bureau of Standards. He had acquired a considerable reputation in his chosen field of metallurgical research and was able to accept many professorships overseas, visiting the universities of Chicago (1951–2),

Ohio State (1962), Witwatersrand (1974), New South Wales (1975), and Queen's University, Ontario (1979). Travel, particularly by sea, always held a deep appeal for him.

His numerous research papers and scientific articles exhibit a distinctive style combining elegance with economy of words. Two of his books, *The Structure of Metals and Alloys* (1944, 1954, 1962), which he wrote with Hume-Rothery, and *An Introduction to the Electron Theory of Metals* (1947, 1988) are classic metallurgical texts. He also wrote *The Physical Metallurgy of Magnesium and its Alloys* (1959), contributed to the *Encyclopaedia Britannica* (1949), and shortly before his death completed with V. G. Rivlin a book on *Phase Equilibria of Iron Ternary Alloys* (1988). Teaching had a special place in his professional life and his lectures, with their intensive and often novel style, were highly regarded by undergraduate students, many of whom joined his research group. These presentations of the latest ideas in alloy research were characterized by precision and a sense of urgency. By temperament, he was gentle, kind-natured and considerate.

Prestigious awards for his scientific achievements included the Beilby memorial award of the Royal Institute of Chemistry and the Institute of Metals (1947), the Rosenhain medal of the Institute of Metals (1951), the Heyn medal of the Deutsche Gesellschaft für Metallkunde (1956), and the Hume-Rothery prize of the Metals Society (1981). Election as FRS (1959) was soon followed by election to the fellowship of the New York Academy of Sciences (1961). Raynor derived special pleasure from his honorary fellowship (1972) at Keble, his old Oxford College, and from the award of a Leverhulme emeritus fellowship (1981). He was simultaneously a fellow of the institutes of Physics, of Chemistry, and of Metallurgists, and he contributed significantly to the national development of metallurgy by serving as vice-president of the Institute of Metals (1953-6) and the Institution of Metallurgists (1963-6, 1977-80), and as president of the Birmingham Metallurgical Society (1965-6).

In 1943 he married Emily Jean, daughter of Dr George Frederick Brockless, musician. They had three sons. Raynor died 20 October 1983 in Birmingham.

[Sir Alan Cottrell in *Biographical Memoirs of Fellows of the Royal Society*, vol. xxx, 1984; R. E. Smallman in *Metals Society World*, January 1984; personal knowledge.]

R. E. Smallman

REDCLIFFE-MAUD, Baron (1906-1982), public servant. [See Maud, John Primatt Redcliffe.]

REDGRAVE, Sir MICHAEL SCUDAMORE (1908-1985), actor, was born 20 March 1908 in theatrical lodgings at St Michael's Hill, Bristol, the only child of George Ellsworthy ('Roy') Redgrave, actor, a specialist in melodrama, and his second wife, Daisy (known later in the theatre as Margaret), actress, daughter of Fortunatus Augustin Scudamore, dramatist. Sixteen months after his birth his mother took him for a short time to Australia where his father was acting. Three years later his parents were divorced. In 1922 his mother, who looked after him, married J. P. Anderson, who had formerly been employed by the Ceylon and Eastern Agency in Ceylon. A half-sister, Peggy, was born.

Michael Redgrave went to Clifton College where he became a competent schoolboy player in male and female parts. His Macbeth, at seventeen, made his mother, who had been opposed to this, think twice about him becoming a professional actor. In 1927 he went to Magdalene College, Cambridge, where he undertook much undergraduate acting and wrote for and edited university magazines. He obtained second classes in both the German section of the medieval and modern languages tripos (part i, 1928) and the English tripos (part i, 1930); in 1931 he gained a third class in part ii of the English tripos. He then went to Cranleigh School, Surrey, as modern languages master. Here, in effect, he was an actor-manager, doing six productions and playing, among other parts, Samson Agonistes, Hamlet, and Lear; moreover he was given work in the semi-professional Guildford repertory company, which was glad to have a recruit so accomplished and personable: he was six feet three and strikingly handsome. Confidently he resigned from Cranleigh and got an audition from Lilian Baylis [q.v.] of the Old Vic, who offered him a contract at three pounds a week. Before deciding, he had an interview with William Armstrong [q.v.], director of Liverpool Playhouse, who persuaded him to go there; between 1934 and 1936 he had a wide variety of parts in the most sympathetic circumstances.

More important to him, he fell in love with Rachel Kempson when they acted together in John van Druten's *The Flowers of the Forest*; two years his junior, she was the daughter of Eric William Edward Kempson, headmaster of the Royal Naval College at Dartmouth. They were married in the college chapel during the spring of 1935 and for another year remained at Liverpool. It was then that (Sir) Tyrone Guthrie [q.v.], becoming one of the principal directors of his time, invited them for a season at the Old Vic where in September 1936 they opened as Ferdinand of Navarre and the Princess of France in *Love's Labour's Lost*. In 1936-

7 Redgrave was Horner in *The Country Wife* (with the American actress, Ruth Gordon) and, to his delight, Orlando in *As You Like It* to the Rosalind of (Dame) Edith Evans [q.v.], forty-eight then but, with her unerring sense of comedy, ready for the adventure. At once they were attracted to each other, an association they sustained during the Old Vic run and a transference for three months to what was then the New Theatre. Before then Redgrave had another rich experience, Laertes to Laurence (later Lord) Olivier's vigorous Hamlet. A daughter was born to Rachel in January 1937 and they named her Vanessa. They later had a son Corin (born 1939) and another daughter Lynn (born 1943). All became well known on the stage.

Even after so brief a time in London, it was clear that Redgrave would be an important player; recognizing this, (Sir) John Gielgud gave him several parts (including Tusenbach in *Three Sisters*) in a season at the Queen's Theatre (1937-8). Work came easily. When he played, surprisingly, Aguecheek during a West End *Twelfth Night* in 1938 the drama critic James Agate [q.v.] called him 'a giddy, witty maypole'. At the Westminster in 1939 he was the first Harry Monchensey in *The Family Reunion* by T. S. Eliot [q.v.], and he had also, inevitably but reluctantly, gone into films: (Sir) Alfred Hitchcock [q.v.] cast him in *The Lady Vanishes* (1939). With the outbreak of war he had to abandon an Old Vic opportunity; instead, during 1940, he was acting and singing Macheath in *The Beggar's Opera* at the Haymarket; later, at a small Kensington theatre and in the West End, he appeared most sensitively as the idealistic recluse of Robert Ardrey's *Thunder Rock*.

His call-up papers reached him in June 1941 in the middle of a film and he found himself, as an ordinary seaman, training in devastated Plymouth. Discharged after a year for medical reasons, he returned to the stage in a sequence of plays, some as successful as Turgenev's *A Month in the Country* (he was Rakitin in 1943) and an American melodrama, *Uncle Harry* (1944), in which he acted with exciting nervous power. Curiously in 1947 he appeared to be out of key in an elaborate production of *Macbeth*. Another good spell was coming: first, the relentless tragedy of Strindberg's *The Father* (1948-9), then a long season with the Old Vic company at the New (its final period before going back to Waterloo Road). Redgrave ended with Hamlet (1950), a performance which lacked only the final quality of excitement: as a disciple of Konstantin Stanislavsky he was apt to concentrate upon a close dissection of the text. Thence he went to Stratford-upon-Avon for a pair of remarkable performances, an intellectually searching Richard II (1951) in

which he did not disguise the man's sexual ambiguity, and a Hotspur, grandly direct, with a precise Northumbrian accent. That summer he also played Prospero and the *Henry V* Chorus. In 1952, at the St James's, he was admirable in Clifford Odets's American drama, *Winter Journey*, though at the time there were awkward differences of opinion with a fellow actor. A good company man, Redgrave never hesitated to speak his mind. Meanwhile he gave an excellent performance in the film *The Browning Version* (1951).

During another Stratford year (1953) he had the unnerving trinity of Shylock, Lear, and the Antony of *Antony and Cleopatra*. As the triumvir at sunset Redgrave reached his Shakespearian height—(Dame) Peggy Ashcroft was Cleopatra—and the play had a London season at the Princes. Films, such as *The Dam Busters* (1954), continued to occupy much of his time. He returned in 1958 to Stratford—his last appearance there—and again, at fifty, acted as Hamlet, a performance mature and deeply considered. His final major work in the theatre was at the Chichester Festival of 1962 (as Uncle Vanya) and at the opening of the National Theatre in 1963, in the Old Vic, when he was an authoritative Claudius to the Hamlet of Peter O'Toole. The next year (1964) he was Solness in *The Master Builder*.

At the opening of the Yvonne Arnaud Theatre, Guildford, in May 1965 he returned to *A Month in the Country* which also had a West End showing. He had become a prodigiously popular film star in such productions as *Kipps* (1941) in which he played the title part, *The Way to the Stars* (1945), *The Importance of Being Earnest* (1952), and *The Quiet American* (1957); one of his last roles was General Wilson in *Oh What a Lovely War* in 1969. Illness was developing: he had Parkinson's disease and he kept to readings, on various international tours, during the ebb of his career. He made his last appearance in Simon Gray's *Close of Play* (National, 1979), a practically silent part during which he sat most of the night in a wheelchair. He died 21 March 1985 in a nursing home at Denham, Buckinghamshire.

Redgrave's publications include *The Actor's Ways and Means* (1953) and *Mask or Face* (1958), and a version of *The Aspern Papers* by Henry James [q.v.], in which he acted, as 'H. J.', at the Queen's in 1959. He published in 1983 an autobiography, *In My Mind's Eye*. He was made a Commander of the Order of Dannebrog (1955) and a D.Litt. of Bristol in 1966. He was appointed CBE in 1952 and knighted in 1959.

[Rachel Kempson, *A. Family and its Fortunes*, 1986; personal knowledge.]

J. C. TREWIN

REDMAYNE, MARTIN, first baronet, and BARON REDMAYNE (1910–1983), Conservative politician and chief whip, was born in Nottingham 16 November 1910, the second son and third child in the family of three sons and three daughters of Leonard Redmayne, civil engineer and farmer, and his wife, Mildred, daughter of Edward Jackson. He was educated at Radley and worked in the family sports business in Nottingham both before and after World War II when he was managing director. During the war he served with the Sherwood Foresters, commanding the 14th battalion in the Italian campaign and later forming and commanding the 66th Infantry brigade. He was appointed to the DSO in 1944. He was Conservative MP for the Rushcliffe division of Nottinghamshire from 1950 until the general election of 1966. He was also a JP for Nottingham (1946–66).

Redmayne's unusual political career was spent almost entirely in the Conservative Whips' Office in a period when the party formed the government from October 1951 to October 1964. With his distinguished war record, it was no surprise to find him in the Whips' Office by 1951 nor, when he became chief whip in 1959, that he should run it like a military headquarters and, off duty, like an officers' mess. He was admitted to the Privy Council in 1959.

He was a shy and reserved man, appearing aloof and even severe to some who served under him. However, he was generally regarded as courteous, loyal, fair, and immensely conscientious in his attention to detail. It came quite naturally to him both to obey commands without question and to expect his orders to be similarly obeyed. He therefore ran an efficient Whips' Office, though not a particularly imaginative one. He might have flourished as chief whip at a relatively uneventful time but 1959–64 was not such a period. Redmayne had courage, stoicism, and durability but he lacked the intuitive 'feel' for a crisis and the vision to handle it successfully.

He has been criticized for not being sufficiently quick to perceive the various serious problems which arose within the party during 1959–64 and adversely affected both the prime minister personally and the government. Nor was he sufficiently influential with Harold Macmillan (later the Earl of Stockton) to be able to alert him in time or to exercise what might be called the 'higher loyalty': putting at his leader's disposal not only his best efforts in carrying out the leader's instructions, but also any intuitive doubts as to the wisdom of what the leader was doing.

Thus Redmayne has been criticized for allowing 'the night of the long knives' in the summer of 1962 to happen, for failing to alert Macmillan much earlier to the dangers of the Profumo affair, and, more important politically, for his part in handling the succession to Macmillan. It is generally agreed that the open leadership battle at the Conservative Party conference at Blackpool in October 1963 was an avoidable blunder and that, if the succession had been decided calmly in London in the normal manner after the conference was over, the Conservatives might well have won the general election of October 1964.

However, Redmayne would have been a most remarkable person to influence all those situations successfully. Indeed he would have deserved—like Edward Heath, his predecessor as chief whip, and William (later Viscount) Whitelaw, his successor—to have gone much further in politics.

After the Conservative defeat in the October 1964 general election, Redmayne was made a baronet and was successively shadow postmaster-general and Conservative spokesman on agriculture and transport. After losing his seat in 1966, he was created a life peer and served in his later years as deputy chairman of the House of Fraser, a director of Boots, and chairman of the Retail Consortium.

In his limited leisure, in addition to some golfing and fishing, he was an enthusiastic water-colourist. He married in 1933 Anne (died 1982), daughter of John Griffiths, coal miner. They had one son, Nicholas John (born 1938), who succeeded to the baronetcy when Redmayne died at King Edward VII Hospital, London, 28 April 1983.

[Private information; personal knowledge.]

MICHAEL FRASER

RENAULT, MARY (pseudonym), writer. [See CHALLANS, EILEEN MARY.]

RENNIE, SIR JOHN OGILVY (1914–1981), diplomat and head of the Secret Intelligence Service (SIS) or MI6, was born 13 January 1914 in Marylebone, London, the only child of Charles Ogilvy Rennie, match manufacturer, and his wife, Agnes Annette Paton. He was educated at Wellington College and Balliol College, Oxford, where he showed precocious talent as a painter, exhibiting at the Royal Academy in 1930 and 1931 and at the Paris Salon in 1932. On leaving Oxford in 1935, with third class honours in modern history, he joined the advertising agency Kenyon & Eckhardt in New York. In 1938 he married a Swiss subject, Anne-Marie Celine Monica, daughter of Charles Godat, of La Chaux-de-Fonds, Switzerland. They had a son.

After the outbreak of war he worked first as a vice-consul in Baltimore. He was then drawn into the British Information Services, which was being organized to combat German propaganda in the United States, and from 1942 to 1946 he worked in New York. It was during this period that another aspect of his intellectual versatility was shown when he took up the study of electronics, which later became a hobby. On returning to London in 1946 he was formally accepted into the Diplomatic Service and was appropriately posted to the Foreign Office's information policy department.

In 1949 he was appointed first secretary commercial in Washington, where he began to establish his reputation in the Service, and he was transferred to Warsaw with similar duties in 1951. It was during the following two years in Poland that he gained firsthand experience of the realities of life behind the Iron Curtain. This enhanced his qualifications for his first senior appointment as counsellor and head of the information research department (IRD) in 1953. IRD had been set up in 1949 at the instigation of Ernest Bevin [q.v.] and Christopher (later Lord) Mayhew, then his parliamentary under-secretary of state. Its main directive was to disseminate anonymously both at home and abroad the factual evidence, largely drawn from Soviet bloc sources, of how communism works, in order to enlighten foreign governments and the media. Rennie soon found his feet in IRD and was widely admired for his skill and ingenuity, deployed over an unusually long five-year tenure of office, in advancing IRD's reputation. He was appointed CMG in 1956.

In 1958 he was promoted commercial minister in Buenos Aires, a post which led to a similarly successful appointment in Washington in 1960. Soon thereafter, his luck turned the other way. His wife fell seriously ill and in 1963 he took a year off in order to be at her side until her death in 1964.

On his return to the Foreign Office in 1964 he was appointed assistant under-secretary for the Americas and headed a special mission to Central America in an effort to resolve the problems between Guatemala and British Honduras. After a short period of secondment in 1966 to the Civil Service Commission as chairman of an interviewing board, he married his second wife, Jennifer Margaret, widow of Julian Miles Wemyss Rycroft, lieutenant in the Royal Navy, and daughter of Lieutenant-Colonel John Gordon Wainwright, of Penn, Buckinghamshire. They had two sons. In 1967 Rennie was promoted deputy under-secretary for defence matters, which involved him in the chairmanship of a number of cabinet committees. Recognition of his performance was shown by his appointment as KCMG in 1967.

Then in 1968, when the post of 'C' became vacant, he was unexpectedly appointed head of MI6, to his surprise, and no less to the surprise of the senior echelons of SIS. The case could have been made that for an outsider, his past service should have provided him with some parallel understanding of what to expect in his new role. But in fact he found it difficult to adjust to his changed responsibilities and, from first to last, he was always overshadowed by his deputy and ultimate successor (Sir) Maurice Oldfield [q.v.]. At no period did he feel at home in SIS, and his personal predicament was greatly worsened by a family tragedy when his son, Charles, and his daughter-in-law were sent to jail on drugs charges in 1973. This incident was widely publicized after the story, involving Rennie's identity, had appeared in the Hamburg magazine *Der Stern*. He retired in 1974 and devoted his time to painting and his lifelong love of sailing, with occasional duties as chairman of the English Speaking Union's current affairs committee. He died 30 September 1981 at St Thomas's Hospital, Lambeth.

[Private information; personal knowledge.]
NIGEL CLIVE

RHYL, BARON (1906-1981), economist and politician. [See BIRCH, (EVELYN) NIGEL (CHETWODE).]

RICHARDS, AUDREY ISABEL (1899-1984), social anthropologist, was born in London 8 July 1899, the second of four daughters (there were no sons) of (Sir) Henry Erle Richards, lawyer, and his wife, Isabel Mary, daughter of Spencer Perceval Butler, of Lincoln's Inn. From 1904 to 1909 her father was legal member of the viceregal council in India and at that period the family lived in Delhi and Simla. In 1911 Sir Henry was appointed Chichele professor of international law and fellow of All Souls College, Oxford, and the family returned to England. Audrey Richards was educated at Downe House School near Newbury and at Newnham College, Cambridge (1918-21). She gained a second class in part i of the natural sciences tripos (1921).

After graduation Audrey Richards taught at her old school for one year and was then an assistant to G. Gilbert A. Murray [q.v.] in Oxford. In 1924 she moved to London where she worked for four years as secretary to the labour department of the League of Nations Union. From 1928 to 1930 she doubled as a teacher of social anthropology at Bedford College and as a student of the subject at the London School of Economics where she registered to work for a Ph.D. (which she obtained in 1931) under the supervision of the professor of ethnology, C. G. Seligman [q.v.]. Her thesis, a slightly modified version of which was later published under the title *Hunger and Work in a Savage Tribe* (1932), was completed in the early spring of 1930. It is based entirely on reading; at that time the author had never visited Africa. But she left immediately afterwards for her first spell of field-work among the Bemba of Rhodesia among whom she worked from May 1930 to July 1931 and again from January 1933 to July 1934. During the academic year 1931-2 she held a lectureship in social anthropology at the London School of Economics and this lectureship was renewed when she returned from the field in the summer of 1934. By this time she had become a devoted follower of Bronislaw Malinowski. The massively detailed *Land, Labour and Diet in Northern Rhodesia* (1939), written under the auspices of the diet committee of the International Institute of African Languages and Cultures of which Audrey Richards was a member, was, in its anthropological aspects, closely modelled on volume i of Malinowski's *Coral Gardens and their Magic* (1935) which was concerned with the food production system of Kiriwina in Melanesia.

At this period Audrey Richards was undoubtedly an outstandingly successful fieldworker which sometimes surprises those who suffered from her haphazard approach to the problems of daily living in later life. She was, however, a member of what has been called 'the intellectual aristocracy'. Through her parents she had many distinguished connections. R. A. Butler (later Lord Butler of Saffron Walden, q.v.), who became chancellor of the Exchequer and master of Trinity College, Cambridge, was her first cousin. Her numerous maternal uncles included a governor of Burma and a governor of the Central Provinces, India. One sister married (Sir) Geoffrey Faber [q.v.], the publisher, who became a fellow of All Souls and another married (Sir) W. Eric Beckett, who was legal adviser to the Foreign Office and also became a fellow of All Souls. With this background it is hardly surprising that Audrey Richards took it for granted that in every situation, even in the middle of a Bemba village, she was a maternal representative of the powers that be. She was certainly eccentric, especially in later life, but adored for her eccentricities. She had a keen sense of humour which included, to a degree that is rare among academics, an ability to laugh at herself.

She continued to teach at the London School of Economics until the spring of 1937 when she moved to the University of Witwatersrand (Johannesburg) in the role of senior lecturer, replacing Winifred Hoernlé. She began further field-work among the Tswana of the Northern Transvaal but broke it off in 1940 to return to England and work in the Colonial Office where she was given the rank of temporary principal. During this period she was an active collaborator of Lord Hailey [q.v.] in designing the Colonial Social Science Research Council and was one of those responsible for planning a post-war programme of anthropological research. From 1946 she held a readership at the London School of Economics but in 1950 she became director of the East African Institute of Social Research which she had helped to plan. The Institute was attached to Makerere College in Uganda which at that time was affiliated to the University of London. Audrey Richards was responsible for organizing a substantial body of research which resulted in a number of important publications, including the symposium *Economic Development and Tribal Change* (1954), a study of immigrant labour in Buganda.

Although administration of the Institute took up much of her time Audrey Richards never forgot that she was an anthropologist. *Chisungu*, a very important study of girls' initiation rites among the Bemba (based on field notes made in 1931), was published in 1956 when the author was still head of the Makerere Institute. The symposium volume *East African Chiefs* was not published until 1960 but all the fifteen contributors were directly linked with the Institute.

She retired from the Makerere Institute in 1956 and took up a fellowship in Newnham College where she was a very active director of studies. She was responsible for founding the University African Studies Centre (an interdisciplinary enterprise) and from 1961 until her final retirement in 1966 she held the Smuts readership at Cambridge. In 1958-9 she was vice-principal of Newnham College and subsequently held various college appointments. From 1966 onwards she was an honorary fellow of the college. Overseas, as visiting lecturer or professor, she taught at Northwestern University, the University of Ghana, Cape Town University, the University of Chicago, and McGill University.

Among her honours were the Wellcome medal and the Rivers memorial medal. She delivered the Munro lectures in Edinburgh (1956), the Mason lectures in Birmingham, and the Jane Harrison lecture in Cambridge (both 1958). She gave a Royal Institution discourse in 1963 and the Frazer lecture in 1965. In 1963-6 she was president of the African Studies Association. She was the first woman to hold the office of president of the Royal Anthropological Institute (1959-61) and also the first woman anthropologist to be elected a fellow of the British Academy (1967). She was appointed CBE in 1955.

Audrey Richards's renown as a social anthropologist derived from the quality of her field research rather than from any notable innovation in theory but her organizational work in the Colonial Office and her directorship of the Makerere Institute did much to enhance the status of her academic field. If one recalls the difficulties that intellectual women faced in the period 1918–21 when she was first at Cambridge and the personal courage that she displayed when she imitated her guru Malinowski by pitching her tent in the middle of the Bemba village of Nkula in 1930, she deserves great respect. Her upper-class background no doubt added to her self-confidence; her reputation for modesty was perhaps deceptive. She was quick to make the most of unexpected opportunities but sometimes authoritarian in her treatment of collaborators. She showed little sympathy for post-functionalist developments in social anthropology.

For a number of years before her final retirement she lived in the Essex village of Elmdon, a twenty-minute drive from Cambridge. Following Malinowski's advice that the only way to learn about interviewing technique is to do it, she had her undergraduate and graduate supervisees carry out interviews with her Elmdon neighbours. The result would have horrified any statistically minded sociologist but it produced some marvellous anthropologists and the research notes, disorganized though they were, were somehow turned into three books. One of these was a short, privately printed affair designed for the people of Elmdon but the other two, one by Jean Robin and the other by Marilyn Strathern, were serious scholarly productions published by Cambridge University Press. Audrey Richards herself contributed to all three volumes. She died at her younger sister's home in Cambridge 29 June 1984. She was unmarried.

[*The Times*, 3 July 1984; obituary by Raymond Firth in *Man*, vol. xx, no. 2, June 1985; *Cambridge Anthropology*, vol. x, no. 1; Helena Wayne, *American Ethnologist*, August 1985; private information; personal knowledge.]

EDMUND LEACH

RICHARDS, OWAIN WESTMACOTT (1901–1984), entomologist, was born 31 December 1901 at Croydon, the second of four sons (there were no daughters) of Harold Meredith Richards MD, then medical officer of health for the district, and his wife, Mary Cecilia, daughter of W. J. Todd, a civil servant from Cumbria. He entered Brasenose College, Oxford, in 1920 from Hereford Cathedral School as an exhibitioner in mathematics, but his boyhood devotion to natural history, shared with his younger brother Paul and encouraged by their mother, led him to abandon mathematics for zoology. He took a first in 1924

and after election as senior Hulme scholar and Christopher Welch scholar he spent three postgraduate years in the Hope department of entomology under Professor E. B. Poulton [q.v.]. In 1927 Richards left Oxford for London as research assistant to J. W. Munro at Imperial College, becoming lecturer then reader in entomology and succeeding Munro in 1953 as professor of zoology and applied entomology and director of the college field station. Even after retirement in 1967 he remained actively associated with his department and continued research at the British Museum (Natural History) working there almost every day until he was over eighty.

Before leaving Oxford Richards had already started publishing in the three fields in which he was to specialize: systematics, ecology, and evolution theory. He was an authority on the Sphaerocerid flies and celebrated internationally for his work on the aculeate Hymenoptera, combining detailed revisionary studies with broader biological, behavioural, and evolutionary considerations, all based on extensive and world-wide field experience. Notable among his very many taxonomic publications were those on the Bethylidae, the Dryinidae, and the genera *Bombus*, *Belonogaster*, and *Mischocyttarus*, together with his books *A Revisional Study of the Masarid Wasps* (1962) and *The Social Wasps of the Americas* (1978).

All this would have been more than enough for a lifetime, but Richards also made major contributions to insect ecology, becoming a leading figure in the quantitative investigation of insect population dynamics. His pioneering study of the butterfly *Pieris rapae* in the 1930s was followed by others on *Ephestia*, *Phytodecta*, and the British Acrididae in collaboration with his colleague Dr Nadia Waloff. An early interest in evolutionary mechanisms led him also to collaborate with the malacologist G. C. Robson. Their joint book, *The Variation of Animals in Nature* (1936), displayed to the full Richards's great powers of critical analysis and his seemingly effortless ability to co-ordinate large bodies of fact, though its sceptical attitude to selectionist theories (a reaction to the views of Sir E. B. Poulton, q.v.) has not stood the test of time, as he later recognized.

Richards was deeply involved in the affairs of the Royal Entomological Society of London, serving as honorary secretary (1937–40) and as president (1957–8). He was almost equally active in the British Ecological Society, being its president in 1944–5 and editor of the *Journal of Animal Ecology* from 1963 to 1967. Both societies elected him to honorary fellowship, as did the Société Entomologique d'Égypte, the Nederlandsche Entomologische Vereeniging, and the Accademia Nazionale Italiana di Entomologia. Fellowship of the Royal Society came in 1959 and presidency of the thirteenth international congress of entomology at London in 1964.

Almost excessively conscientious, Richards never allowed his personal research to overshadow other academic obligations. He was not an exciting teacher, relying on example rather than precept, but his many postgraduate research students, attracted from all over the world, found him patient, helpful, and surprisingly approachable. Visiting professorships at Berkeley and in Ghana and his joint authorship of the two revised editions of A. D. Imms's *General Textbook of Entomology* (1957 and 1977) helped to extend his influence as a university teacher.

Despite his reputation and the great respect in which he was held, Richards was modest to a fault, shunning anything that suggested ostentation or flamboyance. Tall, sparely built, and distinguished in appearance, he could seem austere, remote, or even formidable to those who hardly knew him, an impression enhanced by his capacity for penetrating comment and a reluctance to waste words. But with friends, colleagues, and especially younger people, he was always interesting, stimulating, unobtrusively kind, and gifted with a most characteristic sense of almost boyish humour. He was unusually well read and well informed and he enjoyed travel, the theatre, and the open air.

In 1931 Richards married Maud Jessie, daughter of Captain Colin M. Norris, RN, and herself an entomologist of some note; they had two daughters. She died in 1970 and in 1972 he married Joyce Elinor, daughter of John Morrison McLuckie, minister of the Church of Scotland, and widow of his friend and fellow entomologist Robert Bernard Benson. Richards died at Haslemere 10 November 1984.

[Sir Richard Southwood in *Biographical Memoirs of Fellows of the Royal Society*, vol. xxxiii, 1987; personal knowledge.]

R. G. Davies

RICHARDSON, (FREDERICK) DENYS (1913-1983), metallurgical chemist, was born 17 September 1913 in Streatham, London, the third son of a family of three sons and one daughter of Charles Willerton Richardson, managing director of Asquith & Lord, of Bombay, and his wife, Kate Harriet Bunker, a schoolteacher. He was educated at University School in Hastings, and at University College, London (of which he became a fellow in 1971), where he graduated in chemistry in 1932. (Sir) Charles Goodeve [q.v.], lecturer in chemistry, steered him to carrying out a Ph.D. on the

oxides of chlorine, which he completed in 1936.

In 1937-9 Richardson was a Commonwealth Fund fellow at Princeton, USA. Early in 1940 he joined the Royal Naval Volunteer Reserve to help Goodeve counter the magnetic mine. This partnership found Richardson in the department of miscellaneous weapons development (DMWD) of the Admiralty, where he rose to be deputy director (1943-6) with the rank of commander RNVR (1942). The work he carried out in DMWD was varied and extensive, embracing such topics as 'wiping' ships against magnetic mines, mine-sweeping, radar deception, anti-aircraft weapons, illuminating shells, artificial harbours, rockets, and anti-submarine techniques. The success of this work owed much to his personality, intelligence, and perseverance in the face of difficulties. He gathered a heterogeneous collection of scientists, engineers, and naval personnel, encouraging them to get on with urgent problems, often in the face of official scepticism. The work of DMWD is entertainingly outlined by Gerald Pawle in *The Secret War* (1956).

After the war Richardson joined the British Iron and Steel Research Association under the leadership of Goodeve as head of the chemistry department (1946-50). He organized research into the physical chemistry of iron and steel-making, applying the methods of chemical thermodynamics and kinetics to the high temperature processes hitherto governed by empirical methods. During the next six years his group received wide recognition of his work.

In 1950 he moved as Nuffield fellow to the metallurgy department of Imperial College, London, later becoming professor of extraction metallurgy (1957-76); he remained there until his death. He founded the Nuffield and John Percy research groups which made important contributions in the fields of physical chemistry and process engineering respectively. The reputations of these groups and of their founder was universal and during this period he published about 125 substantial papers with fifty-two collaborators from eight countries. He also published a two-volume book, *The Physical Chemistry of Melts in Metallurgy* (1974).

He travelled widely, delivered many prestigious lectures, and received numerous honours and awards, among them the Sir George Beilby memorial award (1956), Bessemer gold medal (1968), gold medal of the Institution of Mining and Metallurgy (of which he was president in 1975) (1973), gold medal of the American Society of Metals, Peter Tunner medal of the Verein Eisenhütte Österreich (1976), grande medaille de la Société Française de la Métallurgie (1977), Carl Lueg medal of the Verein Deutscher Eisenhüttenleute (1978), and Kelvin medal of the Institute of Civil Engineers

(1983). He had honorary doctorates of the Technische Hochschule, Aachen (1971) and the University of Liège, and was an honorary member of the Association of Engineering of the University of Liège and of the Japanese Iron and Steel Institute. He was a foreign associate of the National Academy of Engineers of the USA. He became FRS in 1968 and F.Eng. in 1976.

Richardson was an outstanding scientist whose energy, tenacity, honesty, and charm won him a special place in the hearts of many of his friends and colleagues. Slight in build, his keenness and quickness of mind together with his spruce appearance marked him out in any discussion, and his ready wit and sense of timing made him a popular speaker at formal and informal occasions. He was a meticulous experimentalist and writer and his colleagues remember wryly the formidable cross-examinations they underwent before their results were accepted. They also remember his generous support and loyalty. He was an accomplished water-colour landscape painter and an enthusiastic gardener.

In 1942 he married Irene Mary, a graduate engineer, the daughter of George Edward Austin, a Lancashire textile manufacturer. They had two sons. Richardson died 8 September 1983 at St George's Hospital, Tooting.

[J. H. E. Jeffes in *Biographical Memoirs of Fellows of the Royal Society*, vol. xxxi, 1985; information from Mrs Irene Richardson; autobiographical notes by F. D. Richardson; personal knowledge.]   J. H. E. Jeffes

RICHARDSON, Sir RALPH DAVID (1902–1983), actor, was born 19 December 1902, the third son and third and youngest child of Arthur Richardson, art master at Cheltenham Ladies' College, and his wife, Lydia Susie, daughter of John Russell, a captain in the merchant navy. When Richardson was four years old, his mother left his father and took him to live with her at Shoreham, Sussex, in a makeshift bungalow constructed out of two old railway carriages. Their allowance from her husband (with whom the two older boys remained) was two pounds and ten shillings a week, and on his own admission Richardson grew up as a 'mother's boy', educated by her at home and at the Xaverian College in Brighton, a seminary for those who intended to be priests from which he soon ran away.

His education thereafter was erratic, and by 1917 he was working as an office boy for the Liverpool and Victoria Insurance Company in Brighton. Two years later, when his grandmother died leaving him £500 in her will, he resigned immediately from the office and enrolled at the Brighton College of Art. Once there, he rapidly discovered that he had no gift for painting; instead, he briefly considered a career in journalism but then, inspired by a touring production of *Hamlet* which had come to Brighton with Sir F. B. (Frank) Benson [q.v.] in the title role, decided that his future lay in the theatre.

He joined a local semi-professional company run by Frank Growcott, who charged him ten shillings a week to learn about acting with the understanding that, once he had learned how to do it, Growcott would in turn pay him the same amount to appear in his company. Richardson seldom saw the colour of Growcott's money but he did make his first stage appearances at Brighton having already created some memorable off-stage sound effects ('I first burst on to the English stage as a bombshell'). He then auditioned successfully for Charles Doran's touring players, with whom he stayed for five seasons while rising through the ranks to such roles as Cassio in *Othello* and Mark Antony in *Julius Caesar*.

In 1924 Richardson married the seventeen-year-old student actress Muriel Bathia Hewitt, daughter of Alfred James Hewitt, a clerk in the Telegraph & Cable Company. They were to have no children. The following year the pair joined the Birmingham Repertory Company, and Richardson made his first London appearance for the Greek Play Society on 10 July 1926 as the Stranger in *Oedipus at Colonus*.

He spent the next four years largely in small West End roles, notably in two plays by Eden Phillpotts [q.v.] (*Yellow Sands* and *The Farmer's Wife*), and at the Royal Court where he spent much of 1928 in H. K. Ayliff's company, which also included a young Laurence (later Lord) Olivier. After touring South Africa in 1929, already aware that his young wife had contracted sleepy sickness (encephalitis lethargica), Richardson returned in 1930 to join the Old Vic Company where he met (Sir) John Gielgud for the first time. Of the three great actor knights of the mid-century (Richardson, Olivier, Gielgud), Richardson was the eldest and the least predictable, the one who looked most like a respectable bank manager possessed of magical powers, and the one who had the most trouble with Shakespeare: the critic James Agate [q.v.] said that his 1932 Iago 'could not hurt a fly' and Richardson soon turned with what seemed a kind of relief to the modern dress of G. B. Shaw, W. Somerset Maugham, and James Bridie [qq.v.] before starting in 1934 (with *Eden End*) an alliance with J. B. Priestley [q.v.] which was to lead to some of his best and most characteristic work.

A year later he was on Broadway for the first

time, playing Mercutio in *Romeo and Juliet* for Katharine Cornell's company, and in 1936 he returned to London for a long-running thriller, *The Amazing Dr Clitterhouse*, in which he was supported by the actress Meriel Forbes whom he married in 1944, two years after the death of his first wife in 1942. She was the daughter of Frank Forbes-Robertson, actor-manager, and grandniece of Sir Johnston Forbes-Robertson [q.v.].

By now, a theory had developed in the theatre that Richardson was at his best playing 'ordinary little men', though as one critic later noted, anyone who believed that could seldom have met many ordinary little men. Those played by Richardson always had an added touch of magic, of something strange, though Agate was still not won over. When Richardson returned to the Old Vic in 1938 to play Othello in a production by (Sir) W. Tyrone Guthrie [q.v.], for which Olivier had elected to play Iago homosexually, much to his partner's horror, Agate simply noted 'the truth is that Nature, which has showered upon this actor the kindly gifts of the comedian, has unkindly refused him any tragic facilities whatever . . . He cannot blaze'.

Richardson returned to Priestley and triumph (*Johnson over Jordan*, 1939) and then, when the war came, rose to the rank of lieutenant-commander in the Royal Naval Volunteer Reserve where he was affectionately known as 'pranger Richardson' on account of the large number of planes which seemed to fall to pieces under his control.

It was in 1944, when he was released to form a directorate of the Old Vic with Olivier and John Burrell, that Richardson reached the height of his considerable form: over four great seasons at the New Theatre with Olivier, Dame A. Sybil Thorndike [q.v.], and Margaret Leighton, he played not only the definitive Falstaff and Peer Gynt of the century but also the title role in Priestley's *An Inspector Calls*, Cyrano de Bergerac, Face in *The Alchemist*, Bluntschli in *Arms and the Man*, and John of Gaunt in *Richard II*, which he unusually also directed.

When the triumvirate was summarily sacked in 1947 by the Old Vic governors who were uneasy about the Olivier and Richardson stardom in what was supposed to be a company of equals, Sir Ralph (it was also the year of his knighthood) returned to the life of a freelance actor, enjoying many more triumphs—as well as another Shakespearian defeat at Stratford-upon-Avon in the title role of *Macbeth* (1952).

The 1960s were highlighted by *Six Characters in Search of an Author* (1963), where in Pirandello he found an ethereal author to satisfy his own other-worldliness, and then in 1969 by a courageous move away from the classics and into the avant-garde as Dr Rance in *What the Butler Saw* by Joe Orton [q.v.]. A year later he was with Gielgud at the Royal Court in David Storey's *Home*, starting a late-life partnership which took them on to Harold Pinter's *No Man's Land* (1975) in the West End and on Broadway, as well as to countless television interviews in which they appeared as two uniquely distinguished but increasingly eccentric brokers' men.

Richardson first joined the National Theatre in 1975, shortly after Olivier left it, as John Gabriel Borkman in the play of that name, and it was there under Sir Peter Hall's administration that he was to do the best of his late work, which culminated a few months before his death in a haunting and characteristic final appearance as Don Alberto in Eduardo de Filippo's *Inner Voices*.

Deeply attached to his second wife, their only child Charles, and a racing motor cycle on which he would speed across Hampstead Heath, Richardson achieved theatrical greatness by turning the ordinary into the extraordinary: on stage as off, he managed to be both unapproachable and instantly accessible, leaving like Priestley's Inspector the impression behind him that perhaps he had not really been there at all, or that if he had, it was only on his way to or from somewhere distinctly unworldly.

He turned somewhat uncertainly to the cinema in the 1930s, at the start of a long contract with (Sir) Alexander Korda (whose notice Richardson later wrote for this Dictionary) which led to such successes as *Things to Come* (1936), *The Four Feathers* (1939), *The Citadel* (1939), *The Fallen Idol* (1948), *An Outcast of the Islands* (1952), and *Richard III* (1955), before he went on to *Long Day's Journey Into Night* (1962), and *A Doll's House* (1973). It was with one of his very last screen roles, however, as the Supreme Being in the 1980 *Time Bandits*, that he achieved the perfect mix of the godly and the homespun that had always been at the heart of his acting. Richardson left an estate valued at just over a million pounds, and the memory of a great and mysterious theatrical wizard. At the National Theatre, and at his suggestion, a rocket is fired from the roof to denote first nights. It is known as Ralph's Rocket. Richardson was awarded an honorary D.Litt. by Oxford (1969) and the Norwegian Order of St Olaf (1950). He died 10 October 1983 in London.

[Harold Hobson, *Ralph Richardson*, 1958; Garry O'Connor, *Ralph Richardson*, 1982; Kenneth Tynan, *Show People*, 1980; personal knowledge.]   SHERIDAN MORLEY

RITCHIE, SIR NEIL METHUEN (1897-

1983), general, was born at Essequibo, British Guiana, 29 July 1897, the youngest of the three children and second son of Dugald MacDougall Ritchie, a sugar plantation manager, and his wife, Anna Catherine Leggatt. He was educated at Lancing College and the Royal Military College, Sandhurst, and in 1914 was commissioned into the Black Watch, joining the 1st battalion in France in May 1915. He was slightly wounded at the battle of Loos in September and, on recovery, was posted to the 2nd battalion in Mesopotamia. He was adjutant of the reformed 2nd battalion of the Black Watch in 1917, and was appointed DSO (1917) and MC (1918).

After the war he attended the Staff College, was promoted major in 1933, brevet lieutenant-colonel in 1936, and lieutenant-colonel in 1938 when he assumed command of the 1st battalion the King's Own Royal Regiment, which he took to Palestine in September of that year to serve in the 7th division. In 1939 he was promoted colonel and became an instructor at the Senior Officer's School at Sheerness; later that year he was appointed brigadier general staff of II Corps, commanded in France by Alan Brooke (later Viscount Alanbrooke, q.v.).

After Dunkirk Ritchie, promoted acting major-general, was given the task of reforming 51st Highland division to replace the one surrounded by the Germans at St Valéry. Early in 1941 he was sent to join Sir A. P. (later Earl) Wavell [q.v.] in North Africa, but soon after he arrived, Wavell was replaced by Sir Claude Auchinleck [q.v.] who decided in November 1941 to relieve Sir Alan Cunningham [q.v.] of the command of Eighth Army and replace him with Ritchie, overriding Ritchie's protests and suggestions that Auchinleck should assume command himself until a more senior officer could arrive from Britain.

Ritchie assumed command when Eighth Army was in a grave situation. After some awkward setbacks, he raised the siege of Tobruk and forced Rommel to abandon Cyrenaica and withdraw to where he had started a year before. If Auchinleck had intended Ritchie's appointment to be a temporary measure, he could not now dismiss him after he had gained a victory. A test of their relationship occurred when Rommel unexpectedly took the offensive at the end of January 1942. Auchinleck had told Ritchie that, if his forward positions, 120 miles south of Benghazi, could not be held, he did not intend 'to try and hold permanently Tobruk or any other locality west of the [Egyptian] frontier'. But when Ritchie acquiesced in the request of the corps commander, (Sir) A. R. Godwin-Austen, to withdraw from Benghazi and Cyrenaica, Auchinleck forced Ritchie to countermand the order. In the confusion which

followed, Ritchie's forces withdrew to the Gazala position, covering Tobruk, and Godwin-Austen resigned his command of XIII Corps.

In the ensuing months (Sir) Winston Churchill pressed Auchinleck to launch an offensive to regain Cyrenaica in order that the RAF could protect convoys sailing to beleaguered Malta, and the unrealistic plan which Ritchie devised to comply with Auchinleck's orders to that effect did much to weaken confidence in him on the part of his corps commanders, C. W. M. (later Lord) Norrie and W. H. E. Gott [q.v.]. When it became clear that Rommel intended to pre-empt the plan, Ritchie was allowed temporarily to adopt a defensive stance, although his dispositions were determined by his offensive designs. Rommel's attack at the end of May 1942 initially threw Eighth Army off balance, and Ritchie's successive attempts to regain the initiative and turn the tables on him failed. This led to another crisis in his relationship with Auchinleck, who, pressed by Churchill, prevaricated over the issue of whether or not Tobruk should be held after Ritchie had lost almost all his tanks. Ritchie tried to carry out Auchinleck's orders and to counter Rommel's advance. Having attempted to hold positions on the Egyptian frontier and been ordered to hold Mersa Matruh, he was dismissed by Auchinleck on 25 June before the town had been attacked. Auchinleck assumed command of Eighth Army himself and promptly reversed the orders he had just given to Ritchie.

On his return to Britain Ritchie was given command of 52nd (Lowland) division, and in 1944 was promoted temporary lieutenant-general in command of XII Corps, which, with three infantry divisions, landed in Normandy. Ritchie commanded the corps throughout the campaign in north-west Europe. He was never given a leading role, but carried out efficiently a series of methodical operations. After the war he was appointed GOC-in-C Scottish Command and in 1947 promoted general, with the post of C-in-C Far East Land Forces, based in Singapore. He left in 1949, after the declaration of an emergency in Malaya, and in 1950 was posted to Washington DC as head of the British Army Staff in the Joint Service Mission to the USA. In 1951 he retired to Canada, where he remained for the rest of his life, taking an active part in commerce, being a director of several companies and chairman of the Mercantile and General Reinsurance Company of Canada and of Macdonald-Buchanan Properties Ltd.

He made no attempt himself, and refused to allow others on his behalf, to refute criticism of his tenure of command of Eighth Army. He was therefore deeply wounded when Auchinleck, to whom he felt that he had been meticulously loyal, not only allowed his biographer, John Connell, to write an account strongly, and Ritchie believed unfairly, criticizing him; but refused to try and get Connell to meet any of Ritchie's objections.

Ritchie was appointed CBE (1940), CB (1944), KBE (1945), KCB (1947), and GBE (1951). He was colonel of the Black Watch from 1950 to 1952 and a member of the Queen's Body Guard for Scotland. He held several foreign decorations and became a Knight of the Order of St John in 1963.

In 1937 he married Catherine Taylor, daughter of James Arnott Minnes, partner in a warehousing firm, of Kingston, Ontario, Canada; they had one son and one daughter. Ritchie died 11 December 1983 in Toronto, Canada.

[Michael Carver, *Tobruk*, 1964, and *Dilemmas of the Desert War*, 1986; Ritchie's personal papers in the Imperial War Museum; private information.]

MICHAEL CARVER

RITCHIE-CALDER, BARON (1906-1982), author and journalist. [See CALDER, PETER RITCHIE.]

ROBBINS, LIONEL CHARLES, BARON ROBBINS (1898-1984), economist, was born 22 November 1898 at Sipson Farm, Harmondsworth, Middlesex, the elder child and only son of Rowland Richard Robbins CBE, market gardener, of Harmondsworth, and his wife, Rosa Marion, daughter of Charles Centurier Harris, a provision merchant, of St John's Wood, London. Rosa died in 1910 and Robbins married her sister, by whom he had another daughter and son. Lionel Robbins was educated at Southall County School, University College, London, and the London School of Economics. Following military service in the Royal Field Artillery in World War I, he graduated from the London School of Economics with a first class honours B.Sc. in 1923 before teaching at New College, Oxford. He returned to the London School of Economics as a lecturer from 1925 to 1927, and was then fellow and lecturer at New College, Oxford, from 1927 to 1929. He was appointed professor of economics at LSE in 1929, at the age of thirty, and held this position until his resignation from the chair in 1961.

He was an inspiring and dedicated teacher. With his great height, and in later years his mane of white hair, he was an impressive and awe-inspiring figure. Yet he devoted himself to his teaching and to his pupils, delivering beautifully organized lectures in his rounded and fine English prose. In the words of one of his distinguished pupils, James Meade: 'The ebullient and exuberant purposefulness of his exposition was infectious . . . I used to describe his performance as combining the qualities of a rowing coach with those of the conductor of a great orchestra.'

His earliest intellectual allegiance was to guild socialism in 1919-20, but after graduating from the LSE he shed any remnants of his one-time socialism. He had come to the conclusion that both personal liberty and economic efficiency were likely to be most prevalent under the dispersed initiatives of a system of markets and private property. He became an economic liberal, but did not belong to any political party: his time in the House of Lords was spent on the cross-benches. At the LSE in the 1930s, the main focus of his intellectual activity was the seminar held with Friedrich von Hayek, which he usually chaired. A great deal of his intellectual effort was devoted to attempts to achieve a consistent view of economics. His first book, *An Essay on the Nature and Significance of Economic Science* (1932), made an immediate mark. In it he did not aspire to produce a system of his own, but to analyse and criticize particular propositions, from the inherited body of economics, for their logic and for the appropriateness of their assumptions. His aim was to enlarge the tradition and to modify it. Above all, as his later work made clear, he wanted to make economics useful for the solution of contemporary problems.

Robbins was a distinguished and meticulous scholar of the history of economic thought, always something of a passion with him. He took this study up again after the war, at a time when he was deeply involved in a multitude of exacting activities. It was for his work on Robert Torrens, published in 1963, that he most wished to be judged as a scholar.

Robbins's interest in applying economic analysis to the problems of the real world first found scope in 1930 when he became involved in the formation of UK economic policy, as a member of the committee of the Economic Advisory Council chaired by J. M. (later Lord) Keynes [q.v.]. He showed his independence in the Council when he was in a minority, against the views of Keynes, in opposing the abandonment of free trade and in being an anti-expansionist where public expenditure was concerned. He came to regret the latter point of view, but never the former. During World War II he was a member, and for the greater part of the time (1941-5) director of the economic section of the war cabinet secretariat. In this capacity he was responsible for advice on the conduct of the war from the economic point of view, and later, on plans for post-war reconstruction. He attended the Hot Springs and Bretton Woods conferences in the United States, and was a member of the team which negotiated the Anglo-American loan agreement in 1945. He was the leader of a brilliant team, and by his clear analysis, his wisdom, and his commanding presence, he succeeded in gaining respect for the value of economic analysis in the conduct of the nation at a critical time. By succeeding in this he laid the foundations for the permanent establishment of organized professional economic advice in the machinery of government.

Robbins himself said that it would be difficult to exaggerate the beneficial effects of this period of public service so far as his own mental and spiritual progress was concerned. Several of his post-war books and essays were concerned with lessons that he derived from his wartime experience. Nevertheless, he returned to the LSE as soon as he could, and took up his teaching and research once more. Robbins's work for, and devotion to, the LSE was an important part of his life. He was chairman of the court of governors from 1968 to 1974, a most fraught period, during which LSE had to be closed for six weeks before order could be restored. He was also chairman of the appeal on behalf of a new building to house the LSE's great library, the British Library of Political and Economic Science. The building was acquired and appropriately named the Lionel Robbins Building.

But Robbins had become too notable a figure not to be called upon for a wide range of public work. The chairmanship of the Robbins committee on higher education (1961–4) was the most notable of these appointments. The committee's demonstration of the need to accommodate increasing numbers of students, with no lowering of standards, was generally accepted and acted upon, although other recommendations (such as the desirability of a unitary as opposed to a binary institutional structure) were rejected. Its analysis led to the establishment of one major new university (Stirling) and the transformation of nine colleges of advanced technology into universities, and dominated national policy towards higher education for the whole of the 1960s and 1970s.

After World War II Robbins was appointed to a number of public positions connected with the arts. He was trustee of the National Gallery (1952–74), and of the Tate Gallery (1953–9 and 1962–7). In 1955 he was appointed a director of the Royal Opera House, Covent Garden, and held this position until 1980. His involvement in the application of economic expertise to the administration of the arts gave him immense pleasure, as did his association with the press. From 1961 to 1970 he was chairman of the *Financial Times*, being also director of the *Economist* and chairman of the Economist Intelligence Unit. He held many of these posts

concurrently, yet his devotion to each of them was such that he seemed to those concerned to be devoting all his time and energies to their particular problems.

Among more academic appointments, Robbins was president of the Royal Economic Society (1954–5) and of the British Academy (1962–7), a member of the planning board of the University of York, and the first chancellor of Stirling University (1968–78). He was a member of the American Academy of Arts and Sciences and several other overseas academies. He received honorary degrees and honorary fellowships from more than a score of universities and colleges at home and abroad. He was elected a fellow of the British Academy in 1942 and was appointed CB in 1944 and CH in 1968. He was created a life peer in 1959.

In 1924 Robbins married Iris Elizabeth, daughter of Alfred George Gardiner [q.v.], formerly editor of the *Daily News*. They had a son and a daughter. Robbins died at Highgate, London, 15 May 1984.

[Lionel Robbins, *Autobiography of an Economist*, 1971; *Who's Who in Economics* (2nd edn.), 1986; *Economist*, 8 December 1984, for tributes in memory of Lord Robbins given on 11 October 1984, at St John's, Smith Square, London SW1; Henry Phelps Brown in *Proceedings* of the British Academy, vol. lxxiii, 1987; personal knowledge.]

AUBREY SILBERSTON

ROBERTSON, SIR JAMES WILSON (1899–1983), overseas civil servant and colonial governor, was born 27 October 1899 in Dundee, the eldest of three sons and the first of six children of James Robertson, jute merchant of Calcutta and Dundee and resident of Broughty Ferry and Edinburgh, and his wife, Margaret Eva, daughter of Adam Wilson, classics master at Dundee High School. He was educated at Merchiston Castle School, Edinburgh, and, after serving with the Black Watch from April 1918 to January 1919, at Balliol College, Oxford, where he gained a blue at rugby football in 1921 and obtained a third class in classical honour moderations (1920) and a second in *literae humaniores* (1922). In 1922 he opted for the Sudan Political Service rather than spend an extra year cramming for the Indian Civil Service. 'John Willie' (his army nickname had stuck) served as district commissioner in four provinces before returning to the White Nile as sub-governor in 1937 and, two years later, to his original province of Blue Nile (Gezira) as deputy and then acting governor.

It was now time for such a successful field officer to acquire secretariat experience. Robertson was transferred to Khartoum as assistant civil secretary in 1941, being promoted to deputy a year later. He was holding this key post when the civil secretary, Sir Douglas Newbold [q.v.], died suddenly at Khartoum in March 1945. Robertson was not immediately seen by the Service as the natural successor: he was junior to several governors, he lacked the headquarters experience of other candidates and, while his rank was equivalent to that of a provincial governor, he had not yet held a substantive governorship. But the governor-general, in whose gift the appointment lay, quickly made up his mind, and five days later confirmed Robertson in the Political Service's top post: his support of Newbold's enthusiasm (by no means shared by all the governors) for the Sudanization of the Service was a likely factor. It fell to him, in his eight years in the office (in the event, he was to be the last in a long line of eminent civil secretaries), to see the Sudan through some of its most difficult political crises, among them the Juba conference of 1947 and the reversal of the so-called 'southern policy', the delicate period following Egypt's decision to take the Sudan question to the United Nations, and the police mutiny of 1951. These were critical years, with the Sudan divided not only culturally between Arab Muslim north and African Christo-animist south, but also politically between the competing claims of the unity of the Nile Valley and the Sudan for the Sudanese. Robertson left Khartoum shortly before the implementation of the 1953 agreement which accepted the right of the Sudanese to self-determination and hence to independence.

On retirement from the Sudan he became a director of the Uganda Company, and in 1953–4 chaired the British Guiana constitutional commission. It was probably the considerable skill he displayed in negotiating the labyrinthine situation there which brought about the unexpected (not least by himself) offer of the governor-generalship of the new federation of Nigeria in 1955. It was left to Robertson as executive architect to build on the foundations so ably laid in Lagos by his predecessor, Sir John Macpherson, and thus consolidate and construct the emerging federal government of Nigeria. In Nigeria as in the Sudan, party unity was in short supply and national integration hard to identify. Dame Margery Perham [q.v.] once likened Robertson to a charioteer, whose task was not to choose the course but by delicate handling to keep his three fiery steeds running in unison. To the unrelenting tensions of a Nigeria already well on its way to independence, Robertson brought—and needed—all the negotiating skills of patient determination and iron geniality that he had learned in the Sudan. To these he added the diplomatic diffidence demanded in a royal representative whose task was to hand over the responsibilities of supreme office to an elected prime minister and cabinet and cede all but the ultimate authority with grace and goodwill, yet who had, while pressing for localization, to retain the confidence of, once again, a still largely expatriate civil service and a sizeable British commercial community. The signal of his success in managing to please most and affront few came when the Nigerian ministers not only asked for his tenure of office to be extended by two years, but also invited him to continue after independence (1 October 1960) as the new nation's first governor-general. Robertson handed over to a Nigerian successor in November 1960.

His second retirement saw little let-up in his involvement in public affairs. After visiting Kenya in 1961 as the commissioner to examine the constitutional issue of the coastal strip, he accepted several directorships (Barclays Bank, the Uganda Company once more), chairmanships (Commonwealth Institute, Royal Over-seas League) and presidentships (Sudan and Overseas Service Pensioners' Associations, Britain–Nigeria Association), and became a governor of Queen Mary College, London University. In 1974 he published his memoirs, *Transition in Africa*.

One of the last generation of Africa's proconsular giants, Robertson succeeded—where not all of them did—in triumphantly bridging the gap, in Khartoum and Lagos alike, between the pre-war style of fundamentally authoritarian ruler and the latter-day lower profile of collaborative constitutional monarch. Innocent of proconsular pomposity, shrewd and intuitive rather than intellectual, the secret of his public persona lay in a mix of imperturbability, integrity, and humour, enhanced by a huge capacity to inspire confidence and often affection in those who worked with him. Robertson was a big man in all senses of the word (his powerful physique and robust attitude inspired a colleague to liken his posting to the civil secretary's office to the bull's arrival in the Khartoum china shop)—and a brave one, too, for when almost eighty he grappled with an armed thief who broke into his home.

Robertson was created KT in 1965. He had been advanced to GCMG in 1957 (KCMG 1953) and KBE in 1948 (MBE 1931) and appointed GCVO in 1956 and KStJ the year before. He was elected an honorary fellow of Balliol College in 1953 and an honorary LLD was conferred on him by Leeds University in 1961, the same year in which the Royal African Society awarded him its Wellcome medal. The Robertsons shared their retirement years between Cholsey, Oxfordshire, where they lived in a converted bakehouse and were much involved in local life, and, in the summer months, a cottage at Killichonan by Loch Rannoch.

He married Nancy, elder daughter of Hugh Smith Walker, woollen manufacturer, of Huddersfield, in 1926. They had a son and a daughter. The Hon. Lord (Ian) Robertson (born 1912), a senator of the College of Justice in Scotland, is a younger brother. Robertson died 23 September 1983 at the Old Vicarage, Moulsford, Berkshire.

[*The Times*, 27 September and 6 October 1983; *West Africa*, 3 October 1983; *Balliol College Annual Record*, 1984; Sir James Robertson, *Transition in Africa*, 1974 (autobiography); the Sudan archive, Durham University; private information; personal knowledge.]   A. H. M. KIRK-GREENE

ROBINSON, JOAN VIOLET (1903-1983), economist, was born in Camberley, Surrey, 31 October 1903, the third child in the family of one son and four daughters of (Major-General Sir) Frederick Barton Maurice [q.v.], a professional soldier who later became principal of East London (later Queen Mary) College, London, and his wife, Helen Margaret, daughter of Frederick Howard Marsh, surgeon and later master of Downing College, Cambridge. Joan Robinson's forebears were upper-middle class English dissenters—her great-grandfather was F. D. Maurice [q.v.], the Christian Socialist, and her father was the central figure in the Maurice debates of 1918. She continued the family tradition with distinction, always a radical with a cause. She was educated at St Paul's Girls School, London, and Girton College, Cambridge, where she was Gilchrist scholar. She obtained second classes (division I) in both parts i and ii of the economics tripos (1924 and 1925). She was appointed an assistant lecturer in economics and politics at Cambridge in 1931, became a university lecturer in 1937, reader in 1949, and professor of economics in 1965.

Her incisive mind made her a powerful critic; her insight and intuition, whereby she provided logical arguments of great penetration (without the help of modern mathematical techniques), allowed her to make significant contributions across the whole spectrum of economic theory. She also made a special study of the socialist countries, especially China, which she frequently visited. She fervently hoped China would create a society in which not only would poverty be vanquished but also the potential of all its citizens would be realized in an environment of co-operation, hard work, and mutual respect and affection—inevitably, as she was to admit, an impossible dream but not the less noble for that.

Her achievements include her contribution to the imperfect (monopolistic) competition 'revolution' of the 1930s, the considerable part she played in helping J. M. (later Lord) Keynes [q.v.] make his revolutionary jump from *A Treatise on Money* (1930) to *The General Theory* (1936), both in discussion and comments and with expository, often original articles and books (her 'told to the children' *Introduction to the Theory of Employment*, 1937, is still the best starting place for understanding Keynes's theory), and her role in putting Marx's insights back on the agenda of modern economics. Her most distinctive personal contribution to economic theory is the special blend of the theories of the classical economists, Marx, Piero Sraffa [q.v.], Keynes, and Michal Kalecki with which she pioneered (along with (Sir) H. Roy Harrod [q.v.], R. F. Kahn, Nicholas (later Lord) Kaldor, and Luigi Pasinetti) a generalization of *The General Theory* to the long period. Joan Robinson thus played a major role in the three main critical movements in economic theory in this century. Her *magnum opus* was *The Accumulation of Capital* (1956).

Her contributions usually arose from criticisms, sometimes hostile, sometimes sympathetic, of the work of others. Her first mentor was Alfred Marshall [q.v.] (together with his faithful protégé, A. C. Pigou, q.v., and interpreter, G. F. Shove). She thoroughly absorbed his insights and methods but angrily resisted his ideology and foxiness, fudges and smokescreens, as she would say. She welcomed Sraffa's attack in the 1920s on Marshallian partial equilibrium theory because it seemed to allow the development of a theory of value, the predictions of which were more in accord with the experiences of firms in United Kingdom industry in the 1920s, and an approach within which to criticize the received theory of distribution, admittedly still within the same neoclassical framework. While writing *The Economics of Imperfect Competition* (1933) she commenced a lifelong intellectual association with Richard Kahn whose contemporaneous study of pricing in the UK textile industry provided some of the important empirical generalizations from which was built her book's theoretical apparatus which exploited the marginal revenue and related concepts. Shove's influence was also crucial.

The urgent stimulus of the great depression led Joan Robinson also to work on the determination of the level of activity in capitalist economies. She increasingly came to see the Keynesian revolution as one of method—the adaptation of Marshallian tools of analysis to the study of the motion of a money-using production economy operating in an uncertain environment, a transition from analysis in terms of logical time to one in terms of historical time, as she put it in the post-war period. By the mid-1930s Joan Robinson became interested in Marxian economics. It was her study of this, resulting in her 1942 *An Essay on Marxian Economics*, together with her friendship with Kalecki, a Polish Marxist economist who independently discovered the principal propositions of *The General Theory*, and her discussions with Maurice Dobb [q.v.] and Sraffa, that revolutionized her own thinking.

In the post-war period she saw the major issue to be a reworking of the grand themes of distribution and accumulation of classical political economy in a framework which incorporated the Keynesian-Kaleckian insights concerning effective demand whereby investment led and saving adjusted to it. Ultimately, though, she was to despair of creating a coherent theory of the long period, arguing that Kalecki's theories of cyclical growth were about as far as economic theory could go. Integral with these developments was an increasing dissatisfaction with the conceptual foundations of neoclassical theory and its method, for which she became most well known if not best understood. She found especially wanting the *meaning* of capital; this was misconstrued by the orthodox to be a complaint about *measurement*. In fact, it was an example of the reasons for her rejection of the supply and demand equilibrium approach and her desire to put in its place, not a rival 'complete' theory, which she considered would be only another 'box of tricks', but an approach that started from institutions, history, and the 'rules of the game', asking what sort of economy was being discussed and trying to model its movements in its own historical setting.

As well as her many books, a stream of challenging evaluative essays, collected together in six volumes, have inspired the young as much as they have irritated their orthodox elders. Never one to mince words, possessor of a civilized wit, sometimes bleakly rude, not always fair but always honest, as hard on herself as on those she criticized, Joan Robinson more than any other economist of this century became a model for progressive radicals, fearlessly following arguments to conclusions no matter how incompatible they proved to be.

She was FBA from 1958 to 1971. She was elected to an unofficial fellowship at Newnham College, Cambridge, in 1962 and to a professorial fellowship in 1965. She became an honorary fellow of Girton in 1965, of Newnham in 1971 when she retired from her chair, and of King's College, Cambridge, in 1979.

In 1926 she married (Sir) (Edward) Austin (Gossage) Robinson, later professor of economics at Cambridge University. They had two daughters. In her later years Joan Robinson, a strict vegetarian, slept all year round in a small unheated hut, open on one side, at the bottom of her garden. In spring the tits would wake her by pecking at her long grey hair for material for their nests. She died 5 August 1983 in Cambridge.

[Obituary in *Annual Report* of King's College, Cambridge, 1984; personal knowledge.]   G. C. HARCOURT

ROBINSON, JOHN ARTHUR THOMAS (1919-1983), bishop, New Testament scholar, and author, was born 15 June 1919 in The Precincts, Canterbury, the elder son and eldest of three children of the Revd Canon Robinson, and his wife, Mary Beatrice Moore. His father married when he was sixty-two and died when Robinson was nine. Both his father and his mother's father were canons of Canterbury. Six of his uncles were ordained, including J. Armitage Robinson [q.v.], former dean of Westminster, and, later, of Wells.

A foundation Scholar of Marlborough College and Rustat exhibitioner of Jesus College, Cambridge (he obtained a second class in part i of the classical tripos in 1940), Robinson won the Burney prize (1941) and the Burney scholarship for 1942-3 to Trinity College, Cambridge. His thesis, 'Thou Who Art', gained him a Ph.D. in 1946. After training for ordination at Westcott House, Cambridge (he gained a first class in part ii of the theology tripos in 1942), Robinson became curate of the inner-city parish of St Matthew, Moorfields, Bristol (1945-8).

Robinson wrote his first book *In the End, God . . .; a Study of the Christian Doctrine of the Last Things* (1950) at Wells Theological College, where he was chaplain from 1948 to 1951. Made fellow and dean of Clare College, Cambridge, in 1951, Robinson pioneered the liturgical revision described in *Liturgy Coming to Life* (1960). His second book, *The Body* (1952), was much acclaimed. His third, *Jesus and His Coming* (1957), was delivered as lectures at Harvard University in 1955. He wrote most of *Twelve New Testament Studies* (1962) at Clare. In 1953 Geoffrey Fisher (later Lord Fisher of Lambeth, q.v.), archbishop of Canterbury, made him examining chaplain and in 1958 one of the six-preachers of Canterbury.

In 1959 A. Mervyn Stockwood, Robinson's vicar in Bristol, became bishop of Southwark and invited Robinson to become his suffragan bishop of Woolwich. Despite contrary advice from fellow academics and from the archbishop of Canterbury, Robinson accepted, and was consecrated on 29 September 1959. He pioneered the Southwark ordination course, a 'theological college without walls'. Elected a proctor in 1960, he spoke in the Convocation of Canterbury and the Church Assembly on the ordination of women, pastoral reorganization, the deployment and payment of the clergy, crown appointments, synodical government, inter-communion, the sharing of churches ecumenically, suicide, capital punishment, and other topics, and was soon recognized as the leading radical of the Church of England. *On Being the Church in the World* (1960) confirmed his stature.

The year 1960 made Robinson notorious. He appeared in court to defend the publication of the unexpurgated edition of *Lady Chatterley's Lover* by D. H. Lawrence [q.v.]. He claimed that Lawrence 'tried to portray this relationship as . . . an act of holy communion'. The archbishop of Canterbury censured Robinson; but he did not regret what he said. He was vociferously supported and equally vociferously attacked.

In 1961 the back trouble began which was to be with Robinson for the rest of his life. Lying on his back he produced *Honest to God* (1963). Preceded by a summary article in the *Observer* headed 'Our Image of God Must Go', it sold over a million copies and was translated into seventeen languages. With liberal quotations from Dietrich Bonhoeffer, Rudolf Bultmann, and Paul Tillich, he argued that the imagery in which God was presented made Him unreal to people of a secular scientific world. Opinions were again violently divided. Robinson received over 4,000 letters. Many who spoke for and against had not read the book. Life for Robinson was never the same after 19 March 1963. There were always those who wanted to meet him and talk with him. The new archbishop of Canterbury, A. Michael Ramsey (later Lord Ramsey of Canterbury), censured Robinson, admitting in *Canterbury Pilgrim* (1974) his 'initial error in reaction'. In October 1963 *The Honest to God Debate* was published, with Robinson as co-editor. He developed the chapter in *Honest to God* entitled 'The New Morality' into three much publicized lectures *Christian Morals Today* (1964). Lectures at Hertford Seminary, Connecticut, and at Cornell University occasioned Robinson's *The New Reformation?* (1965). Lectures at Stanford University, California, afforded further opportunity for *Exploration into God* (1967).

Robinson the radical reformer (the programme of doctrinal, ethical, and pastoral reform he adumbrated, alongside Mervyn Stockwood, was dubbed by the media 'South Bank religion') was at the same time the diligent diocesan pastor, awkwardly shy, incapable of malice, of profound faith and childlike integrity.

Robinson was made BD of Cambridge University in 1962 and DD in 1968. After ten demanding years as bishop of Woolwich, and with the invitation to deliver the Hulsean lectures in 1970—*The Human Face of God* (1973)— Robinson returned to Cambridge in 1969 as fellow, dean of chapel, and lecturer in theology at Trinity College. *Christian Freedom in a Permissive Society* (1970) garnered the fruits of the Woolwich decade. In May 1970 Robinson was present in Westminster Abbey as a translator of the New Testament of the New English Bible. In 1976 he produced a *magnum opus*, *Redating the New Testament*. His New Testament scholarship also led him to concern himself with the Turin shroud.

The appointment to Trinity enabled Robinson to lecture frequently in the USA; in South and Central America in 1971; South Africa in 1975; South Africa, Israel, India, Sri Lanka, Hong Kong, Japan, and the USA in 1977-8. *Truth is Two-Eyed* (1979)—on the Buddhist-Hindu-Christian encounter—was the product. In 1979 he went to New Zealand and Australia to lecture (*The Roots of a Radical*, 1980). In 1980 he was made an honorary DCL of the University of Southern California. In 1981 he visited Israel again to do preparatory work on his second New Testament *magnum opus*—*The Priority of John* (1985), the Bampton lectures for 1984. In 1982 he was able to do further preparation as visiting professor at McMaster University, Hamilton, Ontario, Canada, where he was also able to pursue his long-standing concern with nuclear disarmament.

In 1947 he married Ruth, daughter of Frank Grace, a clerk in a tea merchant's office. Robinson had met Ruth at Cambridge, where she gained a first in the modern languages tripos. They had one son and three daughters. In June 1983 a pancreatic tumour was diagnosed. Robinson lived with cancer courageously for the next six months, preparing *Twelve More New Testament Studies* (1984) for publication and gathering material for *Where Three Ways Meet* (1987). He preached his last sermon, on 'Learning from Cancer' to a packed college chapel on 23 October 1983 and died 5 December 1983 at his home at Arncliffe, Yorkshire. His friend, Professor C. F. D. Moule, delivered Robinson's Bampton lectures *The Priority of John* in 1984.

[Eric James, *A Life of Bishop John A. T. Robinson: Scholar, Pastor, Prophet*, 1987; Alistair Kee, *The Roots of Christian Freedom: The Theology of John A. T. Robinson*, 1988; personal knowledge.]   ERIC JAMES

ROBSON, DAME FLORA (1902-1984), actress, was born in South Shields 28 March

1902, the third of four daughters and the sixth of seven children of David Mather Robson, marine surveyor, and his wife, Eliza, whom he married when second engineer to her father, John McKenzie, a sea captain. The family left their native Scotland, first for Tyneside and later Palmer's Green, London, where Flora was taught singing and elocution to further a talent for speaking poetry when sitting on her father's knee. At her first public appearance, aged six, she recited 'Little Orphan Annie'. In 1919 she left Palmer's Green High School.

Unlike her sisters, she was not good looking and, at five feet eight and a half, was tall for a budding actress. Nevertheless her father sent her to Sir Herbert Beerbohm Tree's (later the Royal) Academy of Dramatic Art. In 1921 she was disappointed to win only the bronze medal. Her first professional engagement was in *Will Shakespeare* at the Shaftesbury Theatre (1921). Advised to acquire more experience, she joined the company at Bristol run by (Sir) P. B. Ben Greet [q.v.]. In 1923, at the Oxford Playhouse, she worked with a younger set of actors including (Sir) John Gielgud and (Sir) Tyrone Guthrie [q.v.], who invited her to his family home in county Monaghan—'a magic and memorable time for both'. Unable to earn her living as an actress, she became welfare officer of the Welgar Shredded Wheat Company at Welwyn Garden City, where her parents now lived. In 1929 Guthrie, the producer at the Festival Theatre, Cambridge, brought her back to acting in Pirandello's *Six Characters in Search of an Author*, with F. Robert Donat [q.v.] as leading man. Her appearance as the stepdaughter was an apparition of innocent corruption voluptuously embodied in an Aeschylean Fury. She and Guthrie became engaged.

However, in 1931 Guthrie married his cousin, Judith. This coincided with his production of *The Anatomist* by James Bridie [q.v.] at the Westminster Theatre, London, in which Flora Robson, playing an Edinburgh prostitute, had only one scene. Her success in the play, as at Cambridge, was overwhelming. The haunting quality of her voice, singing 'My Bonnie Wee Lamb', was remembered long after the play was over.

She conquered the West End, progressing through *Dangerous Corner* (1932, by J. B. Priestley, q.v.) and *All God's Chillun's Got Wings* (1933, by Eugene O'Neill), with Paul Robeson, to the Old Vic where, again under Guthrie, in 1933 she played in several productions, including *Macbeth* (as Lady Macbeth), *Measure for Measure* (as Isabella), *Love for Love* (as Mrs Foresight), and *The Cherry Orchard* (as Varya). She had reached the peak of her achievement, although she herself thought the play *Autumn*, at St Martin's in 1937 (James Agate, q.v., called

it 'Robsonsholm'), was her greatest success.

It was her nature to be immensely happy or unhappy; love could turn to hate and reverse at a moment's notice. She longed so much for Guthrie's children that she lost him; Robert Donat had children of his own; Paul Robeson she loved most but she deeply respected his wife and confided her fear to Bridie that her career would be 'a line of tortured spinsters'. Bridie wrote *Mary Read* (a female pirate) for her and Donat. (Sir) Alexander Korda [q.v.] presented the play in 1934 but hopes of romantic fulfilment through a film version were dashed when he failed to renew her contract. In 1938 she left for America but did not achieve film stardom. She stayed in America during the early years of World War II but returned to England in 1944, when she reappeared at the Lyric, Hammersmith, as Thérèse Raquin in *Guilty*.

But she had missed the wartime Council for the Encouragement of Music and the Arts and ENSA tours which led to reopening the Old Vic and the formation of national companies to which she should have belonged. Her poetry readings remained faultless; but plays like *Black Chiffon* (1949) were pot-boilers and often her passion was diluted into sentimentality. She was appointed CBE in 1952 and DBE in 1960. She was awarded honorary degrees by London (1971), Oxford (1974), Durham, and Wales.

She rejoiced at earning money to support her many relatives, friends, and causes. In her last performance in 1975, as Miss Prism in *The Importance of Being Earnest* by Oscar Wilde [q.v.], she was reborn. A role, usually considered as farcically eccentric, was transformed into a sweet Innocent of high comedy. She died 7 July 1984 at Brighton. She was unmarried.

[Kenneth Barrow, *Flora*, 1981; Janet Dunbar, *Flora Robson*, 1960; personal knowledge.]   MARIUS GORING

ROCHFORT, SIR CECIL CHARLES BOYD- (1887–1983), racehorse trainer. [See BOYD-ROCHFORT.]

ROSEVEARE, SIR MARTIN PEARSON (1898–1985), educationist and civil servant, was born 24 April 1898 in the village of Great Snoring near Walsingham in Norfolk. Roseveare is a common Cornish name but the family, or at least his branch of it, were scattered about the country by absorption into the Victorian professional classes. He was a son of the rectory and lived his life securely in that class. The rectory was a living in the hands of St John's College, Cambridge, where his father, Canon Richard Polgreen Roseveare, had read mathematics before entering the Anglican ministry to become eventually the vicar of Lewisham. Richard and his wife, Minnie Skinner, had seven children. Martin was the third son followed by a daughter and three more sons.

The Roseveares had the characteristic upbringing and schooling of their class and generation. All the boys won scholarships to public secondary schools, Martin to Marlborough and thence, after his father, to read the mathematics tripos at St John's College, Cambridge (of which he became an honorary fellow in 1952), where he emerged in class I from part i (1919) and as a wrangler (b) from part ii (1921). Two of the brothers were killed in action in World War I. Martin himself was wounded in 1918 near Amiens as an artillery officer (he had joined the Royal Field Artillery in 1917). He was mentioned in dispatches and discharged in May 1919.

A common outlook led all five surviving Roseveares into some form of public service, clerical or secular. Martin Roseveare's childhood ambitions were to become a missionary and to go to Africa: the first was turned into secular work as a teacher, the second was eventually and fully realized in Nyasaland (Malawi). His career fell into two stages. He left Cambridge in 1921 to teach mathematics at Repton where the headmaster, Geoffrey Fisher (later Lord Fisher of Lambeth, q.v.), who became archbishop of Canterbury, had previously taught him at Marlborough. His two years at Repton were not sufficiently successful for his probation to be turned into tenure and he moved on to Haileybury in 1923. In 1927, three years after having applied, he was accepted into HM Inspectorate of Schools. He rose to a staff inspectorship in mathematics in 1939 and to the highest office of senior chief inspector from 1944 to 1957. He reorganized the inspectorate in the wake of the 1944 Education Act. However World War II intervened to break his educational career with a period in the Ministry of Food (1939-44 and 1946) in which he distinguished himself by applying his mathematical skill to designing a national scheme for food rationing. He was responsible for the ration book. His meticulous and economic ingenuity in this vital department of administration was mainly responsible for the knighthood conferred on him in 1946.

The second stage of Roseveare's career came late. His call in childhood to Africa was realized. Retiring from his senior chief inspectorship he went to Nyasaland as headmaster of a new secondary school at Mzuzu (1957-63), principal of a teachers' training college, Soche Hill (1964-7), to mathematics teaching at Marymount School (1967-70), and finally to an active retirement in Mzuzu. He thus had an African career of thirteen years after thirty-six years of civil and educational service in England. It completed his life in the sense of taking him closer to his family—one brother was an Anglican priest in Cape Town, another the Anglican bishop of Accra, a son taught in Natal, and a daughter was a missionary-doctor in Zaire. At the same time the move to Africa was also a discontinuity of his family life. In 1921 he had married his first wife, Edith Mary (died 1975), daughter of Arthur Pearce, of Sidcup; she bore him a son and four daughters, one of whom died in 1969. His period in the Ministry of Food had kept him away from home a great deal. He and Edith grew apart and never remade their prewar companionship. They separated in 1954. Then he got to know (Olivia) Margaret Montgomery who was a junior colleague in HM Inspectorate. She was the daughter of Samuel Montgomery, manager-owner of the Irish Glass Bottle Company in Dublin. She followed him to Mzuzu where they married in 1958, after his divorce in the same year, and shared a successful African life.

His privately published autobiography (*Joys, Jobs and Jaunts*, 1985) reveals Roseveare as very much a private man, which is not to say unconvivial; on the contrary his whole life was one of energetic circuit in a complex network of personal and professional friendships. He practised a privacy which avoided politics, ideological movements, and impersonal administration. Thus he was the last senior chief inspector to define himself as the *primus inter pares* of a professional corps rather than the administrative link to the minister. In his own account of African experience over the quarter of a century in which fundamental renegotiation of colonial and Commonwealth relations took place, he never once mentions race or ethnicity or apartheid. He speaks only of kin, friends, acquaintances, and colleagues. His world was of people and places, his interests in teaching, topography, and technical gadgets. There is an evocative photograph of him as the frontispiece to his memoirs—a bespectacled English face by a fireside, smoking a pipe and reading the *Daily Telegraph*. He died in Mzuzu 30 March 1985.

[*The Times*, 2 April 1985; Sir Martin Roseveare, *Joys, Jobs and Jaunts*, Memoirs, 1985; private information.]   A. H. HALSEY

ROSKILL, STEPHEN WENTWORTH (1903-1982), naval officer and historian, was born 1 August 1903 in London, the second of four sons (there were no daughters) of John Henry Roskill, KC and judge of the Salford Hundred Court of Record, and his wife, Sybil

Mary Wentworth, daughter of Ashton Wentworth Dilke, MP for Newcastle upon Tyne. He was educated at the Royal Naval Colleges at Osborne and Dartmouth. In 1921 he was posted as midshipman to his first ship, the cruiser *Durban*, on the China station, where he was fortunate to act as research assistant to Lieutenant-Commander W. Stephen R. (later Lord) King-Hall [q.v.], who was then writing a book on western civilization and the Far East. Here Roskill first learned how, in his own words, 'unremittingly arduous' was the pursuit of history.

In 1927 he began the 'long' gunnery course at the Royal Naval College, Greenwich, and HMS *Excellent*, gunnery being then the élite branch of the navy. He passed out third in the course. The 1930s saw Roskill steadily climb the peacetime ladder of promotion, serving as gunnery officer in the aircraft carrier *Eagle* (1933–5) and the battleship *Warspite* (1936–9), with a spell as instructor at *Excellent* in 1935–6. While in *Warspite* he showed his mettle as a professional who would stand firm for what he believed to be right, refusing to take over the ship's armaments from the dockyards until numerous defects were remedied. He also pioneered the location of fire control beneath the armour instead of exposed aloft.

In March 1939 he was appointed to the Admiralty staff, where he successfully insisted on the Swiss Oerlikon 20 mm. gun instead of an inferior British design, and opposed the proposal of F. A. Lindemann (later Viscount Cherwell, q.v.) that anti-aircraft guns should be replaced by rockets. He later advocated that each main armament turret should have its own fire-control radar—a radical innovation later adopted. However, his unflinching advocacy in these matters brought him into disfavour with more conservative seniors, and in 1941 he saw his posting as executive officer in HMNZS *Leander* in the Pacific as a form of rustication. Here he restored a slack ship's company to a high standard of efficiency and training, so enabling *Leander* to survive a Japanese torpedo hit in 1943. Roskill was reappointed to the ship in command as acting captain, and in 1944 confirmed in rank as captain. In the same year he was awarded the DSC.

In 1944 he was posted as chief staff officer for administration and weapons in the Admiralty delegation in Washington; in 1946 was nominated chief British observer at the Bikini atoll atomic bomb tests; and in 1947 appointed deputy director of naval intelligence. Sadly, increasing deafness caused by exposure to gun detonations at *Excellent* denied him the chance of promotion to flag rank, for in 1948 he was pronounced medically unfit for sea service. While the premature ending of his naval life

was a keen disappointment, Roskill was now to achieve eminence in a new career as historian. In 1949 he was appointed the official naval historian in the Cabinet Office historical section. Although he had little previous experience as a writer and historian, he brought to his new profession the same seamanlike attention to detail, order, and exactitude that he had shown in his naval service and the same sometimes prickly determination to stand firm for what he believed to be right, even in the face of pressure from the most eminent.

It was thanks to him that the official naval history of World War II covered the entire war at sea, and not merely the Atlantic as once envisaged. The three volumes of *The War at Sea* (vol. i 1954, vol. ii 1957, vol. iii pt. I 1960, pt. II 1961), comprehensive, majestic, invested with a sailor's personal knowledge of the men and events as well as the historian's judgement, were to prove only the opening salvo in a prolific career as a writer. Outstanding among Roskill's contributions to twentieth-century history must be accounted his magisterial biography *Hankey, Man of Secrets* (vol. i 1970, vol. ii 1972, vol. iii 1974), his two volumes on *Naval Policy Between the Wars* (1968 and 1976), *Churchill and the Admirals* (1977), and his final work, *The Last Naval Hero: Admiral of the Fleet Earl Beatty; an Intimate Biography* (1980), a penetrating yet sympathetic assessment of Earl Beatty [q.v.] as man, fleet commander, and first sea lord. Roskill's forte as historian consisted in unrivalled professional understanding of naval matters, combined with scholarly thoroughness, mastery of detail, and ability to plumb the complexities of naval policy and strategy, although it could be said that his concern for detail sometimes tended to obscure the main thrust of his narrative. Roskill had no peer among twentieth-century British naval historians, only the American Arthur Marder rivalling him for depth of learning and sheer industry. He wrote several notices for this dictionary.

Roskill's distinction as a historian brought him a senior research fellowship at Churchill College, Cambridge, in 1961; and in 1970 he was made a life fellow. He played a major role in the Churchill Archives Centre, and it was owing to his efforts that many important collections of naval papers were deposited there. Despite worsening deafness, Roskill participated to the full in the life of the college; a much loved colleague and a charming and considerate host to his many guests.

In 1971 he was awarded a Litt.D. by Cambridge, elected a fellow of the British Academy, and appointed CBE. In 1975 he was awarded the Chesney gold medal of the Royal United Services Institute, and made an honorary D.Litt. by Leeds. In 1980 Oxford awarded him an honorary D.Litt., an honour which gave him special delight.

In 1930 he married Elizabeth, daughter of Henry Van den Bergh, margarine manufacturer, from Holland. Her strength of character and devotion to principle matched his own. They had four sons and three daughters. Roskill died 4 November 1982 at his home in Cambridge.

[Personal knowledge.]   CORRELLI BARNETT

ROTHA, PAUL (1907–1984), documentary film producer, director, and film critic, was born Roscoe Treeve Fawcett Thompson 3 June 1907 in Harrow, Middlesex, the only child of Charles John Samuel Thompson from Yorkshire, a writer, medical doctor, and curator of the Wellcome Historical Medical Museum, and his wife, May Tindall, a nurse who came from Lancashire. His parents changed his forename to 'Paul'. Educated at Highgate School and elsewhere, he studied at the Slade School of Art and won an International Design award in 1925. Changing his name by deed poll to the more distinctive 'Paul Rotha' he began work as an artist. His interest in films developed when he discovered the highbrow London Film Society, and he worked briefly at Elstree film studios. In 1930 he established his reputation with the publication of his influential book *The Film Till Now*, the earliest critical history of world cinema, which introduced many to the idea that film could be taken seriously as an art. More books followed down the years, and vigorous film criticism in many different periodicals.

The Empire Marketing Board Film Unit had just started to make officially sponsored promotional factual films, to be called 'documentaries'. Rotha joined the unit but found its austere spirit uncongenial and was quickly sacked. From 1933 onwards he worked independently, finding his own sponsors for the many trenchant films he was to make. He rapidly became a leading figure in the growing documentary movement, itself a significant British contribution to the development of world cinema. His first film *Contact* (1933), made for Imperial Airways, won a gold medal at the Venice film festival of 1934 and *The Face of Britain* (1935) another at the Brussels festival of 1935. Documentary companies proliferated as industry and officialdom turned to films for their prestige publicity.

During World War II the Ministry of Information sponsored hundreds of documentary films to inform the public and sustain morale, many of them by Paul Rotha Productions. *World of Plenty* (1943), a major film about world food production and distribution, was shown at the United Nations food conference. Another one about food politics backing the United Na-

tions Food and Agriculture Organization was *The World is Rich* (1947), which won a British Film Academy award.

For a time after the war Rotha concentrated on writing. In 1951 and 1958 his only two fiction feature films appeared but they aroused little interest. In 1953 another important documentary, *World Without End*, about the future of the underdeveloped countries, was sponsored by Unesco and won a second British Film Academy award. Rotha now turned to television, becoming head of BBC television documentaries from May 1953 to April 1956. His last film, *Das Leben von Adolf Hitler*, was a ninety-minute compilation of archive film made for a German company, and received an award at the Leipzig film festival of 1962.

Rotha was an official adviser to the British government at the inauguration of Unesco in Paris in 1946, a member of the Arts Enquiry 1943–6, a council member of the British Film Academy 1942 and its chairman 1952, president of the jury at the Venice documentary film festival of 1958 and later chairman of the Screen Writers Association. He was also senior Simon research fellow at Manchester University 1967–8. Commemorative seasons of his films were held by the Museum of Modern Art Film Library in New York 1958, the National Film Theatre 1979, and the Oxford Film Makers' Workshop 1982. He was made an honorary member of the Critics' Circle and in 1953 a fellow of the British Film Academy. Honorary membership of the Association of Cinema and Television technicians was awarded posthumously in April 1984.

He was a committed socialist, and his films were remarkable for their vitality, conviction, and relevance to social problems such as unemployment, poverty, food shortages, and the environment. A campaigner rather than a teacher, he eschewed the neutral informational style of many documentarians and used film for forceful and persuasive argument. Lyrical passages, also, reveal an artist's appreciation of the world around him. A man of great integrity, he was endlessly helpful and loyal to those he liked but could be a peppery antagonist to those of whom he disapproved.

In 1930 Rotha married Margaret Louise, daughter of Oscar Stanley Lee, a company director. She died in 1962. In 1973 he married the Irish film actress Constance Mary Smith, formerly Mrs Di Crollalanza, daughter of Sylvester Smith, musician. The marriage was dissolved in 1982. He had no children. His last years were spent in reduced circumstances and poor health, and he died at the John Radcliffe Hospital in Oxford 7 March 1984.

[*The Guardian*, 8 March 1984; *Variety*, 14

March 1984; Paul Rotha, *Documentary Diary* (autobiography), 1973; personal knowledge.]
RACHAEL LOW

RUSSELL, DOROTHY STUART (1895–1983), pathologist, was born 27 June 1895, in Sydney, Australia, the second daughter of Philip Stuart Russell, bank manager, and his wife, Alice Louisa, daughter of William Cave. She was orphaned at the age of eight and, with her sister, was sent to an uncle, Alexander Campbell Yorke, the rector at Fowlmere, near Cambridge, where she attended the Perse High School for Girls, and then Girton College. Graduating from Cambridge in 1918 with first class honours in part i of the natural sciences tripos, she entered the London Hospital Medical College, where she was influenced by Hubert Turnbull (whose notice she later wrote for this Dictionary), professor of morbid anatomy. She qualified MRCS (Eng.), LRCP (Lond.) in 1922 and MB, BS (Lond.) in 1923, with the Sutton prize in pathology, and began work under Turnbull, first as a junior Beit fellow from 1923 to 1926, and later while supported by the Medical Research Council. From 1933 to 1946 she was a member of the MRC scientific staff. Her first major work was 'A Classification of Bright's Disease' (MRC, 1929), which was the basis of her London MD, awarded in 1930, with the University medal.

(Sir) Hugh Cairns [q.v.], the neurosurgeon at the London Hospital, encouraged her into neuropathology and, with a Rockefeller travelling fellowship in 1928–9, she visited Frank B. Mallory in Boston and Wilder Penfield in Montreal, returning in 1930 to the London Hospital. In 1939 she joined Cairns, now Nuffield professor of surgery in Oxford, at the wartime hospital in St Hugh's College, a move partly enforced by the evacuation of many medical departments from London. Pio del Rio-Hortega was briefly a colleague before he departed for Buenos Aires in 1940.

Dorothy Russell's wartime period at Oxford was one of intense productive work, and her 'Observations on the Pathology of Hydrocephalus' (MRC, 1940, 1967) remains a definitive account of the subject. Her major life-work was on brain tumours, their staining and classification, and her masterpiece was the first edition in 1959 of her book, with L. J. Rubinstein, entitled *Pathology of Tumours of the Nervous System*. She gained a Cambridge Sc.D. in 1943. In 1944 she returned from Oxford to the London Hospital and in 1946 succeeded Turnbull as professor of morbid anatomy and director of the Bernhard Baron Institute of Pathology, the first woman to occupy such a prestigious position, in which she remained until her retirement in 1960. These were golden years for her contributions to neuropathology, and to morbid anatomy for she always considered herself a general pathologist. Of greater importance for these subjects were her numerous trainees, who later occupied professorial chairs and founded new departments throughout the world.

Her outstanding achievements brought many honours. She was an honorary fellow of Girton College at Cambridge, St Hugh's College at Oxford, the Royal Microscopical Society, the Royal College of Pathology, and the Royal Society of Medicine. She received honorary doctorates of the universities of Glasgow and McGill, the John Hunter medal and triennial prize of the Royal College of Surgeons, and the Oliver-Sharpey prize of the Royal College of Physicians, of which she was a fellow (MRCP 1943, FRCP 1948). She delivered the Bryce memorial lecture at Somerville College, Oxford, the foundation lecture of the Association of Clinical Pathologists, and the Hugh Cairns and Schorstein lectures at the London Hospital. She was the first woman member of the Medical Research Society.

Her personality had two paradoxical sides. Her penetrating mind, combined with a rigorous training in morbid anatomy, lent her writings and utterances a clarity and thrust which few of her contemporaries equalled. Consequently, in scientific circles, she was regarded with a degree of respect amounting to awe, and in debate was feared because of her customary trenchant exposition of her views. Yet in private she was reserved and shy, sometimes expressing a humble opinion of her own abilities and attainments. She suffered from epilepsy, a disability made generally known, at her request, after her death. To her friends and close colleagues, and to her disciples, with whom she remained in touch, she extended a warm friendship, embracing also their families.

She was active in her retirement at Westcott, near Dorking, in Surrey, for, although she seldom attended meetings, many friends and former pupils visited her and she was in regular correspondence with former colleagues. She died 19 October 1983 in Dorking General Hospital, Dorking, Surrey. She was unmarried.

[*The Times*, 20 October 1983; *Lancet*, 1983, vol. ii, p. 1039; *British Medical Journal*, vol. cclxxxvii, 1983, pp. 1477–8; L. J. Rubinstein in *Munk's Roll*, vol. vii, 1984; Barbara Wootton, *In a World I Never Made*, 1967, p. 33; private information; personal knowledge.]
J. T. HUGHES

RUSSELL, EDWARD FREDERICK LANGLEY, second BARON RUSSELL OF LIVERPOOL (1895–1981), soldier, military lawyer, and writer, was born in Liverpool 10 April 1895, the only child of Richard Henry Langley Russell and his wife, Mabel, daughter of Frederick Younge. His father died when he was four and his mother, a comedy actress, then continued with her profession, entrusting her small son to the care of his paternal grandfather, Sir Edward Richard Russell (later first Baron Russell of Liverpool), the editor of the *Liverpool Daily Post*, whose politics were distinctly Liberal and whose interests were primarily in the stage. His grandfather, who became the dominating influence on his life, preferred to send him to Liverpool College, a day school, rather than to Rugby School for which he was originally destined. In 1913 he entered St John's College, Oxford, to read law. Oxford seems to have made little impression upon him during the one year that he spent there before he entered the army for the period of World War I.

He served in the King's Regiment as an officer. Until gassed in the latter part of the war he spent his whole time in the trenches, an experience which never left him. He was wounded three times and was decorated with the MC (1916) and two bars. Throughout the rest of his life his lungs were affected and he suffered indifferent health. His grandfather died in 1920, while he was still convalescing from his damaged lungs, and he inherited the barony. He remained a further ten years in the army, transferring in 1924 to the 20th Lancers (Indian Army). In India he rode, played polo, engaged in hunting expeditions, and lived to the full the type of life which he afterwards regarded as idyllic. In 1930 the condition of his lungs compelled him to 'send in his papers' on the grounds of poor health.

On leaving the army he felt drawn to the bar. He had admired the legal and political 'giants' whom he had encountered in his grandfather's house during his youth. He was duly called to the bar at Gray's Inn in 1931 and joined the Oxford circuit. He did not acquire any substantial practice and had difficulty in trying to convince his clerk that a member of the aristocracy could have serious ambitions in the profession, particularly in the early 1930s. The result was that after three years he sought and obtained an appointment in the military department of the office of the judge advocate-general as a military prosecutor. This career he followed assiduously, prosecuting and lecturing on military law, at home and abroad, until the outbreak of World War II in 1939.

He remained in the office of the judge advocate-general until 1950, holding important posts throughout the war. He served in France, North Africa, and the Middle East as a senior member of the office, with distinguished members of the bar serving under him. He was ap-

pointed OBE in 1943 and CBE in 1945. He left the army with the rank of brigadier in 1950.

In 1951 he became assistant judge advocate-general but was restless in the appointment, having been actively engaged in 'reviewing' the proceedings of the military courts for the trial of German war criminals in the British occupied zone of Germany from 1946 to 1950. As a result of this experience he wrote *The Scourge of the Swastika* (1954), a detailed exposé of German war criminality, while still holding his judicial appointment. He sent a courtesy copy to the lord chancellor, Lord Simonds [q.v.], before publication. Simonds indicated that if it were published Russell's position would be under consideration, bearing in mind that the author held judicial office. Russell therefore resigned his appointment and published the book, of which over 250,000 copies were sold around the world. It proved to be the first of a series of books, on such topics as Japanese war criminality, lord chancellors, famous murder cases, and his own life. Russell also lectured across the world, engaged in public debates, and introduced into the House of Lords causes that offended his sense of justice, notably that of James Hanratty and the A6 murder. His works showed a high sense of intellectual honesty and a firm attachment to factual accuracy. He had no pretensions to legal erudition.

In 1967, as a result of financial difficulties and differences with the Inland Revenue, he left England to live in France. This precluded his use of 'place and voice' in the Lords, which he much regretted. His life was darkened by the death of his wife in a car accident in France, and of his son and heir, Langley Gordon Haslingden Russell, in England in 1975. He returned to England after his marriage and divorce of a later wife and lived in considerably reduced financial circumstances.

Russell was married four times. In 1920 he married Constance Claudine, daughter of Phillip Cecil Harcourt Gordon, of the RAMC; they had a son and a daughter. This marriage was dissolved in 1933 and in the same year he married Joan Betty, daughter of David Ewart, MD and FRCS; they had one daughter. This marriage was dissolved in 1946 and in the same year he married Alix, widow of Comte Bernard de Richard d'Ivry and daughter of the Marquis de Bréviaire d'Alancourt. She died in 1971 and in 1972 he married Selma (died 1977), formerly wife of A. W. Bradley. Russell suffered a severe stroke while living in Hastings and died in a nursing home there 8 April 1981. He was succeeded in the barony by his grandson, Simon Gordon Jared (born 1952).

[*The Times*, 10 April 1985; Lord Russell of Liverpool, *That Reminds Me* (1959); Michael

Frankel, 'Lord Russell of Liverpool' (doctoral thesis presented to the University of Haute-Bretagne-Rennes II, October 1986); personal knowledge.]   G. I. A. D. DRAPER

RUSSELL, SIR FREDERICK STRATTEN (1897–1984), marine biologist, was born 3 November 1897 at Bridport, Dorset, the younger son and second of three children of William Russell, schoolteacher, of Bridport, and his wife, Lucy Binfield, daughter of Henry Newman, of Liverpool. He was educated at Oundle School and was awarded an open scholarship to Gonville and Caius College, Cambridge, in 1915. His academic career was interrupted by World War I in which he served with distinction as an observer in the Royal Naval Air Service. He was awarded the DSC, DFC, and croix de guerre (with palm). He took up his scholarship to Caius College in 1919 and received a first-class honours degree in natural science (zoology) in 1922.

After a brief period as assistant director of fisheries research for the Egyptian government, he joined the staff of the Marine Biological Association in Plymouth as an assistant naturalist in 1924. In the creative atmosphere of the Plymouth laboratory his exceptional skills as an observer and his incisive ability to unravel the significant patterns from a complex mass of information enabled him to make rapid progress in his research. Some carefully planned and meticulously executed field studies in the English Channel led to the publication of a series of classical papers explaining the vertical distribution of young fish and small animals (zooplankton) in the sea in terms of their sensitivity to light.

In 1928–9 he worked in Australia on the Great Barrier Reef Expedition and, on his return to Plymouth, he began to study the patterns of biological change in the English Channel. Through careful observation he recognized the importance of certain 'indicator' species as heralds of environmental change and he contributed significantly to the establishment of clear seasonal cycles in plankton abundance in the Channel. He also studied the life cycles of the Medusae (jellyfish), in many instances actually culturing these delicate animals in the laboratory with the assistance of W. J. Rees. He had learned the art of water-colour sketching from his father and had also received training as a photographer in World War I. His papers were therefore usually beautifully illustrated and he developed his talents to produce and edit a series of keys, still widely used for the identification of planktonic (drifting) animals. His skills were most eloquently used in the magnificent illustrated volumes on the British Medusae begun at this time.

In 1940–5 he served as a wing commander in air staff intelligence where his contributions were highly praised. Immediately after the war he was invited to return to Plymouth as the director of the laboratory—a challenge which he accepted somewhat reluctantly because of the administrative duties involved. He served as director from 1945 to 1965 and by many this is considered as the 'golden age' of the laboratory. Although he had a relaxed and kindly nature he was energetic and decisive and his confidence in his own abilities enabled him to provide firm direction. His policy of recruiting able and dedicated staff and allowing them to develop their own ideas generated a stimulating atmosphere that proved attractive to a wide range of distinguished scientists including the zoologist O. E. Lowenstein, the neurophysiologists (Sir) A. L. Hodgkin and (Sir) A. F. Huxley, and geologists and geophysicists such as W. B. R. King, W. F. Whittard [qq.v.], and M. N. Hill. During his directorship the laboratory facilities were improved by a carefully planned building programme and by the purchase of an ocean-going research vessel (RV Sarsia). Despite the considerable energy that he devoted to his role as director, Russell continued his research. Volume i (Hydromedusae) of his superbly illustrated work, The Medusae of the British Isles, was published in 1953 and the second volume (Scyphomedusae) in 1970.

He retired from the directorship of the Marine Biological Association in 1965 but retained a room at the laboratory and continued until 1980 his work on the Medusae and on the distribution of young fish and the significance of biological changes in the English Channel. He published The Eggs and Planktonic Stages of British Marine Fishes in 1976. His studies of biological variability continued in collaboration with his colleagues in Plymouth. This work led to the recognition of large periodic changes in the English Channel (the Russell cycle), related to the rise and fall of the herring fishery, that are extremely important for attempts to assess man's impact on the environment.

Russell was elected FRS in 1938. In 1955 he was appointed CBE and in 1957 he was elected an honorary fellow of the Institute of Biology and an honorary LLD at the University of Glasgow. He was awarded honorary D.Sc. degrees at the universities of Exeter (1960), Birmingham (1966), and Bristol (1972). He received the gold medal of the Linnean Society in 1961 and was knighted in 1965. He was elected an honorary fellow of Gonville and Caius College, Cambridge, in 1965. He was a foreign member of the Royal Danish Academy and was an honorary member of the Physiological Society, the Challenger Society, and the Fisheries Society of the British Isles.

In 1923 he married Gweneth, daughter of John Moy Evans, solicitor, of Swansea. They had one son. Although deeply affected by the sudden death of his wife in 1979, Russell continued to work with vigour at the Plymouth laboratory until he moved to a nursing home in Goring-on-Thames in 1980. He died there 5 June 1984.

[E. J. Denton and A. J. Southward in *Biographical Memoirs of Fellows of the Royal Society*, vol. xxxii, 1986; personal knowledge.]   MICHAEL WHITFIELD

RYLE, SIR MARTIN (1918-1984), radio astronomer and Nobel prize-winner, was born at Brighton, Sussex, 27 September 1918, the second son and second child in the family of three sons and two daughters of John Alfred Ryle [q.v.], professor of medicine, and his wife, Miriam Power, daughter of William Charles Scully, civil servant, of Cape Town, who came from a land-owning family in county Tipperary. His uncle was the philosopher Gilbert Ryle [q.v.]. When Ryle was five years old the family home was established at 13 Wimpole Street, London, where he was first educated by a governess. Later he attended Gladstone's preparatory school in Eaton Square and entered Bradfield College at the age of thirteen. He gained an exhibition in the scholarship examination in 1932 and in 1936 entered Christ Church, Oxford, to read physics. He obtained a third class in natural science honour moderations (1937) and a first in physics (1939).

Ryle then joined J. A. Ratcliffe's ionospheric research group at the Cavendish Laboratory, Cambridge. On the outbreak of World War II Ratcliffe joined the Air Ministry Research Establishment (later TRE) and Ryle followed him in May 1940. For two years he worked mainly on the design of aerials and the development of test equipment. In the summer of 1942 he became the leader of a group in the newly formed radio counter-measures division. Ryle played a prominent part in various radar jamming and radio-deception operations amongst which the electronic 'spoof invasion' on D-day, which led the Germans to believe that the invasion was taking place across the Straits of Dover, was particularly important. In this complex field of counter-measures the technical response often had to be immediate. Ryle's extraordinary inventiveness and immediate scientific insight were of great importance in this work and often led him to be intolerant of those not similarly blessed.

At the end of the war Ryle obtained an ICI fellowship and returned to research work at the Cavendish Laboratory. He did not find the ionospheric research sufficiently stimulating and Ratcliffe transferred his interest to an entirely new and challenging research problem. The German jamming of the British radars which enabled the battleships *Scharnhorst* and *Gneisenau* to pass through the English Channel from Brest to Kiel on 12 February 1942 was referred to J. S. Hey of the Army Operational Research Group for investigation. Two weeks later Hey found that an apparently similar case of jamming of the anti-aircraft radars was caused by an intense outburst of radio waves from the sun, associated with a solar flare and large sunspot group. Ratcliffe suggested to Ryle that he should investigate whether in such cases the radio emission came from the region of the sunspots or from the whole solar disc. The solar disc subtends an angle of 0.5 degrees and spots and flares occupy only a very small area of the disc so that the broad beam aerials then available to Ryle were useless for settling this question. He solved this problem by using two simple aerials, which he could move apart, connected by cable to the same receiver. Later, Ryle stated that he did not at first realize that he had invented the radio analogue of the optical interferometer used by A. A. Michelson thirty years earlier to measure the diameter of stars.

This elegant method of obtaining the resolving power equivalent to a very large aerial had far-reaching consequences in astronomy. Ryle's interest soon turned to the study of the cosmic radio waves. He discovered that there were sources of strong radio emission of small angular extent in the universe, and with his colleagues he made a catalogue of the positions and intensities of two thousand of these objects, few of which could be identified optically. He concluded that the number-intensity distribution of these objects provided the evidence that the universe had evolved from a dense and concentrated condition some ten billion years ago. When these results were published in 1955 a historic dispute arose with those astronomers who supported the steady state, continuous creation theory.

In 1954 the idea appears in Ryle's notebooks that he could extend the two aerial interferometers to synthesize a complete large aperture aerial. By moving the two aerials to successive positions and using a digital computer to combine the records he successfully demonstrated the *aperture synthesis* method in 1960-1. He then developed a number of such systems culminating in 1971 with the synthesis of a radio telescope of 5 km. aperture by using eight 13-metre aperture radio telescopes, four fixed and four moving on a railway track. The succession of maps of radio sources with a definition of about an arc second produced by this system revealed the complexity of their structure and raised major problems about the nature of the power sources in the distant galaxies.

An important consequence of the investigations of the radio sources by Ryle's group was the discovery of pulsars (neutron stars) in 1967 by Antony Hewish and Jocelyn Bell. In the joint award of the Nobel prize to Ryle and Hewish in 1974 the development of aperture synthesis was specifically itemized as Ryle's major contribution.

Ryle's early post-war research in Cambridge was carried out at the Grange Road rifle range. In 1956 the research was moved to the site of a former Air Ministry bomb store at Lord's Bridge. On 25 July 1957, in recognition of support from Mullards, the site was named the Mullard Radio-Astronomy Observatory. Ryle was the director from 1957 until his retirement in 1982.

In 1972 Ryle was appointed astronomer royal, the first appointment of an astronomer royal who had not been director of the Royal Greenwich Observatory in the 300 years of its existence. Ryle's appointment coincided with the development of a serious illness and he was unable to exercise the authority which had hitherto been associated with that office, which he relinquished in 1982. In his later years Ryle began experiments on the generation of electricity by wind power. He was deeply concerned about the misuse of science and protested publicly against nuclear armaments.

Ryle's outstanding brilliance as a scientist was widely recognized. He was made lecturer in physics at Cambridge in 1948, became a fellow of Trinity College in 1949, and was appointed to the Cambridge chair of radio astronomy in 1959 (until 1982). He was knighted in 1966, elected FRS (1952), and awarded the Hughes (1954) and Royal (1973) medals of the Royal Society. He was awarded the gold medal of the Royal Astronomical Society in 1964, the Henry Draper medal of the US National Academy of Sciences and the Holweck prize of the Société Française de Physique in 1965, the Popov medal of the USSR Academy of Sciences, the Faraday medal of the IEE, and the Michelson medal of the Franklin Institute of the USA in 1971. He was elected a foreign member of the Royal Danish Academy of Sciences and Letters (1968), of the American Academy of Arts and Sciences (1970), and of the USSR Academy (1971). He was made an honorary D.Sc. of the University of Strathclyde (1968), Oxford (1969), and Torun (1973).

Ryle's chief recreation was sailing and his skill as a craftsman enabled him to construct his own sailing ships. On 19 June 1947 Ryle married Ella Rowena, eighth of the ten children of Reginald Palmer, tradesman, of Mildenhall in Suffolk. There were one son and two daughters of the marriage. Ryle died 14 October 1984 at Cambridge.

[Francis Graham-Smith in *Biographical Memoirs of Fellows of the Royal Society*, vol. xxxii, 1986; Bernard Lovell in *Quarterly Journal of the Royal Astronomical Society*, vol. xxvi, pp. 358-68, 1985; M. Ryle, *Les Prix Nobel en 1974*, pp. 80-99, Stockholm, 1975; personal knowledge.]

BERNARD LOVELL

S

ST OSWALD, fourth BARON (1916–1984), war correspondent, soldier, and politician. [See WINN, ROWLAND DENYS GUY.]

SARONY, LESLIE (1897–1985), song writer and entertainer, was born Leslie Legge Sarony Frye in Surbiton, Surrey, 22 February 1897, the youngest in the family of three sons and three daughters of William Rawston-Frye, portrait painter, of Den Villa, Tolworth, and his wife, Mary Sarony. He first appeared in talent shows at the age of twelve. At fourteen he took his mother's maiden name and became a professional entertainer with Park's Eton Boys and Girton Girls and other juvenile variety acts.

When war began in 1914 he lied about his age, joined the London Scottish Regiment, and saw service on the Somme at Vimy Ridge and in Salonika and Macedonia. He contracted malaria and dysentery and, during recuperation in Malta, wrote his first lyric, a parody of a popular song called 'Three Hundred and Sixty Five Days'. His version was about the surfeit of cheese the troops were served.

On being demobilized he resumed his career in variety, pantomime, revue, and musical comedy. An excellent dancer, in 1926 he played the juvenile lead, Frank, in the original Drury Lane production of *Show Boat*. His West End shows in the 1920s included appearances in *The Peep Show*, *Dover Street to Dixie*, *Brighter London*, and *The Whirl of the World*.

His song-writing and recording careers also started in the 1920s. Between 1926 and 1939 he worked for every recording company in the country. He made over 350 records under his own name and dozens more under assumed ones. Many of his recordings, as featured vocalist, were with Jack Hylton [q.v.] and his band.

In 1935 he formed, with Leslie Holmes, the variety act The Two Leslies. For eleven years they appeared on radio and topped variety bills all over the country. In 1938 they appeared in the royal variety performance. Holmes retired in 1946 and Sarony continued the act for three years with Michael Cole.

He became a solo variety act in the late 1940s. In the 1970s he was much in demand as a character actor, playing everything from Samuel Beckett's *Endgame* and *As You Like It* to film roles in *Chitty Chitty Bang Bang* and *Yanks* as well as, on television, *I Didn't Know You Cared*.

He was a tiny, dynamic, ebullient, aggressively forthright and hard-working, dyed-in-the-wool professional to whom entertaining was a job of work. Yet his songs were special. 'I Lift Up My Finger and I Say Tweet Tweet', 'Jollity Farm', and 'Ain't It Grand To Be Bloomin' Well Dead' were all enormously popular, as was his recording of the folk song 'The Old Sow'. His stirring march 'When the Guards are on Parade' reflected his love of the military life. The titles of a few of his 150 published songs recall the England of the 1930s—'Tune In', 'Teas, Light Refreshments, and Minerals', 'I Like Riding On A Choo Choo Choo', and 'Mucking about in the Garden' (written, incidentally, under the pseudonym Q. Kumber).

Off stage he was a keen golfer and his 'men only' parodies, written and performed for the Vaudeville Golfing Society, are masterpieces of Rabelaisian doggerel.

In the 1980s he was made a member of the Grand Order of Water Rats and president of the Concert Artistes Association, and was presented with the gold badge of merit by the Songwriters Guild of Great Britain. In 1983 he appeared in his second royal variety performance.

In 1939 he married Anita, daughter of Frederick Charles Eaton, a dairy owner, auctioneer, and racehorse owner. They had three sons. The marriage ended in divorce in 1953. Sarony died 12 February 1985 in St George's Hospital, Tooting.

[Private information; personal knowledge.]
ROY HUDD

SAVAGE, SIR (EDWARD) GRAHAM (1886–1981), teacher and educational administrator, was born 31 August 1886 at Upper Sheringham, Norfolk, the eldest in the family of three sons and two daughters, the surviving children of Edward Graham Savage and his wife, Mary Matilda Dewey, whose family were small farmers at Dunhead, near Shaftesbury in Dorset. His parents were the teachers at Upper Sheringham elementary school which he attended before King Edward VI School, Norwich, and Downing College, Cambridge. At Cambridge he obtained firsts in part i of the natural sciences tripos (1905) and part ii of the historical tripos (1906).

Before World War I he held various teaching appointments at Bede College, Durham; St Andrew's College, Toronto; and Tewfikieh School and the Kehdivial Training College, Cairo. He also joined the Territorial Army. On the outbreak of war he was commissioned in the Royal West Kent Regiment, as a captain and later major, fighting at Gallipoli and in France, being severely wounded at Ypres.

In 1919 he became an inspector of schools for the Board of Education and began a career of rapid advancement in the public service. After a stint as a district inspector (1919–27) he became successively staff inspector for science (1927), divisional inspector for the north-west (1931–2), chief inspector of technical schools and colleges (1932), and senior chief inspector from 1933 to 1940.

When E. M. Rich, the education officer to the London county council, retired in 1940, Savage jumped at the opportunity to exchange the role of an HMI for that of senior local authority educational administrator. The London job was a challenge which appealed to him; it was also better paid. Savage's early years in London were spent improvising schooling for children who drifted back to the capital in the lulls between the bombing. It was a period of extreme difficulty when a depleted education service was at full stretch and educational standards inevitably suffered.

By 1943 he was looking ahead to post-war reconstruction. Labour had controlled the LCC since 1934. Soon after taking over, Labour members had been frustrated in attempts to eliminate the sharp distinctions between grammar and senior elementary schools. Savage thought along similar lines. The London School Plan (1947) provided for comprehensive schools, not separate county grammar, technical, and modern schools. He claimed later that his ideas arose from a visit he had paid, as an HMI, to Canada and the United States in the winter of 1925–6. He admired the comprehensive character of the American high schools though much of his subsequent report dwelt on their failure to provide for the differentiated needs of the whole school community.

The London Plan bore Savage's authentic stamp, but he was also a realist and recognized that this was the answer his political masters wanted. He remained ambivalent about many aspects of comprehensive education. He wanted all kinds of pupil together under the same roof, but he expected them to pursue markedly different courses, with a strong practical, technical, and vocational input. The current assumption was that these schools would have to be very large—360–90 children in each year group—to yield a big enough academic sixth form. Moreover, as the law stood, London's fifty-five aided grammar schools remained largely unscathed, to continue alongside the so-called comprehensives.

To the end of his life Savage remained a grammar school man at heart, with a special regard for the direct grant schools, which he described in a *Times* interview in 1965 as 'the scaffolding on which a good state system of secondary education is slowly being built'.

A large man with a large personality, a bald dome of a head, and a face scarred by war wounds with a bristling moustache, Savage was not an easy man to know, even for those who worked closely with him for many years. There was a warmer side to his personality which his family knew, which came out in visits to children and teachers injured in the bombing.

After his retirement in 1951, he acted between 1956 and 1964 as chief assessor to the Industrial Fund for the Advancement of Science Teaching in Schools, channelling money to independent schools for laboratories and equipment. His many other distinctions included being chairman of the League of the Empire (1947–62), the Simplified Spelling Society (1949–68), and the Board of Building Education (1956–66). He was president of the Science Masters' Association (1952–3) and from 1967 to 1971 vice-president of the City and Guilds of London Institute. He was made an honorary fellow of the Institute of Builders in 1966. He was appointed CB in 1935 and knighted in 1947.

In 1911 he married May, daughter of Percy Thwaites, an artist, of Southchurch, Essex. They had two sons and a daughter. Savage died 18 May 1981 in Highgate, north London.

[*The Times*, 20 May 1981; Stuart Maclure, *100 Years of London Education*, 1970; private information.]　　　　　　STUART MACLURE

SCHAPIRO, LEONARD BERTRAM (1908–1983), historian, political scientist, and barrister, was born in Glasgow 22 April 1908, the elder son and second of three children of Max Schapiro and his wife, Leah Levine. His father, educated at the universities of Riga and Glasgow, was the son of a wealthy sawmill owner at Bolderaa near Riga, his mother the daughter of a rabbi and cantor of the Garnett Hill synagogue in Glasgow. His great-uncle, Jacob Shapiro, was a Constitutional Democratic deputy in the second Duma. From 1912 the family resided in Riga, moving in 1915 to Petrograd in wartime conditions and remaining there until 1921, when the father's newly acquired citizenship of recently independent Latvia enabled them to leave and settle in London.

Already fluent in English, German, and Russian, he was educated at St Paul's School and University College, London, where he read law. In 1931, still a student, he married Ynys Mair, an art student and daughter of David Evans, a Newport pharmaceutical chemist. The marriage was dissolved in 1937. Called to the bar (Gray's Inn) in 1932, he practised in London and on the Western circuit, supplementing his income, and gaining his first experience as a teacher, by coaching public school and university entrants.

In 1940 he became a supervisor at the BBC monitoring service at Evesham, where close study, first of German and from 1941 of Soviet news and information, established an intellectual interest and the foundation of a future academic career. There, among a remarkable group of Central European intellectuals, he made friends with G. Katkov and V. Frank who were to share his growing interest in Russian culture and history. Also at Evesham he met and in 1943 married Isabel Margaret, daughter of Salvador de Madariaga [q.v.], King Alfonso XIII professor of Spanish studies in Oxford University, 1929–31. She later became professor of Russian studies in the University of London. Recruited in 1942 into the Intelligence Corps, he was commissioned in early 1943, moved to the general staff at the War Office, and in 1945–6 served in the intelligence division of the German Control Commission, attaining the rank of acting lieutenant-colonel.

He returned to the bar in 1946 and published a number of articles on subjects of international law, but his experience of Soviet military administration in post-war Germany had reinforced a desire to discover the well-spring of Soviet attitudes to law and government practice, and in 1955 he quit law and devoted himself to Soviet studies.

The Origin of the Communist Autocracy (1955, 2nd edn. 1977), a study of the socialist opposition in the early years of Soviet government, in which Schapiro analysed the Bolsheviks' abuse of their political monopoly, established him as a penetrating critic of the Soviet regime and he was offered a lectureship at the London School of Economics, where he inspired and created a generation of Russian historians and political scientists. From 1963 to 1975 he was professor of political science with special reference to Russian studies at LSE. One of his chief contributions to the field, *The Communist Party of the Soviet Union*, was published in 1960 (2nd edn. 1970). As an undogmatic constitutionalist, for Schapiro the acid test of political credibility was respect for the law, a test the Soviet regime consistently failed, based as it was on the same principle of arbitrary rule as tsarism. This was the theme of his Yale lectures, published as *Rationalism and Nationalism in Russian Nineteenth Century Political Thought* (1967). His view that the Stalinist state was the continuation of Lenin's heritage aroused controversy, and he later modified it to allow that Stalin had committed excesses that Lenin might not have countenanced, and still later, in his posthumously published *1917: the Russian Revolutions and the Origins of Present-day Communism* (1984), he conceded that Stalinism was a possible but not a necessary consequence of Leninism.

Born into an anglophile Riga family Schapiro's political ideals were based on his experience as a lawyer; he particularly admired the English common law and he believed that the role of law was essential for the creation of a civilized polity. In Russian culture his ideal was the liberal, cosmopolitan novelist Ivan Turgenev, of whom he wrote a sensitive intellectual biography, *Turgenev: his Life and Times* (1978). Never a Zionist or practising Jew, in his youth he had occasionally frequented East End synagogues for their ethnic colour, but he rejected a perceived nationalism in Judaism; moved by Christianity, as by strong religious feeling in general, but unable to accept the divinity of Christ, he was inclined towards a non-specific spirituality. The Jewish aspect of Schapiro's background was not publicly expressed until relatively late in his life and was as much an outcome of his professional concerns as it was of any 'racial' emanation. As chairman of the editorial board of *Soviet Jewish Affairs* and a member of the Institute of Jewish Affairs, he took an active part in exposing Soviet anti-Semitism.

He was a member of the council of the School of Slavonic and East European Studies from 1956 to 1981. He was elected a fellow of the British Academy in 1971, and in 1980 appointed CBE. He was chairman of the Institute for the Study of Conflict from 1970; a member of the research board of the Institute of Jewish Affairs; a council member of the Institute for Religion and Communism; chairman of the editorial board of *Government and Opposition* from 1965; long-time legal adviser and vice-president from 1976 of the National Council for One-Parent Families; and foreign honorary member of the American Academy of Arts and Sciences (1967).

In 1976, after his second marriage was dissolved, he married (Dorothy) Roma Thewes (née Sherris), a journalist and friend from Evesham days, and daughter of Cyril Sherris, a medical practitioner. Spiritual father to innumerable students, he had no children, but derived great joy from his last wife's grandchildren. He died in London 2 November 1983.

[Peter Reddaway in *Proceedings* of the British Academy, vol. lxx, 1984; S. E. Finer in *Government and Opposition*, no. 1, vol. xix, 1984; Hugh Seton-Watson in *Encounter*, April 1984; private information; personal knowledge.]   HAROLD SHUKMAN

SCHUSTER, SIR GEORGE ERNEST (1881–1982), financial adviser to governments and educationist, was born 25 April 1881 in Hampstead, the elder son (there were no daughters) of Ernest Joseph Schuster, barrister, and his wife, Hilda, daughter of Sir Hermann Weber, MD. His father came from a Jewish family of bankers who had left Frankfurt-on-Main in 1866. One uncle, Felix Schuster [q.v.], was given a baronetcy by H. H. Asquith, and another, Sir Arthur Schuster [q.v.], became a fellow and secretary of the Royal Society. George Schuster was a scholar of Charterhouse and an exhibitioner of New College, Oxford, where he obtained a second class in classical honour moderations in 1901 and a first in *literae humaniores* in 1903. After reading for the bar (Lincoln's Inn, 1905) he accepted a business post with H. R. Merton & Co. Schuster's idealistic and active turn of mind was attracted by politics and in 1911 he was adopted as Liberal candidate for North Cumberland. He was considered a promising recruit but, owing to his German name, had to relinquish any political ambitions in August 1914. He joined the Oxford Yeomanry.

The war suited his mentality and his ideals. In May 1916 he was appointed a chief administrative staff officer at First Army HQ. After the armistice he went to Murmansk, with the rank of lieutenant-colonel, to deal with the financial and currency problems which beset the White Russians.

After the Murmansk operation was wound up he embarked on training as a teacher on the first course at Birmingham University and felt much frustrated when the whole project was abandoned for lack of funds. At just this time he was selected by a city syndicate to report on practical measures for reviving the economy of what remained of the Austro-Hungarian Empire. Though his report to the syndicate contained no effective proposals, it led to a post with Frederick Hull & Co. and another with the finance committee of the League of Nations. He also became a member of the Economic Research Organization of the Labour Party. His growing reputation in economic expertise led to what he styled 'a new adventure'—a request from the Foreign Office that he should go for five years as financial secretary to the Sudan (1922–7). Later he always looked on these years as the most rewarding experience of his life and a 'chance to take part in one of the finest chapters in the history of the British Empire'. This was a disturbed period during which Sir Lee Stack [q.v.], the governor-general, was assassinated, but Schuster won praise not only for his performance of his financial and administrative tasks, but also for his contribution to the handling of political problems.

In October 1927 a request came from the India Office that he would accept the post of finance member of the viceroy's council. The Colonial Office agreed to release him from the Sudan on condition that the appointment be deferred for a year and that he first travel round east and central Africa as a member of the Hilton Young commission, to report on proposals for the closer union of certain African colonies and then act as financial adviser to the Colonial Office on several economic development projects in the area.

Schuster's years in India as finance minister (1928–34) covered an important stage in the transition from British rule to independence. From the first he felt it his prime duty to work for Indian interests. He had to effect large economies in government and military expenditure. It was ironical that he should be the only person wounded, though not seriously, when a bomb was thrown in the assembly, during the introduction of legislation which he himself had suggested should be deferred.

He returned to England in September 1930 to advise the British delegation at the first Round Table conference held in London to plan the political future of the sub-continent. The final effort of his period in India was to be the establishing of a reserve bank as the currency authority for India, independent of the government.

Schuster was fifty-three when he returned from India and still considered himself a young man, with a personal career to make. He was soon invited on to the boards of important companies. Unilever asked him to undertake the chairmanship of Allied Suppliers—the largest retailing group in Britain. His work here brought him to his special interest in relations between management and labour. In 1938 he was elected as Liberal National candidate in the by-election at Walsall. Soon he was working for Sir R. Stafford Cripps [q.v.], an old Oxfordshire friend, in the Ministry of Aircraft Production. He spoke often in the Commons, especially on measures of social policy such as the Beveridge plan, for which he did not share the universal praise, as he regarded as totally inadequate a conception of welfare 'defined in exclusively material terms'. He was a good constituency member and it came as a great shock when he was defeated in 1945.

On his return from India he had joined the governing body of Charterhouse and by 1939 had become deputy to the statutory chairman, the archbishop of Canterbury. He was constantly in demand as a committee member of proven experience in drafting and lobbying. His own interest focused more and more on industrial relations, and he worked closely on this and economic planning with the post-war Labour government. At the request of the government he went in 1950 as financial and economic adviser to Malta. He was convinced the Maltese were justified in asking for financial help from Britain and finally persuaded them to raise more by local taxation. As a result Cripps immediately recommended a grant of £½ million to be met from the Colonial Development Fund.

Early in 1953 Schuster was invited by the

Federation of the Indian Chamber of Commerce to return to their country and advise them on future policy. He advised that the foundation of India's prosperity must be in the development of agricultural production, rather than in large-scale industry and in village communities rather than in vast urban concentrations.

Schuster was never unoccupied. From 1951 to 1963 he was chairman of the Oxford Regional Hospital Board and from 1953 he was for twenty-two years a most active member of the Oxfordshire County Council. He found the work congenial as most of his colleagues were local, voluntary, and non-party. In 1961 came a new 'adventure', the development of Voluntary Service Overseas. These three words summed up his enthusiasms. He became honorary treasurer and eagerly set about creating an orderly administration to support A. G. Dickson's imaginative vision and within two years had received generous support from many private sources plus an annual grant of £500,000 from government funds.

When already over eighty he threw himself into his last great activity, the creation of Atlantic College, which he came to prize more highly than anything else he had done. He was chairman of its board of governors from 1963 to 1973. He was much inspired by discussions with Kurt Hahn and Sir Robert Birley [qq.v.]. The fact that funds were missing merely spurred him on. He drew on every possible source, reasoning rather than wheedling, involving trade-union personalities as well as leaders of finance. He threw himself into every aspect of the project, making it his business to know the pupils and to ensure that staff had frequent opportunities for meeting the governors. His ninetieth birthday was celebrated at St Donats with great display and he remained deeply proud that he had been involved in the college, and also that so constructive a project should have been started in Britain, 'as I do not believe it could have got off the ground in any other country'.

His long life thus ended on a note of triumph. On his ninety-eighth birthday he went up to London to be measured for two new suits. In his hundred and second year he was still writing to headmasters on educational matters. He had published innumerable pamphlets and three books: *India and Democracy* (with G. Wint, 1941), *Christianity and Human Relations in Industry* (1951), and *Private Work and Public Causes* (1979), his autobiography. He listed his recreations in *Who's Who* as 'all country sports'. He was a tall man with an upright and soldierly bearing, with finely marked features (in no way Jewish), and with excellent health. He had a great capacity for work and delighted in it—he nearly always fitted in an hour before breakfast. His financial acumen and his clarity of ex-

position were widely appreciated, as was his gift for warm friendship and for creating a team spirit in any undertaking. He was a rare mixture of intellectual, man of action, and idealist, who never failed the ideal he had set himself as an undergraduate, believing intensely in hard work, personal service, and the values of the British empire. Reading his slim autobiography one gets a vivid picture of the life of a public servant in the closing years of that empire.

Schuster was appointed CBE in 1918, KCMG in 1926, and KCSI in 1931. Oxford University awarded him an honorary DCL in 1964.

In 1908 he married Gwendolen (died 1981), daughter of Robert John Parker, Baron Parker of Waddington [q.v.], judge. They had two sons, one of whom was killed in action in 1941. His marriage was the central happiness of his life. Schuster died 5 June 1982 at Nether Worton House, the seventeenth-century manor he owned near Banbury since 1919.

[Sir George Schuster, *Private Work and Public Causes*, 1979 (autobiography); *The Times*, 8 June 1982; private information.]

OLIVER VAN OSS

SCHWARTZ, GEORGE LEOPOLD (1891-1983), economist and journalist, was born in Brighton 10 February 1891, the elder son and eldest of four children of Adolph George Schwartz, an Austrian Jew who was a wine merchant; his wife, Antonia Held, was Hungarian but not Jewish. He went to Varndean School in Brighton, and after studying at St Paul's training college, Cheltenham, obtained a post as a teacher in 1913 in a school under the aegis of the London County Council. When World War I broke out he enlisted in the army and served at Gallipoli. His brother had by this time emigrated to the USA.

After the war Schwartz studied at the LSE, and in 1923 was appointed secretary of the London Cambridge Economics Service. His first publication, *Output, Employment and Wages in the United Kingdom, 1924*, which appeared in 1928, was mainly a statistical analysis not intended for the general reader. In 1929 Schwartz became an Ernest Cassel lecturer at London University, and, in collaboration with one of his colleagues, Frank W. Paish, published *Insurance Funds and their Investment* (1934).

During the early years of World War II, he assisted (Sir) Arnold Plant in the preparation of a wartime social survey intended to supply the government with information about the reaction of the public to rationing and other emergency measures. Then, from 1944 to 1961 he was deputy city editor of the *Sunday Times* and economic adviser to the Kemsley newspapers. Between 1945 and 1954 he was also editor of the *Bankers' Magazine*, which, when he took it over, was desperately in need of revival, as the war years had seriously affected its circulation. His energy and enthusiasm in this task were amply rewarded; as a result of his nine years' work the circulation of this periodical was materially increased and its reputation substantially enhanced.

After retirement as an editor of the *Sunday Times* at the age of seventy, he continued for another ten years to contribute to the paper a column on economic affairs. A selection of his articles was published in 1959 under the title *Bread and Circuses 1945–1958*.

Schwartz's attitude to politics and economics was summed up in his entry in *Who's Who*, where he described his recreation as 'detesting government'. In the foreword to *Bread and Circuses* he wrote: 'My approach to economic and social problems is governed by the early training I received . . . at the London School of Economics, which a grand and apparently incorrigible popular delusion brands as a hotbed for disaffected unorthodoxy. The unorthodoxy escaped me.' In one of his articles he reiterated his distaste for authority, saying: 'It is a persistent delusion of the human race that Government is composed of men who take a longer view of its destinies than do the governed themselves.' If his ideas are compared with those of his contemporary economists he appears to be most nearly in sympathy with the views of Sir Dennis Robertson [q.v.]; at the other extreme he constantly lambasted the socialist dogma of those, such as Harold Laski and E. Hugh J. N. (later Lord) Dalton [qq.v.], who gave the LSE the reputation which, thought Schwartz, was undeserved.

Schwartz believed passionately in the freedom of the individual and was unsparing of his criticism of any opponent of capitalism from William Temple, archbishop of Canterbury, to Richard Crossman, a controversial figure [qq.v.] in the cabinet of Harold Wilson (later Lord Wilson of Rievaulx). When the archbishop had claimed that capitalism 'doesn't even work' Schwartz, in reply, wrote: 'Such allegations . . . do not emanate appositely from episcopal palaces which have had the milk delivered daily on the doorstep for centuries and into which the necessaries and minor luxuries consonant with plain living and high thinking have flowed uninterruptedly as a result of "uncoordinated and planless human effort".'

To Schwartz economics was never a dismal subject, dry as dust: to him it was a living exciting subject to be explained, not exclusively with theoretical dogma, but with common sense and a pragmatic concern for the welfare of the individual; and these views he succeeded in conveying to a wide circle of readers in simple, realistic prose. He had a genial humour, spiced with sardonic wit, which he used in college common rooms or at the Reform Club. He became an honorary fellow of LSE.

In 1927 he married Rhoda Lomax (died 1966), who was born in Kovno, Russia, and who had previously been married to an artist. They had no children. Schwartz died in London 2 April 1983.

[*The Times*, 6 April 1983; private information.]   H. F. OXBURY

SCOTT, (GUTHRIE) MICHAEL (1907–1983), Anglican clergyman and lifelong champion of the oppressed, particularly in southern Africa, was born at Lowfield Heath, Sussex, 30 July 1907, the third and youngest son (there were no daughters) of Perceval Caleb Scott, a clergyman in the High Church tradition, and his wife, Ethel Maud Burn. The misery of the Southampton slums surrounding his childhood home in Northam made a deep impression on him. He was educated at King's College, Taunton. In 1926 he went to Switzerland and then to South Africa to recuperate from tuberculosis. He studied at St Paul's theological college in Grahamstown, South Africa. After returning to England, he went to Chichester Theological College in 1929. The bishop of Chichester, George Bell [q.v.], ordained him as a deacon in 1930 (he became a priest in 1932). Scott's involvement with political issues began in 1934 when he was a curate at All Souls, Lower Clapton, London, at the time of the hunger marches. He had many contacts with communists during the 1930s.

From 1937 to 1939 he was in India, first as chaplain to the bishop of Bombay and later as chaplain in St Paul's Cathedral, Calcutta. At this time he was continuing to experiment with communist beliefs (which he later abandoned), and his moral and religious questioning led to an inner conflict. He returned to England and, after the outbreak of World War II, chose to enlist in the RAF in 1940, as aircrew rather than as a chaplain. In 1941 he was invalided out with Crohn's disease, which continued to trouble him throughout his life. In 1943 he again went to South Africa. Until 1946 he was assistant priest at St Alban's Coloured Mission on the outskirts of Johannesburg. Appalled by conditions of life for non-white people he worked for the Campaign for Right and Justice. In 1946 he was asked to observe Indian passive resistance in Durban against the Asiatic Land Tenure and Indian Representation Act. Volunteers, who stood on forbidden open land and were attacked by white hooligans, were joined by Scott, who was arrested and sent to prison. On release his parish licence was withdrawn by Bishop Geoffrey H. Clayton, who disapproved of his unauthorized absence from the parish and the extent of his political involvement.

From that time Scott was besieged with requests from oppressed groups who desperately wanted their case to be heard. At the request of African ex-soldiers, he lived in the gang-ruled shanty town called Tobruk outside Johannesburg and was prosecuted for living in a native urban area. He went to Bethal in the eastern Transvaal to investigate the conditions of near slavery in which white farmers held black labourers. His dedication and disregard for his own comfort became legendary. He was a loner but always drew on the help of devoted friends and supporters. His emaciated good looks, height, and shabby crumpled appearance, with pockets bulging with papers, made him conspicuous. He was a poor speaker. He struggled with self-doubt but, having reached a decision, was tenacious in spite of his diffidence. Invited to Bechuanaland (later Botswana) by Tshekedi Khama [q.v.], regent of the Bamangwato people, he met the exiled chief of the Herero who asked him to go to see Chief Hosea Kutako and other chiefs in Windhoek. With them he drew up the Herero petition for the return of the Herero lands and against the incorporation of South-West Africa with South Africa; he was appointed their spokesman to the United Nations in 1947. In 1949 he was the first individual petitioner to be heard by the fourth (trusteeship) committee of the UN and asked for matters pertaining to South-West Africa to be referred to the International Court of Justice. He attended UN sessions for the next thirty years for the International League for Human Rights and kept South African race oppression on the agenda. In 1949 Scott's general licence was withdrawn by Bishop R. Ambrose Reeves on the grounds that he was out of the country. When Bishop George Bell heard of this he gave him a licence.

From 1950, having been refused entry to South Africa, Scott lived in London initially as a guest of the Friends International Centre. When at UN Sessions he lived in New York, often at the General Theological Seminary. He became honorary director of the non-party-political Africa Bureau which in 1952 started to focus attention on issues in British Africa. Scott also undertook projects outside the Bureau's scope, identifying himself with passive resistance in Nyasaland (later Malawi) against a Central African Federation. In 1959 he took part in the peace protests in the Sahara against the French atom bomb and he joined the World Peace Brigade in 1962. From the early 1960s he was associated with Bertrand

(Lord) Russell [q.v.] in the peace movement and became vice-president of the Committee of 100. Russell and he jointly published *Act or Perish* in 1961. Scott served two prison sentences as a result of these activities.

From 1958 representatives of the Naga people, who wanted independence from India, sought his help. In 1962 he brought A. Z. Phizo, president of the Naga National Council, to London to expose the war in Nagaland. Scott consulted Jawaharlal Nehru [q.v.] about a proposal for a cease-fire. In 1964 he was invited by the Nagaland Church Council to be one of three members of a peace mission to the Nagas. In 1964–6 he made strenuous marches in Nagaland, meeting villagers and members of the underground movement. The Indian government deported him from Nagaland in May 1966.

In the 1970s many groups and individual refugees turned to Scott for help. In 1979 he initiated the organizations Rights and Justice and World Wide Research to focus attention on human rights. His great contribution was his identification with the oppressed and his prophetic Christian insights into the seeds of conflict and the need for peaceful change. Africa did not forget him. He was honoured by Zambia in 1968 and in 1975 was made an honorary canon of St George's Cathedral, Windhoek, Namibia. He died 14 September 1983 at 43 King Henry's Road in London. He never married.

[Michael Scott, *A Time to Speak* (autobiography), 1958; personal knowledge.]

TREVOR HUDDLESTON

SCOTT, SIR JOHN ARTHUR GUILLUM (1910–1983), secretary of the Church Assembly and first secretary-general of the General Synod of the Church of England, was born in East Battersea, London, 27 October 1910, the eldest of the three children and elder son of Guy Harden Guillum Scott, a barrister, and his wife, Anne Dorothea Fitzjohn. Scott's paternal grandfather, Sir Arthur Guillum Scott, was accountant-general of the India Office. Guy Guillum Scott became in 1920 assistant secretary of the newly established Church Assembly and in 1939 secretary (jointly with L. G. Dibdin), and was also chancellor of the diocese of Oxford. Scott's younger brother was in government service in Nigeria and his sister Judith was from 1954 to 1971 secretary of the Council for the Care of Churches.

Scott was educated at King's School, Canterbury. In 1929, after a brief period in a City bank, he joined the staff of Queen Anne's Bounty. In the 1930s Scott was a member of the Inns of Court Regiment (TA). He served in World War II, was wounded and mentioned in dispatches, and, when war ended, was on the

Ceylon staff of Lord Louis Mountbatten (later Earl Mountbatten of Burma, q.v.). After demobilization Scott rejoined the Inns of Court Regiment, in which he had a fierce pride, ending as lieutenant-colonel commanding, and being awarded the TD (Territorial Decoration) in 1945. There was much that was soldierly, in the best sense, in Scott's character and bearing; and as a bureaucrat he was always rather more the staff officer than the Whitehall mandarin.

In 1946 he replaced his father on the staff of the Church Assembly, serving for two years under L. G. Dibdin. In 1948 he succeeded to the secretaryship. His primary task was to service the thrice-yearly sessions of the Assembly and its committees. But he was constantly on call to serve as secretary or assessor to a variety of committees dealing with the financial and administrative structures of the church, with the appointment, deployment, and payment of bishops and clergy, with liturgy, with relations with other churches, with church/state relations and much else. He did not see himself as a radical, but he was not against change if it was necessary or desirable. It was often he who, when others had agreed upon aspirations, would give expression to them in terms of positive proposals, leading to acceptable and practicable legislation.

Scott had close, trusting relationships with the two archbishops of Canterbury whom he served. G. F. Fisher (later Lord Fisher of Lambeth, q.v.) appreciated the young staff officer, who helped him to run the Church Assembly with dispatch. With A. Michael Ramsey (later Lord Ramsey of Canterbury) the relationship was more complex. Each man seemed somewhat in awe of the other—Ramsey of the efficient man of business, and Scott of the archbishop whose intellect he found daunting. Recognition came from church (a Lambeth DCL in 1961, shortly before Fisher's retirement) and state (a knighthood in 1964). Church people sensed the depth of Scott's Christian commitment, though he did not wear his religion on his sleeve. He was a man of great personal charm; he was an excellent host; he had a remarkable memory for names and faces. No one enjoyed more the pomp and circumstance of the great church or state occasion. Yet there was a restraint about him which, as the years went by, acquired a touch of the Olympian, placed as he was literally and metaphorically above the battle and seemingly unmoved by it. People liked him, but not many felt that they knew him well. He was, indeed, a private person, on close terms, outside his family, with only a few, such as his secretaries, a close Church House aide of the 1960s, and his successor.

When the General Synod came into being in 1970, replacing the Church Assembly, Scott was its first secretary-general. When it was decided that a younger man should succeed him after two years, he accepted with excellent grace, setting himself to groom his successor with all his own and his father's accumulated professional cunning, for between them they had served the Church Assembly for the whole of its fifty-year life.

In 1938 he married (Muriel) Elizabeth, daughter of James Ross, departmental manager of Fownes, the well-known firm of glovers. They had one daughter.

The Scotts had lived in Buckinghamshire while he worked in London. On his retirement from the Synod in 1972 Scott moved to Chichester, to become the first communar (lay administrator) of the cathedral. He and Lady Scott threw themselves with enthusiasm into the life of the cathedral and its close as he began to establish an administration to meet the pastoral and financial needs of the time, including the vastly increased number of visitors. He retired for a second time in 1979, continuing to live in Chichester. He died there 6 May 1983.

[Church Assembly *Reports of Proceedings*, 1920–70; General Synod *Reports of Proceedings*, 1970–2; *Church Times*, 13 May 1983; personal knowledge.]

DEREK PATTINSON

SCOTT, Sir ROBERT HEATLIE (1905–1982), diplomat and home civil servant, was born at Peterhead, Aberdeenshire, 20 September 1905, the eldest child with two brothers and one sister of Thomas Henderson Scott, civil engineer, of Peterhead, and his wife, Mary Agnes Dixon, teacher. When he was still a boy the family moved to Trinidad following the father's work. There he completed, at Queen's Royal College, his schooling begun in Inverness. He scored a precocious success by winning at the age of fifteen an 'island' scholarship to New College, Oxford. The college refused to accept him at that age and he taught at his school for two years before coming to England.

At Oxford he took some time to settle down. Having tried mathematics (he obtained a third class in moderations in 1924) he switched to law and gained a second class in 1926. He was called to the bar (Gray's Inn, 1927). Eventually he decided to enter the Consular Service in the Far East which was well suited to his adventurous temperament.

In China he met and married another Scot, Rosamond Aeliz, daughter of Robert Nugent Dewar Durie, banker, of the Imperial Bank of Persia. They had a son (who was killed in an accident in 1955 on service with the Royal Marines) and a daughter. Scott served in Japan, Manchuria, Peking, Canton, Chungking,

Shanghai, and Hong Kong. To his Chinese he added Japanese, German, Dutch, French, and some Russian. During his time in China his energy and inquisitiveness attracted the hostile attention of the Japanese authorities. At the outbreak of World War II he was posted to Japan to conduct British propaganda and was subsequently transferred in 1941 to Singapore to open a branch of the Ministry of Information—always equated by the Japanese with intelligence and spying.

As the crisis deepened he was appointed to the governor's war council which consisted of the three Service chiefs—a singular tribute to his talents. Contemporary reports show that he outshone his colleagues in spirit and imagination. When the city fell he attempted to get away on the last boat to leave (his wife had already gone on ahead to Australia). The boat was soon intercepted by a Japanese destroyer. He volunteered to row across to the warship with some others in an attempt to persuade the enemy captain to allow the refugee ship to proceed. The sea was rough and the Japanese captain impatiently opened fire on the refugee ship before the lifeboat could reach the destroyer. There were few survivors. Scott succeeded in reaching Sumatra where he was briefly in hiding until he was returned as a prisoner to Singapore.

After a period in solitary confinement he was put in Changi gaol with other civilian prisoners. Scott was already considered a marked man by the Japanese. However he soon became a leading figure in the camp. The prisoners were slackly administered at first and were able to maintain contact with people outside and from time to time visited the city in the course of administering themselves. But towards the summer of 1943 the Japanese became uneasy and were planning a surprise raid to impose stricter discipline in the camp when a remarkable event occurred. A commando, daringly led by Captain Lyon, reached Singapore from Australia undetected and blew up Japanese tankers in the harbour. The Japanese authorities in Singapore mistakenly thought that the raiders had accomplices within the camp. Scott was assumed to be the ringleader.

For weeks he was terribly beaten and tortured but no confession was ever obtained. He was put on trial and eventually sentenced to six years imprisonment in Outram Road prison. Throughout his ordeal Scott conducted himself calmly and with self-assurance. Drawing on his previous experience of the Japanese, he successfully sought to establish a moral and intellectual ascendancy over his persecutors who found his arguments and predictions most disturbing. During part of this period he was held in solitary confinement at the top of the prison tower where from time to time he could be observed by his fellow prisoners in the yard below. He became known throughout the camp and in the city as 'the man in the tower' and became a symbol for the British and Chinese of defiance and resistance. His wife in Australia was unaware of his fate and drew a widow's pension.

After the Japanese surrender a controversial decision was taken to stage war crimes trials in Singapore. Scott was a principal but somewhat reluctant witness. His testimony was given without rancour and with such fairness that all were astonished—not least the accused. Scott never subsequently showed animosity towards the Japanese. Some years later with the help of the British ambassador in Tokyo he arranged to meet some of his former gaolers who had escaped execution. After a slow start the gathering proved a great success.

After Scott's release he returned to Britain where, following convalescence, he resumed his duties with the Foreign Office. He was appointed CBE (1946), CMG (1950), and in 1950-3 he was under-secretary of state, Foreign Office, with responsibility for Far Eastern affairs. This period included the Korean war and its problems. In 1953-5 Scott served as minister at the British embassy in Washington. In 1954 he was appointed KCMG and in the following year was sent to Singapore as commissioner general for the United Kingdom in South East Asia, a regional post with civil and military responsibilities.

His appointment in South East Asia was greeted with warm approval by the people of the area and by the British colonial officials and Service chiefs with whom he worked in great harmony. He was an invaluable source of good advice locally and helped Whitehall to understand the changing attitudes of old and new countries. He travelled widely and constantly in the area seeing old friends and making new ones amongst the post-war leaders. He became GCMG in 1959.

His war record and his unusual experience of both civilian and Service affairs encouraged the government to appoint him in 1960 as the first civilian commandant of the Imperial Defence College. He was outstandingly successful and moved on to the Ministry of Defence in 1961 as permanent secretary. This move proved to be a disappointment. The management of a huge Civil Service department did not suit his individualistic talents. It was also a period of bitter inter-Service rivalry as the Service chiefs tried to adjust themselves to necessary structural changes. Those who served with him overseas and in Whitehall were impressed by his modesty, originality, peace of mind, ease of manner, and preoccupation with essentials. All his business was conducted with a sense of fun heightened by his habit of drafting directly and inexpertly on to a portable typewriter.

He retired in 1963 and went to live near Peebles on the banks of the Tweed in a cleverly converted railway station which over subsequent years he and his artistic wife made into a charming house and garden. He received an honorary LLD from Dundee in 1972.

His great energy and zest for life ensured that in his retirement in his beloved Scotland he remained active. Absorption with family and with British and foreign friends, counselling young offenders, salmon fishing, his duties as lord lieutenant of Tweeddale—all these and many other interests kept him constantly and happily occupied. He died at Peebles 26 February 1982.

[Personal knowledge.]
GREENHILL OF HARROW

SCOTT, SIR RONALD BODLEY (1906-1982), consultant physician. [See BODLEY SCOTT.]

SEARLE, HUMPHREY (1915-1982), composer and writer on music, was born 26 August 1915 at Oxford, the eldest of the three sons (there were no daughters) of Humphrey Frederic Searle, of Oxford, a commissioner in the Indian Civil Service, and his wife, Charlotte Mathilde Mary ('May'), daughter of Sir William Schlich [q.v.], the pioneer of forestry studies at Oxford. He was educated at Winchester College (1928-33) and New College, Oxford, where he obtained a second class in classical honour moderations (1935) and a third in *literae humaniores* (1937). In 1937 he became Octavia scholar at the Royal College of Music (his teachers were John Ireland, Gordon Jacob [qq.v.], and R. O. Morris), and at the Vienna Conservatorium (1937-8). In Vienna he took private lessons from Anton Webern.

In 1938 Searle joined the BBC music staff. From 1940 to 1946 he served with the Gloucestershire Regiment, Intelligence Corps, and General List, and after the war, when still in Germany, assisted Hugh Trevor-Roper (later Lord Dacre of Glanton) in research for *The Last Days of Hitler* (1947). He resumed producing at the BBC, leaving in 1948 to work free lance. An enthusiast for promoting new music, he was general secretary of the International Society for Contemporary Music (1947-9). In 1951-7 he served as music adviser to Sadler's Wells Ballet.

A distinguished Liszt scholar, he was generous in imparting knowledge to colleagues. He compiled a new catalogue of Liszt's works (*Grove's Dictionary of Music and Musicians*, 1954 edn.; updated in *The New Grove*, 1980), and wrote a seminal book *The Music of Liszt* (1954). He founded the Liszt Society along with his friends Constant Lambert, (Sir) William Walton [qq.v.], (Sir) Sacheverell Sitwell, and others, and was its first honorary secretary (1950-62). Other books included *Twentieth Century Counterpoint* (1954), *Ballet Music* (1958), and *20th Century Composers: Britain, Scandinavia and the Netherlands* (1972). He also edited Schoenberg's *Structural Functions of Harmony* (1977), and translated Josef Rufer's textbook on twelve-note composition (1954) and a selection of Berlioz's letters (1966).

His compositions show the influence of Liszt, Webern, and Schoenberg. This radical continental outlook encouraged a style not then fashionable in Britain, and Searle remained an unfashionable though vigorous, independent, and prolific composer. The neglect of his large output of colourful, powerfully emotional works, written in strongly personal idiom and with a predilection for unusual forms, never seemed to daunt him. *Night Music* (for Webern's sixtieth birthday, 1943) closely approached twelve-note technique. His first truly serial work, *Intermezzo for 11 Instruments* (1946), was written in memory of Webern. Almost all his subsequent compositions use the 12-note method. His finesse in instrumental detail and subtle nuance is the legacy of Webern; his natural romanticism has affinities with the romantic Schoenberg; his fascination with Liszt is seen in the metamorphosis of themes in his Piano Sonata (1951).

A powerful trilogy for speakers, chorus, and orchestra (1949-51) set texts by (Dame) Edith Sitwell [q.v.] and, as centre-piece, *The Riverrun* by James Joyce [q.v.]. Between 1953 and 1964 he produced five symphonies, the Piano Concerto no. 2, three ballets, and two operas: *The Diary of a Madman* (after Gogol; premièred at the Berlin Festival of 1958 and awarded the Unesco Radio Critics prize), and *The Photo of the Colonel* (1964, after Ionesco). He also wrote much other chamber, orchestral, vocal, and incidental music. A BBC production, *The Foundling*, for which he wrote the music, won the Italia prize. His final opera was *Hamlet* (Hamburg 1968, Covent Garden 1969). Many of these works richly deserve revival.

A shy man, Searle's integrity and cosmopolitan outlook won respect among his students. He was composer-in-residence at Stanford University, California (1964-5); professor of composition at the Royal College of Music from 1965; guest composer at the Aspen Music Festival, Colorado (1967); and guest professor at the Staatliche Hochschule für Musik, Karlsruhe (1968-72) and at the University of Southern California, Los Angeles (1976-7).

His deep interest in the spiritual nature of humankind is reflected in his later choral-orchestral works: *Jerusalem*, *Kubla Khan*, *Dr Faustus*, and *Oresteia*. Colour and his love of adventure permeate orchestral works like *Labyrinth* (1971) and *Tamesis* (1979). His humour is seen in splendid settings of Edward Lear's *The Owl and the Pussy Cat* and T. S. Eliot's *Skimbleshanks the Railway Cat*. For all his reserve and professional detachment he loved conviviality, and the memoirs (as yet unpublished) of this fine writer and friend of Cecil Gray, Dylan Thomas [q.v.], Constant Lambert, and the Sitwells are a fascinating record of his times. To this Dictionary he contributed the notices of Sir Eugene Goossens, E. J. Dent, Constant Lambert, and T. F. Dunhill.

He was appointed CBE in 1968; honorary FRCM (1969); and an honorary professorial fellow, University College of Wales, Aberystwyth (1977).

In 1949 Searle married Margaret Gillen ('Lesley'), daughter of John Gray, a cartage contractor. In 1960, three years after her death, he married an actress Fiona Elizabeth Anne, daughter of John Wilfred Nicholson, a forest officer in the service of the Indian government. There were no children of either marriage. Searle died 12 May 1982 in London.

[*The New Grove Dictionary of Music and Musicians*, 1980 (ed. Stanley Sadie); *The Times*, 13 May 1982; private information; personal knowledge.]   DEREK WATSON

SETON-WATSON, (GEORGE) HUGH (NICHOLAS) (1916-1984), historian and political scientist, was born in London 15 February 1916, the eldest of three children (two sons and a daughter) of Robert William Seton-Watson [q.v.], historian, and his wife, Marion ('May') Esther, daughter of Edward Stack, of the Bengal Civil Service. He was educated at Winchester and at New College, Oxford, where he gained a second class in classical honour moderations in 1936 and a first in philosophy, politics, and economics in 1938.

Early in the war he served in the British legations in Belgrade and Bucharest. It was in Yugoslavia that he was recruited into the Special Operations Executive. The *coup d'état* which overthrew the regent Prince Paul on 27 March 1941 was followed by the German invasion and the flight of the British legation staff. Most of them, including Seton-Watson, were repatriated to Britain after internment by the Italians. In August 1941 he was flown out to Cairo, where he served at GHQ Middle East, Special Forces, until 1944. It was during this time that he wrote his first major work, *Eastern Europe between the Wars 1918-1941*, most of it

in Cape Town, to which he had been temporarily evacuated, and where his only source of reference was the public library. The book was finished at the time of the battle of El Alamein but had to wait until 1945 for publication.

After the war his travels in eastern Europe for *The Times* and the *Economist* in 1947 and 1948 had a profound influence on him. His early sympathy with the Soviet Union was justified by his generation on the grounds that Stalin was Hitler's only opponent. These views were now rejected as he saw those who represented ideas of freedom and humanity similar to his own persecuted by the Soviets or by their east European communist allies. Thus what happened then enabled him to form a picture of communism as the ultimate antithesis to all the aspirations of his generation. His experience resulted in *The East European Revolution* (1950) which has provided scholars with a pattern of communist 'take-overs', and a model to think about those events in a systematic fashion.

In spite of an offer by R. M. Barrington-Ward [q.v.] of a job on *The Times* there was never any doubt that he would follow an academic career. The war had prevented him from taking up a lectureship in international politics in Aberystwyth. In 1945 he was appointed praelector in politics at Oxford and elected a fellow of University College. He took up these positions after demobilization in 1946. In 1951 he was appointed to the chair of Russian history at the University of London in the School of Slavonic and East European Studies, and he remained there until his retirement in 1983. In 1952 appeared his first important work on Russian history, *The Decline of Imperial Russia 1855-1914*. It was followed fifteen years later by his monumental study of Russian history in the century before the revolution, *The Russian Empire 1801-1917* (1967). He wrote that he had come to Russian history for three reasons: it was a country which resembled and had always influenced eastern Europe; it was a country which under communism had produced the world communist movement; and thirdly it was the country within which Leninism had been born and which provided the first example of the impact of western ideas and western economy on a backward social and political structure. He stressed that it was the third of these factors which seemed to him to offer the most valuable lessons for our own time.

This set the theme for much of his work. In 1953 came *The Pattern of Communist Revolution: a Historical Analysis* and in 1960 *Neither War nor Peace—The Struggle for Power in the Post-War World*. He had come to the conclusion that the historical analysis of his father's generation had been based on the belief that the destruction of Austria-Hungary had been the last stage of a process of liberal nationalist revolutions which had begun in 1848. His generation, on the other hand, had realized that that destruction had been the first stage of a process which had spread from Europe into Asia and Africa and which had destroyed all the colonial empires except the Soviet. He had always been interested in nationalism and in 1977 appeared his most important work on politics and international relations, *Nations and States. An Enquiry into the Origins of Nations and the Politics of Nationalism*, which analysed the problem of nation building and national movements in every region of the world.

The study of nationalism also brought him back to his father's activity and to the events in eastern Europe which created the Europe of 1918. *The Making of a New Europe. R. W. Seton-Watson and the Last Years of Austria-Hungary* was published in 1981—it was written together with his brother Christopher (at that time a fellow of Oriel College, Oxford).

For more than thirty years Seton-Watson presided over the teaching of Russian and Soviet history in the University of London. His very name and reputation were a magnet for scholars and postgraduate students alike. He took an active part in the work of *The Slavonic and East European Review* and outside the university in the work of the British Academy to which he had been elected a fellow in 1969. From 1952 to 1984 he also served on the council of the Royal Institute of International Affairs.

Seton-Watson was a scholar inspired by travel as much as by books, not merely for pleasure, but to learn and to understand. Eventually his travels took him to all the world's continents, to lecture and to study. Increasingly, contacts with the American academic world became an important element in his life and in his study.

He received a D.Litt. from Oxford in 1974 and an honorary doctorate from the University of Essex in 1983. In 1981 he was appointed CBE. In 1947 he married Mary Hope, daughter of Godfrey Denne Rokeling, of the Ministry of Education. They had three daughters. Seton-Watson died in Washington DC 19 December 1984.

[Dimitri Obolensky in *Proceedings* of the British Academy, vol. lxxiii, 1987; information from wife and brother; personal knowledge.]   HARRY HANAK

SHANKS, MICHAEL JAMES (1927-1984), journalist, economist, and company director, was born 12 April 1927 in London, the only son and eldest of three children of (Alan)James Shanks, managing director in the firm of Moussec drinks, of Mill Hill, and his wife, Margaret

Lee. He was educated at Blundell's School and Balliol College, Oxford, which he attended in Trinity term 1945. He then joined the Royal Artillery and was demobilized as a lieutenant in 1947. He was at Balliol again from 1948 to 1950 and obtained a second class degree in the short course for philosophy, politics, and economics (1950).

In 1950-1 he lectured in economics at Williams College, Massachusetts, USA. He then started on his first career as an economic observer and commentator. During the ten years from 1954 he served successively as leader and feature writer, labour correspondent, and industrial editor on the *Financial Times*. During this period he travelled extensively in western and eastern Europe as well as North America. Other part-time assignments included consultancy jobs with Granada Television and Penguin Books Ltd. In addition, he became one of the best known commentators on current affairs and made frequent appearances on both radio and television. In 1964-5 he was economic correspondent for the *Sunday Times*.

At the beginning of 1965 he started his second career, as a public servant. When the Labour government set up the Department of Economic Affairs, he joined it, becoming industrial policy co-ordinator there, with co-responsibility for industrial policy and thus leading its team of industrial advisers. On leaving the Department in 1967 he started the connections with industry and business which became so prominent later in his life.

He became economic adviser to Leyland Motors and helped to create the British Leyland Motor Corporation for which he worked as director of marketing services and planning from 1968 to 1971. During this time he was also special writer on economic and management topics for the business section of *The Times*. In 1971 he moved to British Oxygen where he was chief executive, finance and planning, and subsequently in January 1973 director of group strategy.

Shanks was always an internationalist; he knew that the world was too small a place for lessons learned in one country to go unheeded in his own. So it was no surprise that, when the United Kingdom joined the European Economic Community, he chose to return to public service. He was appointed in June 1973 as one of the four British directors-general, in charge of employment policy and social affairs. During his period at the commission he inaugurated and implemented the Community's first ever social action programme. French was his working language during this period: he was also able to work in German.

In January 1976 Shanks resigned from his post at the commission in order to return to business interests in Britain. He was a leading consultant at this time and he became a director of a number of companies including BOC International Ltd. in 1976, and P-E Consultants Ltd. in 1977. From 1976 he was a director of Barmel Associates Ltd. and from 1977 of the Henley Centre for Forecasting and Environmental Resources Ltd. At the time of his death he was also a director of Rouger SA, and chairman of Barratt and Co. Ltd. From 1977 to 1982 he was chairman of Datastream PLC, and he became chairman of George Bassett (Holdings) in 1982, having been a director since 1977.

At the same time he continued to devote a considerable portion of his time to public service. He succeeded Michael Young (later Lord Young of Dartington) as chairman of the National Consumer Council in 1977. Founded in 1975, the Council was still young and uncertain of its direction. He consolidated its position and led its attempt to map out a specifically consumer view of the economic world in two major pieces of economic analysis—*Real Money, Real Choice* (1978) and *The Consumer and the State* (1978). He was reappointed twice as chairman and under his leadership the Council achieved a reputation for clear thinking and positive action on behalf of consumers.

He was a prolific author; he wrote a number of books and contributed to many journals, pamphlets, and symposia. Perhaps the most influential of his books was *The Stagnant Society* (1961) which diagnosed Britain's economic ills.

Shanks had a wide circle of friends amongst his colleagues in the many different worlds he inhabited. All remember his friendliness, quickness, and wit, but as well his friends and colleagues valued his deep concern with the fundamental obligations of decency in society—a concern which surfaced clearly in his last book, *What's Wrong with the Modern World* (1978). He could never have held together his different careers—economic commentator, public servant, businessman, and industrialist—without enormous energy.

In 1953 he married Elizabeth Juliet, daughter of Geoffrey Bower Richardson, general practitioner in Penzance and surgeon at the West Cornwall Hospital. They had three sons and one daughter. His wife died in 1972 and in 1973 he married a widow, Patricia Jaffe, who had six children. She was the daughter of Thomas Aspin, schoolmaster. There were no children from this second marriage. Shanks died in hospital in Sheffield 13 January 1984.

[*The Times*, 14 January 1984.]

M. J. MONTAGUE

SHARP, EVELYN ADELAIDE, BARONESS

SHARP (1903-1985), civil servant, was born 25 May 1903 at Ealing, Middlesex, the daughter of the Revd Charles James Sharp, vicar of Ealing, and his wife, Mary Frances Musgrove Harvey. She was the third child, with two older sisters, and a younger sister and younger brother. She was educated at St Paul's Girls' School (where she had athletic distinctions) and at Somerville College, Oxford, where she obtained a second class in modern history (1925). She entered the administrative class of the Home Civil Service in 1926. After service in the Ministry of Health and the Treasury, she became deputy secretary at the Ministry of Town and Country Planning, and subsequently permanent secretary of the Ministry of Housing and Local Government (1955-66).

Even in a generation that included Lords Bridges and Normanbrook [qq.v.], Evelyn Sharp was an outstanding public servant, quite apart from being the first woman permanent secretary. She had an ability to get quickly and decisively to the heart of any matter, however complicated. She also had an extraordinary flair for putting the issues, either in writing or across the table, plainly and summarily. Her manner was short, to the point, and forthright (not for her the ifs and buts or on the one hand, on the other). Ministers found this candid directness compelling.

These formidable professional skills were reinforced in two other ways. Except in the war years she had always worked in the field of housing, planning, and local government. Therefore when she became permanent secretary at the Ministry of Housing and Local Government she had a unique practical knowledge of the problems in this area. Her involvement with the pioneering policy developments of the late 1940s and early 1950s was especially important. Moreover, during that period she fought and won the battle to have housing, planning, and local government together in one central department, a quite remarkable achievement in the circumstances of the time. But above everything else, her specialist knowledge and her professional abilities were very greatly strengthened by her determination to get things done, to improve the lives of ordinary people through better housing, better planning, and better local government. She never hesitated to make her convictions about this absolutely clear, something quite unusual in a civil servant in those days. It was this combination of caring, and the ability to get something done (however radical) that made her so formidable and so respected by ministers, Whitehall, and local authorities alike. Evelyn Sharp pursued her policies resourcefully and with considerable success, a constant and prevailing influence for the public good.

Among her many achievements were major new policies and investment designed to make public housing available to people in need, highly effective land use policies for the constructive protection of the countryside, and an original new towns programme. At the time these were bold steps forward and they have made a lasting contribution to social welfare.

Her greatest personal achievement was to establish a close working relationship between central and local government. Starting with her time in the Ministry of Health, she had always been involved with local authorities (large and small) and had made a point of getting out of Whitehall and seeing for herself. (Even as the head of a busy department she would always make time for this.) Never in any sense a mandarin, she was known up and down the country as a very real person, whose likes and dislikes were pungently expressed whatever the occasion or the company. She believed in strong, independent local government and did everything she could to help it operate effectively (though never slow to point out when it did not). Local authorities had a profound regard and respect for her and trust in her. This gave her considerable influence at the centre of things. To this Dictionary she contributed the notices of Sir William Douglas and Hilda Martindale.

She was more than an unusual public servant. Whether walking in the mountains, cycling, or pottering about at home she was unpredictable, unconventional, never afraid to admit she was wrong, and always herself. She did not allow the strains and tensions of public life to turn her into somebody artificial, false, or contrived. Her oldest friend Lady Meynell put it beautifully in her memorial address. Evelyn Sharp was, she said, 'a very human being. The truest of friends and the best of companions. A great appreciator and giver of enjoyment. In her company there was always laughter.' She had a spontaneous kindness and generosity and an immediate sympathy for anyone in trouble. She never married but her compassion and readiness to respond gave her over the years an 'extended family' of people indebted to her for help and understanding.

She was created DBE in 1948, GBE in 1961, and a life peer in 1966. She was an honorary fellow of Somerville (1955), and was granted an honorary DCL at Oxford (1960), and honorary LLDs at Cambridge (1962), Manchester (1967), and Sussex (1969). A member of the Independent Television Authority (1966-73), she was also president of the London and Quadrant Housing Trust. She died at Lavenham, Suffolk, 1 September 1985.

[Private information; personal knowledge.]

JAMES JONES

SHELDON, SIR WILFRID PERCY HENRY (1901-1983), paediatrician, was born 23 November 1901 at Woodford, Essex, the second of three sons and the fourth of five children of John Joseph Sheldon, a bank clerk in the City of London, and his wife, Marion Squire, daughter of Henry Spring, a Lloyds' underwriter, who insured sailing ships. He was educated at Bancroft's School, Woodford, and King's College, London, where he studied medicine. He qualified MRCS, LRCP in 1923 and graduated MB, BS (Lond.) a year later with honours in anatomy and medicine. He held several resident appointments at his teaching hospital including a house physicianship to (Sir) (George) Frederic Still [q.v.], the well-known British paediatrician. In 1925 he became MRCP (Lond.) and MD (Lond.), both by examination. He became FRCP in 1933.

After holding a medical registrarship at the Royal Free Hospital he was appointed to a key post for an ambitious young paediatrician—that of medical registrar and pathologist to the Hospital for Sick Children, Great Ormond Street, thus beginning an association with that hospital which was to last until his retirement in 1967. He became a consultant there in 1947. He was concurrently consultant paediatrician at King's College Hospital. During World War II Sheldon worked in the Emergency Medical Service organizing the transfer and care of sick children from London to hospitals in Cuckfield and, later, Haywards Heath, Sussex. After the war he was responsible for the successful liaison between Great Ormond Street and Mulago Hospital, Kampala, which lasted until the regime of Idi Amin took control in Uganda.

Like most of his generation in paediatrics Sheldon had no training in the methodology of research and was thus no clinical scientist. But he was an outstanding clinical observer and diagnostician, who relied more on accurate history taking and physical examination than on laboratory findings. He quickly amassed a great store of recollected experience. As a young paediatrician Sheldon published many brief case reports of unusual cases, especially in the *Proceedings* of the Royal Society of Medicine. Later he published full-length papers on a variety of topics relating to child health or disease, but he made no great discoveries or contributions of lasting importance. His *Diseases of Infancy and Childhood* was deservedly the most popular medium-sized textbook of paediatrics during his working life; the first edition appeared in 1936 and the eighth and last in 1962. In the 1940s Sheldon began to take a special interest in coeliac disease and Great Ormond Street became a referral centre for this uncommon but serious disorder. He was an early advocate of a starch-free diet, but he made the false assumption that a diet free of wheat flour was synonymous with this.

Sheldon enjoyed every aspect of his varied career in children's medicine but that which undoubtedly gave him the greatest satisfaction and pleasure was his appointment from 1952 to 1971 as physician/paediatrician to the Queen—the first royal doctor to be accorded that title. He possessed all the qualities of an ideal courtier: he was justifiably self-confident and self-assured, always within the limits of propriety; he was good-mannered and deferential, but never obsequious; he was interesting and amusing and he was discreet, but not to the extent of hindering the escape of some entertaining stories from the royal nurseries. He was rewarded by the friendship of the royal family and his appointment in 1954 as CVO and in 1959 as KCVO. He was also one of the very few London-based paediatricians to make a financial success of private practice in Harley Street.

Sheldon was an inspiring teacher of nurses, medical students, and postgraduates. He was before his time in the importance he attached to effective communication with parents. Students were always present when he talked to parents and parents were always present when he dictated letters to GPs. Early in his career he was consultant paediatrist to the London county council. For many years he was adviser in child health to the Ministry of Health, and wielded much influence in this role. He was involved in a number of published government reports—for example, on welfare foods (1957), cerebral palsy (1960), congenital malformations (1963), child welfare centres (1967), and special care for babies (1971). Each of the two last named was known in its day as the Sheldon report, since he was chairman of the committee reporting. Sheldon was president of the British Paediatric Association (1963-4), and an honorary fellow of both the American Academy of Pediatrics and the Royal Society of Medicine. He became honorary FRCOG (1972) and honorary MMSA (1972), and was a corresponding member of the Portuguese Paediatric Society.

Sheldon was tall, well-built, handsome, and impressive. His appearance and behaviour could and did elicit awe and adoration—not as a rule simultaneously. He was usually kind and urbane in social intercourse, showing to his close friends a genuine and disarming humility. But, if his customary equanimity was threatened—for instance by somebody presuming on a mere acquaintanceship—he could be disagreeably abrupt and arrogant. He was always kind and gentle to children, always thoughtful and considerate to parents, and always at pains to help and befriend a stranger, bewildered and ignored in unfamiliar surroundings. He was an expert, learned, and energetic gardener (his father was FLS).

In 1927 he married Mabel Winifred ('Maithe'), daughter of John Netherway, accountant to Messrs Allen & Hanburys Ltd. They had three daughters, the second of whom, Joanna Sheldon, became FRCP in 1975. She died in 1985. Sheldon died in a private nursing home near his home in Kingston, Surrey, 9 September 1983.

[Valedictory in King's College Hospital *Gazette*, 1967, pp. 172-4; *The Times*, 14 and 21 September 1983; *Lancet*, 1983, vol. ii, p. 749; *British Medical Journal*, 1983, vol. cclxxxvii, pp. 918, 992, and 1146; private information; personal knowledge.] PETER TIZARD

SHEPPARD, SIR RICHARD HERBERT (1910-1982), architect, was born in Bristol 2 July 1910, the eldest of three sons (there were no daughters) of William Herbert Sheppard, commercial traveller, and his wife, Hilda Mabel Kirby-Evans, both of Bristol. Sheppard was educated at Bristol Grammar School, and had begun his professional training at the Royal West of England Academy when at the age of nineteen he was struck down by poliomyelitis. Athletic, extrovert, a born leader, he was hospitalized for two years and lost the use of his legs, but his zest for life seemed undiminished and he moved to the Architectural Association School in London, qualifying as ARIBA in 1936. Here he met his first wife, Jean, daughter of Frank Shufflebotham, doctor of medicine: they were married in 1938 and had a son and a daughter. Both young architects taught at the evacuated AA School (Sheppard as vice-principal) throughout World War II. Thereafter, through Jean's friendship with Henry Morris and other educationists, the Sheppards became heavily involved in the post-war school building programme. In 1958 they took into partnership Geoffrey Robson, an ex-pupil, and the firm of Richard Sheppard, Robson & Partners was founded.

The firm built upwards of eighty schools, but Sheppard had no wish to become a specialist architect and was sceptical of the economy of prefabrication, the monotony of which he deplored. Indeed the first building to demonstrate his own unaffected integrity as a designer was not a school but some shipyard buildings for Swan Hunter on the River Tyne (1950). In the 1960s, like those of other leading architects, the practice was in demand for university and college buildings, notably at Loughborough, Leicester, Brunel, the City of London, Durham, and Newcastle, as well as at Manchester Polytechnic and Imperial College, London. Sheppard's best known work in this field—Churchill College, Cambridge—was won in competition in 1959. In later years the firm designed *inter alia* the major shopping centre at Wood Green, north London, some modest and distinguished office buildings in the City, and not least the delightful conversion of an old warehouse in Camden Town for its own offices. Some twenty buildings designed by the practice received architectural awards during Sheppard's years at its head.

Sheppard always disclaimed the existence of any 'house style'. Therefore it was a tribute to his meticulous yet relaxed leadership that three generations of his partners, each given his head on his own projects, produced work of such consistency. Like others of his generation, he was strongly influenced by Le Corbusier and by welfare-state idealism in his early work—for example in the sadly uncompleted square for Imperial College in Prince's Gardens, London, and in the Brunel University library. But to a person as sensitive as he was to the demands of economics and the nature of the site 'dessicated functionalism', as he called it, had no lasting appeal: he came to prefer warm brickwork, seen at its best in the buildings at Manchester, Newcastle, and Durham, and to lament the decline of craftsmanship. The small house he built in his Hertfordshire garden when he could no longer manage the stairs of his Georgian rectory exemplified his easy mastery in its most appealing form.

Sheppard was a large man in presence and in spirit, who gave himself to the service of his profession and the support of his colleagues, serving on a variety of councils and committees into which he would struggle with a deprecatory grin: experience had taught him fearlessness and irony. A bon viveur, he loved travelling with friends in Europe, was an enthusiastic if irreverent sightseer, and collected the work of artists he admired (he described his recreation in *Who's Who* as 'looking at the work of others'). To this Dictionary he contributed the notice of F. R. S. Yorke.

From 1954 Sheppard served for over two decades on the council of the RIBA, latterly as vice-president, and was chairman of the Association of Consulting Architects until his death. He was appointed CBE in 1964, was elected ARA in 1966 and RA in 1972, and was knighted in 1981. He became an honorary D.Tech. of Brunel in 1972.

His wife Jean died in 1974. The following year on holiday he fell and broke a leg, and subsequently the other, and was from 1966 frustratingly confined to a wheelchair. In 1976 he married Marjorie Head, who had been at his side through these trials. She was the daughter of Stanley Howe Hamilton Head, engineer.

Sheppard died in Hertford General Hospital 18 December 1982.

[Stephen Hitchins, *Sheppard Robson Architects*, 1983; *Building Design*, March 1973; RIBA *Journal*, June 1973 and May 1979; *The Times*, 4 January 1983; private information; personal knowledge.]   ESHER

SHONFIELD, SIR ANDREW AKIBA (1917–1981), political economist, was born 10 August 1917 at Tadworth, Surrey, the fourth of six sons and the fifth of the seven children of Rabbi Victor Schonfeld, of north London, and his wife, Rachel Lea, daughter of Josef Sternberg, of Budapest. He was educated at the Jewish secondary school founded by his Hungarian-born father, at St Paul's School (foundation scholar), and at Magdalen College, Oxford (demy; Chancellor's English essay prize, 1938). He obtained second class honours in philosophy, politics, and economics in 1939. He later changed his name to Shonfield.

In the army (Royal Artillery) from 1940 to 1946, with the final rank of major, General Staff, he was mentioned in dispatches. There followed fifteen years in journalism: with the *Financial Times* (1947–57) as a features writer and from 1950 as foreign editor, and with the *Observer* (1958–61) as economic editor.

In the twenty years of his subsequent career he was director of studies at the Royal Institute of International Affairs (1961–8); chairman of the Social Science Research Council (1969–71); director of the Royal Institute of International Affairs (1972–7); professor of economics at the European University Institute, Florence (from 1978); and consultant to multinational and UK firms. His public activities included lectures and membership of the royal (Donovan) commission on the reform of the trade unions and employers' associations (1965–8); the departmental committee on overseas representation (Duncan committee) (1968–9); the EEC 'Vedel Group' (1972); and of international discussion groups with policy-makers and academics.

His writings included *British Economic Policy since the War* (1958) and *The Attack on World Poverty* (1960). *Modern Capitalism* (1965) surveyed post-1945 changes in the western capitalist system. It examined new approaches to economic management aimed at full employment, improvement for the disadvantaged, and growth, and drew upon comparative European and American evidence. It gave expression to the optimism then so widespread and reinforced perceptions about the directions in which capitalist societies appeared to be moving. His minority report to the Donovan commission stressed that trade unions outside the law could only diminish liberty. *Europe: Journey to an*

Unknown Destination (1973) encompassed his 1972 Reith lectures, and won the Cortina-Ulisse literary prize. It was followed by *International Economic Relations of the Western World 1959-71* (1976). He died whilst engaged in reconsideration of the changed conditions of his earlier assumptions and in the study of political, administrative, and social factors affecting how western democracies and Japan coped with the strain. *The Use of Public Power* (1982) and *In Defence of the Mixed Economy* (1984) posthumously recorded his conclusions. His wide-ranging critique of capitalism in developed democratic societies was influential internationally in the debate about the balance between private and public power.

Shonfield's intellectual leadership derived as much from the qualities of his personality as from those of his mind. His clarity of exposition and ability to remove confusion and reduce something complicated to its simple elements led him sometimes disconcertingly straight to the point. But his was an allusive mind and he would sometimes circle round a subject in order to see it better and reconnoitre before taking a firm position. His charm, generosity of spirit, attractive and distinct voice, and occasional dry comment imparted sheer pleasure to his listeners. He was ready to engage in reasoned discussion with anyone, was tough in debate, and challenged others in order to find out what he really thought himself. At meetings he could change the atmosphere, speaking with energy and emphasis in a constructive tone and in a manner which invited co-operative rather than aggressive responses.

The family influence of orthodox Judaism and Talmudic study had a powerful effect on Shonfield. He was a rebel against tradition who had escaped after great struggle. The environment of the 1930s made him feel a revulsion against totalitarianism and a sense of the needs of others which went beyond mere justice towards active kindness. He described *Modern Capitalism* as an oblique record of his conversion from one view of the world, with strong elements of the cataclysmic, to a more hopeful standpoint. With no pronounced party political stance in his utterances his influence was the wider. His words 'groping for the shape of contemporary history' defined an approach in which beliefs played little part, but guiding principles did: people were seen as ends not means and truth was paramount. He had a love of fiction and a covert passion for creative writing; he published one novel.

He was knighted in 1978, became a fellow of the Imperial College of Science and Technology (1970), and received an honorary D.Litt. from Loughborough University (1972).

In 1942 he married Zuzanna Maria, daughter of Alfred Przeworski, a Polish industrialist. They had one son and one daughter. Shonfield died in St George's Hospital, Tooting, London, 23 January 1981.

[*Financial Times* and *The Times*, 24 January 1981; Alan Bullock in *International Affairs*, spring 1981; private information; personal knowledge.]   ARTHUR KNIGHT

SHOWERING, SIR KEITH STANLEY (1930-1982), chairman and chief executive of Allied-Lyons Ltd. from 1975 to 1982, was born 6 August 1930 in Shepton Mallet, the only son and eldest of the five children of Herbert Marquis Valentine Showering, brewer, and his wife, Ada Agnes Foote. He was educated at Wells Cathedral School and Long Ashton laboratories. He joined the family business of Showerings Ltd., cider makers, in 1947 and became a director in 1951. He first started as a salesman and then worked in all the other departments of Showerings. He took a particular interest in financial transactions.

When he joined the Showering business it was undergoing a period of rapid and unparalleled expansion due to the invention of Babycham, a perry which was marketed strongly through advertising and sales in public houses. It was elegantly packaged with gold foil round the top and it appealed to younger women, who had hitherto had difficulties in finding a suitable drink in public houses, as they did not like port and lemon which was the older women's drink.

As the sales grew Babycham came to be stocked in almost all public houses and the Showering family, led by Keith's father and ably supported by his uncle Francis, the inventor of Babycham, looked for diversification. The discussions leading to new acquisitions were a family matter in which Keith Showering played his full part. Showerings acquired Britvic, Vine Products, and Whiteways of which Keith Showering became deputy chairman in 1964 and chief executive from 1971 to 1975. In 1965 the family decided to break into the upmarket area of imported wine and, after a first unsuccessful bid, by the end of the year had acquired Harveys with its unique position in sherry. Harveys also sold port and fine table wines, including Latour. Keith Showering was chairman of Harveys from 1966 to 1971. In 1968 the family again decided to make a major move and successfully negotiated for Showerings to be merged into Allied Breweries of which Keith Showering became vice-chairman in 1969. In 1975 he became chairman and chief executive of Allied Breweries. Here his capacity for far-sighted and fair leadership was tested and not found wanting. Allied moved into the Scotch whisky business through its acquisition of Teachers in 1976 and then, on Keith Showering's initiative, into food and catering by acquiring the old established business of Lyons in 1979. This was a major step away from reliance on alcoholic beverages and took the company into a new but, in many ways, closely related field. Lyons had been passing through a difficult period, but, as part of Allied-Lyons, as the group came to be called, it was soon restored to profitability.

Aside from the arduous work at Allied-Lyons Showering found time to broaden his business interests, becoming vice-chairman of the Guardian Royal Exchange Assurance Company in 1974 and director of the Midland Bank in 1979. He was also a director of Castlemaine Tooheys Ltd., of Sydney, Australia, and Holland & Holland. He had a keen interest in the arts, being a trustee of the Glyndebourne Arts Trust and the London Philharmonic Orchestra. In addition he was a trustee of the World Wildlife Fund (UK). He was master of the Brewers Company in 1981–2. In 1982 he was appointed high sheriff of Somerset. Perhaps his greatest interest outside his business lay in farming. He was a pioneer in every sense, being amongst the first to import Charollais, Simmentals, and Canadian Holsteins, thus changing the face of British dairy and beef farming. He was also the first UK financier and importer of the ovary transplant fertilization technique.

In his youth he had been a formidable athlete and he remained a first-class shot. He was a generous host and a good friend, though he could also be a hard taskmaster to the second rate and the inefficient. His career covered a period when a small family brewery emerged into the greatest combined force in the United Kingdom in the manufacture and sale of drinks and food. Perhaps his career is best summarized in the words of (Dame) Elisabeth Frink, the sculptor. His wife Marie commissioned a head of him for his fiftieth birthday. When Elisabeth Frink presented it his wife remarked 'but the head is larger than life size', to which the sculptor replied 'but your husband is larger than life size'. Showering was knighted in 1981.

In 1954 Showering married Marie Sadie, daughter of Charles Wesley Golden, a company director; they had four sons and two daughters. Showering died 23 March 1982 at the Bank of England at a meeting of industrial leaders, presided over by the governor.

[Private information; personal knowledge.]
PHILIP SHELBOURNE

SIMPSON, (CEDRIC) KEITH (1907–1985), professor of forensic medicine at Guy's Hospital, was born at Brighton in Sussex 20 July 1907, the younger son and second of three children of George Herbert Simpson, a general medical practitioner, of Brighton, and his wife Mary, a nursing sister, the daughter of Joseph James Bussell, a court uniform and decorations consultant. He was educated at Brighton and Hove Grammar School and in 1924 went to Guy's Hospital Medical School where he had an outstanding academic career, remaining there for the rest of his professional life. As a student, he won the Hilton prize for dissection, the Wooldridge memorial prize in physiology, the Beaney prize in pathology, the Golding-Bird gold medal and scholarship in bacteriology, and the Treasurer's gold medal in clinical surgery. He qualified MRCS, LRCP and MB, BS in 1930.

After qualification he joined the staff of Guy's Hospital as clinical assistant and became Gull scholar and Astley Cooper student in 1932, the year he gained his MD (pathology). In 1932 he became a lecturer in pathology. In 1937 he devoted himself completely to his major interest of forensic medicine and became a lecturer in that subject, subsequently being made reader in 1946. In 1962 he became the first professor of forensic medicine in the University of London.

Simpson was undoubtedly the leading forensic pathologist of the middle decades of the twentieth century. He was at first overshadowed by the ubiquitous figure of Sir Bernard Spilsbury, of whose judgement Simpson was always critical. The death of Spilsbury in 1947 allowed Simpson's expertise to have full rein. He was an expert pathologist before taking up forensic medicine and this background often gave him an advantage over less proficient expert witnesses in many of the hundreds of notorious murder trials in which he was involved. Like Spilsbury, his name became a 'household word', especially after such sensational cases as the Haigh 'acid bath murders' in 1949, the Dobkin Baptist cellar case, and the Neville Heath murders. Simpson was contemporary with R. Donald Teare and Francis Camps [q.v.], the other famous names in London's post-war forensic medicine, and he crossed swords with the latter on many occasions in both the witness box and in print.

Simpson was a superb lecturer and a prolific writer. His textbook *Forensic Medicine* first appeared in 1947 and went into many editions. It won the Swiney prize of the Royal Society of Arts in 1958 for the best work on medical jurisprudence to be published in the preceding ten years. He was also the editor of the twelfth edition of *Taylor's Principles and Practice of Medical Jurisprudence* (1965), the standard work of the time in the English language. He also edited two editions of *Modern Trends in Forensic Medicine* (1953 and 1967), and wrote *A*

Doctor's Guide to Court (1962, 2nd edn. 1967), a textbook for the police, and about 200 papers on forensic medicine. In addition he published cases under the pseudonym 'Guy Bailey', an obvious reference to the two places where he spent much of his working life. One of his most successful books was his autobiography, *Forty Years of Murder* (1978), which had world-wide sales and was in the best-seller list for several months.

He became MRCP (1958) and FRCP (1964), DMJ (Path.) (1962), MCPath. (1964) and later FRCPath. From 1961 to 1973 he was lecturer in forensic medicine at Oxford and was made MA (Oxon.) in 1964 and Havard associate in police science in 1952. He was examiner to seven universities and to the Royal College of Pathologists and the Worshipful Society of Apothecaries. He held the offices of president of the Medico-Legal Society and the British Association of Forensic Medicine, and he was a council member of the Royal College of Pathologists and the Medical Protection Society, being chairman of the cases committee of the latter. He was also a member of the Home Office scientific advisory committee.

Simpson was appointed CBE in 1975 and was awarded the honorary degrees of LLD of Edinburgh (1976) and the MD of Ghent (1973). He travelled widely and was respected amongst the forensic fraternity throughout the world. A linguist and musician, he was a cultured physician of the old school. In 1972 he became emeritus professor on his retirement from Guys' Hospital, but continued to work, lecture and write almost up to his death.

He was married three times, first to Mary McCartney Buchanan in 1932, by whom he had a son and two daughters. After her death in 1955, he married in 1956 Jean Anderson Scott Dunn who died in 1976. In 1982 he married Janet Hazell, MB, Ch.B., LRAM, the widow of Gavin Leonard Bourdas Thurston, CBE, FRCP, coroner, barrister, and lecturer in forensic medicine. Simpson died 21 July 1985 in St Bartholomew's hospital in London.

[*Guys' Hospital Gazette*, August 1985; *Journal of the Forensic Science Society*, vol. v, no. 5, 1985, pp. 403-4; private information; personal knowledge.]   B. H. KNIGHT

SMALLWOOD, NORAH EVELYN (1909-1984), publisher, was born 30 December 1909 in Little Kingshill, Buckinghamshire, the third of the four daughters and the fifth of the eight children of Howard Neville Walford, artist, and his wife, Helen Griffiths. She went to school in Eastbourne and began her publishing career as a secretary in 1936 when she joined Chatto & Windus, the firm to which she devoted the rest of her working life.

She found there in Ian Parsons an exceptionally gifted publisher, who soon involved her in the brilliantly original though short-lived weekly magazine, *Night and Day*. In the war years that followed she proved to the full both her capabilities and her courage: she not only shared the strain of wartime publishing with her very able but hard-pressed senior partner, Harold Raymond, but also had to bear the loss of her husband and a much-loved brother, both killed on active service.

Widowed and without children, she poured her energies into Chatto's, and in 1945 was made a partner. Two years later she was appointed to the board of the Hogarth Press, the firm which Leonard and Virginia Woolf [qq.v.] had founded in 1917 and which had come under Chatto's management in 1946. From then on she worked closely with Leonard Woolf until his death in 1969, doing as much as anyone to foster the separate identity of his list and to care for its authors—Henry Green, William Sansom [qq.v.], (Sir) Laurens van der Post, A. L. Barker, Laurie Lee, and George Mackay Brown among many others.

She was appointed to the board of Chatto's in 1953 when it became a limited company and, with Ian Parsons, effectively ran the firm for the next twenty years before succeeding him as managing director and chairman in 1975 until her own retirement in 1982. She also served for fourteen years on the board of the holding company which was formed when Chatto's merged with Jonathan Cape in 1969, later to be joined by the Bodley Head (1973) and Virago Press (1982).

By now her reputation, both as a dynamic woman publisher in a world still dominated by men and as a vivid, stylish, generous-spirited if sometimes sharp-tongued personality, was widely known. Professionally, she had acquired expertise in many aspects of book production and liked nothing better than turning her hand to typographical design or commissioning work from such gifted artists as Enid Marx, A. Reynolds Stone [q.v.], and John Ward. She also excelled in that essential function of any literary publisher—working creatively with her authors, encouraging and helping them with her quick and incisive reading of their manuscripts. She was especially good with fiction, winning the respect, gratitude and friendship of distinguished writers such as Aldous Huxley, Sir E. M. Compton Mackenzie, Sylvia Townsend Warner [qq.v.], Sir Victor Pritchett, Elizabeth Taylor, Elspeth Huxley, (Dame) J. Iris Murdoch, A. S. Byatt, Amos Oz, and Toni Morrison. She also contributed imaginatively to Chatto's poetry publishing (with Ian Parsons, Cecil Day-Lewis [q.v.], and D. J. Enright), to their children's books, and to their educational and academic list, notably with her early recognition of Sir Moses Finley's stimulating influence in the field of ancient history.

It was her instinct for quality that brought Dirk Bogarde to Chatto's. Her perception of his lively intelligence and sensitive, alert talent soon drew out the writer in him, and led to one of the most rewarding friendships of her latter years. His debt to her is vividly and movingly described in the third and fourth volumes of his autobiography.

Iris Murdoch, eminent among the many authors to whom Norah Smallwood dedicated so much of her life and who in turn dedicated so many of their books to her, remembered her as 'the most rational person I had ever met . . . To us writers she was a combination of comrade, leader, mother, business partner and muse . . . She was always cheerful, always seemed happy, was always joking, was at the same time strong, clever, wise, practical, loyal and absolutely reliable.' That same virtue, the unqualified trust that she inspired in her authors and friends, was given full public expression by Laurens van der Post in the memorial service held in her honour at St Martin-in-the-Fields.

Appointed OBE in 1973, Norah Smallwood received an honorary Litt.D. at Leeds University in 1981. She married Peter Warren Sykes Smallwood, a chartered accountant, in 1938. Son of John Russell Smallwood, from a long line of wine merchants in Birmingham, he was serving as a navigator in the RAF when he was killed over Holland in 1943. Norah Smallwood died in Westminster Hospital, London, 11 October 1984, having for years fought the ill health and crippling arthritic pain with which she was afflicted.

[*The Times*, 12 October 1984; David Holloway, *Daily Telegraph*, 13 October 1984; John Goldsmith, 'Norah Smallwood, Publisher', *Bookseller*, 20 October 1984; Iris Murdoch, 'Norah Smallwood', *Bookseller*, 27 October 1984; Josephine Parfitt, 'Norah Smallwood', *Sphere*, 21 May 1960; Dirk Bogarde, *An Orderly Man*, 1983, and *Backcloth*, 1986; Oliver Warner, *Chatto & Windus: A Brief Account of the Firm's Origin, History and Development*, 1973; private information; personal knowledge.]

JOHN CHARLTON

SNOW, LADY (1912–1981), novelist, dramatist, and critic. [See JOHNSON, PAMELA HANSFORD.]

SOMERSET, HENRY HUGH ARTHUR FITZROY, tenth DUKE OF BEAUFORT (1900–1984), leading figure of equestrian activities, was born at 19 Curzon Street, Mayfair, London, 4 April 1900, the only son and youngest of the three children of Henry Adelbert Wellington FitzRoy Somerset, ninth Duke of Beaufort, and his wife, Louise Emily, widow of Baron Carlo de Tuyll and daughter of William Henry Harford, of Oldown, Almondsbury, Gloucestershire. As Marquess of Worcester he was educated at Eton and the Royal Military College, Sandhurst. He was then commissioned into the Royal Horse Guards, with special leave to hunt two days a week. He regretted that he was too young to have seen active service in the war of 1914–18.

He was given his own pack of harriers at the age of eleven and was known as 'master' by his friends and family for the rest of his life. As master of a great family pack, fox-hunting became 'a permanent love affair . . . the basis of my very existence'. Like his father and grandfather before him he was 'a sportsman by profession', hunting his own pack for forty-seven years, an unparalleled record. His knowledge of every aspect of the sport from kennel management to the correct position of a tie-pin (he regarded the word stock as an unacceptable neologism) and first aid on the hunting field was displayed in his *Fox Hunting* (1980). His long mastership he regarded as a time consuming and expensive duty rewarded by the 'exhilaration and excitement and, what is more, the sense of achievement that is experienced when a successful hunt is concluded'. He was a great breeder and judge of hounds and his proudest possessions were his old established female lines.

Like most of his forebears he took little interest in politics: 'local leadership in country sports and agricultural matters', he wrote, 'took precedence in their lives'. His own leadership was more than local. As chairman of the Master of Foxhounds Association and president of the British Field Sports Association he was the first figure in the hunting world. As such he was the chosen target of those who opposed his sport. In December 1984 a group of people opposed to hunting attempted to dig up his body in order to present his severed head to the daughter of the Queen, Princess Anne (later Princess Royal), since he was considered guilty of introducing the royal family to fox-hunting.

He was a natural and bold rider from childhood. As president of the British Olympic Association he was appalled at the performance of the British team as a 'bunch of amateurs' in the Olympic Games of 1948. It was to improve standards that in 1949 he established at the family home, Badminton, the three-day trials that were to become the most important event in the British equestrian calendar. His other activities—he was president or chairman of some seventy associations and charities—ranged from the chancellorship of Bristol University (1966–70) to the presidency of the

Battersea Dogs Home. He was president of the MCC when England regained the ashes in 1953.

In 1923 he married Lady (Victoria Constance) Mary Cambridge, daughter of Adolphus Charles Alexander Albert Edward George Philip Louis Ladislaus (son of the Prince of Teck), first Marquess of Cambridge and governor and constable of Windsor Castle. She was the favourite niece of King George V and Queen Mary, who spent the years of World War II as a guest at Badminton. The Duke and his Duchess resisted with tact Queen Mary's eccentricities which included the tearing down of ivy and the removal of established trees. The Duke's close personal connection with the royal family was cemented by his appointment as master of the horse, a post that he held longer than any of his predecessors. As the third great officer of the household he was the sovereign's personal attendant on state occasions. He was proud that the same office had been held by his ancestor in the reign of Elizabeth I and that Queen Elizabeth II valued the counsel of his sound common sense.

He regarded his father, whom he succeeded in 1924, as the inspiration of his life. But whereas the ninth Duke was somewhat autocratic his son, apart from the occasional sharp rebuke provoked by bad behaviour in the hunting field, was an amiable man, loved by his family and friends, an admirer of ladies and much admired by them. He took his great possessions and their duties as a matter of course. When it was suggested that his heir's son be called John, he remarked that it was a good family name for, as a Plantagenet, he was a direct, though illegitimate, descendant of John of Gaunt. He did not relish change. He had been lord lieutenant of Gloucestershire and, when Badminton was absorbed in the newly created county of Avon, he protested that he had no desire to live in a four-letter county.

He was appointed GCVO in 1930, admitted to the Privy Council in 1936, and created KG in 1937. He was high steward of Bristol, Gloucester, and Tewkesbury.

He died 5 February 1984 at Badminton. He had no children and was succeeded as Duke of Beaufort by his cousin David Robert Somerset (born 1928), an art dealer and formidable horseman, who had served with the tenth Duke as joint master of the Beaufort.

[Duke of Beaufort, *Memories*, 1981, and *Fox Hunting*, 1980; Sir Osbert Sitwell, *Queen Mary, and Others*, 1974; Barry Campbell, *The Badminton Tradition*, 1978; Raymond Carr, *English Fox Hunting*, 1976; *The Times*, 6 February 1984; private information.]

RAYMOND CARR

SPEAR, (THOMAS GEORGE) PERCIVAL

(1901–1982), historian, was born 2 November 1901 at Bath, the younger son and fourth and last child of Edward Albert Spear, provision merchant, of Bath, and his wife, Lucy Pearce. He was educated at Monkton Combe school and St Catharine's College, Cambridge, where he was an exhibitioner. He read history, and was placed in the upper division of the second class in part i of the tripos in 1921 and also in part ii in 1922. He then took part ii of the theological tripos in 1923, when he was placed in the undivided second class. He went to India in 1924 to teach at St Stephen's College, Delhi, under the auspices of the Cambridge Mission to Delhi. (His elder brother, the Revd Edward Norman Spear, later a canon of Gloucester Cathedral, was also a missionary in India.)

He began at St Stephen's College by teaching British and European history, but he was quickly drawn to the history of India, and of the British connection with it. He later became head of the history department and reader in history in Delhi University. He soon found a highly congenial subject for his own research—English social life in eighteenth-century India. He read widely, both in libraries and in the government archives, which were then in Calcutta, and he continued this research while on leave in England. He successfully completed a Cambridge Ph.D. thesis on this subject in 1931, and it was published in a revised form under the title *The Nabobs. A Study of the Social Life of the English in Eighteenth Century India* (1932). It was an informative survey—witty, anecdotal, but descriptive rather than analytical. It lasted well, and was republished as a paperback in 1963, with an introduction in which Spear outlined the Indian context of the growth of British power in the eighteenth century.

In 1933 he married (Dorothy) Margaret (Gladys), daughter of the Revd Frederick Harry Roberts Perkins, an Anglican clergyman. They had no children. They set up house in Delhi, and quickly established good relations with their Indian neighbours. In their reminiscences, *India Remembered* (1981), Percival and Margaret Spear portrayed their married life in Delhi as a time when they widened and deepened their understanding of Indians of different communities. In these years also Percy was extending his knowledge of Indian, and especially Indo-Muslim, history and culture. Delhi itself always fascinated him. He published a concise but vivid historical survey, *Delhi. An Historical Sketch* (1937), and a guide to its buildings, *Delhi. Its Monuments and History* (1943). The latter book was originally designed for school-children, but a visitor to Delhi today could still learn from it.

During World War II he and his wife joined

the staff of the director-general of information in India, he as a writer, she as librarian. This unit was absorbed into the department of information and broadcasting in 1941, and he was then appointed deputy director of counter-propaganda. In 1943 he became deputy secretary to the government of India in the department of information and broadcasting. He also served for a time in 1944 as government whip in the Legislative Assembly. He was appointed OBE in 1946. Although he had been mildly critical of the aloofness of members of the Indian Civil Service, after his experience in the secretariat he paid tribute to what he had seen of their alertness, energy, and resourcefulness in times of difficulty.

He returned to England after the war, and turned to academic administration when he was appointed bursar and fellow of Selwyn College, Cambridge, in 1945. He also continued his research into the history of Delhi and its surroundings, which culminated in his *Twilight of the Mughuls. Studies in Late Mughul Delhi* (1951), which has become an acknowledged classic of historical writing on India. In the course of it he accepted the argument first put forward in western scholarly writing by F. W. Buckler ('The Political Theory of the Indian Mutiny', *Transactions of the Royal Historical Society* V, 4th series, 1922), that the British had no right, after they had suppressed the mutiny and revolt of 1857, to try the Mogul emperor, or king of Delhi, as a rebel, since he was not a British subject. Spear reinforced this view with an eloquent analysis of Indian attitudes to the fallen Mogul ruler. The book is also remarkable for its portrayal of the life of the people of Delhi and of the villages in the neighbourhood. It is outstanding above all for its literary style and for the sensitive restraint with which Spear expressed a sense of nostalgia for past Mogul splendours.

He rewrote the third part of the *Oxford History of India* in 1958. This was republished as *The Oxford History of Modern India, 1740-1947* (1965). In contrast to Vincent A. Smith [q.v.], the author of the previous *Oxford History of India*, Spear paid considerable attention to socio-economic history and provided a sympathetic account of the nationalist movement. Then he wrote *India. A Modern History* (1961), one of fifteen volumes of the *University of Michigan History of the Modern World*. He also wrote the second volume of *A History of India* published as a paperback by Penguin Books in 1965. He had little more to say of significance in these latter two books, but they occupied much of his time, and distracted him from other projects, such as a study of India in the time of the governor-generalship (1828-35) of Lord William Cavendish Bentinck [q.v.], and a

large-scale history of Delhi, neither of which he completed. To this Dictionary he contributed the notice of L. F. Rushbrook Williams.

He also taught when he could and did what he could to spread a knowledge of Indian history in Cambridge. He resigned office as bursar in 1963, when he was appointed university lecturer in history, a post which he filled with distinction until 1969. Thereafter he continued to take a lively interest in the affairs of the Centre for South Asian Studies at Cambridge, and his services as an external examiner were often sought by other universities. Quiet in manner, and reticent about himself, Spear was always ready to advise and encourage younger scholars. He was alert and witty to the end, especially when the conversation turned to Delhi, as it usually did. He died in Cambridge 16 December 1982.

[India Office Records; private information; personal knowledge.]

KENNETH BALLHATCHET

SPINKS, ALFRED (1917-1982), industrial chemist and biologist, was born in the Fen village of Littleport, Cambridgeshire, 25 February 1917, the only child of Alfred Robert Spinks, manager of a Littleport brewery, and his wife, Ruth Harley. At Soham Grammar School he obtained higher school certificate distinctions in mathematics and botany and with an Isle of Ely major scholarship he went up to University College, Nottingham, in 1935 to read chemistry. In the 1938 University of London B.Sc. external degree examination he gained the highest chemistry marks (external and internal) and was awarded the Neil Arnott studentship for postgraduate studies. He chose to go to Imperial College under (Sir) Ian Heilbron [q.v.], where he worked in association with A. W. Johnson [q.v.] on the vitamin A synthesis, later developed by Roche in Switzerland into the commercial process. He gained his Ph.D. in 1940.

In the same year ICI took him into their employ and from 1942 at Blackley in the dyestuffs division he worked under F. L. Rose in the new medicinal chemistry section on the determination of the concentrations of sulphonamides in animal tissues, pioneering studies in a potentially important field. Over the next eight years he made an intensive study of the fate of organic compounds in animals, devising novel methodology, necessarily on a microscale; consideration of chemical structure-drug absorption relationships became accepted policy and involved him in work with most of the ICI candidate drugs of the period.

Spinks was now self-educated in biochemistry, pharmacology, and physiology and a farsighted divisional research director, Clifford

Paine, sent him to Oxford (Worcester College) in 1950 to study physiology. Taking a degree was not a requisite but, characteristically, he did so and obtained an outstanding first after two years' study. During the vacations he did some enzymic research with J. H. Burn and greatly extended his contacts in the biological world. When he returned to Blackley plans for the independent pharmaceuticals division were well advanced and being put in charge of a new pharmacology section was an exciting responsibility which he grasped wholeheartedly. With his broad background, he became widely involved in many new developments and showed his flair for leadership in the necessarily interdisciplinary discussions. His promotion to research manager of the first biochemistry department in 1961 marked the end of his time as a bench worker, and the beginning of an even more successful career as an administrator. Now in the magnificent establishment at Alderley Park he succeeded W. A. Sexton as divisional research director in 1966 and then, after a brief period as deputy chairman, he joined the ICI main board on the retirement of J. D. Rose [q.v.] and became responsible for research and development in the whole of the company.

Knowing that the scientists at the bench are the driving force in research, Spinks found time to visit and encourage them even though they worked in fields remote from his personal experience. He was particularly sensitive to the problems of chemicals in the environment and established procedures to ascertain the possible mutagenic or carcinogenic risk of compounds made by the company. His more general responsibilities extended to pharmaceuticals, Nobel and agricultural divisions, and to certain countries overseas, particularly South Africa. He believed strongly that ICI's presence there was in the best interests of South Africans as a whole and he took an active part in introducing a unified pay scale for all employees irrespective of colour or race.

He retired from the board in 1979 by which time the pharmaceutical division, which he had done much to develop, was the most profitable in the company. Inevitably his multidisciplinary mastery and wide experience ensured that his advice was widely sought. He was a founder member (1976) of the Advisory Council for Applied Research and Development (ACARD) and under his chairmanship, from 1980 until his death, the council grew both in stature and effectiveness. His impact on science policy was also evident in the Advisory Board for the Research Councils (ABRC) (from 1977) and most notably in the report of a joint working party of ACARD, ABRC, and the Royal Society on the emerging field of biotechnology (the Spinks report). On his retirement he was

appointed to the royal commission on environmental pollution, to the boards of Dunlop and Johnson Matthey, and as chairman of Charter Consolidated. He was involved in the affairs of many learned societies and especially of the Chemical Society of which he became president in 1979–80. He was elected FRS (1977) and appointed CBE (1978). His great natural ability was augmented with vast knowledge and wide experience, and he developed keen interests in music and the arts and no mean skill at photography and gardening. He was a large, kindly, clubbable man, somewhat awesome to those lacking his intellectual gifts.

In 1946 Spinks married Patricia, daughter of Frederick Charles Kilner, an engineer with Manchester Corporation Waterworks; they had two daughters. Spinks died at home in Wilmslow, Cheshire, 11 February 1982.

[*The Times*, 13 February 1982; A. W. Johnson, F. L. Rose, and C. W. Suckling in *Biographical Memoirs of Fellows of the Royal Society*, vol. xxx, 1984; private information; personal knowledge.]   E. R. H. JONES

SPORBORG, HENRY NATHAN (1905–1985), banker and secret organizer, was born 17 September 1905 in Rugby, the elder son of Henry Nathan Sporborg, an electrical engineer, and his wife, Miriam A. Smaith, who were Americans by origin. His younger brother died accidentally in 1928. He went, like his brother, from Rugby School to Emmanuel College, Cambridge; there he took a second class (division II) in history part i (1926) and a third in law (part ii, 1927), and rowed. He was admitted as a solicitor in 1930. He joined Slaughter & May, the City solicitors, and became a partner in 1935. He married in that year Mary, the daughter of Christopher Henry Rowlands, engineer. They had a son and three daughters.

Early in the war of 1939–45 he joined the Ministry of Economic Warfare, and from it soon moved into its secret branch, the Special Operations Executive, when SOE first needed legal advice. Most of his partners joined him; hence the unkind quip about SOE's starting troubles, 'Seems to be all may and no slaughter.' Sporborg first worked, under (Sir) Charles Hambro [q.v.], in the Scandinavian section, where they encountered a mixture of failures and successes. He then took over supervision of several sections working into north-west Europe: one each for Belgium and the Netherlands, two for France, and an escape section to help all four. Some of the foundations for future clandestine work he helped to lay were sound, though troubles shortly developed in the Netherlands. When Hambro became SOE's executive head, Sporborg too was promoted.

Under the cover appointment of principal private secretary to the minister in charge, the third Earl of Selborne [q.v.], Sporborg acted for eighteen months in 1942-3 as SOE's principal liaison officer both with him and with several large departments in Whitehall, none of which wished SOE well: the Foreign Office, the Ministry of Supply, and those for the three other fighting services. He also looked after SOE's relations, sometimes turbulent, with the other secret services. He continued the latter task while, from September 1943 till SOE was disbanded in January 1946, he was deputy to Major-General (Sir) Colin Gubbins [q.v.], its last executive head. Gubbins handled most military business, while he left political affairs to Sporborg. One or other of them was always in London, where Sporborg was prepared to face the chiefs of staff, the foreign secretary, or the prime minister if SOE's needs required it. He saw the permanent under-secretary at the foreign office, Sir Alexander Cadogan [q.v.], almost daily. He became a strong advocate of the role of subversion—SOE's main task—as a tool for unseating the governments of Axis-occupied countries, though he felt the subject too delicate and too secret for him ever to be able to give a publishable account of it. Under Gubbins's and his direction, SOE secured several significant triumphs, particularly in France, Greece, Italy, Norway, Denmark, and Burma, and ended up with a tidy monetary profit as well.

Sporborg then returned to the City, as a businessman. He joined Hambros Bank, of which he was a director for nearly thirty years, and for a time general manager. His main expertise lay in financing take-overs and mergers. His judgement was so widely respected that several large firms took him on as chairman or vice-chairman—Thorn Electrical Industries and the Sun Group of assurance companies among them; and he served on the Port of London Authority for the eight years 1967-75. He was active also in the Fishmongers Company and in various charities; he spent a decade as chairman of St Mary's Hospital, Paddington. He had an active life as a country squire as well, riding to hounds from Upwick Hall near Ware in Hertfordshire; the house where he died of heart failure 6 March 1985.

He had held the wartime rank of lieutenant-colonel on the General List, and was appointed CMG in 1945. He was also a chevalier of the Legion of Honour, and held a French croix de guerre, the order of St Olaf of Norway, the King Christian X liberty medal, and the United States medal of freedom. He was a solidly built, burly man with an affable manner, set off by a drooping eyelid. He died worth over half a million pounds.

[*The Times*, 9 March and 8 August 1985; *Rugby School Register*, 1957, 1171, 1448; private information.]   M. R. D. Foot

SRAFFA, PIERO (1898-1983), economist, was born in Turin 5 August 1898, the only child of Angelo Sraffa, from Pisa, a professor of commercial law at several Italian universities, and his wife, Irma Tivoli, from Piedmont. Both parents came from well-known Jewish families. He was educated in Parma, Milan, and Turin, and at the university of Turin, where he graduated as doctor of law in 1920. He became associate professor of political economy at the University of Perugia in 1924 and then professor at the University of Cagliari (Sardinia) in 1926. After offending Mussolini he migrated to England in 1927 and, through the initiative of J. M. (later Lord) Keynes [q.v.], was appointed to a university lectureship in the Cambridge faculty of economics. In 1930 he resigned his lectureship (he was agonizingly shy about lecturing) and was appointed Marshall librarian and, soon after, assistant director of research to act as mentor to research students. In 1939 he was elected to a fellowship at Trinity College, Cambridge. He was made FBA in 1954 and a reader in economics at Cambridge in 1963. In 1961 he was awarded the prize of the Stockholm Academy of Science which was equivalent to receiving the Nobel prize.

Sraffa had a major influence on the intellectual developments of the twentieth century. He was an intimate friend of Antonio Gramsci, Keynes, and Ludwig Wittgenstein [q.v.], and indeed played an important part in persuading Wittgenstein to change his philosophical views as between the *Tractatus* (1922) and the *Philosophical Investigations* (1953). Sraffa was the most important critic of the orthodox theory of value and distribution in the twentieth century. Yet, though he was to make outstanding contributions to pure theory, about the rigour and coherence of which he had well defined ideas, the object of his theorizing always had a political and social aspect to it.

From his school days he had taken a keen interest in political issues and early on became a socialist. He fought in the Italian army during World War I and, as a result of his experiences, became a pacifist. He opposed Mussolini's Fascist regime; his friendship with Gramsci came about because of this. Even in his earliest economic work—his dissertation 'L'inflazione monetaria in Italia durante e dopo la guerra' (1920)—important political, institutional, and sociological ingredients were already present. And though the analytical structure was then the dominant form of the quantity theory of money, Sraffa's own particular contribution was to integrate the sociological and institutional determinants of wages and employment into this framework. He incurred the wrath of Mussolini by exposing the corrupt practices of the pre-Fascist and Fascist state with regard to the private banking system in his 1922 *Economic Journal* and *Manchester Guardian* articles on the bank crisis in Italy. It was his interest in monetary matters that first attracted Keynes's attention and while they were later to follow different lines of research in economics, their friendship remained as close as ever, not least because they were both passionate bibliophiles. Sraffa translated Keynes's *A Tract on Monetary Reform* into Italian in 1925.

In the mid-1920s Sraffa commenced his critique of the orthodox theory of value and distribution. First, he attacked the partial equilibrium analysis of Alfred Marshall [q.v.] and then the general equilibrium framework, all different examples of the dominant supply and demand theories. As a challenge to these theories, Sraffa spawned the imperfect competition revolution which others developed, in a manner probably not to his liking. Then, changing tack, Sraffa developed a coherent account of the surplus approach of the classical political economists—the contention that the surplus of commodities over those necessary for their reproduction is the core concept of economic theory around which theories of value, distribution, production, employment, and growth may, and should, be set. In Sraffa's view this approach reached its highest form in Marx's work, only to be superseded in mainstream economics by the rise to dominance of the supply and demand theories. By stressing the production interdependencies of the economy as a whole (commodities produced by means of commodities) Sraffa set out a system which gave coherence to the surplus concept, allowed the analysis of the effect of different values of a distributive variable on prices to be examined, and, at the same time, provided a critique of the marginal theories in so far as they were directed to answering classical questions about the origin and size of the rate of profits. The development of these ideas occupied many years. They were published in 1960 as *Production of Commodities by Means of Commodities*.

From 1930 Sraffa also worked on his magnificent eleven-volume edition (1951-73) of the works and correspondence of David Ricardo [q.v.], in later years collaborating with Maurice Dobb [q.v.], who did not share Sraffa's extreme inhibitions against writing for publication. It is one of the finest examples of sustained and meticulous scholarship in the discipline. The arguments of the introduction to volume i, published in 1951, are important complements to those of the 1960 book. In attempting this sustained research programme of both criticism and revival Sraffa may have had in mind Gramsci's injunction to attack at its logical core the very best expression of a rival philosophical system.

Sraffa was a remarkable personality. He had the capacity to evoke great affection and to inspire people to perform at their full potential. He had a subtle original wit and he made wholly unexpected responses to points raised in discussion. He was fluent in four languages. Though he lived in England from 1927 on, he always regarded himself as Italian, reading the Italian papers daily and never changing his nationality (indeed he was interned in the Isle of Man in 1940, until Keynes succeeded in having him returned to Cambridge). Sraffa died in Cambridge 3 September 1983. He was unmarried.

[Obituary in *Annual Report* of King's College, Cambridge, 1984; Nicholas Kaldor in *Proceedings* of the British Academy, vol. lxxi, 1985; personal knowledge.]

G. C. HARCOURT

STORRY, (GEORGE) RICHARD (1913-1982), historian, was born 20 October 1913 in Doncaster, the youngest of three children (all sons) of Frank Storry, manager of the Midland Bank there, and his wife, Kate Roberts. He was educated at Repton and Merton College, Oxford, where he graduated with second class honours in modern history in 1935. At the suggestion of the poet Edmund Blunden [q.v.], a fellow of Merton who had himself taught in Japan, Storry applied for a post teaching English in the College of Commerce at Otaru, a port on the northern Japanese island of Hokkaido. He arrived there in June 1937 and lived in Japan until April 1940. Thus his first contact with Japan, a country to the study of which he was to devote the rest of his life, came at a moment when Anglo-Japanese relations were becoming increasingly difficult; and it was a sign of the warmth and integrity of Storry's character that he maintained good relations with his students, some of whom were to become lifelong friends, without compromising his own position. In his diary and letters from this period he already showed a subtle understanding of Japanese society and politics and a realization of the danger caused by the growing influence of militaristic ideas.

After returning to England he was commissioned in the Intelligence Corps, in spite of poor eyesight, and was posted to Singapore where he arrived late in January 1942, a few days before the siege began. Within three weeks he was evacuated, and after a hazardous and eventful journey eventually reached India. He was subsequently appointed to command a mobile interrogation section with the rank of major, and served through the battle of Imphal, finally arriving in Rangoon in June 1945. The experience of interrogating Japanese prisoners of war increased his understanding of Japanese society: and it was characteristic of him that he noted that the British officers who had lived in Japan, although they had no illusions about the Japanese army, did not consider the Japanese race, as many of the British did, as 'ant-like hordes of subhuman fanatics'. Although there was much in both pre- and post-war Japan which he disliked, he never lost his sometimes humorous but always unpatronizing sympathy and respect for the Japanese.

Storry was demobilized in July 1946, and in 1947 he was awarded a postgraduate studentship by the Australian National University. He started the research—in England, the United States and Japan—which was to result in his first book, *The Double Patriots*, published in 1957. In 1952 he had been made a research fellow of the Australian National University and remained in Canberra till 1955 when he was elected to a Roger Heyworth memorial fellowship at St Antony's College, Oxford. He quickly established a reputation as a leading historian of Japan. *The Double Patriots*, a study of the nationalist societies of the 1930s, was followed by the excellent *A History of Modern Japan* (1960), *The Case of Richard Sorge* (1966) (written in collaboration with (Sir) F. W. Deakin—a fascinating account of a German Communist who became a Soviet spy in Japan and was executed by the Japanese), *The Way of the Samurai* (1978), and *Japan and the Decline of the West in Asia* (1979). These were books in which Storry's scholarship was combined with an admirable prose style which made them attractive to non-specialist readers. He also recorded his impressions of a changing Japan in a number of articles and in his introduction to Kurt Singer's interesting *Mirror, Sword and Jewel* (1973).

In 1970 Storry had taken over from G. F. Hudson the directorship of the Far Eastern Centre at St Antony's and he made a major contribution to the growth of Japanese studies in Britain, training numerous postgraduate students, both British and Japanese. He made several extensive visits to Japan, and it was largely due to his efforts and his influential contacts in Japan that in 1979 the Nissan Motor Company announced a gift of £1½ million to endow an Institute of Japanese Studies in Oxford. By the time the Institute was established Storry was close to the retirement age. However, in recognition of his contribution to Japanese studies, the university granted him the personal title of professor in 1981, and on his retirement later in that year the title of emeritus professor was conferred.

Storry's contacts with Japan were at many levels, from members of the imperial family to ordinary working people: his friends included professors, businessmen, diplomats, and students, in all of whom his warmth, simplicity, cheerfulness, and sincerity found a response. At the same time his romantic conservatism made him sympathetic to many of the traditional elements in Japanese life. He was active in encouraging people to take an interest in Japan and was one of the founders of the British Association for Japanese Studies. He was a true interpreter of the Japanese to the British and of the British to the Japanese. The Japanese government recognized his role by awarding him the Order of the Sacred Treasure (4th class) in 1959, while not long before he died he was given the Japan Foundation Award for his services in developing cultural contacts between the two countries. After his death a group of his Japanese friends endowed a biennial lecture to be delivered in Oxford in his memory.

He married in 1949 Dorothie, daughter of Thomas Edward Morton, of Dublin, a civil servant, and had one son, Terence, born in 1950. Storry died at his home at Woodeaton, near Oxford, 19 February 1982.

[*The Times*, 25 February 1982; Dorothie Storry, '*Second Country': the Story of Richard Storry and Japan 1913–1982*, 1986; private information; personal knowledge.]

JAMES JOLL

STRONG, SIR KENNETH WILLIAM DOBSON (1900–1982), major-general, was born in Montrose, Scotland, 9 September 1900, the only son among the four children of John Strong, rector of Montrose Academy and subsequently professor of education at Leeds University, and his wife, Ethel May, daughter of A. Knapton Dobson. Strong was educated at Montrose Academy, Glenalmond, and the Royal Military College, Sandhurst. He was commissioned into the Royal Scots Fusiliers in July 1920. After being an intelligence staff officer in Germany, he was defence security officer, Malta and Gibraltar, and then assistant military attaché, Berlin (January 1938–August 1939). He was then promoted lieutenant-colonel and GSO1 in MI14 (German section, War Office) until April 1941, when he commanded 4/5 Royal Scots Fusiliers until January 1942. He then became brigadier general staff (Intelligence) Home Forces until he was appointed BGS (Intelligence) Allied Forces HQ Algiers on 11 March 1943. He was promoted major-general in December 1943.

Strong had a distinguished career in Intelligence and became fluent in three foreign languages, particularly German. But the turning point in his career was his appointment as General Dwight Eisenhower's chief of Intelligence in March 1943. This appointment came about as a result of the failure of Intelligence before the Kasserine pass débâcle and subsequent changes amongst the senior staff at AFHQ. Strong brought a new and much needed discipline and direction to the Intelligence services in North Africa. He became a firm friend and confidant of Eisenhower and of his chief of staff, W. Bedell Smith—a friendship which was to last for life.

The Tunisian campaign ended in May 1943. After assisting in the planning of the Sicily landings, Strong's next important role was to accompany Bedell Smith to the Italian armistice negotiations, with the Italian General Castellano, in Lisbon in mid-August 1943. The armistice was signed in Sicily on 3 September by Bedell Smith, with Strong taking an important part in the ceremony and the events preceding it.

Strong then suffered a major disappointment. At the Cairo conference in November 1943 Eisenhower was appointed to take charge of the Allied invasion of north-west Europe. Eisenhower wished to take Strong back to London with him as his chief of Intelligence, but the British chiefs of staff prevented his transfer on the grounds of not wishing to denude AFHQ, although Strong always felt that he was too closely identified with the Americans for the War Office's liking. Eisenhower finally appealed to (Sir) Winston Churchill, and as a result Strong returned to London to join Supreme Headquarters, Allied Expeditionary Force, on 19 May 1944. He was therefore present at and participated in the historic decision to invade Normandy on 6 June 1944.

Strong's close association with Eisenhower and Bedell Smith did not always endear him to Sir B. L. Montgomery (later Viscount Montgomery of Alamein, q.v.) and the War Office. He was not immune from some of the British sniping at Eisenhower's methods and decisions. This was particularly true at the time of the German offensive in the Ardennes (December 1944–January 1945). That the German attack was not foreseen subjected the Intelligence staff in general and Strong in particular to a good deal of criticism. Strong, of course, did a lot of soul-searching to try to discover if he had missed a vital clue. The writer was with him hour by hour during this period and is convinced that the real culprit was lack of air reconnaissance due to bad weather and low cloud. Blame does not attach to Strong and his senior staff.

By 20 December an extremely serious situation had arisen, with the Germans advancing towards the River Meuse. Had they reached the Meuse, they would have driven a wedge through the centre of General Omar Bradley's command, separating his divisions in the north from those in the south. Consequently, Generals Whiteley and Strong felt compelled to recommend to Bedell Smith that the American First Army—Bradley's northern force—be put under Montgomery's command. It says much for the standing of these two British generals that they had the courage to make this very unpopular recommendation and Eisenhower had the wisdom to accept it. In the event, the German offensive was halted, but relations between Montgomery and Generals Bradley and George Patton, never good, went through a bad patch. Relations were not improved by the suggestion emanating from London that there be one ground forces commander, who should be Montgomery.

Montgomery's counter-attack, spearheaded by the US First Army, was launched on 3 January 1945. The capture of the Remagen bridgehead and the crossing of the Rhine by both Bradley and Montgomery put the end of the war finally in sight. Strong played a leading part in organizing the actual surrender ceremony in Rheims on 7 May. He was always mindful of the signing of the armistice by the Germans in 1918, which was done in such a way as to permit the German army to claim it had never been defeated in the field. Strong urged Eisenhower to ensure that this error should not be repeated. Strong participated in the arrest of Wilhelm Keitel and other senior officers at Flensburg and also the final surrender ceremony at Karlshorst, Berlin, on 8 May.

SHAEF was disbanded shortly after the German surrender and Strong had reached another turning point in his career. His first inclination was to continue his military career, which would have meant commanding a brigade, but as Montgomery was about to become CIGS this would probably have been unwise. The post of director-general of MI5 was mooted, but in the end he accepted the appointment of director-general of the political intelligence department of the Foreign Office. This was an interim assignment, as the PID was a wartime organization and was being run down. Strong was then appointed director of the Joint Intelligence Bureau—a job to which he was well suited and of which he made an outstanding success. He retired from the army in May 1947 to become a civil servant. In 1964 there was a reorganization of the Intelligence services and Strong became director-general of Intelligence, Ministry of Defence—a post which he held until his final retirement on 9 May 1966.

Strong subsequently had a career in the City as a director of Eagle Star Insurance Co., Philip

Hill Investment Trust, and other companies. He wrote *Intelligence at the Top* (1968) and *Men of Intelligence* (1970). He was appointed OBE (1942) and CB (1945), was knighted in 1952, and became KBE (1966). He was also an officer of the Legion of Merit (USA), an officer of the Legion of Honour and had the croix de guerre with palms (France), and belonged to the Order of the Red Banner (USSR).

In 1979 he married Brita Charlotta, widow of John Horridge, master of the Supreme Court (King's Bench Division), and daughter of E. S. Persson, engineer, of Malung, Sweden. They had no children. Strong died 11 January 1982 at his home in Eastbourne.

[Sir Kenneth Strong, *Intelligence at the Top*, 1968, and *Men of Intelligence*, 1970; personal knowledge.]   KENNETH KEITH

SUGDEN, SIR (THEODORE) MORRIS (1919–1984), physical chemist, was born at Sowerby Bridge in Yorkshire 31 December 1919, the only child of Frederick Morris Sugden, company secretary, and his wife, Florence Chadwick. He was educated at Sowerby Bridge and District Secondary School from which he won an open scholarship in natural sciences to Jesus College, Cambridge. After obtaining a first class in the preliminary examination in natural sciences in 1939, he decided to read part ii (chemistry) which, because of the war, he took at the end of his second year (1940), obtaining a first. He then began research with W. C. Price in the department of physical chemistry making precise measurements of the ionization potentials of molecules. After eighteen months he was transferred to the group led by R. G. W. Norrish [q.v.] which was investigating methods of suppressing gun flash.

These investigations led to the main themes of his post-war research which were studies of ionization in flames and the chemistry of flames, particularly the behaviour of metal atoms. In this pioneering work he showed that the apparently complex processes could be systematized in terms of equilibrium between some chemical species and an approach to equilibrium by others. His wartime work in the Ministry of Supply involved using the absorption of microwaves to study ionization in gun flash, and in post-war years he extended this to flames and to microwave spectroscopy, building up a large research group in the department of physical chemistry at Cambridge.

In 1945 Sugden was elected to a Stokes studentship at Pembroke College and in the following year he was appointed a university demonstrator being promoted to the Humphrey Owen Jones lectureship in 1950 when Frederick

(later Lord) Dainton accepted the Leeds chair. In 1957 he was elected a fellow of Queens' College, Cambridge (honorary fellow, 1976), and in 1960 he was appointed to an *ad hominem* readership in physical chemistry. He was elected FRS in 1963.

Sugden was a highly gifted pianist and would practice for several hours a day if the opportunity arose, preferring the work of the romantic composers. He was also exceptionally widely read, particularly in the history of thought and in economic theory, some of which he read in the original Italian. His literary interests were reflected in his large circle of friends from many countries and with varied academic interests. They and his remarkable skills as a teacher, an independence of spirit, and a slightly bohemian appearance and disregard for punctuality made Morris seem a natural part of the Cambridge scene. Many of his friends were surprised when in 1963 he accepted Lord Rothschild's invitation to become director of research of the Shell Research Centre at Thornton in Cheshire. There he showed considerable skills in management, particularly in recruiting talented scientists and using their abilities to the full. He rose to be director of Thornton Research Centre (1967) and in 1974 chief executive of Shell Research Limited.

In 1976 he returned to Cambridge as master of Trinity Hall where he and his wife found time to do much entertaining of undergraduates and friends, both in the master's lodge and at their cottage in Elsworth where Morris was able to resume the enthusiasm for gardening which he had developed at Great Barrow near Chester. He was much in demand to serve on committees, being chairman of the faculty board of engineering and of the syndics of the University Press. He became a director and vice-president of the Combustion Institute (1974–82) and was president of the Faraday division and honorary treasurer of the Royal Society of Chemistry before becoming its president in 1978–9. From 1977 he also chaired the government advisory committee on nuclear safety, a responsibility which weighed heavily on his shoulders. He was also increasingly involved in the activities of the Royal Society, becoming physical secretary in 1979, an office which he held at his death. In that post he took a strong interest in the role of the Royal Society and how far it should advise the public as well as the government on scientific issues. He was an excellent chairman of committees, firm, fair, perceptive, and quick-witted; he was always more willing to serve as chairman of a committee rather than as a member. He was appointed CBE in 1975, the year in which he received the Davy medal of the Royal Society, and knighted in 1983.

He had honorary degrees from Bradford, York, Liverpool, and Leeds. His scientific output included over 100 papers and a successful textbook on microwave spectroscopy with C. N. Kenney (1965).

In 1945 he married Marian Florence, daughter of Ernest George Cotton, export manager; they had one son, Andrew, who carried out research in tropical botany before he became an editor of scientific journals. Sugden died 3 January 1984 in Cambridge.

[C. P. Quinn and B. A. Thrush in *Biographical Memoirs of Fellows of the Royal Society*, vol. xxxii, 1986; private information; personal knowledge.]   B. A. THRUSH

SWEET-ESCOTT, BICKHAM ALDRED COWAN (1907-1981), SOE officer, banker, and businessman, was born in Newport, Monmouthshire, 6 June 1907, the eldest of four children (two sons and two daughters) of Aldred Bickham Sweet-Escott, marine engineer of Newport and Bristol, and his wife, Mary Amy, daughter of Michael Waistall Cowan, inspector-general in the Royal Navy. Educated at the choir school of Llandaff Cathedral, and Winchester College (1921-4) he left the latter early owing to family financial difficulties. He was employed by J. S. Fry & Sons, of Bristol (1924-7), and he continued his studies privately. Through the hard work that characterized his career he took a degree at London University and in 1927 won an exhibition to Balliol College, Oxford, taking a first in *literae humaniores* in 1930.

He joined the British Overseas Bank, where he mastered the politics, personalities, and financial structures of most European countries (1930-8). After serving for a year as personal assistant to Courtaulds' chairman he joined the Special Intelligence Service, section D, concerned with organizing sabotage, where he foresaw the importance of establishing section D activities in Europe before Nazi occupation. When section D became part of the Special Operations Executive (SOE) his service with the new organization took him to many regions. In 1941 he was in London concerned with the Balkans and Middle East, becoming regional director. In July 1941 he went to Cairo, as personal assistant to the head of SOE (Sir Frank Nelson, whose notice he later wrote for this Dictionary), returning to London in December to the M (operations) directorate. From July 1942 he was with the SOE mission in Washington, liaising with the Office of Strategic Services (OSS), SOE's opposite number in America. Back in London in 1943 he was briefly head of SOE's Free French section, before returning to Cairo in December, to act as adviser

to SOE's Force 133 until December 1944. From January to December 1945 he was in South East Asia Command as chief of staff to Force 136, becoming acting commander during the commander's absence. He was promoted major in 1941, and colonel in 1945.

In *Baker Street Irregular* (1965) Sweet-Escott gives a reliable and entertaining account of SOE. His experience, outstanding ability, and knowledge of the secrets of the organization repeatedly frustrated his attempts to serve in the field. He was one of the few competent to deal at the top level with SOE's problems, whether in Whitehall, Cairo, Algiers, or Kandy. Whenever and wherever there was a call for staff reinforcement Bickham was the obvious choice.

In 1949 he became general manager of the Ionian Bank which, helped by his wartime involvement with SOE operations in the Balkans, he rebuilt into a banking force in Greece and those parts of the eastern Mediterranean where Greek and British influence was still important. When the Bank to his great regret was taken over in 1958 he joined British Petroleum as group treasurer. BP was at a critical stage in its development, having lost its oil monopoly in Iran, and having 'long lived in its private world of Scottish accountants and Persian oil camps'. It had to find and develop oil elsewhere, and by investing heavily in marketing to reduce its dependence on oil production alone. Sweet-Escott's task was to ensure that financial communities throughout the world 'were aware of BP, understood BP, and trusted BP'. Although at times looked at askance by the old hands, he laid the foundation of much of BP's financial success by his clear-headed guidance, his exceptionally wide banking experience, and his ability to gain the confidence of new colleagues across the world. BP's development during his period of office speaks for his achievement. He retired in 1972.

He wrote several economic and financial surveys of the Balkans and Greece, including two books for the Royal Institute of International Affairs, of which he was an active member. He was a frequent broadcaster on international affairs. He was visiting professor, international finance, at the City University, London (1970-3). He worked hard for charitable causes, especially mental health.

He was a shrewd judge of character and situations, never afraid to state his views, and unfailingly kind, especially to young people planning their careers. A keen sportsman, in his later years he took up hunting, first stag, then fox, 'pursuing both beasts as if they had been storm troopers'.

Sweet-Escott married, first, in 1933, Doris Katharine Mary, daughter of the Revd Percy

George Bulstrode, of Broomfield, Taunton. The marriage was dissolved in 1950 and in the same year he married, secondly, Beryl Mary (died 1984), daughter of Trevor Phelps, businessman, of Botha's Hill, South Africa. There were no children of either marriage. Sweet-Escott died 12 November 1981 at Ipswich.

[*The Times*, 14 November 1981; *BP Shield*, no. 1, 1982; *Winchester College Register, 1901-1946*; SOE adviser; private information.]   CHARLES CRUICKSHANK

SWINNERTON, FRANK ARTHUR (1884-1982), author and publisher's assistant, was born 12 August 1884 at Wood Green, London, the younger son and younger child of Charles Swinnerton, copperplate engraver, and his wife, Rose, a designer, daughter of Richard Pell Cottam, a craftsman printer. Swinnerton's stock, he claimed, derived from the midland English of his father and the Scottishness of his mother. The Swinnertons came from Hanley in Staffordshire, the Cottams from Edinburgh. 'They were all the most modest creatures in the world', claimed Swinnerton, 'but simultaneously they had the utmost indifference to the opinion of any person who was not a Swinnerton or a Cottam.'

Swinnerton's education was perfunctory and spasmodic due to poor health as a child and a lack of means. He attended various educational establishments in the neighbourhood of his home where his childhood was happy and serene. There was little in his background to suggest that he would become a writer of distinction and be found at the intimate centre and in fraternal, almost conspiratorial, association with many of the literary giants of the age.

In his youth Swinnerton read avidly Henry James [q.v.], Ibsen, and Louisa M. Alcott, and was greatly influenced by E. Arnold Bennett, whose notice he later wrote for this Dictionary, as he did that of J. T. ('Frank') Harris. In his fifteenth year he arrived in Fleet Street and was briefly with the *Scottish Cyclist* at a wage of six shillings a week. It could be said that his 'literary life' began in 1901 when he joined the publishing firm of J. M. Dent [q.v.] as a 'clerk receptionist' in which capacity he assisted Hugh Dent in the 1906 launch of Everyman's Library. He had himself started writing at the age of ten, but it was not until 1909, after four abortive novels, that his first, *The Merry Heart*, was published mainly through the advocacy of Arnold Bennett.

As an author his industry and inventiveness never flagged. His calligraphy was exquisite if minute and his publisher was regularly informed that he might soon be receiving the typescript of a new book. There were in all some forty-two novels, fifteen other books of literary interest and purpose, a large output of reviews, comments, and articles, and talks and conversations with recording teams from all over the world both from sound and television.

He enjoyed respect and affection as a writer, but great success, as today measured by sales, eluded him. His short novel *Nocturne* (1917) seems most likely to endure and of his early novels *The Casement* (1911), *September* (1919), and *Young Felix* (1923) were highly regarded. Latterly *The Georgian House* (1922) and *The Two Wives* (1940) were thought to be among his best. His *The Georgian Literary Scene* (1935), *Swinnerton, an Autobiography* (1937), and *Arnold Bennett—A Last Word* (1978) together convey the essences of the era in which he occupied so uniquely observant a vantage point.

After six years with Messrs Dent he joined Chatto & Windus reading for them busily and advising them through the next twenty-five years. Of particular gratification was his discovery of *The Young Visiters* (1919) by Daisy Ashford [q.v.], to which he persuaded Sir J. M. Barrie [q.v.] to write a preface. Other acquisitions to a distinguished list were *Limbo* (1920) by Aldous Huxley [q.v.], *Eminent Victorians* (1918) by G. Lytton Strachey [q.v.], and *The Journal of a Disappointed Man* (1919) by W. N. P. Barbellion [q.v.].

'Swinny'—to all his friends, with the exception of Arnold Bennett, his greatest, who called him Henry—encompassed uniquely an extraordinary era of English writing and authorship and became over his long life its most astute chronicler and the friend and intimate of nearly all the distinguished company which sustained it. Towards the end of his life a visit to his home, Old Tokefield in Cranleigh, Surrey, became almost a pilgrimage for students and literary researchers from all over the world, seeking first-hand knowledge and recollection of so different a scene from any they knew within their modern world of the mass media and communicative gadgetry. As often as not and at least well into his nineties their quarry might be found 'hedging and ditching' the boundaries of his home, but immediately able to launch into vivid recollection and inimitable mimicry of such as Arnold Bennett, H. G. Wells, John Galsworthy, Sir Hugh Walpole, James Barrie [qq.v.], and others who formed the circle they mainly constituted with its centre at the Reform Club of which Swinnerton, in due time, was awarded honorary life membership.

Swinnerton was, above all, a great bookman. He cared deeply for writing and writers. He was the most trusted of men and of a lovable and generous nature. He never faltered for an apt word, a name, or a revealing anecdote. A small, neat figure with a trim beard, twinkling eyes, and small pince-nez glasses (which periodically dropped into his drink), he was a very private man. Once challenged by a friend he admitted with that little gruff grunt which was as distinctive as his chuckle that he had declined appointment as CBE. His own participation in life he rarely referred to but by comment and anecdote about others was able to bring instantly into being a vanished world. He was president of the Royal Literary Fund in 1962-6.

In 1919 he made an unsuccessful marriage. It was whilst at Chatto that he met, working in their office, Mary Dorothy Bennett (unrelated to his lifelong friend); they married in 1924. She was the daughter of George Bennett, journeyman tailor. She died in 1980. They had two daughters, the first of whom survived for only a week. Swinnerton died 6 November 1982 at Cranleigh Village Hospital.

[Frank Swinnerton, *Swinnerton, an Autobiography*, 1937; personal knowledge.]

ROBERT LUSTY

SYKES, SIR CHARLES (1905-1982), physicist and metallurgist, was born 27 February 1905 in Clowne, Derbyshire, the only child of Sam Sykes, manager of the Cresswell Cooperative Society, and his wife, Louisa Webster. With a county scholarship he went to the Staveley Netherthorpe Grammar School, travelling five miles each way daily by train. He thereby became a bridge and chess expert, and indeed he could play chess without a board. He never seemed to work hard and it was said of him in college that he was one of that small band who seemed to succeed without trying. He was good at all sports, gained three distinctions in the higher school certificate, and won scholarships to Leeds, Liverpool, and Sheffield universities. He chose Sheffield and was awarded a first class honours B.Sc. in physics in 1925.

He was awarded a Department of Scientific and Industrial Research grant to do two years' research in physics but after one year the Metropolitan-Vickers Electrical Company of Trafford Park, Manchester, offered him a two-year research studentship to continue work on the alloys of zirconium, which had been begun by one of their other scholars in the metallurgy department of the university. His professor generously allowed Sykes to change horses and thus he entered into a career in applied science in which he remained all his life. He published the work on zirconium with his collaborator and was awarded the degree of Ph.D. in metallurgy in 1928 before being appointed to the staff of the research department of the M-V Co.

In Manchester he continued metallurgical work with a vacuum induction furnace similar to the one he had used in Sheffield, helped to design and construct larger versions, and produced great quantities of steels and other alloys. One aluminium/iron alloy greatly puzzled him; above a certain temperature its coefficient of expansion and its electrical resistivity suddenly changed. X-ray analysis showed that below that critical temperature the atoms were well arranged, while above that temperature the structure became disorganized, and thus Sykes began a new branch of physical metallurgy known as 'order/disorder'. It was his work on this discovery which earned him the fellowship of the Royal Society in 1943.

Sykes then studied metals harder than diamonds, such as tungsten carbide, which were used for making cutting tools. He added other elements to increase hardness and toughness and developed a huge business. His hard metals were put into the tips of the armour-piercing shells used so effectively at the battle of Alamein.

In 1940 he was appointed superintendent of the metallurgy division at the National Physical Laboratory and was shocked at the lax attitude of the staff in wartime, which was so different from the vigour prevailing in the laboratories in Trafford Park. He supervised the work on armour-piercing shells, discovering ways of measuring the hardness inside the tips. The armaments research department at Fort Halstead asked him to take charge of their ballistics department while he continued his NPL work. He accepted the directorship of the Brown-Firth research department in Sheffield in 1944, contributing his great knowledge of armour-piercing steels to the company's products and also to the special steels being developed for the novel high-temperature gas turbines. These were steels which had very low 'creep' so that the blades could run for 100,000 hours or more without failure. Sykes paid great attention to the growth of 'hair-line' cracks which have dominated metallurgical advances ever since and studied closely the part played by hydrogen in the failure of steel castings and forgings. In 1951 he became managing director of Thomas Firth & John Brown Ltd., in 1962 deputy chairman, and in 1964 chairman. He retired in 1967 but remained on the board for another six years.

He was president of the Institute of Physics in 1952-4, chairman (1965-70) of the advisory council on research and development (Ministry of Power), a freeman of the Cutlers' Company of Hallamshire, a Sheffield magistrate, and pro-chancellor of Sheffield University (1967-71), which had given him an honorary D.Met. in 1955. The Iron and Steel Institute awarded him the Bessemer gold medal (1956) and the Institute of Physics the Glazebrook medal

(1967). He was appointed CBE in 1956 and knighted in 1964.

Sykes was one of the very few graduates in pure physics who moved over to an applied science and reached the very top of his new profession; but he remained a dedicated scientist to the end, deceptively slow and careful of his words, critical, and constructive. He was just as thorough in his garden. In 1930 he married Norah, daughter of Joseph Edward Staton, manager of the Clowne Mineral Water Company. They had a son and a daughter. Sykes died 29 January 1982 at his home in Sheffield.

[T. E. Allibone in *Biographical Memoirs of Fellows of the Royal Society*, vol. xxix, 1983; scientific papers in Sheffield University library; papers in Firth Brown library; personal knowledge.] T. E. ALLIBONE

T

TAYLOR, ALEC CLIFTON- (1907-1985), lecturer, broadcaster, and architectural historian. [See CLIFTON-TAYLOR.]

TETLOW, NORMAN (1899-1982), mechanical engineer, was born 7 February 1899 at Oldham, the eldest in the family of three sons and two daughters of James Tetlow, textile manager-owner of the Oldham Velvet Company, and his wife, Emily Goodwin. He was educated at Manchester Secondary School, Oldham, and entered Manchester University as an engineering student in 1915. His studies were interrupted by service with the Royal Flying Corps (1917-19) and he graduated B.Sc. (Hons.) in 1923. Thereafter mechanical engineering was his lifelong career, first in industrial employment and later as a private consultant.

In the 1920s there was an established demand for centrifugal pumps for a great variety of purposes such as irrigation, mine drainage, and boiler feeding. This was subsequently augmented by the growing needs of the petroleum industry in response to the changing pattern of road, sea, and air transport and of power generation. Among the leading manufacturers of such pumps was the long-established firm of Mather & Platt of Manchester (later incorporated within Weir Pumps) whose employ he entered. He became keenly interested in this specialized branch of engineering and was acknowledged as a leading authority on the design and construction of such pumps.

This expertise assumed a new significance during World War II, when German air raids made the transport of petrol and fuel oil by road difficult and dangerous. This led to the construction of an extensive pipeline system linking the principal oil ports with major airfields in Britain. Its operation was dependent on the availability of powerful, high-pressure centrifugal pumps and Tetlow—as estimating and commercial manager of the centrifugal pump department of his firm—became closely involved with the manufacture and installation of these.

This was, however, only the prelude to a project even more critical to the outcome of the war. While planning the invasion of Europe a major problem was how to supply adequate fuel to the vast mechanized forces to be landed in Normandy. In 1943 Lord Louis Mountbatten (later Earl Mountbatten of Burma, q.v.) raised the possibility of laying a submarine pipeline across the Channel. This was referred to A. C. Hartley [q.v.], chief engineer of the Anglo-Iranian Oil Company, who at first pronounced it impossible. However, a successful prototype was laid under the Bristol Channel from Swansea to north Devon and thus was born the highly successful PLUTO project (Pipelines Under the Ocean). The first pipeline—of 3-inch diameter steel pipe welded in 4,000 ft lengths and wound on huge drums—from Sandown to Cherbourg was laid in August 1944 and seventeen others were subsequently laid to supply the Allied armies as they advanced into Flanders and on to the Rhine. The network eventually extended to more than 1,100 miles, and for months delivered more than a million gallons of fuel daily. The success of this project, ultimately the responsibility of Sir Donald Banks, director-general of the petroleum warfare department, depended on many technological factors, but none more vital than the three major pumping stations—with which Tetlow was particularly concerned—at Sandown, Shanklin, and Dungeness.

After the war Tetlow set up as a private consultant, retaining his specialized interest in the pumping and transport of oil. This provided new technological and economic problems as the size of major pipelines increased, up to 36-inch in diameter. He found a particular demand for this service in connection with large new installations in Iran. He retained also his interest in the well-being of the engineering profession. He was a member of all three professional bodies and served on the council of the Institution of Mechanical Engineers (1956) and as chairman of its north-western branch.

His colleagues remember Tetlow not only as a capable and original engineer but also as a man with a pleasing and gifted personality. He had a strong artistic streak, expressed in sketches and water-colours; a keen sense of humour; a ready wit; and a great gift for friendship.

His last years he spent at Holt, Norfolk, and he died in Kelling Hospital 13 February 1982. In 1926 he married Nell, daughter of John Gregson, butcher; they had one son. She died in 1963 and in 1970 he married Sophia, daughter of Hugh Brooks, company director of Oldham Brewery Co. Ltd., and former wife of the Revd Frederick Tattersall, by whom she had one daughter. They had met through their common interest in painting.

[*The Times*, 23 February 1982; Weir Pumps Limited; Manchester University; Institution of Mechanical Engineers.]

TREVOR I. WILLIAMS

TEWSON, SIR (HAROLD) VINCENT (1898-1981), trade-union official and TUC general secretary, was born at Bradford 4 February 1898, the youngest in the family of three sons and two daughters of Edward Tewson, a nursery gardener, and his wife, Harriet Watts. He left elementary school at fourteen and became an office boy at the headquarters of the Amalgamated Society of Dyers (later the Dyers', Bleachers', and Textile Workers' Union). In 1916, when he was eighteen, he joined the army (West Yorkshire regiment) and served on the western front in France where he was promoted to commissioned rank as a lieutenant and won an MC (1917). After demobilization he rejoined the clerical staff of the Dyers' Union where he specialized in matters related to piece-work price lists and the negotiation of piece-work rates. He gained a reputation for meticulous detail in assisting the union membership when the industry was in the midst of slump and large numbers were unemployed.

Tewson joined the Independent Labour Party and became the youngest member of Bradford city council when he was twenty-five years of age. In 1926, following the general strike, he successfully applied for appointment to the new post of secretary to the organization department of the TUC. In its early stages he was largely involved in propaganda campaigns, especially amongst women workers. Experience gained as a result of this work helped him in securing the assistant general secretaryship of the TUC in 1931. He was an industrious and capable deputy to Sir Walter (later Lord) Citrine [q.v.] whom he succeeded as TUC general secretary in 1946. He remained in that important position until his retirement in 1960.

Except for his four years of military service Tewson spent his whole working life in the civil service of the trade-union movement and possibly suffered from lack of direct experience of the hurly-burly of everyday industrial life. He lacked Citrine's drive and imagination and undoubtedly was a hesitant policy-maker. Government leaders with whom he dealt, both Labour and Conservative, expected more from him than he was able to give. Not least, Arthur Deakin [q.v.], the leader of the transport workers, and other members of the TUC general council, were often exasperated by his compromising and cautious nature.

During his initial period as general secretary the TUC leadership was confronted with huge economic problems which had arisen in the aftermath of World War II. Tewson and others were attacked for excessive efforts to support the Labour government with wage restraint policies which resulted in substantial opposition from left-wing trade unionists. Workers were restless because of the changes from wartime to peacetime production, continued rationing, and growing unemployment. Even the na-tionalization of certain industries threw up difficulties of national wage fixation which had not been experienced before.

Although he attracted much criticism and did not have an outstanding record of achievement Tewson was a competent administrator and kept a tidy ship within the TUC itself. During his secretaryship the new TUC headquarters at Congress House was opened and a TUC training college was established in the new building. A production department was created and Tewson helped to shape the trade-union response to automation in industry. He supported efforts towards more amalgamations and closer relationships between trade unions and played a major role in setting up the Anglo-American and the British productivity councils. He also served as a member of the economic planning board from 1947 to 1960.

One of the highlights of Tewson's career was his election as president of the International Confederation of Free Trade Unions, for the two years 1953–5. Previously he had served as secretary of the preparatory commission which led to the formation of the ICFTU in 1949, following the withdrawal of the TUC from the World Federation of Trade Unions. He gave special encouragement to the development of trade unions in British colonies and urged the utmost tolerance. A view he expressed in 1952 is still significant: 'Our organizing work must have regard to the peculiarities and special circumstances of each country . . . Plans must be worked out with the full co-operation of the people on the spot.' He was a part-time member of the London Electricity Board from 1960 to 1968 and a member of the Independent Television Authority from 1964 to 1969. In 1950 he received a knighthood, having been appointed CBE in 1942.

In 1929 Tewson married a fellow employee of the TUC, Florence Elizabeth, the daughter of William Francis Moss, printer. They enjoyed a long and happy, if secluded, family life. His wife became a JP. They had two adopted sons. Tewson died 1 May 1981 in Letchworth, Hertfordshire.

[*The Times*, 2 May 1981; *Free Labour World*, 1981; TUC and ICFTU reports; personal knowledge.]　　　　JACK JONES

THOMAS, GWYN (1913–1981), novelist and playwright, was born 6 July 1913 in Cymmer, Porth, Rhondda, the last of the eight sons and the youngest of the twelve children of Walter Morgan Thomas, colliery ostler, of Porth, and his wife, Ziphorah Davies, of Kenfig Hill, Glamorgan, who died when Gwyn was six. He was educated at Rhondda County Grammar School, Porth, and in 1931 entered St Edmund Hall,

Oxford. He obtained a second class in modern languages in 1934. In 1933 he spent six months at the University of Madrid on a miners' union scholarship. After graduating he returned to south Wales where, with spells of unemployment, he worked in adult education for the Workers' Educational Association and for university extension classes. Work for the social settlement movement in parts of England led to an appointment as an area educational officer for the National Council of Social Service in north-west England and, in 1939, a year's residence in Manchester.

Unfit for war service, in 1940 he was appointed to teach Spanish and French at Cardigan Grammar School. In 1942 he moved to Barry, Glamorgan, as head of Spanish at the Barry County School for Boys. He lived and taught in Barry until he resigned to work as a free-lance writer and broadcaster in 1962. He had already won national fame as a television pundit in the 1950s by his frequent appearances on the Brains Trust programme. Now, for the Welsh commercial station TWW, he made a number of documentary films encapsulating his view of Wales and the world. For almost two decades his distinctive, railing voice and transfixing stare were to be as familiar to listeners and viewers as his unique blend of satire and slapstick had become to many readers. Publicly, his outspoken socialist beliefs and anti-nationalist convictions, at a time when there was a renaissance of Welsh language and nationalist activity in Wales, led to a depiction of him, in some quarters, as a figure out of touch with a changing Wales. He never lacked, however, his champions and admirers.

Gwyn Thomas had begun to write, fluently and prolifically, in the 1930s and had submitted a long, quasi-naturalistic novel about the working-class in unemployment in the Rhondda to a competition organized by Gollancz in 1937. The novel was sympathetically received but rejected (it was published in 1986 in an edited version under its original title, *Sorrow for Thy Sons*) so he did not seek publication again until just after World War II. By this time he had settled on a style (and a voice) that was unique and inimitable. His subject remained the condition of the working-class in south Wales (both during the depression years and as a historical theme) but his new 'sidling, malicious obliquity' turned the topic of countless documentary novels into a black comedy of manners. The wit was poured out in a cascade of metaphor, simile, and hyperbole. The work of this early period—*The Dark Philosophers* (1946), *The Alone to the Alone* (1947), *All Things Betray Thee* (1949)—represents his finest achievement in the novel. He abandoned any traditional structure of narrative, characterization, and even plot to exploit

a rich vein of comic commentary in which a melancholic morality was never far away. In the 1950s his gifts as a comic writer were emphasized at the expense of the earlier, muffled anger with its bitter judgements on all establishment figures. However, the hilarious prose he continued to produce, despite his full-time work as a teacher, marked him out as a notable twentieth-century humorist in novels such as *The World Cannot Hear You* (1951), *Now Lead Us Home* (1952), and *A Frost on my Frolic* (1953). He wrote nine novels in all, exploiting his experience in schools and his knowledge of Spain as in the last one, *The Love Man* (1958), which was based on the Don Juan story. Although he continued to write volumes of short stories, essays, and autobiographical material that culminated in *A Few Selected Exits* (1968), his attention centred on the theatre after the successful London production of his play *The Keep* (1962). He wrote five others (including two musicals) of which the best received, though unpublished, were 'Sap' (1974) and 'The Breakers' (1976). However, his dream, explored in numerous radio plays in the 1950s, of establishing a theatrical tradition in Wales foundered. Ill health, at last, stemmed the flow of words.

Gwyn Thomas was a formidable raconteur whose conversation glittered with epigrams and jests honed for the occasion. This verbal facility acted against a serious appreciation of his mature work in his lifetime but after his death, especially with the reprinting of his neglected early fiction, it became clear that he ranked with his younger contemporaries from Wales, Dylan Thomas [q.v.] and Alun Lewis, as a major writer from within the heart of the urban culture of industrial Wales.

In 1976 he was presented with the Honour (principal prize) of the Welsh Arts Council, awarded for his lifetime achievement.

He married in January 1938 Eiluned (Lyn), daughter of William Thomas, collier. There were no children. He died in Cardiff 13 April 1981.

[*The Times*, 18 April 1981; Gwyn Thomas, *A Few Selected Exits*, 1968 (autobiography); Ian Michael, *Gwyn Thomas*, 1977; private information; personal knowledge.]

D. B. SMITH

THOMPSON, GERTRUDE CATON-(1888-1985), archaeologist and authority on African pre-history. [See CATON-THOMPSON.]

THOMPSON, SIR HAROLD WARRIS (1908-1983), physical chemist, was born in Wombwell, Yorkshire, 15 February 1908, the only son and younger child of William Thompson, chief executive of a colliery, and his wife, Charlotte Emily Warris. He was educated at King Edward VII School in Sheffield and Trinity College, Oxford (of which he became an honorary fellow in 1978), where his tutor was (Sir) C. N. Hinshelwood [q.v.], from whom he derived much inspiration. He graduated with first class honours in chemistry in 1929. The following year he worked in Berlin with Fritz Haber and stayed as a paying guest with Max Planck. He received the degree of Phil.D. from the Humboldt University, and returned to Oxford to take up a fellowship at St John's College (1930).

'Tommy', as he was affectionately known to all his friends, was quickly recognized as one of the outstanding teachers of the university and he went on to inspire many pupils during his lifetime. His first research was concerned with chemical reactions in gases, but his interests soon turned to the effect of light on chemical reactions and then to spectroscopy. During World War II he worked in the Ministry of Supply and the Ministry of Aircraft Production with (Sir) G. B. B. M. Sutherland (Cambridge), J. J. Fox (government chief chemist), and later W. C. Price (ICI) to develop infra-red spectroscopy; this technique was used for the analysis of aviation fuels and many strategic materials. After the war Thompson played a major role in showing how infra-red spectroscopy could be applied to a large range of chemical problems. From 1964 until retirement in 1975 he was professor of chemistry at Oxford, retaining his fellowship at St John's College. He and his colleagues were at the forefront of chemical spectroscopy, making important contributions to the understanding of molecular vibration frequencies, vibration-rotation spectra of gases, intensities of vibrational bands, and the effects of intermolecular forces on vibration frequencies, and latterly on photoelectron spectroscopy. He was elected FRS in 1946. Thompson's scientific work was recognized with many honours, including the Ciamician medal, Bologna (1959), the Davy medal of the Royal Society (1965) and the John Tate gold medal of the American Institute of Physics (1966).

This scientific work was accompanied by remarkable bursts of creative activity in other spheres. In 1928 Thompson had won a blue for soccer at Oxford and he later became treasurer of the university AFC. Soon after the war he was the driving force in the establishment of the joint Oxford and Cambridge Pegasus Football Club which was to have a meteoric existence for eight years and to win the Amateur Cup twice. For the rest of his life he played a major role in national football and was vice-chairman of the Football Association (1967-76), vice-president (1969-80), and chairman (1976-81).

In 1959 Thompson was made a member of the council of the Royal Society, and with only two short gaps of one year each, he remained a member until 1971; this was a period of outstanding service to the society and to international science. He was foreign secretary of the Royal Society from 1965 to 1971 during which he greatly increased the scope of its international activities. He encouraged and expanded scientific exchanges which had already been begun in a modest way with Bulgaria, China, Czechoslovakia, Hungary, Poland, Romania, and the Soviet Union, and initiated a programme of European research conferences and technical exchanges which proved a success. Before his death he saw more than 5,000 exchange visits of short or long duration by European scientists. To this Dictionary he contributed the notices of Sir C. V. Raman, S. Bose, and the footballer W. R. ('Dixie') Dean.

His activities for international science were associated with the international scientific unions. From 1963 to 1966 he was president of the International Council of Scientific Unions, followed by a period as president of the International Union of Pure and Applied Chemistry. On the international scene he had a feel for personal and national sensitivities, combined with an unshakeable insistence on the integrity of science. Many people recognized the considerable charisma that was such an asset to him. His fluent German, passable French, inexhaustible fund of amusing stories, and sometimes astounding powers of concentration and hard work, all impressed and often won over to his point of view the people he was dealing with. It is a testimony to their respect for him that he received world-wide honours, which included being made a chevalier of the Legion of Honour (1971), the Grand Service Cross of Germany (1971), honorary membership of many foreign academies, and honorary degrees at many universities. Thompson was appointed CBE in 1959 and knighted in 1968.

In 1938 Thompson married Grace Penelope, daughter of William Stradling, schoolmaster; they had a son and a daughter. Thompson died in Oxford 31 December 1983.

[Rex Richards in *Biographical Memoirs of Fellows of the Royal Society*, vol. xxxi, 1985; private information; personal knowledge.]

REX RICHARDS

THOMPSON, ROSCOE TREEVE FAWCETT (1907-1984), documentary film producer, director, and film critic. [See ROTHA, PAUL.]

TOYNBEE, (THEODORE) PHILIP (1916–1981), author, critic, and poet, was born at 372 Woodstock Road, Oxford, 25 June 1916, the second in the family of three sons of Arnold Joseph Toynbee [q.v.], historian, and his first wife, Rosalind, daughter of (George) Gilbert (Aimé) Murray [q.v.], scholar, poet, and author, and Lady Mary Howard. Academic celebrity and Castle Howard's baroque splendours played their parts in a long narrative poem later on, but he was more impressed in youth by his Australian Murray ancestors and their perhaps imaginary background of convict hulks and outlawed bushwhackers on the wallaby trail. They fitted the radical yearnings which, at seventeen, made him run away from Rugby (where he played for the first XV) to join Esmond Romilly, who, after a similar flight from Wellington to the London docks, had launched the anti-public-school broadsheet *Out of Bounds*. Expelled, then coached by the monks of Ampleforth, Toynbee won a history scholarship to Christ Church, Oxford, where he took a second class in modern history in 1938 and became the first communist president of the Oxford Union.

His life was now split between left-wing politics, scrapes, the Café Royal and the Gargoyle, the pursuit of 'liberal girls' in London ballrooms, and the beginnings of serious writing: contradictions which he carried off with an engaging assumption of clownish self-mockery. Romilly, meanwhile, back from the Spanish civil war, had eloped with Jessica Mitford and they all lived for a spell hugger-mugger in Rotherhithe. In 1938–9 he was editor of the *Birmingham Town Crier*. He left communism early, but retained a lasting commitment to principles of social justice and felt deeply for the causes he took up. A touch of guilelessness laced his intelligence. His large frame, shabby clothes, and handsome, rugged face gave him the look of an aristocratic stevedore; during his early years, long stretches of country diligence were punctuated by brief London spells of brisk intemperance which may have misled many who were impressed by his gifts; drink, indeed was to remain a bane all his life, but his sensitiveness, good nature, deep honesty, flair, humour, wide reading, urgency in talk, and irresistible sense of the absurd made him widely loved.

In 1939 he married Anne Barbara Denise, daughter of Colonel George Harcourt Powell. They later settled in the Isle of Wight and had two daughters, one of whom, Polly, made a name as a writer. In the war of 1939–45, via the ranks of the Welsh Guards, Toynbee was commissioned and served in the Intelligence Corps. In 1942–4 he was seconded to the Ministry of Economic Warfare; he reached the rank of captain. He was an early contributor to *Horizon*;

his first books, *The Savage Days* (1937), *A School in Private* (1941), and *The Barricades* (1943), were well received and his more experimental *Tea With Mrs Goodman* (1947) was greeted with considerable acclaim. *The Garden to the Sea* (1953) was followed by *Friends Apart* (1954), the spirited double saga of Romilly and Jasper Ridley, both lost in the war. Friendship and discussion were vital to him. At Oxford he had sought out (Sir) Isaiah Berlin and Francis ('Frank') Pakenham (later the Earl of Longford); London friends included (Sir) Stephen Spender, (Sir) A. J. Ayer, Cyril Connolly [q.v.], and, especially, the art historian Benjamin Nicolson [q.v.], with whom he founded a weekly semi-debating luncheon club.

In 1950 his first marriage was dissolved, though both parties always remained friends, and in the same year he married Frances Genevieve ('Sally'), daughter of Charles Stout Smith, oil company executive in the USA. She was a member of the US embassy in Tel Aviv whom he met while he was covering the Levant for the *Observer*. They had a son and two daughters. Toynbee became the leading *Observer* literary critic for the rest of his life and with Connolly and C. Raymond Mortimer [q.v.] of the *Sunday Times*, helped to raise weekly reviewing to a very high level. He lived in the steep woods near Tintern Abbey where he was a devoted fisherman; but a Peacockian experiment at turning their roomy Barn House into a commune turned out a disappointment; meanwhile, to the dismay of fellow agnostics, Toynbee developed the growing intellectual belief in Christian practice which he recounts in *Part of a Journey* (1981) and *End of a Journey* (1988). Visits to London grew rarer; as 'Brother Philip', he became an extramural associate of the nearby Anglican Tymawr Convent, and died 15 June 1981 at Woodroyd Cottage, St Briavels, Lydney, Gloucestershire.

He achieved his avowed longing 'to become a good man' and it may be thought that *Pantaloon* (vol. i, 1961) fulfilled his ambition to write a masterpiece. This astonishing tragi-comic narrative poem in many volumes, the last of them still unpublished, had taken him two decades, and it documents, with sustained imaginative vigour and metrical skill, the trends, lures, revolts, mistakes, rewards, and guilt of a kind of growing up like his own.

[Jessica Mitford, *Faces of Philip*, 1984; personal knowledge.] PATRICK LEIGH FERMOR

TREVELYAN, HUMPHREY, BARON TREVELYAN (1905–1985), diplomat, was born at Hindhead, Surrey, 27 November 1905, the younger son and fifth of six children of the Revd George Philip Trevelyan, rector of Carshalton,

Surrey, and later vicar of St Stephen's, Bournemouth, and his wife, Monica Evelyn Juliet, daughter of the Revd Sidney Phillips, vicar of Kidderminster, and later honorary canon of Worcester. A second cousin of the historian George Macaulay Trevelyan [q.v.] and of Sir Charles P. Trevelyan [q.v.], education minister in Ramsay MacDonald's two Labour administrations, he was educated at Lancing and Jesus College, Cambridge. He obtained a second class in part i (1926) and a first in part ii (1927) of the classical tripos. In 1929 he joined the Indian Civil Service.

From this early age Trevelyan showed that relentless determination to get to the core of the problems facing him, which was to mark his career in the public service and his work in the worlds of commerce and the arts after his retirement. Anxious to plumb the thought processes of Indian nationalism, when a junior district officer in the Madras governorate, he made a personal friend of the local Congress Party representative with whom he used to row at weekends. But such activities were frowned on as too eccentric by the local British community and, after enduring three years of criticism from that quarter, he secured a transfer to the Indian Political Service in 1932 and the more congenial and rewarding work of liaison between New Delhi and the Indian states. There he remained for the next fifteen years, rising to the rank of political agent in the Indian states and joint secretary to the external affairs department of the government of India, for whom he went to Washington in 1944 to prepare the way for the first Indian embassy in the USA. Then, when independence came in 1947, he transferred to the Foreign Service.

From 1948 to 1950 Trevelyan served as counsellor of embassy in Baghdad and from 1951 to 1953 as economic adviser to the UK high commissioner in Germany. In 1953 he was appointed chargé d'affaires in Peking, Britain's first representative to the newly established communist government in China. For the first year the Chinese kept him very much at arm's length, but Trevelyan's tenacity, coupled with the new understanding with China that followed the 1954 Geneva conference, eventually enabled him to enjoy a reasonably businesslike relationship with Zhou Enlai.

In 1955 he became ambassador in Cairo until the breach of diplomatic relations fourteen months later following the Anglo-French attempt to seize the Suez canal by force. Although he never revealed his private abhorrence of his government's action, it was a tribute to the respect in which he was held by the Egyptians that both President Nasser and his foreign minister accepted that he was in no way privy to the conspiracy between Britain, France, and Israel.

After a brief interlude working with Dag Hammarskjöld as under-secretary general at the United Nations—a period he hated owing to Hammarskjöld's inability to delegate work—Trevelyan was sent in 1958 as ambassador to Iraq. Here the unpredictability of the new revolutionary leader, General Kassim, was to tax his patience and diplomatic skills to the utmost. Then in 1962, after a brief spell in the Foreign Office, he became ambassador in Moscow. This was at the end of the Cuban missile crisis and Trevelyan was to witness two years later how this foolish effort of brinkmanship helped to bring about the downfall of the Russian leader, Nikita Khrushchev.

On retirement from the Diplomatic Service in 1965 Trevelyan was offered the post of permanent under-secretary at the Foreign Office, but this he declined, not wishing to block the promotion of someone younger than himself. However he did agree, in 1967, to undertake, as high commissioner in Aden, the thankless but essential task of supervising Britain's withdrawal in the approach to independence.

Retirement from diplomacy did not permit Trevelyan to vegetate. In 1965 he had become president of the Council of Foreign Bondholders. The year 1969 saw him as a trustee of the British Museum and in the following year he became, in succession to Lord Eccles, chairman of the trustees and prime mover in staging at the Museum the exhibition of the Tutankhamun treasures. He was also a director of BP, GEC, and the British Bank of the Middle East. Furthermore, he wrote several memoirs of China, India, Russia, and the Middle East, among them *The India We Left* (1972) and *The Middle East in Revolution* (1970), which will always stand as models of brevity, lucidity, and illumination.

Trevelyan was without doubt the outstanding British diplomat of his time, an ambassador who enjoyed the trust of the governments to which he was accredited no less than his own. Whatever their differences, Jawaharlal Nehru [q.v.], Zhou Enlai, Nasser, Kassim, and Khrushchev held him in the highest esteem as a man of total honour and formidable intellect. Although from a most distinguished clan, he outshone them in his achievements, which found their due reward in the Order of the Garter and a peerage, a distinction unique in modern times for one of his profession.

Possessed of a phenomenal physical and mental energy, he enthused and invigorated those who served with him, especially young people, putting them on their mettle and commanding authority without pomposity by the sheer preeminence of his razor-sharp mind. His recipe for a happy post was 'Not too civilized, too much work, bad climate, a spice of danger.' In his company there was never a dull or an idle moment, but rather always an opportunity to learn from his extraordinary knowledge of politics, art, culture, and music, whether of China, India, or the world of Islam. He was a keen pianist.

He was appointed OBE (1941), CIE (1947), CMG (1951), KCMG (1955), and GCMG (1965). He was made a life baron in 1968 and appointed KG in 1974. He was an honorary fellow of Jesus College and had honorary degrees from Cambridge (LLD, 1970), Durham (DCL, 1973), and Leeds (D.Litt., 1975).

In 1937 he married Violet Margaret ('Peggy'), daughter of General Sir William Henry Bartholomew, CMG, DSO, and later GCB, chief of the general staff in India. They had two daughters. Trevelyan died 8 February 1985 at home in Duchess of Bedford House, London W8.

[Humphrey Trevelyan, *Public and Private*, 1980; personal knowledge.]

ANTHONY NUTTING

TYERMAN, DONALD (1908-1981), journalist and editor of the *Economist*, was born 1 March 1908 in Middlesbrough, the second of the three sons (there were no daughters) of Joseph Tyerman, an accountant, who took jobs in Kenya, and his wife, Catherine Day, who was originally a teacher. At the age of three Tyerman contracted polio and was paralysed from the neck down. He fought through ten years of hospitalization with extraordinary courage and vigour, eventually regaining control of all parts of his body except that he always had to walk with splints. He therefore went late to school but passed through the Friends' School at Great Ayton, Coatham Grammar School (Redcar), and Gateshead Secondary School to Brasenose College, Oxford, with a scholarship. He achieved first class honours in modern history (1929), and from 1930 to 1936 was a lecturer at University College, Southampton.

In 1936 the *Economist* advertised with university appointments boards for a writer on public and economic affairs. As secretary to the appointments board at Southampton Tyerman applied for the job himself, and the *Economist* accepted him with alacrity. Throughout the war of 1939-45 he was one of the most influential journalists in England. He was deputy editor to Geoffrey (later Lord) Crowther [q.v.] at the *Economist* in 1939-44, but, as Crowther was also filling many war jobs, Tyerman was quite often in charge of the paper. In 1943-4 he was

also deputy editor of the *Observer*—frequently putting that paper to bed on the Saturday, after putting the *Economist* to bed on Thursday. Both papers played a major role in preparing the intellectual ground for Britain's post-war welfare state.

In 1944 Tyerman was lured away to be assistant editor of *The Times*, where he infuriated (Sir) Winston Churchill by opposing the government's policy in Greece at the end of the war. His friends hoped Tyerman would become editor of *The Times*, but by now his physical immobility was sapping his boundless energy, and was making it increasingly difficult for him to edit a daily newspaper. Crowther hoped to get him the job as editor of the *News Chronicle*, but he eventually succeeded Crowther as editor of the *Economist* itself (1956-65).

Tyerman's period as editor at the *Economist* cannot have been easy, with Crowther constantly throwing in brilliant new ideas from the chairman's office, and with the young Turks whom Crowther had recruited to the paper after 1945 also being pungent rather than conventional in their opinions. In the last issue before the election of 15 October 1964, Tyerman ended the paper's leading article: 'It does seem to the *Economist* that, on the nicest balance, the riskier choice of Labour—and Mr Wilson—will be the better choice for voters on Thursday.' This was also the view of most of his editorial writers, who thought it crucial that post-Gaitskell Labour should come back for a while into being a party of government, because another spell of opposition during the 1960s could turn it into an old-time socialist party instead of a new-age responsible social democrat party.

Although Crowther and most of the rest of the board of the *Economist* did not agree with this, it is not true that the contretemps over it led to Tyerman's retirement. Tyerman had decided to retire earlier in the year, and before election day in 1964 (Sir) J. W. A. ('Alastair') Burnet had already been approved by all concerned as his successor. Tyerman stayed on as a director of the *Economist*, but he did not have the same post-editorial influence as Crowther did. He devoted more time to his extracurricular activities—on the International Press Institute, the Press Council, the Save the Children Fund, and the council of the University of Sussex. He was appointed CBE for public service in 1978. Throughout his life he was an assiduous trainer of young journalists, with a keen eye for newspaper make-up, and as brave in his integrity as in his long fight against physical disability.

In 1934 he married Margaret Charteris, daughter of Ernest Gray, a businessman who died young. They had three daughters and two sons. Tyerman died 24 April 1981 at his home in Saxmundham, Suffolk.

[Private information; personal knowledge.]
NORMAN MACRAE

V

VAIZEY, JOHN ERNEST, Baron Vaizey (1929-1984), economist, was born 1 October 1929 at 60 Woolwich Road, east Greenwich, the younger child and only son of Ernest Vaizey, wharfinger, and his wife, Lucy Butler Hart. He was educated at Colfe's Grammar School, Lewisham, till in December 1943 he was suddenly struck down by osteomyelitis. Lying on his stomach encased in plaster he suffered acute pain for the next two years. He encapsulated the experience in a minor masterpiece of autobiography, *Scenes from Institutional Life* (1959, reprinted 1986). It is a deeply moving account of the insensitive nursing in an inadequately supplied wartime hospital, varied by touches of humour for he could never be solemn or sad for long. Those painful years gave him, he says, 'a sense of urgency and effort'. It was as if he guessed that he would have a short life and must pack all he could into it.

He completed his school education at Queen Mary's Hospital School, Carshalton, Surrey, and won an open exhibition at Cambridge. ' "Any more for Jesus?" the porters at the station cried evangelically', so he says when he arrived. But he was for Queens' College which he had pricked with a pin on a list in the local library at Lewisham. In 1949 he obtained a second class (division I) in part i of the economics tripos and in 1951 he gained a first class in part ii. After a year as a research officer at the United Nations Economic Commission in Europe he was elected in 1953 to a fellowship at St Catharine's College, Cambridge, together with a five-year university lectureship. In 1956 he moved to Oxford as a university lecturer in economics and economic history. He hated the place and left in 1960 for a research post at the London University Institute of Education. He was there for two years, and in 1961 made a most happy marriage to Marina Stansky, a graduate of Harvard and Cambridge and a distinguished art critic. She was the daughter of Lyman Stansky, lawyer, of New York. They were to have two sons and a daughter.

Vaizey was offered and surprisingly accepted a fellowship at Worcester College, Oxford, in 1962 and became a convert to the Oxford he had once disliked so much. During his two periods in Oxford he produced his major contribution in his chosen field, the virtually unexplored subject of educational finance. *The Costs of Education* (1958) showed that official statistics had been mere surmise. The accepted belief that the slump of 1931 had caused disastrous cuts in educational expenditure was demonstrated as a myth: on the contrary real per capita expenditure rose in the 1930s. He followed this up with numerous books and papers and conference addresses where he expressed a vigorous but unpopular minority view for over twenty years. Although the economics of education was to be his main theme—the subject of some eighteen works written either by himself or in conjunction with others—it was not his only interest. He wrote a history of Guinness's brewery (with Patrick Lynch, 1960) and of the British steel industry (1974); *Capitalism* (1971); *Social Democracy* (1971); and *The Squandered Peace* (1983), a survey of world history from 1945 to 1975. He also produced two light-hearted novels, *Barometer Man* (1966) and *The Sleepless Lunch* (1968). To this Dictionary he contributed the notice of C. P. Snow.

He was of Anglo-Irish descent and had the bubbling wit, the sense of humour, and the power of oratory that often go with that inheritance. He was fascinated by Ireland, also by Portugal where he did much work for the Gulbenkian Foundation. His other favourite country was Australia, but this affection led to an unhappy episode. In 1966 he had become professor of economics at Brunel University and in 1974 head of its school of social sciences which was largely his creation. But in 1975 he suddenly accepted the vice-chancellorship of Monash University, Melbourne. He came, he saw, he withdrew, to the consternation of his friends though they knew that he was volatile and impulsive.

In 1976 he was made a life peer in the resignation honours list of Harold Wilson (later Lord Wilson of Rievaulx). Yet, although he had been in one sense of the word a 'socialist' from early youth, he was becoming increasingly disillusioned with the Labour Party. In 1978 he resigned. He was one of a group of former Labour intellectuals whose faith had been shaken by the manœuvres of the administration of L. James Callaghan (later Lord Callaghan of Cardiff), among them his friends, Hugh Thomas (later Lord Thomas of Swynnerton) and Paul Johnson. He proclaimed his reasons on television in the run-up to the 1979 general election and henceforth took the Conservative whip. He knew and admired Margaret Thatcher, and he was among those whom she consulted. Toward the end of his life he took much interest in the economics of the health service. His *National Health* (1984) is remarkably prophetic of future developments. In 1982 he was appointed principal of the King George VI and Queen Elizabeth Foundation of St Catharine at Cumberland Lodge in the Great Park of Windsor. In two years he rescued its finances and revitalized its staff.

In talk irreverent, amusing, even outrageous ('Trots the lot' was his description of the Labour front bench in the House of Lords) he was a fundamentally serious person who packed an immense amount into a short life—over forty books; consultancies and visiting professorships galore; directorships (mostly unpaid); membership of innumerable councils and committees. He was very religious, describing himself as 'a deeply flawed Puritan', though in fact he was a devout High Anglican. He was a loyal almost passionate friend, but he could also be a formidable enemy especially of those whom he regarded as pompous or priggish.

He received honorary doctorates from the universities of Brunel (1970) and Adelaide (1974), and was appointed to the Order of El Sabio in Spain in 1969. His health was always precarious. In 1980 he had serious heart surgery from which he narrowly recovered and again in 1984, from which he did not. He died in St Thomas's Hospital, London, 19 July 1984.

[*The Times*, 20 and 23 July 1984; Vaizey's own writings mentioned in the text; private information; personal knowledge.]   BLAKE

VICKERS, SIR (CHARLES) GEOFFREY (1894-1982), lawyer and writer, was born 13 October 1894 in Nottingham, the youngest child in the family of two sons and a daughter of Charles Henry Vickers, a lace maker, and his wife, Jessie Lomas. Educated at Oundle School, he gained an exhibition at Merton College, Oxford, and went there to study classics in 1913. When war came a year later he volunteered for service in the army, being gazetted as a second lieutenant in the Sherwood Foresters (7th Robin Hood battalion). He was soon engaged in battle in Flanders and on his twenty-first birthday in 1915 found himself the sole defender of a barricade under heavy attack. In this engagement he showed outstanding courage for which he was awarded the VC.

The wounds incurred on this occasion consigned Vickers to hospital and convalescence for nearly a year. He then returned to France and was again in action in 1918, being awarded the croix de guerre. Leaving the army with the rank of major, he went back to Oxford for two terms and took a pass degree in French, European history, and law in 1919. After qualifying as a solicitor in 1923, he became a partner in the City firm of Slaughter & May in 1926 where he specialized in commercial finance, dealing often with its international ramifications. He enjoyed this work down to 1939 and was later to say that his career in law had been the one undoubted success in his life.

After the outbreak of World War II Vickers joined his old regiment, but was quickly seconded to intelligence work. He went on a mission to South America and on his return in 1941 was put in charge of economic intelligence in the Ministry of Economic Warfare, later becoming deputy director-general. He remained on this work until the war ended and was knighted for his services in 1946. By then Vickers no longer wanted to return to private legal practice and welcomed the chance of joining the new National Coal Board as its legal adviser in late 1945, becoming Board member in charge of personnel and training in 1948. His career as a public servant continued until retirement in 1955. Meanwhile he had become deeply involved in voluntary work in support of medical research, taking a close and informed interest in psychiatry and mental illness. He was an active chairman of the research committee of the Mental Health Research Fund (1951-67) and a member of the Medical Research Council (1952-60).

The years of retirement, which were to stretch over a quarter of a century, were in many respects the most absorbing of his life. It was then that he gained distinction as a writer on action and relationships in complex patterns of social organization. Not that he was any stranger to writing. His literary impulses were always strong, being expressed early on in his World War I letters and unpublished diary as well as in plays and stories for children written in the 1920s, one of which, *The Secret of Tarbury Tor*, was published in 1925. He was also an indefatigable letter writer. Even before 1939 he had gained some reputation for his ideas on social and political questions and was from 1939 to 1942 a visiting fellow of Nuffield College, Oxford, then newly founded. But retirement gave him the chance to concentrate entirely on committing his ideas to paper. Between 1959 and 1980 eight books were published, five in Britain, three in North America. Posthumously two further books appeared, one of which he was still working on shortly before he died. Well over 100 papers, articles, and lectures were also published, many of them in medical and psychiatric journals. Much of this material was later embodied in his books, most of which took shape in this way.

The problem which chiefly preoccupied Vickers was how individuals can best fulfil the requirements of social co-operation in conditions of accelerating economic and scientific change. He came to reject moral and economic individualism and argued that institutions are necessary conditions of satisfactory social co-existence. Influenced by Michael Polanyi [q.v.], he saw the achievement of an adequate understanding of institutions as an epistemological challenge: individuals have to grasp how their actions always involve the regulation of relationships with others, and this occurs only through the exercise of judgement. Consequently much of his work is devoted to the analysis of judgement in terms of what he called 'appreciative behaviour', the most notable contribution being made in *The Art of Judgement* (1965), a study of policy making. Though appreciation and judgement express individual capacities, Vickers never saw the individual as isolated or sovereign, but rather as defined by the relationships he has. He believed that social institutions are best analysed in terms of systems and his published work, notably *Human Systems are Different* (1983), made far-reaching contributions to systems thinking in its applications to human society. These themes, refined, developed, and set within various organizational contexts, recur in all his mature works.

Yet in his later years he was somewhat saddened by what he saw as a certain lack of interest in his ideas in Britain. His work was, however, taken up by the Open University Systems Group where he became a regular contributor to seminars and there were many psychologists and medical scientists who recognized his originality. In the USA and Canada his ideas were warmly received from the start and he became widely known through his frequent visits there. Amongst British social theorists Vickers was unusual in drawing extensively on those experimental sciences concerned with human behaviour, though he never regretted his own humanistic and historical education. The breadth of his reading and knowledge was remarkable, but his sensitivity to the English language enabled him to write about administrative behaviour and organizations with elegance and clarity, keeping footnotes to the minimum and always eschewing jargon. He was that rare combination, a man of action who was also an original thinker.

In 1918 Vickers married Helen Tregoning, daughter of Arthur Henry Newton, a director of Winsor & Newton, makers of water-colour paints and brushes, of Bexhill, Sussex. A son and daughter were born in the early 1920s, but the marriage later broke down and was dissolved in 1934. In 1935 he married (Ethel) Ellen (died 1972), daughter of Henry Richard B. Tweed, solicitor, of Laindon Frith, Billericay, Essex; they had one son. Active and full of intellectual curiosity to the end Vickers died at Goring-on-Thames 16 March 1982. As testimony to his abiding love of poetry a small volume of his poems, *Moods and Tenses*, was published privately in 1983.

[*The Times*, 18, 25, and 29 March 1982; Margaret Blunden, 'Geoffrey Vickers—An In-

tellectual Journey' in *The Vickers Papers*, 1984; private papers and information provided by Mrs R. B. Miller (daughter).]

NEVIL JOHNSON

VOCE, WILLIAM (1909-1984), cricketer, was born at Annesley Woodhouse, Nottinghamshire, 8 August 1909, the eldest of six children (four daughters and two sons) of William Voce, coal miner, of Annesley Woodhouse, and his wife, Kate Leatherland. He was educated at Annesley Woodhouse school to the age of thirteen when, on the death of his father, he assumed responsibility for the family and went to work in the local colliery.

Fred Barratt (1894-1947), the Nottinghamshire and England fast bowler, saw the young Voce playing casual local cricket and was sufficiently impressed to recommend him to Nottinghamshire, who took him on the county staff before his sixteenth birthday. Voce's main ability was as a left-arm bowler, in several different styles at varying periods of his career. Like most lads, he aspired to bowl fast but, on joining the Nottinghamshire staff, under guidance, he bowled orthodox finger-spin from round the wicket well enough to be given a trial for the first XI in June 1927. Still short of his eighteenth birthday, he took five wickets for 36 in the first Gloucestershire innings: and with 36 wickets at 27.16 retained his place in the Nottinghamshire side which finished as runners-up in the county championship. In the following season, however, he changed to fast left-arm-round-the-wicket swing, with some success. Then, in 1929, top of the Nottinghamshire bowling with 107 at 16.03, when the team won the county championship, he was the leading pace bowler in the national averages.

Some experts regretted his change of method, believing he might have become a great slow left-arm bowler. On the other hand, tall and immensely strongly built, he was probably the finest left-arm fast bowler in the world in the 1930s. He was, too, a forcing right-hand batsman who in 1933 scored 1,020 runs; while his powerful, loose left arm made him a valuable and accurate thrower from the deep field.

His penetrative county bowling partnership with Harold Larwood led to them being selected for the unfancied MCC team which D. R. Jardine [q.v.] took to Australia in 1932-3. They were the main instruments of a strategy which employed fast leg theory directed particularly against (Sir) Donald Bradman whom Jardine regarded, correctly, as the likeliest Australian match-winner. In this tactic, Voce bowled fast left-arm over the wicket, took six wickets in the first test match, which England won, and five in the second, which they lost. He was injured for the third test, missed the fourth, and took three in the fifth; England won by four to one what became known as the 'body-line' series. The repercussions of those matches were such that Jardine, Larwood, and Voce did not play in the tests of 1934 with Australia. Larwood and Voce, too, declared themselves unavailable for tests in 1935.

In 1936, however, Voce made himself available; played once against India, and went with G. O. Allen's 1936–7 MCC party to Australia. Reverting to fast left-arm round the wicket, he reached his bowling peak. In the first test he took six for 41 in the first innings; in the second, after rain, he took the first three Australian wickets—O'Brien, Bradman, and McCabe—for one run (in all, four for 16). England won. Voce had seven for 76 in the second test which England also won and five for 169 in the third, which they lost. Unfit, and picked against the captain's wish for the fourth test, Voce took one for 135; three for 123 in the fifth. England lost both tests, and the rubber with them.

Immediately after World War II Voce's four for seven in a test trial took him again to Australia, in 1946–7. Now, though, he was thirty-seven and a knee injury had robbed his bowling of its fire. Unsuccessful in his two tests, in 1947 he reverted to slow left-arm spin and in June retired from county cricket. Strong looking, dark, and ruggedly handsome, he became a valuable and popular coach for Nottinghamshire and the MCC. He was one of the few major England cricketers who played often against Australia—in his case eleven times—but never once in England. Altogether he made 27 test appearances, in which he took 98 wickets at 27.88: in all cricket, he took 1,558 wickets at 23.08 and made 7,583 runs (four centuries) at 19.19.

He married, in 1929, Elsie Emma Soars. They had one son, who died in childhood, and a daughter. Voce died in the University Hospital at Nottingham 6 June 1984.

[Family information; personal knowledge.]
JOHN ARLOTT

W

WAKEFIELD, (WILLIAM) WAVELL, first BARON WAKEFIELD OF KENDAL (1898-1983), sportsman and politician, was born 10 March 1898 at Beckenham, Kent, into an old and respected Westmorland Quaker family, the eldest of four sons (there were no daughters) of Roger William Wakefield, a medical practitioner in Beckenham, and his wife, Ethel Mary, daughter of John Frederick Knott, of Buxton, in Derbyshire. He was educated at the Craig Preparatory School, Windermere, and at Sedbergh, which he left in 1916 to serve in the Royal Naval Air Service and later in the Royal Air Force. He went to Pembroke College, Cambridge, in 1921, on the first RAF course arranged at the university, and took a BA in engineering (1923).

He was a tremendous enthusiast in everything he did, and he was already an England rugby international when he went to Cambridge. He won two blues and was captain of the university in his second year. Typically, he discarded the traditional selection procedures in that year, and combed the colleges for talent. He took players from obscurity, moved them into different positions, and moulded them into such an effective team that they beat Oxford, the hot favourites, by what was then a record score. He went on to win 31 caps for England, which remained a record for forty-two years.

He retired from the RAF as a flight lieutenant in 1923, having reached the rank of captain in his wartime service and having been mentioned in dispatches. On leaving the RAF he joined Boots, the chemists, and qualified as a pharmaceutical chemist during his four years with the firm. In those years he established himself as one of the great players of Rugby Union football and as one of the best-known Englishmen of his time. He played cricket for the MCC and was also an exceptionally gifted athlete. He won the RAF 440 yards championship, and in sprint training at Cambridge he was fast enough to extend Harold Abrahams [q.v.], who went on to win the gold medal for the 100 metres in the 1924 Olympic Games. That speed, then unusual in such a big man, enabled Wakefield to transform forward play in rugby football, and such was his enquiring mind and innovative nature that he was in the forefront of introducing specialization to the various forward positions. With his white scrum cap laced firmly on his fair hair and round his strong, open face, he played for England for eight years and captained his country to the most successful period in its rugby history since the breakaway of the northern clubs which formed the Rugby League in 1892. Wakefield was elected to the

Rugby Union while he was still a player, and he went on to become president in 1950-1 and to represent England on the International Rugby Football Board.

He maintained an active interest in skiing and water-skiing throughout his life. He won the Kandahar gold and became president of the Ski Club of Great Britain. In the summer months he was fond of water-skiing and subaqua diving in his beloved Lake District where he lived and where he had some of his family banking and business interests. After he left the RAF he continued flying as a pilot in the Reserve and in 1939 was recalled to active service for flying duties. He was then transferred to the Air Ministry as parliamentary private secretary and in 1942 was made director of the Air Training Corps.

He stood unsuccessfully for Parliament in a by-election at Swindon in 1934, but a year later, at the general election, he won the seat for the Conservatives. In 1945 he left Swindon and won St Marylebone. He held that seat until 1963, when he was created a baron. He had been knighted for public services in 1944. He served on the committees of the YMCA and the National Playing Fields Association, and was president of various manufacturing and transport associations. He also held a wide range of company directorships, including those of Rediffusion Ltd., Skyways Engineering, and the Portman Building Society.

In 1919 he married Rowena Doris (died 1981), daughter of Llewellyn Lewis, medical practitioner. They had three daughters. Wakefield died in Kendal 12 August 1983. The barony became extinct.

[Family information; personal knowledge.]

JOHN REASON

WALDOCK, SIR (CLAUD) HUMPHREY (MEREDITH) (1904-1981), president of the International Court of Justice, was born 13 August 1904 in Colombo, the fourth son in the family of four sons and one daughter of Frederic William Waldock, tea planter, and his wife, Lizzie Kyd Souter. He was educated at Uppingham and Brasenose College, Oxford, where he represented the university at hockey and played cricket for his college. He obtained second classes in classical honour moderations (1925), jurisprudence (1927), and in the degree of BCL (1928). His professional career began modestly, and there is a measure of uncertainty in the beginnings. He was called to the bar by Gray's Inn in 1928, and practised for a while on the Midland circuit. However, in 1930 election to a tutorial fellowship at Brasenose drew him away from practice.

The war of 1939-45 not only interrupted his academic career but helped to point him away from his interest in land law and equity and towards public international law. In the war years Waldock joined a branch of the Admiralty called military branch I, of which he became the head, attaining the grade of principal assistant secretary in 1944. The work involved the Royal Navy's foreign relations at a time when relations with neutrals and the war at sea generated difficult issues during the critical early phases of the war, issues which were of concern at the highest level of government.

Although Waldock left his duties at the Admiralty in 1945, the experience had marked him out as a man to be relied upon when the stakes were high, and in 1946 he became the United Kingdom member of the commission of experts for the investigation of the Italo-Yugoslav boundary, at a time when the Trieste question was very prominent.

By 1947 the pattern for the future was established. In that year Waldock was elected to the Chichele chair of public international law at Oxford, and combined his university duties with a specialist practice at the bar. He appeared as counsel in leading cases before the International Court of Justice, and provided expert advice to his own and numerous other governments. In 1951 he accompanied the United Nations secretary-general, Dag Hammarskjöld, on a difficult mission to China at a critical point in the Korean war.

Waldock's contribution to books about his subject was modest but by no means insignificant. In 1963 he produced an edition of *The Law of Nations* by J. L. Brierly, whose notice he wrote for this Dictionary. He also wrote a remarkable 'general course' at the Hague Academy of International Law (*Recueil des Cours*, 1962). The latter evidenced his exceptional ability to apply the positivist method to the world of post-colonial diversity and his readiness to assess new trends in state practice objectively. His legal method was always empirical and, though he was not blind to its influence, he did not favour theory.

In the last twenty years of his life Waldock's particular qualities, including his capacity to influence his colleagues, and penchant for hard work carefully done, produced a career which combined personal success with a conspicuous fruitfulness for the practice of the rule of law. The great contributions were to be in the codification of the law and in the field of international adjudication.

Waldock's contribution to codification resulted from his role, in the International Law Commission, as special rapporteur on the law of treaties, and, subsequently, as expert consultant at the United Nations conference on the law of treaties held at Vienna in 1968 and 1969. His role in the field of adjudication began with his membership of the European Commission of Human Rights (1954-61) at a pioneer stage of its activity, and he was president for six years. In 1966 he was elected a judge of the European Court of Human Rights and became president in 1971. In 1973 Waldock laid down his practice and his various offices, not without reluctance, on his election to the International Court of Justice. To no one's surprise, he became president of the court in 1979, and died in office. As a judge he was recognized for what he was, a model of integrity and careful work, attracting the considerable respect of his colleagues.

Apart from his role in the drafting committees of the International Court, and the publications noted earlier, his principal 'published work' is comprised of his six reports on the law of treaties prepared for the International Law Commission, which remain as a major source of the law of treaties. This aside, his other major contribution to the law of nations was institutional and exemplary. He was dedicated to the practical science of the peaceful settlement of disputes between states, and the high standards, and the ability to solve difficult problems, he brought to the sphere of adjudication and arbitration, provide an example which will be long remembered.

Waldock had a quiet manner but was a loyal friend and readily inspired confidence among colleagues: hence his gravitation towards the presidency of any institution of which he was a member. He was a loyal servant of the institutions with which he was associated, and was a bencher of Gray's Inn (1956) and, in 1971, its treasurer. He was appointed OBE (1942) and CMG (1946), became KC (1951), and was knighted (1961).

He enjoyed a very happy home life and his beloved wife, 'Beattie', was a great source of strength. She was Ethel Beatrice, the daughter of James Herbert Williams, shipowner, of the Black Diamond Line, of Wellington, New Zealand. They had a son and a daughter. Waldock died at The Hague 15 August 1981.

[*The Times*, 18 and 24 August 1981; Sir Gerald Fitzmaurice, *Graya*, 1981, p. 54; Sir Francis Vallat, ibid., p. 57; Ian Brownlie, *British Year Book of International Law*, vol. liv, 1983, p. 7; personal knowledge.]

IAN BROWNLIE

WALLACE-HADRILL, (JOHN) MICHAEL (1916-1985), historian, was born at Bromsgrove 29 September 1916, the eldest of the three sons (there were no daughters) of

Frederic Wallace-Hadrill, schoolmaster, and his wife, (Tamsin) Norah White, who came from a family of brewers. From 1930 to 1935 he went to Cheltenham College. From there he won a scholarship to Corpus Christi College, Oxford, to read modern history, in which he gained a second class in 1938. During World War II he served in intelligence and from 1943 to 1945 he was seconded to MI6.

Returning to Corpus in 1945 and elected a junior research fellow, he could now pursue wholeheartedly the Frankish and Anglo-Saxon studies which became his life's work. In 1947 he was elected to a tutorial fellowship at Merton College and a university lectureship soon followed. In 1952 appeared his first book, *The Barbarian West 400-1000*, which surveyed in a spirited manner the German peoples' invasions of the disintegrating western empire, not just in terms of occupation, settlement, and power struggles, but more in search of the Goths', Lombards', and Franks' historical self-awareness after they had become Christians. In 1955 he accepted the chair of medieval history at Manchester where he stayed until 1961 restlessly moving house more than once. The edition of the *Fourth Book of the Chronicle of Fredegar* (1960) was his chief publication during these years, but there were also a number of seminal papers, for example on the Frankish blood feud and the failure of the Visigoths in France. His editing and translation were vigorous, masterly, and full of life. In 1962 appeared *The Long-Haired Kings*, a collection of papers. His study of the Merovingians and the rise of the Arnulfings who replaced them is outstanding in its sensitive use of the foremost historical sources to illuminate the horizons of kings, bishops, and the warrior society they had to try and 'correct'.

The administrative demands of his chair could only be met at the expense of his own scholarly efforts and aims. In 1961, therefore, he gratefully accepted a senior research fellowship, to which his old college, Merton, elected him. Nevertheless in 1965 he took over the onerous editorship of the *English Historical Review*, at first single-handed and from 1967 until 1974 together with his Merton colleague, J. M. Roberts. In 1969–70 he gave the Ford lectures, published in 1971: *Early Germanic Kingship in England and on the Continent*. Never before had Anglo-Saxon, Merovingian, and Carolingian history been looked at so closely together, particularly from the angle of the beliefs and rituals that grew up to sustain kings during the crisis-ridden ninth century. In 1974 he was elected Chichele professor of modern history at Oxford in succession to Geoffrey Barraclough [q.v.] and migrated to All Souls College. He had not coveted the post and was almost a reluctant

professor but as always performed his duties of office vigorously and twice served as an examiner in the final honour school.

During these years his *magnum opus*, *The Frankish Church*, took shape, published though it was shortly after his retirement in 1983. Like all his books, it is headed by a clear statement of intent. Those looking for a textbook systematically setting forth every development would look in vain. His chief themes were the spiritual centres where religious life, ideas, personalities, and learning flourished. He was a man of profound and deep-rooted faith and convictions, which can be sensed by his readers in what he wrote about the great missionaries of the seventh and eighth centuries.

He was elected a fellow of the British Academy in 1969 and received a D.Litt. in the same year. Both Merton in 1974 and Corpus Christi in 1984 elected him to honorary fellowships, and in 1982 he was appointed CBE. Next year his pupils presented him with a Festschrift and in 1984 he became a corresponding fellow of the Medieval Academy of America. His last years were devoted to a new commentary on Bede's *Ecclesiastical History*, which was nearly completed when he died.

He was a sturdy man with a kindly eye. He hated being telephoned and was laconic in his notes, all written in a memorably fine hand. A forceful writer (though sometimes elliptical and allusive), authority seemed to emanate from him and suited him naturally. His advice was much sought after: he served on the councils of the Royal Historical Society and the British Academy for many years. In a world of doubts and equivocation he conveyed a sense of certainty and direction, with his formidable and incorruptible views and criticism. He could be a hard taskmaster to his researchers, expecting punctual work and measured progress. He was all the same accessible and warm-hearted, with an irresistible and indestructible sense of humour. He knew a great deal about flowers, plants, and gardening. During his last months he was well aware of a heart condition but did not allow it to turn him away from his Bede commentary and accustomed pursuits.

In 1950 he married Anne, daughter of Neville Wakefield, schoolmaster. They had two sons. He died 3 November 1985 while working hard in his beloved garden at his home, Reynolds Farm, Cassington, near Oxford.

[*The Times*, 7 November 1985; *English Historical Review*, no. cccc, July 1986; private information; personal knowledge.]

KARL LEYSER

WALTON, SIR WILLIAM TURNER (1902–1983), composer, was born in Oldham, Lancashire, 29 March 1902, the second son in the family of three sons and one daughter of Charles Alexander Walton, the son of an Inland Revenue official, and his wife, Louisa Maria Turner, the daughter of an upholsterer. Both his parents were singing teachers, who instructed pupils at their home. Charles Walton, one of the first enrolments at the Royal Manchester College of Music, had sung oratorio and operatic roles there. As the organist and choirmaster of St John's, Werneth, he had an excellent choir, which included both William and his brother Noel.

In 1912, at the age of nine, Walton took a voice test for a probationer chorister at Christ Church Cathedral Choir School, Oxford, and, on being accepted, became a boarder at the school, remaining there for six years. Then, after being squeezed into Oxford University (Christ Church) at the age of sixteen, without much secondary education, he studied under (Sir) Hugh Allen [q.v.], the professor of music, and on 11 June 1918 passed the first part of the Bachelor of Music examination, but failed responsions at three attempts. He passed the second part of the examination on 8 and 9 June 1920.

While an Oxford undergraduate, Walton completed the writing of a string quartet and a piano quartet. After being revised in the early 1920s, the first of these was performed at the festival of the International Society of Contemporary Music at Salzburg on 4 August 1923; the latter gained a publication award from the Carnegie Trust Fund in 1924. Among his boyhood compositions were three notable works: *Tell Me Where is Fancy Bred* (1916), a Choral Prelude on *Wheatley* for organ, and *A Litany*, a setting of a poem by Phineas Fletcher [q.v.]. Despite early influences acquired from the study of other composers, Walton's own characteristics soon showed in his music.

During this period he made enduring friendships with the poets Siegfried Sassoon and I. Roy Campbell [qq.v.], and the novelist Ronald Firbank. He also met (Sir) Sacheverell Sitwell, who, designating him a musical genius, brought his brother (Sir) Osbert Sitwell [q.v.] to Oxford in February 1919 to meet him and to hear him play his Piano Quartet. Osbert Sitwell was greatly impressed by the music, and as a result the young composer, after leaving Oxford, went to live with the Sitwells at Swan Walk, Chelsea, and at Osbert's house, 2 Carlyle Square, London, in effect becoming another member of the family.

His 'adoption' by the Sitwells, already achieving some literary fame, opened up exciting vistas of opportunity. Combining with Dr Thomas Strong [q.v.], the dean of Christ Church, and with Lord Berners [q.v.] and Sassoon, in guaranteeing him an income of £250 a year, they enabled him to spend his life composing. By introducing him to famous writers, musicians, and painters, they helped to broaden his cultural and social outlook. In the spring of 1920 and on subsequent occasions, he visited Italy with his benefactors; an experience which undoubtedly influenced his music. If he was short of money, his generous friend Sassoon provided it. He even found some for the impoverished Baroness Imma Doernberg, with whom Walton had a romantic association. Born the Princess Imma of Erbach-Schönberg on 11 March 1901, she was the daughter of Alexander, Prince of Erbach-Schönberg. Fortune smiled on the composer in 1932 when Elizabeth Courtauld, wife of Samuel Courtauld [q.v.] the industrialist, died, bequeathing him a life annuity of £500.

Walton had gone to live with the Sitwells in 1919 and he remained with them until 1934, when, having started a romance with Alice, Viscountess Wimborne, he moved into her house. The younger daughter of the second Baron Ebury and a cousin of the Duke of Westminster, she married Ivor Guest, first Viscount Wimborne, heir to the Guest Steel fortune, in 1902. Walton's affair with Imma Doernberg seems, at some time in 1933, to have gone adrift. While staying with the Sitwells, he had composed several works, including *Façade*, *Portsmouth Point*, the *Sinfonia Concertante*, the Viola Concerto, the First Symphony, and *Belshazzar's Feast*.

Drawn to the world of film music by Dallas Bower, he was commissioned by the British and Dominion Film Corporation in 1934 to write a score for *Escape Me Never*, in which the young (Dame) Margot Fonteyn appeared. The fees for this and other commissions helped him to purchase his own house in Eaton Place, London. In 1935 he collaborated with (Sir) C. B. Cochran [q.v.] by writing a short ballet, *The First Shoot*, which formed part of *Follow the Sun* (1935), a spectacular review. After King George V died, Walton was commissioned by the BBC to provide a Symphonic March for the coronation of King George VI in Westminster Abbey on 12 May 1937, and he wrote *Crown Imperial*, which was played for the entry of Queen Mary just before the ceremony began.

He was in America in the spring of 1939, completing the Violin Concerto commissioned by Jascha Heifetz, who gave him some useful technical advice. In the same year Paul Czinner's film *A Stolen Life*, for which Walton had composed music, had its première at the Plaza Theatre, London.

On 28 December 1940, during World War II, he finished work on his comedy-overture *Scapino*, and, about the same time, was attached to the films division of the Ministry of Information as a composer. Walton wrote several film scores for the MOI and these included *The Next of Kin* (1942), *The Foreman Went to France* (1941), *The First of the Few* (1942), and *Went the Day Well?* (1942). Released by the MOI in 1945, he completed his Second String Quartet; then, towards the end of the year, he accepted an invitation from six Scandinavian orchestras to tour Scandinavia, conducting his own music. Imma Doernberg died on 14 March 1947 and Lady Wimborne on 17 April 1948, the latter leaving him £10,000, Lowndes Cottage in Westminster, and various effects. Shortly after settling in a new home on the island of Ischia, off the coast of Italy, in 1949, Walton started work on his opera *Troilus and Cressida*, completing it in 1954. Following the death of King George VI in 1952, he composed a coronation march, *Orb and Sceptre*, and a *Te Deum*: these were performed at the Coronation of Queen Elizabeth II in Westminster Abbey on 2 June 1953.

In the years that followed, Walton wrote many fine works. They included such orchestral pieces as the film music for Laurence (later Lord) Olivier's *Richard III* (1956), the *Johannesburg Festival Overture*, the Cello Concerto, the *Partita* for Orchestra, and the Second Symphony. He composed a one-act opera, or extravaganza, *The Bear*, and among the choral and vocal pieces were *Anon in Love* (1960), *A Song for the Lord Mayor's Table* (1962), a Missa Brevis, and a Magnificat and Nunc Dimittis (1975).

Walton toured the United States in 1955 and appeared there for the first time as a conductor, leading a performance of his *Crown Imperial* at the United Nations. In 1963 he made a return visit, conducting an all-Walton concert at the Lewisohn Stadium, New York.

In January 1957, while travelling along a road near Rome, he was involved in a car crash, sustaining a cracked pelvis and other injuries. He made his first visit to Canada in February 1962 and shared a Canadian Broadcasting Corporation programme with the American composer Aaron Copland and Louis Applebaum from Canada; then in June the same year he was in Los Angeles for the American première of his *Gloria*. He went to Israel in 1963 and conducted three concerts of his own music there, one taking place in July, when the first performance of *Belshazzar's Feast* in Hebrew was given by the Tel Aviv Choir and the Israeli Philharmonic Orchestra in Tel Aviv, Haifa, and Jerusalem. In the first half of 1964 he toured New Zealand and Australia, again acting as an ambassador for, and conductor of, his own compositions. Commissioned by United Artists Ltd. to write music for a new film, *Battle of Britain*, he showed bitter anger when, after he completed

the score in February 1969, it was rejected because there was not enough music to fill a long-playing record.

Walton visited Russia for the first time in 1971, accompanying André Previn and the London Symphony Orchestra, whose performance of his First Symphony before a Moscow audience was described by the *Financial Times* as 'a phenomenal success'. His seventieth birthday was celebrated in Britain by special concerts and tributes from press, radio, and television. Edward Heath, then prime minister, gave a concert for him at 10 Downing Street on 29 May 1972, with the Queen Mother heading the distinguished guests.

Walton's manifestations of ill health surfaced alarmingly in November 1976, when, after attending all the performances of *Troilus and Cressida* at the Royal Opera House, he returned to Ischia, showing definite signs of a stroke, and became a very sick man, unable to work. He made only a partial recovery, but, with characteristic gallantry, visited Britain in March 1977 and 1982, for the London concerts given to celebrate his seventy-fifth and eightieth birthdays. Back on Ischia, a semi-invalid in a wheelchair, with a restricted range of movement and energy, Walton continued to compose in a small way, but the end was near.

To look back over his life is to realize the lofty scale of his achievements. Like Sir Edward Elgar [q.v.], he was largely self-taught as a composer, but, like Elgar, he became a supreme professional. For many people, *Façade*, skittish, catchy, and beautiful by turns, marked him as the brilliant English counterpart of the Parisian playboys of the 1920s, and although this work, which uses strange, capricious poems by (Dame) Edith Sitwell [q.v.], was booed and hissed at its première, it was soon recognized as a concept of rarest originality. The public waited for Walton to repeat the phenomenon. He never did. Between the first performance of *Façade* in 1923 and that of the *Sinfonia Concertante* in 1928 he averaged only one small piece a year, the most impressive of which was *Portsmouth Point* (1926).

The poet in Walton came into his own with the Viola Concerto (1929), when he could with complete conviction write what the critics and music lovers of the day seemed to want. This composition, closely modelled on Elgar's Cello Concerto, has maintained its pre-eminence, like the Violin Concerto (1939), a miracle of delicate, haunting lyricism and elegiac feeling. Between Walton's two concertos came *Belshazzar's Feast* (1931), whose high-flying drama and savage ferocity, offset by passages of songful serenity, broke the bonds of conventional oratorio.

Most of the great masters of the previous generation had avoided the symphony, arguing that it had no relevance to the musical situation of the day, but there was in the early 1930s a definite movement backwards towards this large-scale form. Walton felt sure that he could speak eloquently through the symphony, but his confidence had been shaken by the rather cool response to *Belshazzar's Feast* shown at the 1933 ISCM Festival. Despite doubts and difficulties, his First Symphony proved to be a masterpiece. He was, for a time, unable to find a satisfactory solution to the problem of writing the final movement, and on 3 December 1934 he allowed the symphony to have a première in its unfinished state. Almost a year elapsed before he completed the last movement, and not until 6 November 1935, in London, did the public hear the full score. The tragic nature of the work, with a scherzo marked 'Presto con malizia', reveals the influence of Sibelius.

Walton wrote music for fourteen films, of which *Dreaming Lips* (1937) is not usually included in his list of credits. The score for *Henry V* (1944) is magnificent. He drew a fine line between true 'background' music and those elements of musical pastiche that were needed to evoke the atmosphere of a particular historical period. The most thrilling facet is the Agincourt battle sequence, where the sound effects of horses' hooves, rattling harness, and clinking armour, make the charge of the French knights fearsomely real. Outstanding among the film scores which he wrote for the Ministry of Information during World War II is *The First of the Few* (1942), from which the 'Spitfire Prelude and Fugue' was later published as a separate concert piece.

The comedy-overture *Scapino* (1941), a brittle portrayal of a rascally character of the *commedia dell'arte*, contains the best of all the exhilarating tunes he ever wrote. Both this work and *Portsmouth Point* are superior to the *Johannesburg Festival Overture* (1956). One of his most glorious achievements at this time was the massive and picturesque score which he created for the BBC's radio drama *Christopher Columbus*, used in a transatlantic broadcast on 12 October 1942, the 450th anniversary of the great explorer's first voyage to America.

Walton's String Quartet in A minor (1947) has an air of easy composure, despite the rhythmic vitality of the scherzo and the finale, while the slow movement has a gentle inwardness that emphasizes the Ravelian aspect of much of his art. Like the First Symphony, it was not ready for the publicized première. It seems that, at different stages of his life, unease and doubt over the direction he was taking interrupted his flow of inventiveness.

Uncertainty and discontent haunted him also while he was composing *Troilus and Cressida* (1954), and sometimes irascibility surfaced between him and his librettist, Christopher Hassall [q.v.]. He knew that there was a feeling of disillusionment about English opera, *Gloriana* by Benjamin (later Lord) Britten [q.v.] and *The Midsummer Marriage* by (Sir) Michael Tippett having been poorly received. *Troilus and Cressida*, a love story in the grand manner, with a Chaucerian text, struck some critics as being old-fashioned, but Walton's passionate score, full of gorgeous harmonies and luminous orchestration, inspired much praise after the initial reaction.

His second coronation march *Orb and Sceptre* (1953), enriched by music of coruscating brilliance, proved to be more complex than the previous one, *Crown Imperial* (1937), a work of simple, diatonic grandeur. The other new coronation piece he composed, the *Te Deum* (1953), covered a big expanse of sound and, in its noble utterance, outshone most of the remaining choral works that were sung on this great occasion. Walton's Cello Concerto (1957) also captivated the listener, revealing a warm, Italianate glow and a really fresh invention.

In complete contrast to its mighty predecessor, his Second Symphony (1960) showed him in a more subdued, reflective vein: gone were the truculence, the spiky rhythms, and the restless unease of former years. Because of this, some people complained that he had taken a wrong turning, that he had failed to 'advance' in his style. But after the Cleveland (Ohio) Orchestra, under George Szell, made their splendid recording of the symphony, it was recognized as a feat of orchestral virtuosity which few, if any, British composers could match. The music not only represented a natural development from the preceding period, but introduced entirely fresh ideas that were treated in a quite different way.

Walton's second opera, *The Bear* (1967), has a witty libretto with rhyming lyrics by Paul Dehn based on a short play of the same title by Chekhov, and is a comedy of manners, not of plot. His pungent, high-spirited score never holds up the pace of the merry making, and there is fun for the listener in identifying in the music certain droll parodies of contemporary composers.

He used a spare melodic style of scoring in *Improvisations on an Impromptu of Benjamin Britten* (1970), just one of a stream of finely crafted works too numerous to catalogue fully here. They include the dreamy, skittish miniature *Siesta* (1926), piano pieces *Duets for Children* (1940), a Spenserian ballet *The Quest* (1943), the exquisite setting of the poem by John Masefield [q.v.] *Where Does the Uttered Music Go?* (1946), the *Partita* for Orchestra (1958), the *Gloria* (1961), and the elegant, poetic *Variations on a Theme of Hindemith* (1963).

Walton was always a slow, painstaking perfectionist, who revised a number of his scores after publication.

At a time when a reaction against nineteenth-century romantic music had set in, he had to work extremely hard to find a personal idiom and he did, in fact, create a tone and a rhythm that were unmistakably his own. His pre-1939 compositions were very popular in the concert halls of Europe and America: so that, in winning a reputation abroad, he also acted as an ambassador for his own country. Walton introduced no pioneering techniques into his music, but he demonstrated, most eloquently, that to scale lofty heights a man of genius looks, not necessarily for new forms, but simply for the best means to express his own ideas.

As he grew older, during the post-1945 phase of his career, the inventive quality in his music declined, but he became more prolific, with a greater variety of expression. One example is the *Capriccio Burlesco* (1968), with its sly musical gesticulations and saucy ideas. Walton could not possibly have written this in his so-called *enfant terrible* days.

In the matter of gramophone recordings, Walton conducting Walton became a revelatory, as well as definitive, experience. Sir Eugene Goossens [q.v.], renowned for his brilliance in handling contemporary scores, taught the young composer a technique for mastering the swift nervous changes of rhythm in his own music. Most of Walton's recordings, especially those made for Columbia in the 1950s, are bright, unpretentious, and vibrant with life.

Despite occasional flashes of anger and hostility, there were many appealing facets to his character. He had an endearing habit of self-denigration and once described *Belshazzar's Feast* as 'a beastly noise'. There was no pomposity in his make-up. He loved to tease his friends. A very private person, he often retreated into a haven of brooding silence when questioned about his music or his views. Walton had a curious way of smoking his pipe, balancing it precariously on his lower lip in the centre of his mouth.

He was knighted in 1951, given the freedom of the borough of Oldham in 1961, and admitted to the Order of Merit in 1967. He held honorary doctorates from the universities of Durham (D.Mus. 1937), Oxford (D.Mus. 1942), Dublin (D.Mus. 1948), Manchester (D.Mus. 1952), Cambridge (D.Mus. 1955), London (D.Mus. 1955), and Sussex (D.Litt. 1968). He had a number of honorary fellowships, including those at the Royal College of Music (1937), the Royal Academy of Music (1938), and the Royal Manchester College of Music (1972). Appointed an honorary member of the Royal Swedish Academy of Music (1945) and an Accademico Onorario di Santa Cecilia (Rome) (1962), he was awarded the Benjamin Franklin medal (1972), the gold medal of the Royal Philharmonic Society (1947), and the medal of the Worshipful Company of Musicians (1947).

In 1948 he married Susana Valeria Rose Gil Passo, daughter of Enrique Gil, a prosperous Buenos Aires lawyer. After their marriage the couple settled on the island of Ischia, where 'La Mortella', a beautiful villa with an exotic garden, was built specially for them. Walton died there 8 March 1983. He had no children.

[Neil Tierney, *William Walton: His Life and Music*, 1984; Susana Walton, *William Walton*, 1988; Michael Kennedy, *Portrait of Walton*, 1989; information from Lady Walton and Dr Stewart R. Craggs; private letters and documents.]   NEIL TIERNEY

WANSBROUGH-JONES, SIR OWEN HADDON (1905–1982), chemist and industrialist, was born 25 March 1905 at Attleborough, Norfolk, the youngest of three sons and the fourth and last child of Arthur Wansbrough-Jones, solicitor, of Long Stratton, and his wife, Beatrice Anna, daughter of Thomas Slipper, farmer, of Braydeston Hall. His brothers were regular army officers: Harold, in the Central India Horse, reached the rank of colonel and Llewelyn, a Royal Engineer, that of major-general. He was educated at Gresham's School, Holt, whence he won an open scholarship in 1923 to Trinity Hall, Cambridge. He secured first class honours in parts i and ii of the natural sciences tripos (1925 and 1926). He then became a research student financed by the Goldsmiths' Company, working in colloid science under Professor (Sir) Eric Rideal [q.v.]. In 1930 he took his MA and Ph.D., was elected a fellow of Trinity Hall, and then spent a year working under Fritz Haber in Berlin. On his return in 1932 he was appointed assistant tutor and, in 1934, tutor in his college. These were inspired appointments; while demanding the highest academic standards, Wansbrough-Jones was greatly liked and admired by undergraduates.

When war came he decided that his place was in the army, and he was commissioned in January 1940. He held a number of technical and general staff appointments and, by the time of his demobilization in 1946, had become a brigadier (1945) as director of special weapons and vehicles. He was appointed MBE (1942) and OBE (1946).

He fully intended to return to Cambridge but instead accepted an invitation to be scientific adviser to the Army Council in 1946. With his 'feel' for the army and his knowledge of the War Office, he proved the right man for the job. He established operational research sections in the British Army of the Rhine and Malaya and fought successfully to enhance the standing of the Royal Military College of Science in the academic world. He was appointed CB in 1950. He transferred to the Ministry of Supply in 1951 and became chief scientist in 1953. He believed in the Ministry of Supply and saw his main task as that of deploying its immense scientific potential to maximum advantage. He managed the research programme with skill and tact, and went to endless trouble over senior appointments. He fostered *esprit de corps* among the senior staff, notably by arranging regular informal dinners. He was appointed KBE in 1955.

In 1959 he accepted an invitation from Albright & Wilson to become executive technical director and joined the company at a challenging time of rapid growth. His ability to enthuse research teams in widely different fields of chemistry stood him in good stead and he consistently supported projects where the company had a proven track record. He became vice-chairman in 1965 and was chairman from 1967 until he decided to retire in 1969 after ten years with the company.

Wansbrough-Jones had a lifelong interest in education. He served as a governor of Gresham's from 1938 until 1979 and, in recognition of his distinguished services, the Fishmongers' Company (with which the school is closely associated) made him an honorary freeman in 1973. He became a freeman of the Goldsmiths' Company in 1950 (so enabling him to call himself proudly a 'goldfish'), and was prime warden in 1967-8 and a formidable chairman of the company's education committee from 1967 to 1981. He was treasurer of the Faraday Society (1947-59), a director of BOC (International) in 1960-76, and a governor of Wellington College (1957-75) and of Westminster School (1965-80). He continued to have close links with Trinity Hall which elected him honorary fellow in 1957.

He was throughout a man of uncompromising honesty, integrity, and loyalty, qualities which owed much to the influence of his mother, a remarkable woman of profound faith, who lived to 102. Although shy, he liked people, loved a party, and had a flair for making and keeping friends, many of whom turned instinctively to him for advice. He was a man of infinite kindness. He remained a countryman at heart, proud of his Norfolk roots, and loved a day's shooting. While living in London, he continued to return at weekends to the family home and his beloved garden. He was a recognized authority on silverware, owning perhaps the best Omar Ramsden collection in the country. He was a lover of music, especially of Beethoven string

quartets to which he listened—while following the score—on his ancient gramophone, complete with outsize horn.

He did not enjoy the best of health in his last few years and retired to Long Stratton where he died 10 March 1982. He was unmarried.

[*The Times*, 3 and 19 March 1982; private information; personal knowledge.]

GEORGE LEITCH

WARBURG, SIR SIEGMUND GEORGE (1902-1982), banker, was born 30 September 1902 at Urach, southern Germany, the only child of Georg Siegmund Warburg, landowner, of Urach, and his wife Lucie, daughter of Max Kaulla, a Stuttgart lawyer. He was educated at the Humanistic Gymnasia first of Reutlingen and later of Urach.

Warburg entered the family bank while still seventeen at the invitation of the senior partner, his father's first cousin Max Warburg. He spent three years as a trainee and followed these with prolonged stages at banking and trading houses in London and New York with which there were family or business connections.

M. M. Warburg & Co., established in Hamburg since the eighteenth century, was by the 1920s one of the leading German private banks. It was one of the first German houses to recover from the financial ruin of World War I and the subsequent inflation, and the political leaders of the Weimar Republic frequently looked to Max Warburg for advice. Max encouraged the young Siegmund to meet such men as Walther Rathenau, Gustav Stresemann, and Hjalmar Schacht; he thus developed an early taste for politics and politicians which never left him.

Max made Siegmund a partner in 1930 and deputed him to open an office in Berlin. There he established many connections in banking, industry, and politics. In 1931 the firm got into difficulties. When Hitler came to power in 1933, Jewish bankers suffered growing discrimination and in Siegmund's case all these problems were compounded by an increasing estrangement from Max. In 1933 he took what little capital he could out of Germany and moved to London.

With the backing of Berlin friends and English relations Warburg founded a finance company in the City, the New Trading Co. Ltd. His own financial stake was relatively small. The years to 1939 were spent building up a small team of like-minded bankers, mostly of German-Jewish *émigré* origin, to become the nucleus of a new merchant bank. The beginnings were modest. In 1946 the company adopted the name of S. G. Warburg & Co. Ltd., won banking recognition, and rose over the next thirty-five years to become London's foremost merchant bank. Warburg was naturalized in 1939.

Unable initially to break into the established clientele of the older merchant banks, Warburgs (as it became known) had to make do with financing small companies, many founded by *émigrés* from Germany, experimenting with new techniques such as take-over bids and developing specialized business in the byways of international finance. All the while Warburg travelled unceasingly round western Europe, the United States, and (later) Japan, remaking pre-war connections and establishing new ones.

The first breakthrough came in 1958 with the successful assault on the British Aluminium Co. Ltd. by an Anglo-American group advised by Warburgs and another London house. The prolonged battle attracted extraordinary attention amongst politicians and the general newspaper readership and Warburgs showed great skill in its public relations. The outcome was regarded as a triumph for Warburgs and for Siegmund Warburg over the conservative City establishment.

There followed an ever increasing volume of take-over bids, flotations, capital issues, and creative financial deals involving British and foreign companies and government agencies. The firm trebled in size, moved to new premises and with the careful recruitment of native British and other European talent ceased to be regarded as a purely German-Jewish house.

Warburg's most significant contribution to world finance was the relaunch of international private lending out of Europe, stalled since 1929. The Eurobond market was initiated with a Warburg issue for an Italian government borrower in 1963. With its sister, the Euroloan market, it grew over the next twenty-five years to become the main international source of funds for businesses and governments in the free world.

Warburg was tall (though stooped after middle age) and dark and endowed with classical Jewish good looks. To his friends and business associates he was a man of charm, wisdom, and intelligence. His staff viewed him with pride and fear and many were victims of his ferocious rages. He wrote faultless English but spoke it with a heavy German accent.

Although he made England his home from 1933 till he retired to Switzerland in 1973 he never properly understood the English. Hating inefficiency and complacence, he often mistook English informality for sloppiness and English reticence for self-satisfaction. Consequently some of his judgements on British businessmen and politicians were flawed. He detested social gatherings and was happiest travelling and meeting old friends in health resorts. He wrote letters and telephoned all over the world, even after his nominal retirement.

In 1925 he married Eva Maria, daughter of Mauritz Philipson, of Stockholm, a Swedish banker; they had a son and a daughter. He was knighted in 1966 and received the Japanese Order of the Sacred Treasure in 1978. He died in London 18 October 1982.

[Jacques Attali, *Un homme d'influence*, 1985; Hjalmar Schacht, *76 Jahre meines Lebens*, 1950; personal knowledge.]   IAN FRASER

WARD, BARBARA MARY, BARONESS JACKSON OF LODSWORTH (1914-1981), journalist and broadcaster, was born in York 23 May 1914, the younger child and only daughter of Walter Ward, later solicitor to the port of Ipswich, and his wife, Teresa Burge. Her father was a Quaker and her mother came from a well established Catholic family. She was reared to become a devout Catholic, which she remained all her life, though she also held to a stubborn belief, more often associated with Quakers, that human beings, particularly those adequately financed, should dedicate themselves to creating a better world.

Educated at the Convent of Jesus and Mary at Felixstowe, the Lycée Molière and the Sorbonne in Paris, and in Germany, she then went to Somerville College, Oxford, where she obtained first class honours in philosophy, politics, and economics in 1935. From 1936 to 1939 she was a university extension lecturer, and she became an active member of the Labour Party. Geoffrey (later Lord) Crowther [q.v.], who was building up the fame of the *Economist*, recruited her in 1939 to be assistant editor, then foreign editor. Her exceptionally witty performances in the popular 1940s BBC Brains Trust made her a national institution before she was thirty. Between 1946 and 1950 she was a governor of the BBC; she was also appointed to the boards of Sadler's Wells and the Old Vic (1944-53).

Barbara Ward's rare assortment of talents— literary, oratorical, conceptive, pedagogic, and musical—combined with an encyclopaedic memory, charm, and radiance, enabled her to stamp her thoughts and feelings on a generation. She was a great beauty: in repose her perfectly proportioned face suggested a medieval madonna, though it was frequently enlivened by a gay, sometimes mocking, smile. Her concern over social injustice inclined her to the left but, instead of entering politics, she preferred to exercise what became effective power by influencing the world leaders who made the decisions. Among her personal friends were two presidents of the USA, two popes, several prime ministers, and many editors, scholars, and leading international civil servants. She was equally generous, it should be added, with time and concern for socially unimportant friends.

Her many books (from *The International Share-out*, 1938, to *Progress for a Small Planet*, 1980) are eloquent pleas for her principal causes: the rich countries' obligation to give to the poor ones, the direction of scientific and technological advance towards preserving rather than destroying natural resources, and the supremacy of international over national considerations as the only way of saving the world from the imminent threat of nuclear catastrophe. In 1948, while the Marshall Plan was still being negotiated, she warned that, unless the Americans and Europeans stopped trying to restore the past and co-operated to create a new society, fairer than anything the communists claimed to offer, 'We are for the dark.' On the tenth anniversary of the Marshall Plan, while receiving an honorary degree from Harvard University (1957), she launched her constantly to be repeated plea for a global version of the Plan, in which hard currency would be transferred to the economically backward countries.

In 1950 she married Commander (Sir) Robert Gillman Allen Jackson, who worked for the United Nations and who encouraged her preoccupation with the southern hemisphere. He was the son of Archibald Jackson, an engineer, of Melbourne, Australia. They lived from 1956 to 1961 in Ghana where Jackson was chairman of the Development Commission. They had one son. The marriage did not last and, though Barbara Ward's faith precluded divorce, there was a legal separation in 1966.

Barbara Ward travelled widely in India and Pakistan and shared her husband's conviction that all communities, with adequate financial and technical outside aid, could acquire the prerequisites to sustain their own economic development. She was visiting scholar (1957–68) and Carnegie fellow (1959–67) at Harvard University, where she became a distinguished teacher. In 1967 she was appointed a member of the Pontifical Commission for Justice and Peace, a job which required frequent visits to Rome. From 1968 to 1973 she was the Albert Schweitzer professor of international economic development at Columbia University, New York, and from 1973 president of the International Institute for Environment, to which she insisted on adding 'and Development'. In 1980 she became the Institute's chairman.

Towards the end of her life, increasingly obsessed with what she saw as acute threats to human survival—exhaustion of natural resources, pollution, and overpopulation—she abandoned her earlier preoccupation with individual values, dismissing ideological controversy as irrelevant and arguing that the communist and non-communist systems were converging towards interdependence and active collaboration. She also changed her views on nuclear power: having in 1972 declared that the world's expanding population needed it, she later claimed it could easily be replaced by energy conservation—and installed a solar contraption on her own roof.

She was awarded honorary degrees by many universities, became honorary FRIBA in 1975, and an honorary fellow of LSE in 1976. She won the Jawaharlal Nehru prize (1974) and the Albert medal of the Royal Society of Arts (1980). She was appointed DBE in 1974 and became a life peer in 1976.

In the last twenty-five years of her life she suffered cancer and endured five major operations. But neither unremitting pain nor closeness to death ended her jet-propelled peripatetic life. She died 31 May 1981 in her last and favourite home, in the village of Lodsworth, overlooking the Sussex Downs. She had persuaded Pope John Paul to allow her to be buried at the Anglican parish church and he appropriately sent Cardinal Gantin, a West African, to deliver the papal blessing.

[Obituary by Roland Bird in the *Economist*, 6 June 1981; unpublished autobiography by Irene Hunter (Barbara Ward's secretary); personal knowledge.]   NORA BELOFF

WARD-PERKINS, JOHN BRYAN (1912–1981), archaeologist, was born 3 February 1912 at Bromley, Kent, the elder son (there were no daughters) of Bryan Ward-Perkins, of the Indian Civil Service, and his wife, Winifred Mary Hickman. He was educated at Winchester and New College, Oxford, where he graduated with a first class in *literae humaniores* in 1934, going on to hold the university's Craven travelling fellowship at Magdalen College. The groundwork of his career in archaeology was laid at this time, with excavation and pottery studies in Britain and France. In 1936 he took up a post as assistant at the London Museum under the direction of (Sir) R. E. Mortimer Wheeler [q.v.], producing a notable catalogue of the Museum's medieval collection, besides directing his own excavations at Oldbury hill fort, near Ightham, and at Lockleys Roman villa, near Welwyn Garden City. His work was already remarkable for its range and for the fresh insights which it introduced; the publication of the Lockleys excavation remains a *locus classicus* for the use of archaeology to trace the history of a small Roman farm in south-east Britain.

In 1939 he went to take up a chair in archaeology at the Royal University of Malta but had only been there six months when war broke out, and he returned to England to enlist in the Royal Artillery (Territorial Army). His war service took him to North Africa and ultimately to Italy, where as a lieutenant-colonel, he participated in the Allied invasion. His archaeological expertise and organizing ability were recognized in his appointment as director of the Allied sub-commission for monuments and fine arts in Italy, a post which involved the salvage of artworks dispersed and damaged in the course of the war.

In 1946 Ward-Perkins became the first postwar director of the British School at Rome, where he stayed till his retirement in 1974. His twenty-nine years at the helm of a small, everchanging community of British artists and academics in Italy gave him the opportunity to produce much of his finest work. During the first decade he orchestrated programmes of survey and excavation in Cyrenaica and Tripolitania, writing seminal papers and books on the art and archaeology of a region whose richness in monuments had been revealed by Italian excavations in the 1920s and 1930s. At the same time his concern with Roman architecture, and especially with its technical and organizational aspects, led to important studies on the necropolis under the Vatican basilica (*The Shrine of St Peter and the Vatican Excavations*, with J. M. C. Toynbee, 1956) and on Roman brick construction in *The Great Palace of the Byzantine Emperors* edited by D. Talbot Rice (1958). In the last two decades in Rome his major achievement was the organization of the south Etruria survey, a project for the recording and analysis of remains of all periods within a region where the archaeological evidence was rapidly being destroyed by modern development. This type of systematic field survey, which evolved from the earlier topographical studies of Thomas Ashby [q.v.] to the south and east of Rome, helped to confirm the prestige of the British School as the main centre for research into the historical landscape of Italy.

In his later years Ward-Perkins was involved in the organization of international projects such as the 'Tabula Imperii Romani', a scheme to produce detailed maps for all regions of the Roman empire; he promoted research into the identification and classification of Roman marbles, a vital prerequisite for the understanding of the building industry and of sarcophagus production; and he organized a major exhibition on Pompeii at the Royal Academy in 1976, which later travelled to various cities in America and Australia. His mature thoughts on Roman architecture were set out in his Pelican History of Art volume, originally issued in tandem with Axel Boethius's survey of the Etruscan and early Roman material, but later revised and published separately as *Roman Imperial Architecture* (1981).

Ward-Perkins towered above the Roman archaeologists of his generation. Many of his approaches, as in the use of field survey to study patterns of land use over the ages, and his emphasis on the importance of data on materials and techniques to the understanding of sculpture and architecture, have radically affected the thinking of researchers in the area. He was generous in his encouragement of younger scholars and particularly successful in his recognition of amateur talent. In most of those who knew him he inspired great loyalty and affection; he was a charismatic figure with an expansive personality, and his impatience with those who could not see issues as clearly as he did was readily forgiven. His personal interests—gardening, stamp-collecting, music—were pursued with the same dedication and attention to detail as his professional activities.

His work was rewarded with election to fellowship of the British Academy (1951), honorary degrees from the universities of Birmingham (D.Litt., 1961) and Alberta (LLD, 1969), and appointment as CBE (1955) and CMG (1975). His overseas honours included membership of various academies and election to the presidency of the International Association for Classical Archaeology (1974-9).

In 1943 he married Margaret Sheilah, daughter of Henry William Long, lieutenant-colonel in the Royal Army Medical Corps. They had three sons and a daughter. Ward-Perkins died in Cirencester 28 May 1981.

[*The Times*, 5 June 1981; *Papers of the British School at Rome*, vol. l, 1982; J. J. Wilkes in *Proceedings* of the British Academy, vol. lxix, 1983; private information; personal knowledge.]   ROGER LING

WARNER, JACK (1895-1981), actor, was born Horace John Waters, 24 October 1895 in Bromley, London, the third child and second of four sons among the children of Edward William Waters, master fulling maker and undertaker's warehouseman, and his wife, Maud Mary Best. His two sisters Elsie and Doris Waters became the successful radio and variety comedians 'Gert and Daisy' in the 1930s and 1940s. He was educated at the Coopers' Company School, Mile End Road, and studied automobile engineering for one year at the Northampton Institute, now part of the City University, London. Warner was essentially a practical man, more at home with pistons and people than with books, and left to work in the garage of a firm of undertakers in Balham. In August 1913 he went to work as a mechanic in Paris where unusually for a boy of his background he acquired a working knowledge of French which stood him in good stead throughout his life socially and as an entertainer. An imitation of Maurice Chevalier, in some ways his Parisian opposite number, became a standard part of his repertoire.

During the war he served in France as a driver with the Royal Flying Corps (later the RAF), being awarded the meritorious service medal in 1918. He returned to England and the motor trade in 1919, graduating from hearses to occasional car racing at Brooklands. He was over thirty before he became a professional entertainer, having progressed from choirboy and wartime concert party performer through the Sutton Amateur Dramatic Club to cabaret work, and making his West End début in 1935 in the two-man act of Warner and Darnell. He changed his name to Warner at this point because he did not wish to appear to be resting on the reputation of his sisters. In December 1939 with BBC radio's *Garrison Theatre* he made the transition from cabaret singer to cockney comedian. He epitomized the patient, good hearted, cheeky 'Tommy' of World War I, reborn in 1939 and transferred from the music hall to the broadcasting studio, and later matured into the reliable London bobby as PC Dixon of Dock Green in television. During the war he was a regular performer in radio and stage variety shows. In 1942 he made his first film, *Dummy Talks*.

The Jack Warner father figure emerged in his fourth film, *Holiday Camp* (1947). He played Mr Huggett, with Kathleen Harrison as his wife; they were typical, if romanticized, cockney parents, coping with adversity, often in the shape of their own children. The Huggetts featured in three more films and in a radio show from 1953 to 1962, an everyday story of urban folk. Warner's acting talent and ambitions were limited but although he never aspired to play Hamlet he did hope occasionally to break away from his stereotyped roles. He succeeded in this with more serious films such as *The Captive Heart* (1946), *It Always Rains on Sunday* (1947), and *Against the Wind* (1947). In *The Final Test* (1953) he played a professional cricketer in his last great outing, exchanging his senior NCO role with that of a figure modelled on Sir J. B. (Jack) Hobbs [q.v.]. But after a comparatively small part in *The Blue Lamp* (1949), as a fatherly London policeman shot by a young criminal, Jack Warner was enrolled as the regular screen parental police officer, PC Dixon, in a television series created by E. H. ('Ted', later Lord) Willis that ran from July 1955 until 1976. He was a reassuring traditional officer of the law adapting his pre-war standards and wisdom to the different world of the 1960s. The series altered its style as English society changed. PC Dixon became a less cosy sergeant, but the advent of a harsher view of police life in *Z Cars* brought an end to *Dixon of Dock Green* in 1976.

The series brought Jack Warner fame and financial security, which he enjoyed with Muriel Winifred ('Mollie'), a company secretary whom he married in 1933. She was the daughter of Roberts Peters, of independent means. The Warners had no children. Warner carried on with some stage work until 1980 and died in the Royal Masonic Hospital, Ravenscourt Park, London, 24 May 1981.

Jack Warner was a tall, handsome man who possessed the solid virtues which he portrayed throughout his career. He had no formal training as an actor and performed rather than acted in a style that was ideal for radio, film, and television but was not suited to theatrical work. The character that he developed, a dependable soldier, family man, and policeman, growing from cockney irreverence to maturity, will be interesting for social historians as a picture of the working-class hero of the first half of the twentieth century, romanticized but not unreal. He was appointed OBE in 1965 and was made an honorary D.Litt. by City University in 1975. In 1972 he became a freeman of the City of London.

[Jack Warner, *Jack Of All Trades*, 1975 (autobiography).]　　　　　　D. J. WENDEN

WATERHOUSE, Sir ELLIS KIRKHAM (1905-1985), art historian and gallery director, was born in Epsom, 16 February 1905, the only child of (Percy) Leslie Waterhouse, architect, of Epsom, and his wife, Eleanor, daughter of William Margetson, of Streatham. From Marlborough he won a scholarship in classics to New College, Oxford; he took a first class in classical honour moderations (1925) and a second in *literae humaniores* (1927). His interests had focused irrevocably on art, and as Commonwealth Fund fellow at Princeton (1927-9) he was inducted into the rigorous standards of first-class art-historical scholarship. In contrast, as an assistant in the National Gallery, London, from 1929, he was shocked by the amateurish approach and bureaucracy, and resigned in 1933. Installed as librarian at the British School in Rome till 1936, he sought out neglected baroque paintings wherever they were to be found. His *Baroque Painting in Rome, the Seventeenth Century* (1937, revised 1976), supplemented by *Italian Baroque Painting* (1962), were major pioneering achievements. In England, 1936-9, he worked notably on British art and, for the Royal Academy, on the exhibition of seventeenth-century art in Europe (1938).

A research fellowship at Magdalen College, Oxford (1938-47), was interrupted by war, which caught him in Greece. After some cartographical work there for the British legation, he was commissioned in GHQ Cairo into the Intelligence Corps, and took part in the 1941 campaign from Macedonia to Crete. Later he was seconded to the staff of the British ambassador in Cairo to the Greek government in exile (1943-4). His intelligence, and his diplomatic and tactical skills (though not mentioned, characteristically, by himself) have been given much of the credit for the Lebanon all-party conference of May 1944, and its accord which, though fragile, frustrated the threatened communist take-over in Athens after the German withdrawal. In 1945, now a monuments and fine arts officer, he was enjoyably busy in Holland and Germany, supervising the return of stolen works of art and defending buildings of merit from military vandalism.

In 1946 he was briefly editor of the *Burlington Magazine*, in 1947-8 reader in art history at Manchester University, but in 1949 he accepted the important directorship of the National Galleries of Scotland. However, in Edinburgh, though he made some memorable acquisitions, he endured the bureaucracy of the Scottish Office with difficulty, and in 1952 he resigned to become director of the Barber Institute of Fine Arts in the University of Birmingham.

At Birmingham, amongst congenial colleagues, as professor he blossomed perhaps unexpectedly in academic administration (he was a well-loved dean of arts for three years). He deployed quite generous funds to maximum advantage in acquisitions especially of the British and of the seventeenth-century Italian schools; another special care was the creation of the library. The now remarkable Birmingham School of Byzantine Studies also owes much to his fostering. From Birmingham he was able to lecture widely (notably as Slade professor of fine art at Oxford, 1953-5).

Retiring from Birmingham in 1970, he became London director of the Paul Mellon Centre for Studies in British Art at a critical period in the centre's evolution (1970-3). Throughout the post-war years he accepted various visiting professorships or consultancies in America. His work on the British school—notably his *Sir Joshua Reynolds* (1941), *Painting in Britain 1530-1790* (1953 and subsequent revisions), *Gainsborough* (1958)—and on his beloved Italian school constitute a major achievement that will long hold its value. Amongst other writings there were numerous catalogues, articles, and reviews. His writing, apparently effortless in its classic economy and simple directness of expression, proceeded always from established historical fact and firsthand observation.

He was of independent temperament,

equipped with formidable intellectual powers, supported by a remarkable certainty of principle and of purpose, and driven by inexhaustible energy. He dressed primarily for comfort (though owning a collection of startling neckties); he was no respecter of pretentiousness or pomposity. He could be wilful, and nurtured some idiosyncratic prejudices. His tongue was sharp, and the instrument of a nimble wit unmatched in his profession. He delighted in good company as in good music. Deeper than any critical acerbity, and the wit, were a kindness, an affection for those whom he liked and respected (if not always for their scholarship), and an unstinted generosity with which he drew, in response to any serious enquiry, on his phenomenal memory supported by the superb resources in his own library and archive. He did not personally collect works of art, judging that that might be inimical to a proper scholarly objectivity.

He was appointed FBA (1955), MBE (1943), and CBE (1956). He became an honorary fellow of New College (1975) and was knighted the same year. Nottingham (1968), Leicester (1970), Birmingham (1973), and Oxford (1976) conferred honorary degrees upon him, and he was honoured by the Dutch and Italian governments.

He married in 1949 Helen Thomas, archaeologist and daughter of Frederick William Thomas [q.v.], orientalist. They had two daughters. Through Waterhouse's last decade the distillation and dissemination of his knowledge continued. The companion (seventeenth-century) volume for his *The Dictionary of British Eighteenth Century Painters* (1981) was ready for press when death surprised him, aged eighty, 7 September 1985, at his home on Hinksey Hill above Oxford.

[*The Times*, 9, 14, and 19 September 1985; private information; personal knowledge.]

DAVID PIPER

WATERS, HORACE JOHN (1895-1981), actor. [See WARNER, JACK.]

WATERS, WILLIAM ALEXANDER (1903-1985), organic chemist, was born at 23 Montgomery Street, Cardiff, 8 May 1903, the only child of William Waters and his wife, Elizabeth Annie, daughter of William Roberts, bootmaker. His father was a schoolteacher who eventually became headmaster of Gladstone Council School, Cardiff, and his mother and her sister were also teachers. Waters's educational progress was given every possible assistance, and he recalled that there was always an abundance of books at home that he was encouraged to read. He entered Gladstone Council School in 1908 and went on to Roath Park School in 1911 and then to Cardiff High School for Boys as the top city scholar in 1914. He was persuaded by his father that, rather than follow the school's advice and go to the local Cardiff University College, he should try for Cambridge, and he succeeded in winning one of the scholarships in mathematics and natural sciences that had recently been endowed by Lord Rhondda [q.v.] at Gonville and Caius College. There he obtained first classes in both parts of the natural sciences tripos (1923 and 1924). He went on to a Ph.D. in organic chemistry under the supervision of Hamilton McCombie.

He obtained his Ph.D. in 1927 and in 1928 was appointed to a lectureship at the University of Durham. The department of chemistry there was then very small, with only three lecturers, and although he had opportunities for doing research he felt his prospects would be better elsewhere. However, his applications were unsuccessful. Soon after the start of World War II in 1939 he was seconded as a wartime scientific officer to the Ministry of Supply's Chemical Defence Experimental Station at Porton in Wiltshire. He returned to Durham in 1944 and the following year was elected to a tutorial fellowship at Balliol College, Oxford, which was linked to a university demonstratorship in the Dyson Perrins laboratory. He remained in Oxford until the end of his life. He was appointed to a readership in 1960 and to an *ad hominem* professorship in 1967, when Balliol elected him to a professorial fellowship. He retired formally in 1970, but continued working in the laboratory until about 1977.

Waters's lifelong research interest was in the mechanisms of the reactions of organic compounds in solution. It had been kindled during his Ph.D. work on the mechanisms of chlorination and bromination of aromatic compounds, and soon after he began work in Durham he carried out a survey of the literature at the request of T. M. Lowry [q.v.] to check whether the theory of the ionic nature of organic reactions in solution that was then developing was fully consistent with all the facts. He soon concluded that it was not and that certain reactions took place by way of free-radical intermediates—species which are uncharged and have one unpaired electron. In 1932 he published a speculative article along these lines. Two years later, independently, D. H. Hey at Manchester University published experimental evidence that certain diazo-compounds yield free radicals in solution. Thereafter, Hey and Waters, in consultation, carried out independent research programmes that confirmed the free-radical character of a variety of reactions, and in 1937 they published a seminal

article in *Chemical Reviews* that illustrated the considerable scope of this type of reaction and stimulated work world-wide. Waters's subsequent work focused mainly on radical reactions, including those involving oxidation by metal salts. He was elected FRS for his work in 1954.

Waters was also devoted to teaching organic chemistry. His *Physical Aspects of Organic Chemistry* (1935) was the first text on physical-organic chemistry, and in his *The Chemistry of Free Radicals* (1946) the central theme was the unity of radical reactions in the gas phase and solution. Although never a good lecturer, he was a kind and painstaking tutor with an awesome knowledge of his subject who stimulated in his pupils his own abiding interest; many elected to do research degrees under his supervision and subsequently had distinguished careers in organic chemistry. He was also active in college life, having a spell as estates bursar of Balliol, and he found time, too, for gardening and for playing the piano, chess, and bridge. He was granted an honorary D.Sc. by Warwick University (1977) and was awarded the Chemical Society medal (1973).

In 1932 Waters married his cousin, Elizabeth, younger daughter of William Dougall, foreman steel smelter, of Darlington; they had no children. His shyness was complemented by her gregariousness, and they were fond of entertaining, especially his pupils. They frequently travelled world-wide together, combining conferences and holidays. It was a most happy marriage, and Waters never fully recovered after his wife's death in 1983. He died in Oxford 28 January 1985.

[R. O. C. Norman and J. H. Jones in *Biographical Memoirs of Fellows of the Royal Society*, vol. xxxii, 1986; personal knowledge.]   R. O. C. NORMAN

WATES, NEIL EDWARD (1932–1985), builder and social pioneer, was born 4 February 1932, the eldest son and second child in the family of three boys and three girls of Norman Wates, builder, and his wife, Margot Irene Sidwell. Educated at Stowe, Wates did National Service in the Coldstream Guards. At Brasenose College, Oxford, he read law, was a keen sportsman, and learned to fly. He obtained a second class degree in jurisprudence in 1954.

In the same year he began twenty-one years in Wates Ltd., for the last three as chairman and chief executive. By working on the building sites, he and the men came to know and respect each other. Design, planning, and industrial relations absorbed him and he worked with W. John P. M. Garnett as treasurer of the Industrial Society. His widening interests were stimulated by his wife, (Ann) Jenifer, the daughter of William Guy Weston, of the Ministry of Transport. They married in 1953 after she had taken her degree at Oxford and together they tackled social problems against a background of deep and common religious faith. Later when Wates became a trustee of the Wates Foundation, set up to help the needy, he explored the facts of deprivation, meeting men like Edward D. Berman who fired his imagination.

In 1970 a growing interest in the foreign scene took him to South Africa. He decided that business involvement would be immoral and was persuaded to allow his report to be published. He risked unpopularity, but gained a lasting reputation as a businessman prepared to subordinate expediency to morality.

In 1975, after disagreements, Wates resigned from the company. He and his wife set up at Bore Place in Kent a trust (Commonwork) and centre where they and others could pursue their interests in a diverse farming community. Wates's theme was responsible stewardship—harmonizing the claims and potential of people with the natural environment and resources. All products and residues were used—for methane, for bricks, and for a variety of crafts. The conference centre offered wide-ranging courses from ecology to peacemaking. Wates helped to launch 'Rural'—a forum for the Responsible Use of Resources in Agriculture and on the Land.

By the early 1980s he saw the real seeds of conflict in global environmental deterioration, social injustice, and growing international instability. When an eminent American asked him what he was doing to avert World War III he was ready. A friend, Donald Reeves, was rector of St James's, Piccadilly; there they founded a forum—'Dunamis'—for the discussion of peace and security.

Wates was a peacemaker seeking common ground to resolve enmities by frank discussion. 'Dunamis' considered conflict between east and west and, what Wates thought even more fundamental, between rich north and poor south, with pacifists and leaders of the defence establishment, the orthodox and the revolutionary. Wates believed that problems were not so much technical—weapons or systems—as failures of human understanding. To fellow landowners he showed how respect for the land could be as successful as more ruthless methods. His pack of bloodhounds demonstrated how hunting, which he loved, could avoid the stigma of cruelty. When illness struck he was coordinating all these interests and a new one, holistic medicine.

Fit and dashing on a horse, skis, or hang-glider, he was a natural leader. A sense of drama made his ventures (South Africa or a pep talk at Wates) exciting and momentous; he had a capacity to inspire. In public he was fluent and confident; in private more abrupt and staccato, finding it hard to form close relationships outside the family. A man of paradox, he did not fit easily amongst his peers; a former part-time SAS officer who became a Quaker; very rich but truly abstemious; a bustling entrepreneur whose meetings started with a prayer. He often saw wealth and privilege as handicaps and this, with his admiration of his father, made him strangely anxious to justify himself. His father had made a national reputation for Wates Ltd. and himself by his leadership in the building industry during World War II and in the production of Mulberry harbour. Neil Wates was rich and impetuous enough to put ideas to immediate test. He pioneered projects for others to complete. Often demanding, even imperious, he could be impatient and lose interest if results were not rapid and important.

By 1982 his and his wife's interests diverged. They separated and were divorced in 1984, but remained close and Jenifer continued with Commonwork. Their five children—four boys and a girl—in whose development Wates took delighted interest, were almost grown up. The last phase was not spent alone. Susan Benn, long a supporter of Commonwork, shared his work and developed its artistic side. Wates died 22 September 1985 in the Cromwell Hospital in London.

[*The Times*, 18 and 23 October 1985; private information; personal knowledge.]

HENNIKER

WATSON, (GEORGE) HUGH (NICHOLAS) SETON- (1916-1984), historian and political scientist. [See SETON-WATSON.]

WATSON, JANET VIDA (1923-1985), geologist, was born in Hampstead, London, 1 September 1923, the younger daughter (there were no sons) of David Meredith Seares Watson FRS [q.v.], vertebrate palaeontologist, professor of zoology and comparative anatomy in the University of London, and his wife, Katherine Margarite, daughter of the Revd I. Parker, who was, until marriage, herself active in embryological research. She grew up in a serious, nonconformist, but intellectually lively atmosphere. From South Hampstead High School she went to Reading University, obtaining a first class general honours degree (biology and geology) in 1943.

She then worked on chicken growth and diet in a research institute. Bored, she tried teaching in a girls' school. By the end of the war she had decided to be a geologist, moreover a hard rock geologist, and so she went to Imperial College, emerging, inevitably, with a first class honours degree. She fell under the spell of Professor H. H. Read [q.v.], who proposed that for her Ph.D. project she should attempt, along with fellow student John Sutton, to understand the oldest rocks in the British Isles, the Lewisian gneisses. These, evolved deep in the earth's crust and riddled with veins and dykes, could not be tackled by ordinary geological methods. Janet Watson and John Sutton had to forge their own techniques, in the mists and rock wastes of Sutherlandshire. They were able to show that in the Lewisian gneiss, two components could be separated; the older, Scourian, mapped by Janet Watson, evolved before and the other, Laxfordian, after, the injection of the dyke swarms. These results, presented to the Geological Society of London in 1951, represented a major breakthrough. They were later completely confirmed when radiometric dating showed that the Scourian was hundreds of millions of years older than the Laxfordian.

After her Ph.D. (1949) Janet Watson was awarded a senior 1851 studentship. She continued work on the Lewisian and also took part in a study of similarly ancient rocks in western Tanzania. In 1952 she joined the staff of Imperial College, first as Read's research assistant, and later (1974–80) as professor. Always she was working on, and thinking about, geological problems, especially those of the earth's oldest rocks. The fascination with Scotland lasted: her first, still valuable, Scottish paper was published while an undergraduate, her last, a masterly presidential address to the Geological Society of London, in 1983. But her interests widened, to mineralization, heat flow through the crust, geochemistry, and much else. A visit to Italy in 1983 led her completely to reverse the accepted hypothesis for uranium deposits there. Her last paper, with John Sutton, on 'Lineaments in the Continental Lithosphere' was read to the Royal Society the day before she died.

The whole of her work shows the same clear grasp of essentials, mastery of detail, skilled analysis, and lucid presentation. It adds up to a massive contribution to the understanding of the earth's oldest rocks. She wrote, with H. H. Read, a very successful *Introduction to Geology* (vol. i, 1962; 2nd edn. 1968; vol. ii, 1976); and also *Beginning Geology* (1966).

The value of Janet Watson's work was recognized by awards, by the Geological Society of London of the Lyell Fund, the Bigsby medal (both jointly with John Sutton), and the Lyell medal, and by the Geological Society of Edinburgh of the Clough medal. She was elected FRS (1979), was president of the Geological Society of London (the first woman president) in 1982–4, and a vice-president and member of

council of the Royal Society from 1983 until her death. Outside college she served effectively on the National Water Council (1973–6), the BBC Science Consultative Group, and other bodies.

Janet Watson was like her father: stocky, with the Watsonian jaw and a face full of character. She was reserved, quietly determined, generous, with integrity and above all, warmth. These qualities showed in her teaching. Her lectures were lucid, carefully thought out, penetrating, but it was in sitting down with research students, discussing and sorting out their problems, and giving them ideas, that she excelled.

In 1949 she married John Sutton, geologist, the son of Gerald John Sutton, consulting engineer. They had two daughters, who both died at birth. They continued to work together from Imperial College (John Sutton was appointed to the chair of geology in 1958) throughout her life. After a painful illness, but working and thinking to the end, she died at Ashtead 29 March 1985.

[Private information; personal knowledge.]
ROBERT M. SHACKLETON

WATT, ALEXANDER STUART (1892–1985), plant ecologist, was born at Monquhitter, near Turriff, Aberdeenshire, 21 June 1892, the third son and third of four children of George Watt, a farmer, and his wife, Margaret Jean Stuart. His father and young sister died when Watt was less than three years old but his mother continued to run the farm and bring up the family. His secondary education began at Turriff Secondary School but later he moved to Robert Gordon's College, Aberdeen. From there he entered Aberdeen University in 1910 with a bursary to read for an arts degree but when he found that it was permissible to work for a B.Sc. (agriculture) simultaneously he did this and qualified for both in the minimum time of three years and one month, graduating in 1913. As part of his degree work he had taken a course in forestry which aroused a special interest that was to stay with him all his life. In 1914 he was awarded a Carnegie scholarship and was intending to spend a year in Germany studying forestry. The outbreak of war prevented this and he used the award to work under (Sir) A. G. Tansley [q.v.] at Cambridge on the regeneration of oakwoods.

In 1915 he returned to Aberdeen University as lecturer in forest botany and forest zoology but after a brief spell of teaching he was called up for service in the Royal Engineers. He survived the battle of the Somme but was badly gassed and left with only one functional lung. Despite this he made a good recovery and few people who saw him working in the woods or on the mountains were ever aware of any disability.

He returned to Cambridge in 1919 and completed the requirements for his BA. He then resumed his lectureship at Aberdeen and continued his work on the English beechwoods for his Cambridge Ph.D. (1924) during vacations. As his research developed he became more and more deeply committed to ecology and in 1929 he moved to the Imperial Bureau of Entomology at Farnham Royal, Buckinghamshire, but later in the same year he accepted an invitation to the Gurney lectureship in forestry at the University of Cambridge. In 1933, following a national review of forestry schools, the Cambridge school was closed and Watt was transferred to the botany school.

Watt's early work, published in 1919, on the failure of regeneration in oakwoods attracted immediate attention but it was his studies of the Chiltern beechwoods which clearly marked him out as one of the leaders of plant ecology in Britain. His careful descriptions of the major types of beechwoods and their correlation with distinct soil types set new standards in ecology and played a major role in the integration of the rapidly developing soil sciences into plant ecology.

In 1933 he turned his attention to the characteristic vegetation of the Breckland in East Anglia, an apparently haphazard mosaic of grassland, heath, and bracken. He established a series of permanent observation sites and began careful measurement and recording which continued for almost fifty years even though hindered at times during the war years by tank training in his experimental areas. His elucidation of the way in which the original chalky boulder clay with its grassland cover became progressively leached to form an acid sandy soil dominated by heath and bracken and how local erosion often down to the unchanged clay restarted the process thus giving rise to the mosaic of various developmental stages, will undoubtedly rank as one of the classic ecological studies.

Watt was not a great attender of conferences, nor was he a particularly good lecturer. At times he would seem almost taciturn as he listened to some animated discussion and then with a slight smile on his face and a shrewd penetrating question he would bring the discussion down to earth. The student clerihew, much quoted in the thirties—*Doctor Watt| knows a lot| but confines his speeches| to native beeches*—illustrates how widely it was recognized that when he did speak it was from a basis of deep understanding. He was at his best with a small group in the field whether it was made up of students or well-established ecologists, half reclining on the ground as he dissected a small tuft and pointed out essential features overlooked by less penetrating observers. His observation was me-

ticulous and the value of his guidance in the conduct of field-work is well shown in the important studies of the oakwoods of Killarney and of the vegetation of the Cairngorms undertaken by groups from Cambridge.

His most important contribution to ecology is contained in his presidential address to the British Ecological Society, 'Pattern and Process in the Plant Community', in 1947. His convincing description of the cyclical processes which occur in what appear to be fixed and stable entities threw new light on many features of plant communities not previously recognized or understood.

He was elected FRS in 1957 and awarded the gold medal of the Linnean Society in 1975. He was visiting professor in the University of Colorado in 1963 and the University of Khartoum in 1965. In 1929 he married Annie Constable, daughter of William Kennaway, company director. They had two sons and one daughter. Watt died in Cambridge 2 March 1985.

[Obituary by C. H. Gimmingham in *Journal of Ecology*, 1986, vol. lxxiv, pp. 297–300; John Sheail, *Seventy-five Years in Ecology*, 1987; personal knowledge.]   ALAN BURGES

WAUGH, ALEXANDER RABAN (ALEC) (1898–1981), novelist, was born 8 July 1898, the elder son and elder child of Arthur Waugh and his wife, Catherine Charlotte, daughter of Henry Charles Biddulph Colton Raban, of the Bengal Civil Service. Alec Waugh's younger brother was the writer Evelyn Waugh [q.v.]. Arthur Waugh, who became a publisher and literary critic, was among contributors to the first *Yellow Book*, of April 1894, and a cousin, among others, of Sir Edmund Gosse, the sculptor Thomas Woolner and, through Hunt's marriage to two Misses Waugh in succession, a connection of W. Holman Hunt [qq.v.].

Waugh was educated at Sherborne and Sandhurst, leaving Sherborne under a cloud at the age of seventeen. This episode was later to feature in his first novel, *The Loom of Youth* (1917) which caused a sensation by what then seemed to be its frank treatment of homosexuality in public schools. This was the first time the subject had been mentioned, and Waugh's name was removed from the roll of old boys, much to the distress of his father, another Old Shirburnian, who nevertheless defended his son loyally.

By the time it became apparent that *The Loom of Youth* was a bestseller, Waugh was a prisoner of war near Mainz, having fought at Passchendaele with the Dorset Regiment and been captured as a machine-gunner near Arras. On release he married Barbara, eldest daughter of William Wymark Jacobs [q.v.], the short-story writer, but the marriage was not successful. As he wrote in *The Early Years of Alec Waugh* (1962): 'I had been nicknamed Tank at Sandhurst, yet I could not make my wife a woman.' The marriage was annulled in 1921.

Despite this set-back, despite his small size, and despite emerging from his prisoner-of-war camp, at the age of twenty, rather bald, Waugh enjoyed a considerable success with women. Some of these successes are recounted in his various autobiographies, starting with *Myself When Young* (1923). The chief source for this aspect of his life remains *The Early Years of Alec Waugh*, where, at the age of sixty-four, he is able to record in a memorable sentence, that 'Venus has been kind to me.' He joined his father's publishing firm of Chapman & Hall in September 1919, but left after eight years to be a free-lance writer. He wrote short stories and newspaper essays and embarked upon a world tour.

In 1932 he married as his second wife Joan (died 1969), adopted daughter of Andrew Chirnside, estate owner, of Victoria, Australia. She bore him two sons and a daughter, and, when her father died, was able to provide him with a handsome eighteenth-century home, set in its own park, at Silchester, near Reading. Despite this, Waugh lived a nomadic existence between the wars, as, indeed, he did throughout his entire life, travelling in the Far East, America, and the Caribbean most particularly. His wife returned to Australia with the children. In World War II he rejoined the Dorset Regiment and retired with the rank of major in 1945. His travels inspired him to write about forty books, none particularly successful nor deserving of more than moderate success until, in 1956, he hit the jackpot with a torrid romance, set in the West Indies, called *Island in the Sun*. This touched upon the sensitive subject of another 'forbidden love'—sexual relations across the colour bar—and was made into a successful film.

After this success, in the opinion of his brother Evelyn, Alec Waugh never drew another sober breath, but this was an exaggeration. He lived for much of the year in Tangier, Morocco, where an old age pension from the State of New York enabled him to equip a house with cook, butler, and houseboy; at other times, he lived austerely as writer-in-residence at a mid-Western university, eating his meals from divided, plastic plates in a room above the students' canteen, and emerging from time to time to entertain his friends in London at elegant dinner parties, where he wore immaculately tailored but increasingly eccentric suits.

Within the writing fraternity, Waugh will be remembered as a great survivor, rather than as the author of any particular work of talent. In his later years, he turned his hand to what he called an 'erotic comedy': *A Spy in the Family* (1970) raised a few eyebrows, but by then he was following a fashion set by others rather than breaking new ground as he had with *The Loom of Youth* and, to a lesser extent, with *Island in the Sun*. To this Dictionary he contributed the notice of Ernest Rhys.

Of mild and modest nature, Waugh had few enemies and many friends. He belonged to innumerable clubs and societies in London and New York, and was generally revered in them. He never grudged his younger brother's greater success as a writer, and the two remained on cordial terms until Evelyn's death in April 1966. A year later he published *My Brother Evelyn and Other Profiles*.

Waugh died 3 September 1981 in Tampa, Florida, where he had gone to live with his third wife, Virginia Sorensen, daughter of Claude Eggertsen, of Springfield, Utah, USA, whom he married in 1969.

[Alec Waugh, *The Early Years of Alec Waugh*, 1962, and *The Best Wine Last*, 1978; Evelyn Waugh, *A Little Learning*, 1964; personal knowledge.]   AUBERON WAUGH

WELCHMAN, (WILLIAM) GORDON (1906–1985), cryptanalyst, was born at Fishponds, near Bristol, 15 June 1906, the younger son (the elder was killed in 1914) and youngest of three children of William Welchman, a former missionary who became a country parson and archdeacon of Bristol, and his wife, Elizabeth Marshall, daughter of the Revd Edward Moule Griffith. He went to Marlborough College in 1920 and to Trinity College, Cambridge, in 1925. In the mathematical tripos he obtained a first class in part i (1926) and was a wrangler in part ii (1928). He then went on to teach at Cheltenham for one year before returning to Cambridge where he became a fellow of Sidney Sussex College in 1929 and wrote *Introduction to Algebraic Geometry*, which was published in 1950. He was recruited for service at the Government Communications Headquarters at Bletchley Park in 1938/9 and worked there until 1945.

It was during the early years of World War II that he made a significant contribution to the solving of the Enigma machine cipher which was used extensively by the Germans. He worked with Alan Turing and C. H. O'D. Alexander [q.v.]. Some of his key technical solutions had already been devised by the Poles, and by others at Bletchley, but he instinctively grasped a whole range of problems, possibilities, and solutions which included two vital mathematical constructs as well as a concept of the total process required, from the intercepted German ciphered traffic to passing on significant intelligence implications to the commanders in the field—a highly complex logistical operation for which total secrecy was an added condition.

Welchman, assigned by Dillwyn Knox on arrival at Bletchley Park to comparatively low-level research on call-signs, quickly realized that he and his few colleagues were dealing with an entire communication system that would serve the needs of the German ground and air forces. It was the development of 'traffic analysis' which was his greatest contribution, but in these early months he made two startling breakthroughs in enabling Enigma-coded signals to be read. The first had to do with the indicator setting and indicator of an Enigma message. A long and intricate series of mathematical thought processes resulted in Welchman reinvestigating a system of perforated sheets, ignorant of the fact that the Poles had done this before, and a colleague elsewhere in BP already had production in hand.

Early in 1940 Alan Turing had the idea of making a machine which would test all possible rotor positions of the Enigma to find those at which a given cipher-text could be transformed into a plain-text. Welchman greatly improved on Turing's design by his invention of a device known as a diagonal board, which Turing himself immediately recognized to be invaluable.

These two relevant and vital achievements took place within months of his arrival, and it was not long before Welchman was applying his mind in a wider context. He had practical gifts and a strong personality. Once it was clear that Bletchley Park would be able to read enemy traffic on a massive scale he established the need for increased facilities and close co-operation between the intercepting stations, the cryptographers, the intelligence processors, and the ultimate users. An informed view is clear that the task of converting the original breakthrough into an efficient user of the material was one for which Welchman should receive much of the credit. He himself wrote about his work long after the war in a book for which he was wrongly attacked by the authorities for divulging secrets which might still be of use to a hostile power. His motives however were transparently honourable and the sustained powers of thought and memory evinced in the early chapters of The Hut Six Story (1982), somewhat amended in a subsequent article in Intelligence and National Security, are characteristic not only of his considerable mental powers but also of his deep conviction that there were important lessons to be learned from the breaking of the Enigma secrets, and that governmental refusal to dis-

close such matters in order to learn from them was a matter of overriding public concern. After the publication of his book his accreditation to the Mitre Corporation, which he had joined in America in 1962 and where he concentrated on the development of secure communications systems for the US forces, was withdrawn and the last months of his life, as he was dying of cancer, were marred by the authorities trying to stop him from publishing. He had moved permanently to America in 1948 and became an American citizen in 1962.

Welchman's great achievement took place in 1940-3. At Bletchley Park he became assistant director for mechanization. He was appointed OBE in 1946. After the war he became director of research for the John Lewis Partnership but settled in America in 1948. His wartime experience led him to the computer field and he pioneered developments in digital compiling.

Welchman had an acute analytical mind, boundless drive and enthusiasm, but rather limited imagination. At a crucial moment in World War II he brought together discrete ideas and divergent pieces of evidence to produce a total policy framework. As a man, though not always easy for his colleagues to communicate with, he was admired, trusted, and liked, for his great charm as well as intelligence and kindness.

He married in 1937 a professional musician, Katharine, the daughter of Francis Faith Hodgson, a captain in the 84th Punjabis, Indian Army. They had one son and two daughters. After divorce in 1959 he married Fannie Hillsmith, an artist, the daughter of Clarence Hillsmith, consulting engineer, of New Hampshire. This marriage also ended in divorce (1971) and his later years were made happier by his third marriage, in 1972, to Elisabeth, daughter of his second cousin, Myrtle Octavia Hussey, and her husband, Anton Wilhelm Huber, owner of a sawmill and carpentry contractor, in Aschau in Chiemgau, Bavaria, Germany. She was a physiotherapist. He loved mountains, for climbing and skiing. He was an avid gardener and a keen amateur musician. He died 8 October 1985 at Newburyport, Massachusetts.

[Gordon Welchman, The Hut Six Story, 1982; private information; personal knowledge.]   R. A. DENNISTON

WEST, DAME REBECCA (1892-1983), author, reporter, and literary critic, was born Cicily Isabel in Paddington, London, 21 December 1892, the youngest of three daughters (there were no sons) of Charles Fairfield and his wife, Isabella Campbell Mackenzie. Her father, of Irish parentage, was a largely unsuccessful journalistic and entrepreneurial soldier of fortune, who abandoned his family in 1901 and died in poverty. Her mother, of Scottish stock, was a gifted amateur pianist, who earned her living before her marriage as a governess in a wealthy family, from whom she received an unsolicited and much needed allowance after the marriage broke up.

The fatherless family removed to the mother's native Edinburgh, where Cicily went to George Watson's Ladies' College; her future as a writer was signalled early, when, at the age of fourteen, she won the school's junior essay prize, and in the same year broke into print for the first time with a letter to the *Scotsman* on women's rights. (She was already, with her two sisters, an ardent feminist and socialist.)

Intent first, however, on a stage career, she trained for a year at the Academy of Dramatic Art in London, while beginning to write regularly for a new magazine which might have been created for her, the *Freewoman*. In 1912 she needed a pseudonym for her contributions; she chose the name of one of Henrik Ibsen's proud, rebellious, and independent heroines, and for the rest of her long life, to all the world except her close family, she was Rebecca West.

She made a literary reputation swiftly; her first book, a study of Henry James [q.v.], appeared in 1916. But three years earlier she had sustained a wound that was never to heal, that indeed would bleed afresh after her death; she met H. G. Wells [q.v.], then at the height of his fame as a writer, they became lovers, and she bore his unintentionally conceived child, a son, known as Anthony West.

The relationship with Wells was in any case both stormy and doomed, and inevitably petered out. She had a number of further liaisons both before and after her rather surprising marriage in 1930 to Henry Maxwell Andrews (died 1968), son of Lewis Henry John Andrews, of Wallace Bros., an East India merchant company; he was something of a scholar, and a not notably successful banker, but he clearly provided for her an ordered stability that she might never have found for herself. But nothing ever approached either the intensity or the influence of her love for Wells. She strove to absorb the experience and thus exorcize it, but the task was made more difficult by the deep antipathy that her son developed for her; he felt, not altogether without reason, that she had neglected him. Until she died they were constantly at loggerheads, sometimes legal ones, and after her death he took his revenge by publishing accounts of her, Wells, and himself which make wretched reading, whatever view is taken of the rights and wrongs of the matter.

There remains her literary achievement, much of which is certain to endure. None of her novels is wholly successful as a work of art,

though perhaps *The Fountain Overflows*, published in 1957, comes nearest. It was intended to be the first volume of a trilogy in the form of a family saga; it is clearly based on her own youth, has a spontaneity that rings more true than in any of her other fiction, and is deservedly the most popular of her novels.

Her reportage is of an altogether different quality; at its best it has few equals in all journalism. *The Meaning of Treason* (1949) and *A Train of Powder* (1955) are the fruits of her attendance at such historic assizes as the Nuremberg tribunal and the post-war British treason trials, together with other notable episodes of our troubled world, particularly a gruesome murder. Her account of the case was published as 'Mr Setty and Mr Hume' in *A Train of Powder*. In these writings her omnivorous observation, brilliant sense of colour, and ruthless analysis combine to make a series of unforgettable pictures. But her undoubted masterpiece is *Black Lamb and Grey Falcon*.

It is drawn from a series of visits (telescoped into one for the purpose of the book) to Yugoslavia, or more precisely Serbia, that she made with her husband in the 1930s. She was overwhelmed by the country and its people, and this immense work (a quarter of a million words) is a tribute to both, bursting with their history, architecture, poetry, music, geography, literature, courage, flora, customs, religion, food, loves, hates, philosophy, rulers, and fate; the book was published in New York in 1941, just as Yugoslavia was experiencing the full weight of the Nazi assault.

The post-war tragedy of Yugoslavia's conquest by communism served to cement a hatred of that ideology which she had first felt at the Hitler–Stalin pact of 1939. Thereafter she was as implacable a foe of Soviet totalitarianism as of the Nazi one; her bitterness was accentuated by her knowledge that many of her friends in Yugoslavia who had survived the Nazis had fallen victim to Marshal Tito. But she never became fanatical or obsessed; her wisdom and balance enabled her to control her anger, the better to aim her deadliest darts at all those forces, external and internal alike, which threaten the free society.

Rebecca West was a woman of formidable, even daunting, personality, but with a warmth and understanding that made her company an enriching experience. She did not suffer fools gladly, or indeed at all; her wit could be savage, and the older she grew the less need she felt to restrain it. (Talking one day of a prominent man of undoubted gifts but *outré* beliefs, who affected a decadent air, she delivered a forceful judgement: 'He looks', she said, 'like a dead pimp.')

Her circle of friends was wide and varied;

they gladly endured her quirks and occasional irascibility, and for all her temper she was not naturally quarrelsome, nor did she hug her wrath or bear grudges. Her strength began to ebb at the beginning of the 1980s; her last published article appeared in September 1982, but she fought death as tenaciously as she had battled with the demons all her life. She died 15 March 1983 in London. She was buried, according to her own instructions, in Brookwood cemetery, in Surrey.

She was appointed CBE in 1949 and DBE in 1959. She was made a chevalier of the Legion of Honour in 1957. P. Wyndham Lewis drew a portrait of her when she was forty, which now hangs in the National Portrait Gallery, London.

[Rebecca West, *Family Memories* (ed. Faith Evans), 1987; Victoria Glendinning, *Rebecca West*, 1987; *The Times*, 16 March 1983; personal knowledge.]   BERNARD LEVIN

WETHERED, ROGER HENRY (1899–1983), golfer, was born in Kingston upon Thames, Surrey, 3 January 1899, the only son and elder child of Herbert Newton Wethered, artist, and his wife, Marion Emmeline Lund. Owing to poor health in his younger days, Wethered went to school for a brief period and then was educated by tutors. He was thus introduced to golf earlier than might otherwise have been the case. His father was author and co-author of more than one book on golf, an involvement which almost certainly influenced his son's development as a player. However, Wethered was old enough and fit enough to be commissioned in the Royal Artillery at the end of World War I (1918) and saw service for a few weeks in France.

On leaving the army he went to Christ Church, Oxford, where, for two years, he played golf in the Oxford side with Cyril Tolley [q.v.]. Together they were described as the first and most conspicuous champions of a new generation. As they were also good friends, it was inevitable that their names were repeatedly bracketed together and their achievements compared.

Bernard Darwin [q.v.], who knew them both well, felt that Tolley was 'always and unquestionably the finer driver but that Wethered was equally and beyond question the better iron player'. Wethered was both powerful and accurate, with a capacity for obtaining the maximum backspin. His driving was less than certain although its errant ways established his reputation as an exceptional recovery player. When he won the Amateur championship at Deal in 1922, he drove magnificently.

This feat prompted his sister Joyce to remark 'Why, this is a new Roger.' Joyce Weth-ered (later Lady Heathcoat-Amory) was the best lady golfer of the time and together they stand as the finest brother and sister combination. Wethered's most notable performance was in tieing for the Open Championship at St Andrews in 1921. This was during his last year at Oxford (he obtained his BA in 1921 with the shortened course in English) and followed Tolley's victory in the Amateur the previous summer.

It was the last time that a British amateur came as close to winning the Open; yet Wethered's total of 296 included a penalty stroke in the third round for accidentally treading on his ball. Some assumed it to be logical and incontestable that, if he had not suffered this misfortune, he would have won. It was an unjustifiable assumption but, having been six strokes behind Jock Hutchison, the eventual winner, after the first round, Wethered's final rounds on the last day (72 and 71) were the lowest.

The play-off with Hutchison, a Scotsman who became an American citizen, was a disappointment. Wethered, who had planned to be playing cricket in the south, was beaten by nine strokes but, though he never played quite as well again after his victory at Deal two years later, he remained a formidable, popular, and influential figure in the game. He was a man of engaging modesty and charm, described later by an eminent amateur golfer as one of the two most courteous opponents he had ever faced.

He reached the final of the Amateur twice again, won the President's Putter five times, and was capped six times for Britain in the Walker Cup against America and nine times for England against Scotland. In these later matches, he won all his singles, and eight out of nine foursomes between 1922 and 1931.

As an automatic choice for the Walker Cup he developed a long-standing rivalry and friendship with Bobby Jones, an American, who defeated Wethered in the final of the Amateur at St Andrews in 1930, the first leg of Jones's unique grand slam. That same year Wethered lost to Jones in the Walker Cup singles but his record in that cup did him full justice. Apart from losing his final foursomes in 1934 with his old partner, Tolley, he lost only to Jones, whom he first met in 1922.

By the 1930s Wethered was seldom in good practice, a fact that owed much to his becoming a hard-working stockbroker in London. He worked on committees at the Royal and Ancient and was elected its captain in 1939 although he did not take office until 1946. He thus had plenty of time to contemplate the ordeal of driving himself in.

He continued to play golf and went round Wimbledon in the score of his age when he was seventy-four and, as the senior past captain of the Royal and Ancient gave the address at the memorial service for Bobby Jones at St Andrews in 1972. He was a leading stockbroker and chairman of several investment trusts. In 1925 he married Elizabeth, daughter of Lord Charles Cavendish Cavendish-Bentinck, son of the sixth Duke of Portland. This marriage was dissolved in 1954 and in 1957 he married Marjorie Mitford Campbell Stratford, daughter of Ernest Stubbs, judge. He had no children by either wife. He died at his home in Wimbledon 12 March 1983.

[Bernard Darwin, *Golf Between Two Wars*, 1944.]   DONALD STEEL

WHITE, ERROL IVOR (1901-1985), palacontologist, was born in Woodbury Down in north London 30 June 1901, the younger son of Felix Ernest White, Borough Market merchant, and his second wife, Lilian Emma Daniels, daughter of a wealthy Spitalfields merchant. There were also a son and a daughter from the first marriage. He was educated at Highgate School and entered King's College, London, in October 1918 to read chemistry but he later transferred to the school of honours geology, obtaining a second class B.Sc. in 1921.

White was appointed to the staff of the British Museum (Natural History) in November 1922, in the fossil fish section. In 1925 appeared the first of his 102 publications on all major groups of fossil fishes, ranging throughout their geological history and from every continent. In 1929-30 he participated in a museum expedition to Madagascar and, surmounting many obstacles, made major collections of Triassic fishes and Pleistocene birds. From 1932 White's principal interest centred on the taxonomy, ecology, and stratigraphy of the primitive fish faunas from the Devonian System. In June 1939, and newly appointed as deputy keeper (in 1938), his activities extended into Spitsbergen as a member of a joint Anglo-Norwegian-Swedish expedition to collect from the Old Red Sandstone deposits; but news of the deteriorating international situation forced White's premature return to England where he arrived the day before the declaration of war.

From 1940 to 1945 White was seconded to the Ministry of Health in Reading, where he was responsible for co-ordinating local government emergency administration in southern England. On his return to the museum in April 1945 he was much occupied initially with restoring the collections and the building, but he was happiest in resuming his research which he had nevertheless managed to continue in his own time during his period of wartime secondment. A small collection of Australian Devonian fishes from New South Wales which had been picked up by a sheep farmer and sent just before the war to the museum for comment came to White's attention. The specimens occurred in limestone and it was at this time that a technique of embedding fossils in plastic and with the aid of acetic acid etching them from the surrounding matrix had been devised in the department, an advance that was to revolutionize vertebrate palaeontology. White continued to research on the New South Wales material and new material gathered from north-western Australia long after his retirement in 1966, his work being facilitated by his appointment as a visiting research geologist at Reading University.

The merit of his researches was acknowledged by his being elected a fellow of the Royal Society in 1956. He had been awarded his D.Sc. by the University of London in 1936, having gained a Ph.D. in 1927, and he was made a fellow of King's College, London, in 1958. In 1962 he was awarded the Murchison medal of the Geological Society of London, and in 1970 the gold medal of the Linnean Society. He was appointed CBE in 1960.

Despite his commitment to research White proved to be a conscientious and effective administrator on his appointment in 1955 as keeper of geology (changed to palaeontology in 1956), and he presided over a considerable expansion of his department between 1955 and 1966, fostering the broadening of its activities. Tall, well-built, and invariably well-dressed, he had been in his younger days something of a thorn in the side of the establishment, with a reputation for pugnacity. Yet he was a private and, as he acknowledged himself, a rather shy man and this manifested itself on occasion in apparent brusqueness and peremptoriness. In later life he continued to be forthright and outspoken, but by then he had matured into the epitome of an English gentleman. He had a literary bent: his papers were immaculately phrased and constructed.

Although much preoccupied with his official duties, White found time to contribute to the running of learned societies. As secretary he played a major role in revivifying the Ray Society in the years following the war, and he served as president from 1956 to 1959 when he was made vice-president for life. He served on the councils of the Geological (1949-53, 1956-60, vice-president 1957-60), Zoological (1959-63), and Linnean (1956-9) societies; and he was chairman of the Systematics Association (1955-8). He also served as president of the Linnean Society from 1964 to 1967, and he was greatly moved to be the recipient of the Society's first Festschrift in 1967. To this Dictionary he contributed the notice of T. R. Fox.

In 1933 White married Barbara Gladwyn Christian (died 1969). The marriage, which was childless, was dissolved in 1940 and in 1944 White married Margaret Clare ('Jane'), daughter of Thomas Craven Fawcett, of Bolton Abbey, Yorkshire. They had one son. White died in Wallingford 11 January 1985.

[Sir James Stubblefield in *Biographical Memoirs of Fellows of the Royal Society*, vol. xxxi, 1985; personal knowledge.]   H. W. BALL

WIDGERY, JOHN PASSMORE, BARON WIDGERY (1911–1981), lord chief justice, was born 24 July 1911 at South Molton, Devon, the elder child and only son of Samuel Widgery, house furnisher, and his second wife, Bertha Elizabeth Passmore, JP. Widgery left Queen's College, Taunton, at the age of sixteen to become an articled clerk, and in 1933 he qualified as a solicitor (John Mackrell prizeman). It was not then thought necessary for a solicitor to be a university graduate. One of Widgery's predecessors as lord chief justice, Rufus Isaacs (the Marquess of Reading, q.v.), had never been to a university: but only Charles Russell (first Baron Russell of Killowen, q.v.) had started life as a solicitor.

Perhaps surprisingly, Widgery never practised his profession, but joined the staff of Gibson and Welldon, law tutors in London. He always retained the clarity of expression required of a good teacher. The outbreak of war in 1939 found him, characteristically, adjutant of his Territorial battalion. He saw active service with the Royal Artillery in which he became a lieutenant-colonel in 1942. He was appointed OBE in 1945 and decorated with the croix de guerre and Order of Leopold. In 1952 he was promoted brigadier in the Territorial Army—a position which was held to be both in his favour and against him when in 1972 he thought it his duty as lord chief justice to volunteer to report on civil disorder in Londonderry which led to the deaths of thirteen civilians at the hands of the army.

After war ended in 1945 Widgery did not return to the solicitors' profession. He was called to the bar (Lincoln's Inn) in 1946, and quickly built up a substantial practice on the South-Eastern circuit, mainly in the areas of rating and town planning. In 1958 he became a QC (the most junior of the sixteen new silks but the only one to have been called to the bar since the war). He was recorder of Hastings from 1959 to 1961. In 1961 he was appointed judge of the Queen's Bench division and knighted. (The English bench is traditionally clean-shaven, but Widgery retained his military moustache.) He was the first chairman of the senate of the Inns of Court and the bar.

In 1968 he was promoted to the Court of Appeal, and appointed a privy councillor. In April 1971 he succeeded Lord Parker of Waddington [q.v.] as lord chief justice, with a life peerage.

Widgery was a very good judge for the parties, but hardly a great judge for the jurist. Until struck down by Parkinson's disease, he was always fair and courteous and quick to see the essential point. He was particularly good in a heavy criminal case, being impatient of subtleties, whether forensic or academic, and his summings-up were models of clarity and soldierly brevity. Although he did not have the legal learning or dominant personality of Lord Goddard [q.v.], when his interest was aroused he could write a judgment of quality. Few have improved on his explanation of the reasons for imposing some restraint on the recovery of damages for economic loss (*Weller v Foot & Mouth Disease Research Institute* [1966] 1 QB 569). But normally problems were resolved by common sense. So when the Crown sought an injunction to restrain the publication of the diaries of R. H. S. Crossman [q.v.] some high-flown rhetoric was made to look foolish by Widgery's decision (since approved by a number of appellate tribunals) that after ten years some of the secrets which the ex-cabinet minister had reported with such relish were of no importance to anyone (*Attorney-General* v. *Jonathan Cape Ltd.* [1976] QB 752). Some criticized him for sitting too often in the Divisional Court, but in the 1970s the business of that tribunal had expanded beyond the customary mass of small appeals to include cases of great constitutional importance.

Widgery found that the administrative burden of his office, always considerable, was increased by the great structural changes in the courts proposed by Lord Beeching [q.v.] and adopted by Parliament. There was something frustrating about having to carry into effect the ideas of another man. After some years Widgery was obviously faltering under the strain, and his resignation in April 1980 (the ninth anniversary of his appointment) was not unexpected.

Well-built and ruddy, Widgery in manner and appearance was typical of the west country. He would have finished his game of bowls before dealing with a crisis. Conversely, his devotion to a methodical life meant that at the appointed hour for recreation the papers in even the heaviest case were put aside. Widgery became a bencher of his inn in 1961 and treasurer in 1977. He was awarded honorary doctorates by Exeter (1971) and by Leeds and Columbia (both 1976). He was also, like an ancestor in 1689, a freeman of South Molton (1971).

In 1935 he married Helen Yates, daughter of

Baldwin Walker Peel, estate agent, of Campsall Grange, near Doncaster. This marriage was dissolved in 1946, and in 1948 he married Ann Edith, daughter of William Edwin Kermode, mining engineer, of the Isle of Man, and former wife of David Willett Fleetwood Dillin Paul. There were no children of either marriage. Widgery died at his Chelsea home 26 July 1981.

[*The Times*, 28 July 1981; private information; personal knowledge.]

R. F. V. Heuston

WIGG, GEORGE EDWARD CECIL, Baron Wigg (1900-1983), politician, was born at 139 Uxbridge Road, Ealing, 28 November 1900. He was the eldest of six children (two sons and four daughters) of Edward William Wigg, manager of a dairy business, who declined from a moderately prosperous beginning to losing everything, and of his wife Cecilia Comber whose family had unbroken non-commissioned army service since the time of Sir John Moore's Light Division. Wigg was proud of his Wigg ancestors, once prominent in Hampshire, one of whom founded the Queen Mary's Grammar School, Basingstoke, to which he won a scholarship when twelve, having attended Fairfields Council School. Poverty compelled the end of his formal education when he was fourteen and after a few inconsequential jobs he joined the Hampshire Regiment at eighteen.

Wigg was in the regular army (spending many years in the Middle East) until 1937 and his belief in its virtues never faded though he was prickly with authority when he thought it unjust. The social prejudices of the time unreasonably prevented his being a commissioned officer until he rejoined in 1940 and became a lieutenant-colonel in the Royal Army Education Corps where his enthusiasm for current affairs sessions for the troops prompted the jibe, when in Parliament, that the Education Corps was 'the only unit of the British Army entitled to include the general election of 1945 among its battle honours'.

Wigg's intelligent enquiring mind, added to his bitterness at the world's unfairness for which he blamed and loathed the Tories, impelled him to read widely and to educate himself politically. While working with the Workers' Educational Association he met in 1933 A. D. Lindsay (later Lord Lindsay of Birker, q.v.), master of Balliol. A lasting friendship ensued which greatly influenced Wigg's thinking. Lindsay suggested to Wigg during the war that he should drop his intention of emigrating to Canada in favour of a political career which he began when elected Labour MP for Dudley in 1945.

A man of swirling emotions prone to hero-worship he attached himself passionately to Emanuel (later Lord) Shinwell whose parliamentary private secretary he was when Shinwell was minister of fuel and power (1945-7). Shinwell's replacement by Hugh Gaitskell [q.v.] led Wigg irrationally to dislike Gaitskell and often his associates. His knowledge of and devotion to the army inspired much of his parliamentary work and he was a harrier feared by Tory Service ministers. He could be devastatingly rude. A clash with John Profumo, secretary of state for war, in the autumn of 1962 convinced him of Profumo's deception over army conditions in Kuwait and roused his ire. When the Christine Keeler scandal emerged in 1963 he malevolently pursued Profumo in the Commons until his downfall. An expert in political intrigue (described as wiggery pokery), Wigg was aware of the political advantages to Labour and informed Harold Wilson (later Lord Wilson of Rievaulx) of his activities throughout. Ironically in 1976 Wigg was charged, on the evidence of police officers said by Wigg to be lying, with accosting women from his motor car as he drove slowly near Marble Arch. The Wells Street court magistrate concluded that it was Wigg who was lying but acquitted Wigg solely because he considered that the 'kerb crawling' of which Wigg had been accurately accused did not amount to an offence.

After Gaitskell's death in 1963 Wigg skilfully managed the campaign for Wilson's election as Labour leader using cajolery and threats to Labour MPs relevant to their future under a probable Wilson regime. In 1964 he became paymaster-general with direct access to the prime minister on security and wider political matters. He was admitted to the Privy Council at the same time. Voluble in conspiratorial style whether in person or on the telephone—frequently at unusual hours—he achieved a domination over Wilson which irritated colleagues including Marcia Williams (later Lady Falkender), whose removal from the room he once successfully demanded when he wished to speak confidentially to the prime minister. Eventually Wilson was exhausted by Wigg's constant pummelling and in 1967 adroitly removed him from his presence by making him chairman of the Horserace Betting Levy Board with a seat in the House of Lords as a life peer. Though hipped at this loss of favour and subsequently ungracious about his patron Wigg was also delighted. Wigg loved racing almost as much as he did the army and political intrigue and was intermittently a keen owner of indifferent horses. He had been a member of the Racecourse Betting Control Board (1957-61) and of the Horserace Totalisator Board (1961-4).

As dispenser of the statutory levy collected from bookmakers and the tote Wigg was immediately in conflict with the Jockey Club which considered that he should take their orders as to how the money should be distributed to racing, and told him so at a meeting to which he was summoned. They had mistaken their man. Ferociously Wigg quickened the pace at which the Jockey Club came to realize that their edicts were not as automatically revered as they were at the Club's inception in 1752. By shifting the balance of power to the Levy Board Wigg defused mounting resentment at the Jockey Club's imperious ways. Previously a strong advocate of a tote monopoly, he abruptly reversed his position when he became president of the Betting Office Licensees' Association in 1973, after leaving the Levy Board. This was not surprising to those familiar with his violent lurches.

Though Wigg maintained deep dislikes, rising to hatreds, he was intensely loyal to his friends and unremitting in his zeal for those he felt badly treated. He believed fervently that the poorest and most humble in the land had the same right to identity and self-fulfilment as the richest and most powerful and acted accordingly in his parliamentary and racing career. He employed either considerable charm or hectoring hostility to match his mood or the requirement of the occasion.

Wigg was large and broad-shouldered, with a long and lugubrious face. His beaky nose and wide tall ears were a cartoonist's delight. In 1930 he married Florence Minnie, daughter of William ('Harry') Veal, a stud groom. They had three daughters. There is a bust of Wigg (1984) by Angela Connor placed in the paddock at Epsom racecourse in recognition of his contribution to ensuring into perpetuity the Epsom racecourse and the gallops. Wigg died in London 11 August 1983.

[Lord Wigg, *George Wigg*, 1972; Christopher Hill, *Horse Power: The Politics of the Turf*, 1988; private information; personal knowledge.]   WOODROW WYATT

WILLIAMS, ERIC ERNEST (1911–1983), writer, was born in Golders Green, London, 13 July 1911, the eldest in the family of four sons and one daughter of Ernest Williams, interior decorator and antique dealer, and his wife, Mary Elizabeth Beardmore, a junior-school teacher. Always called 'Bill', except by his family, he was educated at Christ's College, Finchley, which he left at the age of sixteen to become an apprentice to the building trade prior to joining his father's business. Later, he designed and built the décor for some West End theatrical productions. He also enjoyed summer camps as a trooper with the Middlesex Yeomanry.

When his father became bankrupt in the 1929–30 depression Williams went as a management trainee to Waring & Gillows, the furniture firm, and in the evenings wrestled in East End working men's clubs. In 1932 he left London for Liverpool to run the 'design for living' departments in the Lewis's chain of stores. He spent his weekends rock-climbing in north Wales.

On the outbreak of war in 1939 Williams volunteered for the RAF as aircrew. After training in Ontario, Canada, he qualified as an 'observer' (a navigator/bomb-aimer) and was posted to No. 75 (New Zealand) Bomber Squadron based near Cambridge. On 17 December 1942 his Stirling was shot down after a raid on the Ruhr. His mother was told that he was missing, believed killed. In the same year her second and third sons, also in the RAF, were killed. Her eldest son, however, was a prisoner in Stalag-Luft III in Silesia (which later became part of Poland).

A fellow prisoner, Michael Codner, of the Royal Artillery, persuaded Williams to assist in one of the most ingenious escapes of World War II. Escape tunnels had invariably been started from underneath the prisoners' quarters and had soon been discovered by the guards. Codner discussed with Williams the idea of starting a tunnel in the open as near the wire defences as practicable, using a wooden vaulting-horse as a decoy. The escape committee approved the plan; and while other prisoners enthusiastically distracted the guards by leaping over the rudimentary vaulting-horse in the prison compound, Codner and Williams began the exhausting task of digging a tunnel, using the wooden horse as cover. When they realized that the work was beyond their strength they enlisted the aid of Oliver Philpot, another RAF officer. Four and a half months after they started digging the tunnel they reached beyond the wire and all three escaped, Codner and Williams disguised as French workmen and Philpot as a Norwegian salesman. After adventurous journeys all of them reached Britain via Sweden. Williams and his two companions were awarded the MC (1944).

On his return to Britain Williams was sent on a tour of RAF airfields lecturing on what airmen would have to face if they were shot down over enemy territory. A man of his word, he also fulfilled undertakings he had given to his fellow prisoners and visited their families to give them news of their husbands and sons. He then set out to write a novel of his prison-camp experience called *Goon in the Block* (1945) which as he explained in the preface to his later book *The Wooden Horse* (1949, revised 1979) was 'fact thinly disguised as fiction' but without 'details that might have helped the enemy'.

He was demobilized in 1946 and returned to

Lewis's as a buyer for the book departments in their seven stores. At the same time he worked on *The Wooden Horse*, a factual account of his escape from Stalag-Luft III, and after seven publishers had rejected it, Collins decided to publish and found that they had a best seller. 100,000 copies were produced in ten weeks, the film rights were sold, and Williams left Lewis's to assist in the making of the film (1950) with the same title as the book. By 1988 four million copies of the book had been sold, children's versions had become available, and it had been translated into many languages.

After these successes as an author Williams decided to become a free-lance writer, and subsequently published a number of books on the escape theme of which the best known is *The Tunnel* (1951). In 1948 he married as his second wife Sibyl Maud, daughter of Charles Sidney Grain, director of Vedonis hosiery works; she had been an officer in the WRNS and personal secretary to the admiral (submarines) during the war and was awarded the MBE. His first wife Joan Mary, daughter of Richard Owen Roberts, a county court judge, whom he had married in 1940, had been killed in an air raid on Liverpool in 1942. There were no children of either marriage.

After living for a short time in Devon, Williams and Sibyl set off to travel to Hungary, Romania, and Bulgaria, and their experiences were recounted in *Dragoman Pass* (1959). In that year they decided to set out again on a slow journey round the world; they spent some time in Cyprus, eastern Serbia, and Greece, but when they reached Lebanon, they agreed that they wished to explore the islands of the Mediterranean and that for this purpose they needed a boat. Williams himself designed a suitable craft and she was built by a small constructor of Baltic and North Sea fishing boats and named *Escaper*. For the next twenty-one years they sailed together, living on *Escaper*, exploring the eastern shores of the Mediterranean. In 1981 Williams had a heart attack and was flown to England by the RAF for an emergency open-heart operation in Oxford. After the operation he and Sibyl resumed their seafaring life, but two years later he had to undergo further surgery and, though he appeared to have recovered, he died suddenly on *Escaper* at anchor off Portoheli, Greece, 24 December 1983.

[Eric Williams, *The Wooden Horse*, 1949, *The Tunnel*, 1951, and *Complete and Free*, 1957; Oliver Philpot, *Stolen Journey*, 1950; private information.]   H. F. OXBURY

WILSON, PETER CECIL (1913–1984), fine art auctioneer, was born 8 March 1913 in Lon-

don, the third and youngest son (there were no daughters) of Lieutenant-Colonel Sir Mathew Richard Henry ('Scatters') Wilson, fourth baronet, of Eshton Hall, Gargrave, Yorkshire, and his wife, the Hon. Barbara Lister, daughter of the fourth Baron Ribblesdale, a senior official of Queen Victoria's household. Wilson was educated at Eton and New College, Oxford, where he read history, but he left after two years and went to Paris and Geneva to learn French, and later to Hamburg to learn German. There he met a fellow student, (Grace) Helen, daughter of Arthur Ranken, engineer. They married in 1935 and had two sons.

Back in England Wilson worked at first for Spinks, the coin auctioneers. He then moved to Reuters, *Time and Tide*, and later the *Connoisseur*. He was overjoyed when finally an opportunity presented itself in 1936 to take up a junior post in the works of art department of Sotheby's, cataloguing furniture. He applied himself to learning the business with tremendous vigour and determination. His later flair for the instant recognition of objects of quality became legendary. A stepping stone in Wilson's career (he was always known as 'PCW') was his achievement in the detailed cataloguing of an immense collection of antique rings sent to Sotheby's by the descendants of Edward Guilhou of Paris. In December 1938 he was made a director and became a partner in Sotheby's: he had been promoted over his boss's head.

After three happy years at Bond Street, the war broke out and, together with many other members of Sotheby's staff, Wilson was drafted into postal censorship. He worked in Liverpool, Gibraltar, and Bermuda, and was eventually transferred to the counter-espionage side of MI6 at St Albans and then moved to Washington and New York. He specialized initially in the German Intelligence network in South America and then in compiling a 'Who's Who' of the Japanese Intelligence service. He hesitated briefly over the offer of a permanent position in the Intelligence service at the end of hostilities but opted to return to Sotheby's.

The firm began to prosper when currency exchange controls were relaxed, the art market having been assisted by the strong leavening of Jewish refugee dealers who had fled to England before the war and who found the tall, urbane, quick-thinking, aristocratic personality of Wilson particularly attractive to deal with. He became head of Sotheby's picture department in the early 1950s and here formed a tremendously effective working partnership with Carmen Gronau. Wilson saw long before any of his rivals that it was the work of the Impressionist painters that would create the greatest potential interest among collectors. The sale of Jakob

Goldschmidt's seven paintings by Cézanne, Manet, Renoir, and Van Gogh in 1958, for £781,000, is acknowledged as a watershed in the history of the art market and as the beginning of Wilson's two most triumphant decades. He conducted that sale with supreme skill.

He became chairman of Sotheby's in 1957 and when in 1964 he brought off the acquisition of New York's leading auction house, Parke-Bernet, he achieved his longstanding dream of twin centres of sale on either side of the Atlantic. By the time he retired in 1980, he headed a globe-encircling network with three salerooms in London, four in the English provinces, two in New York, one in Los Angeles, and another in Amsterdam. In Europe Sotheby's had twenty-seven offices; in America and Canada there were eleven; there were sizeable outposts in South Africa, Hong Kong, and Australia and strong links with Japan. The annual turnover when he began work at Sotheby's was just over £300,000; when he left the firm it was £200,000,000.

It was an amazing achievement but Wilson had become increasingly autocratic with the years. Old friends were discarded; and because of his relentless determination to have his own way he engendered a great deal of hostility, both inside and outside the firm. He was never a good communicator; his style of management was opportunistic. His sudden departure in 1980 to his beloved estate in Clavary, near Grasse in the South of France, left ominous uncertainty over the future of Sotheby's. His principal weapon had always been the telephone and he thought he could still control the firm's affairs at a distance, but this was not to be. He had long suffered from diabetes. The effects of this, and the general frustration of life away from the centre of events, gave rise to memorably vicious outbursts of temper as he travelled about the world, and towards the end of his life—though he was still incurably active—it was difficult to remember his earlier unfailing courtesy, and the fine and mischievous sense of humour that had charmed everyone for so many years. He was appointed CBE in 1970.

His marriage had been dissolved in 1951 after he had discovered a previous latent homosexuality, although he retained a great affection for his wife after her remarriage, and they continued to meet. He died of leukaemia in the flat of a former colleague in Paris 3 June 1984.

[Frank Herrmann, *Sotheby's: Portrait of an Auction House*, 1980; Peter Wilson, 'Auctioneer Extraordinary' in *Art at Auction*, 1980; Nicholas Faith, *SOLD! A Revolution in the Art Market*, 1985; *Time and Tide*, 12 December 1963; private information; personal knowledge.]   FRANK HERRMANN

WINN, ROWLAND DENYS GUY, fourth BARON ST OSWALD (1916–1984), war correspondent, soldier, and politician, was born 19 September 1916 in London, the elder son (there were no daughters) of Rowland George Winn, third Baron St Oswald, and his wife, Eve Carew, daughter of Charles Greene. He was educated at Stowe, and at Bonn and Freiburg universities. As journalism seemed the quickest route to travel and adventure, he went to Spain for Reuters in 1935 and later covered the civil war for the *Daily Telegraph*. In 1936 he was arrested and condemned by the Republicans in Madrid and spent a number of weeks in a death cell in Barcelona until London secured his release. The *Telegraph* then sent him to the Levant and the Balkans.

At the outbreak of war in 1939 he enlisted in the Coldstream Guards. Commissioned in the 8th King's Royal Irish Hussars the following year, he saw service with them in the Western Desert, then gravitated to Cairo and irregular warfare. Based on Tara, the spirited and unconventional Special Operations Executive household in Zamalek, he was dropped into Yugoslavia and Albania. His adventures and mishaps were later light-heartedly recalled in his novel *Lord Highport Dropped at Dawn* (1949). Dropped behind the Japanese lines in Indo-China, he was recommended for the Legion of Honour, and mentioned in dispatches (1945). After the war, still an ardent hispanophile, he settled near Algeciras and even tried his hand at bullfighting; but when his old regiment was ordered to Korea in 1950, he rejoined it, fought with great courage in several desperate engagements, and returned with the MC (1952), the croix de guerre, and the Belgian Order of Leopold.

Bent on Conservative politics, he stood unsuccessfully for Dearne Valley in 1955, and his next candidature (for Pudsey) in 1957 was cut short by succeeding to the barony on his father's death. He was active in the House of Lords for the rest of his life. He was a lord-in-waiting from 1959 to 1962 in the government led by Harold Macmillan (later the Earl of Stockton) and, from 1962 to 1964, joint parliamentary secretary to the Ministry of Agriculture, Fisheries, and Food. A convinced European, he was a nominated member of the first European Parliament in 1973, but failed to retain his seat (West Yorkshire) in the 1979 election.

He loved entertaining friends at Nostell Priory, the beautiful Adam house he had inherited; it had been made over to the National Trust in 1953 but he kept the magnificent contents, the Chippendale furniture and the hundreds of fine pictures including the vast sixteenth-century copy of the lost Holbein 'Household of Sir Thomas More'. Active in

Yorkshire affairs, he became a deputy lieutenant in 1962 and honorary colonel (1967) of the 150 (Northumbrian) Regiment, Royal Corps of Transport. But his links with Spain remained intact; he worked hard at Anglo-Spanish relations, and he was awarded the grand cross of the Order of Isabel la Catòlica.

He was a sensitive man and not as agile as his career might suggest. Determination and luck got him past the Army Medical Board in 1939 and, later on a wound, and a bad parachute fall in Albania were no help. (Asked how he had got his decorations in Korea, he said: 'A great cloud of Chinese suddenly appeared but, being lame, I couldn't run away.') Fearless, impetuous, and deeply quixotic—he was 'Rowly' at home, but 'Don Rolando' in Andalusia—he sought and attained a cavalier-hidalgo stance to life that sorted well with a nature of great nobility and simplicity; his kindness, generosity, and loyalty knew no limits. There was something boyish about him; his brow was always sunk in urgency, questioningly puckered or disintegrated in laughter; and a typical last glimpse reveals him in rolled-up sleeves in the snow outside his house, loading a truck with food parcels and waving his stick as it set off for Warsaw.

In 1952 he married Laurian, the daughter of Sir (George) Roderick Jones, the head of Reuters, and his wife, Enid Bagnold, writer [qq.v.], but the marriage was dissolved in 1955 and in the same year he married Marie Wanda (died 1981), the daughter of Sigismund Jaxa-Chamiec, banker, of Warsaw; the match set the seal on a devotion to Poland which had begun in Cairo. There were no children of either marriage. The sudden death of his wife in 1981 was a knock he took very hard. He died at his house in London 19 December 1984 and was succeeded in the barony by his younger brother, Captain Derek Edward Anthony Winn (born 1919).

[Personal knowledge.]

PATRICK LEIGH FERMOR

WITTS, LESLIE JOHN (1898–1982), professor of medicine, was born 21 April 1898 at Warrington, the younger son among the four children of Wyndham John Witts, engineer in a soap works, of Warrington, and his wife, Rose Hunt. He was educated at Boteler Grammar School, Warrington, and gained an exhibition in modern languages to Manchester University. From 1916 to 1919 he served in the Royal Field Artillery and was wounded. In 1923 he qualified MB, Ch.B. at Manchester University with first class honours. After house appointments, in 1926 he took his MD at Manchester with commendation. He then received the Dickenson

travelling scholarship, which took him to New York and Prague. He next held the John Lucas Walker postgraduate studentship at Sidney Sussex, Cambridge, where he worked with H. W. (later Lord) Florey [q.v.].

In 1928 he went to work with (Sir) Arthur Ellis [q.v.] on the medical unit at the London Hospital, but a year later was appointed physician to Guy's Hospital, with the Will Edmonds clinical research fellowship. His research was mainly on blood cells, anaemia, and haematemesis, which remained lifelong interests. As well as this specialized work he had an equal interest in general medicine and its psychological aspects. He became FRCP (Lond.) in 1931 (MRCP 1926). In 1937 he published in the *Lancet* an important and practical paper entitled *Ritual Purgation in Modern Medicine* which effectively stopped the senseless purgation of acutely ill patients, an extraordinary survival from past centuries.

In 1933 he was appointed professor of medicine at St Bartholomew's Hospital, where he spent four successful years. In 1938 he became the first Nuffield professor of clinical medicine at Oxford, with a fellowship at Magdalen College. The arrival of four Nuffield professors in the Radcliffe Infirmary inevitably produced problems, chiefly in respect of accommodation and facilities. Witts, despite his own problems, realized those of the non-professorial staff and arranged a weekly meeting (with tea) to discuss matters and problems of mutual interest. Despite the difficulties caused by the prospect of war and the war itself he managed to attract good research workers from Britain and overseas, many of whom later attained distinction. He retired in 1965.

Witts was in demand as a lecturer—he gave the Goulstonian and Lumleian lectures and the Harveian oration at the Royal College of Physicians of London, the Frederick Price lecture in Dublin, the Schorstein memorial lecture at the London Hospital, the Sidney Watson Smith lecture in Edinburgh, the Gwladys and Olwen Williams lecture at Liverpool, the Shepherd lecture in Montreal, the Heath Clark lecture at London University, and the Litchfield lecture at Oxford. He served for two periods on the Medical Research Council, 1938–42 and 1943–7, and was chairman of the Council's leukaemia committee. He was a member of the Ministry of Health committee on the safety of drugs. Another interest was the Association of Physicians of Great Britain and Ireland. He served as honorary secretary and honorary treasurer from 1933 to 1948 and was president of the Association at its meeting in Oxford in 1964. He was second vice-president of the Royal College of Physicians of London in 1965–6.

He published works on *Anaemia and the Al-*

imentary Tract (1956), *The Stomach and An-aemia* (1966), and *Hypochromic Anaemia* (1969), and he edited *Medical Surveys and Clinical Trials* (2nd edn. 1964). As well as his medical books and papers, he wrote delightful essays on the personal, social, and psychological aspects of medicine, notably as 'Doctor Don' in the *Lancet* (1939) and in the series 'Personal Views' in the *British Medical Journal*. To this Dictionary he contributed the notices of Sir Arthur Hurst and T. J. (Lord) Horder. He received honorary degrees from the universities of Dublin, Bristol, Manchester, and Belfast. In 1959 he was appointed CBE. His international reputation was shown by his honorary fellowship of the College of Physicians and Surgeons of Canada, honorary membership of the Association of American Physicians, and honorary membership of the Danish Society of Internal Medicine.

His gentle and courteous manner concealed a strong and principled character. He was unaggressive, of slim build, and later in life had a thick head of white hair. In 1929 he married Nancy Grace, daughter of Louis Francis Salzman, antiquary and historian [q.v.]. It was a happy marriage, with one son, three daughters, and at the time of Witts's death ten grandchildren. He died in Oxford 19 November 1982.

[Information from Mrs Witts; Sir Douglas Black in *Munk's Roll*, vol. vii, 1984; personal knowledge.]   A. M. COOKE

WOLFENDEN, JOHN FREDERICK, BARON WOLFENDEN (1906–1985), schoolmaster and educationist, was born 26 June 1906 at Swindon, the elder son of George Wolfenden, a clerk in the Civil Service, and his wife, Emily Gaukroger, members of large Yorkshire families, who had temporarily emigrated south at the time of their elder son's birth. Two years later they had their second child, also a boy, who died at the age of five, leaving Jack—as he was universally known—an only child. Educated at Wakefield Grammar School, which had been founded in Elizabethan days, in 1924 he won a scholarship to Queen's College, Oxford, where he fell in love with philosophy and the Greeks, passions which were to endure for the rest of his life, and where he also played hockey for the university. He obtained a second class in classical honour moderations (1926) and a first in *literae humaniores* (1928). After a year at Princeton in the USA, in 1929 he was made fellow and tutor in philosophy at Magdalen College, Oxford. While there he graduated from the university hockey xi to play in goal for England in 1930 and the two following years.

When he was only twenty-seven Wolfenden was made headmaster of Uppingham School in 1934 despite formidable competition from a number of other candidates for the job. He remained there for ten years before moving to Shrewsbury School in 1944, also as headmaster, and after six years there he was appointed vice-chancellor of Reading University in 1950, where he stayed for thirteen years. He proved himself to be a man of so many talents that he was widely in demand to serve on various academic committees, charitable trusts, and governmental bodies of one kind or another. For example, during the war he was director of pre-entry training at the Air Ministry, as well as being chairman of the Youth Advisory Council at the Ministry of Education. Thereafter he chaired the Headmasters' Conference (1945, 1946, 1948, 1949), the Secondary Schools Examinations Board Council (1951–7), the National Council of Social Service (1953–60), and a large number of other bodies for greater or lesser periods of time.

But by far the most celebrated body over which he presided, between 1954 and 1957, was the departmental committee on homosexual offences and prostitution, which eventually issued a report which came to be known as the Wolfenden report. At the time these two subjects were highly contentious, especially homosexuality which was proscribed by law and regarded with a mixture of disgust and unhesitating condemnation by many people. Wolfenden, a heterosexual, knew the risk of public obloquy he would run if he were to accept the home secretary's invitation to chair such a committee, and he knew too that some of the mud, which would probably be thrown at him as a result, might well land on members of his family. It was not until after he had consulted them that he accepted the home secretary's invitation. He chaired that celebrated committee with his accustomed tact, skill, and intellectual incisiveness, and eventually produced a document which was years ahead of its time; it displayed great integrity, honesty, and courage, which must have cost him a great deal, and which was typical of him. The report's recommendations led to the legalization of homosexual activity between consenting adults.

In 1963 he stepped down as vice-chancellor of Reading to become chairman of the University Grants Committee, a post he held for five years before being appointed director and principal librarian of the British Museum for four years before retiring in 1973 at the age of sixty-seven.

Wolfenden was far from idle in retirement. Apart from serving in various capacities in the House of Lords, he presided over a number of academic bodies and committees, the Classical Association, and the National Children's Bureau as well as more than one building society, and he was a regular lecturer on Hellenic cruises, in which capacity he became known to a large number of people for the depth and width of his scholarship and knowledge of the ancient world. However, it is probably true to say that few got to know him well as a man, for he was essentially a very private person, sensitive and vulnerable to criticism, shielding himself from prying eyes behind a screen of courtesy, erudition, wit, and civilized urbanity. Among the recreations he listed over the years in *Who's Who* were: innocent (1966), weeding (1978), waiting to cross the A25 on foot (1980), trying to come to terms with arthritis, bifocals, and dentures (1983), and trying to remember (1984).

He was appointed CBE in 1942, knighted in 1956, and made a life peer in 1974. He had honorary degrees from Reading (1963), Hull (1969), Wales (1971), Manchester (1972), York (1973), and Warwick (1977), as well as American honours.

In 1932 he married Eileen Le Messurier, the second daughter of his old headmaster at Wakefield, Alfred John Spilsbury. They had two daughters and two sons, one of whom died in 1965. Wolfenden died after a short illness in hospital at Guildford 18 January 1985.

[Lord Wolfenden, *Turning Points*, 1976 (memoirs); personal knowledge.]

ANTONY BRIDGE

WRIGHT, HELENA ROSA (1887–1982), medical practitioner and pioneer of family planning, was born in Tulse Hill Road, Brixton, London, 17 September 1887, the elder daughter (there were no sons) of Heinz Lowenfeld, a Polish immigrant who had arrived in Britain with the proverbial £5 in his pocket and who later became a property owner and theatre manager, and his wife, Alice Evens, daughter of a naval captain. Her parents separated and were divorced in 1902; her father lived most of his life in Paris. Helena was educated at Cheltenham Ladies' College and the London School of Medicine for Women (later the Royal Free Hospital Medical School). She qualified MRCS (Eng.) and LRCP (Lond.) in 1914 and MB, BS in 1915.

After holding posts at the Hampstead General Hospital and the Hospital for Sick Children, Great Ormond Street, she worked at Bethnal Green Hospital where she met and married a Royal Army Medical Corps surgeon, Henry Wardel Snarey Wright, always known as Peter Wright. He was the son of James Walter Wright, a warehouseman in the postal branch of Somerset House. They had four sons.

After the war Peter and Helena Wright decided to become medical missionaries in China.

They worked at the Shantung Christian University in Tsinan until 1927. On their return to England Helena Wright immediately became involved in the movement for birth control. She was convinced that contraceptive advice should be given by doctors with special training. The North Kensington Women's Welfare Centre in Telford Road, Kensington, had been opened to give advice to poor women in 1924. Helena Wright was appointed medical officer and later became chairman of the medical committee, a position she held for thirty years. She started the first certificate of competence in family planning and the clinic became a training centre for doctors, midwives, nurses, and others, some from overseas. She addressed the Lambeth conference of Anglican bishops in 1930 and obtained limited approval to contraception for married women; in the same year she persuaded the minister of health to allow advice to be given in public welfare clinics.

She set up in private practice as a gynaecologist and was to continue in this until her final retirement in 1975. She published the first of her six books in 1930 and was working on her seventh when she died; the first, *The Sex Factor in Marriage*, was intended for those who were or were about to be married. It sold over one million copies and was the beginning of sex therapy in Britain. During World War II she tried to get contraceptive services organized for the women in the Services, but received little help from those in authority. After the war she turned her attention to the international scene together with Margaret Pyke, secretary to the Family Planning Association, and she organized a conference in Cheltenham in 1948. The twenty-three countries represented there set up the International Committee on Planned Parenthood, later to become the International Planned Parenthood Federation. Helena Wright was first treasurer and later chairman of the medical committee.

In her practice of medicine and in her thinking she was in many ways ahead of her time. This often brought disapproval from the medical establishment. She was prepared to recommend abortion for unplanned pregnancy, thus anticipating the Abortion Act of 1967. Her view was that women should have 'freedom to choose'. She arranged private third-party adoptions for unwanted babies and in doing this she ran counter to the law as in the Adoption Act of 1958; in 1968 she was prosecuted under Section 40 but was given an absolute discharge. She took an interest in prisoners and tried to persuade the governor of Holloway Prison to give contraceptive advice to women prisoners. At Wormwood Scrubs she was active in organizing drama for long-stay men.

In her private life she believed that fidelity

was not essential to a happy marriage and she spent weekends and holidays with men friends, but she and her husband never allowed the marriage to break up. After his death in 1976 she lived alone but was never lonely. She spent holidays with her friend Princess Ceril Birabongse on Lake Garda and spent time painting. As a person she was a little below normal height with piercing eyes accentuated by pince-nez; her personality was evident whenever she entered a room. She could be argumentative, though to good effect, but also gave the impression of great warmth and sympathy. She was a keen traveller, especially to India where she made her last visit alone at the age of ninety. She was a writer, teacher, and propagandist though not a great academic. She died 21 March 1982 in the Royal Free Hospital, London, following an operation for gallstones.

[Barbara Evans, *Freedom to Choose: The Life and Work of Dr Helena Wright*, 1984; *The Times*, 23 March 1982; International Planned Parenthood Federation *Medical Bulletin*, vol. xxi, no. 3, June 1987; personal knowledge.]

JOSEPHINE BARNES

WYCHERLEY, RONALD (1941–1983), singer of popular music. [See FURY, BILLY.]

Y

YATES, Dame FRANCES AMELIA (1899–1981), Renaissance historian, was born in Southsea 28 November 1899, the fourth and by ten years the youngest child of James Alfred Yates, of Portsmouth, naval architect, then chief constructor at Portsmouth dockyard and later at Chatham and on the Clyde, and his wife, Hannah Eliza Malpas. Her early formal schooling was intermittent, but from 1913 to 1917 she was at Birkenhead High School, where the elder of her two sisters then taught. Her only brother was killed in action in 1915. In 1925 the Yates family settled at Claygate, Surrey, in the house which was her home until her death. She never married.

While still at school Frances Yates had determined to be a writer. In 1924, with the aid of part-time study at University College, London, and a correspondence course, she took a first class London external degree in French (1924), following it with an MA on the sixteenth-century French theatre (1926). Her family were devoted Shakespearians, proud of nineteenth-century acting ancestry, as well as francophiles. Their modest resources enabled her to pass the next fifteen years as a private scholar.

A chance find of documents led to her first book, *John Florio: The Life of an Italian in Shakespeare's England* (1934), which won the British Academy's Rose Mary Crawshay prize. *A Study of Love's Labour's Lost* followed in 1936. Many of her later preoccupations—religion and politics in Renaissance England and France, the nature and role of a philosophic academy and the significance of philosophers such as Giordano Bruno—are discernible in these two books. Decisive for their elaboration was her first contact (1936) with the Warburg Institute. She began to frequent the Institute, finding in it the conditions for a career of adventurous scholarship which stands in marked contrast to the calm of her day-to-day existence. She joined its staff part-time in 1941, becoming full-time lecturer and editor of publications in 1944, when the Institute was incorporated in the University of London, reader in the history of the Renaissance in 1956 and, from her retirement in 1967, honorary fellow.

In 1947 the Institute published two works which confirmed her standing among students of the Renaissance. *The French Academies of the Sixteenth Century*, an account of the actual counterparts of the academy which the King of France, in *Love's Labour's Lost*, urges his courtiers to establish, characterized the nexus of political, religious, philosophical, artistic, literary, and musical activity fostered by the academies. 'Queen Elizabeth as Astraea' dealt with the origins and growth of the Protestant imperial-messianic idea and its manifestations in politics, festivals, literature, and art. This seminal essay was republished in 1975, with subsequent essays on related topics, as *Astraea. The Imperial Theme in the Sixteenth Century*. In *The Valois Tapestries* (1959, 2nd edn. 1975) she again used evidence from the arts to throw light on European religious, political, and intellectual history.

In 1949 Frances Yates returned to Bruno by way of the influence exerted on him by Ramon Lull. In *Giordano Bruno and the Hermetic Tradition* (1964) she aroused controversy by insisting that Bruno was a magus in the tradition of Italian Renaissance neo-Platonism. She drew attention to the role of Marsilio Ficino's Latin translation of 'Hermes Trismegistus', arguing also that Bruno's attempt to harness the heavens in the service of religious toleration was important in the development of modern philosophical-scientific thought. *The Art of Memory* (1966) showed how a technical trick devised by Roman orators and later used by preachers for strengthening the natural memory was given a religio-philosophical dimension by Lull, Bruno, and others. In 1965 she received a D.Lit. from London University.

Her four final works were all designed ultimately to shed light on Shakespeare. *Theatre of the World* (1969) aimed at proving influence from the Vitruvian tradition on the building of Elizabethan public theatres. In *The Rosicrucian Enlightenment* (1972) and *Shakespeare's Last Plays: A New Approach* (1975; Northcliffe lectures 1974) she investigated politics and drama in the reign of James I, in relation especially to the origins and diffusion of Rosicrucianism and an attempted revival of the Elizabethan imperial idea. Many of the dominant themes in her later work were drawn together in *The Occult Philosophy in the Elizabethan Age* (1979). Three collections of her essays and reviews (*Lull and Bruno*; *Renaissance and Reform: The Italian Contribution*; and *Ideas and Ideals in the North European Renaissance*) were published posthumously in 1982, 1983, and 1984.

She held honorary D.Litts. from Edinburgh (1969), Oxford (1970), East Anglia (1971), Exeter (1971), and Warwick (1981). She was also an honorary fellow of Lady Margaret Hall, Oxford (1970). In 1970 she was Ford lecturer at Oxford. She won the senior Wolfson history prize (1973) and the Premio Galileo Galilei (1978). Elected FBA in 1967 she was also a foreign member of the American (1975) and Royal Netherlands (1980) Academies of Arts and Sciences. In 1972 she was appointed OBE and in 1977 DBE.

Frances Yates's unselfconsciously magisterial presence was offset by charm and a sort of grand dishevelment. She worked single-mindedly and unremittingly, cared passionately for the people and problems she studied, and firmly defended her strongly individual views. Above all she strove to understand Renaissance thought, especially in its issue in contemporary action and in its continued potency. History was to her an encyclopaedic discipline, concerned as much with ideas and aspirations as with facts and events. She died 19 September 1981 in a nursing home at Surbiton, leaving the bulk of her estate to found research scholarships at the Warburg Institute, where her books and papers are preserved.

['Autobiographical Fragments' and 'List of Writings' in *Ideas and Ideals in the North European Renaissance*, 1984; *Frances A. Yates, 1899-1981*, Warburg Institute memorial booklet, 1982; personal knowledge.]

J. B. TRAPP

YOUNGHUSBAND, DAME EILEEN LOUISE (1902-1981), pioneer of social work, was born in London 1 January 1902, the only daughter and younger child (her brother died in infancy) of (Sir) Francis Edward Younghusband [q.v.], explorer and founder of the World Congress of Faiths, and his wife, Helen Augusta, daughter of Charles Magniac, MP, of Colworth, Bedfordshire. She lived in India for the first seven years of her life before returning to England with her parents. Her unusual father was one of the formative influences on her life and her mother, who was descended from an aristocratic Irish family, wrote a number of books but expected her daughter to lead the life of a debutante and to marry. After a private education Eileen Younghusband attended the London School of Economics, to her mother's dismay. There she gained an external certificate in social studies, followed by the university diploma in sociology (1926).

In 1929 she was appointed half-time tutor at LSE, but four years later she obtained a full-time post there. Her interests lay in the problems of the poor and deprived and much of her time was spent as a voluntary social worker with such organizations as the Citizens' Advice Bureau and the London county council care committees. In 1933 she was appointed a JP in Stepney; her work in the juvenile court meant much to her. She worked in the clubs run by the Bermondsey Settlement and also became involved in courses financed by the British Council for refugee women returning to their own countries after the war.

In 1941 she became principal officer for training and employment for youth leaders for the National Association of Girls' Clubs. Two years later at the request of the National Assistance Board she undertook a survey of the welfare needs of the recipients of benefit, which took her to all parts of England and Wales. In 1945 she undertook for the United Kingdom Carnegie Trust a survey which stressed the need for the provision of training for social workers. This was updated in 1951. In 1955 the Ministry of Health invited her to chair a working party on the role of social workers in the local authority health and welfare services. This covered the elderly, the handicapped, and the mentally ill. So great was the need for social workers that what became known as the Younghusband report recommended that training courses should be set up in the polytechnics and colleges of further education as well as in the universities. As a result the Council for Training in Social Work was set up and the social work certificate initiated. The health visitors formed their own council. There were those who regretted that the child care officers in the children's departments were not included. A further outcome of the Younghusband report was the establishment of the National Institute for Social Work Training in which she was much involved, acting subsequently as its consultant (1961-7).

She was much influenced by social work practice and training in the United States, where she was a frequent visitor. On the basis of the American experience she worked for a course, to be called applied social studies, to be set up in the London School of Economics. Despite considerable disagreement among the staff she remained fiercely adamant; finally with a Carnegie grant and encouragement from Baroness Elliot of Harwood she pioneered in 1954 a generic course, known as the Carnegie experiment, which became the prototype for professional social work training in other universities. The considerable disagreement among the LSE staff led to her resignation in 1957; in 1961 she was awarded an honorary fellowship.

She maintained that there were basic social work principles which applied to all settings of need, but did not deny the need for specialization in particular fields such as child care, the elderly, and the handicapped. She believed in the integration of theory and practice with theory concentrating on the understanding of human growth and development and the use of personal relationships to effect change. She recommended qualified tutors able to knit together theory and practice. She insisted that good social work practice lay not only in acquiring knowledge and skill, but in the possession of personal qualities of a high order.

She was tireless in her work overseas. After World War II she worked for UNRRA and in 1948 went to Geneva as a consultant on the social welfare fellowship programme. In 1952-3 she made a study visit of five weeks to India and Pakistan and in 1954 she followed a six-month course in the USA on a Smith Mundt fellowship. She was a consultant to the School of Social Work in Greece and the Social Welfare Department in Hong Kong and she served as an external examiner at the universities of Hong Kong, Columbia, Nottingham, Khartoum, and at the university college of Makerere. She was president of the International Association of Schools of Social Work, and later an honorary life president; this gave her the opportunity to visit Africa and Asia. To crown her international work, in 1976 she was given the René Sand award, the highest award in the field of international social work.

Eileen Younghusband was a very private person who talked little of herself or her early life. Yet she was humane and had a wide range of deep friendships; many friends visited her in her top-floor flat in Holland Park and to them in her later years she would drop from her window her front door key. She had an intense curiosity about people. Her interests, which were not confined to social work and education, were wide ranging and included the Stock Exchange. She had an abiding love for mountains,

wild flowers, birds, and butterflies. There were aspects of her personality which presented paradoxes; she was on occasion domineering, yet also humble; towards herself she was frugal, towards others she was generous. She was sophisticated and yet was almost childlike in her enjoyment of simple things—food, a journey, a flight. She was a realist, on occasions an opportunist, stubborn and manipulative, but never for herself and only for the things in which she believed. She will be remembered for her contribution to the training of social workers, her belief in social justice, and her influence on social policy. She also had a depth of spirituality, derived perhaps from her father, which was not obtrusive but of which one was aware.

She was appointed MBE (1946), CBE (1955), and DBE (1964). She had honorary degrees from the universities of British Columbia, Nottingham, Bradford, York, and Hong Kong.

Throughout her working life she experienced happiness and stimulation from her many friends in the USA—a country she loved. It was on her way to Raleigh Airport, North Carolina, that she was killed outright in a car accident at Raleigh 27 May 1981. She was unmarried.

[Kathleen Jones, *Eileen Younghusband*, 1984; private information from Robin Huws Jones; personal knowledge.]   LUCY FAITHFULL

Z

ZEC, PHILIP (1909–1983), illustrator and political cartoonist, was born 25 December 1909 on the fringe of London's Bloomsbury, the fourth-born of eleven children (nine daughters and two sons) of Simon Zec, the son of a Russian rabbi and himself a master tailor, who fled with his wife Leah Oistrakh to England to escape Tsarist oppression. His artistic talents, which may have owed something to his maternal grandfather who studied architecture in the Ukraine, were revealed early. A scholarship from the local Stanhope Street elementary school took him to St Martin's School of Art, London, where his education effectively began and where his gifts, notably in portraiture, were rapidly developed. But his vigorous draughtsmanship and flair for illustration pointed more towards commercial art, and at nineteen he set up his own studio.

Working for J. Walter Thompson and other international advertising agencies Zec became one of the leading illustrators of his day. His sculpted heads of prominent political and literary figures of the 1930s extended his range and at the same time exposed what he saw as the superficialities of the world of advertising. His work became widely recognized. One of his early posters, a vivid impression of the Flying Scotsman at speed against a night landscape, is still featured in exhibitions of steam railway memorabilia. But commercial art became too constricting both for his powerful analytical style and a political consciousness spurred by the rise of Hitlerism in Germany. As a socialist and a Jew, the notion of remaining on the sidelines drawing radio valves or coffee labels in Britain's post-Munich era became unthinkable.

Fortuitously, a colleague, Basil Nicholson (creator of the Horlicks 'night starvation' strip) faced a similar dilemma. In 1937 he was hired as features editor of the *Daily Mirror* to help transform it from a genteel picture paper for the wealthy and their servants, to an outspoken radical tabloid. Nicholson recruited Zec as the paper's political cartoonist and also a former copy-writer, (Sir) William Connor [q.v.], who wrote a forthright column under the name 'Cassandra'. Britain at war, and the traumas of Dunkirk and the blitz, offered a compelling landscape for political comment especially by an artist of Zec's passion and boldness of attack. His cartoons, drawn with a starkness and ferocity of line, captured precisely the stoicism and the humour of the British at war. Unlike (Sir) David Low [q.v.], who created satirical stereotypes like Colonel Blimp, or C. R. Giles with his preposterous 'family', Zec aimed at recognizable targets. He presented Hitler,

Goering, and others in the Nazi hierarchy as strutting buffoons. Replacing ridicule with venom, he often drew them in the form of snakes, vultures, toads, or monkeys. Not surprisingly, captured German documents listed Zec's name among those to be arrested immediately England had fallen. The defiance of the population caught in the prolonged air raids; the bravery of the armed services abroad and the civil defence at home, were depicted in the daily cartoons, some sketchily outlined in air-raid shelters.

Zec's reputation as 'the people's cartoonist', however, suffered a blow as a result of one particular cartoon. Drawn in March 1942, it showed a torpedoed sailor adrift on a raft in a dark, empty sea. The caption read: 'The price of petrol has been increased by one penny; *Official*'. It achieved an immediate, but scarcely merited notoriety. Zec had intended to alert the paper's millions of readers to the fact that the petrol they were using, perhaps wasting, cost not only money but men's lives. This was not the view taken by the government. (Sir) Winston Churchill was enraged and endorsed the charge by the home secretary, Herbert Morrison (later Lord Morrison of Lambeth, q.v.), that the cartoon plainly implied that seamen were risking their lives for the profit of the petrol companies. Zec angrily rebutted that suggestion but Morrison warned the paper's proprietors that any further 'transgression' would lead to the paper being shut down. Ironically, the offending caption had not been Zec's but Cassandra's. The columnist considered his friend's original line 'Petrol is dearer now' not strong enough.

Zec continued briefly as a cartoonist after the war but with less zeal, as he became disenchanted with what he saw as growing Left extremism in the Labour Party. In 1950–2 he was editor of the *Sunday Pictorial*. He left the *Mirror* for the *Daily Herald* in 1958. In that year he won an international prize for a cartoon with the greatest political impact, submitted by cartoonists from twenty-four countries. The subject was the crushing of Hungary by the Soviet army. He left the *Herald* in 1961. Zec was a director of the *Jewish Chronicle* for twenty-five years. He also, as a fervent supporter of the Common Market, became editor of *New Europe*. In the last three years of his life he was blind. Nevertheless he continued to proclaim his ideals as passionately and as animatedly as ever.

A tall, genial extrovert, Zec was as fastidious in his dress as he was in his draughtsmanship. Away from the drawing board he was one of the livelier raconteurs in Fleet Street's famous hostelry, El Vino's, during the flourishing years of British journalism. In 1939 he married Betty, daughter of Michael Levy, a tailor. There were no children. Zec died in Middlesex Hospital, London, 14 July 1983.

[Personal knowledge.]   DONALD ZEC

CUMULATIVE INDEX

TO THE BIOGRAPHIES CONTAINED IN THE SUPPLEMENTS

OF THE DICTIONARY OF NATIONAL BIOGRAPHY

1901–1985

Abbey, Edwin Austin	1852–**1911**
Abbey, John Roland	1894–**1969**
Abbott, Edwin Abbott	1838–**1926**
Abbott, Eric Symes	1906–**1983**
Abbott, Evelyn	1843–**1901**
À Beckett, Arthur William	1844–**1909**
Abel, Sir Frederick Augustus	1827–**1902**
Abell, Sir Westcott Stile	1877–**1961**
Aberconway, Baron. See McLaren,	
Charles Benjamin Bright	1850–**1934**
Aberconway, Baron. See McLaren,	
Henry Duncan	1879–**1953**
Abercorn, Duke of. See Hamilton,	
James	1838–**1913**
Abercrombie, Lascelles	1881–**1938**
Abercrombie, Sir (Leslie) Patrick	1879–**1957**
Abercrombie, Michael	1912–**1979**
Aberdare, Baron. See Bruce,	
Clarence Napier	1885–**1957**
Aberdeen and Temair, Marquess	
of. See Gordon, John Campbell	1847–**1934**
Aberdeen and Temair,	
Marchioness of (1857–1939).	
See under Gordon, John	
Campbell	
Aberhart, William	1878–**1943**
Abney, Sir William de Wiveleslie	1843–**1920**
Abraham, Charles John	1814–**1903**
Abraham, William	1842–**1922**
Abrahams, Doris Caroline. See	
Brahms, Caryl	1901–**1982**
Abrahams, Harold Maurice	1899–**1978**
Abramsky, Yehezkel	1886–**1976**
Abu Bakar Tafawa Balewa, Alhaji	
Sir. See Tafawa Balewa	1912–**1966**
Abul Kalam Azad, Maulana. See	
Azad	1888–**1958**
Acland, Sir Arthur Herbert Dyke	1847–**1926**
Acton, Sir Edward	1865–**1945**
Acton, John Adams-. See Adams-	
Acton	1830–**1910**
Acton, Sir John Emerich Edward	
Dalberg, Baron	1834–**1902**
Acworth, Sir William Mitchell	1850–**1925**
Adair, Gilbert Smithson	1896–**1979**
Adam, James	1860–**1907**
Adam Smith, Sir George. See	
Smith	1856–**1942**
Adami, John George	1862–**1926**
Adams, Sir Grantley Herbert	1898–**1971**

Adams, James Williams	1839–**1903**
Adams, Sir John	1857–**1934**
Adams, Sir John Bertram	1920–**1984**
Adams, John Bodkin	1899–**1983**
Adams, John Michael Geoffrey	
Manningham ('Tom')	1931–**1985**
Adams, Mary Grace Agnes	1898–**1984**
Adams, Sir Walter	1906–**1975**
Adams, William Bridges-. See	
Bridges-Adams	1889–**1965**
Adams, William Davenport	1851–**1904**
Adams, William George Stewart	1874–**1966**
Adams-Acton, John	1830–**1910**
Adamson, Sir John Ernest	1867–**1950**
Adamson, Robert	1852–**1902**
Adcock, Sir Frank Ezra	1886–**1968**
Adderley, Charles Bowyer, Baron	
Norton	1814–**1905**
Addison, Christopher, Viscount	1869–**1951**
Adeane, Michael Edward, Baron	1910–**1984**
Adler, Hermann	1839–**1911**
Adrian, Edgar Douglas, Baron	1889–**1977**
Adshead, Stanley Davenport	1868–**1946**
AE, *pseudonym*. See Russell,	
George William	1867–**1935**
Aga Khan, Aga Sultan Sir	
Mohammed Shah	1877–**1957**
Agate, James Evershed	1877–**1947**
Agnew, Sir James Wilson	1815–**1901**
Agnew, Sir William	1825–**1910**
Agnew, Sir William Gladstone	1898–**1960**
Aidé, Charles Hamilton	1826–**1906**
Aikman, George	1830–**1905**
Ainger, Alfred	1837–**1904**
Ainley, Henry Hinchliffe	1879–**1945**
Aird, Sir John	1833–**1911**
Airedale, Baron. See Kitson,	
James	1835–**1911**
Aitchison, Craigie Mason, Lord	1882–**1941**
Aitchison, George	1825–**1910**
Aitken, Alexander Craig	1895–**1967**
Aitken, William Maxwell, Baron	
Beaverbrook	1879–**1964**
Akers, Sir Wallace Alan	1888–**1954**
Akers-Douglas, Aretas, Viscount	
Chilston	1851–**1926**
Akers-Douglas, Aretas, Viscount	
Chilston	1876–**1947**
Alanbrooke, Viscount. See	
Brooke, Alan Francis	1883–**1963**

Albani, Dame Marie Louise Cécilie Emma	1852–1930
Albery, Sir Bronson James	1881–1971
Alcock, Sir John William	1892–1919
Aldenham, Baron. See Gibbs, Henry Hucks	1819–1907
Alderson, Sir Edwin Alfred Hervey	1859–1927
Alderson, Henry James	1834–1909
Aldington, Edward Godfree ('Richard')	1892–1962
Aldrich-Blake, Dame Louisa Brandreth	1865–1925
Aldridge, John Arthur Malcolm	1905–1983
Alexander, Mrs, *pseudonym.* See Hector, Annie French	1825–1902
Alexander, Albert Victor, Earl Alexander of Hillsborough	1885–1965
Alexander, Boyd	1873–1910
Alexander, (Conel) Hugh (O'Donel)	1909–1974
Alexander, Sir George	1858–1918
Alexander, Harold Rupert Leofric George, Earl Alexander of Tunis	1891–1969
Alexander, Samuel	1859–1938
Alexander, William	1824–1911
Alexander-Sinclair, Sir Edwyn Sinclair	1865–1945
Alexandra, Queen	1844–1925
Alexandra Victoria Alberta Edwina Louise Duff, Princess Arthur of Connaught, Duchess of Fife	1891–1959
Alger, John Goldworth	1836–1907
Algeranoff, Harcourt	1903–1967
Alice Mary Victoria Augusta Pauline, Princess of Great Britain and Ireland and Countess of Athlone	1883–1981
Alington, Baron. See Sturt, Henry Gerard	1825–1904
Alington, Cyril Argentine	1872–1955
Alison, Sir Archibald	1826–1907
Allan, Sir William	1837–1903
Allbutt, Sir Thomas Clifford	1836–1925
Allen, Sir Carleton Kemp	1887–1966
Allen, George	1832–1907
Allen, (Herbert) Warner	1881–1968
Allen, Sir Hugh Percy	1869–1946
Allen, Sir James	1855–1942
Allen, John Romilly	1847–1907
Allen, Norman Percy	1903–1972
Allen, Percy Stafford	1869–1933
Allen, Reginald Clifford, Baron Allen of Hurtwood	1889–1939
Allen, Robert Calder	1812–1903
Allenby, Edmund Henry Hynman, Viscount Allenby of Megiddo	1861–1936
Allerton, Baron. See Jackson, William Lawies	1840–1917
Allies, Thomas William	1813–1903

Allingham, Margery Louise	1904–1966
Allman, George Johnston	1824–1904
Alma-Tadema, Sir Lawrence	1836–1912
Almond, Hely Hutchinson	1832–1903
Altham, Harry Surtees	1888–1965
Altrincham, Baron. See Grigg, Edward William Macleay	1879–1955
Alverstone, Viscount. See Webster, Richard Everard	1842–1915
Ambedkar, Bhimrao Ramji	1891–1956
Ameer Ali, Syed	1849–1928
Amery, Leopold Charles Maurice Stennett	1873–1955
Amherst, William Amhurst Tyssen-, Baron Amherst of Hackney	1835–1909
Amoroso, Emmanuel Ciprian	1901–1982
Amory, Derick Heathcoat, Viscount	1899–1981
Amos, Sir (Percy) Maurice (Maclardie) Sheldon	1872–1940
Ampthill, Baron. See Russell, Arthur Oliver Villiers	1869–1935
Ampthill, Baron. See Russell, John Hugo	1896–1973
Amulree, Baron. See Mackenzie, William Warrender	1860–1942
Anderson, Sir Alan Garrett	1877–1952
Anderson, Alexander	1845–1909
Anderson, Sir Donald Forsyth	1906–1973
Anderson, Elizabeth Garrett	1836–1917
Anderson, George	1826–1902
Anderson, Sir Hugh Kerr	1865–1928
Anderson, John, Viscount Waverley	1882–1958
Anderson, Sir Kenneth Arthur Noel	1891–1959
Anderson, Dame Kitty	1903–1979
Anderson (formerly Macarthur), Mary Reid	1880–1921
Anderson, Stanley Arthur Charles	1884–1966
Anderson (formerly Benson), Stella	1892–1933
Anderson, Sir Thomas McCall	1836–1908
Anderson, Sir Warren Hastings	1872–1930
Andrade, Edward Neville da Costa	1887–1971
Andrewes, Sir Frederick William	1859–1932
Andrews, Sir James	1877–1951
Andrews, Thomas	1847–1907
Andrews, Sir (William) Linton	1886–1972
Angell, Sir (Ralph) Norman	1872–1967
Angus, Joseph	1816–1902
Angwin, Sir (Arthur) Stanley	1883–1959
Annandale, Thomas	1838–1907
Anson, Sir William Reynell	1843–1914
Anstey, F., *pseudonym.* See Guthrie, Thomas Anstey	1856–1934
Anstey, Frank	1865–1940
Antal, Frederick	1887–1954
Antrim, Earl of. See McDonnell, Randal John Somerled	1911–1977

Atholstan, Baron. See Graham, Hugh	1848–1938
Atkin, James Richard, Baron	1867–1944
Atkins, Sir Ivor Algernon	1869–1953
Atkinson, Sir Edward Hale Tindal	1878–1957
Atkinson, John, Baron	1844–1932
Atkinson, Robert	1839–1908
Atthill, Lombe	1827–1910
Attlee, Clement Richard, Earl	1883–1967
Attwell, Mabel Lucie	1879–1964
Aubrey, Melbourn Evans	1885–1957
Auchinleck, Sir Claude John Eyre	1884–1981
Auden, Wystan Hugh	1907–1973
Aumonier, James	1832–1911
Austen, Henry Haversham Godwin-. See Godwin-Austen	1834–1923
Austen, Sir William Chandler Roberts-. See Roberts-Austen	1843–1902
Austen Leigh, Augustus	1840–1905
Austin, Alfred	1835–1913
Austin, Herbert, Baron	1866–1941
Austin, John Langshaw	1911–1960
Avebury, Baron. See Lubbock, Sir John	1834–1913
Avon, Earl of. See Eden (Robert) Anthony	1897–1977
Avory, Sir Horace Edmund	1851–1935
Ayerst, William	1830–1904
Aylmer, Sir Felix Edward	1889–1979
Aylward, Gladys May	1902–1970
Ayrton, Michael	1921–1975
Ayrton, William Edward	1847–1908
Azad, Maulana Abul Kalam	1888–1958
Azariah, Samuel Vedanayakam	1874–1945
Babington Smith, Sir Henry. See Smith	1863–1923
Bacharach, Alfred Louis	1891–1966
Backhouse, Sir Edmund Trelawny	1873–1944
Backhouse, Sir Roger Roland Charles	1878–1939
Bacon, Sir Edmund Castell	1903–1982
Bacon, John Mackenzie	1846–1904
Bacon, Sir Reginald Hugh Spencer	1863–1947
Badcock, Sir Alexander Robert	1844–1907
Baddeley, Mountford John Byrde	1843–1906
Badeley, Henry John Fanshawe, Baron	1874–1951
Baden-Powell, Olave St. Clair, Lady	1889–1977
Baden-Powell, Robert Stephenson Smyth, Baron	1857–1941
Bader, Sir Douglas Robert Steuart	1910–1982
Bagnold, Enid Algerine, Lady Jones	1889–1981
Bagrit, Sir Leon	1902–1979
Bailey, Sir Abe	1864–1940
Bailey, Cyril	1871–1957
Bailey, Sir Edward Battersby	1881–1965

Bailey, Frederick Marshman	1882–1967
Bailey, Sir George Edwin	1879–1965
Bailey, John Cann	1864–1931
Bailey, Kenneth	1909–1963
Bailey, Mary, Lady	1890–1960
Bailey, Philip James	1816–1902
Bailhache, Sir Clement Meacher	1856–1924
Baillie, Charles Wallace Alexander Napier Ross Cochrane-, Baron Lamington	1860–1940
Baillie, Dame Isobel	1895–1983
Baillie, Sir James Black	1872–1940
Bain, Alexander	1818–1903
Bain, Francis William	1863–1940
Bain, Sir Frederick William	1889–1950
Bain, Robert Nisbet	1854–1909
Bainbridge, Francis Arthur	1874–1921
Baines, Frederick Ebenezer	1832–1911
Baird, Andrew Wilson	1842–1908
Baird, John Logie	1888–1946
Bairnsfather, Charles Bruce	1888–1959
Bairstow, Sir Edward Cuthbert	1874–1946
Bairstow, Sir Leonard	1880–1963
Bajpai, Sir Girja Shankar	1891–1954
Baker, Sir Benjamin	1840–1907
Baker, Sir Geoffrey Harding	1912–1980
Baker, Sir George Gillespie	1910–1984
Baker, Henry Frederick	1866–1956
Baker, Sir Herbert	1862–1946
Baker, Herbert Brereton	1862–1935
Baker, James Franklin Bethune-. See Bethune-Baker	1861–1951
Baker, John Fleetwood, Baron	1901–1985
Baker, Philip John Noel-, Baron Noel-Baker. See Noel-Baker	1889–1982
Baker, Shirley Waldemar	1835–1903
Balcarres, Earl of. See Lindsay, David Alexander Robert	1900–1975
Balcon, Sir Michael Elias	1896–1977
Baldwin, Stanley, Earl Baldwin of Bewdley	1867–1947
Baldwin Brown, Gerard. See Brown	1849–1932
Balewa, Alhaji Sir Abu Bakar Tafawa. See Tafawa Balewa	1912–1966
Balfour, Sir Andrew	1873–1931
Balfour, Arthur, Baron Riverdale	1873–1957
Balfour, Arthur James, Earl of Balfour	1848–1930
Balfour, Lady Frances	1858–1931
Balfour, George William	1823–1903
Balfour, Gerald William, Earl of Balfour	1853–1945
Balfour, Henry	1863–1939
Balfour, Sir Isaac Bayley	1853–1922
Balfour, John Blair, Baron Kinross	1837–1905
Balfour, Sir Thomas Graham	1858–1929
Balfour of Burleigh, Baron. See Bruce, Alexander Hugh	1849–1921
Balfour-Browne, William Alexander Francis	1874–1967

Bateson, Mary	1865–1906	Bedson, Sir Samuel Phillips	1886–1969
Bateson, William	1861–1926	Beecham, Thomas	1820–1907
Bathurst, Charles, Viscount		Beecham, Sir Thomas	1879–1961
Bledisloe	1867–1958	Beeching, Henry Charles	1859–1919
Batsford, Harry	1880–1951	Beeching, Richard, Baron	1913–1985
Batten, Edith Mary	1905–1985	Beerbohm, Sir Henry Maximilian	
Battenberg, Prince Louis		(Max)	1872–1956
Alexander of. See Mountbatten	1854–1921	Beevor, Charles Edward	1854–1908
Bauerman, Hilary	1835–1909	Bégin, Louis Nazaire	1840–1925
Bawden, Sir Frederick Charles	1908–1972	Beilby, Sir George Thomas	1850–1924
Bax, Sir Arnold Edward Trevor	1883–1953	Beit, Alfred	1853–1906
Baxter, Lucy, 'Leader Scott'	1837–1902	Beit, Sir Otto John	1865–1930
Bayley, Sir Steuart Colvin	1836–1925	Beith, John Hay, 'Ian Hay'	1876–1952
Baylis, Lilian Mary	1874–1937	Belcher, John	1841–1913
Baylis, Thomas Henry	1817–1908	Belisha, (Isaac) Leslie Hore-,	
Bayliss, Sir William Maddock	1860–1924	Baron Hore-Belisha. See Hore-	
Bayliss, Sir Wyke	1835–1906	Belisha	1893–1957
Bayly, Ada Ellen, 'Edna Lyall'	1857–1903	Bell, Alexander Graham	1847–1922
Bayly, Sir Lewis	1857–1938	Bell, (Arthur) Clive (Heward)	1881–1964
Baynes, Norman Hepburn	1877–1961	Bell, Sir Charles Alfred	1870–1945
Beach, Sir Michael Edward Hicks,		Bell, Charles Frederic Moberly	1847–1911
Earl St. Aldwyn. See Hicks Beach	1837–1916	Bell, Sir Francis Henry Dillon	1851–1936
Beachcomber, *pseudonym*. See		Bell, George Kennedy Allen	1883–1958
Morton, John Cameron Andrieu		Bell, Gertrude Margaret Lowthian	1868–1926
Bingham Michael	1893–1975	Bell, Sir (Harold) Idris	1879–1967
Beadle, Sir (Thomas) Hugh		Bell, Sir Henry Hesketh Joudou	1864–1952
(William)	1905–1980	Bell, Horace	1839–1903
Beaglehole, John Cawte	1901–1971	Bell, Sir Isaac Lowthian	1816–1904
Beale, Dorothea	1831–1906	Bell, James	1824–1908
Beale, Lionel Smith	1828–1906	Bell, Sir Thomas	1865–1952
Beardmore, William, Baron		Bell, Valentine Graeme	1839–1908
Invernairn	1856–1936	Bell, Vanessa	1879–1961
Bearsted, Viscount. See Samuel,		Bellamy, James	1819–1909
Marcus	1853–1927	Bellew, Harold Kyrle	1855–1911
Beaton, Sir Cecil Walter Hardy	1904–1980	Bellman, Sir (Charles) Harold	1886–1963
Beatrice Mary Victoria Feodore,		Bello, Sir Ahmadu, Sardauna of	
princess of Great Britain	1857–1944	Sokoto	1910–1966
Beattie-Brown, William	1831–1909	Belloc, Joseph Hilaire Pierre René	1870–1953
Beatty, Sir (Alfred) Chester	1875–1968	Bellows, John	1831–1902
Beatty, David, Earl	1871–1936	Beloe, Robert	1905–1984
Beatty, Sir Edward Wentworth	1877–1943	Bemrose, William	1831–1908
Beauchamp, Earl. See Lygon,		Bendall, Cecil	1856–1906
William	1872–1938	Benham, William	1831–1910
Beaufort, Duke of. See Somerset,		Benn, Sir Ernest John Pickstone	1875–1954
Henry Hugh Arthur FitzRoy	1900–1984	Benn, William Wedgwood,	
Beaumont, Hughes Griffiths	1908–1973	Viscount Stansgate	1877–1960
Beaver, Sir Hugh Eyre Campbell	1890–1967	Bennet-Clark, Thomas Archibald	1903–1975
Beaver, Stanley Henry	1907–1984	Bennett, Alfred William	1833–1902
Beaverbrook, Baron. See Aitken,		Bennett, Edward Hallaran	1837–1907
William Maxwell	1879–1964	Bennett, (Enoch) Arnold	1867–1931
Beazley, Sir John Davidson	1885–1970	Bennett, George Macdonald	1892–1959
Beckett, Sir Edmund, Baron		Bennett, Jack Arthur Walter	1911–1981
Grimthorpe	1816–1905	Bennett, Sir John Wheeler	
Beddoe, John	1826–1911	Wheeler-. See Wheeler-	
Bedford, Duke of. See Russell,		Bennett	1902–1975
Herbrand Arthur	1858–1940	Bennett, Peter Frederick Blaker,	
Bedford, Duchess of (1865–1937).		Baron Bennett of Edgbaston	1880–1957
See under Russell, Herbrand		Bennett, Richard Bedford,	
Arthur		Viscount	1870–1947
Bedford, William Kirkpatrick		Bennett, Sir Thomas Penberthy	1887–1980
Riland	1826–1905	Benson, Arthur Christopher	1862–1925

Blackman, Geoffrey Emett	1903–1980	Bodley Scott, Sir Ronald	1906–1982
Blackman, Vernon Herbert	1872–1967	Body, George	1840–1911
Blackwell, Sir Basil Henry	1889–1984	Boldero, Sir Harold Esmond	
Blackwell, Elizabeth	1821–1910	Arnison	1889–1960
Blackwell, Richard	1918–1980	Bols, Sir Louis Jean	1867–1930
Blackwood, Algernon Henry	1869–1951	Bomberg, David Garshen	1890–1957
Blackwood, Frederick Temple		Bompas, Henry Mason (1836–	
Hamilton-Temple, Marquess of		1909). See under Bompas,	
Dufferin and Ava	1826–1902	William Carpenter	
Blair, David	1932–1976	Bompas, William Carpenter	1834–1906
Blair, Eric Arthur, 'George		Bonar, James	1852–1941
Orwell'	1903–1950	Bonar Law, Andrew. See Law	1858–1923
Blake, Edward	1833–1912	Bond, Sir (Charles) Hubert	1870–1945
Blake, Dame Louisa Brandreth		Bond, Sir Robert	1857–1927
Aldrich-. See Aldrich-Blake	1865–1925	Bond, William Bennett	1815–1906
Blakenham, Viscount. See Hare,		Bondfield, Margaret Grace	1873–1953
John Hugh	1911–1982	Bone, James	1872–1962
Blakiston, Herbert Edward		Bone, Sir Muirhead	1876–1953
Douglas	1862–1942	Bone, Stephen	1904–1958
Blamey, Sir Thomas Albert	1884–1951	Bone, William Arthur	1871–1938
Bland, Edith (E. Nesbit)	1858–1924	Bonham-Carter, Sir Edgar	1870–1956
Bland, John Otway Percy	1863–1945	Bonham Carter, (Helen) Violet,	
Bland-Sutton, Sir John. See		Baroness Asquith of Yarnbury	1887–1969
Sutton	1855–1936	Bonney, Thomas George	1833–1923
Blandford, George Fielding	1829–1911	Bonney, (William Francis) Victor	1872–1953
Blanesburgh, Baron. See Younger,		Bonwick, James	1817–1906
Robert	1861–1946	Boosey, Leslie Arthur	1887–1979
Blaney, Thomas	1823–1903	Boot, Henry Albert Howard	1917–1983
Blanford, William Thomas	1832–1905	Boot, Jesse, Baron Trent	1850–1931
Blatchford, Robert Peel Glanville	1851–1943	Booth, Charles	1840–1916
Blaydes, Frederick Henry Marvell	1818–1908	Booth, Hubert Cecil	1871–1955
Bledisloe, Viscount. See Bathurst,		Booth, Paul Henry Gore-, Baron	
Charles	1867–1958	Gore-Booth. See Gore-Booth	1909–1984
Blennerhassett, Sir Rowland	1839–1909	Booth, William ('General' Booth)	1829–1912
Blind, Karl	1826–1907	Booth, William Bramwell	1856–1929
Bliss, Sir Arthur Edward		Boothby, Guy Newell	1867–1905
Drummond	1891–1975	Boothman, Sir John Nelson	1901–1957
Blogg, Henry George	1876–1954	Borden, Sir Robert Laird	1854–1937
Blomfield, Sir Reginald Theodore	1856–1942	Borg Olivier, Giorgio (George).	
Blood, Sir Bindon	1842–1940	See Olivier	1911–1980
Blood, Sir Hilary Rudolph Robert	1893–1967	Borthwick, Algernon, Baron	
Bloomfield, Georgiana, Lady	1822–1905	Glenesk	1830–1908
Blouet, Léon Paul, 'Max O'Rell'	1848–1903	Bosanquet, Bernard	1848–1923
Blount, Sir Edward Charles	1809–1905	Bosanquet, Sir Frederick Albert	1837–1923
Blumenfeld, Ralph David	1864–1948	Bosanquet, Robert Carr	1871–1935
Blumenthal, Jacques (Jacob)	1829–1908	Bose, Satyendranath	1894–1974
Blunden, Edmund Charles	1896–1974	Boswell, John James	1835–1908
Blunt, Lady Anne Isabella Noel		Boswell, Percy George Hamnall	1886–1960
(1837–1917). See under Blunt,		Bosworth Smith, Reginald. See	
Wilfrid Scawen		Smith	1839–1908
Blunt, Anthony Frederick	1907–1983	Botha, Louis	1862–1919
Blunt, Wilfrid Scawen	1840–1922	Bottomley, Gordon	1874–1948
Blythswood, Baron. See Campbell,		Bottomley, Horatio William	1860–1933
Archibald Campbell	1835–1908	Boucherett, Emilia Jessie	1825–1905
Blyton, Enid Mary	1897–1968	Boucicault, Dion, the younger	1859–1929
Boase, Thomas Sherrer Ross	1898–1974	Boughton, George Henry	1833–1905
Bodda Pyne, Louisa Fanny	1832–1904	Boughton, Rutland	1878–1960
Bodington, Sir Nathan	1848–1911	Boult, Sir Adrian Cedric	1889–1983
Bodkin, Sir Archibald Henry	1862–1957	Boulting, John	1913–1985
Bodkin, Thomas Patrick	1887–1961	Bourchier, Arthur	1863–1927
Bodley, George Frederick	1827–1907	Bourchier, James David	1850–1920

Bridges, Sir (George) Tom (Molesworth)	1871–1939
Bridges, John Henry	1832–1906
Bridges, Robert Seymour	1844–1930
Bridges, Sir William Throsby	1861–1915
Bridges-Adams, William	1889–1965
Bridie, James, *pseudonym*. See Mavor, Osborne Henry	1888–1951
Bridson, (Douglas) Geoffrey	1910–1980
Brierly, James Leslie	1881–1955
Briggs, John	1862–1902
Bright, Gerald Walcan-, 'Geraldo'. See Walcan-Bright	1904–1974
Bright, James Franck	1832–1920
Bright, William	1824–1901
Brightman, Frank Edward	1856–1932
Brightwen, Eliza	1830–1906
Brind, Sir (Eric James) Patrick	1892–1963
Brise, Sir Evelyn John Ruggles-. See Ruggles-Brise	1857–1935
Brittain, Sir Henry Ernest (Harry)	1873–1974
Brittain, Vera Mary	1893–1970
Britten, (Edward) Benjamin, Baron	1913–1976
Broad, Sir Charles Noel Frank	1882–1976
Broad, Charlie Dunbar	1887–1971
Broadbent, Sir William Henry	1835–1907
Broadhurst, Henry	1840–1911
Brock, Sir Osmond de Beauvoir	1869–1947
Brock, Russell Claude, Baron	1903–1980
Brock, Sir Thomas	1847–1922
Brodetsky, Selig	1888–1954
Brodie, Sir Israel	1895–1979
Brodribb, Charles William	1878–1945
Brodribb, William Jackson	1829–1905
Brodrick, George Charles	1831–1903
Brodrick, (William) St. John (Fremantle), Earl of Midleton	1856–1942
Brogan, Sir Denis William	1900–1974
Bromby, Charles Hamilton (1843–1904). See under Bromby, Charles Henry	
Bromby, Charles Henry	1814–1907
Bronowski, Jacob	1908–1974
Broodbank, Sir Joseph Guinness	1857–1944
Brook, Norman Craven, Baron Normanbrook	1902–1967
Brooke, Alan England	1863–1939
Brooke, Alan Francis, Viscount Alanbrooke	1883–1963
Brooke, Basil Stanlake, Viscount Brookeborough	1888–1973
Brooke, Sir Charles Anthony Johnson	1829–1917
Brooke, Rupert Chawner	1887–1915
Brooke, Stopford Augustus	1832–1916
Brooke, Zachary Nugent	1883–1946
Brookeborough, Viscount. See Brooke, Basil Stanlake	1888–1973
Brooke-Popham, Sir (Henry) Robert (Moore)	1878–1953

Brooking Rowe, Joshua. See Rowe	1837–1908
Broom, Robert	1866–1951
Brotherhood, Peter	1838–1902
Brough, Bennett Hooper	1860–1908
Brough, Lionel	1836–1909
Brough, Robert	1872–1905
Broughton, Rhoda	1840–1920
Brown, (Alfred) Ernest	1881–1962
Brown, Sir Arthur Whitten	1886–1948
Brown, Douglas Clifton, Viscount Ruffside	1879–1958
Brown, Ernest William	1866–1938
Brown, Frederick	1851–1941
Brown, George Alfred, Baron George-Brown	1914–1985
Brown, George Douglas, 'George Douglas'	1869–1902
Brown, Sir (George) Lindor	1903–1971
Brown, Sir George Thomas	1827–1906
Brown, Gerard Baldwin	1849–1932
Brown, Horatio Robert Forbes	1854–1926
Brown, Ivor John Carnegie	1891–1974
Brown, Sir John	1880–1958
Brown, Joseph	1809–1902
Brown, Oliver Frank Gustave	1885–1966
Brown, Peter Hume	1849–1918
Brown, Spencer Curtis. See Curtis Brown, Spencer	1906–1980
Brown, Thomas Graham	1882–1965
Brown, Sir Walter Langdon Langdon-. See Langdon-Brown	1870–1946
Brown, William	1888–1975
Brown, William Francis	1862–1951
Brown, William Haig. See Haig Brown	1823–1907
Brown, William John	1894–1960
Brown, William Michael Court	1918–1968
Browne, Edward Granville	1862–1926
Browne, George Forrest	1833–1930
Browne, Sir James Crichton-	1840–1938
Browne, Sir James Frankfort Manners	1823–1910
Browne, John Francis Archibald, sixth Baron Kilmaine	1902–1978
Browne, Sir Samuel James	1824–1901
Browne, Sir Stewart Gore-. See Gore-Browne	1883–1967
Browne, Thomas	1870–1910
Browne, William Alexander Francis Balfour-. See Balfour- Browne	1874–1967
Browning, Sir Frederick Arthur Montague	1896–1965
Browning, Sir Montague Edward	1863–1947
Browning, Oscar	1837–1923
Bruce, Alexander Hugh, Baron Balfour of Burleigh	1849–1921
Bruce, Charles Granville	1866–1939
Bruce, Clarence Napier, Baron Aberdare	1885–1957
Bruce, Sir David	1855–1931
Bruce, Sir George Barclay	1821–1908

Bury, John Bagnell 1861–1927
Bush, Eric Wheler 1899–1985
Bushell, Stephen Wootton 1844–1908
Busk, Rachel Harriette 1831–1907
Bustamante, Sir William
   Alexander 1884–1977
Butcher, Samuel Henry 1850–1910
Butler, Arthur Gray 1831–1909
Butler, Arthur John 1844–1910
Butler, Edward Joseph Aloysius
   (Dom Cuthbert) 1858–1934
Butler, Elizabeth Southerden,
   Lady 1846–1933
Butler, Frank Hedges 1855–1928
Butler, Sir (George) Geoffrey
   (Gilbert) 1887–1929
Butler, Sir Harold Beresford 1883–1951
Butler, Henry Montagu 1833–1918
Butler, Sir James Ramsay
   Montagu 1889–1975
Butler, Josephine Elizabeth 1828–1906
Butler, Sir Montagu Sherard
   Dawes 1873–1952
Butler, Reginald Cotterell 1913–1981
Butler, Richard Austen, Baron
   Butler of Saffron Walden 1902–1982
Butler, Sir Richard Harte
   Keatinge 1870–1935
Butler, Samuel 1835–1902
Butler, Sir (Spencer) Harcourt 1869–1938
Butler, Sir William Francis 1838–1910
Butlin, Sir Henry Trentham 1845–1912
Butlin, Sir William Heygate
   Edmund Colborne 1899–1980
Butt, Dame Clara Ellen 1872–1936
Butterfield, Sir Herbert 1900–1979
Butterworth, George Sainton
   Kaye 1885–1916
Buxton, Noel Edward Noel-,
   Baron Noel-Buxton. See Noel-
   Buxton 1869–1948
Buxton, Patrick Alfred 1892–1955
Buxton, Sydney Charles, Earl 1853–1934
Buxton, Sir Thomas Fowell 1837–1915
Buzzard, Sir (Edward) Farquhar 1871–1945
Byers, (Charles) Frank, Baron 1915–1984
Byng, Julian Hedworth George,
   Viscount Byng of Vimy 1862–1935
Byrne, Sir Edmund Widdrington 1844–1904
Byron, Robert 1905–1941
Bywater, Ingram 1840–1914

Cable, (Alice) Mildred 1878–1952
Cadbury, George 1839–1922
Cadman, John, Baron 1877–1941
Cadogan, Sir Alexander George
   Montagu 1884–1968
Cadogan, George Henry, Earl 1840–1915
Caillard, Sir Vincent Henry
   Penalver 1856–1930

Caine, Sir (Thomas Henry) Hall 1853–1931
Caine, William Sproston 1842–1903
Caird, Edward 1835–1908
Caird, George Bradford 1917–1984
Caird, Sir James 1864–1954
Cairnes, William Elliot 1862–1902
Cairns, David Smith 1862–1946
Cairns, Sir Hugh William Bell 1896–1952
Caldecote, Viscount. See Inskip,
   Thomas Walker Hobart 1876–1947
Caldecott, Sir Andrew 1884–1951
Calder, Peter Ritchie, Baron
   Ritchie-Calder 1906–1982
Calderon, George 1868–1915
Calkin, John Baptiste 1827–1905
Callaghan, Sir George Astley 1852–1920
Callendar, Hugh Longbourne 1863–1930
Callender, Sir Geoffrey Arthur
   Romaine 1875–1946
Callow, William 1812–1908
Callwell, Sir Charles Edward 1859–1928
Calman, William Thomas 1871–1952
Calthorpe, Baron. See Gough-
   Calthorpe, Augustus
   Cholmondeley 1829–1910
Calthorpe, Sir Somerset Arthur
   Gough- 1864–1937
Cam, Helen Maud 1885–1968
Cambridge, Duke of. See George
   William Frederick Charles 1819–1904
Cambridge, Alexander Augustus
   Frederick William Alfred
   George, Earl of Athlone 1874–1957
Cameron, Sir David Young 1865–1945
Cameron, Sir Donald Charles 1872–1948
Cameron, Sir (Gordon) Roy 1899–1966
Cameron, (Mark) James (Walter) 1911–1985
Cameron, Neil, Baron Cameron of
   Balhousie 1920–1985
Camm, Sir Sydney 1893–1966
Campbell, Archibald Campbell,
   Baron Blythswood 1835–1908
Campbell, Beatrice Stella (Mrs
   Patrick Campbell) 1865–1940
Campbell, Sir David 1889–1978
Campbell, Frederick Archibald
   Vaughan, Earl Cawdor 1847–1911
Campbell, Gordon 1886–1953
Campbell, (Ignatius) Royston
   Dunnachie (Roy) 1901–1957
Campbell, James Henry Mussen,
   Baron Glenavy 1851–1931
Campbell, Sir James Macnabb 1846–1903
Campbell, Dame Janet Mary 1877–1954
Campbell, John Charles 1894–1942
Campbell, John Douglas
   Sutherland, Duke of Argyll 1845–1914
Campbell, Lewis 1830–1908
Campbell, Sir Malcolm 1885–1948
Campbell, Patrick Gordon, Baron
   Glenavy 1913–1980

Cecil, James Edward Hubert Gascoyne-, Marquess of Salisbury	1861–1947	
Cecil, Robert Arthur James Gascoyne-, Marquess of Salisbury	1893–1972	
Cecil, Robert Arthur Talbot Gascoyne-, Marquess of Salisbury	1830–1903	
Centlivres, Albert van de Sandt	1887–1966	
Chads, Sir Henry	1819–1906	
Chadwick, Hector Munro	1870–1947	
Chadwick, Sir James	1891–1974	
Chadwick, Roy	1893–1947	
Chain, Sir Ernst Boris	1906–1979	
Challans, (Eileen) Mary, 'Mary Renault'	1905–1983	
Chalmers, James	1841–1901	
Chalmers, Sir Mackenzie Dalzell	1847–1927	
Chalmers, Robert, Baron	1858–1938	
Chamberlain, (Arthur) Neville	1869–1940	
Chamberlain, Sir Crawford Trotter	1821–1902	
Chamberlain, Houston Stewart	1855–1927	
Chamberlain, Joseph	1836–1914	
Chamberlain, Sir (Joseph) Austen	1863–1937	
Chamberlain, Sir Neville Bowles	1820–1902	
Chamberlin, Peter Hugh Girard	1919–1978	
Chambers, Dorothea Katharine	1878–1960	
Chambers, Sir Edmund Kerchever	1866–1954	
Chambers, Raymond Wilson	1874–1942	
Chambers, Sir (Stanley) Paul	1904–1981	
Chamier, Stephen Henry Edward	1834–1910	
Champneys, Basil	1842–1935	
Champneys, Sir Francis Henry	1848–1930	
Chance, Sir James Timmins	1814–1902	
Chancellor, Sir John Robert	1870–1952	
Chandos, Viscount. See Lyttelton, Oliver	1893–1972	
Channell, Sir Arthur Moseley	1838–1928	
Channer, George Nicholas	1842–1905	
Chaplin, Sir Charles Spencer	1889–1977	
Chaplin, Henry, Viscount	1840–1923	
Chapman, (Anthony) Colin (Bruce)	1928–1982	
Chapman, David Leonard	1869–1958	
Chapman, Edward John	1821–1904	
Chapman, Frederick Spencer	1907–1971	
Chapman, Robert William	1881–1960	
Chapman, Sir Ronald Ivelaw-. See Ivelaw-Chapman	1899–1978	
Chapman, Sydney	1888–1970	
Chapman, Sir Sydney John	1871–1951	
Charles, James	1851–1906	
Charles, Robert Henry	1855–1931	
Charlesworth, Martin Percival	1895–1950	
Charley, Sir William Thomas	1833–1904	
Charlot, André Eugene Maurice	1892–1956	
Charnley, Sir John	1911–1982	
Charnwood, Baron. See Benson, Godfrey Rathbone	1864–1945	
Charoux, Siegfried Joseph	1896–1967	
Charrington, Frederick Nicholas	1850–1936	
Charteris, Archibald Hamilton	1835–1908	
Chase, Drummond Percy	1820–1902	
Chase, Frederic Henry	1853–1925	
Chase, Marian Emma	1844–1905	
Chase, William St. Lucian	1856–1908	
Chatfield, Alfred Ernle Montacute, Baron	1873–1967	
Chatterjee, Sir Atul Chandra	1874–1955	
Chauvel, Sir Henry George	1865–1945	
Chavasse, Christopher Maude	1884–1962	
Chavasse, Francis James	1846–1928	
Cheadle, Walter Butler	1835–1910	
Cheatle, Arthur Henry	1866–1929	
Cheesman, Robert Ernest	1878–1962	
Cheetham, Samuel	1827–1908	
Chelmsford, Baron. See Thesiger, Frederic Augustus	1827–1905	
Chelmsford, Viscount. See Thesiger, Frederic John Napier	1868–1933	
Chenevix-Trench, Anthony	1919–1979	
Chermside, Sir Herbert Charles	1850–1929	
Cherry-Garrard, Apsley George Benet	1886–1959	
Cherwell, Viscount. See Lindemann, Frederick Alexander	1886–1957	
Cheshire, Geoffrey Chevalier	1886–1978	
Chesterton, Gilbert Keith	1874–1936	
Chetwode, Sir Philip Walhouse, Baron	1869–1950	
Chevalier, Albert	1861–1923	
Cheylesmore, Baron. See Eaton, Herbert Francis	1848–1925	
Cheylesmore, Baron. See Eaton, William Meriton	1843–1902	
Cheyne, Thomas Kelly	1841–1915	
Cheyne, Sir (William) Watson	1852–1932	
Chichester, Sir Francis Charles	1901–1972	
Chick, Dame Harriette	1875–1977	
Chifley, Joseph Benedict	1885–1951	
Child, Harold Hannyngton	1869–1945	
Child, Thomas	1839–1906	
Child-Villiers, Margaret Elizabeth, Countess of Jersey. See Villiers	1849–1945	
Child-Villiers, Victor Albert George, Earl of Jersey. See Villiers	1845–1915	
Childe, Vere Gordon	1892–1957	
Childers, Robert Erskine	1870–1922	
Childs, William Macbride	1869–1939	
Chilston, Viscount. See Akers-Douglas, Aretas	1851–1926	
Chilston, Viscount. See Akers-Douglas, Aretas	1876–1947	
Chirol, Sir (Ignatius) Valentine	1852–1929	
Chisholm, Hugh	1866–1924	
Cholmondeley, Hugh, Baron Delamere	1870–1931	

Cochrane, Douglas Mackinnon
Baillie Hamilton, Earl of
Dundonald 1852–1935
Cochrane, Sir Ralph Alexander 1895–1977
Cochrane-Baillie, Charles
Wallace Alexander Napier Ross,
Baron Lamington. See Baillie 1860–1940
Cockburn, (Francis) Claud 1904–1981
Cockcroft, Sir John Douglas 1897–1967
Cockerell, Douglas Bennett 1870–1945
Cockerell, Sir Sydney Carlyle 1867–1962
Cocks, Arthur Herbert Tennyson
Somers-, Baron Somers. See
Somers-Cocks 1887–1944
Codner, Maurice Frederick 1888–1958
Coghill, Nevill Henry Kendal
Aylmer 1899–1980
Coghlan, Sir Charles Patrick John 1863–1927
Cohen, Sir Andrew Benjamin 1909–1968
Cohen, Arthur 1829–1914
Cohen, Harriet 1896–1967
Cohen, Henry, Baron Cohen of
Birkenhead 1900–1977
Cohen, Sir John Edward (Jack) 1898–1979
Cohen, Lionel Leonard, Baron 1888–1973
Cohen, Sir Robert Waley 1877–1952
Coia, Jack Antonio 1898–1981
Coillard, François 1834–1904
Cokayne, George Edward 1825–1911
Coke, Thomas William, Earl of
Leicester 1822–1909
Coker, Ernest George 1869–1946
Cole, Cecil Jackson-. See Jackson-
Cole 1901–1979
Cole, George Douglas Howard 1889–1959
Cole, George James, Baron 1906–1979
Cole, Dame Margaret Isabel 1893–1980
Colebrook, Leonard 1883–1967
Coleman, William Stephen 1829–1904
Coleraine, Baron. See Law,
Richard Kidston 1901–1980
Coleridge, Bernard John Seymour,
Baron 1851–1927
Coleridge, Mary Elizabeth 1861–1907
Coleridge, Stephen William
Buchanan 1854–1936
Coleridge-Taylor, Samuel 1875–1912
Coles, Charles Edward, Coles
Pasha 1853–1926
Coles, Vincent Stuckey Stratton 1845–1929
Collen, Sir Edwin Henry Hayter 1843–1911
Colles, Henry Cope 1879–1943
Collett, Sir Henry 1836–1901
Collie, John Norman 1859–1942
Collier, John 1850–1934
Collings, Jesse 1831–1920
Collingwood, Cuthbert 1826–1908
Collingwood, Sir Edward Foyle 1900–1970
Collingwood, Robin George 1889–1943
Collins, John Churton 1848–1908
Collins, Josephine (José) 1887–1958

Collins, (Lewis) John 1905–1982
Collins, Michael 1890–1922
Collins, Norman Richard 1907–1982
Collins, Richard Henn, Baron 1842–1911
Collins, Sir William Alexander
Roy 1900–1976
Collins, William Edward 1867–1911
Colnaghi, Martin Henry 1821–1908
Colomb, Sir John Charles Ready 1838–1909
Colquhoun, Robert 1914–1962
Colton, Sir John 1823–1902
Colvile, Sir Henry Edward 1852–1907
Colville, David John, Baron
Clydesmuir 1894–1954
Colville, Sir Stanley Cecil James 1861–1939
Colvin, Sir Auckland 1838–1908
Colvin, Ian Duncan 1877–1938
Colvin, Sir Sidney 1845–1927
Colvin, Sir Walter Mytton. See
under Colvin, Sir Auckland
Commerell, Sir John Edmund 1829–1901
Common, Andrew Ainslie 1841–1903
Comper, Sir (John) Ninian 1864–1960
Compton, Lord Alwyne Frederick 1825–1906
Compton, Fay 1894–1978
Compton-Burnett, Dame Ivy 1884–1969
Comrie, Leslie John 1893–1950
Conder, Charles 1868–1909
Conder, Claude Reignier 1848–1910
Conesford, Baron. See Strauss,
Henry George 1892–1974
Congreve, Sir Walter Norris 1862–1927
Coningham, Sir Arthur 1895–1948
Connard, Philip 1875–1958
Connaught and Strathearn, Duke
of. See Arthur William Patrick
Albert 1850–1942
Connell, Amyas Douglas 1901–1980
Connemara, Baron. See Bourke,
Robert 1827–1902
Connolly, Cyril Vernon 1903–1974
Connor, Ralph, *pseudonym*. See
Gordon, Charles William 1860–1937
Connor, Sir William Neil 1909–1967
Conquest, George Augustus 1837–1901
Conrad, Joseph 1857–1924
Constable, William George 1887–1976
Constant, Hayne 1904–1968
Constantine, Learie Nicholas,
Baron Constantine 1901–1971
Conway, Robert Seymour 1864–1933
Conway, William Martin, Baron
Conway of Allington 1856–1937
Conybeare, Frederick Cornwallis 1856–1924
Conyngham, Sir Gerald Ponsonby
Lenox-. See Lenox-
Conyngham 1866–1956
Cook, Arthur Bernard 1868–1952
Cook, Arthur James 1883–1931
Cook, Sir Basil Alfred Kemball-.
See Kemball-Cook 1876–1949

Craigie, Pearl Mary Teresa, 'John
   Oliver Hobbes'   1867–1906
Craigie, Sir Robert Leslie   1883–1959
Craigie, Sir William Alexander   1867–1957
Craigmyle, Baron. See Shaw,
   Thomas   1850–1937
Craik, Sir Henry   1846–1927
Cranbrook, Earl of. See Gathorne-
   Hardy, Gathorne   1814–1906
Crane, Walter   1845–1915
Cranko, John Cyril   1927–1973
Crankshaw, Edward   1909–1984
Crathorne, Baron. See Dugdale,
   William Lionel   1897–1977
Craven, Hawes   1837–1910
Craven, Henry Thornton   1818–1905
Crawford, Earl of. See Lindsay,
   David Alexander Edward   1871–1940
Crawford, Earl of. See Lindsay,
   David Alexander Robert   1900–1975
Crawford, Earl of. See Lindsay,
   James Ludovic   1847–1913
Crawford, Osbert Guy Stanhope 1886–1957
Crawfurd, Oswald John Frederick 1834–1909
Crawfurd, Sir Raymond Henry
   Payne   1865–1938
Crawley, Leonard George   1903–1981
Creagh, Sir Garrett O'Moore   1848–1923
Creagh, William   1828–1901
Creasy, Sir George Elvey   1895–1972
Creech Jones, Arthur. See Jones 1891–1964
Creed, John Martin   1889–1940
Creed, Sir Thomas Percival   1897–1969
Creedy, Sir Herbert James   1878–1973
Cremer, Robert Wyndham
   Ketton-. See Ketton-Cremer 1906–1969
Cremer, Sir William Randal   1838–1908
Crew-Milnes, Robert Offley
   Ashburton, Marquess of Crewe 1858–1945
Crichton-Browne, Sir James. See
   Browne   1840–1938
Cripps, Charles Alfred, Baron
   Parmoor   1852–1941
Cripps, Dame Isobel   1891–1979
Cripps, Sir (Richard) Stafford   1889–1952
Cripps, Wilfred Joseph   1841–1903
Crispin, Edmund, *pseudonym*. See
   Montgomery, Robert Bruce   1921–1978
Crocker, Henry Radcliffe-. See
   Radcliffe-Crocker   1845–1909
Crockett, Samuel Rutherford   1860–1914
Croft, Henry Page, Baron   1881–1947
Croft, John   1833–1905
Crofts, Ernest   1847–1911
Croke, Thomas William   1824–1902
Cromer, Earl of. See Baring, Evelyn 1841–1917
Cromer, Earl of. See Baring,
   Rowland Thomas   1877–1953
Crompton, Henry   1836–1904
Crompton, Richmal. See
   Lamburn, Richmal Crompton 1890–1969

Crompton, Rookes Evelyn Bell   1845–1940
Cronin, Archibald Joseph   1896–1981
Crookes, Sir William   1832–1919
Crooks, William   1852–1921
Crookshank, Harry Frederick
   Comfort, Viscount Crookshank 1893–1961
Crosland, (Charles) Anthony (Raven) 1918–1977
Cross, Sir (Alfred) Rupert (Neale) 1912–1980
Cross, Charles Frederick   1855–1935
Cross, Kenneth Mervyn
   Baskerville   1890–1968
Cross, Richard Assheton, Viscount 1823–1914
Crossman, Richard Howard
   Stafford   1907–1974
Crossman, Sir William   1830–1901
Crosthwaite, Sir Charles Haukes
   Todd   1835–1915
Crowdy, Dame Rachel Eleanor   1884–1964
Crowe, Sir Edward Thomas
   Frederick   1877–1960
Crowe, Eyre   1824–1910
Crowe, Sir Eyre Alexander Barby
   Wichart   1864–1925
Crowther, Geoffrey, Baron   1907–1972
Crozier, William Percival   1879–1944
Cruikshank, Robert James   1898–1956
Crum, Walter Ewing   1865–1944
Crump, Charles George   1862–1935
Cruttwell, Charles Robert
   Mowbray Fraser   1887–1941
Cruttwell, Charles Thomas   1847–1911
Cubitt, William George   1835–1903
Cudlipp, Percival Thomas James 1905–1962
Cullen, William   1867–1948
Cullingworth, Charles James   1841–1908
Cullis, Winifred Clara   1875–1956
Cumberlege, Geoffrey Fenwick
   Jocelyn   1891–1979
Cummings, Arthur John   1882–1957
Cummings, Bruce Frederick, 'W.
   N. P. Barbellion'   1889–1919
Cuningham, James McNabb   1829–1905
Cunliffe-Lister, Philip, Earl of
   Swinton   1884–1972
Cunningham, Sir Alan Gordon 1887–1983
Cunningham, Andrew Browne,
   Viscount Cunningham of
   Hyndhope   1883–1963
Cunningham, Daniel John   1850–1909
Cunningham, Sir George   1888–1963
Cunningham, Sir John Henry
   Dacres   1885–1962
Cunningham, William   1849–1919
Cunninghame Graham, Robert
   Bontine. See Graham   1852–1936
Currie, Sir Arthur William   1875–1933
Currie, Sir Donald   1825–1909
Currie, Sir James   1868–1937
Currie (formerly Singleton), Mary
   Montgomerie, Lady, 'Violet
   Fane'   1843–1905

Davies, William Henry	1871–1940	De la Ramée, Marie Louise,	
Davies, William John Abbott	1890–1967	'Ouida'	1839–1908
Davies, Sir William (Llewelyn)	1887–1952	De la Rue, Sir Thomas Andros	1849–1911
D'Avigdor-Goldsmid, Sir Henry		De László, Philip Alexius. See	
Joseph	1909–1976	László de Lombos	1869–1937
Davis, Charles Edward	1827–1902	De La Warr, Earl. See Sackville,	
Davis, Henry William Carless	1874–1928	Herbrand Edward Dundonald	
Davis, Joseph	1901–1978	Brassey	1900–1976
Davitt, Michael	1846–1906	Delevingne, Sir Malcolm	1868–1950
Dawber, Sir (Edward) Guy	1861–1938	Delius, Frederick	1862–1934
Dawkins, Richard McGillivray	1871–1955	Dell, Ethel Mary. See Savage	1881–1939
Dawkins, Sir William Boyd	1837–1929	Deller, Sir Edwin	1883–1936
Dawson, Bertrand Edward,		Delmer, (Denis) Sefton	1904–1979
Viscount Dawson of Penn	1864–1945	De Madariaga, Salvador. See	
Dawson, (George) Geoffrey	1874–1944	Madariaga	1886–1978
Dawson, George Mercer	1849–1901	Demant, Vigo Auguste	1893–1983
Dawson, John	1827–1903	De Montmorency, James Edward	
Dawtry, Frank Dalmeny	1902–1968	Geoffrey	1866–1934
Day, Sir John Charles Frederic		De Montmorency, Raymond	
Sigismund	1826–1908	Harvey, Viscount Frankfort de	
Day, Lewis Foreman	1845–1910	Montmorency	1835–1902
Day, William Henry	1823–1908	De Morgan, William Frend	1839–1917
Day-Lewis, Cecil	1904–1972	Dempsey, Sir Miles Christopher	1896–1969
Deacon, Sir George Edward		Denman, Gertrude Mary, Lady	1884–1954
Raven	1906–1984	Denney, James	1856–1917
Deacon, George Frederick	1843–1909	Denning, Sir Norman Egbert	1904–1979
Deakin, Alfred	1856–1919	Denniston, Alexander Guthrie	
Deakin, Arthur	1890–1955	(Alastair)	1881–1961
Dean, Basil Herbert	1888–1978	Denniston, John Dewar	1887–1949
Dean, Sir Maurice Joseph	1906–1978	Denny, Sir Archibald	1860–1936
Dean, William Ralph ('Dixie')	1907–1980	Denny, Sir Maurice Edward	1886–1955
Deane, Sir James Parker	1812–1902	Denny, Sir Michael Maynard	1896–1972
Dearmer, Percy	1867–1936	Dent, Charles Enrique	1911–1976
De Baissac, (Marc) Claude (de		Dent, Edward Joseph	1876–1957
Boucherville)	1907–1974	Dent, Joseph Malaby	1849–1926
De Beer, Sir Gavin Rylands	1899–1972	Derby, Earl of. See Stanley,	
Debenham, Frank	1883–1965	Edward George Villiers	1865–1948
De Bunsen, Sir Maurice William		Derby, Earl of. See Stanley,	
Ernest	1852–1932	Frederick Arthur	1841–1908
De Burgh, William George	1866–1943	D'Erlanger, Sir Gerard John Regis	
De Burgh Canning, Hubert		Leo	1906–1962
George, Marquess of		De Robeck, Sir John Michael	1862–1928
Clanricarde. See Burgh		De Saulles, George William	1862–1903
Canning	1832–1916	Desborough, Baron. See Grenfell,	
De Chair, Sir Dudley Rawson		William Henry	1855–1945
Stratford	1864–1958	De Selincourt, Ernest. See	
Dee, Philip Ivor	1904–1983	Selincourt	1870–1943
Deedes, Sir Wyndham Henry	1883–1956	De Soissons, Louis Emmanuel	
De Ferranti, Sebastian Ziani. See		Jean Guy de Savoie-Carignan	1890–1962
Ferranti	1864–1930	De Stein, Sir Edward Sinauer	1887–1965
De Guingand, Sir Francis Wilfred	1900–1979	De Syllas, Stelios Messinesos	
De Havilland, Sir Geoffrey	1882–1965	(Leo)	1917–1964
De Havilland, Geoffrey Raoul	1910–1946	Des Voeux, Sir (George) William	1834–1909
De la Bedoyere, Count Michael		Detmold, Charles Maurice	1883–1908
Anthony Maurice Huchet	1900–1973	De Valera, Eamon	1882–1975
Delafield, E. M., pseudonym. See		De Vere, Aubrey Thomas	1814–1902
Dashwood, Edmée Elizabeth		De Vere, Sir Stephen Edward	1812–1904
Monica	1890–1943	Deverell, Sir Cyril John	1874–1947
De la Mare, Walter John	1873–1956	De Villiers, John Henry, Baron	1842–1914
Delamere, Baron. See		Devine, George Alexander	
Cholmondeley, Hugh	1870–1931	Cassady	1910–1966

Douglas, Sir (Henry) Percy	1876–1939	Du Cros, Sir Arthur Philip	1871–1955
Douglas, William Sholto, Baron		Dudgeon, Leonard Stanley	1876–1938
Douglas of Kirtleside	1893–1969	Dudgeon, Robert Ellis	1820–1904
Douglas, Sir William Scott	1890–1953	Dudley, Earl of. See Ward,	
Douglas-Home, Charles		William Humble	1867–1932
Cospatrick	1937–1985	Duff, Sir Alexander Ludovic	1862–1933
Douglas-Pennant, George Sholto		Duff, Sir Beauchamp	1855–1918
Gordon, Baron Penrhyn	1836–1907	Duff, Sir James Fitzjames	1898–1970
Douglas-Scott-Montagu, John		Duff, Sir Lyman Poore	1865–1955
Walter Edward, Baron Montagu		Duff, Sir Mountstuart Elphinstone	
of Beaulieu	1866–1929	Grant. See Grant Duff	1829–1906
Dove, Dame (Jane) Frances	1847–1942	Dufferin and Ava, Marquess of.	
Dove, John	1872–1934	See Blackwood, Frederick	
Dover Wilson, John. See Wilson	1881–1969	Temple Hamilton-Temple	1826–1902
Dowden, Edward	1843–1913	Duffy, Sir Charles Gavan	1816–1903
Dowden, John	1840–1910	Duffy, Sir Frank Gavan	1852–1936
Dowding, Hugh Caswall		Duffy, Patrick Vincent	1836–1909
Tremenheere, Baron	1882–1970	Duffy, Terence	1922–1985
Dowie, John Alexander	1847–1907	Dugdale, William Lionel, Baron	
Downey, Richard Joseph	1881–1953	Crathorne	1897–1977
Dowty, Sir George Herbert	1901–1975	Duke, Sir Frederick William	1863–1924
Doyle, Sir Arthur Conan	1859–1930	Duke, Henry Edward, Baron	
Doyle, John Andrew	1844–1907	Merrivale	1855–1939
D'Oyly Carte, Dame Bridget	1908–1985	Duke-Elder, Sir (William) Stewart	1898–1978
D'Oyly Carte, Richard	1844–1901	Dukes, Ashley	1885–1959
Drax, Sir Reginald Aylmer		Dulac, Edmund	1882–1953
Ranfurly Plunkett-Ernle-Erle-.		Du Maurier, Sir Gerald Hubert	
See Plunkett	1880–1967	Edward Busson	1873–1934
Drayton, Harold Charles Gilbert		Duncan, Sir Andrew Rae	1884–1952
(Harley)	1901–1966	Duncan, George Simpson	1884–1965
Dredge, James	1840–1906	Duncan, Sir John Norman Valette	
Dreschfeld, Julius	1846–1907	(Val)	1913–1975
Drew, Sir Thomas	1838–1910	Duncan, Sir Patrick	1870–1943
Dreyer, Sir Frederic Charles	1878–1956	Dundas, Lawrence John Lumley,	
Dreyer, Georges	1873–1934	Marquess of Zetland	1876–1961
Dreyer, John Louis Emil	1852–1926	Dundonald, Earl of. See	
Driberg, Thomas Edward Neil,		Cochrane, Douglas Mackinnon	
Baron Bradwell	1905–1976	Baillie Hamilton	1852–1935
Drinkwater, John	1882–1937	Dunedin, Viscount. See Murray,	
Driver, Sir Godfrey Rolles	1892–1975	Andrew Graham	1849–1942
Driver, Samuel Rolles	1846–1914	Dunhill, Thomas Frederick	1877–1946
Druce, George Claridge	1850–1932	Dunhill, Sir Thomas Peel	1876–1957
Drummond, Sir George Alexander	1829–1910	Dunlop, Sir Derrick Melville	1902–1980
Drummond, Sir Jack Cecil	1891–1952	Dunlop, John Boyd	1840–1921
Drummond, James	1835–1918	Dunmore, Earl of. See Murray,	
Drummond, James Eric, Earl of		Charles Adolphus	1841–1907
Perth	1876–1951	Dunne, Sir Laurence Rivers	1893–1970
Drummond, Sir Peter Roy		Dunphie, Charles James	1820–1908
Maxwell	1894–1945	Dunraven and Mount-Earl, Earl	
Drummond, William Henry	1854–1907	of. See Quin, Windham Thomas	
Drury, Sir Alan Nigel	1889–1980	Wyndham-	1841–1926
Drury, (Edward) Alfred (Briscoe)	1856–1944	Dunrossil, Viscount. See	
Drury-Lowe, Sir Drury Curzon	1830–1908	Morrison, William Shepherd	1893–1961
Dryland, Alfred	1865–1946	Dunsany, Baron of. See Plunkett,	
Drysdale, Charles Vickery	1874–1961	Edward John Moreton Drax	1878–1957
Drysdale, Learmont	1866–1909	Dunstan, Sir Wyndham Rowland	1861–1949
Du Cane, Sir Edmund Frederick	1830–1903	Du Parcq, Herbert, Baron	1880–1949
Duckett, Sir George Floyd	1811–1902	Dupré, August	1835–1907
Duckworth, Sir Dyce	1840–1928	Durand, Sir Henry Mortimer	1850–1924
Duckworth, Wynfrid Laurence		Durnford, Sir Walter	1847–1926
Henry	1870–1956	Dutt, (Rajani) Palme	1896–1974

Elphinstone, Sir (George) Keith		Evans, Sir Trevor Maldwyn	1902–1981
(Buller)	1865–1941	Evans, Sir (Worthington) Laming	
Elsie, Lily	1886–1962	Worthington-	1868–1931
Elsmie, George Robert	1838–1909	Evans-Pritchard, Sir Edward Evan	1902–1973
Elton, Sir Arthur Hallam Rice	1906–1973	Evatt, Herbert Vere	1894–1965
Elton, Godfrey, Baron	1892–1973	Eve, Sir Harry Trelawney	1856–1940
Elton, Oliver	1861–1945	Everard, Harry Stirling Crawfurd	1848–1909
Elvin, Sir (James) Arthur	1899–1957	Everett, Joseph David	1831–1904
Elwes, Gervase Henry Cary-	1866–1921	Everett, Sir William	1844–1908
Elwes, Henry John	1846–1922	Evershed, (Francis) Raymond,	
Elwes, Simon Edmund Vincent		Baron	1899–1966
Paul	1902–1975	Evershed, John	1864–1956
Elworthy, Frederick Thomas	1830–1907	Eversley, Baron. See Shaw-	
Embry, Sir Basil Edward	1902–1977	Lefevre, George John	1831–1928
Emery, Richard Gilbert (Dick)	1915–1983	Eves, Reginald Grenville	1876–1941
Emery, (Walter) Bryan	1903–1971	Evill, Sir Douglas Claude	
Emery, William	1825–1910	Strathern	1892–1971
Emmott, Alfred, Baron	1858–1926	Ewart, Alfred James	1872–1937
Empson, Sir William	1906–1984	Ewart, Charles Brisbane	1827–1903
Engledow, Sir Frank Leonard	1890–1985	Ewart, Sir John Alexander	1821–1904
Ensor, Sir Robert Charles		Ewart, Sir John Spencer	1861–1930
Kirkwood	1877–1958	Ewer, William Norman	1885–1977
Entwistle, William James	1895–1952	Ewing, Sir (James) Alfred	1855–1935
Epstein, Sir Jacob	1880–1959	Ewins, Arthur James	1882–1957
Erdélyi, Arthur	1908–1977	Exeter, Marquess of. See Cecil,	
Erith, Raymond Charles	1904–1973	David George Brownlow	1905–1981
Ernle, Baron. See Prothero,		Eyre, Edward John	1815–1901
Rowland Edmund	1851–1937	Eyston, George Edward Thomas	1897–1979
Ervine, (John) St. John (Greer)	1883–1971		
Escott, Bickham Aldred Cowan			
Sweet-. See Sweet-Escott	1907–1981	Faber, Sir Geoffrey Cust	1889–1961
Esdaile, Katharine Ada	1881–1950	Faber, Oscar	1886–1956
Esher, Viscount. See Brett,		Fachiri, Adila Adrienne	
Reginald Baliol	1852–1930	Adalbertina Maria	1886–1962
Esmond, Henry Vernon	1869–1922	Faed, John	1819–1902
Etheridge, Robert	1819–1903	Fagan, James Bernard	1873–1933
Euan-Smith, Sir Charles Bean	1842–1910	Fagan, Louis Alexander	1845–1903
Eumorfopoulos, George	1863–1939	Fairbairn, Andrew Martin	1838–1912
Eva, pseudonym. See under		Fairbairn, Stephen	1862–1938
O'Doherty, Kevin Izod	1823–1905	Fairbridge, Kingsley Ogilvie	1885–1924
Evan-Thomas, Sir Hugh	1862–1928	Fairey, Sir (Charles) Richard	1887–1956
Evans, Sir Arthur John	1851–1941	Fairfield, Baron. See Greer,	
Evans, (Benjamin) Ifor, Baron		(Frederick) Arthur	1863–1945
Evans of Hungershall	1899–1982	Fairfield, Cicily Isabel. See West,	
Evans, Sir Charles Arthur Lovatt	1884–1968	Dame Rebecca	1892–1983
Evans, Daniel Silvan	1818–1903	Fairley, Sir Neil Hamilton	1891–1966
Evans, Sir David Gwynne	1909–1984	Falcke, Isaac	1819–1909
Evans, Dame Edith Mary	1888–1976	Falconer, Lanoe, pseudonym. See	
Evans, Edmund	1826–1905	Hawker, Mary Elizabeth	1848–1908
Evans, Edward Ratcliffe Garth		Falconer, Sir Robert Alexander	1867–1943
Russell, Baron Mountevans	1880–1957	Falkiner, Caesar Litton	1863–1908
Evans, Sir (Evan) Vincent	1851–1934	Falkiner, Sir Frederick Richard	1831–1908
Evans, George Essex	1863–1909	Falkner, John Meade	1858–1932
Evans, Sir Guildhaume Myrddin-.		Falls, Cyril Bentham	1888–1971
See Myrddin-Evans	1894–1964	Fane, Violet, pseudonym. See	
Evans, Horace, Baron	1903–1963	Currie, Mary Montgomerie,	
Evans, Sir John	1823–1908	Lady	1843–1905
Evans, John Gwenogvryn	1852–1930	Fanshawe, Sir Edward Gennys	1814–1906
Evans, Meredith Gwynne	1904–1952	Faringdon, Baron. See	
Evans, Sir Samuel Thomas	1859–1918	Henderson, (Alexander) Gavin	1902–1977
Evans, Sebastian	1830–1909	Farjeon, Benjamin Leopold	1838–1903

FitzGerald, George Francis	1851–1901	Forbes-Sempill, William Francis,	
FitzGerald, Sir Thomas Naghten	1838–1908	Baron Sempill	1893–1965
FitzGibbon, Gerald	1837–1909	Ford, Edward Onslow	1852–1901
FitzGibbon, (Robert Louis)		Ford, Ford Madox (formerly Ford	
Constantine (Lee-Dillon)	1919–1983	Hermann Hueffer)	1873–1939
Fitzmaurice, Baron. See Petty-		Ford, Patrick	1837–1913
Fitzmaurice, Edmond George	1846–1935	Ford, William Justice	1853–1904
Fitzmaurice, Sir Gerald Gray	1901–1982	Fordham, Sir Herbert George	1854–1929
Fitzmaurice, Sir Maurice	1861–1924	Forester, Cecil Scott	1899–1966
Fitzmaurice-Kelly, James	1857–1923	Forestier-Walker, Sir Frederick	
Fitzpatrick, Sir Dennis	1837–1920	William Edward Forestier	1844–1910
FitzPatrick, Sir (James) Percy	1862–1931	Forman, Alfred William. See	
FitzRoy, Edward Algernon	1869–1943	Forman, Henry Buxton	
Flanagan, Bud	1896–1968	Forman, Henry Buxton	1842–1917
Flanders, Allan David	1910–1973	Formby, George	1904–1961
Flanders, Michael Henry	1922–1975	Forrest, Sir George William David	
Fleay, Frederick Gard	1831–1909	Starck	1845–1926
Fleck, Alexander, Baron	1889–1968	Forrest, John, Baron	1847–1918
Flecker, Herman Elroy (James		Forster, Edward Morgan	1879–1970
Elroy)	1884–1915	Forster, Hugh Oakeley Arnold-.	
Fleming, Sir Alexander	1881–1955	See Arnold-Forster	1855–1909
Fleming, Sir Arthur Percy Morris	1881–1960	Forster, Sir Martin Onslow	1872–1945
Fleming, David Hay	1849–1931	Forsyth, Andrew Russell	1858–1942
Fleming, David Pinkerton, Lord	1877–1944	Fortes, Meyer	1906–1983
Fleming, George	1833–1901	Fortescue, George Knottesford	1847–1912
Fleming, Ian Lancaster	1908–1964	Fortescue, Hugh, Earl	1818–1905
Fleming, James	1830–1908	Fortescue, Sir John William	1859–1933
Fleming, Sir (John) Ambrose	1849–1945	Foss, Hubert James	1899–1953
Fleming, (Robert) Peter	1907–1971	Foster, Sir Clement Le Neve	1841–1904
Fleming, Sir Sandford	1827–1915	Foster, Sir George Eulas	1847–1931
Fletcher, Sir Banister Flight	1866–1953	Foster, Sir Harry Braustyn Hylton	
Fletcher, Charles Robert Leslie	1857–1934	Hylton-. See Hylton-Foster	1905–1965
Fletcher, Sir Frank	1870–1954	Foster, Sir Idris Llewelyn	1911–1984
Fletcher, James	1852–1908	Foster, Sir John Galway	1903/4–1982
Fletcher, Reginald Thomas		Foster, Joseph	1844–1905
Herbert, Baron Winster	1885–1961	Foster, Sir Michael	1836–1907
Fletcher, Sir Walter Morley	1873–1933	Foster, Sir (Thomas) Gregory	1866–1931
Flett, Sir John Smith	1869–1947	Fotheringham, John Knight	1874–1936
Fleure, Herbert John	1877–1969	Fougasse, *pseudonym*. See Bird,	
Flint, Robert	1838–1910	(Cyril) Kenneth	1887–1965
Flint, Sir William Russell	1880–1969	Foulkes, Isaac	1836–1904
Florey, Howard Walter, Baron	1898–1968	Fowle, Thomas Welbank	1835–1903
Flower, Sir Cyril Thomas	1879–1961	Fowler, Alfred	1868–1940
Flower, Robin Ernest William	1881–1946	Fowler, Ellen Thorneycroft. See	
Floyer, Ernest Ayscoghe	1852–1903	Felkin	1860–1929
Fluck, Diana Mary. See Dors,		Fowler, Henry Hartley, Viscount	
Diana	1931–1984	Wolverhampton	1830–1911
Flux, Sir Alfred William	1867–1942	Fowler, Henry Watson	1858–1933
Foakes Jackson, Frederick John.		Fowler, Sir James Kingston	1852–1934
See Jackson	1855–1941	Fowler, Sir Ralph Howard	1889–1944
Fogerty, Elsie	1865–1945	Fowler, Thomas	1832–1904
Folley, (Sydney) John	1906–1970	Fowler, William Warde	1847–1921
Foot, Sir Dingle Mackintosh	1905–1978	Fox, Sir Cyril Fred	1882–1967
Foot, Isaac	1880–1960	Fox, Douglas Gerard Arthur	1893–1978
Forbes, Sir Charles Morton	1880–1960	Fox, Dame Evelyn Emily Marian	1874–1955
Forbes, George William	1869–1947	Fox, Sir Francis	1844–1927
Forbes, James Staats	1823–1904	Fox, Harold Munro	1889–1967
Forbes, (Joan) Rosita	1890–1967	Fox, Sir Lionel Wray	1895–1961
Forbes, Stanhope Alexander	1857–1947	Fox, Samson	1838–1903
Forbes-Robertson, Sir Johnston.		Fox, Terence Robert Corelli	1912–1962
See Robertson	1853–1937	Fox, Uffa	1898–1972

Gainford, Baron. See Pease,		Gaskell, Walter Holbrook	1847–1914
Joseph Albert	1860–1943	Gasquet, Francis Neil	1846–1929
Gairdner, James	1828–1912	Gaster, Moses	1856–1939
Gairdner, Sir William Tennant	1824–1907	Gatacre, Sir William Forbes	1843–1906
Gaitskell, Hugh Todd Naylor	1906–1963	Gatenby, James Brontë	1892–1960
Galbraith, Vivian Hunter	1889–1976	Gater, Sir George Henry	1886–1963
Gale, Frederick	1823–1904	Gates, Reginald Ruggles	1882–1962
Gale, Sir Humfrey Myddelton	1890–1971	Gathorne-Hardy, Gathorne, Earl	
Gale, Sir Richard Nelson	1896–1982	of Cranbrook	1814–1906
Gallacher, William	1881–1965	Gatty, Alfred	1813–1903
Galloway, Sir Alexander	1895–1977	Gauvain, Sir Henry John	1878–1945
Galloway, Sir William	1840–1927	Geddes, Auckland Campbell,	
Gallwey, Peter	1820–1906	Baron	1879–1954
Galsworthy, John	1867–1933	Geddes, Sir Eric Campbell	1875–1937
Galton, Sir Francis	1822–1911	Geddes, Sir Patrick	1854–1932
Game, Sir Philip Woolcott	1876–1961	Gedye, (George) Eric (Rowe)	1890–1970
Gamgee, Arthur	1841–1909	Gee, Samuel Jones	1839–1911
Gandhi, Indira Priyadarshani	1917–1984	Geikie, Sir Archibald	1835–1924
Gandhi, Mohandas Karamchand	1869–1948	Geikie, John Cunningham	1824–1906
Gann, Thomas William Francis	1867–1938	Gell, Sir James	1823–1905
Garbett, Cyril Forster	1875–1955	Gellibrand, Sir John	1872–1945
García, Manuel Patricio		Genée, Dame Adeline	1878–1970
Rodríguez	1805–1906	George V, King	1865–1936
Gardiner, Sir Alan Henderson	1879–1963	George VI, King	1895–1952
Gardiner, Alfred George	1865–1946	George Edward Alexander	
Gardiner, Henry Balfour	1877–1950	Edmund, Duke of Kent	1902–1942
Gardiner, Samuel Rawson	1829–1902	George William Frederick	
Gardiner, Sir Thomas Robert	1883–1964	Charles, Duke of Cambridge	1819–1904
Gardner, Ernest Arthur (1862–		George, David Lloyd, Earl Lloyd-	
1939). See under Gardner,		George of Dwyfor. See Lloyd	
Percy		George	1863–1945
Gardner, Percy	1846–1937	George, Sir Ernest	1839–1922
Gargan, Denis	1819–1903	George, Frances Louise Lloyd,	
Garner, (Joseph John) Saville,		Countess Lloyd-George of	
Baron	1908–1983	Dwyfor. See Lloyd George	1888–1972
Garner, Thomas	1839–1906	George, Gwilym Lloyd-, Viscount	
Garner, William Edward	1889–1960	Tenby. See Lloyd-George	1894–1967
Garnett, Constance Clara	1861–1946	George, Hereford Brooke	1838–1910
Garnett, David	1892–1981	George, Lady Megan Lloyd. See	
Garnett, James Clerk Maxwell	1880–1958	Lloyd George	1902–1966
Garnett, Richard	1835–1906	George, Thomas Neville	1904–1980
Garran (formerly Gamman),		George-Brown, Baron. See	
Andrew	1825–1901	Brown, George Alfred	1914–1985
Garrard, Apsley George Benet		Geraldo, *pseudonym*. See Walcan-	
Cherry-. See Cherry-Garrard	1886–1959	Bright, Gerald	1904–1974
Garrett, Fydell Edmund	1865–1907	Gerard (afterwards de Laszowska),	
Garrett Anderson, Elizabeth. See		(Jane) Emily	1849–1905
Anderson	1836–1917	Gerard, Sir Montagu Gilbert	1842–1905
Garrod, Sir Alfred Baring	1819–1907	Gere, Charles March	1869–1957
Garrod, Sir (Alfred) Guy (Roland)	1891–1965	Gerhardie, William Alexander	1895–1977
Garrod, Sir Archibald Edward	1857–1936	Gérin, Winifred Eveleen	1901–1981
Garrod, Heathcote William	1878–1960	German, Sir Edward	1862–1936
Garrod, Lawrence Paul	1895–1979	Gertler, Mark	1891–1939
Garstang, John	1876–1956	Gibb, Sir Alexander	1872–1958
Garstin, Sir William Edmund	1849–1925	Gibb, Sir Claude Dixon	1898–1959
Garth, Sir Richard	1820–1903	Gibb, Elias John Wilkinson	1857–1901
Garvie, Alfred Ernest	1861–1945	Gibb, Sir Hamilton Alexander	
Garvin, James Louis	1868–1947	Rosskeen	1895–1971
Gaselee, Sir Alfred	1844–1918	Gibberd, Sir Frederick Ernest	1908–1984
Gaselee, Sir Stephen	1882–1943	Gibbings, Robert John	1889–1958
Gask, George Ernest	1875–1951	Gibbins, Henry de Beltgens	1865–1907

Gordon, Charles William, 'Ralph
  Connor'   1860–1937
Gordon, George Stuart   1881–1942
Gordon (formerly Marjoribanks),
  Ishbel Maria, Marchioness of
  Aberdeen and Temair (1857–
  1939). See under Gordon, John
  Campbell
Gordon, James Frederick Skinner   1821–1904
Gordon, John Campbell,
  Marquess of Aberdeen and
  Temair   1847–1934
Gordon, Sir John James Hood   1832–1908
Gordon, John Rutherford   1890–1974
Gordon, Mervyn Henry   1872–1953
Gordon, Sir Thomas Edward   1832–1914
Gordon-Lennox, Charles Henry,
  Duke of Richmond and Gordon   1818–1903
Gordon-Taylor, Sir Gordon   1878–1960
Gordon Walker, Patrick
  Chrestien, Baron Gordon-
  Walker   1907–1980
Gore, Albert Augustus   1840–1901
Gore, Charles   1853–1932
Gore, George   1826–1908
Gore, John Ellard   1845–1910
Gore, (William) David Ormsby,
  Baron Harlech. See Ormsby
  Gore   1918–1985
Gore, William George Arthur
  Ormsby-, Baron Harlech. See
  Ormsby-Gore   1885–1964
Gore-Booth, Paul Henry, Baron   1909–1984
Gore-Browne, Sir Stewart   1883–1967
Gorell, Baron. See Barnes, John
  Gorell   1848–1913
Gorer, Peter Alfred Isaac   1907–1961
Gorst, Sir John Eldon   1835–1916
Gorst, Sir (John) Eldon   1861–1911
Gort, Viscount. See Vereker, John
  Standish Surtees Prendergast   1886–1946
Goschen, George Joachim,
  Viscount   1831–1907
Gosling, Harry   1861–1930
Gossage, Sir (Ernest) Leslie   1891–1949
Gosse, Sir Edmund William   1849–1928
Gosselin, Sir Martin le Marchant
  Hadsley   1847–1905
Gosset, William Sealy, 'Student'   1876–1937
Gotch, John Alfred   1852–1942
Gott, John   1830–1906
Gott, William Henry Ewart   1897–1942
Goudge, Elizabeth de Beauchamp   1900–1984
Gough, Sir Charles John Stanley   1832–1912
Gough, Herbert John   1890–1965
Gough, Sir Hubert de la Poer   1870–1963
Gough, Sir Hugh Henry   1833–1909
Gough, John Edmond   1871–1915
Gough-Calthorpe, Augustus
  Cholmondeley, Baron
  Calthorpe   1829–1910

Gough-Calthorpe, Sir Somerset
  Arthur. See Calthorpe   1864–1937
Gould, Sir Francis Carruthers   1844–1925
Gould, Nathaniel   1857–1919
Goulding, Frederick   1842–1909
Gouzenko, Igor Sergeievich   1919–1982
Gower, (Edward) Frederick
  Leveson-. See Leveson-Gower   1819–1907
Gower, Sir Henry Dudley
  Gresham Leveson   1873–1954
Gowers, Sir Ernest Arthur   1880–1966
Gowers, Sir William Richard   1845–1915
Gowrie, Earl of. See Hore-Ruthven,
  Alexander Gore Arkwright   1872–1955
Grace, Edward Mills   1841–1911
Grace, William Gilbert   1848–1915
Graham, Henry Grey   1842–1906
Graham, Hugh, Baron Atholstan   1848–1938
Graham, John Anderson   1861–1942
Graham, Robert Bontine
  Cunninghame   1852–1936
Graham, Sir Ronald William   1870–1949
Graham, Thomas Alexander
  Ferguson   1840–1906
Graham, William   1839–1911
Graham, William   1887–1932
Graham Brown, Thomas. See
  Brown   1882–1965
Graham-Harrison, Sir William
  Montagu   1871–1949
Graham-Little, Sir Ernest Gordon
  Graham   1867–1950
Grahame, Kenneth   1859–1932
Grahame-White, Claude   1879–1959
Granet, Sir (William) Guy   1867–1943
Grant, Sir (Alfred) Hamilton   1872–1937
Grant, Sir Charles (1836–1903).
  See under Grant, Sir Robert
Grant, Duncan James Corrowr   1885–1978
Grant, George Monro   1835–1902
Grant, Sir Robert   1837–1904
Grant Duff, Sir Mountstuart
  Elphinstone   1829–1906
Grantham, Sir William   1835–1911
Granville-Barker, Harley Granville   1877–1946
Graves, Alfred Perceval   1846–1931
Graves, George Windsor   1873?–1949
Graves, Robert Ranke   1895–1985
Gray, Sir Alexander   1882–1968
Gray, Sir Archibald Montague
  Henry   1880–1967
Gray, Benjamin Kirkman   1862–1907
Gray, George Buchanan   1865–1922
Gray, George Edward Kruger   1880–1943
Gray, Herbert Branston   1851–1929
Gray, Sir James   1891–1975
Gray, (Kathleen) Eileen (Moray)   1879–1976
Gray, Louis Harold   1905–1965
Greame, Philip Lloyd-, Earl of
  Swinton. See Cunliffe-Lister   1884–1972
Greaves, Walter   1846–1930

Gully, William Court, Viscount Selby	1835–1909	Haldane, John Burdon Sanderson	1892–1964
Gunn, Battiscombe George	1883–1950	Haldane, John Scott	1860–1936
Gunn, Sir James	1893–1964	Haldane, Richard Burdon,	
Günther, Albert Charles Lewis Gotthilf	1830–1914	Viscount	1856–1928
		Hale-White, Sir William	1857–1949
		Halford, Frank Bernard	1894–1955
Gunther, Robert William Theodore	1869–1940	Haliburton, Arthur Lawrence, Baron	1832–1907
Gurney, Sir Henry Lovell Goldsworthy	1898–1951	Halifax, Viscount. See Wood, Charles Lindley	1839–1934
Gurney, Henry Palin	1847–1904	Halifax, Earl of. See Wood,	
Guthrie, Sir James	1859–1930	Edward Frederick Lindley	1881–1959
Guthrie, Thomas Anstey, 'F. Anstey'	1856–1934	Hall, Sir (Alfred) Daniel	1864–1942
		Hall, Arthur Henry	1876–1949
Guthrie, William	1835–1908	Hall, Sir Arthur John	1866–1951
Guthrie, (William) Keith (Chambers)	1906–1981	Hall, Christopher Newman	1816–1902
		Hall, Sir Edward Marshall	1858–1927
Guthrie, Sir (William) Tyrone	1900–1971	Hall, FitzEdward	1825–1901
Gutteridge, Harold Cooke	1876–1953	Hall, Harry Reginald Holland	1873–1930
Guttmann, Sir Ludwig	1899–1980	Hall, Hubert	1857–1944
Guy, Sir Henry Lewis	1887–1956	Hall, Sir John	1824–1907
Gwatkin, Henry Melvill	1844–1916	Hall, Philip	1904–1982
Gwyer, Sir Maurice Linford	1878–1952	Hall, William George Glenvil	1887–1962
Gwynn, John	1827–1917	Hall, Sir (William) Reginald	1870–1943
Gwynn, Stephen Lucius	1864–1950	Hall, (William) Stephen (Richard) King-, Baron King-Hall. See	
Gwynne, Howell Arthur	1865–1950	King-Hall	1893–1966
Gwynne-Jones, Allan	1892–1982	Hall-Patch, Sir Edmund Leo	1896–1975
Gwynne-Vaughan, Dame Helen Charlotte Isabella	1879–1967	Hallé (formerly Norman-Neruda), Wilma Maria Francisca, Lady	1839–1911
		Hallett, John Hughes-. See Hughes-Hallett	1901–1972
Hacker, Arthur	1858–1919	Halliburton, William Dobinson	1860–1931
Hacking, Sir John	1888–1969	Halliday, Edward Irvine	1902–1984
Haddon, Alfred Cort	1855–1940	Halliday, Sir Frederick James	1806–1901
Haddow, Sir Alexander	1907–1976	Halliley, John Elton. See Le Mesurier, John	1912–1983
Haden, Sir Francis Seymour	1818–1910		
Hadfield, Sir Robert Abbott	1858–1940	Halsbury, Earl of. See Giffard, Hardinge Stanley	1823–1921
Hadley, Patrick Arthur Sheldon	1899–1973		
Hadley, William Waite	1866–1960	Halsey, Sir Lionel	1872–1949
Hadow, Grace Eleanor	1875–1940	Hambleden, Viscount. See Smith, William Frederick Danvers	1868–1928
Hadow, Sir (William) Henry	1859–1937		
Hadrill, (John) Michael Wallace-. See Wallace-Hadrill	1916–1985	Hamblin Smith, James. See Smith	1829–1901
		Hambourg, Mark	1879–1960
Haggard, Sir Henry Rider	1856–1925	Hambro, Sir Charles Jocelyn	1897–1963
Hahn, Kurt Matthias Robert Martin	1886–1974	Hamidullah, Nawab of Bhopal. See Bhopal	1894–1960
Haig, Douglas, Earl	1861–1928	Hamilton, Charles Harold St. John, 'Frank Richards'	1876–1961
Haig Brown, William	1823–1907		
Haigh, Arthur Elam	1855–1905	Hamilton, David James	1849–1909
Hailes, Baron. See Buchan-Hepburn, Patrick George Thomas	1901–1974	Hamilton, Sir Edward Walter	1847–1908
		Hamilton, Eugene Jacob Lee-. See Lee-Hamilton	1845–1907
Hailey, (William) Malcolm, Baron	1872–1969		
Hailsham, Viscount. See Hogg, Douglas McGarel	1872–1950	Hamilton, Sir Frederick Hew George Dalrymple-. See Dalrymple-Hamilton	1890–1974
Hailwood, (Stanley) Michael (Bailey)	1940–1981	Hamilton, Lord George Francis	1845–1927
Haines, Sir Frederick Paul	1819–1909	Hamilton, Sir Ian Standish Monteith	1853–1947
Haking, Sir Richard Cyril Byrne	1862–1945		
Halcrow, Sir William Thomson	1883–1958	Hamilton, James, Duke of Abercorn	1838–1913
Haldane, Elizabeth Sanderson	1862–1937		

Harris, Sir Percy Wyn-. See Wyn-
Harris   1903–1979
Harris, Thomas Lake   1823–1906
Harris, Tomás   1908–1964
Harris, Sir William Henry   1883–1973
Harrison, Frederic   1831–1923
Harrison, Henry   1867–1954
Harrison, Jane Ellen   1850–1928
Harrison, Mary St. Leger, 'Lucas
Malet'   1852–1931
Harrison, Reginald   1837–1908
Harrison, Sir William Montagu
Graham-. See Graham-
Harrison   1871–1949
Harrisson, Thomas Harnett
(Tom)   1911–1976
Harrod, Sir (Henry) Roy (Forbes)   1900–1978
Hart, Sir Basil Henry Liddell   1895–1970
Hart, Sir Raymund George   1899–1960
Hart, Sir Robert   1835–1911
Hartington, Marquess of. See
Cavendish, Spencer Compton   1833–1908
Hartley, Arthur Clifford   1889–1960
Hartley, Sir Charles Augustus   1825–1915
Hartley, Sir Harold Brewer   1878–1972
Hartley, Leslie Poles   1895–1972
Hartnell, Sir Norman Bishop   1901–1979
Hartog, Sir Philip(pe) Joseph   1864–1947
Hartree, Douglas Rayner   1897–1958
Hartshorn, Vernon   1872–1931
Hartshorne, Albert   1839–1910
Harty, Sir (Herbert) Hamilton   1879–1941
Harvey, Hildebrand Wolfe   1887–1970
Harvey, Sir John Martin Martin-.
See Martin-Harvey   1863–1944
Harvey, Sir Oliver Charles, Baron
Harvey of Tasburgh   1893–1968
Harwood, Basil   1859–1949
Harwood, Sir Henry Harwood   1888–1950
Haskell, Arnold Lionel David   1903–1980
Haslett, Dame Caroline Harriet   1895–1957
Hassall, Christopher Vernon   1912–1963
Hassall, John   1868–1948
Hastie, William   1842–1903
Hastings, Anthea Esther   1924–1981
Hastings, James   1852–1922
Hastings, Sir Patrick Gardiner   1880–1952
Hatry, Clarence Charles   1888–1965
Hatton, Harold Heneage Finch-.
See Finch-Hatton   1856–1904
Hatton, Joseph   1841–1907
Hatton, Sir Ronald George   1886–1965
Havelock, Sir Arthur Elibank   1844–1908
Havelock, Sir Thomas Henry   1877–1968
Haverfield, Francis John   1860–1919
Havilland, Sir Geoffrey de. See de
Havilland   1882–1965
Haweis, Hugh Reginald   1838–1901
Haweis, Mary (d. 1898). See under
Haweis, Hugh Reginald
Hawke, Sir (Edward) Anthony   1895–1964

Hawke, Sir (John) Anthony   1869–1941
Hawke, Martin Bladen, Baron
Hawke of Towton   1860–1938
Hawker, Mary Elizabeth, 'Lanoe
Falcner'   1848–1908
Hawkins, Sir Anthony Hope,
'Anthony Hope'   1863–1933
Hawkins, Henry, Baron Brampton   1817–1907
Hawkins, Herbert Leader   1887–1968
Haworth, Sir (Walter) Norman   1883–1950
Hawthorn, John Michael   1929–1959
Hawtrey, Sir Charles Henry   1858–1923
Hawtrey, Sir Ralph George   1879–1975
Hay, Sir Harley Hugh Dalrymple-   1861–1940
Hay, Ian, pseudonym. See Beith,
John Hay   1876–1952
Hayes, Edwin   1819–1904
Hayman, Henry   1823–1904
Hayne, Charles Hayne Seale-. See
Seale-Hayne   1833–1903
Hayward, Sir Isaac James   1884–1976
Hayward, John Davy   1905–1965
Hayward, Robert Baldwin   1829–1903
Hazlitt, William Carew   1834–1913
Head, Antony Henry, Viscount   1906–1983
Head, Barclay Vincent   1844–1914
Head, Sir Henry   1861–1940
Headlam, Arthur Cayley   1862–1947
Headlam, Walter George   1866–1908
Headlam-Morley, Sir James
Wycliffe   1863–1929
Heal, Sir Ambrose   1872–1959
Healy, John Edward   1872–1934
Healy, Timothy Michael   1855–1931
Hearn, Mary Anne, 'Marianne
Farningham'   1834–1909
Heath, Christopher   1835–1905
Heath, Sir (Henry) Frank   1863–1946
Heath, Sir Leopold George   1817–1907
Heath, Sir Thomas Little   1861–1940
Heath Robinson, William. See
Robinson   1872–1944
Heathcoat Amory, Derick,
Viscount Amory. See Amory   1899–1981
Heathcote, John Moyer   1834–1912
Heaton, Sir John Henniker   1848–1914
Heaviside, Oliver   1850–1925
Hector, Annie French, 'Mrs
Alexander'   1825–1902
Hector, Sir James   1834–1907
Heenan, John Carmel   1905–1975
Heffer, Reuben George   1908–1985
Heilbron, Sir Ian Morris   1886–1959
Heinemann, William   1863–1920
Heitler, Walter Heinrich   1904–1981
Hele-Shaw, Henry Selby   1854–1941
Helena Victoria, Princess   1870–1948
Hellmuth, Isaac   1817–1901
Hely-Hutchinson, Richard Walter
John, Earl of Donoughmore   1875–1948
Hemming, George Wirgman   1821–1905

Hilton, James 1900–1954
Hilton, Roger 1911–1975
Hind, Arthur Mayger 1880–1957
Hind, Henry Youle 1823–1908
Hind, Richard Dacre Archer-. See
Archer-Hind 1849–1910
Hindley, Sir Clement Daniel
Maggs 1874–1944
Hindley, John Scott, Viscount
Hyndley 1883–1963
Hingeston-Randolph (formerly
Hingston), Francis Charles 1833–1910
Hingley, Sir Benjamin 1830–1905
Hingston, Sir William Hales 1829–1907
Hinks, Arthur Robert 1873–1945
Hinkson (formerly Tynan),
Katharine 1861–1931
Hinshelwood, Sir Cyril Norman 1897–1967
Hinsley, Arthur 1865–1943
Hinton, Christopher, Baron
Hinton of Bankside 1901–1983
Hipkins, Alfred James 1826–1903
Hirst, Sir Edmund Langley 1898–1975
Hirst, Francis Wrigley 1873–1953
Hirst, George Herbert 1871–1954
Hirst, Hugo, Baron 1863–1943
Hitchcock, Sir Alfred Joseph 1899–1980
Hitchcock, Sir Eldred Frederick 1887–1959
Hitchens, (Sydney) Ivon 1893–1979
Hives, Ernest Walter, Baron 1886–1965
Hoare, Joseph Charles 1851–1906
Hoare, Sir Reginald Hervey 1882–1954
Hoare, Sir Samuel John Gurney,
Viscount Templewood 1880–1959
Hobart, Sir Percy Cleghorn
Stanley 1885–1957
Hobbes, John Oliver, *pseudonym*.
See Craigie, Pearl Mary Teresa 1867–1906
Hobbs, Sir John Berry (Jack) 1882–1963
Hobday, Sir Frederick Thomas
George 1869–1939
Hobhouse, Arthur, Baron 1819–1904
Hobhouse, Edmund 1817–1904
Hobhouse, Henry 1854–1937
Hobhouse, Sir John Richard 1893–1961
Hobhouse, Leonard Trelawny 1864–1929
Hobson, Ernest William 1856–1933
Hobson, Geoffrey Dudley 1882–1949
Hobson, John Atkinson 1858–1940
Hobson, Sir John Gardiner
Sumner 1912–1967
Hocking, Joseph (1860–1937).
See under Hocking, Silas Kitto
Hocking, Silas Kitto 1850–1935
Hodge, John 1855–1937
Hodge, Sir William Vallance
Douglas 1903–1975
Hodgetts, James Frederick 1828–1906
Hodgkin, Thomas 1831–1913
Hodgkins, Frances Mary 1869–1947
Hodgson, Ralph Edwin 1871–1962

Hodgson, Richard Dacre. See
Archer-Hind 1849–1910
Hodgson, Sir Robert MacLeod 1874–1956
Hodgson, Shadworth Hollway 1832–1912
Hodsoll, Sir (Eric) John 1894–1971
Hodson, (Francis Lord) Charlton,
Baron 1895–1984
Hodson (afterward Labouchere),
Henrietta 1841–1910
Hoey, Frances Sarah (Mrs Cashel
Hoey) 1830–1908
Hofmeyr, Jan Hendrik 1845–1909
Hofmeyr, Jan Hendrik 1894–1948
Hogarth, David George 1862–1927
Hogben, Lancelot Thomas 1895–1975
Hogg, Douglas McGarel, Viscount
Hailsham 1872–1950
Hogg, Quintin 1845–1903
Holden, Charles Henry 1875–1960
Holden, Henry Smith 1887–1963
Holden, Luther 1815–1905
Holder, Sir Frederick William 1850–1909
Holderness, Sir Thomas William 1849–1924
Holdich, Sir Thomas Hungerford 1843–1929
Holdsworth, Sir William Searle 1871–1944
Hole, Samuel Reynolds 1819–1904
Holford, William Graham, Baron 1907–1975
Holiday, Henry 1839–1927
Hollams, Sir John 1820–1910
Holland, Sir Eardley Lancelot 1879–1967
Holland, Sir (Edward) Milner 1902–1969
Holland, Henry Scott 1847–1918
Holland, Sir Henry Thurstan,
Viscount Knutsford 1825–1914
Holland, Sir Henry Tristram 1875–1965
Holland, Sir Sidney George 1893–1961
Holland, Sydney George, Viscount
Knutsford 1855–1931
Holland, Sir Thomas Erskine 1835–1926
Holland, Sir Thomas Henry 1868–1947
Holland-Martin, Sir Douglas Eric
(Deric) 1906–1977
Hollinghurst, Sir Leslie Norman 1895–1971
Hollingshead, John 1827–1904
Hollingworth, Sydney Ewart 1899–1966
Hollis, Sir Leslie Chasemore 1897–1963
Hollis, (Maurice) Christopher 1902–1977
Hollis, Sir Roger Henry 1905–1973
Holloway, Stanley Augustus 1890–1982
Hollowell, James Hirst 1851–1909
Holman Hunt, William. See Hunt 1827–1910
Holme, Charles 1848–1923
Holmes, Arthur 1890–1965
Holmes, Augusta Mary Anne 1847–1903
Holmes, Sir Charles John 1868–1936
Holmes, Sir Gordon Morgan 1876–1965
Holmes, Sir Richard Rivington 1835–1911
Holmes, Thomas 1846–1918
Holmes, Thomas Rice Edward 1855–1933
Holmes, Timothy 1825–1907
Holmes, Sir Valentine 1888–1956

Hudson, William Henry	1841–1922	Hutchinson, Francis Ernest	1871–1947
Hueffer, Ford Hermann. See		Hutchinson, Horatio Gordon	
Ford, Ford Madox	1873–1939	(Horace)	1859–1932
Hügel, Friedrich von, Baron of the		Hutchinson, John	1884–1972
Holy Roman Empire. See Von		Hutchinson, Sir Jonathan	1828–1913
Hügel	1852–1925	Hutchinson, Richard Walter John	
Hugessen, Sir Hughe		Hely-, Earl of Donoughmore.	
Montgomery Knatchbull-. See		See Hely-Hutchinson	1875–1948
Knatchbull-Hugessen	1886–1971	Hutchison, Sir Robert	1871–1960
Huggins, Godfrey Martin,		Hutchison, Sir William Oliphant	1889–1970
Viscount Malvern	1883–1971	Huth, Alfred Henry	1850–1910
Huggins, Sir William	1824–1910	Hutton, Alfred	1839–1910
Hughes, Arthur	1832–1915	Hutton, Frederick Wollaston	1836–1905
Hughes, Edward	1832–1908	Hutton, George Clark	1825–1908
Hughes, Edward David	1906–1963	Hutton, William Holden	1860–1930
Hughes, Hugh Price	1847–1902	Huxley, Aldous Leonard	1894–1963
Hughes, John	1842–1902	Huxley, Sir Julian Sorell	1887–1975
Hughes, Richard Arthur Warren	1900–1976	Huxley, Leonard	1860–1933
Hughes, Sir Sam	1853–1921	Hwfa Môn. See Williams,	
Hughes, William Morris	1862–1952	Rowland	1823–1905
Hughes-Hallett, John	1901–1972	Hyde, Douglas	1860–1949
Hulbert, John Norman (Jack)	1892–1978	Hyde, Sir Robert Robertson	1878–1967
Hulme, Frederick Edward	1841–1909	Hylton, Jack	1892–1965
Hulton, Sir Edward	1869–1925	Hylton-Foster, Sir Harry Braustyn	
Hume, Allan Octavian	1829–1912	Hylton	1905–1965
Hume, Martin Andrew Sharp	1843–1910	Hyndley, Viscount. See Hindley,	
Hume-Rothery, William	1899–1968	John Scott	1883–1963
Humphrey, Sir Andrew Henry	1921–1977	Hyndman, Henry Mayers	1842–1921
Humphrey, Herbert Alfred	1868–1951		
Humphreys, Leslie Alexander			
Francis Longmore	1904–1976	Ibbetson, Sir Denzil Charles Jelf	1847–1908
Humphreys, Sir (Richard Somers)		Ibbetson, Henry John Selwin-,	
Travers (Christmas)	1867–1956	Baron Rookwood. See Selwin-	
Humphreys, (Travers) Christmas	1901–1983	Ibbetson	1826–1902
Hunt, Dame Agnes Gwendoline	1866–1948	Ignatius, Father. See Lyne, Joseph	
Hunt, Arthur Surridge	1871–1934	Leycester	1837–1908
Hunt, George William (1829?–		Ilbert, Sir Courtenay Peregrine	1841–1924
1904). See under Macdermott,		Ilchester, Earl of. See Fox-	
Gilbert Hastings	1845–1901	Strangways, Giles Stephen	
Hunt, William	1842–1931	Holland	1874–1959
Hunt, William Holman	1827–1910	Iliffe, Edward Mauger, Baron	1877–1960
Hunter, Sir Archibald	1856–1936	Illing, Vincent Charles	1890–1969
Hunter, Colin	1841–1904	Image, Selwyn	1849–1930
Hunter, Donald	1898–1978	Imms, Augustus Daniel	1880–1949
Hunter, Sir Ellis	1892–1961	Ince, Sir Godfrey Herbert	1891–1960
Hunter, Sir George Burton	1845–1937	Ince, William	1825–1910
Hunter, Leslie Stannard	1890–1983	Inchcape, Earl of. See Mackay,	
Hunter, Philip Vassar	1883–1956	James Lyle	1852–1932
Hunter, Sir Robert	1844–1913	Inderwick, Frederick Andrew	1836–1904
Hunter, Sir William Guyer	1827–1902	Ing, (Harry) Raymond	1899–1974
Hunter-Weston, Sir Aylmer		Inge, William Ralph	1860–1954
Gould. See Weston	1864–1940	Ingham, Albert Edward	1900–1967
Huntington, George	1825–1905	Inglis, Sir Charles Edward	1875–1952
Hurcomb, Cyril William, Baron	1883–1975	Inglis, Sir Claude Cavendish	1883–1974
Hurlstone, William Yeates	1876–1906	Inglis, Elsie Maud	1864–1917
Hurst, Sir Arthur Frederick	1879–1944	Ingold, Sir Christopher Kelk	1893–1970
Hurst, Sir Cecil James Barrington	1870–1963	Ingram, Arthur Foley	
Husband, Sir (Henry) Charles	1908–1983	Winnington-. See Winnington-	
Hussey, Christopher Edward		Ingram	1858–1946
Clive	1899–1970	Ingram, Sir Bruce Stirling	1877–1963
Hutchinson, Arthur	1866–1937	Ingram, John Kells	1823–1907

Jennings, Sir (William) Ivor	1903–1965	Jones, Lady. See Bagnold, Enid	
Jephcott, Sir Harry	1891–1978	Algerine	1889–1981
Jephson, Arthur Jermy Mounteney	1858–1908	Jones, Adrian	1845–1938
Jerome, Jerome Klapka	1859–1927	Jones, (Alfred) Ernest	1879–1958
Jerram, Sir (Thomas Henry) Martyn	1858–1933	Jones, Sir Alfred Lewis	1845–1909
Jerrold, Douglas Francis	1893–1964	Jones, Allan Gwynne-. See	
Jersey, Countess of. See Villiers,		Gwynne-Jones	1892–1982
Margaret Elizabeth Child-	1849–1945	Jones, Arnold Hugh Martin	1904–1970
Jersey, Earl of. See Villiers, Victor		Jones, Arthur Creech	1891–1964
Albert George Child-	1845–1915	Jones, Sir (Bennett) Melvill	1887–1975
Jessop, Gilbert Laird	1874–1955	Jones, Bernard Mouat	1882–1953
Jessopp, Augustus	1823–1914	Jones, David	1895–1974
Jeune, Francis Henry, Baron St.		Jones, (Frederic) Wood	1879–1954
Helier	1843–1905	Jones, Sir (George) Roderick	1877–1962
Jex-Blake, Sophia Louisa	1840–1912	Jones, Sir Harold Spencer	1890–1960
Jex-Blake, Thomas William	1832–1915	Jones, Sir Henry	1852–1922
Jinnah, Mahomed Ali	1876–1948	Jones, Henry Arthur	1851–1929
Joachim, Harold Henry	1868–1938	Jones, Henry Cadman	1818–1902
Joad, Cyril Edwin Mitchinson	1891–1953	Jones, Sir Henry Stuart-	1867–1939
Joel, Jack Barnato (1862–1940).		Jones, (James) Sidney	1861–1946
See under Joel, Solomon		Jones, John Daniel	1865–1942
Barnato		Jones, Sir John Edward Lennard-.	
Joel, Solomon Barnato	1865–1931	See Lennard-Jones	1894–1954
John, Augustus Edwin	1878–1961	Jones, Sir John Morris-. See	
John, Sir Caspar	1903–1984	Morris-Jones	1864–1929
John, Sir William Goscombe	1860–1952	Jones, John Viriamu	1856–1901
Johns, Claude Hermann Walter	1857–1920	Jones, Owen Thomas	1878–1967
Johns, William Earl	1893–1968	Jones, Sir Robert	1857–1933
Johnson, Alan Woodworth	1917–1982	Jones, Sir Robert Armstrong-. See	
Johnson, Alfred Edward Webb-,		Armstrong-Jones	1857–1943
Baron Webb-Johnson. See		Jones, Thomas	1870–1955
Webb-Johnson	1880–1958	Jones, Thomas Rupert	1819–1911
Johnson, Amy	1903–1941	Jones, William West	1838–1908
Johnson, Dame Celia	1908–1982	Jordan, (Heinrich Ernst) Karl	1861–1959
Johnson, Charles	1870–1961	Jordan, Sir John Newell	1852–1925
Johnson, Harry Gordon	1923–1977	Jordan Lloyd, Dorothy. See Lloyd	1889–1946
Johnson, Hewlett	1874–1966	Joseph, Horace William Brindley	1867–1943
Johnson, John de Monins	1882–1956	Joseph, Sir Maxwell	1910–1982
Johnson, Lionel Pigot	1867–1902	Joubert, de la Ferté, Sir Philip	
Johnson, Sir Nelson King	1892–1954	Bennet	1887–1965
Johnson, Pamela Hansford, Lady		Jourdain, Francis Charles Robert	1865–1940
Snow	1912–1981	Jowitt, William Allen, Earl	1885–1957
Johnson, William Ernest	1858–1931	Joyce, James Augustine	1882–1941
Johnson, William Percival	1854–1928	Joyce, Sir Matthew Ingle	1839–1930
Johnson-Marshall, Sir Stirrat		Joynson-Hicks, William, Viscount	
Andrew William	1912–1981	Brentford. See Hicks	1865–1932
Johnston, Christopher Nicholson,		Julius, Sir George Alfred	1873–1946
Lord Sands	1857–1934		
Johnston, Edward	1872–1944		
Johnston, George Lawson, Baron			
Luke	1873–1943	Kahn-Freund, Sir Otto	1900–1979
Johnston, Sir Harry Hamilton	1858–1927	Kane, Robert Romney	1842–1902
Johnston, Sir Reginald Fleming	1874–1938	Kapitza, Piotr Leonidovich	1894–1984
Johnston, Thomas	1881–1965	Karloff, Boris	1887–1969
Johnston, William	1829–1902	Kay, (Sydney Francis) Patrick	
Joicey, James, Baron	1846–1954	(Chippindall Healey). See	
Jolowicz, Herbert Felix	1890–1954	Dolin, Sir Anton	1904–1983
Joly, Charles Jasper	1864–1906	Kearley, Hudson Ewbanke,	
Joly, John	1857–1933	Viscount Devonport	1856–1934
Joly de Lotbinière, Sir Henry		Keating, Thomas Patrick	1917–1984
Gustave	1829–1908	Keay, John Seymour	1839–1909

Kindersley, Hugh Kenyon
Molesworth, Baron   1899–1976
Kindersley, Robert Molesworth,
Baron   1871–1954
King, Earl Judson   1901–1962
King, Edward   1829–1910
King, Sir (Frederic) Truby   1858–1938
King, Sir George   1840–1909
King, Harold   1887–1956
King, Haynes   1831–1904
King, William Bernard Robinson   1889–1963
King, William Lyon Mackenzie   1874–1950
King-Hall, (William) Stephen
(Richard), Baron   1893–1966
Kingdon-Ward, Francis (Frank)   1885–1958
Kingsburgh, Lord. See
Macdonald, John Hay Athole   1836–1919
Kingscote, Sir Robert Nigel
Fitzhardinge   1830–1908
Kingsford, Charles Lethbridge   1862–1926
Kingston, Charles Cameron   1850–1908
Kinnear, Alexander Smith, Baron   1833–1917
Kinnear, Sir Norman Boyd   1882–1957
Kinns, Samuel   1826–1903
Kinross, Baron. See Balfour, John
Blair   1837–1905
Kipling, (Joseph) Rudyard   1865–1936
Kipping, Frederic Stanley   1863–1949
Kipping, Sir Norman Victor   1901–1979
Kirk, Sir John   1832–1922
Kirk, Sir John   1847–1922
Kirk, Kenneth Escott   1886–1954
Kirk, Norman Eric   1923–1974
Kirkbride, Sir Alec Seath   1897–1978
Kirkman, Sir Sidney Chevalier   1895–1982
Kirkpatrick, Sir Ivone Augustine   1897–1964
Kirkwood, David, Baron   1872–1955
Kitchener, Horatio Herbert, Earl   1850–1916
Kitchin, George William   1827–1912
Kitson, James, Baron Airedale   1835–1911
Kitson Clark, George Sidney
Roberts   1900–1975
Kitton, Frederick George   1856–1904
Klein, Melanie   1882–1960
Klugmann, Norman John ('James')   1912–1977
Knatchbull-Hugessen, Sir Hughe
Montgomery   1886–1971
Knight, (George) Wilson   1897–1985
Knight, Harold   1874–1961
Knight, Joseph   1829–1907
Knight, Joseph   1837–1909
Knight, Dame Laura (1877–
1970). See under Knight, Harold
Knollys, Edward George William
Tyrwhitt, Viscount   1895–1966
Knollys, Francis, Viscount   1837–1924
Knott, Ralph   1878–1929
Knowles, Dom David. See
Knowles, Michael Clive   1896–1974
Knowles, Sir Francis Gerald
William   1915–1974

Knowles, Sir James Thomas   1831–1908
Knowles, Michael Clive (Dom
David)   1896–1974
Knox, Edmund Arbuthnott   1847–1937
Knox, Edmund George Valpy   1881–1971
Knox, Sir Geoffrey George   1884–1958
Knox, Sir George Edward   1845–1922
Knox (formerly Craig), Isa   1831–1903
Knox, Ronald Arbuthnott   1888–1957
Knox, Wilfred Lawrence   1886–1950
Knox-Little, William John   1839–1918
Knutsford, Viscount. See Holland,
Sir Henry Thurstan   1825–1914
Knutsford, Viscount. See Holland,
Sydney George   1855–1931
Koestler, Arthur   1905–1983
Kokoschka, Oskar   1886–1980
Komisarjevsky, Theodore   1882–1954
Kompfner, Rudolf   1909–1977
Korda, Sir Alexander   1893–1956
Kotzé, Sir John Gilbert   1849–1940
Krebs, Sir Hans Adolf   1900–1981
Kronberger, Hans   1920–1970
Kruger Gray, George Edward. See
Gray   1880–1943
Küchemann, Dietrich   1911–1976
Kuczynski, Robert Rene   1876–1947
Kylsant, Baron. See Philipps,
Owen Cosby   1863–1937
Kynaston (formerly Snow),
Herbert   1835–1910

Labouchere, Henrietta. See
Hodson   1841–1910
Labouchere, Henry Du Pré   1831–1912
Lacey, Thomas Alexander   1853–1931
Lachmann, Gustav Victor   1896–1966
Lack, David Lambert   1910–1973
Lafont, Eugène   1837–1908
Laidlaw, Anna Robena   1819–1901
Laidlaw, John   1832–1906
Laidlaw, Sir Patrick Playfair   1881–1940
Laird, John   1887–1946
Lake, Kirsopp   1872–1946
Lake, Sir Percy Henry Noel   1855–1940
Lamb, Henry Taylor   1883–1960
Lamb, Sir Horace   1849–1934
Lamb, Lynton Harold   1907–1977
Lambart, Frederick Rudolph, Earl
of Cavan   1865–1946
Lambe, Sir Charles Edward   1900–1960
Lambert, Brooke   1834–1901
Lambert, Constant   1905–1951
Lambert, George   1842–1915
Lambert, George, Viscount   1866–1958
Lambert, Maurice   1901–1964
Lambourne, Baron. See
Lockwood, Amelius Mark
Richard   1847–1928
Lamburn, Richmal Crompton-   1890–1969

Ledward, Gilbert	1888–1960	Leno, Dan	1860–1904
Ledwidge, Francis	1891–1917	Lenox-Conyngham, Sir Gerald	
Lee, Sir (Albert) George	1879–1967	Ponsonby	1866–1956
Lee, Arthur Hamilton, Viscount		Leon, Henry Cecil, 'Henry Cecil'	1902–1976
Lee of Fareham	1868–1947	Le Sage, Sir John Merry	1837–1926
Lee, Sir Frank Godbould	1903–1971	Leslie, Sir Bradford	1831–1926
Lee, Frederick George	1832–1902	Leslie, Sir John Randolph	
Lee, Rawdon Briggs	1845–1908	('Shane')	1885–1971
Lee, Robert Warden	1868–1958	Lester, Sean (John Ernest)	1888–1959
Lee, Sir Sidney	1859–1926	Le Strange, Guy	1854–1933
Lee, Vernon, *pseudonym*. See		Lethaby, William Richard	1857–1931
Paget, Violet	1856–1935	Lett, Sir Hugh	1876–1964
Lee-Hamilton, Eugene Jacob	1845–1907	Lever, Sir (Samuel) Hardman	1869–1947
Lee-Warner, Sir William	1846–1914	Lever, William Hesketh, Viscount	
Lees, George Martin	1898–1955	Leverhulme	1851–1925
Leese, Sir Oliver William		Leverhulme, Viscount. See Lever,	
Hargreaves	1894–1978	William Hesketh	
Leeson, Spencer Stottesbery		Leveson-Gower, (Edward)	
Gwatkin	1892–1956	Frederick	1819–1907
Le Fanu, Sir Michael	1913–1970	Leveson Gower, Sir Henry Dudley	
Lefroy, William	1836–1909	Gresham. See Gower	1873–1954
Le Gallienne, Richard Thomas	1866–1947	Levick, George Murray	1876–1956
Legg, John Wickham	1843–1921	Levy, Benn Wolfe	1900–1973
Legh, Thomas Wodehouse, Baron		Levy, Hyman	1889–1975
Newton	1857–1942	Levy-Lawson, Edward, Baron	
Legros, Alphonse	1837–1911	Burnham	1833–1916
Le Gros Clark, Frederick. See		Levy-Lawson, Harry Lawson	
Clark	1892–1977	Webster, Viscount Burnham.	
Le Gros Clark, Sir Wilfrid		See Lawson	1862–1933
Edward. See Clark	1895–1971	Lewin, (George) Ronald	1914–1984
Lehmann, Rudolf	1819–1905	Lewis, Agnes	1843–1926
Leicester, Earl of. See Coke,		Lewis, Sir Anthony Carey	1915–1983
Thomas William	1822–1909	Lewis, Sir Aubrey Julian	1900–1975
Leigh, Vivien	1913–1967	Lewis, Bunnell	1824–1908
Leigh-Mallory, Sir Trafford Leigh	1892–1944	Lewis, Cecil Day-. See Day-Lewis	1904–1972
Leighton, Stanley	1837–1901	Lewis, Clive Staples	1898–1963
Leiningen, Prince Ernest Leopold		Lewis, David (1814–1895). See	
Victor Charles Auguste Joseph		under Lewis, Evan	
Emich	1830–1904	Lewis, Evan	1818–1901
Leiper, Robert Thomson	1881–1969	Lewis, Sir George Henry	1833–1911
Leishman, Thomas	1825–1904	Lewis, John Spedan	1885–1963
Leishman, Sir William Boog	1865–1926	Lewis, John Travers	1825–1901
Leitch, Charlotte Cecilia Pitcairn		Lewis, Percy Wyndham	1882–1957
(Cecil)	1891–1977	Lewis, Richard	1821–1905
Leith-Ross, Sir Frederick William	1887–1968	Lewis, Rosa	1867–1952
Lejeune, Caroline Alice	1897–1973	Lewis, Sir Thomas	1881–1945
Le Jeune, Henry	1819–1904	Lewis, Sir Wilfrid Hubert Poyer	1881–1950
Lemass, Sean Francis	1899–1971	Lewis, William Cudmore	
Le Mesurier, John	1912–1983	McCullagh	1885–1956
Lemmens-Sherrington, Helen	1834–1906	Lewis, William Thomas, Baron	
Lemon, Sir Ernest John Hutchings	1884–1954	Merthyr	1837–1914
Lempriere, Charles	1818–1901	Lewis, Sir Willmott Harsant	1877–1950
Leng, Sir John	1828–1906	Ley, Henry George	1887–1962
Leng, Sir William Christopher	1825–1902	Leyel, Hilda Winifred Ivy (Mrs C.	
Lennard-Jones, Sir John Edward	1894–1954	F. Leyel)	1880–1957
Lennon, John Winston	1940–1980	Liaqat Ali Khan	1895–1951
Lennox, Charles Henry Gordon-,		Liberty, Sir Arthur Lasenby	1843–1917
Duke of Richmond and Gordon.		Liddell, Edward George Tandy	1895–1981
See Gordon-Lennox	1818–1903	Liddell Hart, Sir Basil Henry. See	
Lennox-Boyd, Alan Tindal,		Hart	1895–1970
Viscount Boyd of Merton	1904–1983	Lidderdale, William	1832–1902

Lombard, Adrian Albert	1915–1967	Lowther, James	1840–1904
London, Heinz	1907–1970	Lowther, James William, Viscount	
Londonderry, Marquess of. See		Ullswater	1855–1949
Vane-Tempest-Stewart,		Löwy, Albert or Abraham	1816–1908
Charles Stewart	1852–1915	Loyd-Lindsay, Robert James,	
Londonderry, Marquess of. See		Baron Wantage. See Lindsay	1832–1901
Vane-Tempest-Stewart,		Luard, Sir William Garnham	1820–1910
Charles Stewart Henry	1878–1949	Lubbock, Sir John, Baron Avebury	1834–1913
Long, Walter Hume, Viscount		Lubbock, Percy	1879–1965
Long of Wraxall	1854–1924	Luby, Thomas Clarke	1821–1901
Longhurst, Henry Carpenter	1909–1979	Lucas, Baron. See Herbert,	
Longhurst, William Henry	1819–1904	Auberon Thomas	1876–1916
Longmore, Sir Arthur Murray	1885–1970	Lucas, Sir Charles Prestwood	1853–1931
Longstaff, Tom George	1875–1964	Lucas, Edward Verrall	1868–1938
Lonsdale, Earl of. See Lowther,		Lucas, Frank Laurence	1894–1967
Hugh Cecil	1857–1944	Lucas, Keith	1879–1916
Lonsdale, Frederick	1881–1954	Luckock, Herbert Mortimer	1833–1909
Lonsdale, Dame Kathleen	1903–1971	Lucy, Sir Henry William	1843–1924
Lopes, Sir Lopes Massey	1818–1908	Ludlow, John Malcolm Forbes	1821–1911
Lopokova, Lydia Vasilievna, Lady		Ludlow-Hewitt, Sir Edgar Rainey	1886–1973
Keynes	1892–1981	Lugard, Frederick John Dealtry,	
Loraine, Sir Percy Lyham	1880–1961	Baron	1858–1945
Loraine, Violet Mary	1886–1956	Luke, Baron. See Johnston,	
Lord, Thomas	1808–1908	George Lawson	1873–1943
Loreburn, Earl. See Reid, Robert		Luke, Sir Harry Charles	1884–1969
Threshie	1846–1923	Luke, Jemima	1813–1906
Lorimer, Sir Robert Stodart	1864–1929	Lukin, Sir Henry Timson	1860–1925
Lotbinière, Sir Henry Gustave Joly		Lumley, Lawrence Roger, Earl of	
de. See Joly de Lotbinière	1829–1908	Scarbrough	1896–1969
Lothian, Marquess of. See Kerr,		Lunn, Sir Arnold Henry Moore	1888–1974
Philip Henry	1882–1940	Lunn, Sir Henry Simpson	1859–1939
Louise Caroline Alberta, princess		Lupton, Joseph Hirst	1836–1905
of Great Britain	1848–1939	Lush, Sir Charles Montague	1853–1930
Louise Victoria Alexandra		Lusk, Sir Andrew	1810–1909
Dagmar, Princess Royal of		Luthuli, Albert John	1898?–1967
Great Britain	1867–1931	Lutyens, (Agnes) Elisabeth	1906–1983
Lovat, Baron. See Fraser, Simon		Lutyens, Sir Edwin Landseer	1869–1944
Joseph	1871–1933	Lutz, (Wilhelm) Meyer	1829–1903
Lovatt Evans, Sir Charles Arthur.		Luxmoore, Sir (Arthur) Fairfax	
See Evans	1884–1968	(Charles Coryndon)	1876–1944
Love, Augustus Edward Hough	1863–1940	Lyall, Sir Alfred Comyn	1835–1911
Lovelace, Earl of. See Milbanke,		Lyall, Sir Charles James	1845–1920
Ralph Gordon Noel King	1839–1906	Lyall, Edna, pseudonym. See Bayly,	
Lovett, Richard	1851–1904	Ada Ellen	1857–1903
Low, Alexander, Lord	1845–1910	Lygon, William, Earl Beauchamp	1872–1938
Low, Sir David Alexander Cecil	1891–1963	Lyle, Charles Ernest Leonard,	
Low, Sir Robert Cunliffe	1838–1911	Baron Lyle of Westbourne	1882–1954
Low, Sir Sidney James Mark	1857–1932	Lynch, Arthur Alfred	1861–1934
Lowe, Sir Drury Curzon Drury-.		Lynd, Robert Wilson	1879–1949
See Drury-Lowe	1830–1908	Lyne, Joseph Leycester (Father	
Lowe, Eveline Mary	1869–1956	Ignatius)	1837–1908
Lowke, Wenman Joseph Bassett-.		Lyne, Sir William John	1844–1913
See Bassett-Lowke	1877–1953	Lynskey, Sir George Justin	1888–1957
Lowry, Clarence Malcolm	1909–1957	Lyon, Claude George Bowes-,	
Lowry, Henry Dawson	1869–1906	Earl of Strathmore and	
Lowry, Laurence Stephen	1887–1976	Kinghorne. See Bowes-Lyon	1855–1944
Lowry, Thomas Martin	1874–1936	Lyons, Sir Algernon McLennan	1833–1908
Lowson, Sir Denys Colquhoun		Lyons, (Francis Stewart) Leland	1923–1983
Flowerdew	1906–1975	Lyons, Sir Henry George	1864–1944
Lowther, Hugh Cecil, Earl of		Lyons, Joseph Aloysius	1879–1939
Lonsdale	1857–1944	Lyons, Sir William	1901–1985

Mackay, Donald James, Baron Reay	1839–1921	Maclean, Donald Duart	1913–1983
Mackay, James Lyle, Earl of Inchcape	1852–1932	Maclean, Sir Harry Aubrey de Vere	1848–1920
Mackay, Mary, 'Marie Corelli'	1855–1924	Maclean, James Mackenzie	1835–1906
McKechnie, William Sharp	1863–1930	McLean, Norman	1865–1947
McKenna, Reginald	1863–1943	Maclear, George Frederick	1833–1902
Mackennal, Alexander	1835–1904	Maclear, John Fiot Lee Pearse	1838–1907
Mackennal, Sir (Edgar) Bertram	1863–1931	McLennan, Sir John Cunningham	1867–1935
Mackenzie, Sir Alexander	1842–1902	Macleod, Fiona, pseudonym. See	
McKenzie, Alexander	1869–1951	Sharp, William	1855–1905
Mackenzie, Sir Alexander Campbell	1847–1935	Macleod, Henry Dunning	1821–1902
		Macleod, Iain Norman	1913–1970
Mackenzie, Sir (Edward Montague) Compton	1883–1972	McLeod, (James) Walter	1887–1978
		Macleod, John James Rickard	1876–1935
Mackenzie, Sir George Sutherland	1844–1910	McLintock, Sir William	1873–1947
		McLintock, William Francis Porter	1887–1960
Mackenzie, Sir James	1853–1925	Maclure, Edward Craig	1833–1906
M'Kenzie, Sir John	1836–1901	Maclure, Sir John William (1835–	
MacKenzie, John Stuart	1860–1935	1901). See under Maclure,	
McKenzie, (Robert) Tait	1867–1938	Edward Craig	
McKenzie, Robert Trelford	1917–1981	McMahon, Sir (Arthur) Henry	1862–1949
Mackenzie, Sir Stephen	1844–1909	McMahon, Charles Alexander	1830–1904
Mackenzie, Sir William	1849–1923	MacMahon, Percy Alexander	1854–1929
Mackenzie, William Warrender, Baron Amulree	1860–1942	MacMichael, Sir Harold Alfred	1882–1969
		Macmillan, Sir Frederick Orridge	1851–1936
Mackenzie King, William Lyon. See King	1874–1950	Macmillan, Hugh	1833–1903
		Macmillan, Hugh Pattison, Baron	1873–1952
McKerrow, Ronald Brunlees	1872–1940	McMillan, Margaret	1860–1931
McKie, Douglas	1896–1967	McMillan, William	1887–1977
McKie, Sir William Neil	1901–1984	McMurrich, James Playfair	1859–1939
Mackinder, Sir Halford John	1861–1947	Macnaghten, Sir Edward, Baron	1830–1913
MacKinlay, Antoinette. See Sterling	1843–1904	McNair, Arnold Duncan, Baron	1885–1975
		McNair, John Frederick Adolphus	1828–1910
Mackinnon, Sir Frank Douglas	1871–1946	MacNalty, Sir Arthur Salusbury	1880–1969
Mackinnon, Sir William Henry	1852–1929	Macnamara, Thomas James	1861–1931
Mackintosh, Sir Alexander	1858–1948	McNaughton, Andrew George Latta	1887–1966
Mackintosh, Charles Rennie	1868–1928		
Mackintosh, Harold Vincent, Viscount Mackintosh of Halifax	1891–1964	McNee, Sir John William	1887–1984
		MacNeice, (Frederick) Louis	1907–1963
Mackintosh, Hugh Ross	1870–1936	McNeil, Hector	1907–1955
Mackintosh, James Macalister	1891–1966	McNeile, (Herman) Cyril, 'Sapper'	1888–1937
Mackintosh, John	1833–1907		
Mackintosh, John Pitcairn	1929–1978	McNeill, James	1869–1938
Mackworth-Young, Gerard	1884–1965	McNeill, Sir James McFadyen	1892–1964
McLachlan, Robert	1837–1904	MacNeill, John (otherwise Eoin)	1867–1945
Maclagan, Christian	1811–1901	McNeill, Sir John Carstairs	1831–1904
Maclagan, Sir Eric Robert Dalrymple	1879–1951	MacNeill, John Gordon Swift	1849–1926
		McNeill, Ronald John, Baron Cushendun	1861–1934
Maclagan, William Dalrymple	1826–1910		
Maclaren, Alexander	1826–1910	Macphail, Sir (John) Andrew	1864–1938
MacLaren, Archibald Campbell	1871–1944	Macpherson, (James) Ian, Baron Strathcarron	1880–1937
McLaren, Charles Benjamin Bright, Baron Aberconway	1850–1934		
		Macpherson, Sir John Molesworth	1853–1914
McLaren, Henry Duncan, Baron Aberconway	1879–1953	McQueen, Sir John Withers	1836–1909
		Macqueen-Pope, Walter James	1888–1960
Maclaren, Ian, pseudonym. See Watson, John	1850–1907	Macready, Sir (Cecil Frederick) Nevil	1862–1946
McLaren, John, Lord	1831–1910	Macrorie, William Kenneth	1831–1905
Maclay, Joseph Paton, Baron	1857–1951	M'Taggart, John M'Taggart Ellis	1866–1925
Maclean, Sir Donald	1864–1932	McTaggart, William	1835–1910

Martel, Sir Giffard Le Quesne	1889–1958
Marten, Sir (Clarence) Henry (Kennett)	1872–1948
Martin, Alexander	1857–1946
Martin, (Basil) Kingsley	1897–1969
Martin, Sir Charles James	1866–1955
Martin, Sir David Christie	1914–1976
Martin, Sir Douglas Eric (Deric) Holland-. See Holland-Martin	1906–1977
Martin, Herbert Henry	1881–1954
Martin, Hugh	1890–1964
Martin, Sir Theodore	1816–1909
Martin, Sir Thomas Acquin	1850–1906
Martin, Violet Florence, 'Martin Ross'	1862–1915
Martin, William Keble	1877–1969
Martin-Harvey, Sir John Martin	1863–1944
Martindale, Cyril Charlie	1879–1963
Martindale, Hilda	1875–1952
Marwick, Sir James David	1826–1908
Mary, Queen	1867–1953
Masefield, John Edward	1878–1967
Masham, Baron. See Lister, Samuel Cunliffe	1815–1906
Maskelyne, Mervyn Herbert Nevil Story-. See Story-Maskelyne	1823–1911
Mason, Alfred Edward Woodley	1865–1948
Mason, Arthur James	1851–1928
Mason, James Neville	1909–1984
Mason-MacFarlane, Sir (Frank) Noel	1889–1953
Massey, (Charles) Vincent	1887–1967
Massey, Gerald	1828–1907
Massey, Sir Harrie Stewart Wilson	1908–1983
Massey, William Ferguson	1856–1925
Massingberd, Sir Archibald Armar Montgomery-. See Montgomery-Massingberd	1871–1947
Massingham, Harold John	1888–1952
Massingham, Henry William	1860–1924
Masson, David	1822–1907
Masson, Sir David Orme	1858–1937
Massy, William Godfrey Dunham	1838–1906
Masterman, Charles Frederick Gurney	1874–1927
Masterman, Sir John Cecil	1891–1977
Masters, John	1914–1983
Masters, Maxwell Tylden	1833–1907
Matheson, George	1842–1906
Mathew, Anthony (1905–1976). See under Mathew, David James	
Mathew, David James	1902–1975
Mathew, Gervase (1905–1976). See under Mathew, David James	
Mathew, Sir James Charles	1830–1908
Mathew, Theobald	1866–1939
Mathew, Sir Theobald	1898–1964
Mathews, Basil Joseph	1879–1951

Mathews, Charles Edward	1834–1905
Mathews, Sir Charles Willie	1850–1920
Mathews, Sir Lloyd William	1850–1901
Mathews, Dame Vera (Elvira Sibyl Maria) Laughton	1888–1959
Mathieson, William Law	1868–1938
Matthew, Sir Robert Hogg	1906–1975
Matthews, Alfred Edward	1869–1960
Matthews, Henry, Viscount Llandaff	1826–1913
Matthews, Jessie Margaret	1907–1981
Matthews, Walter Robert	1881–1973
Matthews, Sir William	1844–1922
Maturin, Basil William	1847–1915
Maud Charlotte Mary Victoria, Queen of Norway	1869–1938
Maud, John Primatt Redcliffe, Baron Redcliffe-Maud	1906–1982
Maude, Aylmer	1858–1938
Maude, Sir (Frederick) Stanley	1864–1917
Maudling, Reginald	1917–1979
Maufe, Sir Edward Brantwood	1883–1974
Maugham, Frederic Herbert, Viscount	1866–1958
Maugham, William Somerset	1874–1965
Maurice, Sir Frederick Barton	1871–1951
Maurice, Sir John Frederick	1841–1912
Mavor, Osborne Henry, 'James Bridie'	1888–1951
Mawdsley, James	1848–1902
Mawer, Sir Allen	1879–1942
Mawson, Sir Douglas	1882–1958
Maxim, Sir Hiram Stevens	1840–1916
Maxse, Sir (Frederick) Ivor	1862–1958
Maxse, Leopold James	1864–1932
Maxton, James	1885–1946
Maxwell, Sir Alexander	1880–1963
Maxwell, Gavin	1914–1969
Maxwell, Sir Herbert Eustace	1845–1937
Maxwell, Sir John Grenfell	1859–1929
Maxwell (formerly Braddon), Mary Elizabeth	1837–1915
Maxwell Fyfe, David Patrick, Earl of Kilmuir. See Fyfe	1900–1967
Maxwell Lyte, Sir Henry Churchill. See Lyte	1848–1940
May, George Ernest, Baron	1871–1946
May, Philip William (Phil)	1864–1903
May, Sir William Henry	1849–1930
Maybury, Sir Henry Percy	1864–1943
Mayer, Sir Robert	1879–1985
Mayor, John Eyton Bickersteth	1825–1910
Meade, Richard James, Earl of Clanwilliam	1832–1907
Meade-Fetherstonhaugh, Sir Herbert	1875–1964
Meakin, James Edward Budgett	1866–1906
Meath, Earl of. See Brabazon, Reginald	1841–1929
Medd, Peter Goldsmith	1829–1908
Medlicott, Henry Benedict	1829–1905

Mitchell, John Murray	1815–1904	Montagu-Douglas-Scott, Lord	
Mitchell, Leslie Scott Falconer	1905–1985	Charles Thomas. See Scott	1839–1911
Mitchell, Sir Peter Chalmers	1864–1945	Montagu-Douglas-Scott, Lord	
Mitchell, Sir Philip Euen	1890–1964	Francis George. See Scott	1879–1952
Mitchell, Reginald Joseph	1895–1937	Montague, Charles Edward	1867–1928
Mitchell, Sir William Gore		Montague, Francis Charles	1858–1935
Sutherland	1888–1944	Monteath, Sir James	1847–1929
Mitford, Algernon Bertram		Montefiore, Claude Joseph	
Freeman-, Baron Redesdale	1837–1916	Goldsmid-	1858–1938
Mitford, Nancy Freeman-	1904–1973	Montgomerie, Robert Archibald	
Moberly, Robert Campbell	1845–1903	James	1855–1908
Moberly, Sir Walter Hamilton	1881–1974	Montgomery, Bernard Law,	
Mocatta, Frederic David	1828–1905	Viscount Montgomery of	
Mockler-Ferryman, Eric Edward	1896–1978	Alamein	1887–1976
Möens, William John Charles	1833–1904	Montgomery, (Robert) Bruce,	
Moeran, Ernest John	1894–1950	'Edmund Crispin'	1921–1978
Moffatt, James	1870–1944	Montgomery-Massingberd, Sir	
Moir, Frank Lewis	1852–1904	Archibald Armar	1871–1947
Mollison, Amy. See Johnson	1903–1941	Montmorency, James Edward	
Mollison, James Allan	1905–1959	Geoffrey de. See de	
Molloy, Gerald	1834–1906	Montmorency	1866–1934
Molloy, James Lynam	1837–1909	Montmorency, Raymond Harvey	
Molloy, Joseph FitzGerald	1858–1908	de, Viscount Frankfort de	
Molony, Sir Thomas Francis	1865–1949	Montmorency. See de	
Molyneux, Sir Robert Henry		Montmorency	1835–1902
More-. See More-Molyneux	1838–1904	Monypenny, William Flavelle	1866–1912
'Môn, Hwfa', *pseudonym*. See		Moody, Harold Arundel	1882–1947
Williams, Rowland	1823–1905	Moor, Sir Frederick Robert	1853–1927
Monash, Sir John	1865–1931	Moor, Sir Ralph Denham	
Monckton, Walter Turner,		Rayment	1860–1909
Viscount Monckton of		Moore, Arthur William	1853–1909
Brenchley	1891–1965	Moore, Edward	1835–1916
Moncreiff, Henry James, Baron	1840–1909	Moore, George Augustus	1852–1933
Moncrieff, Sir Alexander	1829–1906	Moore, George Edward	1873–1958
Moncreiffe of that Ilk, Sir (Rupert)		Moore, Mary. See Wyndham,	
Iain (Kay),	1919–1985	Mary, Lady	1861–1931
Mond, Alfred Moritz, Baron		Moore, Stuart Archibald	1842–1907
Melchett	1868–1930	Moore, Temple Lushington	1856–1920
Mond, Julian Edward Alfred,		Moore-Brabazon, John Theodore	
Baron Melchett	1925–1973	Cuthbert, Baron Brabazon of	
Mond, Ludwig	1839–1909	Tara. See Brabazon	1884–1964
Mond, Sir Robert Ludwig	1867–1938	Moorehead, Alan McCrae	1910–1983
Monkhouse, William Cosmo	1840–1901	Moorhouse, James	1826–1915
Monnington, Sir (Walter) Thomas	1902–1976	Moran, Baron. See Wilson,	
Monro, Sir Charles Carmichael	1860–1929	Charles McMoran	1882–1977
Monro, Charles Henry	1835–1908	Moran, Patrick Francis	1830–1911
Monro, David Binning	1836–1905	Morant, Geoffrey Miles	1899–1964
Monro, Harold Edward	1879–1932	Morant, Sir Robert Laurie	1863–1920
Monro, Sir Horace Cecil	1861–1949	Mordell, Louis Joel	1888–1972
Monro, Matt	1930–1985	More-Molyneux, Sir Robert	
Monsarrat, Nicholas John Turney	1910–1979	Henry	1838–1904
Monson, Sir Edmund John	1834–1909	Morecambe, Eric	1926–1984
Montagu of Beaulieu, Baron. See		Moresby, John	1830–1922
Douglas-Scott-Montagu, John		Morfill, William Richard	1834–1909
Walter Edward	1866–1929	Morgan, Charles Langbridge	1894–1958
Montagu, Edwin Samuel	1879–1924	Morgan, Conwy Lloyd	1852–1936
Montagu, Ewen Edward Samuel	1901–1985	Morgan, Edward Delmar	1840–1909
Montagu, Ivor Goldsmid Samuel	1904–1984	Morgan, Sir Frederick Edgworth	1894–1967
Montagu, Lord Robert	1825–1902	Morgan, Sir Gilbert Thomas	1872–1940
Montagu, Samuel, Baron		Morgan, John Hartman	1876–1955
Swaythling	1832–1911	Morgan, Sir Morien Bedford	1912–1978

Murray, Sir James Wolfe	1853–1919	Nettleship, Edward	1845–1913
Murray, Sir John	1841–1914	Nettleship, John Trivett	1841–1902
Murray, Sir John	1851–1928	Neubauer, Adolf	1832–1907
Murray, John	1879–1964	Neville, Henry	1837–1910
Murray, Sir (John) Hubert		Nevinson, Christopher Richard	
(Plunkett)	1861–1940	Wynne	1889–1946
Murray, Margaret Alice	1863–1963	Nevinson, Henry Woodd	1856–1941
Murray, Sir Oswyn Alexander		Newall (formerly Phillpotts),	
Ruthven	1873–1936	Dame Bertha Surtees	1877–1932
Murry, John Middleton	1889–1957	Newall, Cyril Louis Norton, Baron	1886–1963
Murry, Kathleen, 'Katherine		Newall, Hugh Frank	1857–1944
Mansfield'	1888–1923	Newberry, Percy Edward	1869–1949
Musgrave, Sir James	1826–1904	Newbold, Sir Douglas	1894–1945
Muybridge, Eadweard	1830–1904	Newbolt, Sir Henry John	1862–1938
Myers, Charles Samuel	1873–1946	Newbolt, William Charles	
Myers, Ernest James	1844–1921	Edmund	1844–1930
Myers, Leopold Hamilton	1881–1944	Newitt, Dudley Maurice	1894–1980
Myrddin-Evans, Sir Guildhaume	1894–1964	Newman, Ernest	1868–1959
Myres, Sir John Linton	1869–1954	Newman, Sir George	1870–1948
Mysore, Sir Shri Krishnaraja		Newman, Maxwell Herman,	
Wadiyar Bahadur, Maharaja of	1884–1940	Alexander	1897–1984
		Newman, William Lambert	1834–1923
		Newmarch, Charles Henry	1824–1903
Nabarro, Sir Gerald David Nunes	1913–1973	Newnes, Sir George	1851–1910
Naipaul, Shivadhar Srinivasa		Newsam, Sir Frank Aubrey	1893–1964
(Shiva)	1945–1985	Newsholme, Sir Arthur	1857–1943
Nair, Sir Chettur Sankaran. See		Newsom, Sir John Hubert	1910–1971
Sankaran Nair	1857–1934	Newton, Baron. See Legh,	
Nairne, Alexander	1863–1936	Thomas Wodehouse	1857–1942
Namier, Sir Lewis Bernstein	1888–1960	Newton, Alfred	1829–1907
Narbeth, John Harper	1863–1944	Newton, Ernest	1856–1922
Nares, Sir George Strong	1831–1915	Nichol Smith, David. See Smith	1875–1962
Nash, John Northcote	1893–1977	Nicholls, Frederick William	1889–1974
Nash, Paul	1889–1946	Nichols, Robert Malise Bowyer	1893–1944
Nash, Sir Walter	1882–1968	Nicholson, Sir Charles	1808–1903
Nathan, Harry Louis, Baron	1889–1963	Nicholson, Sir Charles Archibald	1867–1949
Nathan, Sir Matthew	1862–1939	Nicholson, Charles Ernest	1868–1954
Nawanagar, Maharaja Shri		Nicholson, Edward William Byron	1849–1912
Ranjitsinhji Vibhaji, Maharaja		Nicholson, George	1847–1908
Jam Saheb of	1872–1933	Nicholson, Joseph Shield	1850–1927
Neale, Sir John Ernest	1890–1975	Nicholson, Reynold Alleyne	1868–1945
Neave, Airey Middleton Sheffield	1916–1979	Nicholson, Sir Sydney Hugo	1875–1947
Neel, (Louis) Boyd	1905–1981	Nicholson, William Gustavus,	
Nehru, Jawaharlal	1889–1964	Baron	1845–1918
Nehru, Pandit Motilal	1861–1931	Nicholson, Sir William Newzam	
Neil, Robert Alexander	1852–1901	Prior	1872–1949
Neil, Samuel	1825–1901	Nickalls, Guy	1866–1935
Neill, Alexander Sutherland	1883–1973	Nicol, Erskine	1825–1904
Neill, Stephen Charles	1900–1984	Nicoll, (John Ramsay) Allardyce	1894–1976
Neilson, George	1858–1923	Nicoll, Sir William Robertson	1851–1923
Neilson, Julia Emilie	1868–1957	Nichols, (John) Beverley	1898–1983
Nelson, Eliza (1827–1908). See		Nicolson, Adela Florence,	
under Craven, Henry Thornton	1818–1905	'Laurence Hope'	1865–1904
Nelson, Sir Frank	1883–1966	Nicolson, Sir Arthur, Baron	
Nelson, George Horatio, Baron		Carnock	1849–1928
Nelson of Stafford	1887–1962	Nicholson, Benjamin Lauder (Ben)	1894–1982
Nelson, Sir Hugh Muir	1835–1906	Nicolson, Sir Harold George	1886–1968
Nemon, Oscar	1906–1985	Nicolson, (Lionel) Benedict	1914–1978
Neruda, Wilma Maria Francisca.		Nicolson, Malcolm Hassels	
See Hallé, Lady	1839–1911	(1843–1904). See under	
Nesbit, Edith. See Bland	1858–1924	Nicolson, Adela Florence	

Oliver, Sir Geoffrey Nigel	1898–1980	Ouless, Walter William	1848–1933
Oliver, Sir Henry Francis	1865–1965	Overton, John Henry	1835–1903
Oliver, Samuel Pasfield	1838–1907	Overtoun, Baron. See White, John	
Oliver, Sir Thomas	1853–1942	Campbell	1843–1908
Olivier, Giorgio Borg (George)	1911–1980	Owen, Sir (Arthur) David (Kemp)	1904–1970
Olivier, Sydney Haldane, Baron	1859–1943	Owen, (Humphrey) Frank	1905–1979
Olpherts, Sir William	1822–1902	Owen, John	1854–1926
Olsson, Julius	1864–1942	Owen, Robert	1820–1902
Oman, Sir Charles William		Owen, Sir (William) Leonard	1897–1971
Chadwick	1860–1946	Oxford and Asquith, Countess of.	
Oman, John Wood	1860–1939	See Asquith, Emma Alice	
Ommanney, Sir Erasmus	1814–1904	Margaret (Margot)	1864–1945
Ommanney, George Druce		Oxford and Asquith, Earl of. See	
Wynne	1819–1902	Asquith, Herbert Henry	1852–1928
Onions, Charles Talbut	1873–1965		
Onslow, Sir Richard George	1904–1975		
Onslow, William Hillier, Earl of		Page, Sir Archibald	1875–1949
Onslow	1853–1911	Page, Sir Denys Lionel	1908–1978
Opie, Peter Mason	1918–1982	Page, Sir Frederick Handley	1885–1962
Oppé, Adolph Paul	1878–1957	Page, H. A., pseudonym. See Japp,	
Oppenheim, Edward Phillips	1866–1946	Alexander Hay	1837–1905
Oppenheim, Lassa Francis		Page, Sir Leo Francis	1890–1951
Lawrence	1858–1919	Page, Thomas Ethelbert	1850–1936
Oppenheimer, Sir Ernest	1880–1957	Page, William	1861–1934
Orage, Alfred Richard	1873–1934	Paget, Sir Bernard Charles Tolver	1887–1961
Oram, Sir Henry John	1858–1939	Paget, Edward Francis	1886–1971
Orchardson, Sir William Quiller	1832–1910	Paget, Francis	1851–1911
Orczy, Emma Magdalena Rosalia		Paget, Dame (Mary) Rosalind	1855–1948
Marie Josepha Barbara, Baroness	1865–1947	Paget, Lady Muriel Evelyn Vernon	1876–1938
Ord, Bernhard (Boris)	1897–1961	Paget, Sir Richard Arthur Surtees	1869–1955
Ord, William Miller	1834–1902	Paget, Sidney Edward	1860–1908
Orde, Cuthbert Julian	1888–1968	Paget, Stephen	1855–1926
O'Rell, Max, pseudonym. See		Paget, Violet, 'Vernon Lee'	1856–1935
Blouet, Léon Paul	1848–1903	Pain, Barry Eric Odell	1864–1928
Ormerod, Eleanor Anne	1828–1901	Paine, Charles Hubert Scott-. See	
Ormsby Gore, (William) David,		Scott-Paine	1891–1954
Baron Harlech	1918–1985	Pakenham, Sir Francis John	1832–1905
Ormsby-Gore, William George		Pakenham, Sir William	
Arthur, Baron Harlech	1885–1964	Christopher	1861–1933
Orpen, Sir William Newenham		Palairet, Sir (Charles) Michael	1882–1956
Montague	1878–1931	Palgrave, Sir Reginald Francis	
Orr, Alexandra Sutherland	1828–1903	Douce	1829–1904
Orr, John Boyd, Baron Boyd Orr	1880–1971	Palles, Christopher	1831–1920
Orr, William McFadden	1866–1934	Palmer, Sir Arthur Power	1840–1904
Orton, Charles William Previté-.		Palmer, Sir Charles Mark	1822–1907
See Previté-Orton	1877–1947	Palmer, Sir Elwin Mitford	1852–1906
Orton, John Kingsley (Joe)	1933–1967	Palmer, George Herbert	1846–1926
Orwell, George, pseudonym. See		Palmer, George William	1851–1913
Blair, Eric Arthur	1903–1950	Palmer, Roundell Cecil, Earl of	
Orwin, Charles Stewart	1876–1955	Selborne	1887–1971
Osborn, Sir Frederic James	1885–1978	Palmer, William Waldegrave, Earl	
Osborne, Walter Frederick	1859–1903	of Selborne	1859–1942
O'Shea, John Augustus	1839–1905	Paneth, Friedrich Adolf	1887–1958
O'Shea, William Henry	1840–1905	Pankhurst, Dame Christabel	
Osler, Abraham Follett	1808–1903	Harriette	1880–1958
Osler, Sir William	1849–1919	Pankhurst, Emmeline	1858–1928
O'Sullivan, Cornelius	1841–1907	Pantin, Carl Frederick Abel	1899–1967
Otté, Elise	1818–1903	Pares, Sir Bernard	1867–1949
Ottley, Sir Charles Langdale	1858–1932	Paris, Sir Archibald	1861–1937
Ouida, pseudonym. See De la		Parish, William Douglas	1833–1904
Ramée, Marie Louise	1839–1908	Park, Sir Keith Rodney	1892–1975

Primrose, (Albert Edward) Harry
(Mayer Archibald), Earl of
Rosebery — 1882–1974
Primrose, Archibald Philip, Earl of
Rosebery — 1847–1929
Primrose, Sir Henry William — 1846–1923
Pringle, John William Sutton — 1912–1982
Pringle, Mia Lilly Kellmer — 1920–1983
Pringle, William Mather
Rutherford — 1874–1928
Pringle-Pattison, Andrew Seth.
See Pattison — 1856–1931
Prinsep, Valentine Cameron (Val) — 1838–1904
Prior, Melton — 1845–1910
Pritchard, Sir Charles Bradley — 1837–1903
Pritchard, Sir Edward Evan
Evans-. See Evans-Pritchard — 1902–1973
Pritchett, Robert Taylor — 1828–1907
Pritt, Denis Nowell — 1887–1972
Prittie, Terence Cornelius Farmer — 1913–1985
Probert, Lewis — 1841–1908
Procter, Francis — 1812–1905
Proctor, Robert George Collier — 1868–1903
Propert, John Lumsden — 1834–1902
Prothero, Sir George Walter — 1848–1922
Prothero, Rowland Edmund,
Baron Ernle — 1851–1937
Proudman, Joseph — 1888–1975
Prout, Ebenezer — 1835–1909
Pryde, James Ferrier — 1866–1941
Prynne, George Rundle — 1818–1903
Puddicombe, Anne Adalisa, 'Allen
Raine' — 1836–1908
Pudney, John Sleigh — 1909–1977
Pugh, Sir Arthur — 1870–1955
Pullen, Henry William — 1836–1903
Pumphrey, Richard Julius — 1906–1967
Purcell, Albert Arthur William — 1872–1935
Purse, Benjamin Ormond — 1874–1950
Purser, Louis Claude — 1854–1932
Purvis, Arthur Blaikie — 1890–1941
Pye, Sir David Randall — 1886–1960
Pym, Barbara Mary Crampton — 1913–1980
Pyne, Louisa Fanny Bodda. See
Bodda Pyne — 1832–1904

Quarrier, William — 1829–1903
Quick, Sir John — 1852–1932
Quick, Oliver Chase — 1885–1944
Quickswood, Baron. See Cecil,
Hugh Richard Heathcote
Gascoyne- — 1869–1956
Quiller-Couch, Sir Arthur
Thomas ('Q') — 1863–1944
Quilter, Harry — 1851–1907
Quilter, Roger Cuthbert — 1877–1953
Quilter, Sir William Cuthbert — 1841–1911
Quin, Windham Thomas
Wyndham-, Earl of Dunraven
and Mount-Earl — 1841–1926

Race, Robert Russell — 1907–1984
Rackham, Arthur — 1867–1939
Rackham, Bernard — 1876–1964
Radcliffe, Cyril John, Viscount — 1899–1977
Radcliffe-Crocker, Henry — 1845–1909
Radhakrishnan, Sir Sarvepalli — 1888–1975
Rae, William Fraser — 1835–1905
Raggi, Mario — 1821–1907
Raikes, Humphrey Rivaz — 1891–1955
Railton, Herbert — 1858–1910
Raine, Allen, pseudonym. See
Puddicombe, Anne Adalisa — 1836–1908
Raines, Sir Julius Augustus Robert — 1827–1909
Rainy, Adam Rolland (1862–
1911). See under Rainy, Robert
Rainy, Robert — 1826–1906
Raistrick, Harold — 1890–1971
Rait, Sir Robert Sangster — 1874–1936
Rajagopalachari, Chakravarti — 1878–1972
Raleigh, Sir Walter Alexander — 1861–1922
Ralston, James Layton — 1881–1948
Ram, Sir (Lucius Abel John)
Granville — 1885–1952
Raman, Sir (Chandrasekhara)
Venkata — 1888–1970
Ramberg, Cyvia Myriam. See
Rambert, Dame Marie — 1888–1982
Rambert, Dame Marie — 1888–1982
Ramé, Marie Louise, 'Ouida'. See
De la Ramée — 1839–1908
Ramsay, Alexander — 1822–1909
Ramsay, Sir Bertram Home — 1883–1945
Ramsay, Sir James Henry — 1832–1925
Ramsay, Lady (Victoria) Patricia
(Helena Elizabeth) — 1886–1974
Ramsay, Sir William — 1852–1916
Ramsay, Sir William Mitchell — 1851–1939
Ramsay-Steel-Maitland, Sir
Arthur Herbert Drummond.
See Steel-Maitland — 1876–1935
Ramsbottom, John — 1885–1974
Ramsden, Omar — 1873–1939
Ramsey, Ian Thomas — 1915–1972
Randall, Sir John Turton — 1905–1984
Randall, Richard William — 1824–1906
Randall-MacIver, David — 1873–1945
Randegger, Alberto — 1832–1911
Randles, Marshall — 1826–1904
Randolph, Francis Charles
Hingeston-. See Hingeston-
Randolph — 1833–1910
Randolph, Sir George Granville — 1818–1907
Ranjitsinhji, Maharaja Jam Saheb
of Nawanagar. See Nawanagar — 1872–1933
Rank, (Joseph) Arthur, Baron — 1888–1972
Rankeillour, Baron. See Hope,
James Fitzalan — 1870–1949
Rankin, Sir George Claus — 1877–1946
Ransom, William Henry — 1824–1907
Ransome, Arthur Michell — 1884–1967
Raper, Robert William — 1842–1915

Richards, Owain Westmacott	1901–1984	Robeck, Sir John Michael De. See	
Richardson, Alan	1905–1975	De Robeck	1862–1928
Richardson, Sir Albert Edward	1880–1964	Roberts, Alexander	1826–1901
Richardson, Ethel Florence		Roberts, Frederick Sleigh, Earl	1832–1914
Lindesay, 'Henry Handel		Roberts, George Henry	1869–1928
Richardson'	1870–1946	Roberts, Isaac	1829–1904
Richardson, (Frederick) Denys	1913–1983	Roberts, Robert Davies	1851–1911
Richardson, Henry Handel. See		Roberts, Thomas d'Esterre	1893–1976
Richardson, Ethel Florence		Roberts, William Patrick	1895–1980
Lindesay	1870–1946	Roberts-Austen, Sir William	
Richardson, Lewis Fry	1881–1953	Chandler	1843–1902
Richardson, Sir Owen Willans	1879–1959	Robertson, Alexander	1896–1970
Richardson, Sir Ralph David	1902–1983	Robertson, Andrew	1883–1977
Richey, James Ernest	1886–1968	Robertson, Archibald	1853–1931
Richmond, Sir Bruce Lyttelton	1871–1964	Robertson, Brian Hubert, Baron	
Richmond, Sir Herbert William	1871–1946	Robertson of Oakbridge	1896–1974
Richmond, Sir Ian Archibald	1902–1965	Robertson, Sir Charles Grant	1869–1948
Richmond, Sir William Blake	1842–1921	Robertson, Sir Dennis Holme	1890–1963
Richmond and Gordon, Duke of.		Robertson, Donald Struan	1885–1961
See Gordon-Lennox, Charles		Robertson, Douglas Moray	
Henry	1818–1903	Cooper Lamb Argyll	1837–1909
Ricketts, Charles de Sousy	1866–1931	Robertson, George Matthew	1864–1932
Riddell, Charles James Buchanan	1817–1903	Robertson, Sir George Scott	1852–1916
Riddell, Charlotte Eliza Lawson		Robertson, Sir Howard Morley	1888–1963
(Mrs J. H. Riddell), 'F. G.		Robertson, James Patrick	
Trafford'	1832–1906	Bannerman, Baron	1845–1909
Riddell, George Allardice, Baron	1865–1934	Robertson, Sir James Wilson	1899–1983
Ridding, George	1828–1904	Robertson, John Mackinnon	1856–1933
Rideal, Sir Eric Keightley	1890–1974	Robertson, Sir Johnston Forbes-	1853–1937
Ridgeway, Sir Joseph West	1844–1930	Robertson, Sir Robert	1869–1949
Ridgeway, Sir William	1853–1926	Robertson, Sir William Robert	1860–1933
Ridley, Henry Nicholas	1855–1956	Robertson Scott, John William	1866–1962
Ridley, Sir Matthew White,		Robey, Sir George Edward	1869–1954
Viscount	1842–1904	Robins, Thomas Ellis, Baron	1884–1962
Rieu, Charles Pierre Henri	1820–1902	Robinson, (Esmé Stuart) Lennox	1886–1958
Rieu, Emile Victor	1887–1972	Robinson, Frederick William	1830–1901
Rigby, Sir John	1834–1903	Robinson, George Frederick	
Rigg, James Harrison	1821–1909	Samuel, Marquess of Ripon	1827–1909
Rigg, James McMullen	1855–1926	Robinson, (George) Geoffrey. See	
Ringer, Sydney	1835–1910	Dawson	1874–1944
Ripon, Marquess of. See		Robinson, Henry Wheeler	1872–1945
Robinson, George Frederick		Robinson, Joan Violet	1903–1983
Samuel	1827–1909	Robinson, Sir John	1839–1903
Risley, Sir Herbert Hope	1851–1911	Robinson, John Arthur Thomas	1919–1983
Ritchie, Anne Isabella, Lady		Robinson, Sir John Charles	1824–1913
(1837–1919). See under		Robinson, Sir John Richard	1828–1903
Ritchie, Sir Richmond		Robinson, Joseph Armitage	1858–1933
Thackeray Willoughby		Robinson, Sir Joseph Benjamin	1840–1929
Ritchie, Charles Thomson, Baron		Robinson, Philip Stewart (Phil)	1847–1902
Ritchie of Dundee	1838–1906	Robinson, Sir Robert	1886–1975
Ritchie, David George	1853–1903	Robinson, Roy Lister, Baron	1883–1952
Ritchie, Sir John Neish	1904–1977	Robinson, Vincent Joseph	1829–1910
Ritchie, Sir Neil Methuen	1897–1983	Robinson, Sir (William) Arthur	1874–1950
Ritchie, Sir Richmond Thackeray		Robinson, William Heath	1872–1944
Willoughby	1854–1912	Robinson, William Leefe	1895–1918
Ritchie-Calder, Baron. See Calder	1906–1982	Robison, Robert	1883–1941
Rivaz, Sir Charles Montgomery	1845–1926	Robson, Dame Flora	1902–1984
Riverdale, Baron. See Balfour,		Robson, William Alexander	1895–1980
Arthur	1873–1957	Robson, William Snowdon, Baron	1852–1918
Riviere, Briton	1840–1920	Roby, Henry John	1830–1915
Robbins, Lionel Charles, Baron	1898–1984	Roche, Alexander Adair, Baron	1871–1956

Rushton, William Albert Hugh	1901–1980
Russell, Arthur Oliver Villiers, Baron Ampthill	1869–1935
Russell, Bertrand Arthur William, Earl	1872–1970
Russell, Sir Charles	1863–1928
Russell, Dorothy Stuart	1895–1983
Russell, Edward Frederick Langley, Baron Russell of Liverpool	1895–1981
Russell, Sir (Edward) John	1872–1965
Russell, Edward Stuart	1887–1954
Russell, Francis Xavier Joseph (Frank), Baron Russell of Killowen	1867–1946
Russell, Sir Frederick Stratten	1897–1984
Russell, George William, 'AE'	1867–1935
Russell, Sir Guy Herbrand Edward	1898–1977
Russell, Henry Chamberlaine	1836–1907
Russell, Herbrand Arthur, Duke of Bedford	1858–1940
Russell, John Hugo, Baron Ampthill	1896–1973
Russell, Mary Annette, Countess	1866–1941
Russell, Mary du Caurroy, Duchess of Bedford (1865–1937). See under Russell, Herbrand Arthur	
Russell, Sir (Sydney) Gordon	1892–1980
Russell, Thomas O'Neill	1828–1908
Russell, Sir Thomas Wentworth, Russell Pasha	1879–1954
Russell, Sir Walter Westley	1867–1949
Russell, William Clark	1844–1911
Russell, Sir William Howard	1820–1907
Russell, William James	1830–1909
Russell, (William) Ritchie	1903–1980
Russell Flint, Sir William. See Flint	1880–1969
Rutherford, Ernest, Baron Rutherford of Nelson	1871–1937
Rutherford, Dame Margaret	1892–1972
Rutherford, Mark, *pseudonym*. See White, William Hale	1831–1913
Rutherford, William Gunion	1853–1907
Rutland, Duke of. See Manners, (Lord) John James Robert	1818–1906
Ruttledge, Hugh	1884–1961
Ryan, Elizabeth Montague	1892–1979
Ryde, John Walter	1898–1961
Ryder, Charles Henry Dudley	1868–1945
Rye, Maria Susan	1829–1903
Rye, William Brenchley	1818–1901
Ryle, Gilbert	1900–1976
Ryle, Herbert Edward	1856–1925
Ryle, John Alfred	1889–1950
Ryle, Sir Martin	1918–1984
Ryrie, Sir Granville de Laune	1865–1937
Sachs, Sir Eric Leopold Otho	1898–1979
Sackville, Herbrand Edward Dundonald Brassey, Earl De La Warr	1900–1976
Sackville-West, Edward Charles, Baron Sackville	1901–1965
Sackville-West, Lionel Sackville, Baron Sackville	1827–1908
Sackville-West, Victoria Mary	1892–1962
Sadleir, Michael Thomas Harvey	1888–1957
Sadler, Sir Michael Ernest	1861–1943
Saha, Meghnad. See Meghnad Saha	1893–1956
St Aldwyn, Earl. See Hicks Beach, Sir Michael Edward	1837–1916
St Davids, Viscount. See Philipps, Sir John Wynford	1860–1938
St Helier, Baron. See Jeune, Francis Henry	1843–1905
St John, Sir Spenser Buckingham	1825–1910
St John, Vane Ireton Shaftesbury (1839–1911). See under St John, Sir Spenser Buckingham	
St Just, Baron. See Grenfell, Edward Charles	1870–1941
St Laurent, Louis Stephen	1882–1973
St Oswald, Baron. See Winn, Rowland Denys Guy	1916–1984
Saintsbury, George Edward Bateman	1845–1933
Saklatvala, Shapurji	1874–1936
Saladin, *pseudonym*. See Ross, William Stewart	1844–1906
Salaman, Charles Kensington	1814–1901
Salaman, Julia. See Goodman	1812–1906
Salaman, Redcliffe Nathan	1874–1955
Salisbury, Marquess of. See Cecil, James Edward Hubert Gascoyne-	1861–1947
Salisbury, Marquess of. See Cecil, Robert Arthur James Gascoyne-	1893–1972
Salisbury, Marquess of. See Cecil, Robert Arthur Talbot Gascoyne-	1830–1903
Salisbury, Sir Edward James	1886–1978
Salisbury, Francis Owen (Frank)	1874–1962
Salmon, Sir Eric Cecil Heygate	1896–1946
Salmon, George	1819–1904
Salmond, Sir John Maitland	1881–1968
Salmond, Sir (William) Geoffrey (Hanson)	1878–1933
Salomons, Sir Julian Emanuel	1835–1909
Salt, Dame Barbara	1904–1975
Salter, Sir Arthur Clavell	1859–1928
Salter, Herbert Edward	1863–1951
Salter, (James) Arthur, Baron	1881–1975
Salting, George	1835–1909
Salvidge, Sir Archibald Tutton James	1863–1928
Salvin, Francis Henry	1817–1904
Salzman, Louis Francis	1878–1971
Sambourne, Edward Linley	1844–1910

Scott, Paul Mark	1920–1978	Sexton, Sir James	1856–1938
Scott, Sir Percy Moreton	1853–1924	Sexton, Thomas	1848–1932
Scott, Robert Falcon	1868–1912	Seymour, Sir Edward Hobart	1840–1929
Scott, Sir Robert Heatlie	1905–1982	Shackleton, Sir David James	1863–1938
Scott, Sir Ronald Bodley. See		Shackleton, Sir Ernest Henry	1874–1922
Bodley Scott	1906–1982	Shadwell, Charles Lancelot	1840–1919
Scott-Ellis, Thomas Evelyn, Baron		Shand (afterwards Burns),	
Howard de Walden	1880–1946	Alexander, Baron	1828–1904
Scott-James, Rolfe Arnold	1878–1959	Shand, Alexander Innes	1832–1907
Scott-Paine, Charles Hubert	1891–1954	Shandon, Baron. See O'Brien,	
Scrutton, Sir Thomas Edward	1856–1934	Ignatius John	1857–1930
Seago, Edward Brian	1910–1974	Shanks, Michael James	1927–1984
Seale-Hayne, Charles Hayne	1833–1903	Shannon, Charles Haslewood	1863–1937
Seaman, Sir Owen	1861–1936	Shannon, Sir James Jebusa	1862–1923
Searle, Humphrey	1915–1982	Sharp, Cecil James	1859–1924
Seccombe, Thomas	1866–1923	Sharp, Evelyn Adelaide, Baroness	1903–1985
Seddon, Richard John	1845–1906	Sharp, Thomas Wilfred	1901–1978
Sedgwick, Adam	1854–1913	Sharp, William, 'Fiona Macleod'	1855–1905
See, Sir John	1844–1907	Sharpe, Richard Bowdler	1847–1909
Seebohm, Frederic	1833–1912	Sharpey-Schafer, Sir Edward	
Seeley, Harry Govier	1839–1909	Albert. See Schafer	1850–1935
Seely, John Edward Bernard,		Shattock, Samuel George	1852–1924
Baron Mottistone	1868–1947	Shaughnessy, Thomas George,	
Selbie, William Boothby	1862–1944	Baron	1853–1923
Selborne, Earl of. See Palmer,		Shaw, Alfred	1842–1907
Roundell Cecil	1887–1971	Shaw, Sir Eyre Massey	1830–1908
Selborne, Earl of. See Palmer,		Shaw, George Bernard	1856–1950
William Waldegrave	1859–1942	Shaw, Henry Selby Hele-. See	
Selby, Viscount. See Gully,		Hele-Shaw	1854–1941
William Court	1835–1909	Shaw, James Johnston	1845–1910
Selby, Thomas Gunn	1846–1910	Shaw, John Byam Lister	1872–1919
Selfridge, Harry Gordon	1858–1947	Shaw, Richard Norman	1831–1912
Seligman, Charles Gabriel	1873–1940	Shaw, Thomas, Baron Craigmyle	1850–1937
Selincourt, Ernest de	1870–1943	Shaw, Thomas	1872–1938
Sellers, Richard Henry ('Peter')	1925–1980	Shaw, William Arthur	1865–1943
Selous, Frederick Courteney	1851–1917	Shaw, Sir (William) Napier	1854–1945
Selwin-Ibbetson, Henry John,		Shaw-Lefevre, George John,	
Baron Rookwood	1826–1902	Baron Eversley	1831–1928
Selwyn, Alfred Richard Cecil	1824–1902	Shearman, Sir Montague	1857–1930
Selwyn-Lloyd, Baron. See Lloyd,		Sheepshanks, Sir Thomas Herbert	1895–1964
John Selwyn Brooke	1904–1978	Sheffield, Earl of. See Holroyd,	
Semon, Sir Felix	1849–1921	Henry North	1832–1909
Sempill, Baron. See Forbes-		Sheffield, Baron. See Stanley,	
Sempill, William Francis	1893–1965	Edward Lyulph	1839–1925
Senanayake, Don Stephen	1884–1952	Sheldon, Sir Wilfrid Percy Henry	1901–1983
Sendall, Sir Walter Joseph	1832–1904	Shelford, Sir William	1834–1905
Sequeira, James Harry	1865–1948	Shenstone, William Ashwell	1850–1908
Sergeant, (Emily Frances) Adeline	1851–1904	Shepard, Ernest Howard	1879–1976
Sergeant, Lewis	1841–1902	Shepherd, George Robert, Baron	1881–1954
Service, Robert William	1874–1958	Sheppard, Hugh Richard Lawrie	1880–1937
Seth, Andrew. See Pattison,		Sheppard, Philip Macdonald	1921–1976
Andrew Seth Pringle-	1856–1931	Sheppard, Sir Richard Herbert	1910–1982
Seton, George	1822–1908	Sherborn, Charles William	1831–1912
Seton-Watson, (George) Hugh		Sherek, (Jules) Henry	1900–1967
(Nicholas)	1916–1984	Sheridan, Clare Consuelo	1885–1970
Seton-Watson, Robert William	1879–1951	Sherriff, George	1898–1967
Severn, Walter	1830–1904	Sherriff, Robert Cedric	1896–1975
Seward, Sir Albert Charles	1863–1941	Sherrington, Sir Charles Scott	1857–1952
Sewell, Elizabeth Missing	1815–1906	Sherrington, Helen Lemmens-.	
Sewell, James Edwards	1810–1903	See Lemmens-Sherrington	1834–1906
Sewell, Robert Beresford Seymour	1880–1964	Shields, Frederic James	1833–1911

Smith, Henry Spencer	1812–1901	Somervell, Donald Bradley, Baron	
Smith, Herbert	1862–1938	Somervell of Harrow	1889–1960
Smith, Sir Hubert Llewellyn	1864–1945	Somervell, (Theodore) Howard	1890–1975
Smith, James Hamblin	1829–1901	Somerville, Edith Anna Œnone	1858–1949
Smith, John Alexander	1863–1939	Somerville, Sir James Fownes	1882–1949
Smith, (Lloyd) Logan Pearsall	1865–1946	Somerville, Mary	1897–1963
Smith, Lucy Toulmin	1838–1911	Somerville, Sir William	1860–1932
Smith, Sir Matthew Arnold Bracy	1879–1959	Sonnenschein, Edward Adolf	1851–1929
Smith, Reginald Bosworth	1839–1908	Sorabji, Cornelia	1866–1954
Smith, Reginald John	1857–1916	Sorby, Henry Clifton	1826–1908
Smith, Rodney	1860–1947	Sorley, Sir Ralph Squire	1898–1974
Smith, Sir Ross Macpherson	1892–1922	Sorley, William Ritchie	1855–1935
Smith, Samuel	1836–1906	Soskice, Frank, Baron Stow Hill	1902–1979
Smith, Sarah, 'Hesba Stretton'	1832–1911	Sotheby, Sir Edward Southwell	1831–1902
Smith, Stevie. See Smith,		Soutar, Ellen. See Farren	1848–1904
Florence Margaret	1902–1971	Southborough, Baron. See	
Smith, Sir Sydney Alfred	1883–1969	Hopwood, Francis John	
Smith, Thomas	1817–1906	Stephens	1860–1947
Smith, Sir Thomas	1833–1909	Southesk, Earl of. See Carnegie,	
Smith, Thomas	1883–1969	James	1827–1905
Smith, Thomas Roger	1830–1903	Southey, Sir Richard	1808–1901
Smith, Vincent Arthur	1848–1920	Southward, John	1840–1902
Smith, Vivian Hugh, Baron Bicester	1867–1956	Southwell, Sir Richard Vynne	1888–1970
Smith, Walter Chalmers	1824–1908	Southwell, Thomas	1831–1909
Smith, William Frederick Danvers,		Southwood, Viscount. See Elias,	
Viscount Hambleden	1868–1928	Julius Salter	1873–1946
Smith, William Saumarez	1836–1909	Souttar, Sir Henry Sessions	1875–1964
Smith-Dorrien, Sir Horace		Spare, Austin Osman	1886–1956
Lockwood	1858–1930	Speaight, Robert William	1904–1976
Smith-Rose, Reginald Leslie	1894–1980	Spear, (Thomas George) Percival	1901–1982
Smithells, Arthur	1860–1939	Spearman, Charles Edward	1863–1945
Smuts, Jan Christian	1870–1950	Spears, Sir Edward Louis	1886–1974
Smyly, Sir Philip Crampton	1838–1904	Spence, Sir Basil Urwin	1907–1976
Smyth, Dame Ethel Mary	1858–1944	Spence, Sir James Calvert	1892–1954
Smyth, Sir Henry Augustus	1825–1906	Spencer, Gilbert	1892–1979
Smythe, Francis Sydney	1900–1949	Spencer, Sir Henry Francis	1892–1964
Snedden, Sir Richard	1900–1970	Spencer, Herbert	1820–1903
Snell, Henry, Baron	1865–1944	Spencer, John Poyntz, Earl	
Snell, Sir John Francis Cleverton	1869–1938	Spencer	1835–1910
Snelus, George James	1837–1906	Spencer, Leonard James	1870–1959
Snow, Lady. See Johnson, Pamela		Spencer, Sir Stanley	1891–1959
Hansford	1912–1981	Spencer, Sir Walter Baldwin	1860–1929
Snow, Charles Percy, Baron	1905–1980	Spencer-Churchill, Baroness. See	
Snow, Sir Frederick Sydney	1899–1976	Churchill, Clementine Ogilvy	
Snow, Herbert. See Kynaston	1835–1910	Spencer-	1885–1977
Snow, Sir Thomas D'Oyly	1858–1940	Spender, John Alfred	1862–1942
Snowden, Philip, Viscount	1864–1937	Spens, Sir William (Will)	1882–1962
Soddy, Frederick	1877–1956	Spens, (William) Patrick, Baron	1885–1973
Soissons, Louis Emmanuel Jean		Speyer, Sir Edgar	1862–1932
Guy de Savoie-Carignan de.		Spiers, Richard Phené	1838–1916
See de Soissons	1890–1962	Spilsbury, Sir Bernard Henry	1877–1947
Sollas, William Johnson	1849–1936	Spinks, Alfred	1917–1982
Solomon, Sir Richard	1850–1913	Spofforth, Frederick Robert	1853–1926
Solomon, Simeon	1840–1905	Spooner, William Archibald	1844–1930
Solomon, Solomon Joseph	1860–1927	Sporborg, Henry Nathan	1905–1985
Somers-Cocks, Arthur Herbert		Sprengel, Hermann Johann	
Tennyson, Baron Somers	1887–1944	Philipp	1834–1906
Somerset, Henry Hugh Arthur		Sprigg, Sir John Gordon	1830–1913
FitzRoy, Duke of Beaufort	1900–1984	Sprigge, Sir (Samuel) Squire	1860–1937
Somerset, Lady Isabella Caroline		Spring, (Robert) Howard	1889–1965
(Lady Henry Somerset)	1851–1921	Spring-Rice, Sir Cecil Arthur	1859–1918

Stirling, Sir James	1836–1916	Holland Fox-, Earl of Ilchester.	
Stirling, James Hutchison	1820–1909	See Fox-Strangways	1874–1959
Stirling, Walter Francis	1880–1958	Strathalmond, Baron. See Fraser,	
Stockdale, Sir Frank Arthur	1883–1949	William	1888–1970
Stocks, John Leofric	1882–1937	Strathcarron, Baron. See	
Stocks, Mary Danvers, Baroness	1891–1975	Macpherson, (James) Ian	1880–1937
Stoddart, Andrew Ernest	1863–1915	Strathclyde, Baron. See Ure,	
Stokes, Adrian	1887–1927	Alexander	1853–1928
Stokes, Adrian Durham	1902–1972	Strathcona, Baron. See Smith,	
Stokes, Sir Frederick Wilfrid		Donald Alexander	1820–1914
Scott	1860–1927	Strathmore and Kinghorne, Earl	
Stokes, Sir George Gabriel	1819–1903	of. See Bowes-Lyon, Claude	
Stokes, Sir John	1825–1902	George	1855–1944
Stokes, Whitley	1830–1909	Stratton, Frederick John Marrian	1881–1960
Stokowski, Leopold Anthony	1882–1977	Strauss, Henry George, Baron	
Stoll, Sir Oswald	1866–1942	Conesford	1892–1974
Stone, (Alan) Reynolds	1909–1979	Street, Arthur George	1892–1966
Stone, Darwell	1859–1941	Street, Sir Arthur William	1892–1951
Stoner, Edmund Clifton	1899–1968	Streeter, Burnett Hillman	1874–1937
Stoney, Bindon Blood	1828–1909	Stretton, Hesba, *pseudonym*. See	
Stoney, George Gerald	1863–1942	Smith, Sarah	1832–1911
Stoney, George Johnstone	1826–1911	Strickland, Gerald, Baron	1861–1940
Stoop, Adrian Dura	1883–1957	Strijdom, Johannes Gerhardus	1893–1958
Stopes, Marie Charlotte		Strong, Eugénie	1860–1943
Carmichael	1880–1958	Strong, Sir Kenneth William	
Stopford, Sir Frederick William	1854–1929	Dobson	1900–1982
Stopford, John Sebastian Bach,		Strong, Leonard Alfred George	1896–1958
Baron Stopford of Fallowfield	1888–1961	Strong, Sir Samuel Henry	1825–1909
Stopford, Robert Wright	1901–1976	Strong, Sandford Arthur	1863–1904
Storrs, Sir Ronald Henry Amherst	1881–1955	Strong, Thomas Banks	1861–1944
Storry, (George) Richard	1913–1982	Struthers, Sir John	1857–1925
Story, Robert Herbert	1835–1907	Strutt, Edward Gerald	1854–1930
Story-Maskelyne, Mervyn		Strutt, John William, Baron	
Herbert Nevil	1823–1911	Rayleigh	1842–1919
Stout, George Frederick	1860–1944	Strutt, Robert John, Baron	
Stout, Sir Robert	1844–1930	Rayleigh	1875–1947
Stow Hill, Baron. See Soskice,		Stuart, Sir Campbell Arthur	1885–1972
Frank	1902–1979	Stuart, James Gray, Viscount	
Strachan, Douglas	1875–1950	Stuart of Findhorn	1897–1971
Strachan, John	1862–1907	Stuart, Sir John Theodosius	
Strachan-Davidson, James Leigh	1843–1916	Burnett-. See Burnett-Stuart	1875–1958
Strachey, Sir Arthur (1858–1901).		Stuart-Jones, Sir Henry. See Jones	1867–1939
See under Strachey, Sir John		Stubbs, Sir Reginald Edward	1876–1947
Strachey, Christopher	1916–1975	Stubbs, William	1825–1901
Strachey, Sir Edward	1812–1901	Studd, Sir (John Edward)	
Strachey, Sir Edward, Baron		Kynaston	1858–1944
Strachie	1858–1936	Sturdee, Sir Frederick Charles	
Strachey, (Evelyn) John (St. Loe)	1901–1963	Doveton	1859–1925
Strachey, (Giles) Lytton	1880–1932	Sturgis, Julian Russell	1848–1904
Strachey, Sir John	1823–1907	Sturt, George	1863–1927
Strachey, John St. Loe	1860–1927	Sturt, Henry Gerard, Baron	
Strachey, Sir Richard	1817–1908	Alington	1825–1904
Strachie, Baron. See Strachey, Sir		Sueter, Sir Murray Frazer	1872–1960
Edward	1858–1936	Sugden, Sir (Theodore) Morris	1919–1984
Stradling, Sir Reginald Edward	1891–1952	Sullivan, Alexander Martin	1871–1959
Straight, Whitney Willard	1912–1979	Summerskill, Edith Clara,	
Strakosch, Sir Henry	1871–1943	Baroness	1901–1980
Strang, William	1859–1921	Sumner, Viscount. See Hamilton,	
Strang, William, Baron	1893–1978	John Andrew	1859–1934
Strangways, Arthur Henry Fox	1859–1948	Sumner, Benedict Humphrey	1893–1951
Strangways, Giles Stephen		Sutcliffe, Herbert William	1894–1978

Temple, Sir Richard	1826–1902	Thomas, Sir William Beach	1868–1957
Temple, Sir Richard Carnac	1850–1931	Thomas, William Moy	1828–1910
Temple, William	1881–1944	Thompson, Alexander Hamilton	1873–1952
Templer, Sir Gerald Walter		Thompson, D'Arcy Wentworth	1829–1902
Robert	1898–1979	Thompson, Sir D'Arcy	
Templewood, Viscount. See		Wentworth	1860–1948
Hoare, Sir Samuel John Gurney	1880–1959	Thompson, Edmund Symes-. See	
Tenby, Viscount. See Lloyd-		Symes-Thompson	1837–1906
George, Gwilym	1894–1967	Thompson, Edward John	1886–1946
Tennant, Sir Charles	1823–1906	Thompson, Sir Edward Maunde	1840–1929
Tennant, Sir David	1829–1905	Thompson, Francis	1859–1907
Tennant, Margaret Mary Edith		Thompson, Gertrude Caton-. See	
(May)	1869–1946	Caton-Thompson	1888–1985
Tenniel, Sir John	1820–1914	Thompson, Sir Harold Warris	1908–1983
Tennyson-d'Eyncourt, Sir		Thompson, Sir Henry	1820–1904
Eustace Henry William	1868–1951	Thompson, Sir (Henry Francis)	
Terry, Dame (Alice) Ellen	1847–1928	Herbert	1859–1944
Terry, Charles Sanford	1864–1936	Thompson, Henry Yates	1838–1928
Terry, Fred	1863–1933	Thompson, James Matthew	1878–1956
Terry, Sir Richard Runciman	1865–1938	Thompson, Sir John Eric Sidney	1898–1975
Tertis, Lionel	1876–1975	Thompson, Lydia	1836–1908
Tetlow, Norman	1899–1982	Thompson, Reginald Campbell	1876–1941
Tewson, Sir (Harold) Vincent	1898–1981	Thompson, Roscoe Treeve	
Teyte, Dame Margaret (Maggie)	1888–1976	Fawcett. See Rotha, Paul	1907–1984
Thankerton, Baron. See Watson,		Thompson, Silvanus Phillips	1851–1916
William	1873–1948	Thompson, William Marcus	1857–1907
Thesiger, Frederic Augustus,		Thomson, Arthur	1858–1935
Baron Chelmsford	1827–1905	Thomson, Sir (Arthur)	
Thesiger, Frederic John Napier,		Landsborough	1890–1977
Viscount Chelmsford	1868–1933	Thomson, Sir Basil Home	1861–1939
Thirkell, Angela Margaret	1890–1961	Thomson, Christopher Birdwood,	
Thiselton-Dyer, Sir William		Baron	1875–1930
Turner	1843–1928	Thomson, Sir George Paget	1892–1975
Thoday, David	1883–1964	Thomson, Sir George Pirie	1887–1965
Thomas, Bertram Sidney	1892–1950	Thomson, George Reid, Lord	1893–1962
Thomas, David Alfred, Viscount		Thomson, Hugh	1860–1920
Rhondda	1856–1918	Thomson, Jocelyn Home	1859–1908
Thomas, Dylan Marlais	1914–1953	Thomson, John	1856–1926
Thomas, Forest Frederic Edward		Thomson, Sir Joseph John	1856–1940
Yeo-. See Yeo-Thomas	1902–1964	Thomson, Roy Herbert, Baron	
Thomas, Frederick William	1867–1956	Thomson of Fleet	1894–1976
Thomas, Freeman Freeman-,		Thomson, William, Baron Kelvin	1824–1907
Marquess of Willingdon. See		Thomson, Sir William	1843–1909
Freeman-Thomas	1866–1941	Thorndike, Dame (Agnes) Sybil	1882–1976
Thomas, Sir George Alan	1881–1972	Thorne, William James (Will)	1857–1946
Thomas, George Holt	1869–1929	Thornton, Alfred Henry Robinson	1863–1939
Thomas, Gwyn	1913–1981	Thornton, Sir Edward	1817–1906
Thomas, Sir Henry	1878–1952	Thornycroft, Sir John Isaac	1843–1928
Thomas, Herbert Henry	1876–1935	Thornycroft, Sir (William) Hamo	1850–1925
Thomas, Sir Hugh Evan-. See		Thorpe, Sir Thomas Edward	1845–1925
Evan-Thomas	1862–1928	Threlfall, Sir Richard	1861–1932
Thomas, Hugh Hamshaw	1885–1962	Thring, Godfrey	1823–1903
Thomas, James Henry	1874–1949	Thring, Henry, Baron	1818–1907
Thomas, James Purdon Lewes,		Thrupp, George Athelstane	1822–1905
Viscount Cilcennin	1903–1960	Thuillier, Sir Henry Edward	
Thomas, Margaret Haig,		Landor	1813–1906
Viscountess Rhondda	1883–1958	Thursfield, Sir James	1840–1923
Thomas, Meirion	1894–1977	Thurso, Viscount. See Sinclair, Sir	
Thomas, (Philip) Edward	1878–1917	Archibald Henry Macdonald	1890–1970
Thomas, Sir (Thomas) Shenton		Thurston (formerly Madden),	
(Whitelegge)	1879–1962	Katherine Cecil	1875–1911

Tynan, Katharine. See Hinkson	1861–1931
Tynan, Kenneth Peacock	1927–1980
Tyndale-Biscoe, Cecil Earle	1863–1949
Tyndall, Arthur Mannering	1881–1961
Tyrrell, George	1861–1909
Tyrrell, Robert Yelverton	1844–1914
Tyrrell, William George, Baron	1866–1947
Tyrwhitt, Sir Reginald Yorke	1870–1951
Tyrwhitt-Wilson, Sir Gerald Hugh, Baron Berners	1883–1950
Ullswater, Viscount. See Lowther, James William	1855–1949
Underhill, Edward Bean	1813–1901
Underhill, Evelyn (Mrs Stuart Moore)	1875–1941
Underwood, (George Claude) Leon	1890–1975
Unwin, Sir Raymond	1863–1940
Unwin, Sir Stanley	1884–1968
Unwin, William Cawthorne	1838–1933
Upjohn, Gerald Ritchie, Baron	1903–1971
Ure, Alexander, Baron Strathclyde	1853–1928
Urwick, William	1826–1905
Uthwatt, Augustus Andrewes, Baron	1879–1949
Uttley, Alice Jane (Alison)	1884–1976
Uvarov, Sir Boris Petrovitch	1889–1970
Uwins, Cyril Frank	1896–1972
Vachell, Horace Annesley	1861–1955
Vaizey, John Ernest, Baron	1929–1984
Vallance, Gerald Aylmer	1892–1955
Vallance, William Fleming	1827–1904
Vanbrugh, Dame Irene	1872–1949
Vanbrugh, Violet	1867–1942
Vandam, Albert Dresden	1843–1903
Vane-Tempest-Stewart, Charles Stewart, Marquess of Londonderry	1852–1915
Vane-Tempest-Stewart, Charles Stewart Henry, Marquess of Londonderry	1878–1949
Van Horne, Sir William Cornelius	1843–1915
Vansittart, Edward Westby	1818–1904
Vansittart, Robert Gilbert, Baron	1881–1957
Vaughan, Bernard John	1847–1922
Vaughan, David James	1825–1905
Vaughan, Dame Helen Charlotte Isabella Gwynne-. See Gwynne-Vaughan	1879–1967
Vaughan, Herbert Alfred	1832–1903
Vaughan, (John) Keith	1912–1977
Vaughan, Kate	1852?–1903
Vaughan, William Wyamar	1865–1938
Vaughan Williams, Ralph	1872–1958
Veale, Sir Douglas	1891–1973
Veitch, Sir Harry James	1840–1924
Veitch, James Herbert	1868–1907

Venables, Sir Percy Frederick Ronald ('Peter')	1904–1979
Venn, John	1834–1923
Ventris, Michael George Francis	1922–1956
Verdon-Roe, Sir (Edwin) Alliott Verdon	1877–1958
Vereker, John Standish Surtees Prendergast, Viscount Gort	1886–1946
Verney, Ernest Basil	1894–1967
Verney, Margaret Maria, Lady	1844–1930
Vernon-Harcourt, Leveson Francis	1839–1907
Verrall, Arthur Woollgar	1851–1912
Vestey, William, Baron	1859–1940
Vezin, Hermann	1829–1910
Vezin (formerly Mrs Charles Young), Jane Elizabeth	1827–1902
Vian, Sir Philip Louis	1894–1968
Vickers, Sir (Charles) Geoffrey	1894–1982
Vickers, Kenneth Hotham	1881–1958
Vicky. See Weisz, Victor	1913–1966
Victoria Adelaide Mary Louise, Princess Royal of Great Britain and German Empress	1840–1901
Victoria Alexandra Alice Mary, Princess Royal of Great Britain	1897–1965
Victoria Alexandra Olga Mary, princess of Great Britain	1868–1935
Victoria Eugénie Julia Ena, Queen of Spain	1887–1969
Villiers, George Herbert Hyde, Earl of Clarendon	1877–1955
Villiers, John Henry De, Baron. See De Villiers	1842–1914
Villiers, Margaret Elizabeth Child-, Countess of Jersey	1849–1945
Villiers, Victor Albert George Child-, Earl of Jersey	1845–1915
Vincent, Sir (Charles Edward) Howard	1849–1908
Vincent, Sir Edgar, Viscount D'Abernon	1857–1941
Vincent, James Edmund	1857–1909
Vines, Sydney Howard	1849–1934
Vinogradoff, Sir Paul Gavrilovitch	1854–1925
Voce, William	1909–1984
Voigt, Frederick Augustus	1892–1957
Von Hügel, Friedrich, Baron of the Holy Roman Empire	1852–1925
Voyce, (Anthony) Thomas	1897–1980
Voysey, Charles	1828–1912
Voysey, Charles Francis Annesley	1857–1941
Wace, Henry	1836–1924
Waddell, Helen Jane	1889–1965
Waddell, Lawrence Augustine (later Austine)	1854–1938
Waddington, Conrad Hal	1905–1975
Wade, Sir Willoughby Francis	1827–1906
Wadsworth, Alfred Powell	1891–1956

Wilcox, Herbert Sydney 1890–1977
Wild, (John Robert) Francis 1873–1939
Wilde, Johannes (János) 1891–1970
Wilde, William James (Jimmy) 1892–1969
Wilding, Anthony Frederick 1883–1915
Wilkie, Sir David Percival Dalbreck 1882–1938
Wilkins, Augustus Samuel 1843–1905
Wilkins, Sir (George) Hubert 1888–1958
Wilkins, William Henry 1860–1905
Wilkinson, Ellen Cicely 1891–1947
Wilkinson, George Howard 1833–1907
Wilkinson, (Henry) Spenser 1853–1937
Wilkinson, Sir Nevile Rodwell 1869–1940
Wilkinson, Norman 1882–1934
Wilks, Sir Samuel 1824–1911
Will, John Shiress 1840–1910
Willcocks, Sir James 1857–1926
Willcox, Sir William Henry 1870–1941
Willes, Sir George Ommanney 1823–1901
Willett, William 1856–1915
William Henry Andrew Frederick,
    prince of Great Britain (1941–
    1972). See under Henry William
    Frederick Albert, Duke of
    Gloucester
Williams, Alfred 1832–1905
Williams, Alwyn Terrell Petre 1888–1968
Williams, (Arthur Frederic) Basil 1867–1950
Williams, Charles 1838–1904
Williams, Charles Hanson
    Greville 1829–1910
Williams, Charles Walter Stansby 1886–1945
Williams, Edward Francis, Baron
    Francis-Williams 1903–1970
Williams, Sir Edward Leader 1828–1910
Williams, Ella Gwendolen Rees,
    'Jean Rhys' 1890?–1979
Williams, Eric Ernest 1911–1983
Williams, Sir Frederic Calland 1911–1977
Williams, Sir George 1821–1905
Williams, Sir Harold Herbert 1880–1964
Williams, Hugh 1843–1911
Williams, Ivy 1877–1966
Williams, John Carvell 1821–1907
Williams, Sir John Coldbrook
    Hanbury-. See Hanbury-
    Williams 1892–1965
Williams, Sir John Fischer 1870–1947
Williams, (Laurence Frederick)
    Rushbrook 1890–1978
Williams, Norman Powell 1883–1943
Williams, Ralph Vaughan. See
    Vaughan Williams 1872–1958
Williams, (Richard) Tecwyn 1909–1979
Williams, Sir Roland Bowdler
    Vaughan 1838–1916
Williams, Rowland, 'Hwfa Môn' 1823–1905
Williams, Thomas, Baron
    Williams of Barnburgh 1888–1967
Williams, Watkin Hezekiah,
    'Watcyn Wyn' 1844–1905

Williams, Sir William Emrys 1896–1977
Williams-Ellis, Sir (Bertram)
    Clough 1883–1978
Williams-Freeman, John Peere 1858–1943
Williamson, Alexander William 1824–1904
Williamson, Henry 1895–1977
Williamson, John Thoburn 1907–1958
Willingdon, Marquess of. See
    Freeman-Thomas, Freeman 1866–1941
Willink, Sir Henry Urmston 1894–1973
Willis, Sir Algernon Usborne 1889–1976
Willis, Henry 1821–1901
Wills, Leonard Johnston 1884–1979
Willis, William 1835–1911
Willock, Henry Davis 1830–1903
Willoughby, Digby 1845–1901
Wills, Sir George Alfred 1854–1928
Wills, William Henry, Baron
    Winterstoke 1830–1911
Wilmot, John, Baron Wilmot of
    Selmeston 1895–1964
Wilmot, Sir Sainthill Eardley- 1852–1929
Wilshaw, Sir Edward 1879–1968
Wilson, Sir Arnold Talbot 1884–1940
Wilson, Arthur (1836–1909). See
    under Wilson, Charles Henry,
    Baron Nunburnholme
Wilson, Sir Arthur Knyvet 1842–1921
Wilson, Charles Henry, Baron
    Nunburnholme 1833–1907
Wilson, Charles McMoran, Baron
    Moran 1882–1977
Wilson, Sir Charles Rivers 1831–1916
Wilson, Charles Robert 1863–1904
Wilson, Charles Thomson Rees 1869–1959
Wilson, Sir Charles William 1836–1905
Wilson, Edward Adrian 1872–1912
Wilson, Eleanora Mary Carus-.
    See Carus-Wilson 1897–1977
Wilson, Frank Percy 1889–1963
Wilson, George Fergusson 1822–1902
Wilson, Sir Gerald Hugh
    Tyrwhitt-, Baron Berners. See
    Tyrwhitt-Wilson 1883–1950
Wilson, Sir Henry Hughes 1864–1922
Wilson, Henry Maitland, Baron 1881–1964
Wilson, Henry Schütz 1824–1902
Wilson, Herbert Wrigley 1866–1940
Wilson, Sir Horace John 1882–1972
Wilson, Sir Jacob 1836–1905
Wilson, James Maurice 1836–1931
Wilson, Sir (James) Steuart 1889–1966
Wilson, John Cook 1849–1915
Wilson, John Dove 1833–1908
Wilson, John Dover 1881–1969
Wilson, John Gideon 1876–1963
Wilson, (John) Leonard 1897–1970
Wilson, Joseph Havelock 1858–1929
Wilson, Peter Cecil 1913–1984
Wilson, Samuel Alexander
    Kinnier 1874–1937

Wright, Edward Perceval	1834–1910	Yeo-Thomas, Forest Frederic	
Wright, Helena Rosa	1887–1982	Edward	1902–1964
Wright, Joseph	1855–1930	Yerbury, Francis Rowland (Frank)	1885–1970
Wright, Sir Norman Charles	1900–1970	Yonge, Charlotte Mary	1823–1901
Wright, Robert Alderson, Baron	1869–1964	Yorke, Albert Edward Philip	
Wright, Sir Robert Samuel	1839–1904	Henry, Earl of Hardwicke	1867–1904
Wright, Whitaker	1845–1904	Yorke, Francis Reginald Stevens	1906–1962
Wright, William Aldis	1831–1914	Yorke, Henry Vincent, 'Henry	
Wright, Sir (William) Charles	1876–1950	Green'	1905–1973
Wrong, Sir George Mackinnon	1860–1948	Yorke, Warrington	1883–1943
Wroth, Warwick William	1858–1911	Youl, Sir James Arndell	1811–1904
Wrottesley, Sir Frederic John	1880–1948	Young, Sir Allen William	1827–1915
Wrottesley, George	1827–1909	Young, Mrs Charles. See Vezin,	
Wycherley, Ronald. See Fury, Billy	1941–1983	Jane Elizabeth	1827–1902
Wyld, Henry Cecil Kennedy	1870–1945	Young, Edward Hilton, Baron	
Wylie, Charles Hotham Montagu		Kennet	1879–1960
Doughty-. See Doughty-Wylie	1868–1915	Young, Francis Brett	1884–1954
Wylie, Sir Francis James	1865–1952	Young, Geoffrey Winthrop	1876–1958
Wyllie, Sir William Hutt Curzon	1848–1909	Young, George, Lord	1819–1907
Wyllie, William Lionel	1851–1931	Young, Sir George	1837–1930
Wyndham, Sir Charles	1837–1919	Young, George Malcolm	1883–1959
Wyndham, George	1863–1913	Young, Gerard Mackworth-. See	
Wyndham, John. See Harris, John		Mackworth-Young	1884–1965
Wyndham Parkes Lucas Beynon	1903–1969	Young, Sir Hubert Winthrop	1885–1950
Wyndham, John Edward Reginald,		Young, Sir Robert Arthur	1871–1959
Baron Egremont and Baron		Young, Sydney	1857–1937
Leconfield	1920–1972	Young, William Henry	1863–1942
Wyndham (formerly Moore),		Young, Sir William Mackworth	1840–1924
Mary, Lady	1861–1931	Younger, George, Viscount	
Wyndham-Quin, Windham		Younger of Leckie	1851–1929
Thomas, Earl of Dunraven and		Younger, Sir Kenneth Gilmour	1908–1976
Mount-Earl. See Quin	1841–1926	Younger, Robert, Baron	
Wyn-Harris, Sir Percy	1903–1979	Blanesburgh	1861–1946
Wynn-Carrington, Charles		Younghusband, Dame Eileen	
Robert, Baron Carrington and		Louise	1902–1981
Marquess of Lincolnshire	1843–1928	Younghusband, Sir Francis	
Wynne-Edwards, Sir Robert		Edward	1863–1942
Meredydd	1897–1974	Yoxall, Sir James Henry	1857–1925
Wynyard, Diana	1906–1964	Ypres, Earl of. See French, John	
Wyon, Allan	1843–1907	Denton Pinkstone	1852–1925
		Yule, George Udny	1871–1951
Yapp, Sir Arthur Keysall	1869–1936		
Yarrow, Sir Alfred Fernandez	1842–1932	Zangwill, Israel	1864–1926
Yate, Sir Charles Edward	1849–1940	Zec, Philip	1909–1983
Yates, Dornford, *pseudonym*. See		Zetland, Marquess of. See	
Mercer, Cecil William	1885–1960	Dundas, Lawrence John	
Yates, Dame Frances Amelia	1899–1981	Lumley	1876–1961
Yeats, Jack Butler	1871–1957	Zimmern, Sir Alfred Eckhard	1879–1957
Yeats, William Butler	1865–1939	Zulueta, Francis de (Francisco	
Yeo, Gerald Francis	1845–1909	María José)	1878–1958